KU-023-548

Overarching themes

The following three overarching themes have been fully integrated throughout the Pearson Edexcel AS and A level Mathematics series, so they can be applied alongside your learning and practice.

1. Mathematical argument, language and proof

- Rigorous and consistent approach throughout
- Notation boxes explain key mathematical language and symbols
- Dedicated sections on mathematical proof explain key principles and strategies
- Opportunities to critique arguments and justify methods

2. Mathematical problem solving

- Hundreds of problem-solving questions, fully integrated into the main exercises
- Problem-solving boxes provide tips and strategies
- Structured and unstructured questions to build confidence
- Challenge boxes provide extra stretch

The Mathematical Problem-solving cycle

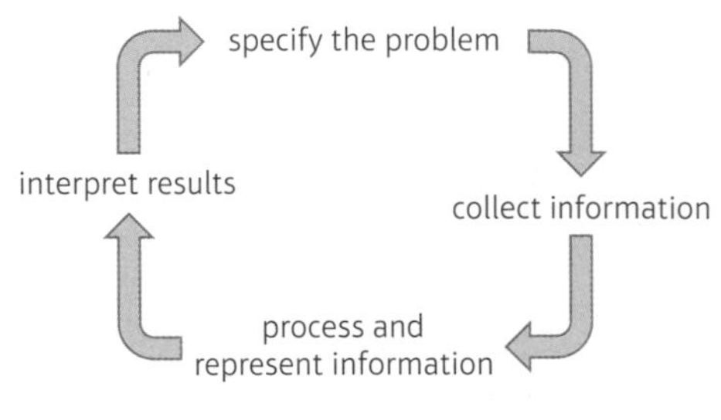

3. Mathematical modelling

- Dedicated modelling sections in relevant topics provide plenty of practice where you need it
- Examples and exercises include qualitative questions that allow you to interpret answers in the context of the model
- Dedicated chapter in Statistics & Mechanics Year 1/AS explains the principles of modelling in mechanics

Finding your way around the book

Access an online digital edition using the code at the front of the book.

Each chapter starts with a list of objectives

The *Prior knowledge check* helps make sure you are ready to start the chapter

3 Equations and inequalities

Objectives

After completing this chapter you should be able to:

- Solve linear simultaneous equations using elimination or substitution → pages 39 – 40
- Solve simultaneous equations: one linear and one quadratic → pages 41 – 42
- Interpret algebraic solutions of equations graphically → pages 42 – 45
- Solve linear inequalities → pages 46 – 48
- Solve quadratic inequalities → pages 48 – 51
- Interpret inequalities graphically → pages 51 – 53
- Represent linear and quadratic inequalities graphically → pages 53 – 55

Prior knowledge check

1 A = {factors of 12}
B = {factors of 20}
Write down the numbers in each of these sets:
a $A \cap B$ b $(A \cup B)'$
← GCSE Mathematics

2 Simplify these expressions.
a $\sqrt{75}$ b $\frac{2\sqrt{45} + 3\sqrt{32}}{6}$
← Section 1.5

3 Match the equations to the correct graph. Label the points of intersection with the axes and the coordinates of the turning point.
a $y = 9 - x^2$ b $y = (x - 2)^2 + 4$
c $y = (x - 7)(2x + 5)$
i ii iii
← Section 2.4

Food scientists use **regions** on graphs to optimise athletes' nutritional intake and ensure they satisfy the minimum dietary requirements for calories and vitamins.

38

The real world applications of the maths you are about to learn are highlighted at the start of the chapter with links to relevant questions in the chapter

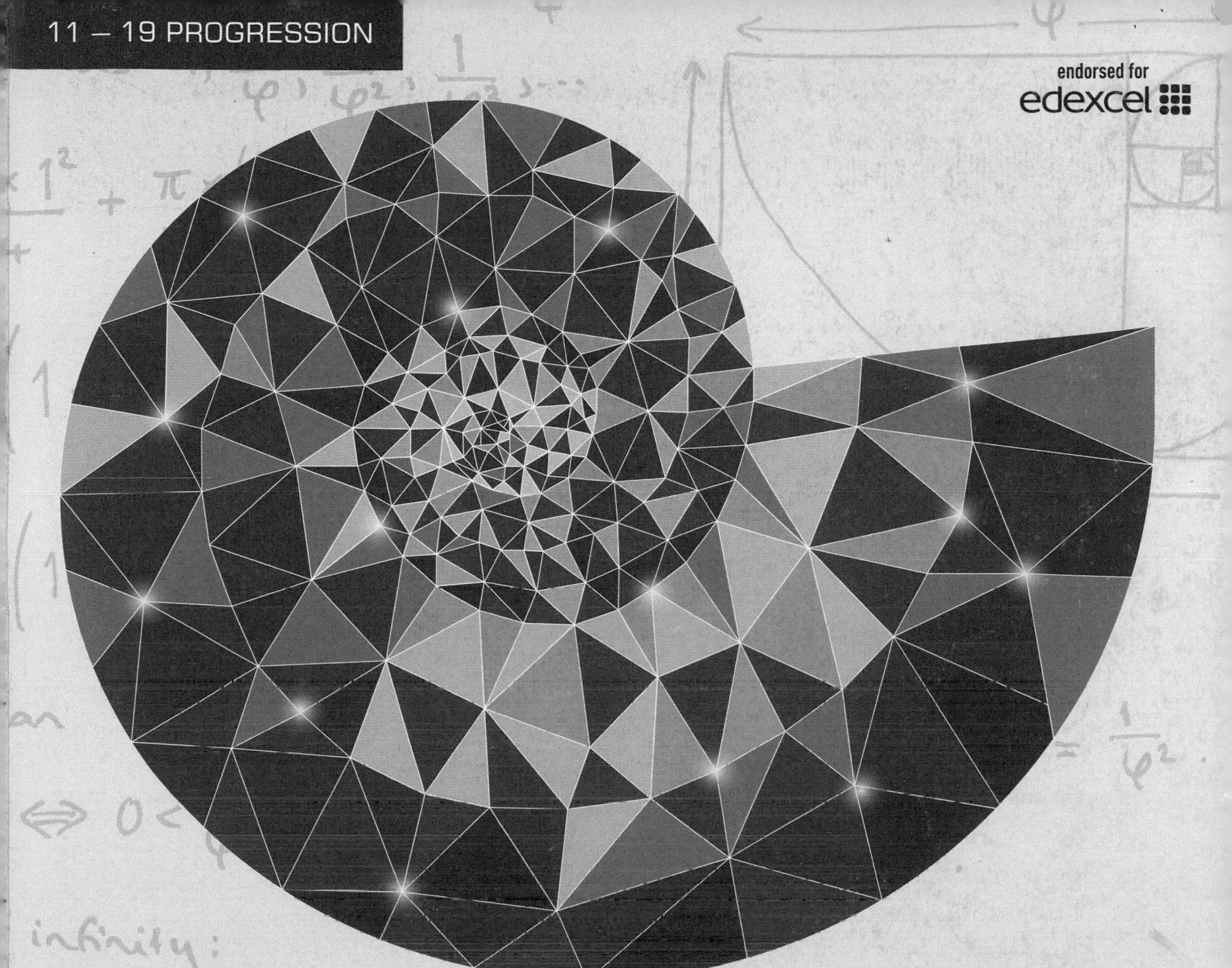

Edexcel AS and A level Mathematics

Pure Mathematics Year 1/AS

Series Editor: Harry Smith

Authors: Greg Attwood, Jack Barraclough, Ian Bettison, Alistair Macpherson, Bronwen Moran, Su Nicholson, Diane Oliver, Joe Petran, Keith Pledger, Harry Smith, Geoff Staley, Robert Ward-Penny, Dave Wilkins

Pearson

Contents

Exercise questions are carefully graded so they increase in difficulty and gradually bring you up to exam standard

Challenge boxes give you a chance to tackle some more difficult questions

Each section begins with explanation and key learning points

Step-by-step worked examples focus on the key types of questions you'll need to tackle

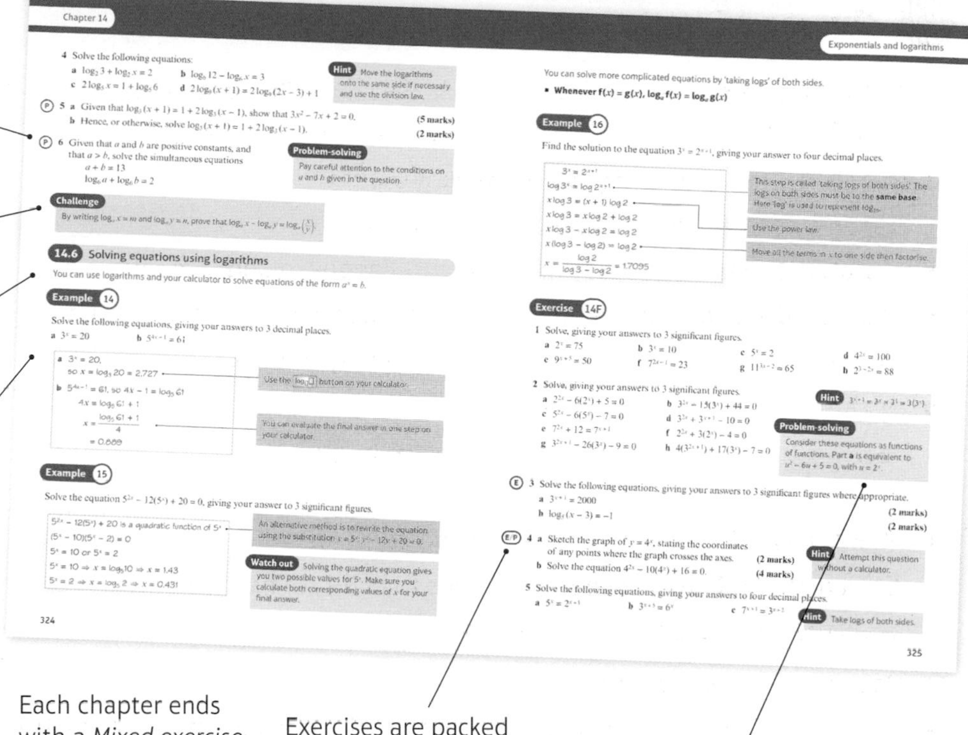

Exam-style questions are flagged with Ⓔ

Problem-solving questions are flagged with Ⓟ

Each chapter ends with a *Mixed exercise* and a *Summary of key points*

Exercises are packed with exam-style questions to ensure you are ready for the exams

Problem-solving boxes provide hints, tips and strategies, and *Watch out* boxes highlight areas where students often lose marks in their exams

Every few chapters a *Review exercise* helps you consolidate your learning with lots of exam-style questions

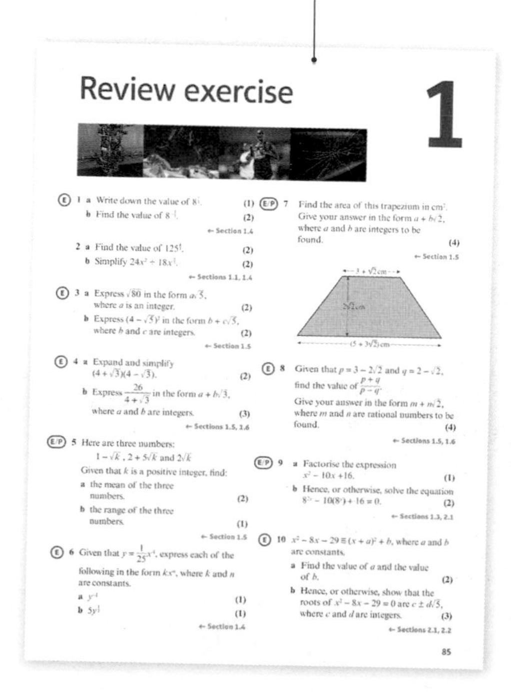

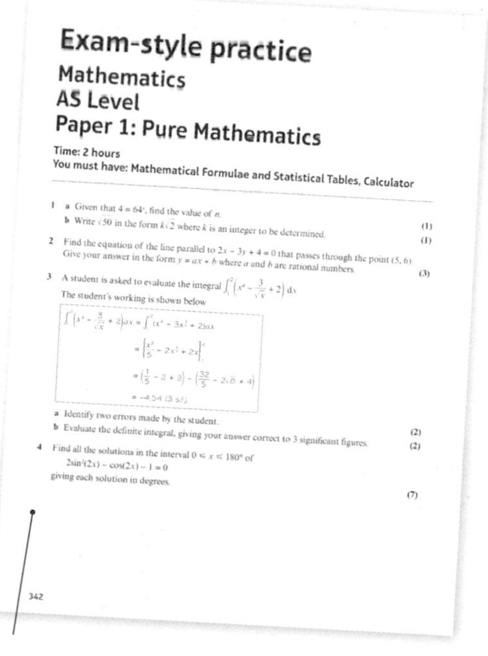

A full AS level practice paper at the back of the book helps you prepare for the real thing

Extra online content

Whenever you see an *Online* box, it means that there is extra online content available to support you.

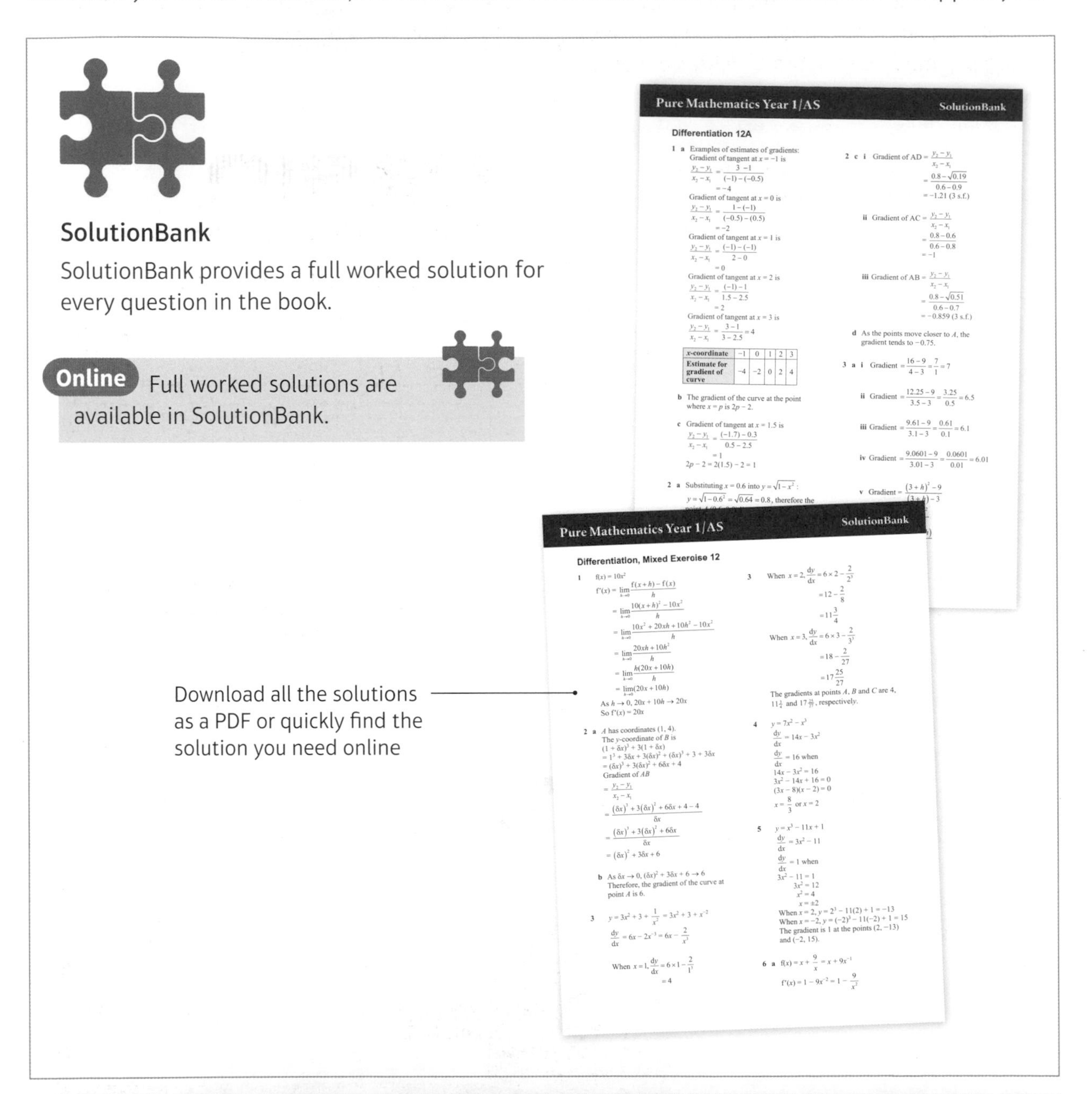

SolutionBank

SolutionBank provides a full worked solution for every question in the book.

Online Full worked solutions are available in SolutionBank.

Download all the solutions as a PDF or quickly find the solution you need online

Access all the extra online content for free at:

www.pearsonschools.co.uk/p1maths

You can also access the extra online content by scanning this QR code:

Use of technology

Explore topics in more detail, visualise problems and consolidate your understanding. Use pre-made GeoGebra activities or Casio resources for a graphic calculator.

Online Find the point of intersection graphically using technology.

GeoGebra

GeoGebra-powered interactives

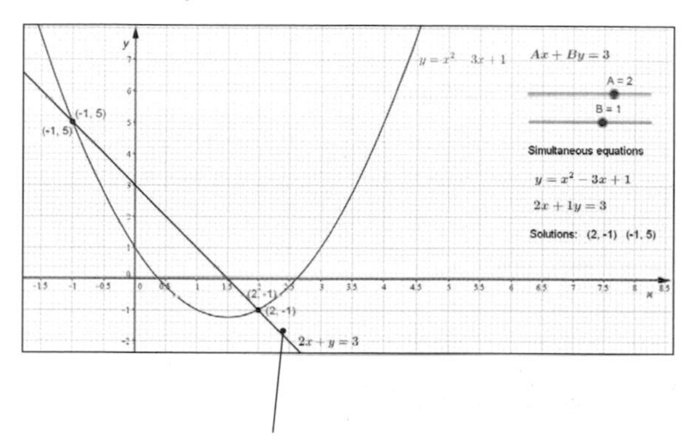

Interact with the maths you are learning using GeoGebra's easy-to-use tools

CASIO

Graphic calculator interactives

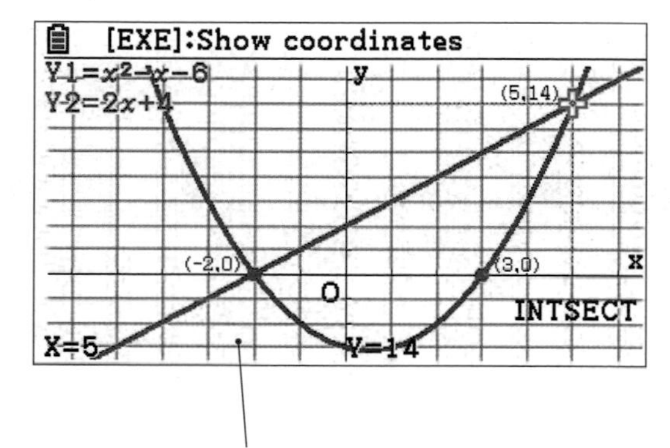

Explore the maths you are learning and gain confidence in using a graphic calculator

Calculator tutorials

Our helpful video tutorials will guide you through how to use your calculator in the exams. They cover both Casio's scientific and colour graphic calculators.

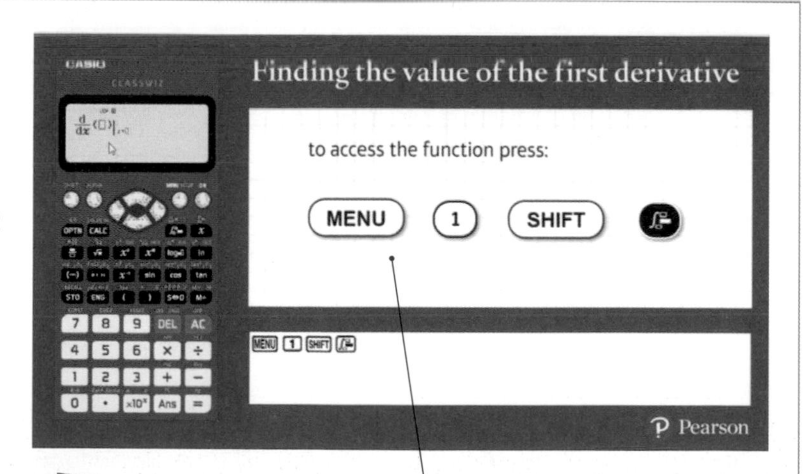

Step-by-step guide with audio instructions on exactly which buttons to press and what should appear on your calculator's screen

Online Work out each coefficient quickly using the ^nC_r and power functions on your calculator.

Published by Pearson Education Limited, 80 Strand, London WC2R 0RL.

www.pearsonschoolsandfecolleges.co.uk

Copies of official specifications for all Pearson qualifications may be found on the website: qualifications.pearson.com

Edited by Tech-Set Ltd, Gateshead
Typeset by Tech-Set Ltd, Gateshead

Cover illustration Marcus@kja-artists

First published 2017

20 19 18
10 9 8 7 6

British Library Cataloguing in Publication Data
A catalogue record for this book is available from the British Library

ISBN 978 1 292 18339 8

Printed and bound in Great Britain by Bell and Bain Ltd, Glasgow

Acknowledgements
The authors and publisher would like to thank the following individuals and organisations for permission to reproduce photographs:

(Key: b-bottom; c-centre; l-left; r-right; t-top)

123RF.com: David Acosta Allely 287, 338cr; Alamy Images: Utah Images 113, 226l, Xinhua 38, 85cr, ZUMA Press, Inc. 311, 338r; Fotolia.com: Kajano 137, 226cl, sborisov 173, 226r, Thaut Images 202, 226tr; Getty Images: Graiki 255, 338cl, Henglein and Steets 18, 85c, Jeff Schultz 230, 338l, mviamonte 1, 85l, Steve Dunwell 158, 226cr; Science Photo Library Ltd: CMS EXPERIMENT, CERN 59, 85; Shutterstock.com: vladimir salman 89, 226tl

Pearson has robust editorial processes, including answer and fact checks, to ensure the accuracy of the content in this publication, and every effort is made to ensure this publication is free of errors. We are, however, only human, and occasionally errors do occur. Pearson is not liable for any misunderstandings that arise as a result of errors in this publication, but it is our priority to ensure that the content is accurate. If you spot an error, please do contact us at resourcescorrections@pearson.com so we can make sure it is corrected.

A note from the publisher
In order to ensure that this resource offers high-quality support for the associated Pearson qualification, it has been through a review process by the awarding body. This process confirms that this resource fully covers the teaching and learning content of the specification or part of a specification at which it is aimed. It also confirms that it demonstrates an appropriate balance between the development of subject skills, knowledge and understanding, in addition to preparation for assessment.

Endorsement does not cover any guidance on assessment activities or processes (e.g. practice questions or advice on how to answer assessment questions), included in the resource nor does it prescribe any particular approach to the teaching or delivery of a related course.

While the publishers have made every attempt to ensure that advice on the qualification and its assessment is accurate, the official specification and associated assessment guidance materials are the only authoritative source of information and should always be referred to for definitive guidance.

Pearson examiners have not contributed to any sections in this resource relevant to examination papers for which they have responsibility.

Examiners will not use endorsed resources as a source of material for any assessment set by Pearson.

Endorsement of a resource does not mean that the resource is required to achieve this Pearson qualification, nor does it mean that it is the only suitable material available to support the qualification, and any resource lists produced by the awarding body shall include this and other appropriate resources.

Algebraic expressions

1

Objectives

After completing this chapter you should be able to:

- Multiply and divide integer powers → pages 2–3
- Expand a single term over brackets and collect like terms → pages 3–4
- Expand the product of two or three expressions → pages 4–6
- Factorise linear, quadratic and simple cubic expressions → pages 6–9
- Know and use the laws of indices → pages 9–11
- Simplify and use the rules of surds → pages 12–13
- Rationalise denominators → pages 13–16

Computer scientists use indices to describe very large numbers. A quantum computer with 1000 qubits (quantum bits) can consider 2^{1000} values simultaneously. This is greater than the number of particles in the observable universe.

Prior knowledge check

1 Simplify:

a $4m^2n + 5mn^2 - 2m^2n + mn^2 - 3mn^2$

b $3x^2 - 5x + 2 + 3x^2 - 7x - 12$

← GCSE Mathematics

2 Write as a single power of 2:

a $2^5 \times 2^3$ **b** $2^6 \div 2^2$

c $(2^3)^2$

← GCSE Mathematics

3 Expand:

a $3(x + 4)$ **b** $5(2 - 3x)$

c $6(2x - 5y)$

← GCSE Mathematics

4 Write down the highest common factor of:

a 24 and 16 **b** $6x$ and $8x^2$

c $4xy^2$ and $3xy$

← GCSE Mathematics

5 Simplify:

a $\frac{10x}{5}$ **b** $\frac{20x}{2}$ **c** $\frac{40x}{24}$

← GCSE Mathematics

1.1 Index laws

■ **You can use the laws of indices to simplify powers of the same base.**

- $a^m \times a^n = a^{m+n}$
- $a^m \div a^n = a^{m-n}$
- $(a^m)^n = a^{mn}$
- $(ab)^n = a^n b^n$

Notation

x^5 — x: This is the **base**. 5: This is the **index**, **power** or **exponent**.

Example 1

Simplify these expressions:

a $x^2 \times x^5$ **b** $2r^2 \times 3r^3$ **c** $\frac{b^7}{b^4}$ **d** $6x^5 \div 3x^3$ **e** $(a^3)^2 \times 2a^2$ **f** $(3x^2)^3 \div x^4$

a $x^2 \times x^5 = x^{2+5} = x^7$

Use the rule $a^m \times a^n = a^{m+n}$ to simplify the index.

b $2r^2 \times 3r^3 = 2 \times 3 \times r^2 \times r^3$
$= 6 \times r^{2+3} = 6r^5$

Rewrite the expression with the numbers together and the r terms together.
$2 \times 3 = 6$
$r^2 \times r^3 = r^{2+3}$

c $\frac{b^7}{b^4} = b^{7-4} = b^3$

Use the rule $a^m \div a^n = a^{m-n}$ to simplify the index.

d $6x^5 \div 3x^3 = \frac{6}{3} \times \frac{x^5}{x^3}$
$= 2 \times x^2 = 2x^2$

$x^5 \div x^3 = x^{5-3} = x^2$

e $(a^3)^2 \times 2a^2 = a^6 \times 2a^2$
$= 2 \times a^6 \times a^2 = 2a^8$

Use the rule $(a^m)^n = a^{mn}$ to simplify the index.

$a^6 \times a^2 = a^{6+2} = a^8$

f $\frac{(3x^2)^3}{x^4} = 3^3 \times \frac{(x^2)^3}{x^4}$
$= 27 \times \frac{x^6}{x^4} = 27x^2$

Use the rule $(ab)^n = a^n b^n$ to simplify the numerator.
$(x^2)^3 = x^{2 \times 3} = x^6$

$\frac{x^6}{x^4} = x^{6-4} = x^2$

Example 2

Expand these expressions and simplify if possible:

a $-3x(7x - 4)$ **b** $y^2(3 - 2y^3)$
c $4x(3x - 2x^2 + 5x^3)$ **d** $2x(5x + 3) - 5(2x + 3)$

Watch out A minus sign outside brackets changes the sign of every term inside the brackets.

a $-3x(7x - 4) = -21x^2 + 12x$

b $y^2(3 - 2y^3) = 3y^2 - 2y^5$

c $4x(3x - 2x^2 + 5x^3)$
$= 12x^2 - 8x^3 + 20x^4$

d $2x(5x + 3) - 5(2x + 3)$
$= 10x^2 + 6x - 10x - 15$
$= 10x^2 - 4x - 15$

$-3x \times 7x = -21x^{1+1} = -21x^2$
$-3x \times (-4) = +12x$

$y^2 \times (-2y^3) = -2y^{2+3} = -2y^5$

Remember a minus sign outside the brackets changes the signs within the brackets.

Simplify $6x - 10x$ to give $-4x$.

Example 3

Simplify these expressions:

a $\dfrac{x^7 + x^4}{x^3}$ b $\dfrac{3x^2 - 6x^5}{2x}$ c $\dfrac{20x^7 + 15x^3}{5x^2}$

a $\dfrac{x^7 + x^4}{x^3} = \dfrac{x^7}{x^3} + \dfrac{x^4}{x^3}$
$= x^{7-3} + x^{4-3} = x^4 + x$

b $\dfrac{3x^2 - 6x^5}{2x} = \dfrac{3x^2}{2x} - \dfrac{6x^5}{2x}$
$= \dfrac{3}{2}x^{2-1} - 3x^{5-1} = \dfrac{3x}{2} - 3x^4$

c $\dfrac{20x^7 + 15x^3}{5x^2} = \dfrac{20x^7}{5x^2} + \dfrac{15x^3}{5x^2}$
$= 4x^{7-2} + 3x^{3-2} = 4x^5 + 3x$

Divide each term of the numerator by x^3.

x^1 is the same as x.

Divide each term of the numerator by $2x$.

Simplify each fraction:
$\dfrac{3x^2}{2x} = \dfrac{3}{2} \times \dfrac{x^2}{x} = \dfrac{3}{2} \times x^{2-1}$
$-\dfrac{6x^5}{2x} = -\dfrac{6}{2} \times \dfrac{x^5}{x} = -3 \times x^{5-1}$

Divide each term of the numerator by $5x^2$.

Exercise 1A

1 Simplify these expressions:

a $x^3 \times x^4$ **b** $2x^3 \times 3x^2$ **c** $\dfrac{k^3}{k^2}$

d $\dfrac{4p^3}{2p}$ **e** $\dfrac{3x^3}{3x^2}$ **f** $(y^2)^5$

g $10x^5 \div 2x^3$ **h** $(p^3)^2 \div p^4$ **i** $(2a^3)^2 \div 2a^3$

j $8p^4 \div 4p^3$ **k** $2a^4 \times 3a^5$ **l** $\dfrac{21a^3b^7}{7ab^4}$

m $9x^2 \times 3(x^2)^3$ **n** $3x^3 \times 2x^2 \times 4x^6$ **o** $7a^4 \times (3a^4)^2$

p $(4y^3)^3 \div 2y^3$ **q** $2a^3 \div 3a^2 \times 6a^5$ **r** $3a^4 \times 2a^5 \times a^3$

2 Expand and simplify if possible:

a $9(x - 2)$ **b** $x(x + 9)$ **c** $-3y(4 - 3y)$

d $x(y + 5)$ **e** $-x(3x + 5)$ **f** $-5x(4x + 1)$

g $(4x + 5)x$ **h** $-3y(5 - 2y^2)$ **i** $-2x(5x - 4)$

j $(3x - 5)x^2$ **k** $3(x + 2) + (x - 7)$ **l** $5x - 6 - (3x - 2)$

m $4(c + 3d^2) - 3(2c + d^2)$ **n** $(r^2 + 3t^2 + 9) - (2r^2 + 3t^2 - 4)$

o $x(3x^2 - 2x + 5)$ **p** $7y^2(2 - 5y + 3y^2)$ **q** $-2y^2(5 - 7y + 3y^2)$

r $7(x - 2) + 3(x + 4) - 6(x - 2)$ **s** $5x - 3(4 - 2x) + 6$

t $3x^2 - x(3 - 4x) + 7$ **u** $4x(x + 3) - 2x(3x - 7)$ **v** $3x^2(2x + 1) - 5x^2(3x - 4)$

3 Simplify these fractions:

a $\dfrac{6x^4 + 10x^6}{2x}$ **b** $\dfrac{3x^5 - x^7}{x}$ **c** $\dfrac{2x^4 - 4x^2}{4x}$

d $\dfrac{8x^3 + 5x}{2x}$ **e** $\dfrac{7x^7 + 5x^2}{5x}$ **f** $\dfrac{9x^5 - 5x^3}{3x}$

1.2 Expanding brackets

To find the **product** of two expressions you **multiply** each term in one expression by each term in the other expression.

Multiplying each of the 2 terms in the first expression by each of the 3 terms in the second expression gives $2 \times 3 = 6$ terms.

$x \times$

$$(x + 5)(4x - 2y + 3) = x(4x - 2y + 3) + 5(4x - 2y + 3)$$
$$= 4x^2 - 2xy + 3x + 20x - 10y + 15$$
$$= 4x^2 - 2xy + 23x - 10y + 15$$

$5 \times$

Simplify your answer by collecting like terms.

Example 4

Expand these expressions and simplify if possible:

a $(x + 5)(x + 2)$ **b** $(x - 2y)(x^2 + 1)$ **c** $(x - y)^2$ **d** $(x + y)(3x - 2y - 4)$

a $(x + 5)(x + 2)$
$= x^2 + 2x + 5x + 10$
$= x^2 + 7x + 10$

Multiply x by $(x + 2)$ and then multiply 5 by $(x + 2)$.

Simplify your answer by collecting like terms.

b $(x - 2y)(x^2 + 1)$
$= x^3 + x - 2x^2y - 2y$

$-2y \times x^2 = -2x^2y$

There are no like terms to collect.

c $(x - y)^2$
$= (x - y)(x - y)$
$= x^2 - xy - xy + y^2$
$= x^2 - 2xy + y^2$

$(x - y)^2$ means $(x - y)$ multiplied by itself.

$-xy - xy = -2xy$

d $(x + y)(3x - 2y - 4)$
$= x(3x - 2y - 4) + y(3x - 2y - 4)$
$= 3x^2 - 2xy - 4x + 3xy - 2y^2 - 4y$
$= 3x^2 + xy - 4x - 2y^2 - 4y$

Multiply x by $(3x - 2y - 4)$ and then multiply y by $(3x - 2y - 4)$.

Example 5

Expand these expressions and simplify if possible:

a $x(2x + 3)(x - 7)$ **b** $x(5x - 3y)(2x - y + 4)$ **c** $(x - 4)(x + 3)(x + 1)$

a $x(2x + 3)(x - 7)$
$= (2x^2 + 3x)(x - 7)$
$= 2x^3 - 14x^2 + 3x^2 - 21x$
$= 2x^3 - 11x^2 - 21x$

Start by expanding one pair of brackets: $x(2x + 3) = 2x^2 + 3x$

You could also have expanded the second pair of brackets first: $(2x + 3)(x - 7) = 2x^2 - 11x - 21$ Then multiply by x.

b $x(5x - 3y)(2x - y + 4)$
$= (5x^2 - 3xy)(2x - y + 4)$
$= 5x^2(2x - y + 4) - 3xy(2x - y + 4)$
$= 10x^3 - 5x^2y + 20x^2 - 6x^2y + 3xy^2 - 12xy$
$= 10x^3 - 11x^2y + 20x^2 + 3xy^2 - 12xy$

Be careful with minus signs. You need to change every sign in the second pair of brackets when you multiply it out.

c $(x - 4)(x + 3)(x + 1)$
$= (x^2 - x - 12)(x + 1)$
$= x^2(x + 1) - x(x + 1) - 12(x + 1)$
$= x^3 + x^2 - x^2 - x - 12x - 12$
$= x^3 - 13x - 12$

Choose one pair of brackets to expand first, for example:
$(x - 4)(x + 3) = x^2 + 3x - 4x - 12$
$= x^2 - x - 12$

You multiplied together three linear terms, so the final answer contains an x^3 term.

Exercise 1B

1 Expand and simplify if possible:

a $(x + 4)(x + 7)$ **b** $(x - 3)(x + 2)$ **c** $(x - 2)^2$
d $(x - y)(2x + 3)$ **e** $(x + 3y)(4x - y)$ **f** $(2x - 4y)(3x + y)$
g $(2x - 3)(x - 4)$ **h** $(3x + 2y)^2$ **i** $(2x + 8y)(2x + 3)$
j $(x + 5)(2x + 3y - 5)$ **k** $(x - 1)(3x - 4y - 5)$ **l** $(x - 4y)(2x + y + 5)$
m $(x + 2y - 1)(x + 3)$ **n** $(2x + 2y + 3)(x + 6)$ **o** $(4 - y)(4y - x + 3)$
p $(4y + 5)(3x - y + 2)$ **q** $(5y - 2x + 3)(x - 4)$ **r** $(4y - x - 2)(5 - y)$

2 Expand and simplify if possible:

a $5(x + 1)(x - 4)$ **b** $7(x - 2)(2x + 5)$ **c** $3(x - 3)(x - 3)$

d $x(x - y)(x + y)$ **e** $x(2x + y)(3x + 4)$ **f** $y(x - 5)(x + 1)$

g $y(3x - 2y)(4x + 2)$ **h** $y(7 - x)(2x - 5)$ **i** $x(2x + y)(5x - 2)$

j $x(x + 2)(x + 3y - 4)$ **k** $y(2x + y - 1)(x + 5)$ **l** $y(3x + 2y - 3)(2x + 1)$

m $x(2x + 3)(x + y - 5)$ **n** $2x(3x - 1)(4x - y - 3)$ **o** $3x(x - 2y)(2x + 3y + 5)$

p $(x + 3)(x + 2)(x + 1)$ **q** $(x + 2)(x - 4)(x + 3)$ **r** $(x + 3)(x - 1)(x - 5)$

s $(x - 5)(x - 4)(x - 3)$ **t** $(2x + 1)(x - 2)(x + 1)$ **u** $(2x + 3)(3x - 1)(x + 2)$

v $(3x - 2)(2x + 1)(3x - 2)$ **w** $(x + y)(x - y)(x - 1)$ **x** $(2x - 3y)^3$

(P) **3** The diagram shows a rectangle with a square cut out.
The rectangle has length $3x - y + 4$ and width $x + 7$.
The square has length $x - 2$.
Find an expanded and simplified expression for the shaded area.

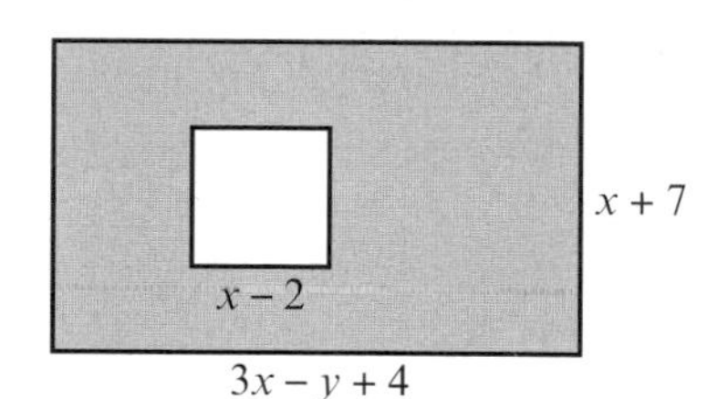

Problem-solving

Use the same strategy as you would use if the lengths were given as numbers:

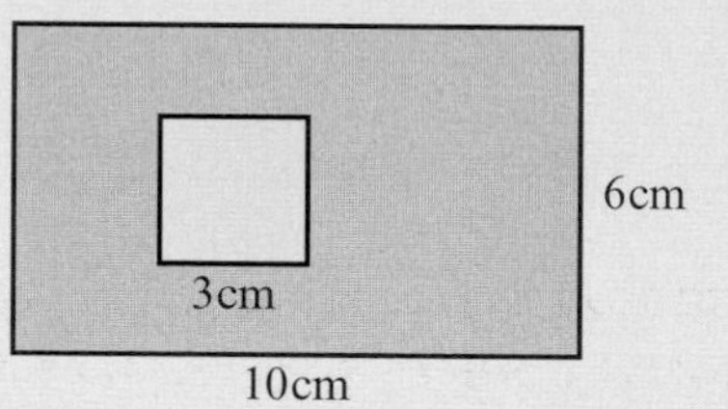

(P) **4** A cuboid has dimensions $x + 2$ cm, $2x - 1$ cm and $2x + 3$ cm.
Show that the volume of the cuboid is $4x^3 + 12x^2 + 5x - 6$ cm³.

(E/P) **5** Given that $(2x + 5y)(3x - y)(2x + y) = ax^3 + bx^2y + cxy^2 + dy^3$, where a, b, c and d are constants, find the values of a, b, c and d. **(2 marks)**

Challenge

Expand and simplify $(x + y)^4$.

Links You can use the binomial expansion to expand expressions like $(x + y)^4$ quickly. → Section 8.3

1.3 Factorising

You can write expressions as a **product of their factors**.

- **Factorising is the opposite of expanding brackets.**

Expanding brackets →

$$4x(2x + y) = 8x^2 + 4xy$$
$$(x + 5)^3 = x^3 + 15x^2 + 75x + 125$$
$$(x + 2y)(x - 5y) = x^2 - 3xy - 10y^2$$

← Factorising

Example 6

Factorise these expressions completely:

a $3x + 9$ **b** $x^2 - 5x$ **c** $8x^2 + 20x$ **d** $9x^2y + 15xy^2$ **e** $3x^2 - 9xy$

a $3x + 9 = 3(x + 3)$ — 3 is a common factor of $3x$ and 9.

b $x^2 - 5x = x(x - 5)$ — x is a common factor of x^2 and $-5x$.

c $8x^2 + 20x = 4x(2x + 5)$ — 4 and x are common factors of $8x^2$ and $20x$. So take $4x$ outside the brackets.

d $9x^2y + 15xy^2 = 3xy(3x + 5y)$ — 3, x and y are common factors of $9x^2y$ and $15xy^2$. So take $3xy$ outside the brackets.

e $3x^2 - 9xy = 3x(x - 3y)$ — x and $-3y$ have no common factors so this expression is completely factorised.

- **A quadratic expression has the form $ax^2 + bx + c$ where a, b and c are real numbers and $a \neq 0$.**

Notation Real numbers are all the positive and negative numbers, or zero, including fractions and surds.

To factorise a quadratic expression:

- Find two factors of ac that add up to b — For the expression $2x^2 + 5x - 3$, $ac = -6 = -1 \times 6$ and $-1 + 6 = 5 = b$.
- Rewrite the b term as a sum of these two factors — $2x^2 - x + 6x - 3$
- Factorise each pair of terms — $= x(2x - 1) + 3(2x - 1)$
- Take out the common factor — $= (x + 3)(2x - 1)$

- **$x^2 - y^2 = (x + y)(x - y)$**

Notation An expression in the form $x^2 - y^2$ is called the **difference** of two squares.

Example 7

Factorise:

a $x^2 - 5x - 6$ **b** $x^2 + 6x + 8$ **c** $6x^2 - 11x - 10$ **d** $x^2 - 25$ **e** $4x^2 - 9y^2$

a $x^2 - 5x - 6$

$ac = -6$ and $b = -5$

So $x^2 - 5x - 6 = x^2 + x - 6x - 6$

$= x(x + 1) - 6(x + 1)$

$= (x + 1)(x - 6)$

Here $a = 1$, $b = -5$ and $c = -6$.

① Work out the two factors of $ac = -6$ which add to give you $b = -5$. $-6 + 1 = -5$

② Rewrite the b term using these two factors.

③ Factorise first two terms and last two terms.

④ $x + 1$ is a factor of both terms, so take that outside the brackets. This is now completely factorised.

b $x^2 + 6x + 8$
$= x^2 + 2x + 4x + 8$ — $ac = 8$ and $2 + 4 = 6 = b$.
$= x(x + 2) + 4(x + 2)$ — Factorise.
$= (x + 2)(x + 4)$

c $6x^2 - 11x - 10$
$= 6x^2 - 15x + 4x - 10$ — $ac = -60$ and $4 - 15 = -11 = b$.
$= 3x(2x - 5) + 2(2x - 5)$ — Factorise.
$= (2x - 5)(3x + 2)$

d $x^2 - 25$ — This is the difference of two squares as the two terms are x^2 and 5^2.
$= x^2 - 5^2$
$= (x + 5)(x - 5)$ — The two x terms, $5x$ and $-5x$, cancel each other out.

e $4x^2 - 9y^2$ — This is the same as $(2x)^2 - (3y)^2$.
$= 2^2x^2 - 3^2y^2$
$= (2x + 3y)(2x - 3y)$

Example 8

Factorise completely:

a $x^3 - 2x^2$ **b** $x^3 - 25x$ **c** $x^3 + 3x^2 - 10x$

a $x^3 - 2x^2 = x^2(x - 2)$ — You can't factorise this any further.

b $x^3 - 25x = x(x^2 - 25)$ — x is a common factor of x^3 and $-25x$. So take x outside the brackets. $x^2 - 25$ is the difference of two squares.
$= x(x^2 - 5^2)$
$= x(x + 5)(x - 5)$

c $x^3 + 3x^2 - 10x = x(x^2 + 3x - 10)$ — Write the expression as a product of x and a quadratic factor.
$= x(x + 5)(x - 2)$ — Factorise the quadratic to get three linear factors.

Exercise 1C

1 Factorise these expressions completely:

a $4x + 8$ **b** $6x - 24$ **c** $20x + 15$
d $2x^2 + 4$ **e** $4x^2 + 20$ **f** $6x^2 - 18x$
g $x^2 - 7x$ **h** $2x^2 + 4x$ **i** $3x^2 - x$
j $6x^2 - 2x$ **k** $10y^2 - 5y$ **l** $35x^2 - 28x$
m $x^2 + 2x$ **n** $3y^2 + 2y$ **o** $4x^2 + 12x$
p $5y^2 - 20y$ **q** $9xy^2 + 12x^2y$ **r** $6ab - 2ab^2$
s $5x^2 - 25xy$ **t** $12x^2y + 8xy^2$ **u** $15y - 20yz^2$
v $12x^2 - 30$ **w** $xy^2 - x^2y$ **x** $12y^2 - 4yx$

2 Factorise:

a $x^2 + 4x$ **b** $2x^2 + 6x$ **c** $x^2 + 11x + 24$
d $x^2 + 8x + 12$ **e** $x^2 + 3x - 40$ **f** $x^2 - 8x + 12$
g $x^2 + 5x + 6$ **h** $x^2 - 2x - 24$ **i** $x^2 - 3x - 10$
j $x^2 + x - 20$ **k** $2x^2 + 5x + 2$ **l** $3x^2 + 10x - 8$
m $5x^2 - 16x + 3$ **n** $6x^2 - 8x - 8$
o $2x^2 + 7x - 15$ **p** $2x^4 + 14x^2 + 24$
q $x^2 - 4$ **r** $x^2 - 49$
s $4x^2 - 25$ **t** $9x^2 - 25y^2$ **u** $36x^2 - 4$
v $2x^2 - 50$ **w** $6x^2 - 10x + 4$ **x** $15x^2 + 42x - 9$

Hint For part **n**, take 2 out as a common factor first. For part **p**, let $y = x^2$.

3 Factorise completely:

a $x^3 + 2x$ **b** $x^3 - x^2 + x$ **c** $x^3 - 5x$
d $x^3 - 9x$ **e** $x^3 - x^2 - 12x$ **f** $x^3 + 11x^2 + 30x$
g $x^3 - 7x^2 + 6x$ **h** $x^3 - 64x$ **i** $2x^3 - 5x^2 - 3x$
j $2x^3 + 13x^2 + 15x$ **k** $x^3 - 4x$ **l** $3x^3 + 27x^2 + 60x$

(P) **4** Factorise completely $x^4 - y^4$. **(2 marks)**

Problem-solving
Watch out for terms that can be written as a function of a function: $x^4 = (x^2)^2$

(E) **5** Factorise completely $6x^3 + 7x^2 - 5x$. **(2 marks)**

Challenge
Write $4x^4 - 13x^2 + 9$ as the product of four linear factors.

1.4 Negative and fractional indices

Indices can be negative numbers or fractions.

$x^{\frac{1}{2}} \times x^{\frac{1}{2}} = x^{\frac{1}{2}+\frac{1}{2}} = x^1 = x,$

similarly $\underbrace{x^{\frac{1}{n}} \times x^{\frac{1}{n}} \times \ldots \times x^{\frac{1}{n}}}_{n \text{ terms}} = x^{\frac{1}{n}+\frac{1}{n}+\ldots+\frac{1}{n}} = x^1 = x$

Notation **Rational** numbers are those that can be written as $\frac{a}{b}$ where a and b are integers.

- **You can use the laws of indices with any rational power.**
  - $a^{\frac{1}{m}} = \sqrt[m]{a}$
  - $a^{\frac{n}{m}} = \sqrt[m]{a^n}$
  - $a^{-m} = \frac{1}{a^m}$
  - $a^0 = 1$

Notation $a^{\frac{1}{2}} = \sqrt{a}$ is the positive square root of a. For example $9^{\frac{1}{2}} = \sqrt{9} = 3$ but $9^{\frac{1}{2}} \neq -3$.

Example 9

Simplify:

a $\frac{x^3}{x^{-3}}$ **b** $x^{\frac{1}{2}} \times x^{\frac{3}{2}}$ **c** $(x^3)^{\frac{2}{3}}$ **d** $2x^{1.5} \div 4x^{-0.25}$ **e** $\sqrt[3]{125x^6}$ **f** $\frac{2x^2 - x}{x^5}$

a $\frac{x^3}{x^{-3}} = x^{3-(-3)} = x^6$

Use the rule $a^m \div a^n = a^{m-n}$.

b $x^{\frac{1}{2}} \times x^{\frac{3}{2}} = x^{\frac{1}{2}+\frac{3}{2}} = x^2$

This could also be written as $\sqrt{x}$.
Use the rule $a^m \times a^n = a^{m+n}$.

c $(x^3)^{\frac{2}{3}} = x^{3 \times \frac{2}{3}} = x^2$

Use the rule $(a^m)^n = a^{mn}$.

d $2x^{1.5} \div 4x^{-0.25} = \frac{1}{2}x^{1.5-(-0.25)} = \frac{1}{2}x^{1.75}$

Use the rule $a^m \div a^n = a^{m-n}$.
$1.5 - (-0.25) = 1.75$

e $\sqrt[3]{125x^6} = (125x^6)^{\frac{1}{3}}$

Using $a^{\frac{1}{m}} = \sqrt[m]{a}$.

$= (125)^{\frac{1}{3}}(x^6)^{\frac{1}{3}} = \sqrt[3]{125}(x^{6 \times \frac{1}{3}}) = 5x^2$

f $\frac{2x^2 - x}{x^5} = \frac{2x^2}{x^5} - \frac{x}{x^5}$

Divide each term of the numerator by x^5.

$= 2 \times x^{2-5} - x^{1-5} = 2x^{-3} - x^{-4}$

$= \frac{2}{x^3} - \frac{1}{x^4}$

Using $a^{-m} = \frac{1}{a^m}$

Example 10

Evaluate:

a $9^{\frac{1}{2}}$ **b** $64^{\frac{1}{3}}$ **c** $49^{\frac{3}{2}}$ **d** $25^{-\frac{3}{2}}$

a $9^{\frac{1}{2}} = \sqrt{9} = 3$

Using $a^{\frac{1}{m}} = \sqrt[m]{a}$. $9^{\frac{1}{2}} = \sqrt{9}$

b $64^{\frac{1}{3}} = \sqrt[3]{64} = 4$

This means the cube root of 64.

c $49^{\frac{3}{2}} = (\sqrt{49})^3$

Using $a^{\frac{n}{m}} = \sqrt[m]{a^n}$.
This means the square root of 49, cubed.

$= 7^3 = 343$

d $25^{-\frac{3}{2}} = \frac{1}{25^{\frac{3}{2}}} = \frac{1}{(\sqrt{25})^3}$

Using $a^{-m} = \frac{1}{a^m}$

$= \frac{1}{5^3} = \frac{1}{125}$

Online Use your calculator to enter negative and fractional powers.

Example 11

Given that $y = \frac{1}{16}x^2$ express each of the following in the form kx^n, where k and n are constants.

a $y^{\frac{1}{2}}$ **b** $4y^{-1}$

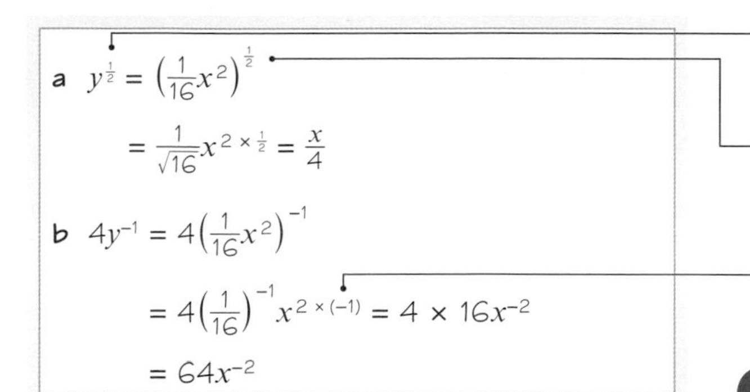

Substitute $y = \frac{1}{16}x^2$ into $y^{\frac{1}{2}}$.

$\left(\frac{1}{16}\right)^{\frac{1}{2}} = \frac{1}{\sqrt{16}}$ and $(x^2)^{\frac{1}{2}} = x^{2 \times \frac{1}{2}}$

$\left(\frac{1}{16}\right)^{-1} = 16$ and $x^{2 \times -1} = x^{-2}$

Problem-solving

Check that your answers are in the correct form. If k and n are constants they could be positive or negative, and they could be integers, fractions or surds.

Exercise 1D

1 Simplify:

a $x^3 \div x^{-2}$ **b** $x^5 \div x^7$ **c** $x^{\frac{3}{2}} \times x^{\frac{5}{2}}$

d $(x^2)^{\frac{3}{2}}$ **e** $(x^3)^{\frac{5}{3}}$ **f** $3x^{0.5} \times 4x^{-0.5}$

g $9x^{\frac{2}{3}} \div 3x^{\frac{1}{6}}$ **h** $5x^{\frac{7}{5}} \div x^{\frac{2}{5}}$ **i** $3x^4 \times 2x^{-5}$

j $\sqrt{x} \times \sqrt[3]{x}$ **k** $(\sqrt{x})^3 \times (\sqrt[3]{x})^4$ **l** $\frac{(\sqrt[3]{x})^2}{\sqrt{x}}$

2 Evaluate:

a $25^{\frac{1}{2}}$ **b** $81^{\frac{3}{2}}$ **c** $27^{\frac{1}{3}}$

d 4^{-2} **e** $9^{-\frac{1}{2}}$ **f** $(-5)^{-3}$

g $\left(\frac{3}{4}\right)^0$ **h** $1296^{\frac{3}{4}}$ **i** $\left(\frac{25}{16}\right)^{\frac{3}{2}}$

j $\left(\frac{27}{8}\right)^{\frac{2}{3}}$ **k** $\left(\frac{6}{5}\right)^{-1}$ **l** $\left(\frac{343}{512}\right)^{-\frac{2}{3}}$

3 Simplify:

a $(64x^{10})^{\frac{1}{2}}$ **b** $\frac{5x^3 - 2x^2}{x^5}$ **c** $(125x^{12})^{\frac{1}{3}}$ **d** $\frac{x + 4x^3}{x^3}$

e $\frac{2x + x^2}{x^4}$ **f** $\left(\frac{4}{9}x^4\right)^{\frac{3}{2}}$ **g** $\frac{9x^2 - 15x^5}{3x^3}$ **h** $\frac{5x + 3x^2}{15x^3}$

(E) **4** **a** Find the value of $81^{\frac{1}{4}}$. **(1 mark)**

b Simplify $x(2x^{-\frac{1}{3}})^4$. **(2 marks)**

(E) **5** Given that $y = \frac{1}{8}x^3$ express each of the following in the form kx^n, where k and n are constants.

a $y^{\frac{1}{3}}$ **(2 marks)**

b $\frac{1}{2}y^{-2}$ **(2 marks)**

1.5 Surds

If n is an integer that is **not** a square number, then any multiple of $\sqrt{n}$ is called a surd.
Examples of surds are $\sqrt{2}$, $\sqrt{19}$ and $5\sqrt{2}$.

Surds are examples of **irrational numbers**.
The decimal expansion of a surd is never-ending and never repeats, for example $\sqrt{2} = 1.414213562\ldots$

Notation Irrational numbers cannot be written in the form $\frac{a}{b}$ where a and b are integers.
Surds are examples of **irrational numbers**.

You can use surds to write exact answers to calculations.

- **You can manipulate surds using these rules:**
  - $\sqrt{ab} = \sqrt{a} \times \sqrt{b}$
  - $\sqrt{\frac{a}{b}} = \frac{\sqrt{a}}{\sqrt{b}}$

Example 12

Simplify:

a $\sqrt{12}$ **b** $\frac{\sqrt{20}}{2}$ **c** $5\sqrt{6} - 2\sqrt{24} + \sqrt{294}$

a $\sqrt{12} = \sqrt{(4 \times 3)}$

$= \sqrt{4} \times \sqrt{3} = 2\sqrt{3}$

Look for a factor of 12 that is a square number. Use the rule $\sqrt{ab} = \sqrt{a} \times \sqrt{b}$. $\sqrt{4} = 2$

b $\frac{\sqrt{20}}{2} = \frac{\sqrt{4} \times \sqrt{5}}{2}$

$\sqrt{20} = \sqrt{4} \times \sqrt{5}$

$\sqrt{4} = 2$

$= \frac{2 \times \sqrt{5}}{2} = \sqrt{5}$

Cancel by 2.

c $5\sqrt{6} - 2\sqrt{24} + \sqrt{294}$

$= 5\sqrt{6} - 2\sqrt{6}\sqrt{4} + \sqrt{6} \times \sqrt{49}$

$\sqrt{6}$ is a common factor.

$= \sqrt{6}(5 - 2\sqrt{4} + \sqrt{49})$

Work out the square roots $\sqrt{4}$ and $\sqrt{49}$.

$= \sqrt{6}(5 - 2 \times 2 + 7)$

$5 - 4 + 7 = 8$

$= \sqrt{6}(8)$

$= 8\sqrt{6}$

Example 13

Expand and simplify if possible:

a $\sqrt{2}(5 - \sqrt{3})$ **b** $(2 - \sqrt{3})(5 + \sqrt{3})$

a $\sqrt{2}(5 - \sqrt{3})$
$= 5\sqrt{2} - \sqrt{2}\sqrt{3}$
$= 5\sqrt{2} - \sqrt{6}$

b $(2 - \sqrt{3})(5 + \sqrt{3})$
$= 2(5 + \sqrt{3}) - \sqrt{3}(5 + \sqrt{3})$
$= 10 + 2\sqrt{3} - 5\sqrt{3} - \sqrt{9}$
$= 7 - 3\sqrt{3}$

$\sqrt{2} \times 5 - \sqrt{2} \times \sqrt{3}$

Using $\sqrt{a} \times \sqrt{b} = \sqrt{ab}$

Expand the brackets completely before you simplify.

Collect like terms: $2\sqrt{3} - 5\sqrt{3} = -3\sqrt{3}$

Simplify any roots if possible: $\sqrt{9} = 3$

Exercise 1E

1 Do not use your calculator for this exercise. Simplify:

a $\sqrt{28}$ **b** $\sqrt{72}$ **c** $\sqrt{50}$
d $\sqrt{32}$ **e** $\sqrt{90}$ **f** $\frac{\sqrt{12}}{2}$
g $\frac{\sqrt{27}}{3}$ **h** $\sqrt{20} + \sqrt{80}$ **i** $\sqrt{200} + \sqrt{18} - \sqrt{72}$
j $\sqrt{175} + \sqrt{63} + 2\sqrt{28}$ **k** $\sqrt{28} - 2\sqrt{63} + \sqrt{7}$ **l** $\sqrt{80} - 2\sqrt{20} + 3\sqrt{45}$
m $3\sqrt{80} - 2\sqrt{20} + 5\sqrt{45}$ **n** $\frac{\sqrt{44}}{\sqrt{11}}$ **o** $\sqrt{12} + 3\sqrt{48} + \sqrt{75}$

2 Expand and simplify if possible:

a $\sqrt{3}(2 + \sqrt{3})$ **b** $\sqrt{5}(3 - \sqrt{3})$ **c** $\sqrt{2}(4 - \sqrt{5})$
d $(2 - \sqrt{2})(3 + \sqrt{5})$ **e** $(2 - \sqrt{3})(3 - \sqrt{7})$ **f** $(4 + \sqrt{5})(2 + \sqrt{5})$
g $(5 - \sqrt{3})(1 - \sqrt{3})$ **h** $(4 + \sqrt{3})(2 - \sqrt{3})$ **i** $(7 - \sqrt{11})(2 + \sqrt{11})$

E **3** Simplify $\sqrt{75} - \sqrt{12}$ giving your answer in the form $a\sqrt{3}$, where a is an integer. **(2 marks)**

1.6 Rationalising denominators

If a fraction has a surd in the denominator, it is sometimes useful to **rearrange** it so that the denominator is a **rational** number. This is called rationalising the denominator.

- **The rules to rationalise denominators are:**
  - **For fractions in the form $\frac{1}{\sqrt{a}}$, multiply the numerator and denominator by $\sqrt{a}$.**
  - **For fractions in the form $\frac{1}{a + \sqrt{b}}$, multiply the numerator and denominator by $a - \sqrt{b}$.**
  - **For fractions in the form $\frac{1}{a - \sqrt{b}}$, multiply the numerator and denominator by $a + \sqrt{b}$.**

Example 14

Rationalise the denominator of:

a $\frac{1}{\sqrt{3}}$ **b** $\frac{1}{3+\sqrt{2}}$ **c** $\frac{\sqrt{5}+\sqrt{2}}{\sqrt{5}-\sqrt{2}}$ **d** $\frac{1}{(1-\sqrt{3})^2}$

a $\frac{1}{\sqrt{3}} = \frac{1 \times \sqrt{3}}{\sqrt{3} \times \sqrt{3}}$ — Multiply the numerator and denominator by $\sqrt{3}$.

$= \frac{\sqrt{3}}{3}$ — $\sqrt{3} \times \sqrt{3} = (\sqrt{3})^2 = 3$

b $\frac{1}{3+\sqrt{2}} = \frac{1 \times (3-\sqrt{2})}{(3+\sqrt{2})(3-\sqrt{2})}$ — Multiply numerator and denominator by $(3-\sqrt{2})$.

$= \frac{3-\sqrt{2}}{9-3\sqrt{2}+3\sqrt{2}-2}$ — $\sqrt{2} \times \sqrt{2} = 2$

$= \frac{3-\sqrt{2}}{7}$ — $9-2=7$, $-3\sqrt{2}+3\sqrt{2}=0$

c $\frac{\sqrt{5}+\sqrt{2}}{\sqrt{5}-\sqrt{2}} = \frac{(\sqrt{5}+\sqrt{2})(\sqrt{5}+\sqrt{2})}{(\sqrt{5}-\sqrt{2})(\sqrt{5}+\sqrt{2})}$ — Multiply numerator and denominator by $\sqrt{5}+\sqrt{2}$.

$= \frac{5+\sqrt{5}\sqrt{2}+\sqrt{2}\sqrt{5}+2}{5-2}$ — $-\sqrt{2}\sqrt{5}$ and $\sqrt{5}\sqrt{2}$ cancel each other out.

$= \frac{7+2\sqrt{10}}{3}$ — $\sqrt{5}\sqrt{2} = \sqrt{10}$

d $\frac{1}{(1-\sqrt{3})^2} = \frac{1}{(1-\sqrt{3})(1-\sqrt{3})}$ — Expand the brackets.

$= \frac{1}{1-\sqrt{3}-\sqrt{3}+\sqrt{9}}$ — Simplify and collect like terms. $\sqrt{9} = 3$

$= \frac{1}{4-2\sqrt{3}}$

$= \frac{1 \times (4+2\sqrt{3})}{(4-2\sqrt{3})(4+2\sqrt{3})}$ — Multiply the numerator and denominator by $4+2\sqrt{3}$.

$= \frac{4+2\sqrt{3}}{16+8\sqrt{3}-8\sqrt{3}-12}$ — $\sqrt{3} \times \sqrt{3} = 3$

$= \frac{4+2\sqrt{3}}{4} = \frac{2+\sqrt{3}}{2}$ — $16-12=4$, $8\sqrt{3}-8\sqrt{3}=0$

Exercise 1F

1 Simplify:

a $\frac{1}{\sqrt{5}}$ b $\frac{1}{\sqrt{11}}$ c $\frac{1}{\sqrt{2}}$ d $\frac{\sqrt{3}}{\sqrt{15}}$

e $\frac{\sqrt{12}}{\sqrt{48}}$ f $\frac{\sqrt{5}}{\sqrt{80}}$ g $\frac{\sqrt{12}}{\sqrt{156}}$ h $\frac{\sqrt{7}}{\sqrt{63}}$

2 Rationalise the denominators and simplify:

a $\frac{1}{1+\sqrt{3}}$ b $\frac{1}{2+\sqrt{5}}$ c $\frac{1}{3-\sqrt{7}}$ d $\frac{4}{3-\sqrt{5}}$ e $\frac{1}{\sqrt{5}-\sqrt{3}}$

f $\frac{3-\sqrt{2}}{4-\sqrt{5}}$ g $\frac{5}{2+\sqrt{5}}$ h $\frac{5\sqrt{2}}{\sqrt{8}-\sqrt{7}}$ i $\frac{11}{3+\sqrt{11}}$ j $\frac{\sqrt{3}-\sqrt{7}}{\sqrt{3}+\sqrt{7}}$

k $\frac{\sqrt{17}-\sqrt{11}}{\sqrt{17}+\sqrt{11}}$ l $\frac{\sqrt{41}+\sqrt{29}}{\sqrt{41}-\sqrt{29}}$ m $\frac{\sqrt{2}-\sqrt{3}}{\sqrt{3}-\sqrt{2}}$

3 Rationalise the denominators and simplify:

a $\frac{1}{(3-\sqrt{2})^2}$ b $\frac{1}{(2+\sqrt{5})^2}$ c $\frac{4}{(3-\sqrt{2})^2}$

d $\frac{3}{(5+\sqrt{2})^2}$ e $\frac{1}{(5+\sqrt{2})(3-\sqrt{2})}$ f $\frac{2}{(5-\sqrt{3})(2+\sqrt{3})}$

4 Simplify $\frac{3-2\sqrt{5}}{\sqrt{5}-1}$ giving your answer in the form $p + q\sqrt{5}$, where p and q are rational numbers. **(4 marks)**

Problem-solving

You can check that your answer is in the correct form by writing down the values of p and q and checking that they are rational numbers.

Mixed exercise 1

1 Simplify:

a $y^3 \times y^5$ b $3x^2 \times 2x^5$ c $(4x^2)^3 \div 2x^5$ d $4b^2 \times 3b^3 \times b^4$

2 Expand and simplify if possible:

a $(x + 3)(x - 5)$ b $(2x - 7)(3x + 1)$ c $(2x + 5)(3x - y + 2)$

3 Expand and simplify if possible:

a $x(x + 4)(x - 1)$ b $(x + 2)(x - 3)(x + 7)$ c $(2x + 3)(x - 2)(3x - 1)$

4 Expand the brackets:

a $3(5y + 4)$ b $5x^2(3 - 5x + 2x^2)$ c $5x(2x + 3) - 2x(1 - 3x)$ d $3x^2(1 + 3x) - 2x(3x - 2)$

5 Factorise these expressions completely:

a $3x^2 + 4x$ **b** $4y^2 + 10y$ **c** $x^2 + xy + xy^2$ **d** $8xy^2 + 10x^2y$

6 Factorise:

a $x^2 + 3x + 2$ **b** $3x^2 + 6x$ **c** $x^2 - 2x - 35$ **d** $2x^2 - x - 3$

e $5x^2 - 13x - 6$ **f** $6 - 5x - x^2$

7 Factorise:

a $2x^3 + 6x$ **b** $x^3 - 36x$ **c** $2x^3 + 7x^2 - 15x$

8 Simplify:

a $9x^3 \div 3x^{-3}$ **b** $(4^{\frac{3}{2}})^{\frac{1}{3}}$ **c** $3x^{-2} \times 2x^4$ **d** $3x^{\frac{1}{3}} \div 6x^{\frac{2}{3}}$

9 Evaluate:

a $\left(\frac{8}{27}\right)^{\frac{2}{3}}$ **b** $\left(\frac{225}{289}\right)^{\frac{3}{2}}$

10 Simplify:

a $\frac{3}{\sqrt{63}}$ **b** $\sqrt{20} + 2\sqrt{45} - \sqrt{80}$

11 a Find the value of $35x^2 + 2x - 48$ when $x = 25$.

b By factorising the expression, show that your answer to part **a** can be written as the product of two prime factors.

12 Expand and simplify if possible:

a $\sqrt{2}(3 + \sqrt{5})$ **b** $(2 - \sqrt{5})(5 + \sqrt{3})$ **c** $(6 - \sqrt{2})(4 - \sqrt{7})$

13 Rationalise the denominator and simplify:

a $\frac{1}{\sqrt{3}}$ **b** $\frac{1}{\sqrt{2} - 1}$ **c** $\frac{3}{\sqrt{3} - 2}$ **d** $\frac{\sqrt{23} - \sqrt{37}}{\sqrt{23} + \sqrt{37}}$ **e** $\frac{1}{(2 + \sqrt{3})^2}$ **f** $\frac{1}{(4 - \sqrt{7})^2}$

14 a Given that $x^3 - x^2 - 17x - 15 = (x + 3)(x^2 + bx + c)$, where b and c are constants, work out the values of b and c.

b Hence, fully factorise $x^3 - x^2 - 17x - 15$.

(E) **15** Given that $y = \frac{1}{64}x^3$ express each of the following in the form kx^n, where k and n are constants.

a $y^{\frac{1}{3}}$ **(1 mark)**

b $4y^{-1}$ **(1 mark)**

(E/P) **16** Show that $\frac{5}{\sqrt{75} - \sqrt{50}}$ can be written in the form $\sqrt{a} + \sqrt{b}$, where a and b are integers. **(5 marks)**

(E) **17** Expand and simplify $(\sqrt{11} - 5)(5 - \sqrt{11})$. **(2 marks)**

(E) **18** Factorise completely $x - 64x^3$. **(3 marks)**

(E/P) **19** Express $27^{2x + 1}$ in the form 3^y, stating y in terms of x. **(2 marks)**

(E/P) **20** Solve the equation $8 + x\sqrt{12} = \frac{8x}{\sqrt{3}}$

Give your answer in the form $a\sqrt{b}$ where a and b are integers. **(4 marks)**

(P) **21** A rectangle has a length of $(1 + \sqrt{3})$ cm and area of $\sqrt{12}$ cm^2.
Calculate the width of the rectangle in cm.
Express your answer in the form $a + b\sqrt{3}$, where a and b are integers to be found.

(E) **22** Show that $\frac{(2 - \sqrt{x})^2}{\sqrt{x}}$ can be written as $4x^{-\frac{1}{2}} - 4 + x^{\frac{1}{2}}$. **(2 marks)**

(E/P) **23** Given that $243\sqrt{3} = 3^a$, find the value of a. **(3 marks)**

(E/P) **24** Given that $\frac{4x^3 + x^{\frac{5}{2}}}{\sqrt{x}}$ can be written in the form $4x^a + x^b$, write down the value of a and the value of b. **(2 marks)**

Challenge

a Simplify $(\sqrt{a} + \sqrt{b})(\sqrt{a} - \sqrt{b})$.

b Hence show that $\frac{1}{\sqrt{1} + \sqrt{2}} + \frac{1}{\sqrt{2} + \sqrt{3}} + \frac{1}{\sqrt{3} + \sqrt{4}} + \ldots + \frac{1}{\sqrt{24} + \sqrt{25}} = 4$

Summary of key points

1 You can use the laws of indices to simplify powers of the **same base**.

- $a^m \times a^n = a^{m+n}$
- $a^m \div a^n = a^{m-n}$
- $(a^m)^n = a^{mn}$
- $(ab)^n = a^n b^n$

2 Factorising is the opposite of expanding brackets.

3 A quadratic expression has the form $ax^2 + bx + c$ where a, b and c are real numbers and $a \neq 0$.

4 $x^2 - y^2 = (x + y)(x - y)$

5 You can use the laws of indices with any rational power.

- $a^{\frac{1}{m}} = \sqrt[m]{a}$
- $a^{\frac{n}{m}} = \sqrt[m]{a^n}$
- $a^{-m} = \frac{1}{a^m}$
- $a^0 = 1$

6 You can manipulate surds using these rules:

- $\sqrt{ab} = \sqrt{a} \times \sqrt{b}$
- $\sqrt{\frac{a}{b}} = \frac{\sqrt{a}}{\sqrt{b}}$

7 The rules to rationalise denominators are:

- Fractions in the form $\frac{1}{\sqrt{a}}$, multiply the numerator and denominator by $\sqrt{a}$.
- Fractions in the form $\frac{1}{a + \sqrt{b}}$, multiply the numerator and denominator by $a - \sqrt{b}$.
- Fractions in the form $\frac{1}{a - \sqrt{b}}$, multiply the numerator and denominator by $a + \sqrt{b}$.

2 Quadratics

Objectives

After completing this chapter you should be able to:

- Solve quadratic equations using factorisation, the quadratic formula and completing the square → pages 19 – 24
- Read and use $\mathrm{f}(x)$ notation when working with functions → pages 25 – 27
- Sketch the graph and find the turning point of a quadratic function → pages 27 – 30
- Find and interpret the discriminant of a quadratic expression → pages 30 – 32
- Use and apply models that involve quadratic functions → pages 32 – 35

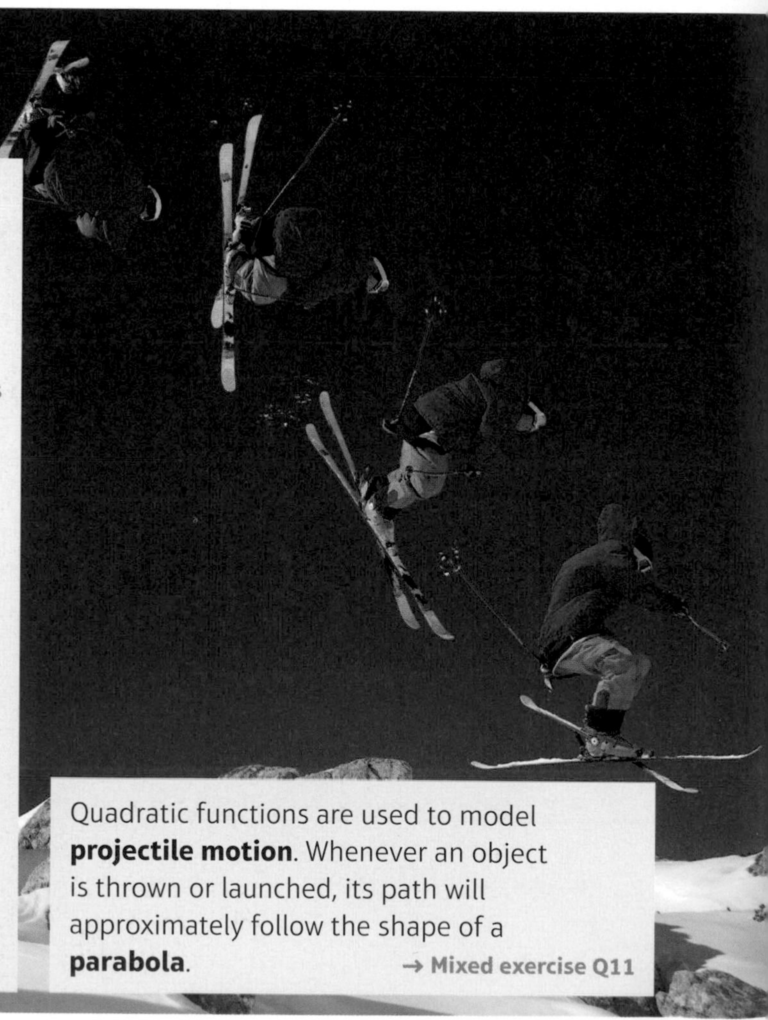

Prior knowledge check

1 Solve the following equations:

a $3x + 6 = x - 4$

b $5(x + 3) = 6(2x - 1)$

c $4x^2 = 100$

d $(x - 8)^2 = 64$ ← GCSE Mathematics

2 Factorise the following expressions:

a $x^2 + 8x + 15$ **b** $x^2 + 3x - 10$

c $3x^2 - 14x - 5$ **d** $x^2 - 400$

← Section 1.3

3 Sketch the graphs of the following equations, labelling the points where each graph crosses the axes:

a $y = 3x - 6$ **b** $y = 10 - 2x$

c $x + 2y = 18$ **d** $y = x^2$

← GCSE Mathematics

4 Solve the following inequalities:

a $x + 8 < 11$ **b** $2x - 5 \geqslant 13$

c $4x - 7 \leqslant 2(x - 1)$ **d** $4 - x < 11$

← GCSE Mathematics

Quadratic functions are used to model **projectile motion**. Whenever an object is thrown or launched, its path will approximately follow the shape of a **parabola**. → Mixed exercise Q11

2.1 Solving quadratic equations

A quadratic equation can be written in the form $ax^2 + bx + c = 0$, where a, b and c are real constants, and $a \neq 0$. Quadratic equations can have one, two, or no real solutions.

- **To solve a quadratic equation by factorising:**
  - **Write the equation in the form $ax^2 + bx + c = 0$**
  - **Factorise the left-hand side**
  - **Set each factor equal to zero and solve to find the value(s) of x**

Notation The solutions to an equation are sometimes called the **roots** of the equation.

Example 1

Solve the following equations:

a $x^2 - 2x - 15 = 0$ **b** $x^2 = 9x$

c $6x^2 + 13x - 5 = 0$ **d** $x^2 - 5x + 18 = 2 + 3x$

a $x^2 - 2x - 15 = 0$

$(x + 3)(x - 5) = 0$ — Factorise the quadratic. ← Section 1.3

Then either $x + 3 = 0 \Rightarrow x = -3$ — If the product of the factors is zero, one of the factors must be zero.

or $x - 5 = 0 \Rightarrow x = 5$

Notation The symbol ⇒ means 'implies that'. This statement says 'If $x + 3 = 0$, then $x = -3$'.

So $x = -3$ and $x = 5$ are the two solutions of the equation. — A quadratic equation with two distinct factors has two distinct solutions.

Watch out The signs of the solutions are **opposite** to the signs of the constant terms in each factor.

b $x^2 = 9x$

$x^2 - 9x = 0$ — Be careful not to divide both sides by x, since x may have the value 0. Instead, rearrange into the form $ax^2 + bx + c = 0$.

$x(x - 9) = 0$ — Factorise.

Then either $x = 0$

or $x - 9 = 0 \Rightarrow x = 9$

The solutions are $x = 0$ and $x = 9$.

c $6x^2 + 13x - 5 = 0$

$(3x - 1)(2x + 5) = 0$ — Factorise.

Then either $3x - 1 = 0 \Rightarrow x = \frac{1}{3}$ — Solutions to quadratic equations do not have to be integers. The quadratic equation $(px + q)(rx + s) = 0$ will have solutions $x = -\frac{q}{p}$ and $x = -\frac{s}{r}$.

or $2x + 5 = 0 \Rightarrow x = -\frac{5}{2}$

The solutions are $x = \frac{1}{3}$ and $x = -\frac{5}{2}$

d $x^2 - 5x + 18 = 2 + 3x$

$x^2 - 8x + 16 = 0$ — Rearrange into the form $ax^2 + bx + c = 0$.

$(x - 4)(x - 4) = 0$ — Factorise.

Then either $x - 4 = 0 \Rightarrow x = 4$

or $x - 4 = 0 \Rightarrow x = 4$

$\Rightarrow x = 4$

Notation When a quadratic equation has exactly one root it is called a **repeated root**. You can also say that the equation has two equal roots.

In some cases it may be more straightforward to solve a quadratic equation without factorising.

Example 2

Solve the following equations

a $(2x - 3)^2 = 25$ **b** $(x - 3)^2 = 7$

Notation The symbol ± lets you write two statements in one line of working. You say 'plus or minus'.

a $(2x - 3)^2 = 25$

$2x - 3 = \pm 5$ — Take the square root of both sides. Remember $5^2 = (-5)^2 = 25$.

$2x = 3 \pm 5$ — Add 3 to both sides.

Then either $2x = 3 + 5 \Rightarrow x = 4$

or $2x = 3 - 5 \Rightarrow x = -1$

The solutions are $x = 4$ and $x = -1$

b $(x - 3)^2 = 7$ — Take square roots of both sides.

$x - 3 = \pm\sqrt{7}$

$x = 3 \pm \sqrt{7}$

The solutions are $x = 3 + \sqrt{7}$ and $x = 3 - \sqrt{7}$ — You can leave your answer in surd form.

Exercise 2A

1 Solve the following equations using factorisation:

a $x^2 + 3x + 2 = 0$ **b** $x^2 + 5x + 4 = 0$ **c** $x^2 + 7x + 10 = 0$ **d** $x^2 - x - 6 = 0$

e $x^2 - 8x + 15 = 0$ **f** $x^2 - 9x + 20 = 0$ **g** $x^2 - 5x - 6 = 0$ **h** $x^2 - 4x - 12 = 0$

2 Solve the following equations using factorisation:

a $x^2 = 4x$ **b** $x^2 = 25x$ **c** $3x^2 = 6x$ **d** $5x^2 = 30x$

e $2x^2 + 7x + 3 = 0$ **f** $6x^2 - 7x - 3 = 0$ **g** $6x^2 - 5x - 6 = 0$ **h** $4x^2 - 16x + 15 = 0$

3 Solve the following equations:

a $3x^2 + 5x = 2$ **b** $(2x - 3)^2 = 9$ **c** $(x - 7)^2 = 36$ **d** $2x^2 = 8$ **e** $3x^2 = 5$

f $(x - 3)^2 = 13$ **g** $(3x - 1)^2 = 11$ **h** $5x^2 - 10x^2 = -7 + x + x^2$

i $6x^2 - 7 = 11x$ **j** $4x^2 + 17x = 6x - 2x^2$

(P) **4** This shape has an area of $44\,m^2$. Find the value of x.

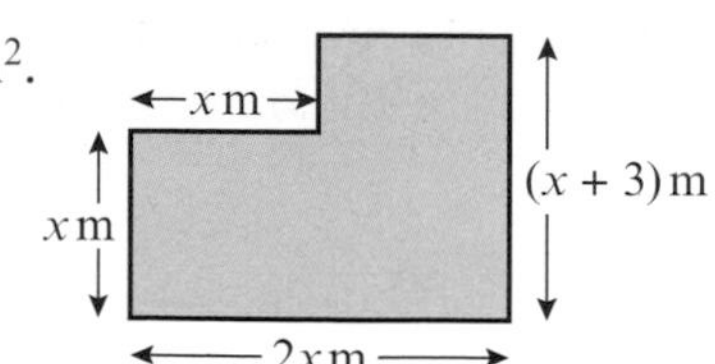

Problem-solving

Divide the shape into two sections:

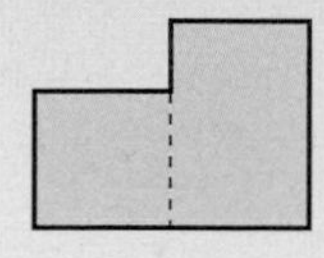

(P) **5** Solve the equation $5x + 3 = \sqrt{3x + 7}$.

Some equations cannot be easily factorised. You can also solve quadratic equations using the **quadratic formula**.

- **The solutions of the equation $ax^2 + bx + c = 0$ are given by the formula:**

$$x = \frac{-b \pm \sqrt{b^2 - 4ac}}{2a}$$

Watch out You need to rearrange the equation into the form $ax^2 + bx + c = 0$ before reading off the coefficients.

Notation In $ax^2 + bx + c = 0$, the constants a, b and c are called **coefficients**.

Example 3

Solve $3x^2 - 7x - 1 = 0$ by using the formula.

$$x = \frac{-(-7) \pm \sqrt{(-7)^2 - 4(3)(-1)}}{2 \times 3}$$

$a = 3$, $b = -7$ and $c = -1$.

Put brackets around any negative values.

$$x = \frac{7 \pm \sqrt{49 + 12}}{6}$$

$-4 \times 3 \times (-1) = +12$

$$x = \frac{7 \pm \sqrt{61}}{6}$$

Then $x = \frac{7 + \sqrt{61}}{6}$ or $x = \frac{7 - \sqrt{61}}{6}$

Or $x = 2.47$ (3 s.f.) or $x = -0.135$ (3 s.f.)

Exercise 2B

1 Solve the following equations using the quadratic formula.
Give your answers exactly, leaving them in surd form where necessary.

a $x^2 + 3x + 1 = 0$ **b** $x^2 - 3x - 2 = 0$ **c** $x^2 + 6x + 6 = 0$ **d** $x^2 - 5x - 2 = 0$
e $3x^2 + 10x - 2 = 0$ **f** $4x^2 - 4x - 1 = 0$ **g** $4x^2 - 7x = 2$ **h** $11x^2 + 2x - 7 = 0$

2 Solve the following equations using the quadratic formula.
Give your answers to three significant figures.

a $x^2 + 4x + 2 = 0$ **b** $x^2 - 8x + 1 = 0$ **c** $x^2 + 11x - 9 = 0$ **d** $x^2 - 7x - 17 = 0$
e $5x^2 + 9x - 1 = 0$ **f** $2x^2 - 3x - 18 = 0$ **g** $3x^2 + 8 = 16x$ **h** $2x^2 + 11x = 5x^2 - 18$

3 For each of the equations below, choose a suitable method and find all of the solutions.
Where necessary, give your answers to three significant figures.

a $x^2 + 8x + 12 = 0$ **b** $x^2 + 9x - 11 = 0$
c $x^2 - 9x - 1 = 0$ **d** $2x^2 + 5x + 2 = 0$
e $(2x + 8)^2 = 100$ **f** $6x^2 + 6 = 12x$
g $2x^2 - 11 = 7x$ **h** $x = \sqrt{8x - 15}$

Hint You can use any method you are confident with to solve these equations.

(P) **4** This trapezium has an area of $50\,\text{m}^2$.
Show that the height of the trapezium is equal to $5(\sqrt{5} - 1)\,\text{m}$.

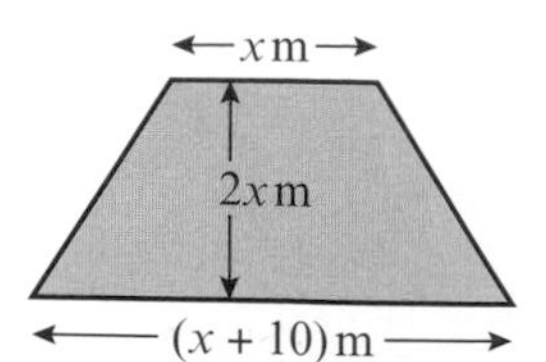

Problem-solving

Height must be positive. You will have to discard the negative solution of your quadratic equation.

Challenge

Given that x is positive, solve the equation
$$\frac{1}{x} + \frac{1}{x+2} = \frac{28}{195}$$

Hint Write the equation in the form $ax^2 + bx + c = 0$ before using the quadratic formula or factorising.

2.2 Completing the square

It is frequently useful to rewrite quadratic expressions by **completing the square:**

- $\boldsymbol{x^2 + bx = \left(x + \frac{b}{2}\right)^2 - \left(\frac{b}{2}\right)^2}$

You can draw a diagram of this process when x and b are positive:

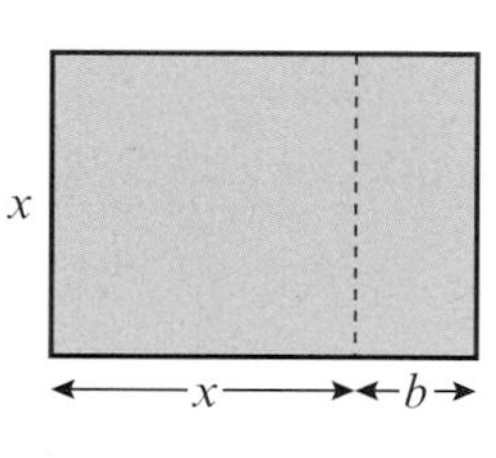

The original rectangle has been rearranged into the shape of a square with a smaller square missing. The two areas shaded blue are the same.

$$x^2 + bx \quad = \quad \left(x + \frac{b}{2}\right)^2 - \left(\frac{b}{2}\right)^2$$

Example 4

Complete the square for the expressions:

a $x^2 + 8x$ **b** $x^2 - 3x$ **c** $2x^2 - 12x$

Notation A quadratic expression in the form $p(x + q)^2 + r$ where p, q and r are real constants is in **completed square form**.

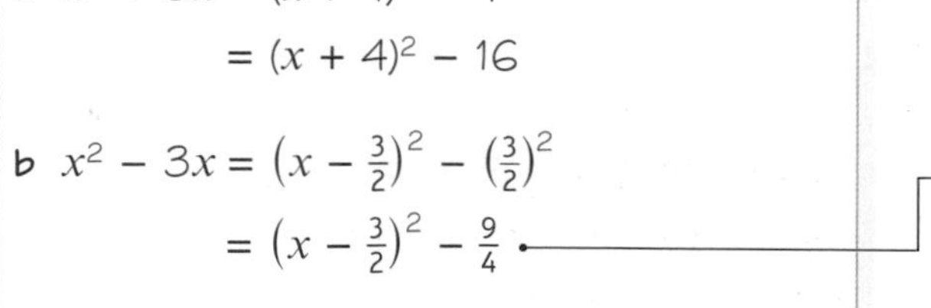

a $x^2 + 8x = (x + 4)^2 - 4^2$
$= (x + 4)^2 - 16$

Begin by halving the coefficient of x. Using the rule given above, $b = 8$ so $\frac{b}{2} = 4$.

b $x^2 - 3x = \left(x - \frac{3}{2}\right)^2 - \left(\frac{3}{2}\right)^2$
$= \left(x - \frac{3}{2}\right)^2 - \frac{9}{4}$

Be careful if $\frac{b}{2}$ is a fraction. Here $\left(\frac{3}{2}\right)^2 = \frac{3^2}{2^2} = \frac{9}{4}$.

c $2x^2 - 12x = 2(x^2 - 6x)$
$= 2((x - 3)^2 - 3^2)$
$= 2((x - 3)^2 - 9)$
$= 2(x - 3)^2 - 18$

Here the coefficient of x^2 is 2, so take out a factor of 2. The other factor is in the form $(x^2 + bx)$ so you can use the rule to complete the square.

Expand the outer bracket by multiplying 2 by 9 to get your answer in this form.

- $ax^2 + bx + c = a\left(x + \frac{b}{2a}\right)^2 + \left(c - \frac{b^2}{4a}\right)$

Example 5

Write $3x^2 + 6x + 1$ in the form $p(x + q)^2 + r$, where p, q and r are integers to be found.

$3x^2 + 6x + 1$
$= 3(x^2 + 2x) + 1$
$= 3((x + 1)^2 - 1^2) + 1$
$= 3(x + 1)^2 - 3 + 1$
$= 3(x + 1)^2 - 2$
So $p = 3$, $q = 1$ and $r = -2$.

Watch out This is an **expression**, so you can't divide every term by 3 without changing its value. Instead, you need to take a factor of 3 out of $3x^2 + 6x$.

You could also use the rule given above to complete the square for this expression, but it is safer to learn the method shown here.

Exercise 2C

Hint In question **3d**, write the expression as $-4x^2 - 16x + 10$ then take a factor of -4 out of the first two terms to get $-4(x^2 + 4x) + 10$.

1 Complete the square for the expressions:
a $x^2 + 4x$ **b** $x^2 - 6x$ **c** $x^2 - 16x$ **d** $x^2 + x$ **e** $x^2 - 14x$

2 Complete the square for the expressions:
a $2x^2 + 16x$ **b** $3x^2 - 24x$ **c** $5x^2 + 20x$ **d** $2x^2 - 5x$ **e** $8x - 2x^2$

3 Write each of these expressions in the form $p(x + q)^2 + r$, where p, q and r are constants to be found:
a $2x^2 + 8x + 1$ **b** $5x^2 - 15x + 3$ **c** $3x^2 + 2x - 1$ **d** $10 - 16x - 4x^2$ **e** $2x - 8x^2 + 10$

(E) **4** Given that $x^2 + 3x + 6 = (x + a)^2 + b$, find the values of the constants a and b. **(2 marks)**

(E) **5** Write $2 + 0.8x - 0.04x^2$ in the form $A - B(x + C)^2$, where A, B and C are constants to be determined. **(3 marks)**

Example 6

Solve the equation $x^2 + 8x + 10 = 0$ by completing the square.
Give your answers in surd form.

$x^2 + 8x + 10 = 0$ — Check coefficient of $x^2 = 1$.
$x^2 + 8x = -10$ — Subtract 10 to get the LHS in the form $x^2 + bx$.
$(x + 4)^2 - 4^2 = -10$ — Complete the square for $x^2 + 8x$.
$(x + 4)^2 = -10 + 16$ — Add 4^2 to both sides.
$(x + 4)^2 = 6$
$(x + 4) = \pm\sqrt{6}$ — Take square roots of both sides.
$x = -4 \pm \sqrt{6}$ — Subtract 4 from both sides.
So the solutions are
$x = -4 + \sqrt{6}$ and $x = -4 - \sqrt{6}$. — Leave your answer in surd form.

Example 7

Solve the equation $2x^2 - 8x + 7 = 0$. Give your answers in surd form.

$2x^2 - 8x + 7 = 0$

$x^2 - 4x + \frac{7}{2} = 0$

$x^2 - 4x = -\frac{7}{2}$

$(x - 2)^2 - 2^2 = -\frac{7}{2}$ — Complete the square for $x^2 - 4x$.

$(x - 2)^2 = -\frac{7}{2} + 4$ — Add 2^2 to both sides.

$(x - 2)^2 = \frac{1}{2}$

$x - 2 = \pm\sqrt{\frac{1}{2}}$ — Take square roots of both sides.

$x = 2 \pm \frac{1}{\sqrt{2}}$ — Add 2 to both sides.

So the roots are

$x = 2 + \frac{1}{\sqrt{2}}$ and $x = 2 - \frac{1}{\sqrt{2}}$

Problem-solving

This is an **equation** so you can divide every term by the same constant. Divide by 2 to get x^2 on its own. The right-hand side is 0 so it is unchanged.

Online Use your calculator to check solutions to quadratic equations quickly.

Exercise 2D

1 Solve these quadratic equations by completing the square. Leave your answers in surd form.

a $x^2 + 6x + 1 = 0$ **b** $x^2 + 12x + 3 = 0$ **c** $x^2 + 4x - 2 = 0$ **d** $x^2 - 10x = 5$

2 Solve these quadratic equations by completing the square. Leave your answers in surd form.

a $2x^2 + 6x - 3 = 0$ **b** $5x^2 + 8x - 2 = 0$ **c** $4x^2 - x - 8 = 0$ **d** $15 - 6x - 2x^2 = 0$

(E) **3** $x^2 - 14x + 1 = (x + p)^2 + q$, where p and q are constants.

a Find the values of p and q. **(2 marks)**

b Using your answer to part **a**, or otherwise, show that the solutions to the equation $x^2 - 14x + 1 = 0$ can be written in the form $r \pm s\sqrt{3}$, where r and s are constants to be found. **(2 marks)**

(E/P) **4** By completing the square, show that the solutions to the equation $x^2 + 2bx + c = 0$ are given by the formula $x = -b \pm \sqrt{b^2 - c}$. **(4 marks)**

Problem-solving

Follow the same steps as you would if the coefficients were numbers.

Challenge

a Show that the solutions to the equation $ax^2 + 2bx + c = 0$ are given by $x = -\frac{b}{a} \pm \sqrt{\frac{b^2 - ac}{a^2}}$.

b Hence, or otherwise, show that the solutions to the equation $ax^2 + bx + c = 0$ can be written as $x = \frac{-b \pm \sqrt{b^2 - 4ac}}{2a}$.

Hint Start by dividing the whole equation by a.

Links You can use this method to prove the quadratic formula. → Section 7.4

2.3 Functions

A function is a mathematical relationship that maps each value of a set of inputs to a single output. The notation f(x) is used to represent a function of x.

- **The set of possible inputs for a function is called the domain.**
- **The set of possible outputs of a function is called the range.**

This diagram shows how the function $f(x) = x^2$ maps five values in its domain to values in its range.

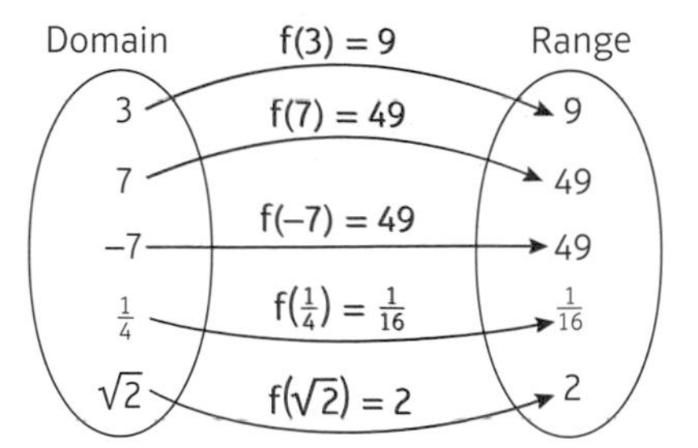

- **The roots of a function are the values of x for which f(x) = 0.**

Example 8

The functions f and g are given by $f(x) = 2x - 10$ and $g(x) = x^2 - 9$, $x \in \mathbb{R}$.

a Find the values of f(5) and g(10).

b Find the value of x for which f(x) = g(x).

a $f(5) = 2(5) - 10 = 10 - 10 = 0$
$g(10) = (10)^2 - 9 = 100 - 9 = 91$

To find f(5), substitute $x = 5$ into the function f(x).

b
$$f(x) = g(x)$$
$$2x - 10 = x^2 - 9$$
$$x^2 - 2x + 1 = 0$$
$$(x - 1)^2 = 0$$
$$x = 1$$

Set f(x) equal to g(x) and solve for x.

Notation If the input of a function, x, can be any real number the domain can be written as $x \in \mathbb{R}$.
The symbol $\in$ means 'is a member of' and the symbol $\mathbb{R}$ represents the real numbers.

Example 9

The function f is defined as $f(x) = x^2 + 6x - 5$, $x \in \mathbb{R}$.

a Write f(x) in the form $(x + p)^2 + q$.

b Hence, or otherwise, find the roots of f(x), leaving your answers in surd form.

c Write down the minimum value of f(x), and state the value of x for which it occurs.

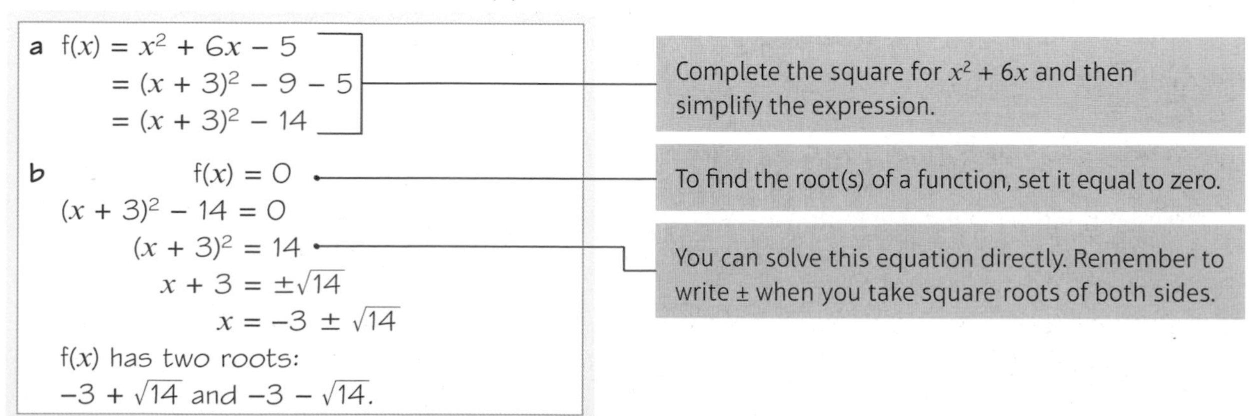

a $f(x) = x^2 + 6x - 5$
$= (x + 3)^2 - 9 - 5$
$= (x + 3)^2 - 14$

Complete the square for $x^2 + 6x$ and then simplify the expression.

b
$$f(x) = 0$$
$$(x + 3)^2 - 14 = 0$$
$$(x + 3)^2 = 14$$
$$x + 3 = \pm\sqrt{14}$$
$$x = -3 \pm \sqrt{14}$$

To find the root(s) of a function, set it equal to zero.

You can solve this equation directly. Remember to write ± when you take square roots of both sides.

f(x) has two roots:
$-3 + \sqrt{14}$ and $-3 - \sqrt{14}$.

c $(x+3)^2 \geqslant 0$
So the minimum value of $f(x)$ is -14.
This occurs when $(x+3)^2 = 0$,
so when $x = -3$

A squared value must be greater than or equal to 0.

$(x+3)^2 \geqslant 0$ so $(x+3)^2 - 14 \geqslant -14$

Example 10

Find the roots of the function $f(x) = x^6 + 7x^3 - 8$, $x \in \mathbb{R}$.

$f(x) = 0$
$x^6 + 7x^3 - 8 = 0$
$(x^3)^2 + 7(x^3) - 8 = 0$
$(x^3 - 1)(x^3 + 8) = 0$
So $x^3 = 1$ or $x^3 = -8$
$x^3 = 1 \Rightarrow x = 1$
$x^3 = -8 \Rightarrow x = -2$
The roots of $f(x)$ are 1 and -2.

Alternatively, let $u = x^3$.
$f(x) = x^6 + 7x^3 - 8$
$= (x^3)^2 + 7(x^3) - 8$
$= u^2 + 7u - 8$
$= (u - 1)(u + 8)$
So when $f(x) = 0$, $u = 1$ or $u = -8$.
If $u = 1 \Rightarrow x^3 = 1 \Rightarrow x = 1$
If $u = -8 \Rightarrow x^3 = -8 \Rightarrow x = -2$
The roots of $f(x)$ are 1 and -2.

Problem-solving
$f(x)$ can be written as a function of a function. The only powers of x in $f(x)$ are 6, 3 and 0 so you can write it as a quadratic function of x^3.

Treat x^3 as a single variable and factorise.

Solve the quadratic equation to find two values for x^3, then find the corresponding values of x.

You can simplify this working with a substitution.

Replace x^3 with u and solve the quadratic equation in u.

Watch out The solutions to the quadratic equation will be values of u. Convert back to values of x using your substitution.

Exercise 2E

1 Using the functions $f(x) = 5x + 3$, $g(x) = x^2 - 2$ and $h(x) = \sqrt{x+1}$, find the values of:

a $f(1)$ **b** $g(3)$ **c** $h(8)$ **d** $f(1.5)$ **e** $g(\sqrt{2})$

f $h(-1)$ **g** $f(4) + g(2)$ **h** $f(0) + g(0) + h(0)$ **i** $\dfrac{g(4)}{h(3)}$

(P) **2** The function $f(x)$ is defined by $f(x) = x^2 - 2x$, $x \in \mathbb{R}$.
Given that $f(a) = 8$, find two possible values for a.

Problem-solving
Substitute $x = a$ into the function and set the resulting expression equal to 8.

3 Find all of the roots of the following functions:

a $f(x) = 10 - 15x$ **b** $g(x) = (x + 9)(x - 2)$ **c** $h(x) = x^2 + 6x - 40$

d $j(x) = 144 - x^2$ **e** $k(x) = x(x + 5)(x + 7)$ **f** $m(x) = x^3 + 5x^2 - 24x$

4 The functions p and q are given by $p(x) = x^2 - 3x$ and $q(x) = 2x - 6, x \in \mathbb{R}$.
Find the two values of x for which $p(x) = q(x)$.

5 The functions f and g are given by $f(x) = 2x^3 + 30x$ and $g(x) = 17x^2,\ x \in \mathbb{R}$.
Find the three values of x for which $f(x) = g(x)$.

(E) **6** The function f is defined as $f(x) = x^2 - 2x + 2, x \in \mathbb{R}$.

a Write $f(x)$ in the form $(x + p)^2 + q$, where p and q are constants to be found. **(2 marks)**

b Hence, or otherwise, explain why $f(x) > 0$ for all values of x, and find the minimum value of $f(x)$. **(1 mark)**

homework

7 Find all roots of the following functions:

a $f(x) = x^6 + 9x^3 + 8$

b $g(x) = x^4 - 12x^2 + 32$

c $h(x) = 27x^6 + 26x^3 - 1$

d $j(x) = 32x^{10} - 33x^5 + 1$

e $k(x) = x - 7\sqrt{x} + 10$

f $m(x) = 2x^{\frac{2}{3}} + 2x^{\frac{1}{3}} - 12$

Hint The function in part **b** has four roots.

(E/P) **8** The function f is defined as $f(x) = 3^{2x} - 28(3^x) + 27, x \in \mathbb{R}$.

a Write $f(x)$ in the form $(3^x - a)(3^x - b)$, where a and b are real constants. **(2 marks)**

b Hence find the two roots of $f(x)$. **(2 marks)**

Problem-solving

Consider $f(x)$ as a function of a function.

2.4 Quadratic graphs

When $f(x) = ax^2 + bx + c$, the graph of $y = f(x)$ has a curved shape called a parabola.

You can sketch a quadratic graph by identifying key features.

The coefficient of x^2 determines the overall shape of the graph.

When a is positive the parabola will have this shape: ⋃

When a is negative the parabola will have this shape: ⋂

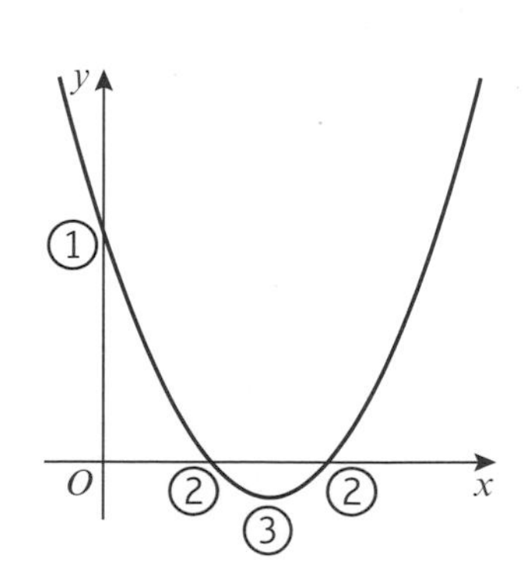

① The graph crosses the y-axis when $x = 0$. The y-coordinate is equal to c.

② The graph crosses the x-axis when $y = 0$. The x-coordinates are roots of the function $f(x)$.

③ Quadratic graphs have one turning point. This can be a minimum or a maximum. Since a parabola is symmetrical, the turning point and line of symmetry are half-way between the two roots.

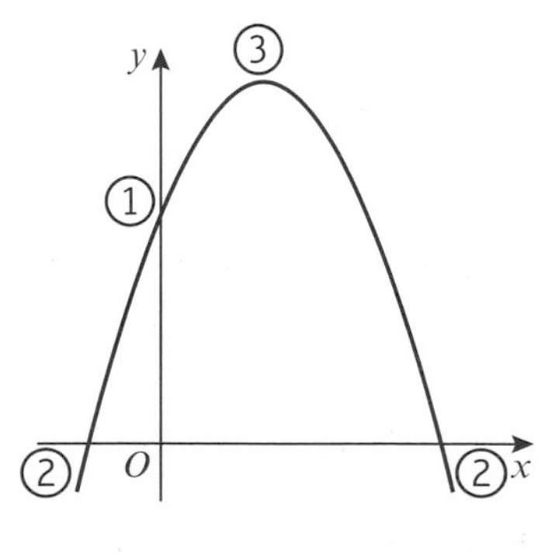

- **You can find the coordinates of the turning point of a quadratic graph by completing the square. If $f(x) = a(x + p)^2 + q$, the graph of $y = f(x)$ has a turning point at $(-p, q)$.**

Links The graph of $y = a(x + p)^2 + q$ is a translation of the graph of $y = ax^2$ by $\begin{pmatrix} -p \\ q \end{pmatrix}$. → Section 4.5

Example 11

Sketch the graph of $y = x^2 - 5x + 4$, and find the coordinates of its turning point.

As $a = 1$ is positive, the graph has a $\vee$ shape and a minimum point.

When $x = 0$, $y = 4$, so the graph crosses the y-axis at $(0, 4)$.

When $y = 0$,

$$x^2 - 5x + 4 = 0$$

$$(x - 1)(x - 4) = 0$$

$x = 1$ or $x = 4$, so the graph crosses the x-axis at $(1, 0)$ and $(4, 0)$.

Completing the square:

$$x^2 - 5x + 4 = \left(x - \frac{5}{2}\right)^2 - \frac{25}{4} + 4$$

$$= \left(x - \frac{5}{2}\right)^2 - \frac{9}{4}$$

So the minimum point has coordinates $\left(\frac{5}{2}, -\frac{9}{4}\right)$.

Alternatively, the minimum occurs when x is half-way between 1 and 4,

$$\text{so } x = \frac{1 + 4}{2} = \frac{5}{2}$$

$$y = \left(\frac{5}{2}\right)^2 - 5 \times \left(\frac{5}{2}\right) + 4 = -\frac{9}{4}$$

so the minimum has coordinates $\left(\frac{5}{2}, -\frac{9}{4}\right)$.

The sketch of the graph is:

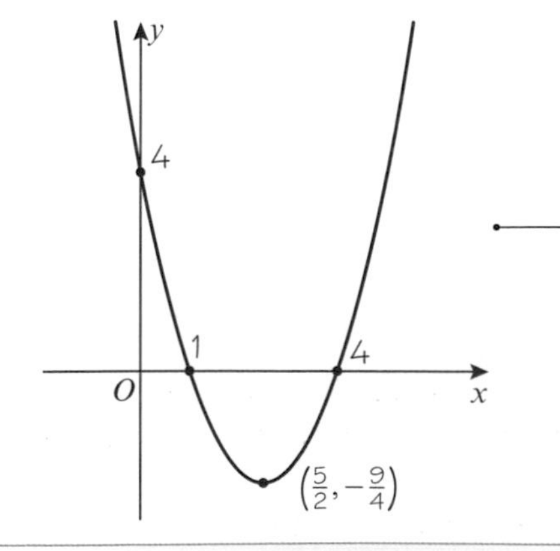

Use the coefficient of x^2 to determine the general shape of the graph.

This example factorises, but you may need to use the quadratic formula or complete the square.

Complete the square to find the coordinates of the turning point.

Watch out If you use symmetry to find the x-coordinate of the minimum point, you need to substitute this value into the equation to find the y-coordinate of the minimum point.

You could use a graphic calculator or substitute some values to check your sketch.

When $x = 5$, $y = 5^2 - 5 \times 5 + 4 = 4$.

Online Explore how the graph of $y = (x + p)^2 + q$ changes as the values of p and q change using technology.

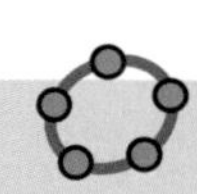

Example 12

Sketch the graph of $y = 4x - 2x^2 - 3$. Find the coordinates of its turning point and write down the equation of its line of symmetry.

As $a = -2$ is negative, the graph has a ∩ shape and a maximum point.

When $x = 0$, $y = -3$, so the graph crosses the y-axis at $(0, -3)$.

When $y = 0$,

$-2x^2 + 4x - 3 = 0$

Using the quadratic formula,

$$x = \frac{-4 \pm \sqrt{4^2 - 4(-2)(-3)}}{2 \times (-2)}$$

$$x = \frac{-4 \pm \sqrt{-8}}{-4}$$

There are no real solutions, so the graph does not cross the x-axis.

Completing the square:

$-2x^2 + 4x - 3$

$= -2(x^2 - 2x) - 3$

$= -2((x - 1)^2 - 1) - 3$

$= -2(x - 1)^2 + 2 - 3$

$= -2(x - 1)^2 - 1$

So the maximum point has coordinates $(1, -1)$.

The line of symmetry is vertical and goes through the maximum point. It has the equation $x = 1$.

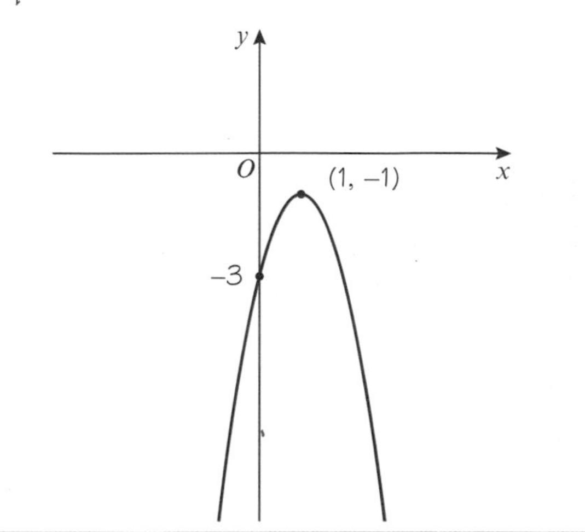

It's easier to see that $a < 0$ if you write the equation in the form $y = -2x^2 + 4x - 3$.

$a = -2$, $b = 4$ and $c = -3$

You would need to square root a negative number to evaluate this expression. Therefore this equation has no real solutions.

The coefficient of x^2 is -2 so take out a factor of -2

Watch out A sketch graph does not need to be plotted exactly or drawn to scale. However you should:

- draw a smooth curve by hand
- identify any relevant key points (such as intercepts and turning points)
- label your axes.

Exercise 2F

1 Sketch the graphs of the following equations. For each graph, show the coordinates of the point(s) where the graph crosses the coordinate axes, and write down the coordinate of the turning point and the equation of the line of symmetry.

a $y = x^2 - 6x + 8$ **b** $y = x^2 + 2x - 15$ **c** $y = 25 - x^2$ **d** $y = x^2 + 3x + 2$

e $y = -x^2 + 6x + 7$ **f** $y = 2x^2 + 4x + 10$ **g** $y = 2x^2 + 7x - 15$ **h** $y = 6x^2 - 19x + 10$

i $y = 4 - 7x - 2x^2$ **j** $y = 0.5x^2 + 0.2x + 0.02$

(P) 2 These sketches are graphs of quadratic functions of the form $ax^2 + bx + c$.
Find the values of a, b and c for each function.

a

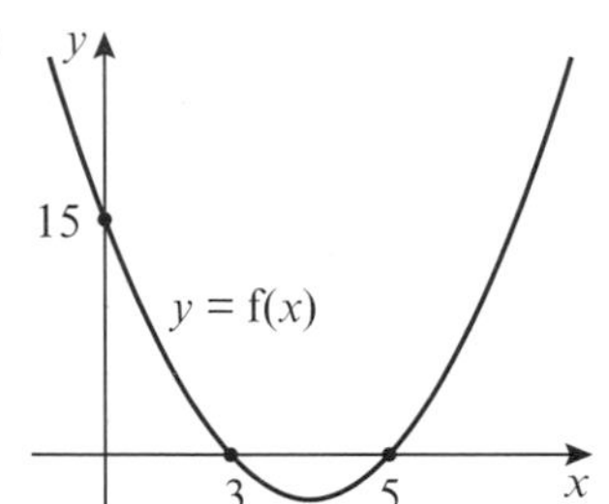

b

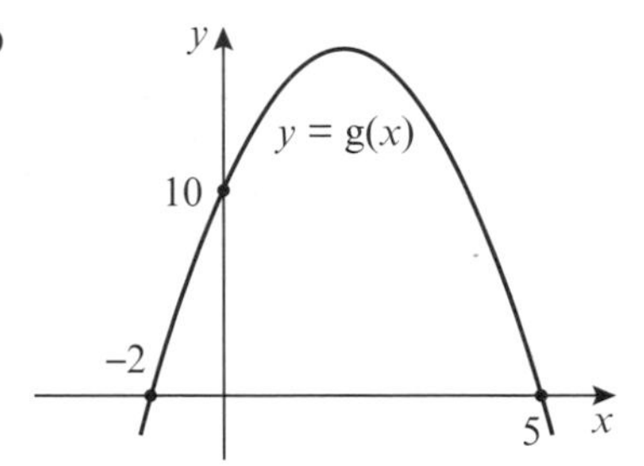

Problem-solving

Check your answers by substituting values into the function. In part **c** the graph passes through (0, −18), so h(0) should be −18.

c

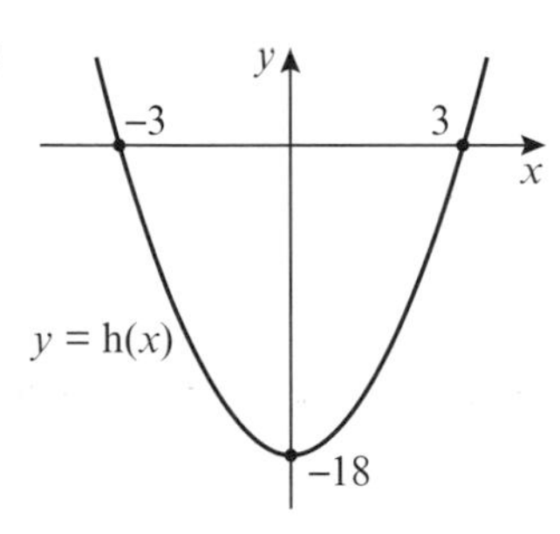

d

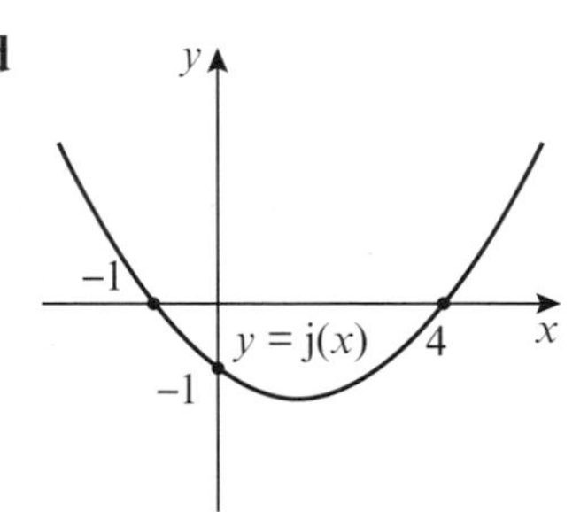

(E/P) 3 The graph of $y = ax^2 + bx + c$ has a minimum at $(5, -3)$ and passes through $(4, 0)$.
Find the values of a, b and c. **(3 marks)**

2.5 The discriminant

If you square any real number, the result is greater than or equal to 0. This means that if y is negative, $\sqrt{y}$ cannot be a real number. Look at the quadratic formula:

$$x = \frac{-b \pm \sqrt{b^2 - 4ac}}{2a}$$

If the value under the square root sign is negative, x cannot be a real number and there are no real solutions. If the value under the square root is equal to 0, both solutions will be the same.

- **For the quadratic function $f(x) = ax^2 + bx + c$, the expression $b^2 - 4ac$ is called the discriminant. The value of the discriminant shows how many roots f(x) has:**
  - **If $b^2 - 4ac > 0$ then f(x) has two distinct real roots.**
  - **If $b^2 - 4ac = 0$ then f(x) has one repeated root.**
  - **If $b^2 - 4ac < 0$ then f(x) has no real roots.**

You can use the discriminant to check the shape of sketch graphs.

Below are some graphs of $y = \text{f}(x)$ where $\text{f}(x) = ax^2 + bx + c$.

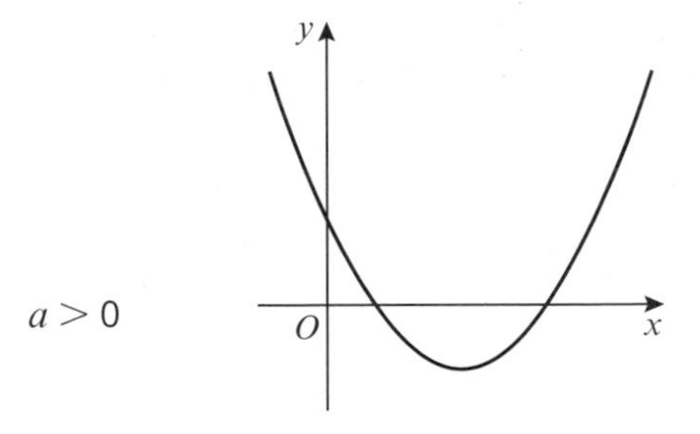

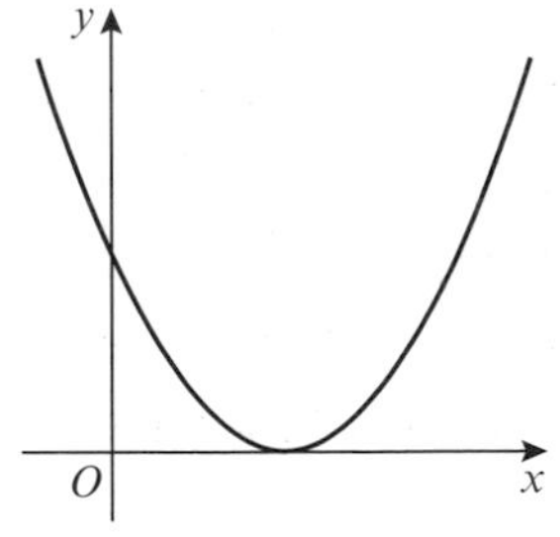

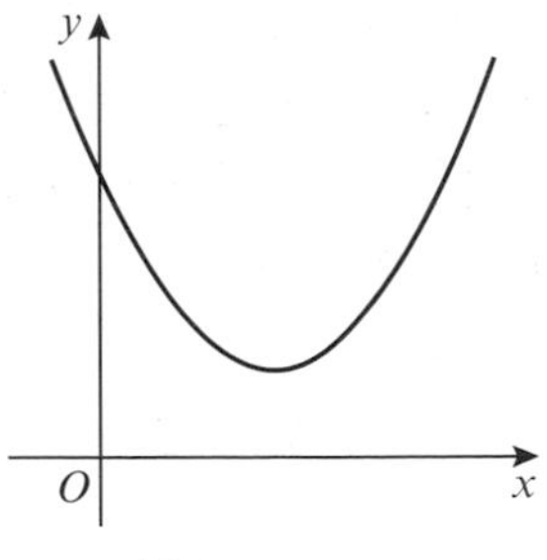

$b^2 - 4ac > 0$
Two distinct real roots

$b^2 - 4ac = 0$
One repeated root

$b^2 - 4ac < 0$
No real roots

$a < 0$

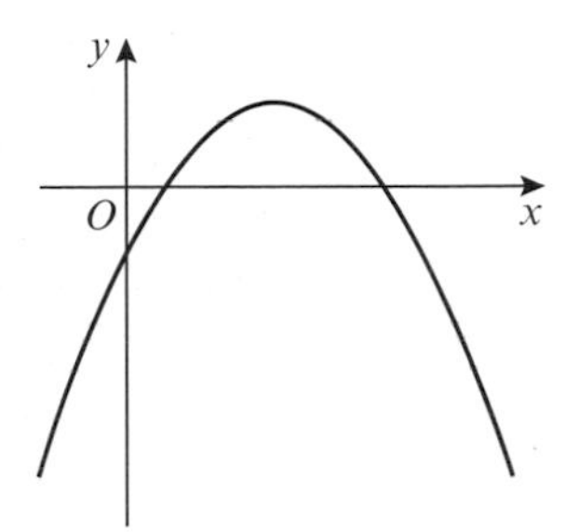

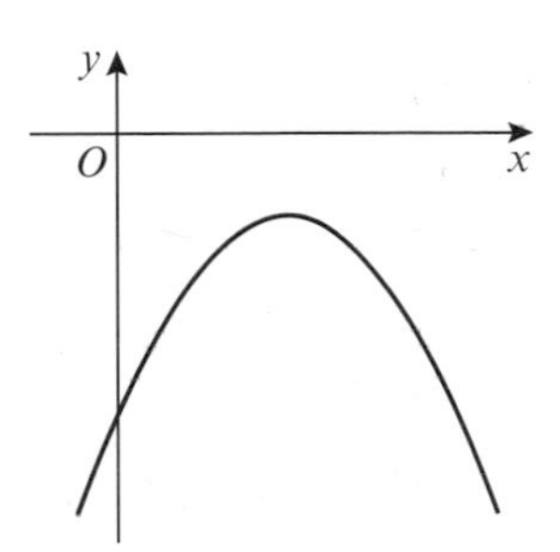

Example 13

Find the values of k for which $\text{f}(x) = x^2 + kx + 9$ has equal roots.

$x^2 + kx + 9 = 0$
Here $a = 1$, $b = k$ and $c = 9$
For equal roots, $b^2 - 4ac = 0$
$k^2 - 4 \times 1 \times 9 = 0$
$k^2 - 36 = 0$
$k^2 = 36$
so $k = \pm 6$

Problem-solving

Use the condition given in the question to write a statement about the discriminant.

Substitute for a, b and c to get an equation with one unknown.

Solve to find the values of k.

Example 14

Find the range of values of k for which $x^2 + 4x + k = 0$ has two distinct real solutions.

$x^2 + 4x + k = 0$
Here $a = 1$, $b = 4$ and $c = k$.
For two real solutions, $b^2 - 4ac > 0$
$4^2 - 4 \times 1 \times k > 0$
$16 - 4k > 0$
$16 > 4k$
$4 > k$
So $k < 4$

This statement involves an inequality, so your answer will also be an inequality.

For any value of k less than 4, the equation will have 2 distinct real solutions.

Online Explore how the value of the discriminant changes with k using technology.

Exercise 2G

1 a Calculate the value of the discriminant for each of these five functions:

i $f(x) = x^2 + 8x + 3$ **ii** $g(x) = 2x^2 - 3x + 4$ **iii** $h(x) = -x^2 + 7x - 3$

iv $j(x) = x^2 - 8x + 16$ **v** $k(x) = 2x - 3x^2 - 4$

b Using your answers to part **a**, match the same five functions to these sketch graphs.

i
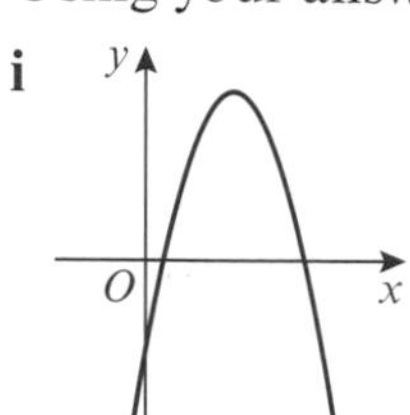

ii
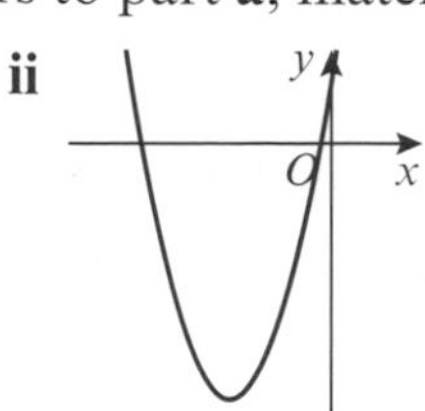

iii
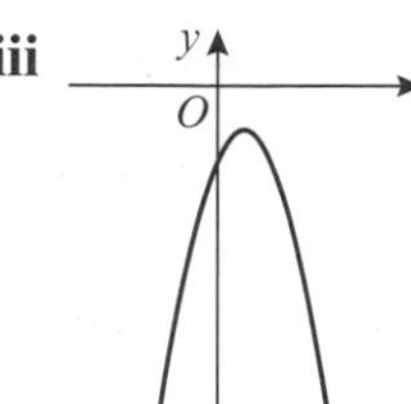

iv
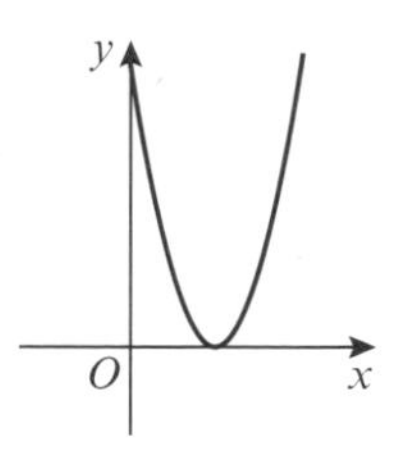

v
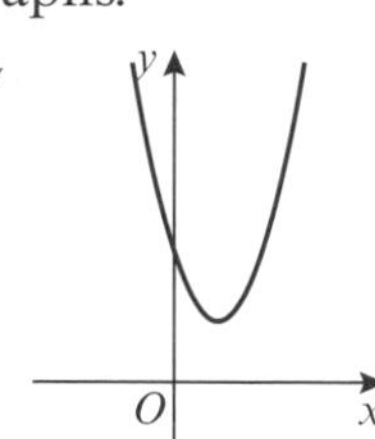

E/P **2** Find the values of k for which $x^2 + 6x + k = 0$ has two real solutions. **(2 marks)**

E/P **3** Find the value of t for which $2x^2 - 3x + t = 0$ has exactly one solution. **(2 marks)**

E/P **4** Given that the function $f(x) = sx^2 + 8x + s$ has equal roots, find the value of the positive constant s. **(2 marks)**

E/P **5** Find the range of values of k for which $3x^2 - 4x + k = 0$ has no real solutions. **(2 marks)**

E/P **6** The function $g(x) = x^2 + 3px + (14p - 3)$, where p is an integer, has two equal roots.

a Find the value of p. **(2 marks)**

b For this value of p, solve the equation $x^2 + 3px + (14p - 3) = 0$. **(2 marks)**

E/P **7** $h(x) = 2x^2 + (k + 4)x + k$, where k is a real constant.

a Find the discriminant of $h(x)$ in terms of k. **(3 marks)**

b Hence or otherwise, prove that $h(x)$ has two distinct real roots for all values of k. **(3 marks)**

Problem-solving

If a question part says 'hence or otherwise' it is usually easier to use your answer to the previous question part.

Challenge

a Prove that, if the values of a and c are given and non-zero, it is always possible to choose a value of b so that $f(x) = ax^2 + bx + c$ has distinct real roots.

b Is it always possible to choose a value of b so that $f(x)$ has equal roots? Explain your answer.

2.6 Modelling with quadratics

A **mathematical model** is a mathematical description of a real-life situation. Mathematical models use the language and tools of mathematics to represent and explore real-life patterns and relationships, and to predict what is going to happen next.

Models can be simple or complicated, and their results can be approximate or exact. Sometimes a model is only valid under certain circumstances, or for a limited range of inputs. You will learn more about how models involve simplifications and assumptions in Statistics and Mechanics.

Quadratic functions can be used to model and explore a range of practical contexts, including projectile motion.

Example 15

A spear is thrown over level ground from the top of a tower.
The height, in metres, of the spear above the ground after t seconds is modelled by the function:

$$h(t) = 12.25 + 14.7t - 4.9t^2, \ t \geqslant 0$$

a Interpret the meaning of the constant term 12.25 in the model.

b After how many seconds does the spear hit the ground?

c Write $h(t)$ in the form $A - B(t - C)^2$, where A, B and C are constants to be found.

d Using your answer to part **c** or otherwise, find the maximum height of the spear above the ground, and the time at which this maximum height is reached.

a The tower is 12.25 m tall, since this is the height at time 0.

b When the spear hits the ground, the height is equal to 0.

$$12.25 + 14.7t - 4.9t^2 = 0$$

Using the formula, where $a = -4.9$, $b = 14.7$ and $c = 12.25$,

$$t = \frac{-14.7 \pm \sqrt{14.7^2 - 4(-4.9)(12.25)}}{(2 \times -4.9)}$$

$$t = \frac{-14.7 \pm \sqrt{456.19}}{-9.8}$$

$t = -0.679$ or $t = 3.68$ (to 3 s.f.)

As $t \geqslant 0$, $t = 3.68$ seconds (to 3 s.f.).

c $12.25 + 14.7t - 4.9t^2$

$= -4.9(t^2 - 3t) + 12.25$

$= -4.9((t - 1.5)^2 - 2.25) + 12.25$

$= -4.9(t - 1.5)^2 + 11.025 + 12.25$

$= 23.275 - 4.9(t - 1.5)^2$

So $A = 23.275$, $B = 4.9$ and $C = 1.5$.

d The maximum height of the spear is 23.275 metres, 1.5 seconds after the spear is thrown.

Problem-solving

Read the question carefully to work out the meaning of the constant term in the **context of the model**. Here, $t = 0$ is the time the spear is thrown.

To solve a quadratic, factorise, use the quadratic formula, or complete the square.

Give any non-exact numerical answers correct to 3 significant figures unless specified otherwise.

Always interpret your answers in the context of the model. t is the time after the spear was thrown so it must be positive.

$4.9(t - 1.5)^2$ must be positive or 0, so $h(t) \leqslant 23.275$ for all possible values of t.

The turning point of the graph of this function would be at (1.5, 23.275). You may find it helpful to draw a sketch of the function when working through modelling questions.

Online Explore the trajectory of the spear using technolgy.

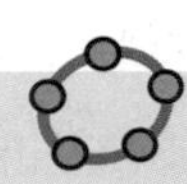

Exercise 2H

E/P **1** The diagram shows a section of a suspension bridge carrying a road over water.

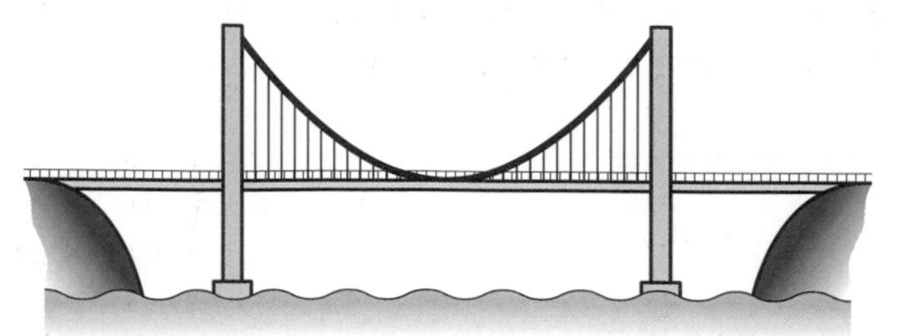

> **Problem-solving**
> For part **a**, make sure your answer is in the context of the model.

The height of the cables above water level in metres can be modelled by the function $h(x) = 0.000\,12x^2 + 200$, where x is the displacement in metres from the centre of the bridge.

a Interpret the meaning of the constant term 200 in the model. **(1 mark)**

b Use the model to find the two values of x at which the height is 346 m. **(3 marks)**

c Given that the towers at each end are 346 m tall, use your answer to part **b** to calculate the length of the bridge to the nearest metre. **(1 mark)**

E/P **2** A car manufacturer uses a model to predict the fuel consumption, y miles per gallon (mpg), for a specific model of car travelling at a speed of x mph.

$$y = -0.01x^2 + 0.975x + 16, x > 0$$

a Use the model to find two speeds at which the car has a fuel consumption of 32.5 mpg. **(3 marks)**

b Rewrite y in the form $A - B(x - C)^2$, where A, B and C are constants to be found. **(3 marks)**

c Using your answer to part **b**, find the speed at which the car has the greatest fuel efficiency. **(1 mark)**

d Use the model to calculate the fuel consumption of a car travelling at 120 mph. Comment on the validity of using this model for very high speeds. **(2 marks)**

E/P **3** A fertiliser company uses a model to determine how the amount of fertiliser used, f kilograms per hectare, affects the grain yield g, measured in tonnes per hectare.

$$g = 6 + 0.03f - 0.000\,06f^2$$

a According to the model, how much grain would each hectare yield without any fertiliser? **(1 mark)**

b One farmer currently uses 20 kilograms of fertiliser per hectare. How much more fertiliser would he need to use to increase his grain yield by 1 tonne per hectare? **(4 marks)**

E/P **4** A football stadium has 25 000 seats. The football club know from past experience that they will sell only 10 000 tickets if each ticket costs £30. They also expect to sell 1000 more tickets every time the price goes down by £1.

a The number of tickets sold t can be modelled by the linear equation $t = M - 1000p$, where £p is the price of each ticket and M is a constant. Find the value of M. **(1 mark)**

The total revenue, £r, can be calculated by multiplying the number of tickets sold by the price of each ticket. This can be written as $r = p(M - 1000p)$.

b Rearrange r into the form $A - B(p - C)^2$, where A, B and C are constants to be found. **(3 marks)**

c Using your answer to part **b** or otherwise, work out how much the football club should charge for each ticket if they want to make the maximum amount of money. **(2 marks)**

Challenge

Accident investigators are studying the stopping distance of a particular car.
When the car is travelling at 20 mph, its stopping distance is 6 feet.
When the car is travelling at 30 mph, its stopping distance is 14 feet.
When the car is travelling at 40 mph, its stopping distance is 24 feet.

The investigators suggest that the stopping distance in feet, d, is a quadratic function of the speed in miles per hour, s.

a Given that $\text{d}(s) = as^2 + bs + c$, find the values of the constants a, b and c.

b At an accident scene a car has left behind a skid that is 20 feet long.
Use your model to calculate the speed that this car was going at before the accident.

Hint Start by setting up three simultaneous equations. Combine two different pairs of equations to eliminate c. Use the results to find the values of a and b first.

Mixed exercise 2

1 Solve the following equations without a calculator. Leave your answers in surd form where necessary.

a $y^2 + 3y + 2 = 0$ **b** $3x^2 + 13x - 10 = 0$ **c** $5x^2 - 10x = 4x + 3$ **d** $(2x - 5)^2 = 7$

2 Sketch graphs of the following equations:

a $y = x^2 + 5x + 4$ **b** $y = 2x^2 + x - 3$ **c** $y = 6 - 10x - 4x^2$ **d** $y = 15x - 2x^2$

(E) **3** $\text{f}(x) = x^2 + 3x - 5$ and $\text{g}(x) = 4x + k$, where k is a constant.

a Given that $\text{f}(3) = \text{g}(3)$, find the value of k. **(3 marks)**

b Find the values of x for which $\text{f}(x) = \text{g}(x)$. **(3 marks)**

4 Solve the following equations, giving your answers correct to 3 significant figures:

a $k^2 + 11k - 1 = 0$ **b** $2t^2 - 5t + 1 = 0$ **c** $10 - x - x^2 = 7$ **d** $(3x - 1)^2 = 3 - x^2$

5 Write each of these expressions in the form $p(x + q)^2 + r$, where p, q and r are constants to be found:

a $x^2 + 12x - 9$ **b** $5x^2 - 40x + 13$ **c** $8x - 2x^2$ **d** $3x^2 - (x + 1)^2$

(E) **6** Find the value k for which the equation $5x^2 - 2x + k = 0$ has exactly one solution. **(2 marks)**

(E) **7** Given that for all values of x:

$$3x^2 + 12x + 5 = p(x + q)^2 + r$$

a find the values of p, q and r. **(3 marks)**

b Hence solve the equation $3x^2 + 12x + 5 = 0$. **(2 marks)**

(E/P) **8** The function f is defined as $f(x) = 2^{2x} - 20(2^x) + 64, x \in \mathbb{R}$.

a Write f(x) in the form $(2^x - a)(2^x - b)$, where a and b are real constants. **(2 marks)**

b Hence find the two roots of f(x). **(2 marks)**

9 Find, as surds, the roots of the equation:

$$2(x + 1)(x - 4) - (x - 2)^2 = 0.$$

10 Use algebra to solve $(x - 1)(x + 2) = 18$.

(E/P) **11** A diver launches herself off a springboard. The height of the diver, in metres, above the pool t seconds after launch can be modelled by the following function:

$$h(t) = 5t - 10t^2 + 10, t \geqslant 0$$

a How high is the springboard above the water? **(1 mark)**

b Use the model to find the time at which the diver hits the water. **(3 marks)**

c Rearrange h(t) into the form $A - B(t - C)^2$ and give the values of the constants A, B and C. **(3 marks)**

d Using your answer to part **c** or otherwise, find the maximum height of the diver, and the time at which this maximum height is reached. **(2 marks)**

(E/P) **12** For this question, $f(x) = 4kx^2 + (4k + 2)x + 1$, where k is a real constant.

a Find the discriminant of f(x) in terms of k. **(3 marks)**

b By simplifying your answer to part **a** or otherwise, prove that f(x) has two distinct real roots for all non-zero values of k. **(2 marks)**

c Explain why f(x) cannot have two distinct real roots when $k = 0$. **(1 mark)**

(E/P) **13** Find all of the roots of the function $r(x) = x^8 - 17x^4 + 16$. **(5 marks)**

(E/P) **14** Lynn is selling cushions as part of an enterprise project. On her first attempt, she sold 80 cushions at the cost of £15 each. She hopes to sell more cushions next time. Her adviser suggests that she can expect to sell 10 more cushions for every £1 that she lowers the price.

a The number of cushions sold c can be modelled by the equation $c = 230 - Hp$, where £p is the price of each cushion and H is a constant. Determine the value of H. **(1 mark)**

To model her total revenue, £r, Lynn multiplies the number of cushions sold by the price of each cushion. She writes this as $r = p(230 - Hp)$.

b Rearrange r into the form $A - B(p - C)^2$, where A, B and C are constants to be found. **(3 marks)**

c Using your answer to part **b** or otherwise, show that Lynn can increase her revenue by £122.50 through lowering her prices, and state the optimum selling price of a cushion. **(2 marks)**

Challenge

a The ratio of the lengths $a:b$ in this line is the same as the ratio of the lengths $b:c$.

Show that this ratio is $\frac{1+\sqrt{5}}{2}:1$.

b Show also that the infinite square root

$$\sqrt{1+\sqrt{1+\sqrt{1+\sqrt{1+\sqrt{1+\dots}}}}} = \frac{1+\sqrt{5}}{2}$$

Summary of key points

1 To solve a quadratic equation by factorising:
- Write the equation in the form $ax^2 + bx + c = 0$
- Factorise the left-hand side
- Set each factor equal to zero and solve to find the value(s) of x

2 The solutions of the equation $ax^2 + bx + c = 0$ where $a \neq 0$ are given by the formula:

$$x = \frac{-b \pm \sqrt{b^2 - 4ac}}{2a}$$

3 $x^2 + bx = \left(x + \frac{b}{2}\right)^2 - \left(\frac{b}{2}\right)^2$

4 $ax^2 + bx + c = a\left(x + \frac{b}{2a}\right)^2 + \left(c - \frac{b^2}{4a}\right)$

5 The set of possible inputs for a function is called the **domain**.
The set of possible outputs of a function is called the **range**.

6 The **roots** of a function are the values of x for which $f(x) = 0$.

7 You can find the coordinates of a **turning point** of a quadratic graph by completing the square. If $f(x) = a(x + p)^2 + q$, the graph of $y = f(x)$ has a turning point at $(-p, q)$.

8 For the quadratic function $f(x) = ax^2 + bx + c = 0$, the expression $b^2 - 4ac$ is called the **discriminant**. The value of the discriminant shows how many roots $f(x)$ has:
- If $b^2 - 4ac > 0$ then a quadratic function has two distinct real roots.
- If $b^2 - 4ac = 0$ then a quadratic function has one repeated real root.
- If $b^2 - 4ac < 0$ then a quadratic function has no real roots

9 Quadratics can be used to model real-life situations.

3 Equations and inequalities

Objectives

After completing this chapter you should be able to:

Prior knowledge check

1 A = {factors of 12}
B = {factors of 20}
Write down the numbers in each of these sets:

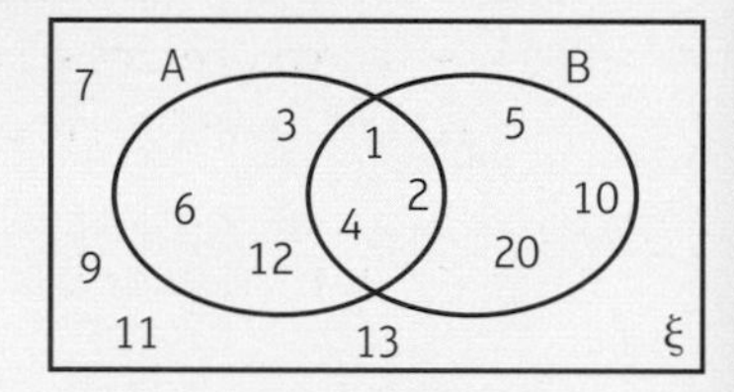

a $A \cap B$ **b** $(A \cup B)'$

← GCSE Mathematics

2 Simplify these expressions.

a $\sqrt{75}$ **b** $\frac{2\sqrt{45} + 3\sqrt{32}}{6}$

← Section 1.5

3 Match the equations to the correct graph. Label the points of intersection with the axes and the coordinates of the turning point.

a $y = 9 - x^2$ **b** $y = (x - 2)^2 + 4$
c $y = (x - 7)(2x + 5)$

i

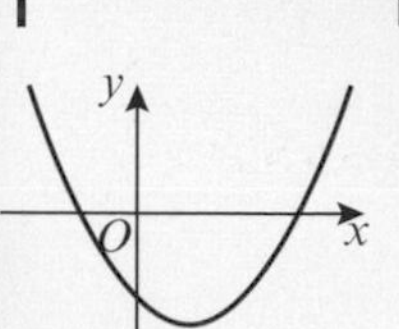

ii

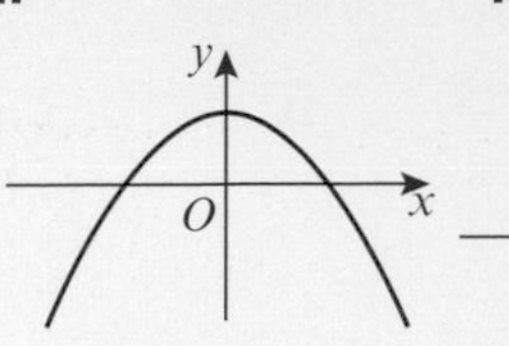

iii

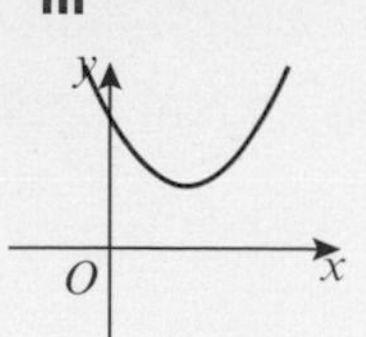

← Section 2.4

Food scientists use **regions** on graphs to optimise athletes' nutritional intake and ensure they satisfy the minimum dietary requirements for calories and vitamins.

3.1 Linear simultaneous equations

Linear simultaneous equations in two unknowns have **one set of values** that will make a pair of equations true at the same time.

The solution to this pair of simultaneous equations is $x = 5$, $y = 2$

$x + 3y = 11$ (1) — $5 + 3(2) = 5 + 6 = 11$ ✓

$4x - 5y = 10$ (2) — $4(5) - 5(2) = 20 - 10 = 10$ ✓

- **Linear simultaneous equations can be solved using elimination or substitution.**

Example 1

Solve the simultaneous equations:

a $2x + 3y = 8$
$3x - y = 23$

b $4x - 5y = 4$
$6x + 2y = 25$

a $2x + 3y = 8$ (1)

$3x - y = 23$ (2) — First look for a way to eliminate x or y. Multiply equation (2) by 3 to get $3y$ in each equation.

$9x - 3y = 69$ (3) — Number this new equation (3).

$11x = 77$ — Then add equations (1) and (3), since the $3y$ terms have different signs and y will be eliminated.

$x = 7$

$14 + 3y = 8$ — Substitute $x = 7$ into equation (1) to find y.

$3y = 8 - 14$

$y = -2$

The solution is $x = 7$, $y = -2$. — Remember to check your solution by substituting into equation (2). $3(7) - (-2) = 21 + 2 = 23$ ✓ Note that you could also multiply equation (1) by 3 and equation (2) by 2 to get $6x$ in both equations. You could then subtract to eliminate x.

b $4x - 5y = 4$ (1) — Multiply equation (1) by 3 and multiply equation (2) by 2 to get $12x$ in each equation.

$6x + 2y = 25$ (2)

$12x - 15y = 12$ (3)

$12x + 4y = 50$ (4)

$-19y = -38$ — Subtract, since the $12x$ terms have the same sign (both positive).

$y = 2$

$4x - 10 = 4$ — Substitute $y = 2$ into equation (1) to find x.

$4x = 14$

$x = 3\frac{1}{2}$

The solution is $x = 3\frac{1}{2}$, $y = 2$.

Example 2

Solve the simultaneous equations:

$2x - y = 1$
$4x + 2y = -30$

$2x - y = 1$ (1)
$4x + 2y = -30$ (2)
$y = 2x - 1$

Rearrange an equation, in this case equation (1), to get either $x = \ldots$ or $y = \ldots$ (here $y = \ldots$).

$4x + 2(2x - 1) = -30$

Substitute this into the other equation (here into equation (2) in place of y).

$4x + 4x - 2 = -30$
$8x = -28$
$x = -3\frac{1}{2}$

Solve for x.

$y = 2(-3\frac{1}{2}) - 1 = -8$

Substitute $x = -3\frac{1}{2}$ into equation (1) to find the value of y.

The solution is $x = -3\frac{1}{2}$, $y = -8$.

Remember to check your solution in equation (2).
$4(-3.5) + 2(-8) = -14 - 16 = -30$ ✓

Exercise 3A

1 Solve these simultaneous equations by elimination:

a $2x - y = 6$
$4x + 3y = 22$

b $7x + 3y = 16$
$2x + 9y = 29$

c $5x + 2y = 6$
$3x - 10y = 26$

d $2x - y = 12$
$6x + 2y = 21$

e $3x - 2y = -6$
$6x + 3y = 2$

f $3x + 8y = 33$
$6x = 3 + 5y$

2 Solve these simultaneous equations by substitution:

a $x + 3y = 11$
$4x - 7y = 6$

b $4x - 3y = 40$
$2x + y = 5$

c $3x - y = 7$
$10x + 3y = -2$

d $2y = 2x - 3$
$3y = x - 1$

3 Solve these simultaneous equations:

a $3x - 2y + 5 = 0$
$5(x + y) = 6(x + 1)$

b $\frac{x - 2y}{3} = 4$
$2x + 3y + 4 = 0$

c $3y = 5(x - 2)$
$3(x - 1) + y + 4 = 0$

Hint First rearrange both equations into the same form e.g. $ax + by = c$.

(E/P) **4** $3x + ky = 8$
$x - 2ky = 5$
are simultaneous equations where k is a constant.

a Show that $x = 3$. **(3 marks)**

b Given that $y = \frac{1}{2}$ determine the value of k. **(1 mark)**

Problem-solving
k is a constant, so it has the same value in both equations.

(E/P) **5** $2x - py = 5$
$4x + 5y + q = 0$
are simultaneous equations where p and q are constants.
The solution to this pair of simultaneous equations is $x = q$, $y = -1$.
Find the value of p and the value of q. **(5 marks)**

3.2 Quadratic simultaneous equations

You need to be able to solve simultaneous equations where one equation is linear and one is quadratic.

To solve simultaneous equations involving one linear equation and one quadratic equation, you need to use a substitution method from the linear equation into the quadratic equation.

- **Simultaneous equations with one linear and one quadratic equation can have up to two pairs of solutions. You need to make sure the solutions are paired correctly.**

The solutions to this pair of simultaneous equations are $x = 4$, $y = -3$ and $x = 5.5$, $y = -1.5$.

$x - y = 7$ (1) — $4 - (-3) = 7$ ✓ and $5.5 - (-1.5) = 7$ ✓

$y^2 + xy + 2x = 5$ (2) — $(-3)^2 + (4)(-3) + 2(4) = 9 - 12 + 8 = 5$ ✓ and $(-1.5)^2 + (5.5)(-1.5) + 2(5.5) = 2.25 - 8.25 + 11 = 5$ ✓

Example 3

Solve the simultaneous equations:

$x + 2y = 3$
$x^2 + 3xy = 10$

$x + 2y = 3$ (1)
$x^2 + 3xy = 10$ (2) — The quadratic equation can contain terms involving y^2 and xy.

$x = 3 - 2y$ — Rearrange linear equation (1) to get $x = \ldots$ or $y = \ldots$ (here $x = \ldots$).

$(3 - 2y)^2 + 3y(3 - 2y) = 10$ — Substitute this into quadratic equation (2) (here in place of x).

$9 - 12y + 4y^2 + 9y - 6y^2 = 10$ — $(3 - 2y)^2$ means $(3 - 2y)(3 - 2y)$ ← Section 1.2

$-2y^2 - 3y - 1 = 0$

$2y^2 + 3y + 1 = 0$

$(2y + 1)(y + 1) = 0$ — Solve for y using factorisation.

$y = -\frac{1}{2}$ or $y = -1$

So $x = 4$ or $x = 5$ — Find the corresponding x-values by substituting the y-values into linear equation (1), $x = 3 - 2y$.

Solutions are $x = 4$, $y = -\frac{1}{2}$ and $x = 5$, $y = -1$. — There are two solution pairs for x and y.

Exercise 3B

1 Solve the simultaneous equations:

a $x + y = 11$
$xy = 30$

b $2x + y = 1$
$x^2 + y^2 = 1$

c $y = 3x$
$2y^2 - xy = 15$

d $3a + b = 8$
$3a^2 + b^2 = 28$

e $2u + v = 7$
$uv = 6$

f $3x + 2y = 7$
$x^2 + y = 8$

2 Solve the simultaneous equations:

a $2x + 2y = 7$
$x^2 - 4y^2 = 8$

b $x + y = 9$
$x^2 - 3xy + 2y^2 = 0$

c $5y - 4x = 1$
$x^2 - y^2 + 5x = 41$

3 Solve the simultaneous equations, giving your answers in their simplest surd form:

a $x - y = 6$
$xy = 4$

b $2x + 3y = 13$
$x^2 + y^2 = 78$

Watch out Use brackets when you are substituting an expression into an equation.

(E/P) 4 Solve the simultaneous equations:

$x + y = 3$
$x^2 - 3y = 1$

(6 marks)

(E/P) 5 a By eliminating y from the equations

$y = 2 - 4x$
$3x^2 + xy + 11 = 0$

show that $x^2 - 2x - 11 = 0$. **(2 marks)**

b Hence, or otherwise, solve the simultaneous equations

$y = 2 - 4x$
$3x^2 + xy + 11 = 0$

giving your answers in the form $a \pm b\sqrt{3}$, where a and b are integers. **(5 marks)**

(P) 6 One pair of solutions for the simultaneous equations

$y = kx - 5$
$4x^2 - xy = 6$

is $(1, p)$ where k and p are constants.

a Find the values of k and p.

b Find the second pair of solutions for the simultaneous equations.

Problem-solving

If $(1, p)$ is a solution, then $x = 1$, $y = p$ satisfies both equations.

Challenge

$y - x = k$
$x^2 + y^2 = 4$

Given that the simultaneous equations have exactly one pair of solutions, show that

$k = \pm 2\sqrt{2}$

3.3 Simultaneous equations on graphs

You can represent the solutions of simultaneous equations graphically. As every point on a line or curve satisfies the equation of that line or curve, the points of intersection of two lines or curves satisfy both equations simultaneously.

- **The solutions to a pair of simultaneous equations represent the points of intersection of their graphs.**

Example 4

a On the same axes, draw the graphs of:

$2x + 3y = 8$
$3x - y = 23$

b Use your graph to write down the solutions to the simultaneous equations.

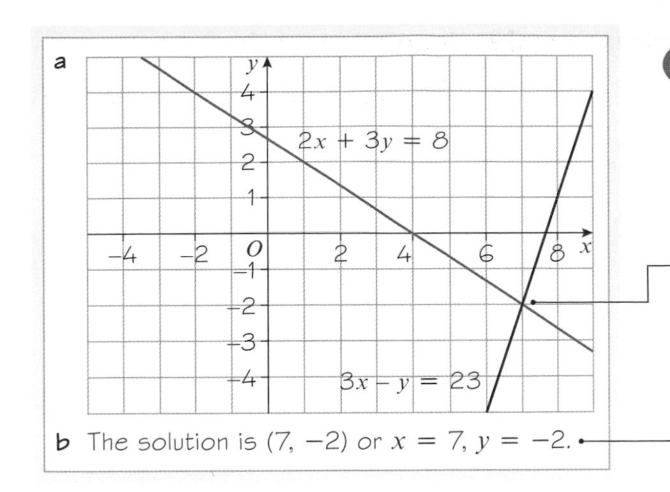

Online Find the point of intersection graphically using technology.

The point of intersection is the solution to the simultaneous equations

$2x + 3y = 8$

$3x - y = 23$

This solution matches the algebraic solution to the simultaneous equations.

Example 5

a On the same axes, draw the graphs of:

$2x + y = 3$

$y = x^2 - 3x + 1$

b Use your graph to write down the solutions to the simultaneous equations.

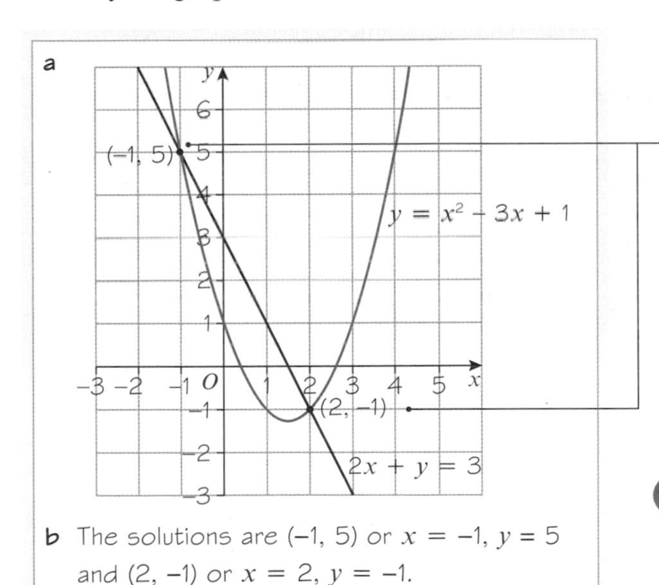

There are **two** solutions. Each solution will have an x-value and a y-value.

Check your solutions by substituting into both equations.

$2(-1) + (5) = -2 + 5 = 3$ ✓ and

$5 = (-1)^2 - 3(-1) + 1 = 1 + 3 + 1 = 5$ ✓

$2(2) + (-1) = 4 - 1 = 3$ ✓ and

$-1 = (2)^2 - 3(2) + 1 = 4 - 6 + 1 = -1$ ✓

Online Plot the curve and the line using technology to find the two points of intersection.

The graph of a linear equation and the graph of a quadratic equation can either:

- intersect twice
- intersect once
- not intersect

After substituting, you can use the **discriminant** of the resulting quadratic equation to determine the number of points of intersection.

- **For a pair of simultaneous equations that produce a quadratic equation of the form $ax^2 + bx + c = 0$:**
  - **$b^2 - 4ac > 0$ two real solutions**
  - **$b^2 - 4ac = 0$ one real solution**
  - **$b^2 - 4ac < 0$ no real solutions**

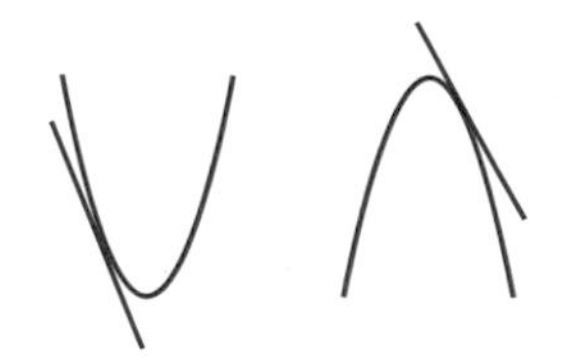
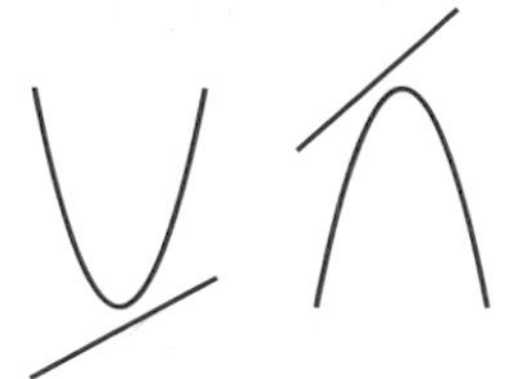

Example 6

The line with equation $y = 2x + 1$ meets the curve with equation $kx^2 + 2y + (k - 2) = 0$ at exactly one point. Given that k is a positive constant

a find the value of k

b for this value of k, find the coordinates of the point of intersection.

Online Explore how the value of k affects the line and the curve using technology.

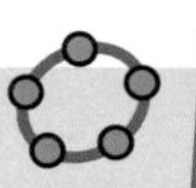

a

$$y = 2x + 1 \quad (1)$$
$$kx^2 + 2y + (k - 2) = 0 \quad (2)$$
$$kx^2 + 2(2x + 1) + (k - 2) = 0$$
$$kx^2 + 4x + 2 + k - 2 = 0$$
$$kx^2 + 4x + k = 0$$

Substitute $y = 2x + 1$ into equation (2) and simplify the quadratic equation. The resulting quadratic equation is in the form $ax^2 + bx + c = 0$ with $a = k$, $b = 4$ and $c = k$.

$$4^2 - 4 \times k \times k = 0$$
$$16 - 4k^2 = 0$$
$$k^2 - 4 = 0$$
$$(k - 2)(k + 2) = 0$$

Problem-solving

You are told that the line meets the curve at exactly one point, so use the discriminant of the resulting quadratic. There will be exactly one solution, so $b^2 - 4ac = 0$.

Factorise the quadratic to find the values of k.

$k = 2$ or $k = -2$

So $k = 2$

The solution is $k = +2$ as k is a positive constant.

b

$$2x^2 + 4x + 2 = 0$$
$$x^2 + 2x + 1 = 0$$
$$(x + 1)(x + 1) = 0$$
$$x = -1$$

Substitute $k = +2$ into the quadratic equation $kx^2 + 4x + k = 0$. Simplify and factorise to find the x-coordinate.

$$y = 2(-1) + 1 = -1$$

Substitute $x = -1$ into linear equation (1) to find the y-coordinate.

Point of intersection is $(-1, -1)$.

Check your answer by substituting into equation (2):

$$2x^2 + 2y = 0$$
$$2(-1)^2 + 2(-1) = 2 - 2 = 0 \checkmark$$

Exercise 3C

1 In each case:

i draw the graphs for each pair of equations on the same axes

ii find the coordinates of the point of intersection.

a $y = 3x - 5$
$y = 3 - x$

b $y = 2x - 7$
$y = 8 - 3x$

c $y = 3x + 2$
$3x + y + 1 = 0$

2 **a** Use graph paper to draw accurately the graphs of $2y = 2x + 11$ and $y = 2x^2 - 3x - 5$ on the same axes.

b Use your graph to find the coordinates of the points of intersection.

c Verify your solutions by substitution.

3 **a** On the same axes sketch the curve with equation $x^2 + y = 9$ and the line with equation $2x + y = 6$.

b Find the coordinates of the points of intersection.

c Verify your solutions by substitution.

4 **a** On the same axes sketch the curve with equation $y = (x - 2)^2$ and the line with equation $y = 3x - 2$.

b Find the coordinates of the point of intersection.

Hint You need to use algebra in part **b** to find the coordinates.

5 Find the coordinates of the points at which the line with equation $y = x - 4$ intersects the curve with equation $y^2 = 2x^2 - 17$.

6 Find the coordinates of the points at which the line with equation $y = 3x - 1$ intersects the curve with equation $y^2 = xy + 15$.

(P) **7** Determine the number of points of intersection for these pairs of simultaneous equations.

a $y = 6x^2 + 3x - 7$
$y = 2x + 8$

b $y = 4x^2 - 18x + 40$
$y = 10x - 9$

c $y = 3x^2 - 2x + 4$
$7x + y + 3 = 0$

(E/P) **8** Given the simultaneous equations

$$2x - y = 1$$
$$x^2 + 4ky + 5k = 0$$

where k is a non-zero constant

a show that $x^2 + 8kx + k = 0$. **(2 marks)**

Given that $x^2 + 8kx + k = 0$ has equal roots,

b find the value of k **(3 marks)**

c for this value of k, find the solution of the simultaneous equations. **(3 marks)**

(E/P) **9** A swimmer dives into a pool. Her position, p m, underwater can be modelled in relation to her horizontal distance, x m, from the point she entered the water as a quadratic equation $p = \frac{1}{2}x^2 - 3x$.

The position of the bottom of the pool can be modelled by the linear equation $p = 0.3x - 6$.

Determine whether this model predicts that the swimmer will touch the bottom of the pool. **(5 marks)**

p

x

3.4 Linear inequalities

You can solve linear inequalities using similar methods to those for solving linear equations.

- **The solution of an inequality is the set of all real numbers x that make the inequality true.**

Example 7

Find the set of values of x for which:

a $5x + 9 \geqslant x + 20$ b $12 - 3x < 27$

c $3(x - 5) > 5 - 2(x - 8)$

a $5x + 9 \geqslant x + 20$

$4x + 9 \geqslant 20$

$4x \geqslant 11$

$x \geqslant 2.75$

Rearrange to get $x \geqslant \ldots$

Notation You can write the solution to this inequality using set notation as $\{x : x \geqslant 2.75\}$. This means the set of all values x for which x is greater than or equal to 2.75.

b $12 - 3x < 27$

$-3x < 15$

$x > -5$

Subtract 12 from both sides.

Divide both sides by −3. (You therefore need to turn round the inequality sign.)

In set notation $\{x : x > -5\}$.

c $3(x - 5) > 5 - 2(x - 8)$

$3x - 15 > 5 - 2x + 16$

$5x > 5 + 16 + 15$

$5x > 36$

$x > 7.2$

Multiply out (note: $-2 \times -8 = +16$).

Rearrange to get $x > \ldots$

In set notation $\{x : x > 7.2\}$.

You may sometimes need to find the set of values for which **two** inequalities are true together. Number lines can be useful to find your solution.

Notation In set notation $x > -2$ **and** $x \leqslant 4$ is written $\{x : -2 < x \leqslant 4\}$ or alternatively $\{x : x > -2\} \cap \{x : x \leqslant 4\}$
$x \leqslant -1$ **or** $x > 3$ is written $\{x : x \leqslant -1\} \cup \{x : x > 3\}$

For example, in the number line below the solution set is $x > -2$ **and** $x \leqslant 4$.

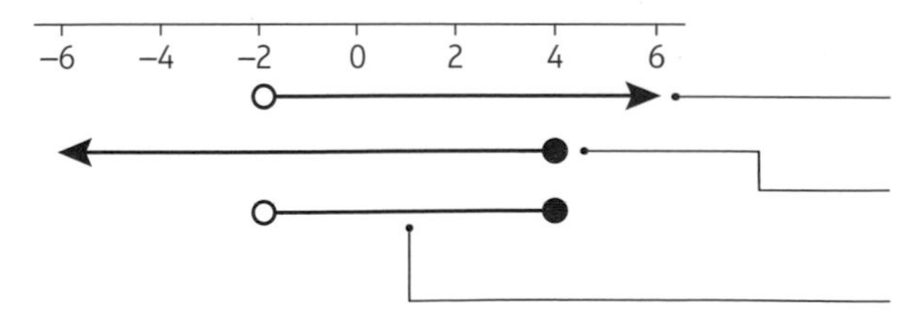

○ is used for < and > and means the end value is not included.

● is used for ⩽ and ⩾ and means the end value is included.

These are the only real values that satisfy both equalities simultaneously so the solution is $-2 < x \leqslant 4$.

Here the solution sets are $x \leqslant -1$ **or** $x > 3$.

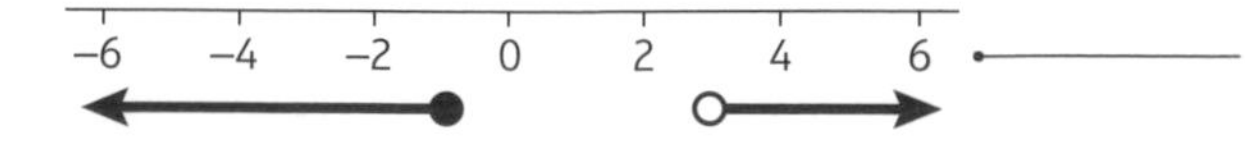

Here there is no overlap and the two inequalities have to be written separately as $x \leqslant -1$ or $x > 3$.

Example 8

Find the set of values of x for which:

a $3x - 5 < x + 8$ and $5x > x - 8$

b $x - 5 > 1 - x$ or $15 - 3x > 5 + 2x$.

a $3x - 5 < x + 8$ $\qquad$ $5x > x - 8$

$2x - 5 < 8$ $\qquad$ $4x > -8$

$2x < 13$ $\qquad$ $x > -2$

$x < 6.5$

−4 −2 0 2 4 6 8

$x < 6.5$

$x > -2$

So the required set of values is $-2 < x < 6.5$.

Draw a number line to illustrate the two inequalities.

The two sets of values overlap (intersect) where $-2 < x < 6.5$.

Notice here how this is written when x lies between two values.

In set notation this can be written as $\{x : -2 < x < 6.5\}$.

b $x - 5 > 1 - x$ $\qquad$ $15 - 3x > 5 + 2x$

$2x - 5 > 1$ $\qquad$ $10 - 3x > 2x$

$2x > 6$ $\qquad$ $10 > 5x$

$x > 3$ $\qquad$ $2 > x$

$x < 2$

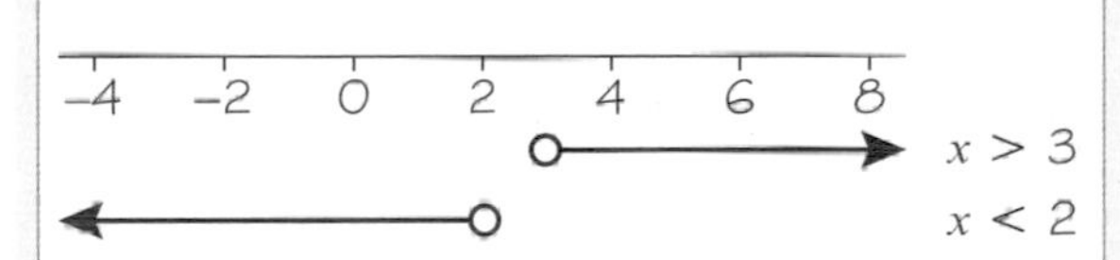

The solution is $x > 3$ or $x < 2$.

Draw a number line. Note that there is no overlap between the two sets of values.

In set notation this can be written as $\{x : x < 2\} \cup \{x : x > 3\}$.

Exercise 3D

1 Find the set of values of x for which:

a $2x - 3 < 5$

b $5x + 4 \geqslant 39$

c $6x - 3 > 2x + 7$

d $5x + 6 \leqslant -12 - x$

e $15 - x > 4$

f $21 - 2x > 8 + 3x$

g $1 + x < 25 + 3x$

h $7x - 7 < 7 - 7x$

i $5 - 0.5x \geqslant 1$

j $5x + 4 > 12 - 2x$

2 Find the set of values of x for which:

a $2(x - 3) \geqslant 0$
b $8(1 - x) > x - 1$
c $3(x + 7) \leqslant 8 - x$
d $2(x - 3) - (x + 12) < 0$
e $1 + 11(2 - x) < 10(x - 4)$
f $2(x - 5) \geqslant 3(4 - x)$
g $12x - 3(x - 3) < 45$
h $x - 2(5 + 2x) < 11$
i $x(x - 4) \geqslant x^2 + 2$
j $x(5 - x) \geqslant 3 + x - x^2$
k $3x + 2x(x - 3) \leqslant 2(5 + x^2)$
l $x(2x - 5) \leqslant \dfrac{4x(x + 3)}{2} - 9$

3 Use set notation to describe the set of values of x for which:

a $3(x - 2) > x - 4$ and $4x + 12 > 2x + 17$
b $2x - 5 < x - 1$ and $7(x + 1) > 23 - x$
c $2x - 3 > 2$ and $3(x + 2) < 12 + x$
d $15 - x < 2(11 - x)$ and $5(3x - 1) > 12x + 19$
e $3x + 8 \leqslant 20$ and $2(3x - 7) \geqslant x + 6$
f $5x + 3 < 9$ or $5(2x + 1) > 27$
g $4(3x + 7) \leqslant 20$ or $2(3x - 5) \geqslant \dfrac{7 - 6x}{2}$

Challenge

$A = \{x : 3x + 5 > 2\}$ $\quad B = \left\{x : \dfrac{x}{2} + 1 \leqslant 3\right\}$ $\quad C = \{x : 11 < 2x - 1\}$

Given that $A \cap (B \cup C) = \{x : p < x \leqslant q\} \cup \{x : x > r\}$, find the values of p, q and r.

3.5 Quadratic inequalities

- **To solve a quadratic inequality:**
  - **Rearrange so that the right-hand side of the inequality is 0**
  - **Solve the corresponding quadratic equation to find the critical values**
  - **Sketch the graph of the quadratic function**
  - **Use your sketch to find the required set of values.**

The sketch shows the graph of $f(x) = x^2 - 4x - 5$
$= (x + 1)(x - 5)$

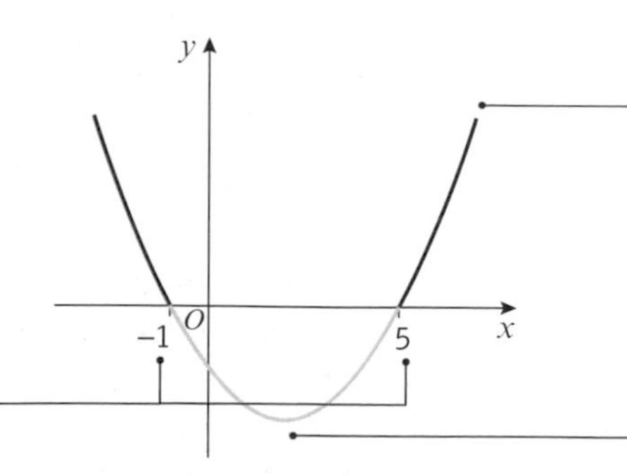

The solutions to $f(x) = 0$ are $x = -1$ and $x = 5$. These are called the critical values.

The solutions to the quadratic inequality $x^2 - 4x - 5 > 0$ are the x-values when the curve is **above** the x-axis (the darker part of the curve). This is when $x < -1$ or $x > 5$. In set notation the solution is $\{x : x < -1\} \cup \{x : x > 5\}$.

The solutions to the quadratic inequality $x^2 - 4x - 5 < 0$ are the x-values when the curve is **below** the x-axis (the lighter part of the curve). This is when $x > -1$ and $x < 5$ or $-1 < x < 5$. In set notation the solution is $\{x : -1 < x < 5\}$.

Example 9

Find the set of values of x for which:

$3 - 5x - 2x^2 < 0$.

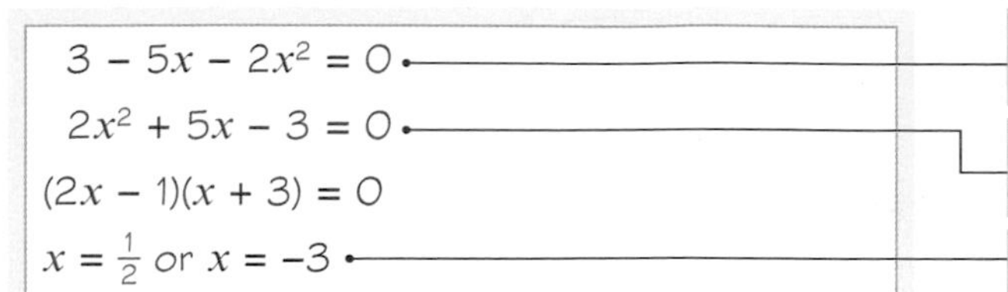

$3 - 5x - 2x^2 = 0$ — Quadratic equation.

$2x^2 + 5x - 3 = 0$ — Multiply by −1 (so it's easier to factorise).

$(2x - 1)(x + 3) = 0$

$x = \frac{1}{2}$ or $x = -3$ — $\frac{1}{2}$ and −3 are the critical values.

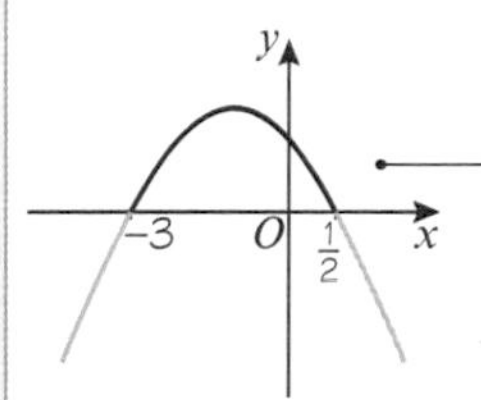

Draw a sketch to show the shape of the graph and the critical values.

Since the coefficient of x^2 is negative, the graph is 'upside-down ∪-shaped'. It crosses the x-axis at −3 and $\frac{1}{2}$. ← Section 2.4

So the required set of values is

$x < -3$ or $x > \frac{1}{2}$.

$3 - 5x - 2x^2 < 0$ ($y < 0$) for the outer parts of the graph, below the x-axis, as shown by the paler parts of the curve.

In set notation this can be written as

$\{x : x < -3\} \cup \{x : x > \frac{1}{2}\}$.

Example 10

a Find the set of values of x for which $12 + 4x > x^2$.

b Hence find the set of values for which $12 + 4x > x^2$ and $5x - 3 > 2$.

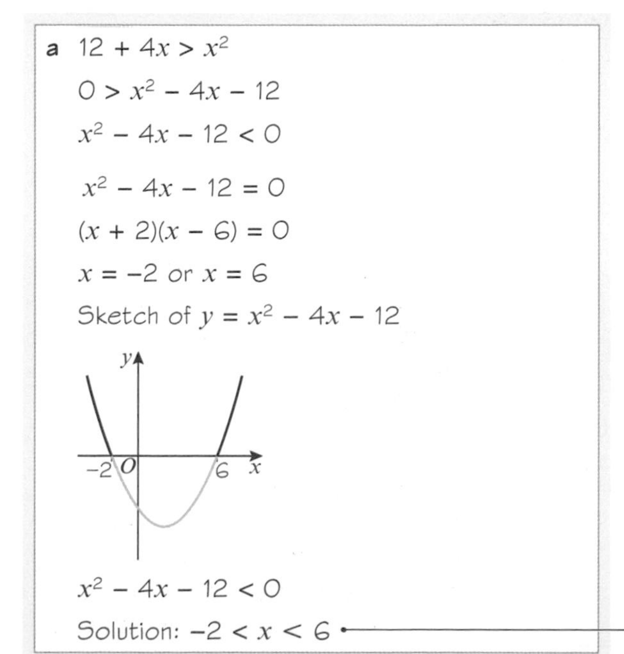

a $12 + 4x > x^2$

$0 > x^2 - 4x - 12$

$x^2 - 4x - 12 < 0$

$x^2 - 4x - 12 = 0$

$(x + 2)(x - 6) = 0$

$x = -2$ or $x = 6$

Sketch of $y = x^2 - 4x - 12$

$x^2 - 4x - 12 < 0$

Solution: $-2 < x < 6$

You can use a table to check your solution.

$-2 < x < 6$

Use the critical values to split the real number line into sets.

	$x < -2$	$-2 < x < 6$	$x > 6$
$x + 2$	−	+	+
$x - 6$	−	−	+
$(x + 2)(x - 6)$	+	−	+

For each set, check whether the set of values makes the value of the bracket positive or negative. For example, if $x < -2$, $(x + 2)$ is negative, $(x - 6)$ is negative, and $(x + 2)(x - 6)$ is (neg) × (neg) = positive.

In set notation the solution is $\{x : -2 < x < 6\}$.

b Solving $12 + 4x > x^2$ gives $-2 < x < 6$.

Solving $5x - 3 > 2$ gives $x > 1$.

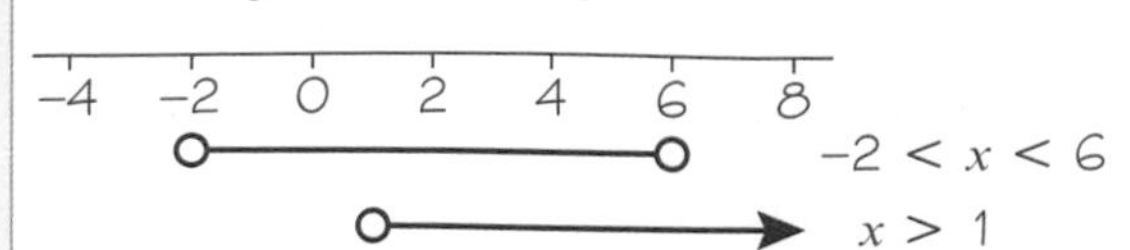

The two sets of values overlap where $1 < x < 6$.

So the solution is $1 < x < 6$.

In set notation this can be written as $\{x : 1 < x < 6\}$.

Problem-solving

This question is easier if you represent the information in more than one way. Use a sketch graph to solve the quadratic inequality, and use a number line to combine it with the linear inequality.

Example 11

Find the set of values for which $\frac{6}{x} > 2$, $x \neq 0$

$$\frac{6}{x} > 2$$

$$6x > 2x^2$$

$$6x - 2x^2 > 0$$

$$6x - 2x^2 = 0$$

Solve the corresponding quadratic equation to find the critical values.

$$x(6 - 2x) = 0$$

$$x = 0 \text{ or } x = 3$$

$x = 0$ can still be a critical value even though $x \neq 0$. But it would not be part of the solution set, even if the inequality was $\geqslant$ rather than $>$.

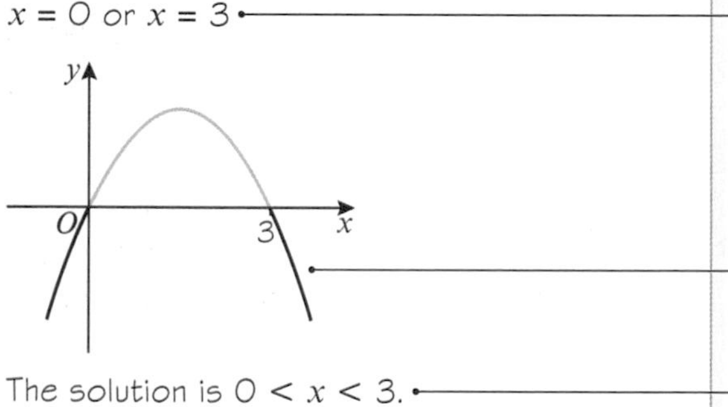

Sketch $y = x(6 - 2x)$. You are interested in the values of x where the graph is above the x-axis.

The solution is $0 < x < 3$.

In set notation this can be written as $\{x : 0 < x < 3\}$.

Watch out x could be either positive or negative, so you can't multiply both sides of this inequality by x. Instead, multiply both sides by x^2. Because x^2 is never negative, and $x \neq 0$ so $x^2 \neq 0$, the inequality sign stays the same.

Exercise 3E

1 Find the set of values of x for which:

a $x^2 - 11x + 24 < 0$ **b** $12 - x - x^2 > 0$ **c** $x^2 - 3x - 10 > 0$

d $x^2 + 7x + 12 \geqslant 0$ **e** $7 + 13x - 2x^2 > 0$ **f** $10 + x - 2x^2 < 0$

g $4x^2 - 8x + 3 \leqslant 0$ **h** $-2 + 7x - 3x^2 < 0$ **i** $x^2 - 9 < 0$

j $6x^2 + 11x - 10 > 0$ **k** $x^2 - 5x > 0$ **l** $2x^2 + 3x \leqslant 0$

2 Find the set of values of x for which:

a $x^2 < 10 - 3x$ **b** $11 < x^2 + 10$

c $x(3 - 2x) > 1$ **d** $x(x + 11) < 3(1 - x^2)$

3 Use set notation to describe the set of values of x for which:

a $x^2 - 7x + 10 < 0$ and $3x + 5 < 17$

b $x^2 - x - 6 > 0$ and $10 - 2x < 5$

c $4x^2 - 3x - 1 < 0$ and $4(x + 2) < 15 - (x + 7)$

d $2x^2 - x - 1 < 0$ and $14 < 3x - 2$

e $x^2 - x - 12 > 0$ and $3x + 17 > 2$

f $x^2 - 2x - 3 < 0$ and $x^2 - 3x + 2 > 0$

(P) 4 Given that $x \neq 0$, find the set of values of x for which:

a $\frac{2}{x} < 1$

b $5 > \frac{4}{x}$

c $\frac{1}{x} + 3 > 2$

d $6 + \frac{5}{x} > \frac{8}{x}$

e $25 > \frac{1}{x^2}$

f $\frac{6}{x^2} + \frac{7}{x} \leqslant 3$

5 **a** Find the range of values of k for which the equation $x^2 - kx + (k + 3) = 0$ has no real roots.

b Find the range of values of p for which the roots of the equation $px^2 + px - 2 = 0$ are real.

Hint The quadratic equation $ax^2 + bx + c = 0$ has real roots if $b^2 - 4ac \geqslant 0$. ← Section 2.5

(E) 6 Find the set of values of x for which $x^2 - 5x - 14 > 0$. **(4 marks)**

(E) 7 Find the set of values of x for which

a $2(3x - 1) < 4 - 3x$ **(2 marks)**

b $2x^2 - 5x - 3 < 0$ **(4 marks)**

c **both** $2(3x - 1) < 4 - 3x$ **and** $2x^2 - 5x - 3 < 0$. **(2 marks)**

(E/P) 8 Given that $x \neq 3$, find the set of values for which $\frac{5}{x - 3} < 2$. **(6 marks)**

Problem-solving

Multiply both sides of the inequality by $(x - 3)^2$.

(E/P) 9 The equation $kx^2 - 2kx + 3 = 0$, where k is a constant, has no real roots. Prove that k satisfies the inequality $0 \leqslant k < 3$. **(4 marks)**

3.6 Inequalities on graphs

You may be asked to interpret graphically the solutions to inequalities by considering the graphs of functions that are related to them.

- **The values of x for which the curve $y = f(x)$ is below the curve $y = g(x)$ satisfy the inequality $f(x) < g(x)$.**
- **The values of x for which the curve $y = f(x)$ is above the curve $y = g(x)$ satisfy the inequality $f(x) > g(x)$.**

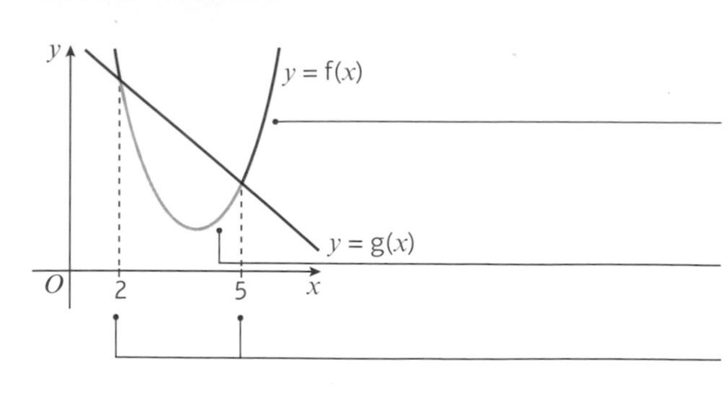

f(x) is above g(x) when $x < 2$ and when $x > 5$. These values of x satisfy f(x) > g(x).

f(x) is below g(x) when $2 < x < 5$. These values of x satisfy f(x) < g(x).

The solutions to f(x) = g(x) are $x = 2$ and $x = 5$.

Example 12

L_1 has equation $y = 12 + 4x$.

L_2 has equation $y = x^2$.

The diagram shows a sketch of L_1 and L_2 on the same axes.

a Find the coordinates of P_1 and P_2, the points of intersection.

b Hence write down the solution to the inequality $12 + 4x > x^2$.

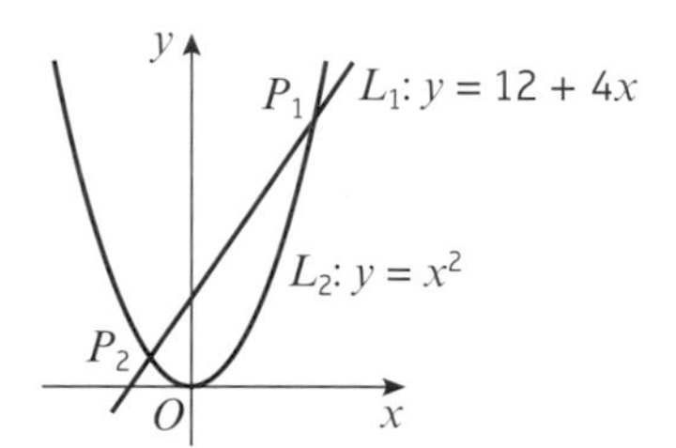

a $x^2 = 12 + 4x$

$x^2 - 4x - 12 = 0$

$(x - 6)(x + 2) = 0$

$x = 6$ and $x = -2$

substitute into $y = x^2$

when $x = 6$, $y = 36$ P_1 (6, 36)

when $x = -2$, $y = 4$ P_2 (−2, 4)

b $12 + 4x > x^2$ when the graph of L_1 is above the graph of L_2

$-2 < x < 6$

Equate to find the points of intersection, then rearrange to solve the quadratic equation.

Factorise to find the x-coordinates at the points of intersection.

This is the range of values of x for which the graph of $y = 12 + 4x$ is above the graph of $y = x^2$ i.e. between the two points of intersection. In set notation this is $\{x : -2 < x < 6\}$.

Exercise 3F

1 L_1 has equation $2y + 3x = 6$.

L_2 has the equation $x - y = 5$.

The diagram shows a sketch of L_1 and L_2.

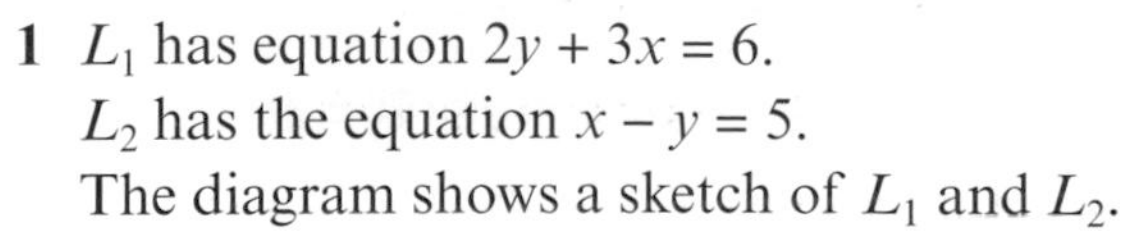

a Find the coordinates of P, the point of intersection.

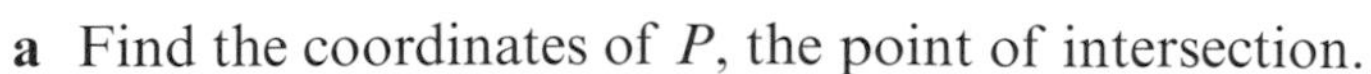

b Hence write down the solution to the inequality $2y + 3x > x - y$.

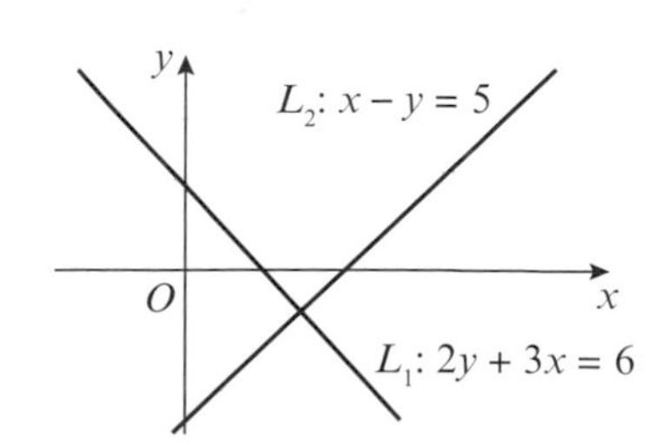

2 For each pair of functions:

i Sketch the graphs of $y = f(x)$ and $y = g(x)$ on the same axes.

ii Find the coordinates of any points of intersection.

iii Write down the solutions to the inequality $f(x) \leqslant g(x)$.

a $f(x) = 3x - 7$
$g(x) = 13 - 2x$

b $f(x) = 8 - 5x$
$g(x) = 14 - 3x$

c $f(x) = x^2 + 5$
$g(x) = 5 - 2x$

d $f(x) = 3 - x^2$
$g(x) = 2x - 12$

e $f(x) = x^2 - 5$
$g(x) = 7x + 13$

f $f(x) = 7 - x^2$
$g(x) = 2x - 8$

(P) **3** Find the set of values of x for which the curve with equation $y = f(x)$ is below the line with equation $y = g(x)$.

a $f(x) = 3x^2 - 2x - 1$
$g(x) = x + 5$

b $f(x) = 2x^2 - 4x + 1$
$g(x) = 3x - 2$

c $f(x) = 5x - 2x^2 - 4$
$g(x) = -2x - 1$

d $f(x) = \frac{2}{x}, x \neq 0$
$g(x) = 1$

e $f(x) = \frac{3}{x^2} - \frac{4}{x}, x \neq 0$
$g(x) = -1$

f $f(x) = \frac{2}{x+1}, x \neq -1$
$g(x) = 8$

Challenge

The sketch shows the graphs of

$f(x) = x^2 - 4x - 12$
$g(x) = 6 + 5x - x^2$

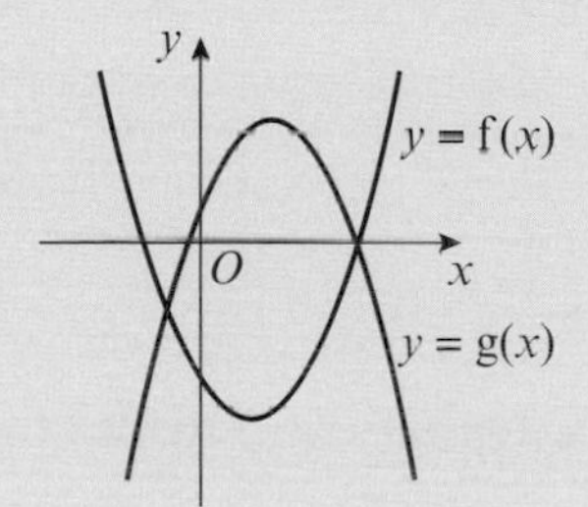

a Find the coordinates of the points of intersection.

b Find the set of values of x for which $f(x) < g(x)$.
Give your answer in set notation.

3.7 Regions

You can use shading on graphs to identify regions that satisfy linear and quadratic inequalities.

- **$y < f(x)$ represents the points on the coordinate grid below the curve $y = f(x)$.**
- **$y > f(x)$ represents the points on the coordinate grid above the curve $y = f(x)$.**

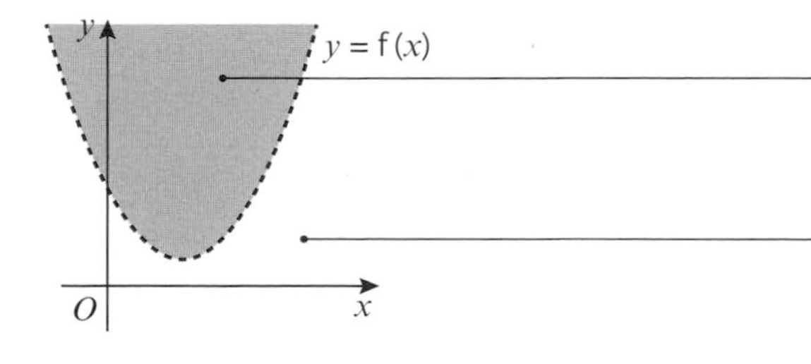

All the shaded points in this region satisfy the inequality $y > f(x)$.

All the unshaded points in this region satisfy the inequality $y < f(x)$.

- **If $y > f(x)$ or $y < f(x)$ then the curve $y = f(x)$ is not included in the region and is represented by a dotted line.**
- **If $y \geqslant f(x)$ or $y \leqslant f(x)$ then the curve $y = f(x)$ is included in the region and is represented by a solid line.**

Example 13

On graph paper, shade the region that satisfies the inequalities:

$y \geqslant -2$, $x < 5$, $y \leqslant 3x + 2$ and $x > 0$.

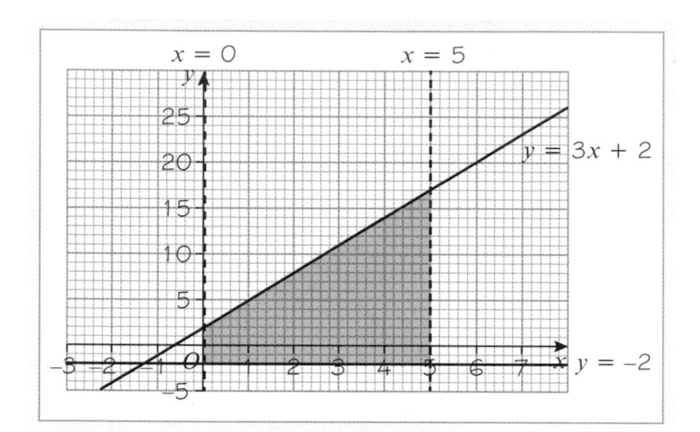

Draw dotted lines for $x = 0$, $x = 5$.

Shade the required region.
Test a point in the region. Try (1, 2).
For $x = 1$: $1 < 5$ and $1 > 0$ ✓
For $y = 2$: $2 \geqslant -2$ and $2 \leqslant 3 + 2$ ✓

Draw solid lines for $y = -2$, $y = 3x + 2$.

Example 14

On graph paper, shade the region that satisfies the inequalities:

$2y + x < 14$

$y \geqslant x^2 - 3x - 4$

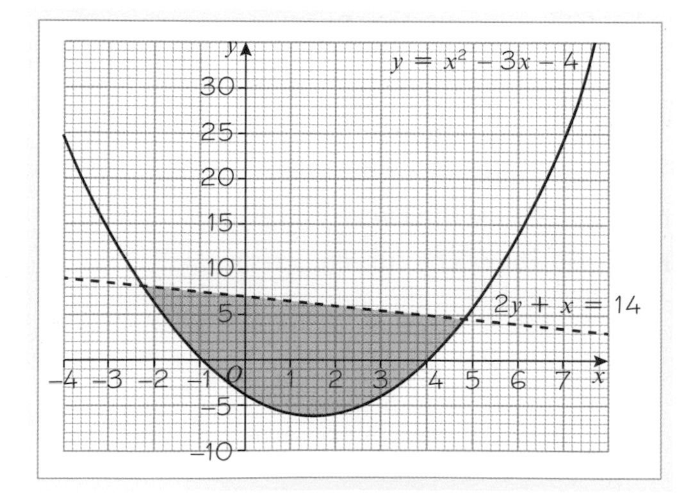

Online Explore which regions on the graph satisfy which inequalities using technology.

Draw a dotted line for $2y + x = 14$ and a solid line for $y = x^2 - 3x - 4$.

Shade the required region.
Test a point in the region. Try (0, 0).
$0 + 0 < 14$ and $0 > 0 - 0 - 4$ ✓

Exercise 3G

1 On a coordinate grid, shade the region that satisfies the inequalities:

$y > x - 2$, $y < 4x$ and $y \leqslant 5 - x$.

2 On a coordinate grid, shade the region that satisfies the inequalities:

$x \geqslant -1$, $y + x < 4$, $2x + y \leqslant 5$ and $y > -2$.

3 On a coordinate grid, shade the region that satisfies the inequalities:

$y < (3 - x)(2 + x)$ and $y + x \geqslant 3$.

4 On a coordinate grid, shade the region that satisfies the inequalities:

$y > x^2 - 2$ and $y \leqslant 9 - x^2$.

5 On a coordinate grid, shade the region that satisfies the inequalities:

$y > (x - 3)^2$, $y + x \geqslant 5$ and $y < x - 1$.

6 The sketch shows the graphs of the straight lines with equations:

$y = x + 1$, $y = 7 - x$ and $x = 1$.

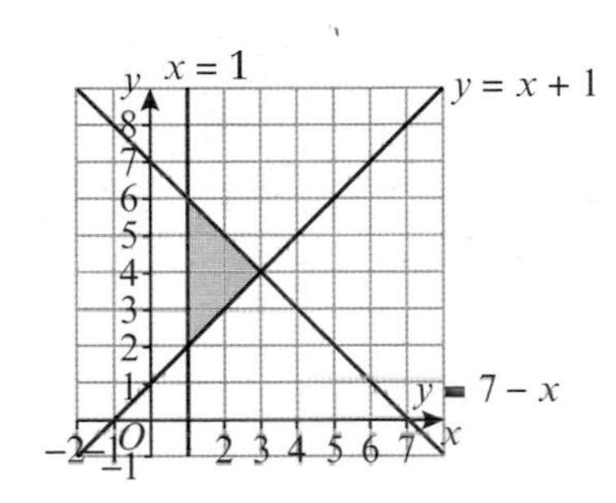

a Work out the coordinates of the points of intersection of the functions.

b Write down the set of inequalities that represent the shaded region shown in the sketch.

7 The sketch shows the graphs of the curves with equations:

$y = 2 - 5x - x^2$, $2x + y = 0$ and $x + y = 4$.

Write down the set of inequalities that represent the shaded region shown in the sketch.

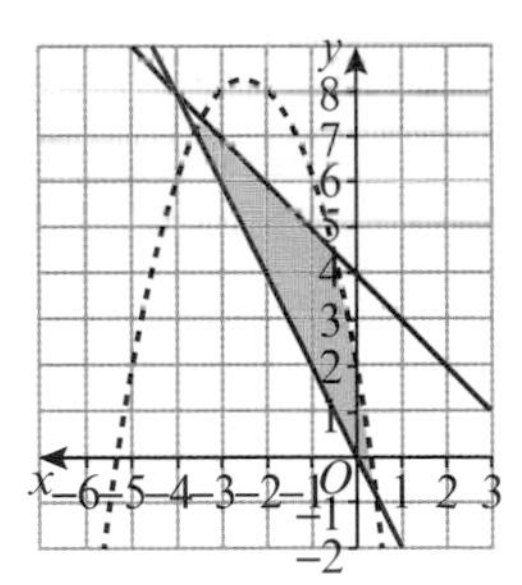

(P) **8** **a** On a coordinate grid, shade the region that satisfies the inequalities

$y < x + 4$, $y + 5x + 3 \geqslant 0$, $y \geqslant -1$ and $x < 2$.

b Work out the coordinates of the vertices of the shaded region.

c Which of the vertices lie within the region identified by the inequalities?

d Work out the area of the shaded region.

Problem-solving

A vertex is only included if both intersecting lines are included.

Mixed exercise 3

(E) **1** $2kx - y = 4$
$4kx + 3y = -2$

are two simultaneous equations, where k is a constant.

a Show that $y = -2$. **(3 marks)**

b Find an expression for x in terms of the constant k. **(1 mark)**

(E) **2** Solve the simultaneous equations:

$x + 2y = 3$
$x^2 - 4y^2 = -33$ **(7 marks)**

(E) **3** Given the simultaneous equations

$x - 2y = 1$
$3xy - y^2 = 8$

a Show that $5y^2 + 3y - 8 = 0$. **(2 marks)**

b Hence find the pairs (x, y) for which the simultaneous equations are satisfied. **(5 marks)**

(E) **4 a** By eliminating y from the equations

$x + y = 2$
$x^2 + xy - y^2 = -1$

show that $x^2 - 6x + 3 = 0$. **(2 marks)**

b Hence, or otherwise solve the simultaneous equations

$x + y = 2$
$x^2 + xy - y^2 = -1$

giving x and y in the form $a \pm b\sqrt{6}$, where a and b are integers. **(5 marks)**

(E) **5 a** Given that $3^x = 9^{y-1}$, show that $x = 2y - 2$. **(1 mark)**

b Solve the simultaneous equations:

$x = 2y - 2$
$x^2 = y^2 + 7$ **(6 marks)**

(E) **6** Solve the simultaneous equations:

$x + 2y = 3$
$x^2 - 2y + 4y^2 = 18$ **(7 marks)**

(E/P) **7** The curve and the line given by the equations

$kx^2 - xy + (k + 1)x = 1$
$-\frac{k}{2}x + y = 1$

where k is a non-zero constant, intersect at a single point.

a Find the value of k. **(5 marks)**

b Give the coordinates of the point of intersection of the line and the curve. **(3 marks)**

E/P **8** A person throws a ball in a sports hall. The height of the ball, h m, can be modelled in relation to the horizontal distance from the point it was thrown from by the quadratic equation:

$$h = -\frac{3}{10}x^2 + \frac{5}{2}x + \frac{3}{2}$$

The hall has a sloping ceiling which can be modelled with equation $h = \frac{15}{2} - \frac{1}{5}x$.

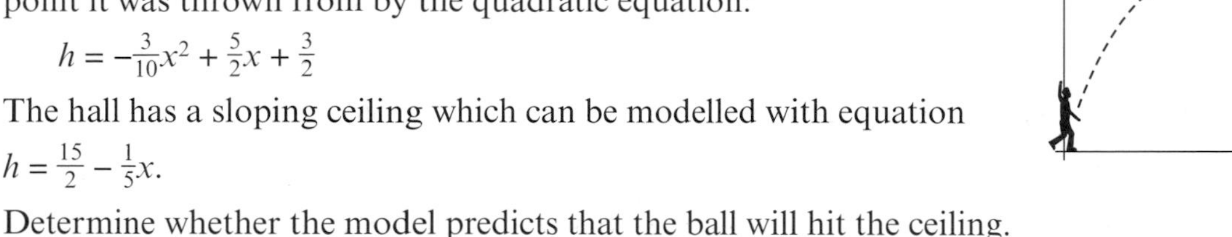

Determine whether the model predicts that the ball will hit the ceiling. **(5 marks)**

E **9** Give your answers in set notation.

a Solve the inequality $3x - 8 > x + 13$. **(2 marks)**

b Solve the inequality $x^2 - 5x - 14 > 0$. **(4 marks)**

E **10** Find the set of values of x for which $(x - 1)(x - 4) < 2(x - 4)$. **(6 marks)**

E **11** **a** Use algebra to solve $(x - 1)(x + 2) = 18$. **(2 marks)**

b Hence, or otherwise, find the set of values of x for which $(x - 1)(x + 2) > 18$. Give your answer in set notation. **(2 marks)**

12 Find the set of values of x for which:

a $6x - 7 < 2x + 3$ **(2 marks)**

b $2x^2 - 11x + 5 < 0$ **(4 marks)**

c $5 < \frac{20}{x}$ **(4 marks)**

d both $6x - 7 < 2x + 3$ and $2x^2 - 11x + 5 < 0$. **(2 marks)**

E **13** Find the set of values of x that satisfy $\frac{8}{x^2} + 1 \leqslant \frac{9}{x}$, $x \neq 0$ **(5 marks)**

E **14** Find the values of k for which $kx^2 + 8x + 5 = 0$ has real roots. **(3 marks)**

E/P **15** The equation $2x^2 + 4kx - 5k = 0$, where k is a constant, has no real roots. Prove that k satisfies the inequality $-\frac{5}{2} < k < 0$. **(3 marks)**

E **16** **a** Sketch the graphs of $y = \mathrm{f}(x) = x^2 + 2x - 15$ and $\mathrm{g}(x) = 6 - 2x$ on the same axes. **(4 marks)**

b Find the coordinates of any points of intersection. **(3 marks)**

c Write down the set of values of x for which $\mathrm{f}(x) > \mathrm{g}(x)$. **(1 mark)**

E **17** Find the set of values of x for which the curve with equation $y = 2x^2 + 3x - 15$ is below the line with equation $y = 8 + 2x$. **(5 marks)**

E **18** On a coordinate grid, shade the region that satisfies the inequalities:

$y > x^2 + 4x - 12$ and $y < 4 - x^2$. **(5 marks)**

E/P **19** **a** On a coordinate grid, shade the region that satisfies the inequalities

$y + x < 6$, $y < 2x + 9$, $y > 3$ and $x > 0$. **(6 marks)**

b Work out the area of the shaded region. **(2 marks)**

Challenge

1 Find the possible values of k for the quadratic equation $2kx^2 + 5kx + 5k - 3 = 0$ to have real roots.

2 A straight line has equation $y = 2x - k$ and a parabola has equation $y = 3x^2 + 2kx + 5$ where k is a constant. Find the range of values of k for which the line and the parabola do not intersect.

Summary of key points

1 Linear simultaneous equations can be solved using elimination or substitution.

2 Simultaneous equations with one linear and one quadratic equation can have up to two pairs of solutions. You need to make sure the solutions are paired correctly.

3 The solutions of a pair of simultaneous equations represent the points of intersection of their graphs.

4 For a pair of simultaneous equations that produce a quadratic equation of the form $ax^2 + bx + c = 0$:

- $b^2 - 4ac > 0$ two real solutions
- $b^2 - 4ac = 0$ one real solution
- $b^2 - 4ac < 0$ no real solutions

5 The solution of an inequality is the set of all real numbers x that make the inequality true.

6 To solve a quadratic inequality:

- Rearrange so that the right-hand side of the inequality is 0
- Solve the corresponding quadratic equation to find the critical values
- Sketch the graph of the quadratic function
- Use your sketch to find the required set of values.

7 The values of x for which the curve $y = f(x)$ is **below** the curve $y = g(x)$ satisfy the inequality $f(x) < g(x)$.

The values of x for which the curve $y = f(x)$ is **above** the curve $y = g(x)$ satisfy the inequality $f(x) > g(x)$.

8 $y < f(x)$ represents the points on the coordinate grid below the curve $y = f(x)$.

$y > f(x)$ represents the points on the coordinate grid above the curve $y = f(x)$.

9 If $y > f(x)$ or $y < f(x)$ then the curve $y = f(x)$ is not included in the region and is represented by a dotted line.

If $y \geqslant f(x)$ or $y \leqslant f(x)$ then the curve $y = f(x)$ is included in the region and is represented by a solid line.

Graphs and transformations

4

Objectives

After completing this chapter you should be able to:

Many complicated functions can be understood by transforming simpler functions using stretches, reflections and translations. Particle physicists compare observed results with transformations of known functions to determine the nature of subatomic particles.

Prior knowledge check

1 Factorise these quadratic expressions:

a $x^2 + 6x + 5$ **b** $x^2 - 4x + 3$

← GCSE Mathematics

2 Sketch the graphs of the following functions:

a $y = (x + 2)(x - 3)$ **b** $y = x^2 - 6x - 7$

← Section 2.4

3 **a** Copy and complete the table of values for the function $y = x^3 + x - 2$.

x	−2	−1.5	−1	−0.5	0	0.5	1	1.5	2
y	−12	−6.875			−2	−1.375			

b Use your table of values to draw the graph of $y = x^3 + x - 2$.

← GCSE Mathematics

4 Solve each pair of simultaneous equations:

a $y = 2x$
$x + y = 6$

b $y = x^2$
$y = 2x - 1$

← Sections 3.1, 3.2

4.1 Cubic graphs

A **cubic function** has the form $f(x) = ax^3 + bx^2 + cx + d$, where a, b, c and d are real numbers and a is non-zero.

The graph of a cubic function can take several different forms, depending on the exact nature of the function.

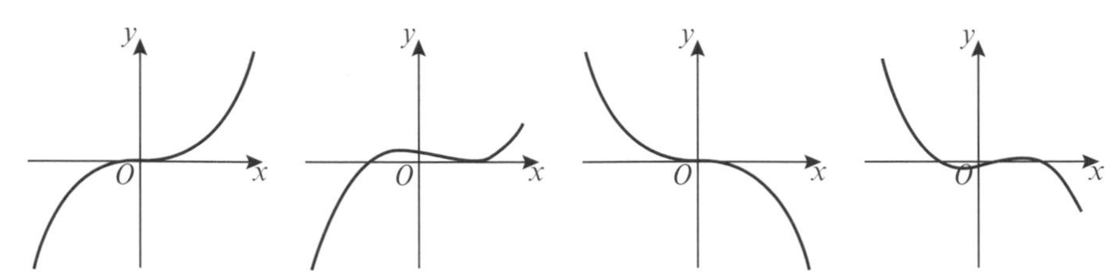

For these two functions a is positive. For these two functions a is negative.

- **If p is a root of the function $f(x)$, then the graph of $y = f(x)$ touches or crosses the x-axis at the point $(p, 0)$.**

You can sketch the graph of a cubic function by finding the roots of the function.

Example 1

Sketch the curves with the following equations and show the points where they cross the coordinate axes.

a $y = (x - 2)(1 - x)(1 + x)$ **b** $y = x(x + 1)(x + 2)$

Online Explore the graph of $y = (x - p)(x - q)(x - r)$ where p, q and r are constants using technology.

a $y = (x - 2)(1 - x)(1 + x)$

$0 = (x - 2)(1 - x)(1 + x)$ — Put $y = 0$ and solve for x.

So $x = 2$, $x = 1$ or $x = -1$

So the curve crosses the x-axis at $(2, 0)$, $(1, 0)$ and $(-1, 0)$.

When $x = 0$, $y = -2 \times 1 \times 1 = -2$ — Find the value of y when $x = 0$.

So the curve crosses the y-axis at $(0, -2)$.

$x \to \infty, y \to -\infty$

$x \to -\infty, y \to \infty$ — Check what happens to y for large positive and negative values of x.

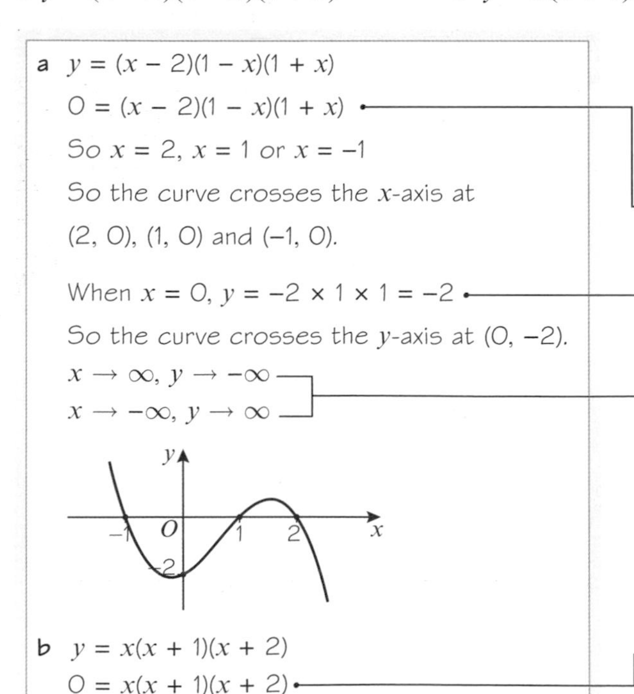

The x^3 term in the expanded function would be $x \times (-x) \times x = -x^3$ so the curve has a negative x^3 coefficient.

b $y = x(x + 1)(x + 2)$

$0 = x(x + 1)(x + 2)$ — Put $y = 0$ and solve for x.

So $x = 0$, $x = -1$ or $x = -2$

So the curve crosses the x-axis at $(0, 0)$, $(-1, 0)$ and $(-2, 0)$.

You know that the curve crosses the x-axis at (0, 0) so you don't need to calculate the y-intercept separately.

$x \to \infty, y \to \infty$
$x \to -\infty, y \to -\infty$

Check what happens to y for large positive and negative values of x.

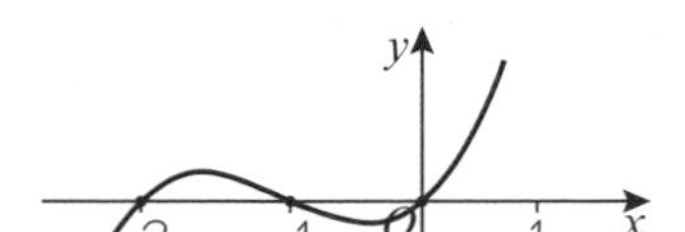

The x^3 term in the expanded function would be $x \times x \times x = x^3$ so the curve has a positive x^3 coefficient.

Example 2

Sketch the following curves.

a $y = (x - 1)^2(x + 1)$ **b** $y = x^3 - 2x^2 - 3x$ **c** $y = (x - 2)^3$

a $y = (x - 1)^2(x + 1)$
$0 = (x - 1)^2(x + 1)$

Put $y = 0$ and solve for x.

So $x = 1$ or $x = -1$

So the curve crosses the x-axis at $(-1, 0)$ and touches the x-axis at $(1, 0)$.

$(x - 1)$ is squared so $x = 1$ is a 'double' repeated root. This means that the curve just touches the x-axis at (1, 0).

When $x = 0$, $y = (-1)^2 \times 1 = 1$
So the curve crosses the y-axis at $(0, 1)$.

Find the value of y when $x = 0$.

$x \to \infty, y \to \infty$
$x \to -\infty, y \to -\infty$

Check what happens to y for large positive and negative values of x.

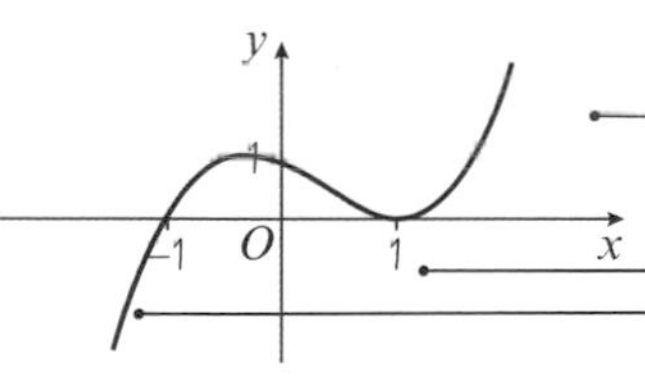

$x \to \infty, y \to \infty$

$x = 1$ is a 'double' repeated root.

$x \to -\infty, y \to -\infty$

b $y = x^3 - 2x^3 - 3x$
$= x(x^2 - 2x - 3)$
$= x(x - 3)(x + 1)$

First factorise.

$0 = x(x - 3)(x + 1)$
So $x = 0$, $x = 3$ or $x = -1$
So the curve crosses the x-axis at $(0, 0)$, $(3, 0)$ and $(-1, 0)$.

$x \to \infty, y \to \infty$
$x \to -\infty, y \to -\infty$

Check what happens to y for large positive and negative values of x.

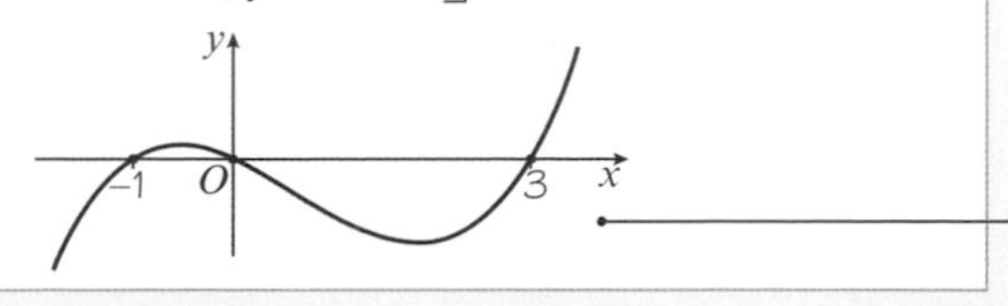

This is a cubic curve with a positive coefficient of x^3 and three distinct roots.

c $y = (x - 2)^3$

$0 = (x - 2)^3$

So $x = 2$ and the curve crosses the x-axis at $(2, 0)$ only.

When $x = 0$, $y = (-2)^3 = -8$

So the curve crosses the y-axis at $(0, -8)$.

$x \to \infty, y \to \infty$

$x \to -\infty, y \to -\infty$

Check what happens to y for large positive and negative values of x.

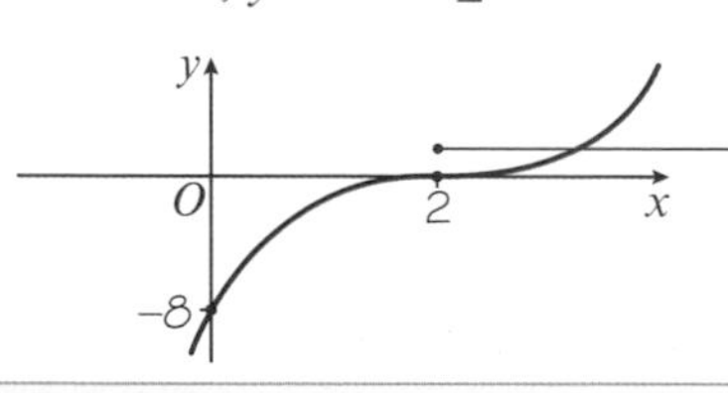

$x = 2$ is a 'triple' repeated root.

Example 3

Sketch the curve with equation $y = (x - 1)(x^2 + x + 2)$.

$y = (x - 1)(x^2 + x + 2)$

$0 = (x - 1)(x^2 + x + 2)$

So $x = 1$ only and the curve crosses the x-axis at $(1, 0)$.

The quadratic factor $x^2 + x + 2$ gives no solutions since the discriminant $b^2 - 4ac = (1)^2 - 4(1)(2) = -7$.

← Section 2.5

Watch out A cubic graph could intersect the x-axis at 1, 2 or 3 points.

When $x = 0$, $y = (-1)(2) = -2$

So the curve crosses the y-axis at $(0, -2)$.

$x \to \infty, y \to \infty$

$x \to -\infty, y \to -\infty$

Check what happens to y for large positive and negative values of x.

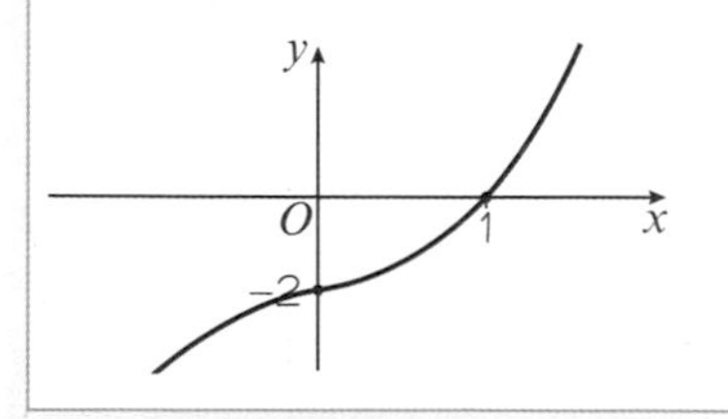

You haven't got enough information to know the exact shape of the graph. It could also be shaped like this:

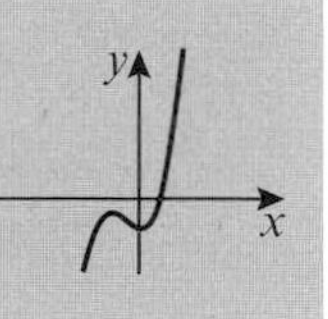

Exercise 4A

1 Sketch the following curves and indicate clearly the points of intersection with the axes:

a $y = (x - 3)(x - 2)(x + 1)$ **b** $y = (x - 1)(x + 2)(x + 3)$

c $y = (x + 1)(x + 2)(x + 3)$ **d** $y = (x + 1)(1 - x)(x + 3)$

e $y = (x - 2)(x - 3)(4 - x)$ **f** $y = x(x - 2)(x + 1)$

g $y = x(x + 1)(x - 1)$ **h** $y = x(x + 1)(1 - x)$

i $y = (x - 2)(2x - 1)(2x + 1)$ **j** $y = x(2x - 1)(x + 3)$

2 Sketch the curves with the following equations:

a $y = (x + 1)^2(x - 1)$ **b** $y = (x + 2)(x - 1)^2$ **c** $y = (2 - x)(x + 1)^2$

d $y = (x - 2)(x + 1)^2$ **e** $y = x^2(x + 2)$ **f** $y = (x - 1)^2x$

g $y = (1 - x)^2(3 + x)$ **h** $y = (x - 1)^2(3 - x)$ **i** $y = x^2(2 - x)$

j $y = x^2(x - 2)$

3 Factorise the following equations and then sketch the curves:

a $y = x^3 + x^2 - 2x$ **b** $y = x^3 + 5x^2 + 4x$ **c** $y = x^3 + 2x^2 + x$

d $y = 3x + 2x^2 - x^3$ **e** $y = x^3 - x^2$ **f** $y = x - x^3$

g $y = 12x^3 - 3x$ **h** $y = x^3 - x^2 - 2x$ **i** $y = x^3 - 9x$

j $y = x^3 - 9x^2$

4 Sketch the following curves and indicate the coordinates of the points where the curves cross the axes:

a $y = (x - 2)^3$ **b** $y = (2 - x)^3$ **c** $y = (x - 1)^3$ **d** $y = (x + 2)^3$

e $y = -(x + 2)^3$ **f** $y = (x + 3)^3$ **g** $y = (x - 3)^3$ **h** $y = (1 - x)^3$

i $y = -(x - 2)^3$ **j** $y = -\left(x - \frac{1}{2}\right)^3$

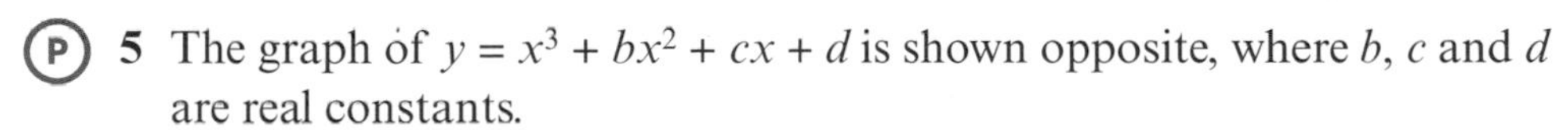

(P) **5** The graph of $y = x^3 + bx^2 + cx + d$ is shown opposite, where b, c and d are real constants.

a Find the values of b, c and d. **(3 marks)**

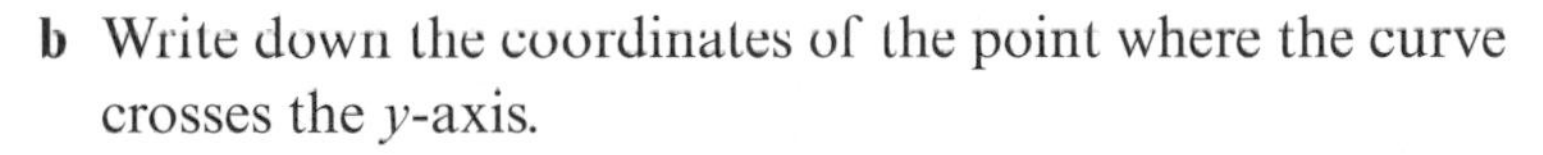

b Write down the coordinates of the point where the curve crosses the y-axis. **(1 mark)**

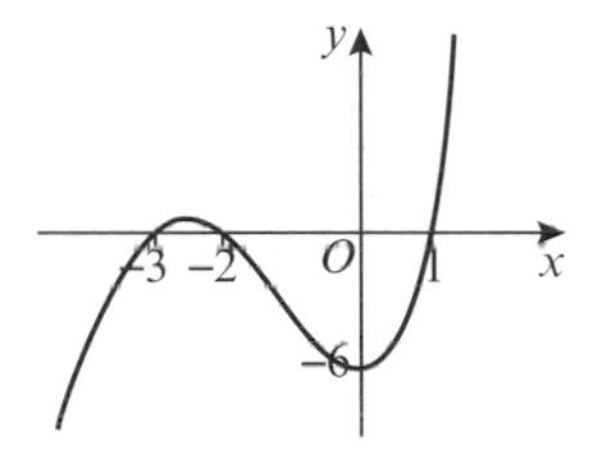

Problem-solving

Start by writing the equation in the form $y = (x - p)(x - q)(x - r)$.

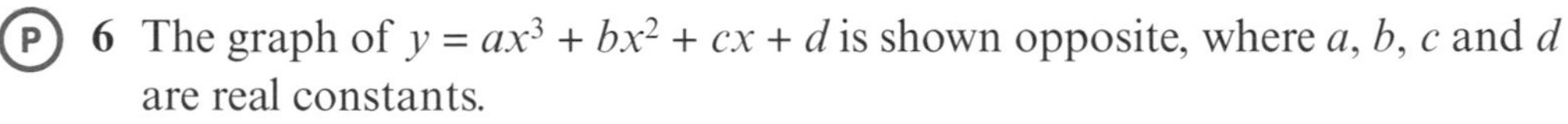

(P) **6** The graph of $y = ax^3 + bx^2 + cx + d$ is shown opposite, where a, b, c and d are real constants.
Find the values of a, b, c and d. **(4 marks)**

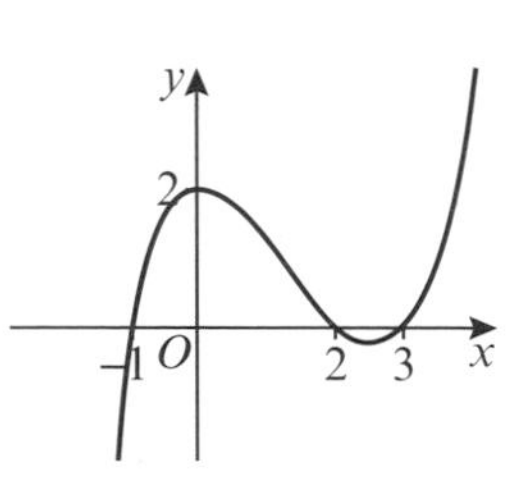

(E) **7** Given that $f(x) = (x - 10)(x^2 - 2x) + 12x$

a Express $f(x)$ in the form $x(ax^2 + bx + c)$ where a, b and c are real constants. **(3 marks)**

b Hence factorise $f(x)$ completely. **(2 marks)**

c Sketch the graph of $y = f(x)$ showing clearly the points where the graph intersects the axes. **(3 marks)**

4.2 Quartic graphs

A **quartic function** has the form $f(x) = ax^4 + bx^3 + cx^2 + dx + e$, where a, b, c, d and e are real numbers and a is non-zero.

The graph of a quartic function can take several different forms, depending on the exact nature of the function.

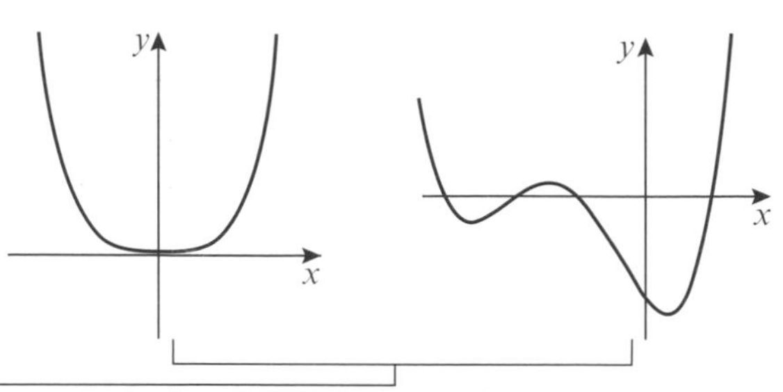

For these two functions a is positive.

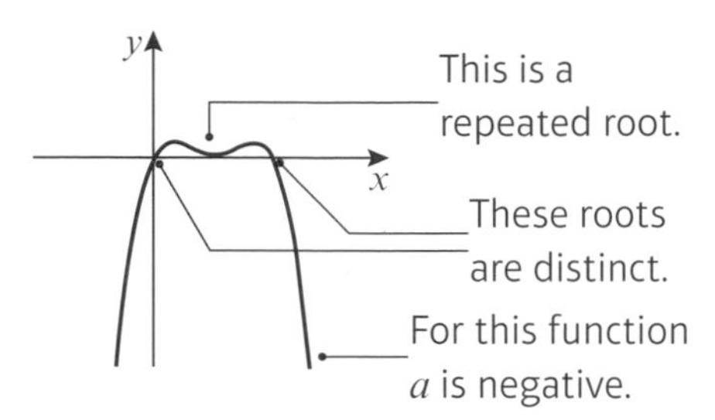

This is a repeated root.

These roots are distinct.

For this function a is negative.

You can sketch the graph of a quartic function by finding the roots of the function.

Example 4

Sketch the following curves:

a $y = (x + 1)(x + 2)(x - 1)(x - 2)$ **b** $y = x(x + 2)^2(3 - x)$ **c** $y = (x - 1)^2(x - 3)^2$

a $y = (x + 1)(x + 2)(x - 1)(x - 2)$

$0 = (x + 1)(x + 2)(x - 1)(x - 2)$

So $x = -1, -2, 1$ or 2

The curve cuts the x-axis at $(-2, 0)$, $(-1, 0)$, $(1, 0)$ and $(2, 0)$.

When $x = 0$, $y = 1 \times 2 \times (-1) \times (-2) = 4$.

So the curve cuts the y-axis at $(0, 4)$.

$x \to \infty, y \to \infty$

$x \to -\infty, y \to \infty$

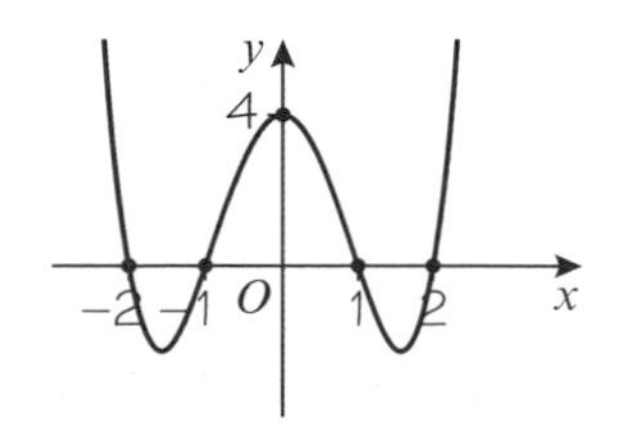

Online Explore the graph of $y = (x - p)(x - q)(x - r)(x - s)$ where p, q, r and s are constants using technology.

Set $y = 0$ and solve to find the roots of the function.

Substitute $x = 0$ into the function to find the coordinates of the y-intercept.

Check what happens to y for large positive and negative values of x.

We know the general shape of the quartic graph so we can draw a smooth curve through the points.

b $y = x(x + 2)^2(3 - x)$
$0 = x(x + 2)^2(3 - x)$
So $x = 0, -2$ or 3
The curve cuts the x-axis at (0, 0), (−2, 0) and (3, 0)

$x \to \infty, y \to -\infty$
$x \to -\infty, y \to -\infty$

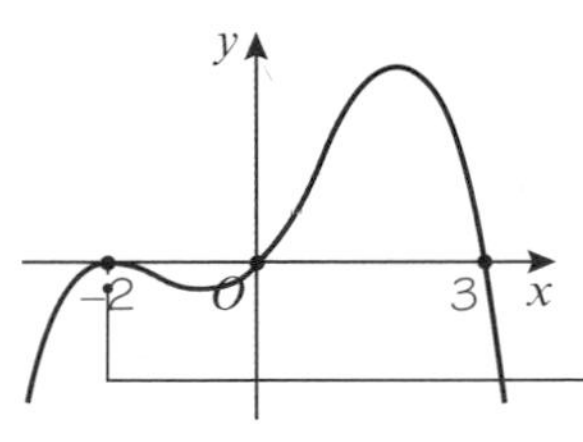

The coefficient of x^4 in the expanded function will be negative so you know the general shape of the curve.

There is a 'double' repeated root at $x = -2$ so the graph just touches the x-axis at this point.

c $y = (x - 1)^2(x - 3)^2$
$0 = (x - 1)^2(x - 3)^2$
So $x = 1$ or 3

The curve touches the x-axis at (1, 0) and (3, 0).
When $x = 0$, $y = 9$.
So the curve cuts the y-axis at (0, 9).

These are both 'double' repeated roots, so the curve will just touch the x-axis at these points.

$x \to \infty, y \to \infty$
$x \to -\infty, y \to \infty$

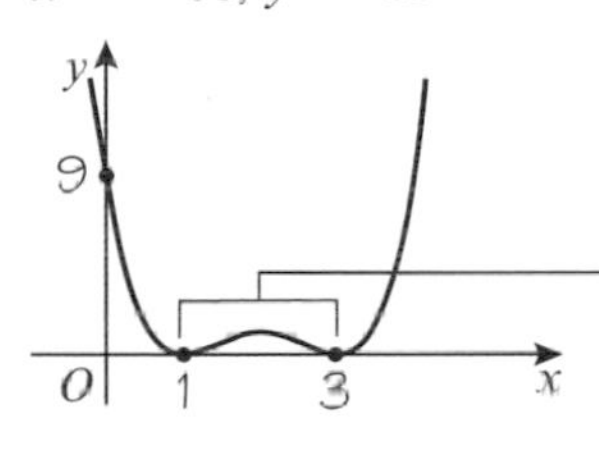

The coefficient of x^4 in the expanded function will be positive.

There are two 'double' repeated roots.

Exercise 4B

1 Sketch the following curves and indicate clearly the points of intersection with the axes:

a $y = (x + 1)(x + 2)(x + 3)(x + 4)$ **b** $y = x(x - 1)(x + 3)(x - 2)$
c $y = x(x + 1)^2(x + 2)$ **d** $y = (2x - 1)(x + 2)(x - 1)(x - 2)$
e $y = x^2(4x + 1)(4x - 1)$ **f** $y = -(x - 4)^2(x - 2)^2$
g $y = (x - 3)^2(x + 1)^2$ **h** $y = (x + 2)^3(x - 3)$
i $y = -(2x - 1)^3(x + 5)$ **j** $y = (x + 4)^4$

Hint In part **f** the coefficient of x^4 will be negative.

2 Sketch the following curves and indicate clearly the points of intersection with the axes:

a $y = (x + 2)(x - 1)(x^2 - 3x + 2)$ **b** $y = (x + 3)^2(x^2 - 5x + 6)$
c $y = (x - 4)^2(x^2 - 11x + 30)$ **d** $y = (x^2 - 4x - 32)(x^2 + 5x - 36)$

Hint Factorise the quadratic factor first.

(E/P) **3** The graph of $y = x^4 + bx^3 + cx^2 + dx + e$ is shown opposite, where b, c, d and e are real constants.

a Find the coordinates of point P. **(2 marks)**

b Find the values of b, c, d and e. **(3 marks)**

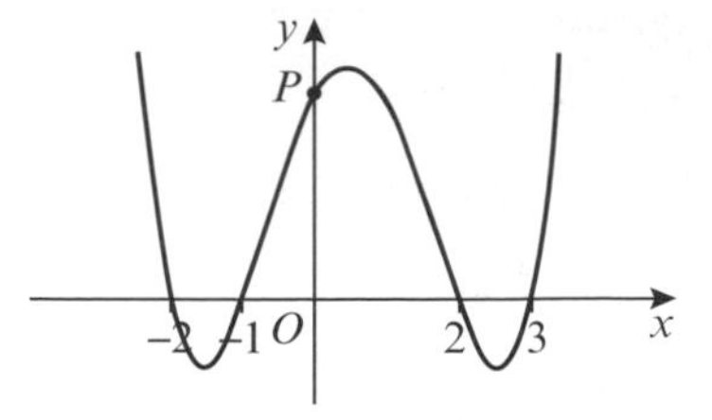

(E/P) **4** Sketch the graph of $y = (x + 5)(x - 4)(x^2 + 5x + 14)$. **(3 marks)**

Problem-solving

Consider the discriminant of the quadratic factor.

Challenge

The graph of $y = ax^4 + bx^3 + cx^2 + dx + e$ is shown, where a, b, c, d and e are real constants.

Find the values of a, b, c, d and e.

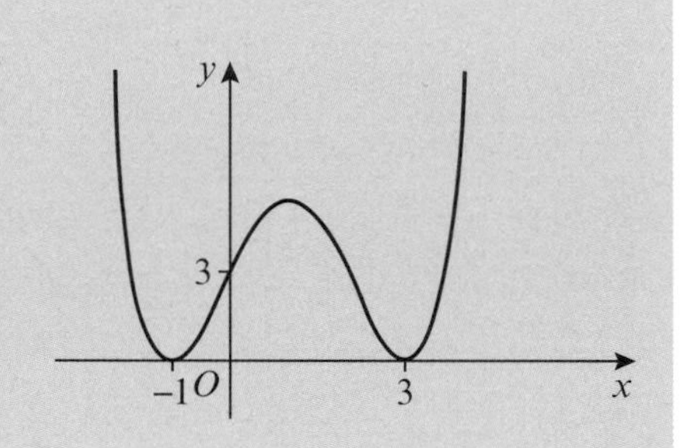

4.3 Reciprocal graphs

You can sketch graphs of **reciprocal functions** such as $y = \frac{1}{x}$, $y = \frac{1}{x^2}$ and $y = -\frac{2}{x}$ by considering their asymptotes.

- **The graphs of $y = \frac{k}{x}$ and $y = \frac{k}{x^2}$, where k is a real constant, have asymptotes at $x = 0$ and $y = 0$.**

Notation An **asymptote** is a line which the graph approaches but never reaches.

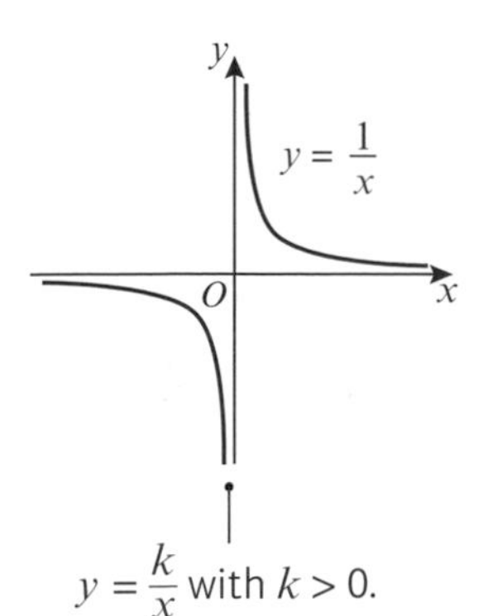

$y = \frac{k}{x}$ with $k > 0$.

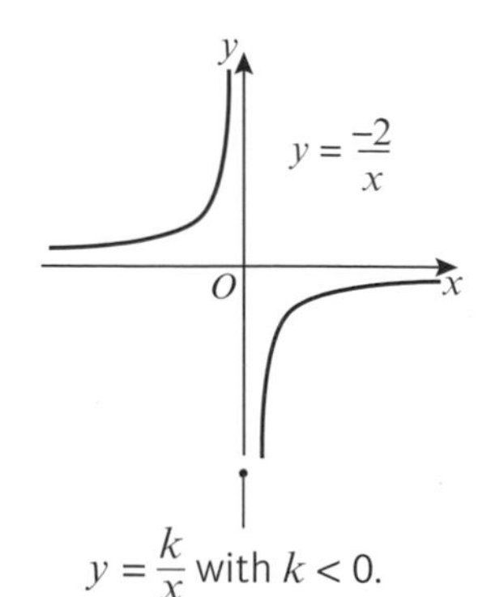

$y = \frac{k}{x}$ with $k < 0$.

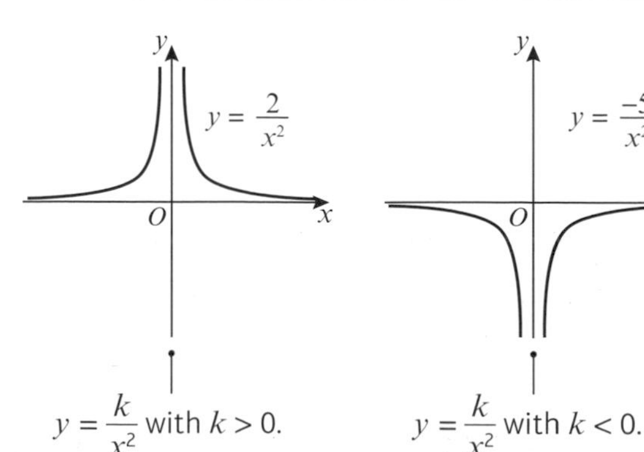

$y = \frac{k}{x^2}$ with $k > 0$.

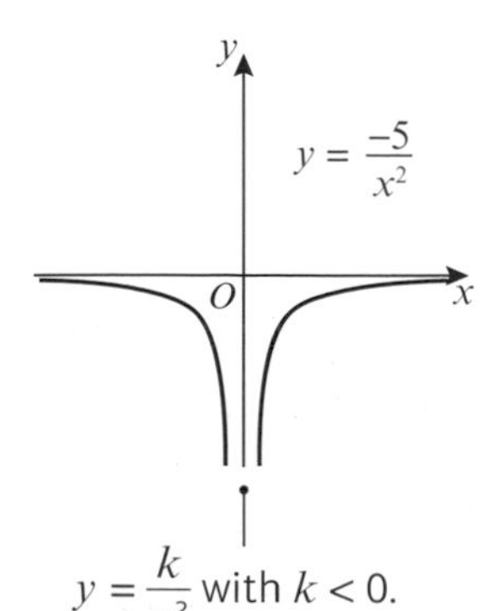

$y = \frac{k}{x^2}$ with $k < 0$.

Example 5

Sketch on the same diagram:

a $y = \frac{4}{x}$ and $y = \frac{12}{x}$ **b** $y = -\frac{1}{x}$ and $y = -\frac{3}{x}$ **c** $y = \frac{4}{x^2}$ and $y = \frac{10}{x^2}$

Online Explore the graph of $y = \frac{a}{x}$ for different values of a using technology.

This is a $y = \frac{k}{x}$ graph with $k > 0$

In this quadrant, $x > 0$ so for any values of x: $\frac{12}{x} > \frac{4}{x}$

In this quadrant, $x < 0$ so for any values of x: $\frac{12}{x} < \frac{4}{x}$

This is a $y = \frac{k}{x}$ graph with $k < 0$

In this quadrant, $x < 0$ so for any values of x: $-\frac{3}{x} > -\frac{1}{x}$

In this quadrant, $x > 0$ so for any values of x: $-\frac{3}{x} < -\frac{1}{x}$

This is a $y = \frac{k}{x^2}$ graph with $k > 0$.
x^2 is always positive and $k > 0$ so the y-values are all positive.

Exercise 4C

1 Use a separate diagram to sketch each pair of graphs.

a $y = \frac{2}{x}$ and $y = \frac{4}{x}$ **b** $y = \frac{2}{x}$ and $y = -\frac{2}{x}$ **c** $y = -\frac{4}{x}$ and $y = -\frac{2}{x}$

d $y = \frac{3}{x}$ and $y = \frac{8}{x}$ **e** $y = -\frac{3}{x}$ and $y = -\frac{8}{x}$

2 Use a separate diagram to sketch each pair of graphs.

a $y = \frac{2}{x^2}$ and $y = \frac{5}{x^2}$ **b** $y = \frac{3}{x^2}$ and $y = -\frac{3}{x^2}$ **c** $y = -\frac{2}{x^2}$ and $y = -\frac{6}{x^2}$

4.4 Points of intersection

You can sketch curves of functions to show points of intersection and solutions to equations.

- **The x-coordinate(s) at the points of intersection of the curves with equations $y = f(x)$ and $y = g(x)$ are the solution(s) to the equation $f(x) = g(x)$.**

Example 6

a On the same diagram sketch the curves with equations $y = x(x - 3)$ and $y = x^2(1 - x)$.

b Find the coordinates of the points of intersection.

a

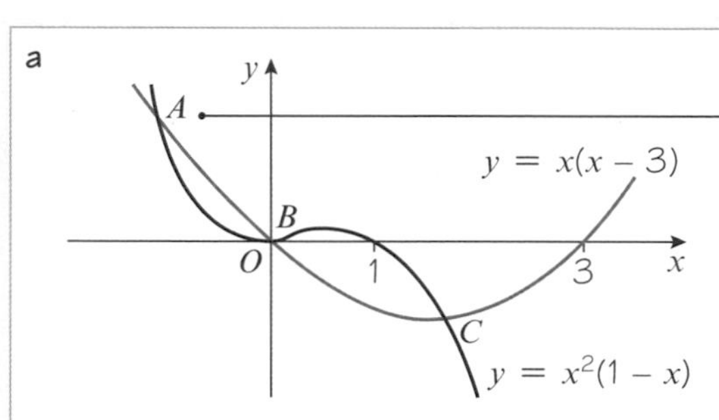

A cubic curve will eventually get steeper than a quadratic curve, so the graphs will intersect for some negative value of x.

b From the graph there are three points where the curves cross, labelled A, B and C. The x-coordinates are given by the solutions to the equation.

$$x(x - 3) = x^2(1 - x)$$

$$x^2 - 3x = x^2 - x^3$$

$$x^3 - 3x = 0$$

$$x(x^2 - 3) = 0$$

So $x = 0$ or $x^2 = 3$

So $x = -\sqrt{3}, 0, \sqrt{3}$

There are three points of intersection so the equation $x(x - 3) = x^2(1 - x)$ has three real roots.

Multiply out brackets.
Collect terms on one side.
Factorise.

Substitute into $y = x^2(1 - x)$

The graphs intersect for these values of x, so you can substitute into either equation to find the y-coordinates.

The points of intersection are:

$A(-\sqrt{3}, 3 + 3\sqrt{3})$

$B(0, 0)$

$C(\sqrt{3}, 3 - 3\sqrt{3})$

Leave your answers in surd form.

Example 7

a On the same diagram sketch the curves with equations $y = x^2(3x - a)$ and $y = \frac{b}{x}$, where a and b are positive constants.

b State, giving a reason, the number of real solutions to the equation $x^2(3x - a) - \frac{b}{x} = 0$

a

$3x - a = 0$ when $x = \frac{1}{3}a$, so the graph of $y = x^2(3x - a)$ touches the x-axis at (0, 0) and intersects it at $\left(\frac{1}{3}a, 0\right)$

b From the sketch there are only two points of intersection of the curves. This means there are only two values of x where

$$x^2(3x - a) = \frac{b}{x}$$

or $x^2(3x - a) - \frac{b}{x} = 0$

So this equation has two real solutions.

Problem-solving

You can sketch curves involving unknown constants. You should give any points of intersection with the coordinate axes in terms of the constants where appropriate.

You only need to state the **number** of solutions. You don't need to find the solutions.

Example 8

a Sketch the curves $y = \frac{4}{x^2}$ and $y = x^2(x - 3)$ on the same axes.

b Using your sketch, state, with a reason, the number of real solutions to the equation $x^4(x - 3) - 4 = 0$.

a

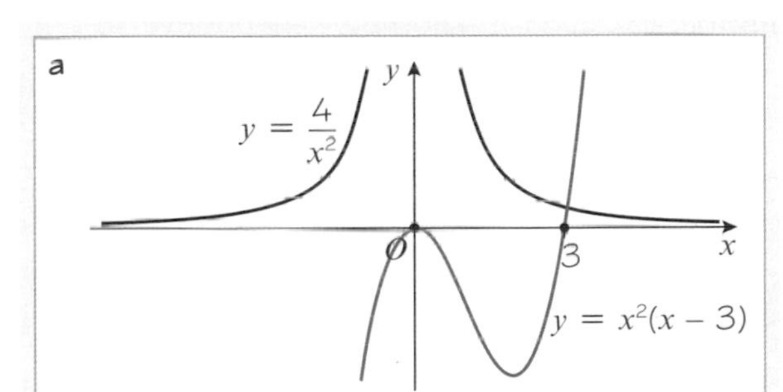

b There is a single point of intersection so the equation $x^2(x - 3) = \frac{4}{x^2}$ has one real solution.

Rearranging:

$x^4(x - 3) = 4$

$x^4(x - 3) - 4 = 0$

So this equation has one real solution.

Problem-solving

Set the functions equal to each other to form an equation with one real solution, then rearrange the equation into the form given in the question.

You would not be expected to solve this equation in your exam.

Exercise 4D

1 In each case:

i sketch the two curves on the same axes

ii state the number of points of intersection

iii write down a suitable equation which would give the x-coordinates of these points. (You are not required to solve this equation.)

a $y = x^2, y = x(x^2 - 1)$ **b** $y = x(x + 2), y = -\frac{3}{x}$ **c** $y = x^2, y = (x + 1)(x - 1)^2$

d $y = x^2(1 - x), y = -\frac{2}{x}$ **e** $y = x(x - 4), y = \frac{1}{x}$ **f** $y = x(x - 4), y = -\frac{1}{x}$

g $y = x(x - 4), y = (x - 2)^3$ **h** $y = -x^3, y = -\frac{2}{x}$ **i** $y = -x^3, y = x^2$

j $y = -x^3, y = -x(x + 2)$ **k** $y = 4, y = x(x - 1)(x + 2)^2$ **l** $y = x^3, y = x^2(x + 1)^2$

2 **a** On the same axes sketch the curves given by $y = x^2(x - 3)$ and $y = \frac{2}{x}$

b Explain how your sketch shows that there are only two real solutions to the equation $x^3(x - 3) = 2$.

3 **a** On the same axes sketch the curves given by $y = (x + 1)^3$ and $y = 3x(x - 1)$.

b Explain how your sketch shows that there is only one real solution to the equation $x^3 + 6x + 1 = 0$.

4 **a** On the same axes sketch the curves given by $y = \frac{1}{x}$ and $y = -x(x - 1)^2$.

b Explain how your sketch shows that there are no real solutions to the equation $1 + x^2(x - 1)^2 = 0$.

E/P **5** **a** On the same axes sketch the curves given by $y = x^2(x - a)$ and $y = \frac{b}{x}$ where a and b are both positive constants. **(5 marks)**

b Using your sketch, state, giving a reason, the number of real solutions to the equation $x^4 - ax^3 - b = 0$. **(1 mark)**

Problem-solving

Even though you don't know the values of a and b, you know they are positive, so you know the shapes of the graphs. You can label the point a on the x-axis on your sketch of $y = x^2(x - a)$.

E **6** **a** On the same set of axes sketch the graphs of $y = \frac{4}{x^2}$ and $y = 3x + 7$. **(3 marks)**

b Write down the number of real solutions to the equation $\frac{4}{x^2} = 3x + 7$. **(1 mark)**

c Show that you can rearrange the equation to give $(x + 1)(x + 2)(3x - 2) = 0$. **(2 marks)**

d Hence determine the exact coordinates of the points of intersection. **(3 marks)**

7 **a** On the same axes sketch the curve $y = x^3 - 3x^2 - 4x$ and the line $y = 6x$.

b Find the coordinates of the points of intersection.

P **8** **a** On the same axes sketch the curve $y = (x^2 - 1)(x - 2)$ and the line $y = 14x + 2$.

b Find the coordinates of the points of intersection.

P **9** **a** On the same axes sketch the curves with equations $y = (x - 2)(x + 2)^2$ and $y = -x^2 - 8$.

b Find the coordinates of the points of intersection.

10 **a** Sketch the graphs of $y = x^2 + 1$ and $2y = x - 1$. **(3 marks)**

b Explain why there are no real solutions to the equation $2x^2 - x + 3 = 0$. **(2 marks)**

c Work out the range of values of a such that the graphs of $y = x^2 + a$ and $2y = x - 1$ have two points of intersection. **(5 marks)**

11 **a** Sketch the graphs of $y = x^2(x - 1)(x + 1)$ and $y = \frac{1}{3}x^3 + 1$. **(5 marks)**

b Find the number of real solutions to the equation $3x^2(x - 1)(x + 1) = x^3 + 3$. **(1 mark)**

4.5 Translating graphs

You can transform the graph of a function by altering the function. Adding or subtracting a constant 'outside' the function translates a graph vertically.

- **The graph of $y = f(x) + a$ is a translation of the graph $y = f(x)$ by the vector $\begin{pmatrix} 0 \\ a \end{pmatrix}$.**

Adding or subtracting a constant 'inside' the function translates the graph horizontally.

- **The graph of $y = f(x + a)$ is a translation of the graph $y = f(x)$ by the vector $\begin{pmatrix} -a \\ 0 \end{pmatrix}$.**

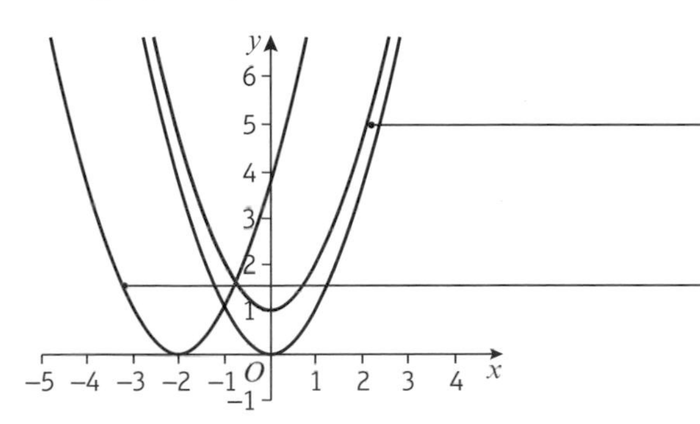

$y = f(x) + 1$ is a translation $\begin{pmatrix} 0 \\ 1 \end{pmatrix}$, or 1 unit in the direction of the positive y-axis.

$y = f(x + 2)$ is a translation $\begin{pmatrix} -2 \\ 0 \end{pmatrix}$, or 2 units in the direction of the negative x-axis.

Example 9

Sketch the graphs of:

a $y = x^2$ **b** $y = (x - 2)^2$ **c** $y = x^2 + 2$

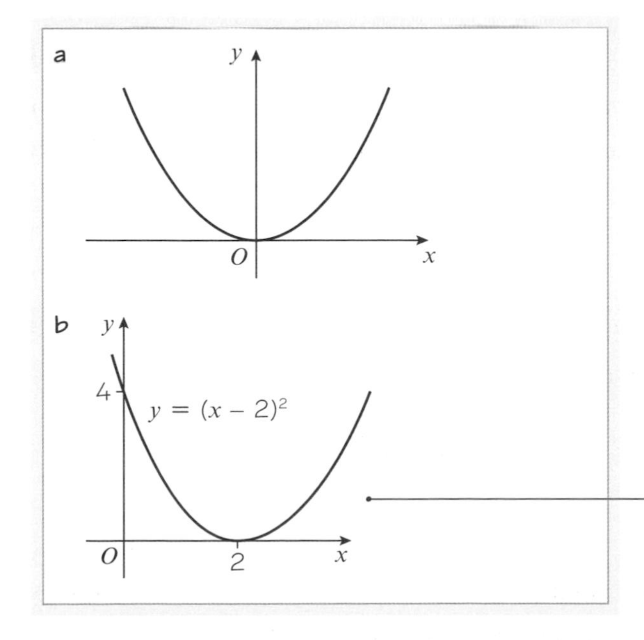

This is a translation by vector $\begin{pmatrix} 2 \\ 0 \end{pmatrix}$.

Remember to mark on the intersections with the axes.

c $y = x^2 + 2$

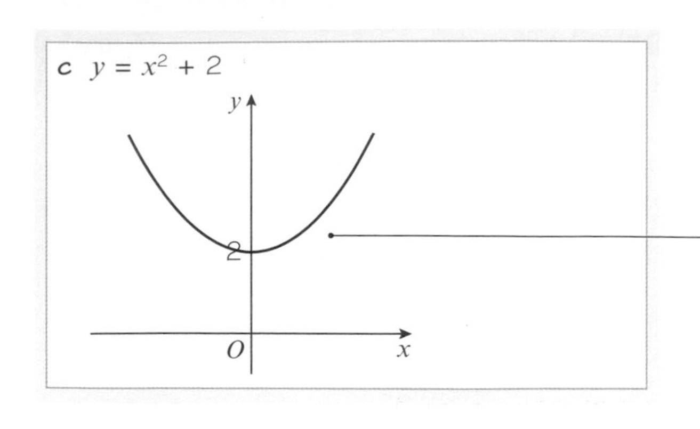

This is a translation by vector $\begin{pmatrix} 0 \\ 2 \end{pmatrix}$.

Remember to mark on the y-axis intersection.

Example 10

$f(x) = x^3$

$g(x) = x(x - 2)$

Sketch the following graphs, indicating any points where the curves cross the axes:

a $y = f(x + 1)$

b $y = g(x + 1)$

a The graph of $f(x)$ is

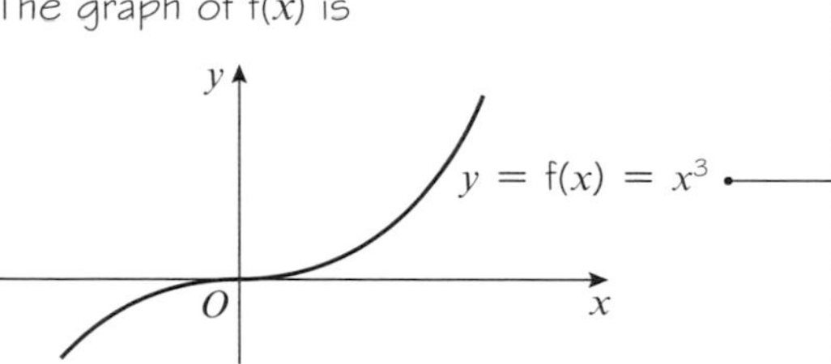

First sketch $y = f(x)$.

So the graph of $y = f(x + 1)$ is

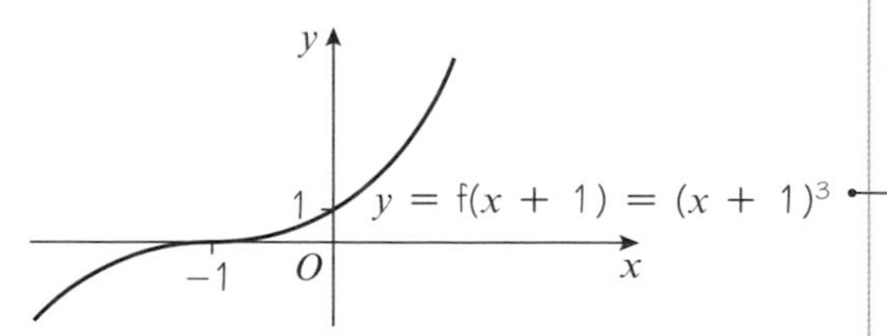

This is a translation of the graph of $y = f(x)$ by vector $\begin{pmatrix} -1 \\ 0 \end{pmatrix}$.
You could also write out the equation as $y = (x + 1)^3$ and sketch the graph directly.

Online Explore translations of the graph of $y = x^3$ using technology.

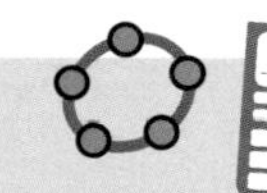

b $g(x) = x(x - 2)$

The curve is $y = x(x - 2)$

$0 = x(x - 2)$

Put $y = 0$ to find where the curve crosses the x-axis.

So $x = 0$ or $x = 2$

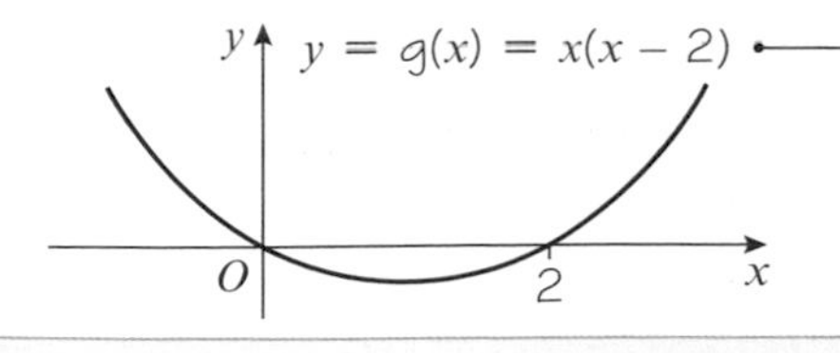

First sketch $g(x)$.

So the graph of $y = g(x + 1)$ is

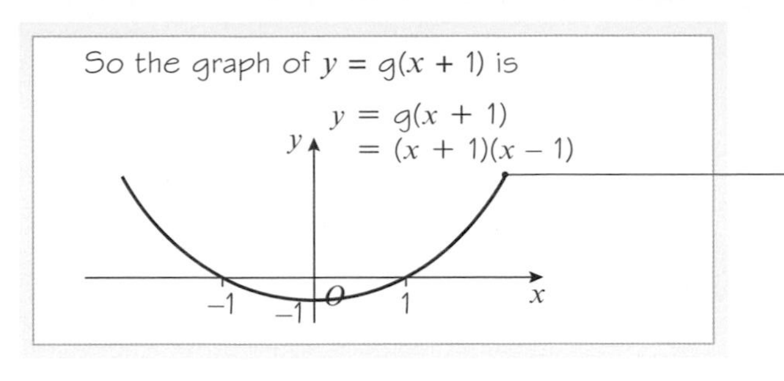

This is a translation of the graph of $y = g(x)$ by vector $\begin{pmatrix} -1 \\ 0 \end{pmatrix}$.

You could also write out the equation and sketch the graph directly:

$y = g(x + 1)$
$= (x + 1)(x + 1 - 2)$
$= (x + 1)(x - 1)$

- **When you translate a function, any asymptotes are also translated.**

Example 11

Given that $h(x) = \frac{1}{x}$, sketch the curve with equation $y = h(x) + 1$ and state the equations of any asymptotes and intersections with the axes.

The graph of $y = h(x)$ is

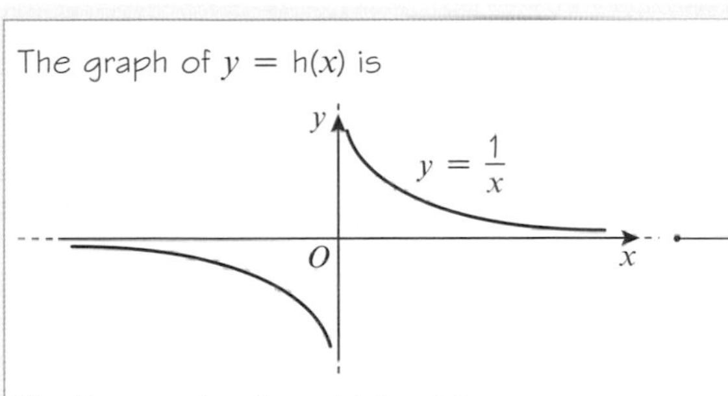

First sketch $y = h(x)$.

So the graph of $y = h(x) + 1$ is

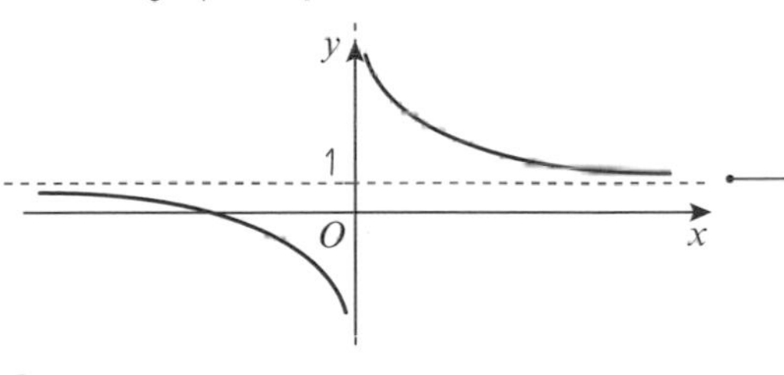

The curve is translated by vector $\begin{pmatrix} 0 \\ 1 \end{pmatrix}$ so the asymptote is translated by the same vector.

The curve crosses the x-axis once.

$$y = h(x) + 1 = \frac{1}{x} + 1$$

$$0 = \frac{1}{x} + 1$$

Put $y = 0$ to find where the curve crosses the x-axis.

$$-1 = \frac{1}{x}$$

$$x = -1$$

So the curve intersects the x-axis at $(-1, 0)$.

The horizontal asymptote is $y = 1$.

The vertical asymptote is $x = 0$.

Remember to write down the equation of the vertical asymptote as well. It is the y-axis so it has equation $x = 0$.

Exercise 4E

1 Apply the following transformations to the curves with equations $y = f(x)$ where:

i $f(x) = x^2$ **ii** $f(x) = x^3$ **iii** $f(x) = \frac{1}{x}$

In each case state the coordinates of points where the curves cross the axes and in **iii** state the equations of the asymptotes.

a $f(x + 2)$ **b** $f(x) + 2$ **c** $f(x - 1)$

d $f(x) - 1$ **e** $f(x) - 3$ **f** $f(x - 3)$

2 a Sketch the curve $y = f(x)$ where $f(x) = (x - 1)(x + 2)$.

b On separate diagrams sketch the graphs of **i** $y = f(x + 2)$ **ii** $y = f(x) + 2$.

c Find the equations of the curves $y = f(x + 2)$ and $y = f(x) + 2$, in terms of x, and use these equations to find the coordinates of the points where your graphs in part **b** cross the y-axis.

3 a Sketch the graph of $y = f(x)$ where $f(x) = x^2(1 - x)$.

b Sketch the curve with equation $y = f(x + 1)$.

c By finding the equation $f(x + 1)$ in terms of x, find the coordinates of the point in part **b** where the curve crosses the y-axis.

4 a Sketch the graph of $y = f(x)$ where $f(x) = x(x - 2)^2$.

b Sketch the curves with equations $y = f(x) + 2$ and $y = f(x + 2)$.

c Find the coordinates of the points where the graph of $y = f(x + 2)$ crosses the axes.

5 a Sketch the graph of $y = f(x)$ where $f(x) = x(x - 4)$.

b Sketch the curves with equations $y = f(x + 2)$ and $y = f(x) + 4$.

c Find the equations of the curves in part **b** in terms of x and hence find the coordinates of the points where the curves cross the axes.

6 a Sketch the graph of $y = f(x)$ where $f(x) = x^2(x - 1)(x - 2)$.

b Sketch the curves with equations $y = f(x + 2)$ and $y = f(x) - 1$.

(E) **7** The point $P(4, -1)$ lies on the curve with equation $y = f(x)$.

a State the coordinates that point P is transformed to on the curve with equation $y = f(x - 2)$. **(1 mark)**

b State the coordinates that point P is transformed to on the curve with equation $y = f(x) + 3$. **(1 mark)**

(E/P) **8** The graph of $y = f(x)$ where $f(x) = \frac{1}{x}$ is translated so that the asymptotes are at $x = 4$ and $y = 0$. Write down the equation for the transformed function in the form $y = \frac{1}{x + a}$ **(3 marks)**

(P) **9 a** Sketch the graph of $y = x^3 - 5x^2 + 6x$, marking clearly the points of intersection with the axes.

b Hence sketch $y = (x - 2)^3 - 5(x - 2)^2 + 6(x - 2)$.

(P) **10** **a** Sketch the graph of $y = x^2(x - 3)(x + 2)$, marking clearly the points of intersection with the axes.

b Hence sketch $y = (x + 2)^2(x - 1)(x + 4)$.

(E/P) **11** **a** Sketch the graph of $y = x^3 + 4x^2 + 4x$. **(6 marks)**

b The point with coordinates $(-1, 0)$ lies on the curve with equation $y = (x + a)^3 + 4(x + a)^2 + 4(x + a)$ where a is a constant. Find the two possible values of a. **(3 marks)**

Problem-solving

Look at your sketch and picture the curve sliding to the left or right.

12 **a** Sketch the graph of $y = x(x + 1)(x + 3)^2$. **(4 marks)**

b Find the possible values of b such that the point $(2, 0)$ lies on the curve with equation $y = (x + b)(x + b + 1)(x + b + 3)^2$. **(3 marks)**

Challenge

1 Sketch the graph of $y = (x - 3)^3 + 2$ and determine the coordinates of the point of inflection. → Section 12.9

2 The point $Q(-5, -7)$ lies on the curve with equation $y = \mathrm{f}(x)$.

a State the coordinates that point Q is transformed to on the curve with equation $y = \mathrm{f}(x + 2) - 5$.

b The coordinates of the point Q on a transformed curve are $(-3, -6)$. Write down the transformation in the form $y = \mathrm{f}(x + a) - b$.

4.6 Stretching graphs

Multiplying by a constant 'outside' the function stretches the graph vertically.

- **The graph of $y = a\mathrm{f}(x)$ is a stretch of the graph $y = \mathrm{f}(x)$ by a scale factor of a in the vertical direction.**

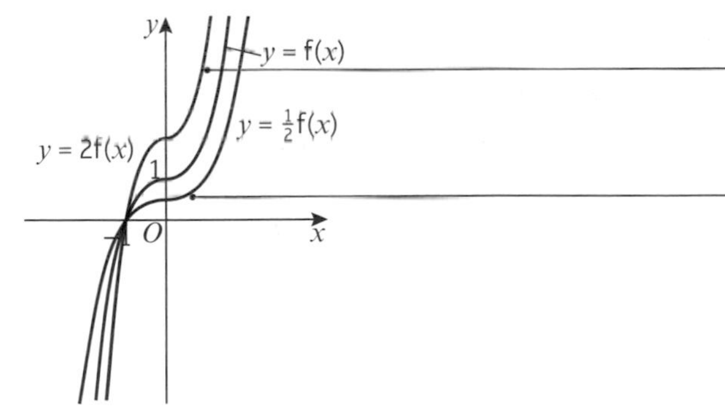

$2\mathrm{f}(x)$ is a stretch with scale factor 2 in the y-direction. All y-coordinates are doubled.

$\frac{1}{2}\mathrm{f}(x)$ is a stretch with scale factor $\frac{1}{2}$ in the y-direction. All y-coordinates are halved.

Multiplying by a constant 'inside' the function stretches the graph horizontally.

- **The graph of $y = \mathrm{f}(ax)$ is a stretch of the graph $y = \mathrm{f}(x)$ by a scale factor of $\frac{1}{a}$ in the horizontal direction.**

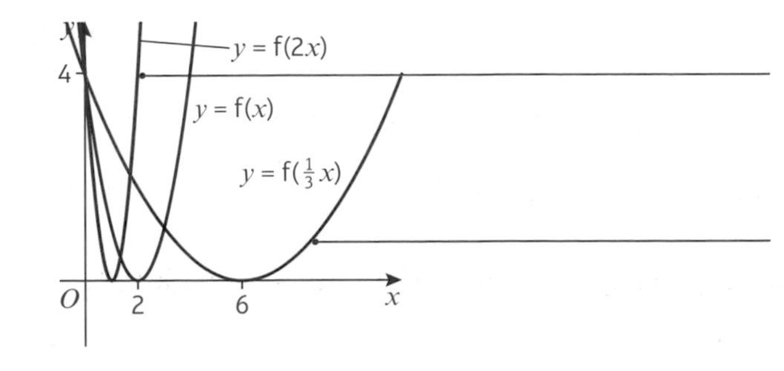

$y = \mathrm{f}(2x)$ is a stretch with scale factor $\frac{1}{2}$ in the x-direction. All x-coordinates are halved.

$y = \mathrm{f}(\frac{1}{3}x)$ is a stretch with scale factor 3 in the x-direction. All x-coordinates are tripled.

Example 12

Given that $f(x) = 9 - x^2$, sketch the curves with equations:

a $y = f(2x)$ **b** $y = 2f(x)$

a $f(x) = 9 - x^2$

So $f(x) = (3 - x)(3 + x)$

You can factorise the expression.

The curve is $y = (3 - x)(3 + x)$

$0 = (3 - x)(3 + x)$

Put $y = 0$ to find where the curve crosses the x-axis.

So $x = 3$ or $x = -3$

So the curve crosses the x-axis at (3, 0) and (−3, 0).

When $x = 0$, $y = 3 \times 3 = 9$

So the curve crosses the y-axis at (0, 9).

Put $x = 0$ to find where the curve crosses the y-axis.

The curve $y = f(x)$ is

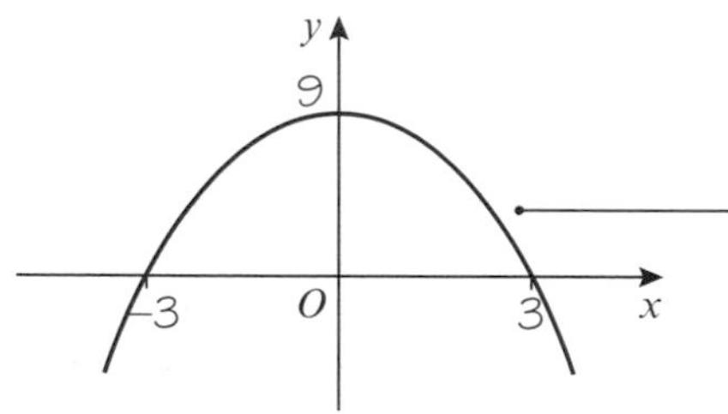

First sketch $y = f(x)$.

$y = f(2x)$ so the curve is

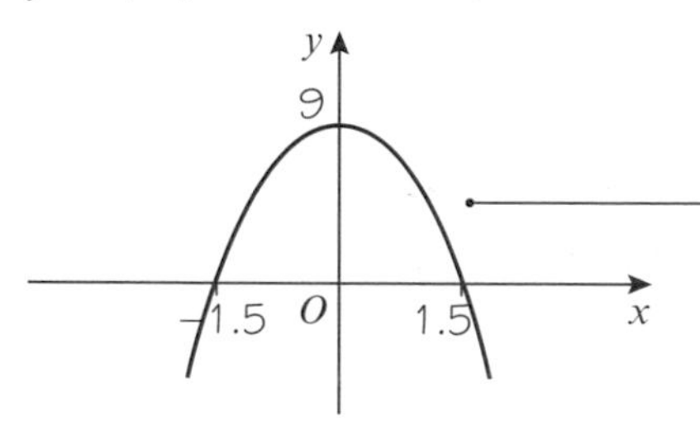

$y = f(ax)$ where $a = 2$ so it is a horizontal stretch with scale factor $\frac{1}{2}$.

Check: The curve is $y = f(2x)$.

So $y = (3 - 2x)(3 + 2x)$.

When $y = 0$, $x = -1.5$ or $x = 1.5$.

So the curve crosses the x-axis at (−1.5, 0) and (1.5, 0).

When $x = 0$, $y = 9$.

So the curve crosses the y-axis at (0, 9).

b $y = 2f(x)$ so the curve is

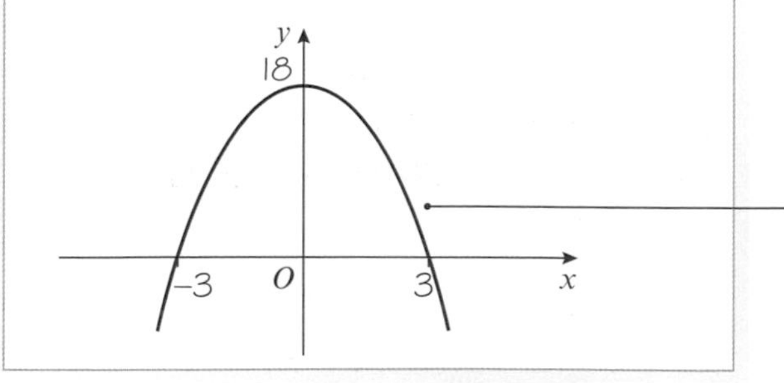

$y = af(x)$ where $a = 2$ so it is a vertical stretch with scale factor 2.

Check: The curve is $y = 2f(x)$.

So $y = 2(3 - x)(3 + x)$.

When $y = 0$, $x = 3$ or $x = -3$.

So the curve crosses the x-axis at (−3, 0) and (3, 0).

When $x = 0$, $y = 2 \times 9 = 18$.

So the curve crosses the y-axis at (0, 18).

Example 13

a Sketch the curve with equation $y = x(x - 2)(x + 1)$.

b On the same axes, sketch the curves $y = 2x(2x - 2)(2x + 1)$ and $y = -x(x - 2)(x + 1)$.

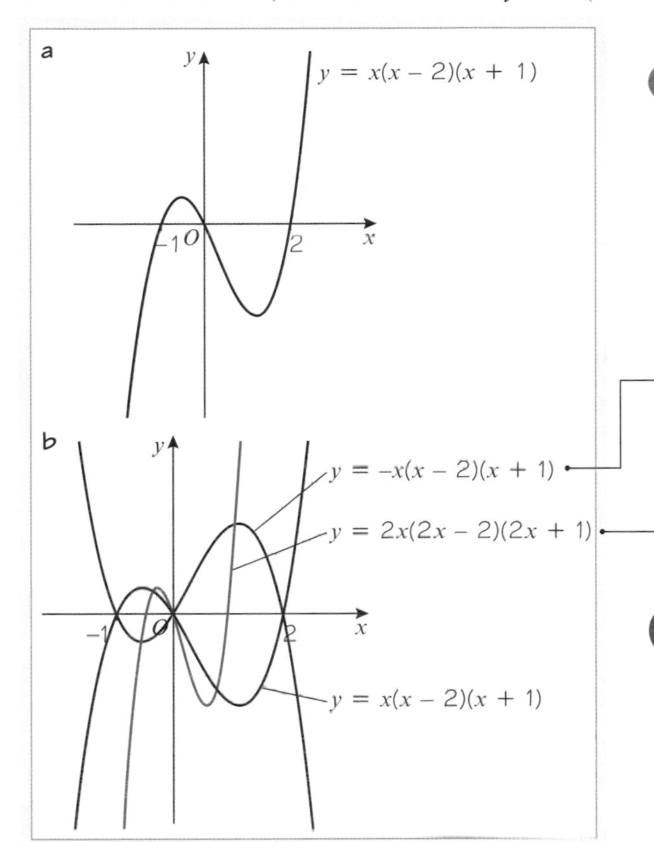

Online Explore stretches of the graph of $y = x(x - 2)(x + 1)$ using technology.

$y = -x(x - 2)(x + 1)$ is a stretch with scale factor -1 in the y-direction. Notice that this stretch has the effect of reflecting the curve in the x-axis.

$y = 2x(2x - 2)(2x + 1)$ is a stretch with scale factor $\frac{1}{2}$ in the x-direction.

Problem-solving

You need to work out the relationship between each new function and the original function.
If $x(x - 2)(x + 1) = f(x)$ then
$2x(2x - 2)(2x + 1) = f(2x)$, and
$-x(x - 2)(x + 1) = -f(x)$.

- **The graph of $y = -f(x)$ is a reflection of the graph of $y = f(x)$ in the x-axis.**
- **The graph of $y = f(-x)$ is a reflection of the graph of $y = f(x)$ in the y-axis.**

Example 14

On the same axes sketch the graphs of $y = f(x)$, $y = f(-x)$ and $y = -f(x)$ where $f(x) = x(x + 2)$.

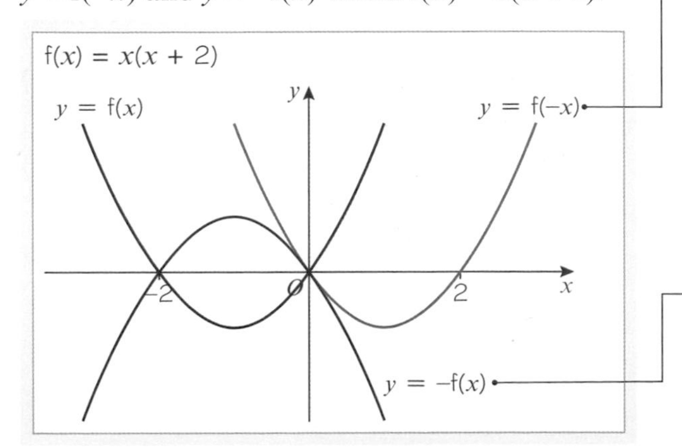

$y = f(-x)$ is $y = (-x)(-x + 2)$ which is $y = x^2 - 2x$ or $y = x(x - 2)$ and this is a reflection of the original curve in the y-axis.
Alternatively multiply each x-coordinate by -1 and leave the y-coordinates unchanged.
This is the same as a stretch parallel to the x-axis scale factor -1.

$y = -f(x)$ is $y = -x(x + 2)$ and this is a reflection of the original curve in the x-axis.
Alternatively multiply each y-coordinate by -1 and leave the x-coordinates unchanged.
This is the same as a stretch parallel to the y-axis scale factor -1.

Exercise 4F

1 Apply the following transformations to the curves with equations $y = f(x)$ where:

i $f(x) = x^2$ **ii** $f(x) = x^3$ **iii** $f(x) = \frac{1}{x}$

In each case show both f(x) and the transformation on the same diagram.

a $f(2x)$ **b** $f(-x)$ **c** $f(\frac{1}{2}x)$ **d** $f(4x)$ **e** $f(\frac{1}{4}x)$

f $2f(x)$ **g** $-f(x)$ **h** $4f(x)$ **i** $\frac{1}{2}f(x)$ **j** $\frac{1}{4}f(x)$

2 **a** Sketch the curve with equation $y = f(x)$ where $f(x) = x^2 - 4$.

b Sketch the graphs of $y = f(4x)$, $\frac{1}{3}y = f(x)$, $y = f(-x)$ and $y = -f(x)$.

Hint For part **b**, rearrange the second equation into the form $y = 3f(x)$.

3 **a** Sketch the curve with equation $y = f(x)$ where $f(x) = (x - 2)(x + 2)x$.

b Sketch the graphs of $y = f(\frac{1}{2}x)$, $y = f(2x)$ and $y = -f(x)$.

(P) **4** **a** Sketch the curve with equation $y = x^2(x - 3)$.

b On the same axes, sketch the curves with equations:

i $y = (2x)^2(2x - 3)$ **ii** $y = -x^2(x - 3)$

Problem-solving

Let $f(x) = x^2(x - 3)$ and try to write each of the equations in part **b** in terms of f(x).

5 **a** Sketch the curve $y = x^2 + 3x - 4$.

b On the same axes, sketch the graph of $5y = x^2 + 3x - 4$.

6 **a** Sketch the graph of $y = x^2(x - 2)^2$.

b On the same axes, sketch the graph of $3y = -x^2(x - 2)^2$.

(E) **7** The point $P(2, -3)$ lies on the curve with equation $y = f(x)$.

a State the coordinates that point P is transformed to on the curve with equation $y = f(2x)$. **(1 mark)**

b State the coordinates that point P is transformed to on the curve with equation $y = 4f(x)$. **(1 mark)**

(E) **8** The point $Q(-2, 8)$ lies on the curve with equation $y = f(x)$.

State the coordinates that point Q is transformed to on the curve with equation $y = f(\frac{1}{2}x)$. **(1 mark)**

(E/P) **9** **a** Sketch the graph of $y = (x - 2)(x - 3)^2$. **(4 marks)**

b The graph of $y = (ax - 2)(ax - 3)^2$ passes through the point (1, 0). Find two possible values for a. **(3 marks)**

Challenge

1 The point $R(4, -6)$ lies on the curve with equation $y = f(x)$. State the coordinates that point R is transformed to on the curve with equation $y = \frac{1}{3}f(2x)$.

2 The point $S(-4, 7)$ is transformed to a point $S'(-8, 1.75)$. Write down the transformation in the form $y = af(bx)$.

4.7 Transforming functions

You can apply transformations to unfamiliar functions by considering how specific points and features are transformed.

Example 15

The following diagram shows a sketch of the curve f(x) which passes through the origin.
The points $A(1, 4)$ and $B(3, 1)$ also lie on the curve.

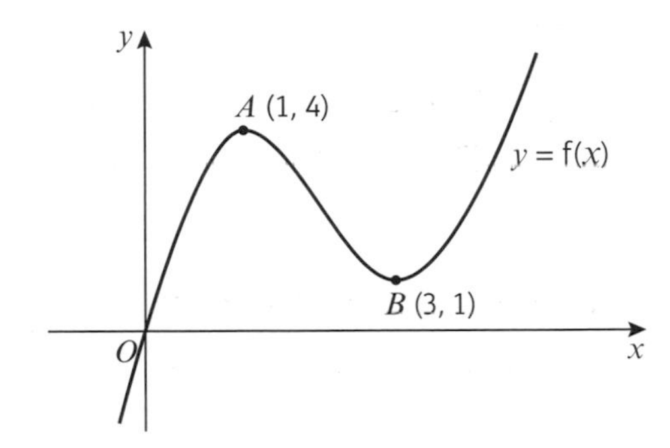

Sketch the following:

a $y = f(x + 1)$ **b** $y = f(x - 1)$ **c** $y = f(x) - 4$

d $2y = f(x)$ **e** $y - 1 = f(x)$

In each case you should show the positions of the images of the points O, A and B.

a f(x + 1)

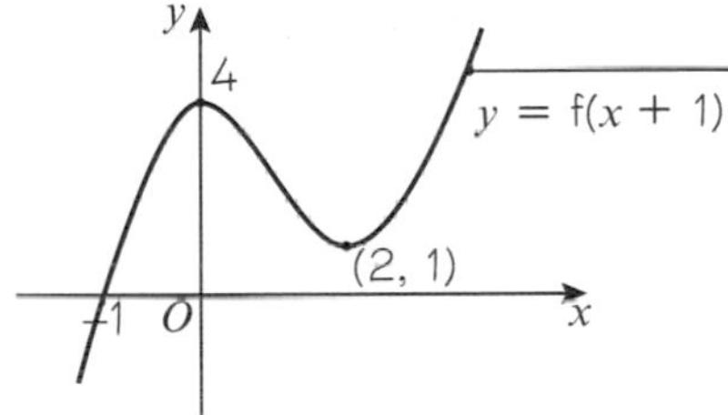

Translate f(x) 1 unit in the direction of the negative x-axis.

b f(x − 1)

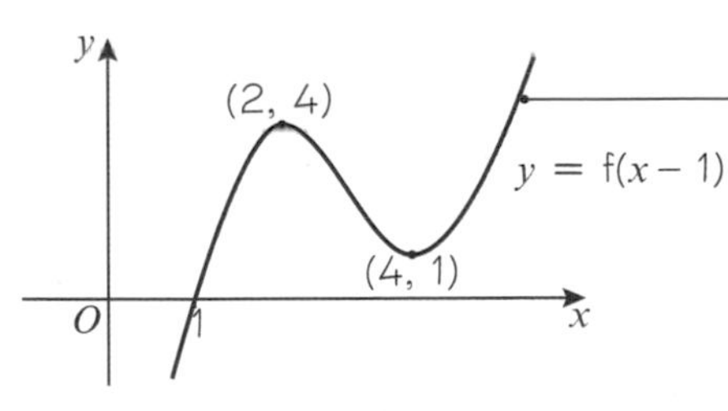

Translate f(x) 1 unit in the direction of the positive x-axis.

c f(x) − 4

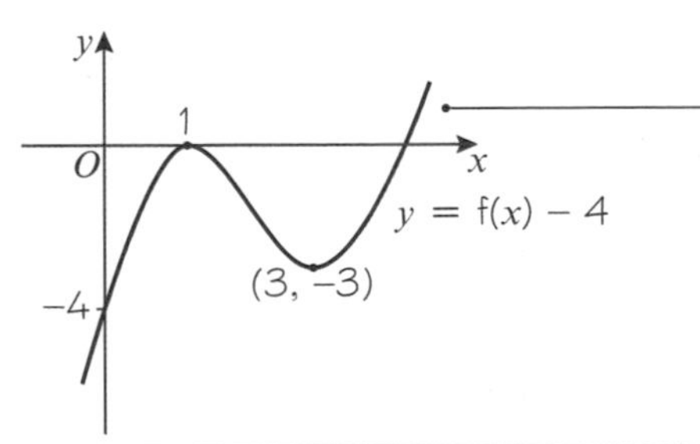

Translate f(x) 4 units in the direction of the negative y-axis.

d $2y = f(x)$ so $y = \frac{1}{2}f(x)$

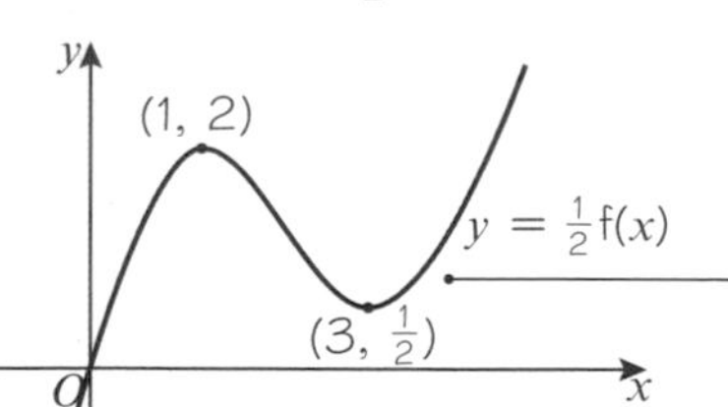

Rearrange in the form $y = \ldots$

Stretch $f(x)$ by scale factor $\frac{1}{2}$ in the y-direction.

e $y - 1 = f(x)$ so $y = f(x) + 1$

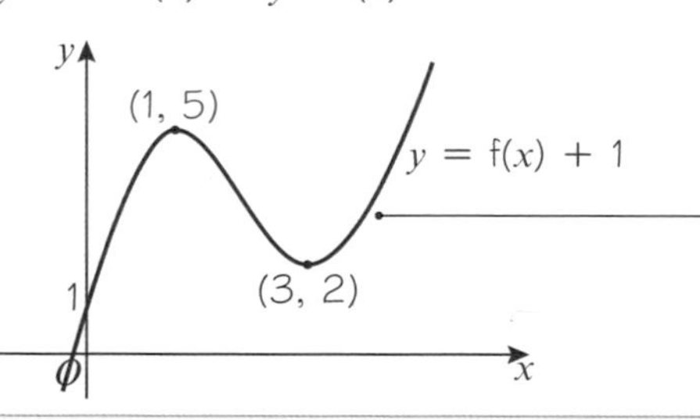

Rearrange in the form $y = \ldots$

Translate $f(x)$ 1 unit in the direction of the positive y-axis.

Exercise 4G

1 The following diagram shows a sketch of the curve with equation $y = f(x)$. The points $A(0, 2)$, $B(1, 0)$, $C(4, 4)$ and $D(6, 0)$ lie on the curve.

Sketch the following graphs and give the coordinates of the points, A, B, C and D after each transformation:

a $f(x + 1)$ **b** $f(x) - 4$ **c** $f(x + 4)$

d $f(2x)$ **e** $3f(x)$ **f** $f(\frac{1}{2}x)$

g $\frac{1}{2}f(x)$ **h** $f(-x)$

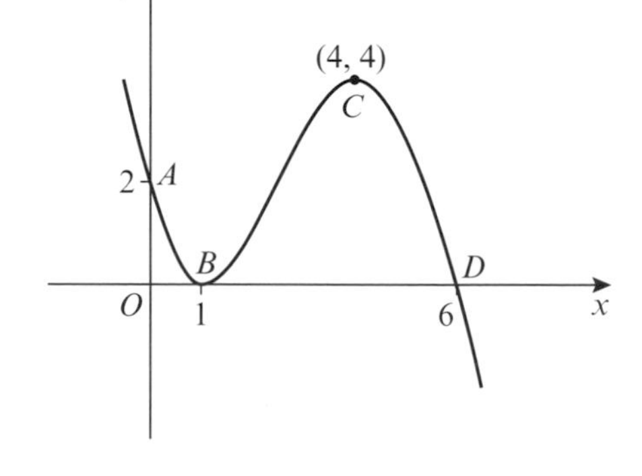

2 The curve $y = f(x)$ passes through the origin and has horizontal asymptote $y = 2$ and vertical asymptote $x = 1$, as shown in the diagram.

Sketch the following graphs. Give the equations of any asymptotes and give the coordinates of intersections with the axes after each transformation.

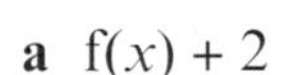

a $f(x) + 2$ **b** $f(x + 1)$ **c** $2f(x)$

d $f(x) - 2$ **e** $f(2x)$ **f** $f(\frac{1}{2}x)$

g $\frac{1}{2}f(x)$ **h** $-f(x)$

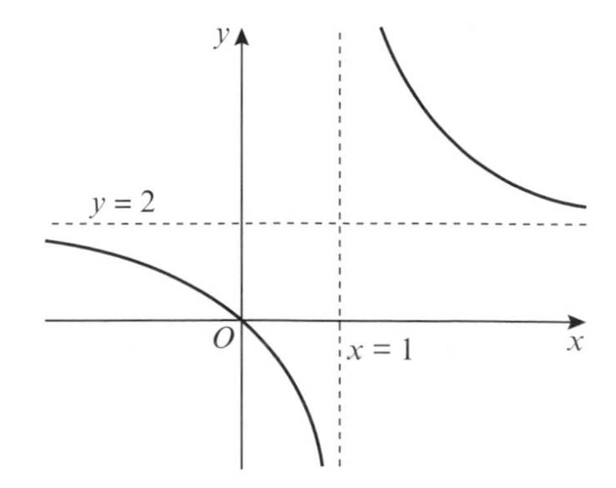

3 The curve with equation $y = \mathrm{f}(x)$ passes through the points $A(-4, -6)$, $B(-2, 0)$, $C(0, -3)$ and $D(4, 0)$ as shown in the diagram.

Sketch the following and give the coordinates of the points A, B, C and D after each transformation.

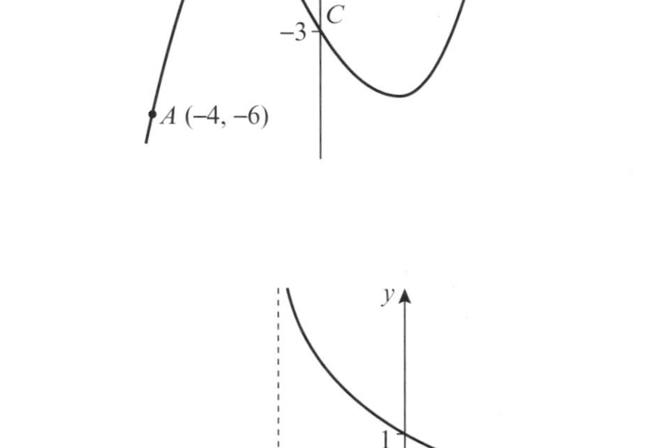

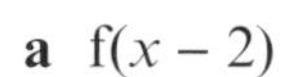

a $\mathrm{f}(x - 2)$ **b** $\mathrm{f}(x) + 6$ **c** $\mathrm{f}(2x)$

d $\mathrm{f}(x + 4)$ **e** $\mathrm{f}(x) + 3$ **f** $3\mathrm{f}(x)$

g $\frac{1}{3}\mathrm{f}(x)$ **h** $\mathrm{f}(\frac{1}{4}x)$ **i** $-\mathrm{f}(x)$

j $\mathrm{f}(-x)$

4 A sketch of the curve $y = \mathrm{f}(x)$ is shown in the diagram. The curve has a vertical asymptote with equation $x = -2$ and a horizontal asymptote with equation $y = 0$. The curve crosses the y-axis at $(0, 1)$.

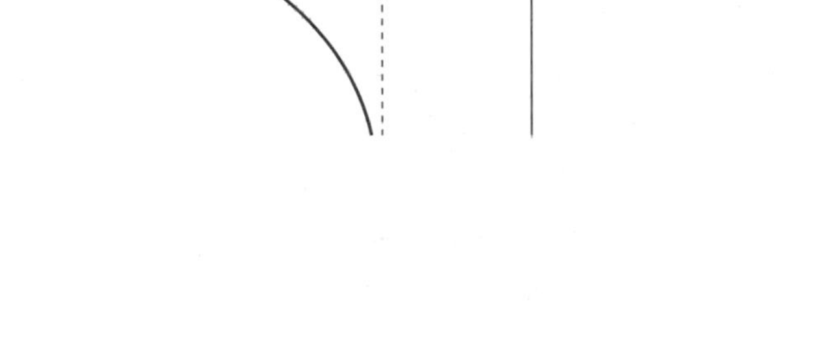

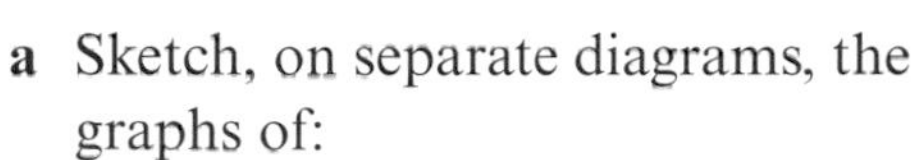

a Sketch, on separate diagrams, the graphs of:

i $2\mathrm{f}(x)$ **ii** $\mathrm{f}(2x)$ **iii** $\mathrm{f}(x - 2)$

iv $\mathrm{f}(x) - 1$ **v** $\mathrm{f}(-x)$ **vi** $-\mathrm{f}(x)$

In each case state the equations of any asymptotes and, if possible, points where the curve cuts the axes.

b Suggest a possible equation for $\mathrm{f}(x)$.

(E/P) **5** The point $P(2, 1)$ lies on the graph with equation $y = \mathrm{f}(x)$.

a On the graph of $y = \mathrm{f}(ax)$, the point P is mapped to the point $Q(4, 1)$. Determine the value of a. **(1 mark)**

b Write down the coordinates of the point to which P maps under each transformation

i $\mathrm{f}(x - 4)$ **ii** $3\mathrm{f}(x)$ **iii** $\frac{1}{2}\mathrm{f}(x) - 4$ **(3 marks)**

(P) **6** The diagram shows a sketch of a curve with equation $y = \mathrm{f}(x)$. The points $A(-1, 0)$, $B(0, 2)$, $C(1, 2)$ and $D(2, 0)$ lie on the curve.

Sketch the following graphs and give the coordinates of the points A, B, C and D after each transformation:

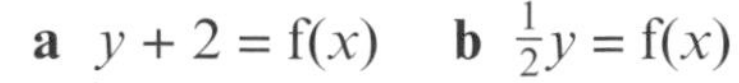

a $y + 2 = \mathrm{f}(x)$ **b** $\frac{1}{2}y = \mathrm{f}(x)$

c $y - 3 = \mathrm{f}(x)$ **d** $3y = \mathrm{f}(x)$

e $2y - 1 = \mathrm{f}(x)$

Problem-solving

Rearrange each equation into the form $y = \ldots$

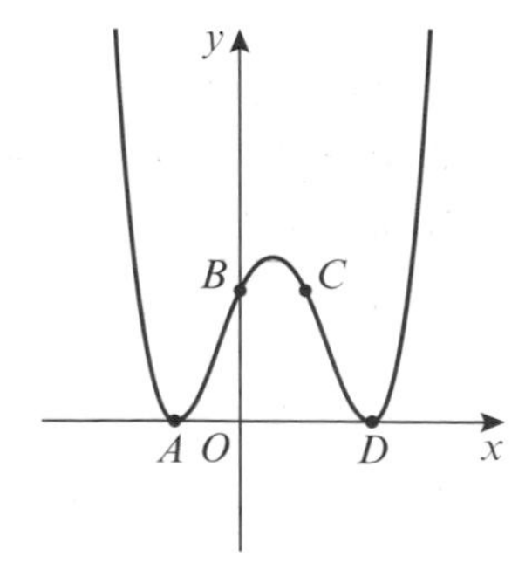

Mixed exercise 4

1 **a** On the same axes sketch the graphs of $y = x^2(x-2)$ and $y = 2x - x^2$.

b By solving a suitable equation find the points of intersection of the two graphs.

(P) 2 **a** On the same axes sketch the curves with equations $y = \frac{6}{x}$ and $y = 1 + x$.

b The curves intersect at the points A and B. Find the coordinates of A and B.

c The curve C with equation $y = x^2 + px + q$, where p and q are integers, passes through A and B. Find the values of p and q.

d Add C to your sketch.

3 The diagram shows a sketch of the curve $y = f(x)$. The point $B(0, 0)$ lies on the curve and the point $A(3, 4)$ is a maximum point. The line $y = 2$ is an asymptote.

Sketch the following and in each case give the coordinates of the new positions of A and B and state the equation of the asymptote:

a $f(2x)$ **b** $\frac{1}{2}f(x)$ **c** $f(x) - 2$

d $f(x + 3)$ **e** $f(x - 3)$ **f** $f(x) + 1$

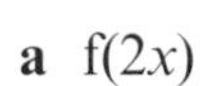

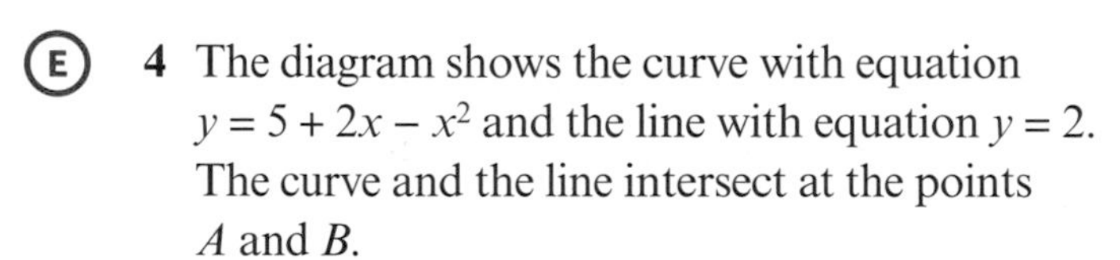

(E) 4 The diagram shows the curve with equation $y = 5 + 2x - x^2$ and the line with equation $y = 2$. The curve and the line intersect at the points A and B.

Find the x-coordinates of A and B. **(4 marks)**

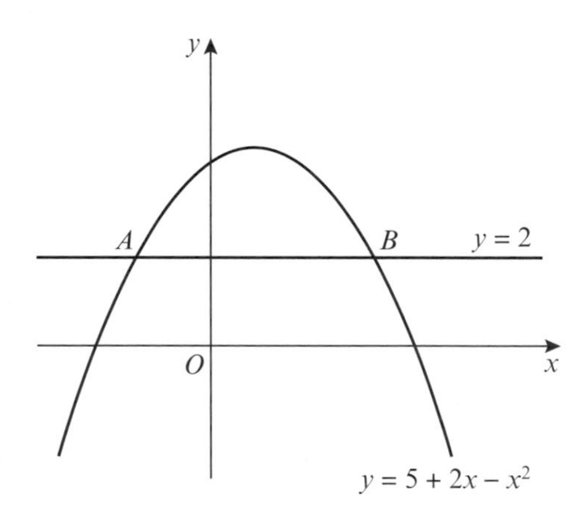

(E/P) 5 $f(x) = x^2(x-1)(x-3)$.

a Sketch the graph of $y = f(x)$. **(2 marks)**

b On the same axes, draw the line $y = 2 - x$. **(2 marks)**

c State the number of real solutions to the equation $x^2(x-1)(x-3) = 2 - x$. **(1 mark)**

d Write down the coordinates of the point where the graph with equation $y = f(x) + 2$ crosses the y-axis. **(1 mark)**

(E) 6 The figure shows a sketch of the curve with equation $y = f(x)$.

On separate axes sketch the curves with equations:

a $y = f(-x)$ **(2 marks)**

b $y = -f(x)$ **(2 marks)**

Mark on each sketch the x-coordinate of any point, or points, where the curve touches or crosses the x-axis.

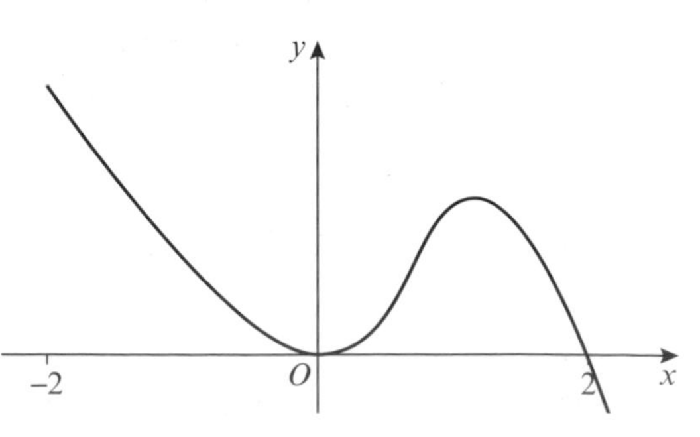

(E/P) **7** The diagram shows the graph of the quadratic function $f(x)$. The graph meets the x-axis at $(1, 0)$ and $(3, 0)$ and the minimum point is $(2, -1)$.

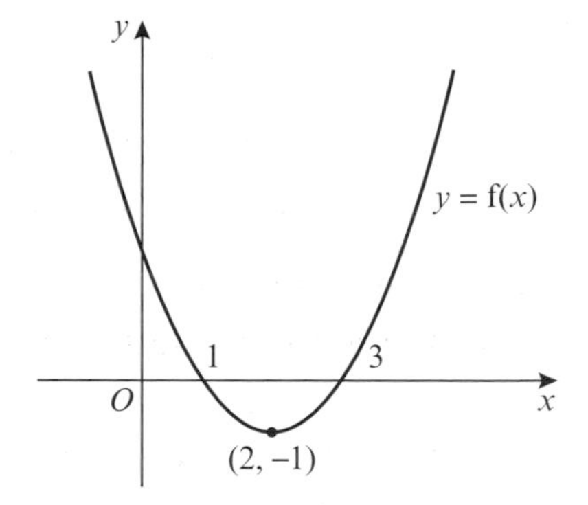

a Find the equation of the graph in the form $y = ax^2 + bx + c$ **(2 marks)**

b On separate axes, sketch the graphs of
i $y = f(x + 2)$ **ii** $y = f(2x)$. **(2 marks)**

c On each graph label the coordinates of the points at which the graph meets the x-axis and label the coordinates of the minimum point.

(E/P) **8** $f(x) = (x - 1)(x - 2)(x + 1)$.

a State the coordinates of the point at which the graph $y = f(x)$ intersects the y-axis. **(1 mark)**

b The graph of $y = af(x)$ intersects the y-axis at $(0, -4)$. Find the value of a. **(1 mark)**

c The graph of $y = f(x + b)$ passes through the origin. Find three possible values of b. **(3 marks)**

(P) **9** The point $P(4, 3)$ lies on a curve $y = f(x)$.

a State the coordinates of the point to which P is transformed on the curve with equation:
i $y = f(3x)$ **ii** $\frac{1}{2}y = f(x)$ **iii** $y = f(x - 5)$ **iv** $-y = f(x)$ **v** $2(y + 2) = f(x)$

b P is transformed to point $(2, 3)$. Write down two possible transformations of $f(x)$.

c P is transformed to point $(8, 6)$. Write down a possible transformation of $f(x)$ if
i $f(x)$ is translated only **ii** $f(x)$ is stretched only.

(E/P) **10** The curve C_1 has equation $y = -\frac{a}{x^2}$ where a is a positive constant. The curve C_2 has the equation $y = x^2(3x + b)$ where b is a positive constant.

a Sketch C_1 and C_2 on the same set of axes, showing clearly the coordinates of any point where the curves touch or cross the axes. **(4 marks)**

b Using your sketch state, giving reasons, the number of solutions to the equation $x^4(3x + b) + a = 0$. **(2 marks)**

(E/P) **11** **a** Factorise completely $x^3 - 6x^2 + 9x$. **(2 marks)**

b Sketch the curve of $y = x^3 - 6x^2 + 9x$ showing clearly the coordinates of the points where the curve touches or crosses the axes. **(4 marks)**

c The point with coordinates $(-4, 0)$ lies on the curve with equation $y = (x - k)^3 - 6(x - k)^2 + 9(x - k)$ where k is a constant.
Find the two possible values of k. **(3 marks)**

(E) **12** $f(x) = x(x - 2)^2$

Sketch on separate axes the graphs of:

a $y = f(x)$ **(2 marks)**

b $y = f(x + 3)$ **(2 marks)**

Show on each sketch the coordinates of the points where each graph crosses or meets the axes.

(E) **13** Given that $f(x) = \frac{1}{x}, x \neq 0$,

a Sketch the graph of $y = f(x) - 2$ and state the equations of the asymptotes. **(3 marks)**

b Find the coordinates of the point where the curve $y = f(x) - 2$ cuts a coordinate axis. **(2 marks)**

c Sketch the graph of $y = f(x + 3)$. **(2 marks)**

d State the equations of the asymptotes and the coordinates of the point where the curve cuts a coordinate axis. **(2 marks)**

Challenge

The point $R(6, -4)$ lies on the curve with equation $y = f(x)$. State the coordinates that point R is transformed to on the curve with equation $y = f(x + c) - d$.

Summary of key points

1 If p is a root of the function $f(x)$, then the graph of $y = f(x)$ touches or crosses the x-axis at the point $(p, 0)$.

2 The graphs of $y = \frac{k}{x}$ and $y = \frac{k}{x^2}$, where k is a real constant, have asymptotes at $x = 0$ and $y = 0$.

3 The x-coordinate(s) at the points of intersection of the curves with equations $y = f(x)$ and $y = g(x)$ are the solution(s) to the equation $f(x) = g(x)$.

4 The graph of $y = f(x) + a$ is a translation of the graph $y = f(x)$ by the vector $\begin{pmatrix} 0 \\ a \end{pmatrix}$.

5 The graph of $y = f(x + a)$ is a translation of the graph $y = f(x)$ by the vector $\begin{pmatrix} -a \\ 0 \end{pmatrix}$.

6 When you translate a function, any asymptotes are also translated.

7 The graph of $y = af(x)$ is a stretch of the graph $y = f(x)$ by a scale factor of a in the vertical direction.

8 The graph of $y = f(ax)$ is a stretch of the graph $y = f(x)$ by a scale factor of $\frac{1}{a}$ in the horizontal direction.

9 The graph of $y = -f(x)$ is a reflection of the graph of $y = f(x)$ in the x-axis.

10 The graph of $y = f(-x)$ is a reflection of the graph of $y = f(x)$ in the y-axis.

Review exercise

(E) **1 a** Write down the value of $8^{\frac{1}{3}}$. **(1)**

b Find the value of $8^{-\frac{2}{3}}$. **(2)**

← Section 1.4

2 a Find the value of $125^{\frac{4}{3}}$. **(2)**

b Simplify $24x^2 \div 18x^{\frac{4}{3}}$. **(2)**

← Sections 1.1, 1.4

(E) **3 a** Express $\sqrt{80}$ in the form $a\sqrt{5}$, where a is an integer. **(2)**

b Express $(4 - \sqrt{5})^2$ in the form $b + c\sqrt{5}$, where b and c are integers. **(2)**

← Section 1.5

(E) **4 a** Expand and simplify $(4 + \sqrt{3})(4 - \sqrt{3})$. **(2)**

b Express $\dfrac{26}{4 + \sqrt{3}}$ in the form $a + b\sqrt{3}$, where a and b are integers. **(3)**

← Sections 1.5, 1.6

(E/P) **5** Here are three numbers:

$1 - \sqrt{k}$, $2 + 5\sqrt{k}$ and $2\sqrt{k}$

Given that k is a positive integer, find:

a the mean of the three numbers. **(2)**

b the range of the three numbers. **(1)**

← Section 1.5

(E) **6** Given that $y = \dfrac{1}{25}x^4$, express each of the following in the form kx^n, where k and n are constants.

a y^{-1} **(1)**

b $5y^{\frac{1}{2}}$ **(1)**

← Section 1.4

(E/P) **7** Find the area of this trapezium in cm^2. Give your answer in the form $a + b\sqrt{2}$, where a and b are integers to be found. **(4)**

← Section 1.5

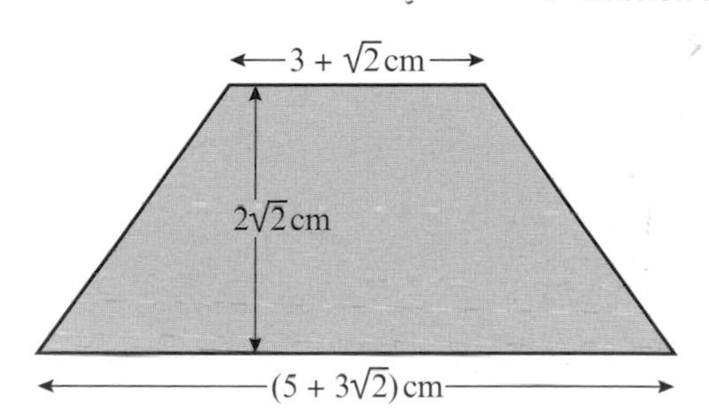

(E) **8** Given that $p = 3 - 2\sqrt{2}$ and $q = 2 - \sqrt{2}$, find the value of $\dfrac{p+q}{p-q}$.

Give your answer in the form $m + n\sqrt{2}$, where m and n are rational numbers to be found. **(4)**

← Sections 1.5, 1.6

(E/P) **9 a** Factorise the expression $x^2 - 10x + 16$. **(1)**

b Hence, or otherwise, solve the equation $8^{2y} - 10(8^y) + 16 = 0$. **(2)**

← Sections 1.3, 2.1

(E) **10** $x^2 - 8x - 29 \equiv (x + a)^2 + b$, where a and b are constants.

a Find the value of a and the value of b. **(2)**

b Hence, or otherwise, show that the roots of $x^2 - 8x - 29 = 0$ are $c \pm d\sqrt{5}$, where c and d are integers. **(3)**

← Sections 2.1, 2.2

E/P **11** The functions f and g are defined as $f(x) = x(x - 2)$ and $g(x) = x + 5, x \in \mathbb{R}$. Given that $f(a) = g(a)$ and $a > 0$, find the value of a to three significant figures. **(3)**

← Sections 2.1, 2.3

P **12** An athlete launches a shot put from shoulder height. The height of the shot put, in metres, above the ground t seconds after launch, can be modelled by the following function:

$h(t) = 1.7 + 10t - 5t^2 \qquad t \geqslant 0$

a Give the physical meaning of the constant term 1.7 in the context of the model.

b Use the model to calculate how many seconds after launch the shot put hits the ground.

c Rearrange $h(t)$ into the form $A - B(t - C)^2$ and give the values of the constants A, B and C.

d Using your answer to part **c** or otherwise, find the maximum height of the shot put, and the time at which this maximum height is reached.

← Section 2.6

E/P **13** Given that $f(x) = x^2 - 6x + 18, x \geqslant 0$,

a express $f(x)$ in the form $(x - a)^2 + b$, where a and b are integers. **(2)**

The curve C with equation $y = f(x)$, $x \geqslant 0$, meets the y-axis at P and has a minimum point at Q.

b Sketch the graph of C, showing the coordinates of P and Q. **(3)**

The line $y = 41$ meets C at the point R.

c Find the x-coordinate of R, giving your answer in the form $p + q\sqrt{2}$, where p and q are integers. **(2)**

← Sections 2.2, 2.4

E **14** The function $h(x) = x^2 + 2\sqrt{2}x + k$ has equal roots.

a Find the value of k. **(1)**

b Sketch the graph of $y = h(x)$, clearly labelling any intersections with the coordinate axes. **(3)**

← Sections 1.5, 2.4, 2.5

E/P **15** The function $g(x)$ is defined as

$g(x) = x^9 - 7x^6 - 8x^3, x \in \mathbb{R}$.

a Write $g(x)$ in the form $x^3(x^3 + a)(x^3 + b)$, where a and b are integers. **(1)**

b Hence find the three roots of $g(x)$. **(1)**

← Section 2.3

E/P **16** Given that

$x^2 + 10x + 36 \equiv (x + a)^2 + b$,

where a and b are constants,

a find the value of a and the value of b. **(2)**

b Hence show that the equation $x^2 + 10x + 36 = 0$ has no real roots. **(2)**

The equation $x^2 + 10x + k = 0$ has equal roots.

c Find the value of k. **(2)**

d For this value of k, sketch the graph of $y = x^2 + 10x + k$, showing the coordinates of any points at which the graph meets the coordinate axes. **(3)**

← Sections 2.2, 2.4, 2.5

E/P **17** Given that $x^2 + 2x + 3 \equiv (x + a)^2 + b$,

a find the value of the constants a and b **(2)**

b Sketch the graph of $y = x^2 + 2x + 3$, indicating clearly the coordinates of any intersections with the coordinate axes. **(3)**

c Find the value of the discriminant of $x^2 + 2x + 3$. Explain how the sign of the discriminant relates to your sketch in part **b**. **(2)**

The equation $x^2 + kx + 3 = 0$, where k is a constant, has no real roots.

d Find the set of possible values of k, giving your answer in surd form. **(2)**

← Section 2.2, 2.4, 2.5

E **18** **a** By eliminating y from the equations:

$y = x - 4,$
$2x^2 - xy = 8,$

show that

$x^2 + 4x - 8 = 0.$ **(2)**

b Hence, or otherwise, solve the simultaneous equations:

$y = x - 4,$
$2x^2 - xy = 8,$

giving your answers in the form $a \pm b\sqrt{3}$, where a and b are integers. **(4)**

← Section 3.2

E **19** Find the set of values of x for which:

a $3(2x + 1) > 5 - 2x,$ **(2)**

b $2x^2 - 7x + 3 > 0,$ **(3)**

c **both** $3(2x + 1) > 5 - 2x$ **and** $2x^2 - 7x + 3 > 0.$ **(1)**

← Sections 3.4, 3.5

E/P **20** The functions p and q are defined as $p(x) = -2(x + 1)$ and $q(x) = x^2 - 5x + 2$, $x \in \mathbb{R}$. Show algebraically that there is no value of x for which $p(x) = q(x)$. **(3)**

← Sections 2.3, 2.5

E **21** **a** Solve the simultaneous equations:

$y + 2x = 5$
$2x^2 - 3x - y = 16.$ **(5)**

b Hence, or otherwise, find the set of values of x for which:

$2x^2 - 3x - 16 > 5 - 2x.$ **(2)**

← Sections 3.2, 3.5

E/P **22** The equation $x^2 + kx + (k + 3) = 0$, where k is a constant, has different real roots.

a Show that $k^2 - 4k - 12 > 0.$ **(2)**

b Find the set of possible values of k. **(2)**

← Sections 2.5, 3.5

E **23** Find the set of values for which

$\frac{6}{x + 5} < 2, x \neq -5.$ **(6)**

← Section 3.5

E **24** The functions f and g are defined as $f(x) = 9 - x^2$ and $g(x) = 14 - 6x$, $x \in \mathbb{R}$.

a On the same set of axes, sketch the graphs of $y = f(x)$ and $y = g(x)$. Indicate clearly the coordinates of any points where the graphs intersect with each other or the coordinate axes. **(5)**

b On your sketch, shade the region that satisfies the inequalities $y > 0$ and $f(x) > g(x)$. **(1)**

← Sections 3.2, 3.3, 3.7

E/P **25** **a** Factorise completely $x^3 - 4x$. **(1)**

b Sketch the curve with equation $y = x^3 - 4x$, showing the coordinates of the points where the curve crosses the x-axis. **(2)**

c On a separate diagram, sketch the curve with equation

$y = (x - 1)^3 - 4(x - 1)$

showing the coordinates of the points where the curve crosses the x-axis. **(2)**

← Sections 1.3, 4.1, 4.5

E **26**

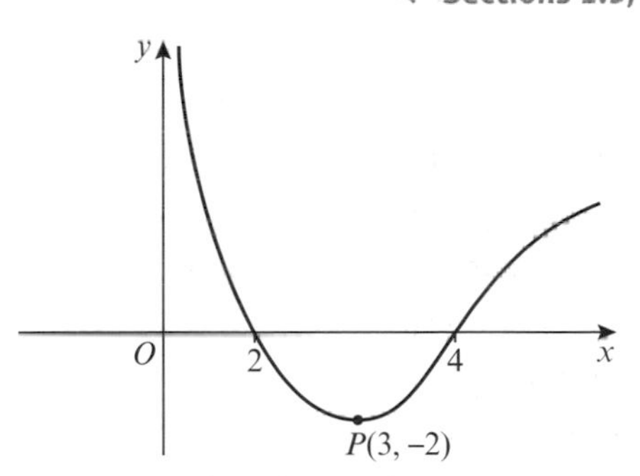

The figure shows a sketch of the curve with equation $y = f(x)$. The curve crosses the x-axis at the points (2, 0) and (4, 0). The minimum point on the curve is $P(3, -2)$.

In separate diagrams, sketch the curves with equation

a $y = -f(x)$ **(2)**

b $y = f(2x)$ **(2)**

On each diagram, give the coordinates of the points at which the curve crosses the x-axis, and the coordinates of the image of P under the given transformation.

← Sections 4.6, 4.7

E **27**

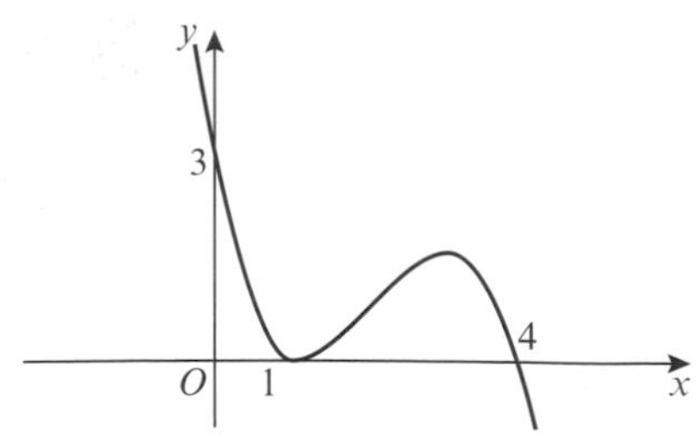

The figure shows a sketch of the curve with equation $y = f(x)$. The curve passes through the points (0, 3) and (4, 0) and touches the x-axis at the point (1, 0).

On separate diagrams, sketch the curves with equations

a $y = f(x + 1)$ **(2)**

b $y = 2f(x)$ **(2)**

c $y = f\left(\frac{1}{2}x\right)$ **(2)**

On each diagram, show clearly the coordinates of all the points where the curve meets the axes.

← Sections 4.5, 4.6, 4.7

E **28** Given that $f(x) = \frac{1}{x}, x \neq 0$,

a sketch the graph of $y = f(x) + 3$ and state the equations of the asymptotes **(2)**

b find the coordinates of the point where $y = f(x) + 3$ crosses a coordinate axis. **(2)**

← Sections 4.3, 4.5

E **29** The quartic function t is defined as $t(x) = (x^2 - 5x + 2)(x^2 - 5x + 4), x \in \mathbb{R}$.

a Find the four roots of $t(x)$, giving your answers to 3 significant figures where necessary. **(3)**

b Sketch the graph of $y = t(x)$, showing clearly the coordinates of all the points where the curve meets the axes. **(2)**

← Sections 4.2, 2.1

E **30** The point (6, −8) lies on the graph of $y = f(x)$. State the coordinates of the point to which P is transformed on the graph with equation:

a $y = -f(x)$ **(1)**

b $y = f(x - 3)$ **(1)**

c $2y = f(x)$ **(1)**

← Section 4.7

E/P **31** The curve C_1 has equation $y = -\frac{a}{x}$, where a is a positive constant.

The curve C_2 has equation $y = (x - b)^2$, where b is a positive constant.

a Sketch C_1 and C_2 on the same set of axes. Label any points where either curve meets the coordinate axes, giving your coordinates in terms of a and b. **(4)**

b Using your sketch, state the number of real solutions to the equation $x(x - 5)^2 = -7$. **(1)**

← Sections 4.3, 4.4

E/P **32** **a** Sketch the graph of $y = \frac{1}{x^2} - 4$, showing clearly the coordinates of the points where the curve crosses the coordinate axes and stating the equations of the asymptotes. **(4)**

b The curve with $y = \frac{1}{(x + k)^2} - 4$ passes through the origin. Find the two possible values of k. **(2)**

← Sections 4.1, 4.5, 4.7

Challenge

1 a Solve the equation $x^2 - 10x + 9 = 0$

b Hence, or otherwise, solve the equation $3^{x-2}(3^x - 10) = -1$ ← Sections 1.1, 1.3, 2.1

2 A rectangle has an area of 6 cm² and a perimeter of $8\sqrt{2}$ cm. Find the dimensions of the rectangle, giving your answers as surds in their simplest form. ← Sections 1.5, 2.2

3 Show algebraically that the graphs of $y = 3x^3 + x^2 - x$ and $y = 2x(x - 1)(x + 1)$ have only one point of intersection, and find the coordinates of this point. ← Section 3.3

4 The quartic function $f(x) = (x^2 + x - 20)(x^2 + x - 2)$ has three roots in common with the function $g(x) = f(x - k)$, where k is a constant. Find the two possible values of k. ← Sections 4.2, 4.5, 4.7

Straight line graphs

5

Objectives

After completing this unit you should be able to:

- Calculate the gradient of a line joining a pair of points → pages 90 – 91
- Understand the link between the equation of a line, and its gradient and intercept → pages 91 – 93
- Find the equation of a line given (i) the gradient and one point on the line or (ii) two points on the line → pages 93 – 95
- Find the point of intersection for a pair of straight lines → pages 95 – 96
- Know and use the rules for parallel and perpendicular gradients → pages 97 – 100
- Solve length and area problems on coordinate grids → pages 100 – 103
- Use straight line graphs to construct mathematical models → pages 103 – 108

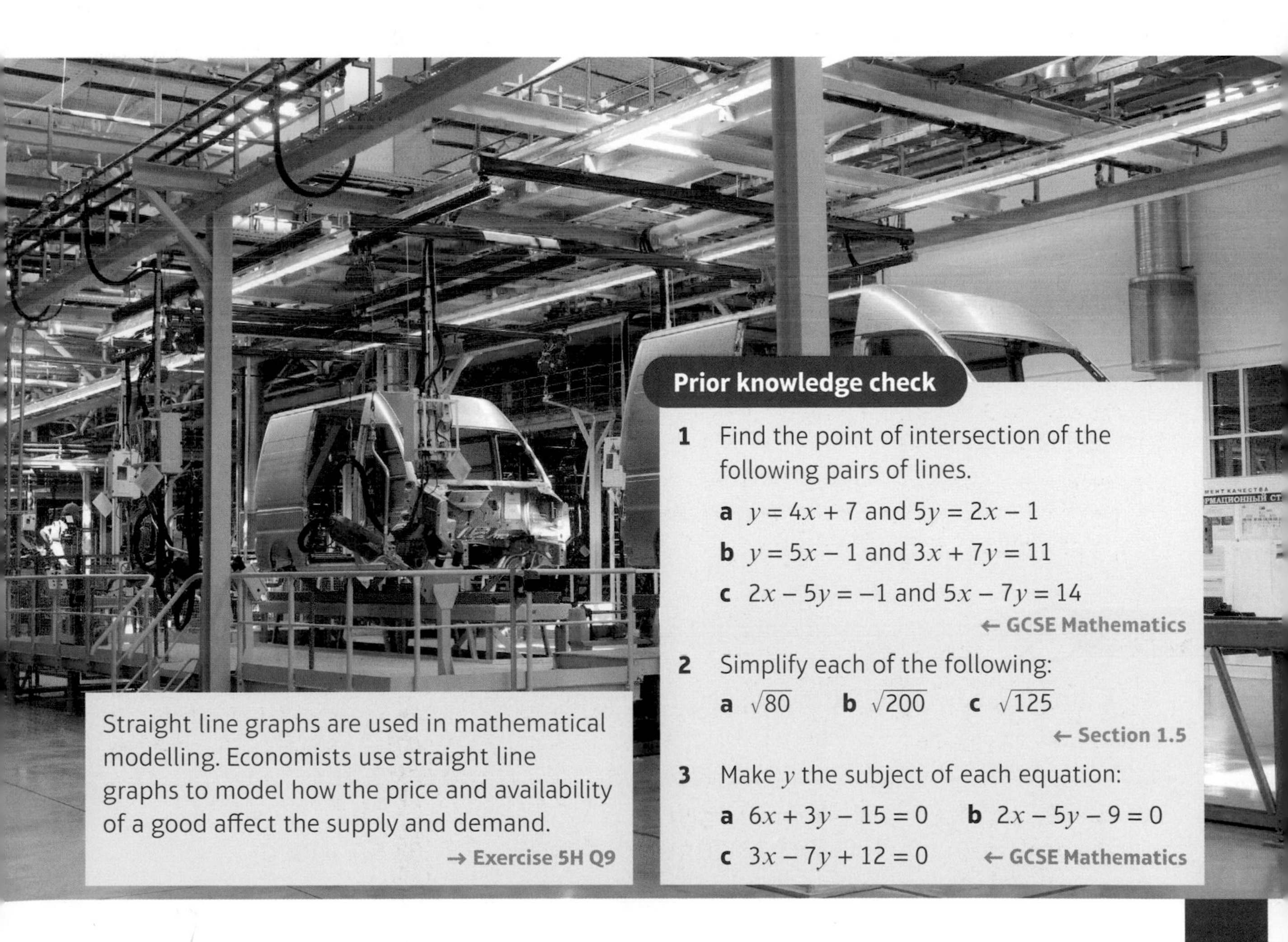

Straight line graphs are used in mathematical modelling. Economists use straight line graphs to model how the price and availability of a good affect the supply and demand.

→ Exercise 5H Q9

Prior knowledge check

1 Find the point of intersection of the following pairs of lines.

a $y = 4x + 7$ and $5y = 2x - 1$

b $y = 5x - 1$ and $3x + 7y = 11$

c $2x - 5y = -1$ and $5x - 7y = 14$

← GCSE Mathematics

2 Simplify each of the following:

a $\sqrt{80}$ **b** $\sqrt{200}$ **c** $\sqrt{125}$

← Section 1.5

3 Make y the subject of each equation:

a $6x + 3y - 15 = 0$ **b** $2x - 5y - 9 = 0$

c $3x - 7y + 12 = 0$

← GCSE Mathematics

5.1 $y = mx + c$

You can find the gradient of a straight line joining two points by considering the vertical distance and the horizontal distance between the points.

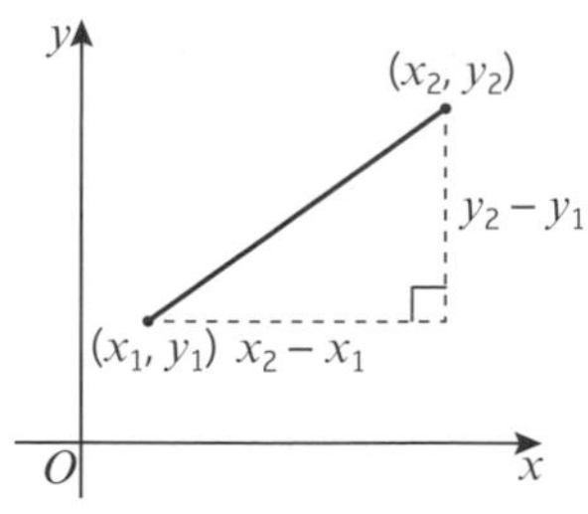

- **The gradient m of a line joining the point with coordinates (x_1, y_1) to the point with coordinates (x_2, y_2) can be calculated using the formula $m = \frac{y_2 - y_1}{x_2 - x_1}$**

Online Explore the gradient formula using GeoGebra.

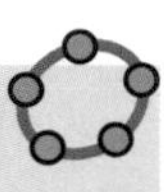

Example 1

Work out the gradient of the line joining (−2, 7) and (4, 5)

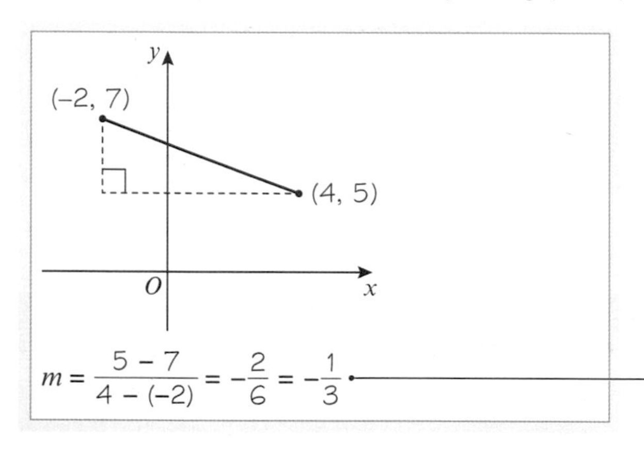

$m = \frac{5 - 7}{4 - (-2)} = -\frac{2}{6} = -\frac{1}{3}$

Use $m = \frac{y_2 - y_1}{x_2 - x_1}$. Here $(x_1, y_1) = (-2, 7)$ and $(x_2, y_2) = (4, 5)$

Example 2

The line joining (2, −5) to (4, a) has gradient −1. Work out the value of a.

$\frac{a - (-5)}{4 - 2} = -1$

Use $m = \frac{y_2 - y_1}{x_2 - x_1}$. Here $m = -1$, $(x_1, y_1) = (2, -5)$ and $(x_2, y_2) = (4, a)$.

So $\frac{a + 5}{2} = -1$

$a + 5 = -2$

$a = -7$

Exercise 5A

1 Work out the gradients of the lines joining these pairs of points:

a (4, 2), (6, 3) **b** (−1, 3), (5, 4) **c** (−4, 5), (1, 2)

d (2, −3), (6, 5) **e** (−3, 4), (7, −6) **f** (−12, 3), (−2, 8)

g (−2, −4), (10, 2) **h** $(\frac{1}{2}, 2), (\frac{3}{4}, 4)$ **i** $(\frac{1}{4}, \frac{1}{2}), (\frac{1}{2}, \frac{2}{3})$

j (−2.4, 9.6), (0, 0) **k** (1.3, −2.2), (8.8, −4.7) **l** $(0, 5a), (10a, 0)$

m $(3b, -2b), (7b, 2b)$ **n** $(p, p^2), (q, q^2)$

2 The line joining $(3, -5)$ to $(6, a)$ has a gradient 4. Work out the value of a.

3 The line joining $(5, b)$ to $(8, 3)$ has gradient -3. Work out the value of b.

4 The line joining $(c, 4)$ to $(7, 6)$ has gradient $\frac{3}{4}$. Work out the value of c.

5 The line joining $(-1, 2d)$ to $(1, 4)$ has gradient $-\frac{1}{4}$. Work out the value of d.

6 The line joining $(-3, -2)$ to $(2e, 5)$ has gradient 2. Work out the value of e.

7 The line joining $(7, 2)$ to $(f, 3f)$ has gradient 4. Work out the value of f.

8 The line joining $(3, -4)$ to $(-g, 2g)$ has gradient -3. Work out the value of g.

(P) **9** Show that the points $A(2, 3)$, $B(4, 4)$ and $C(10,7)$ can be joined by a straight line.

Problem-solving Find the gradient of the line joining the points A and B and the line joining the points A and C.

(E/P) **10** Show that the points $A(-2a, 5a)$, $B(0, 4a)$ and points $C(6a, a)$ are collinear. **(3 marks)**

Notation Points are collinear if they all lie on the same straight line.

- **The equation of a straight line can be written in the form $y = mx + c$, where m is the gradient and c is the y-intercept.**
- **The equation of a straight line can also be written in the form $ax + by + c = 0$, where a, b and c are integers.**

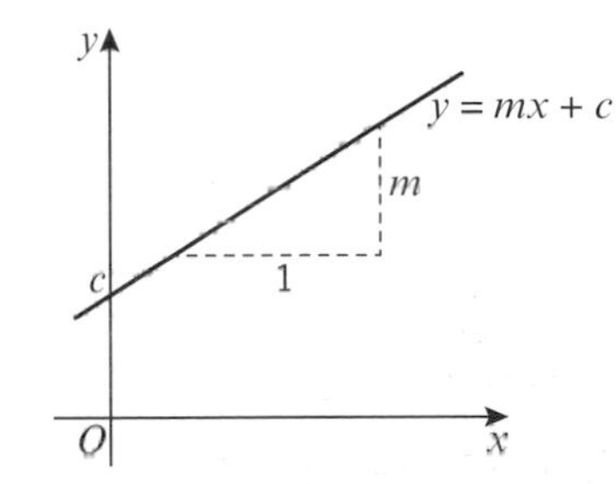

Example 3

Write down the gradient and y-intercept of these lines:

a $y = -3x + 2$ **b** $4x - 3y + 5 = 0$

a Gradient $= -3$ and y-intercept $= (0, 2)$.

Compare $y = -3x + 2$ with $y = mx + c$. From this, $m = -3$ and $c = 2$.

b $y = \frac{4}{3}x + \frac{5}{3}$

Gradient $= \frac{4}{3}$ and y-intercept $= (0, \frac{5}{3})$.

Rearrange the equation into the form $y = mx + c$. From this $m = \frac{4}{3}$ and $c = \frac{5}{3}$

Watch out Use fractions rather than decimals in coordinate geometry questions.

Example 4

Write these lines in the form $ax + by + c = 0$

a $y = 4x + 3$ **b** $y = -\frac{1}{2}x + 5$

a $4x - y + 3 = 0$ — Rearrange the equation into the form $ax + by + c = 0$

b $\frac{1}{2}x + y - 5 = 0$ — Collect all the terms on one side of the equation.

$x + 2y - 10 = 0$

Example 5

The line $y = 4x - 8$ meets the x-axis at the point P. Work out the coordinates of P.

$4x - 8 = 0$ — The line meets the x-axis when $y = 0$, so substitute $y = 0$ into $y = 4x - 8$.

$4x = 8$ — Rearrange the equation for x.

$x = 2$

So P has coordinates (2, 0) — Always write down the coordinates of the point.

Exercise 5B

1 Work out the gradients of these lines:

a $y = -2x + 5$ **b** $y = -x + 7$ **c** $y = 4 + 3x$

d $y = \frac{1}{3}x - 2$ **e** $y = -\frac{2}{3}x$ **f** $y = \frac{5}{4}x + \frac{2}{3}$

g $2x - 4y + 5 = 0$ **h** $10x - 5y + 1 = 0$ **i** $-x + 2y - 4 = 0$

j $-3x + 6y + 7 = 0$ **k** $4x + 2y - 9 = 0$ **l** $9x + 6y + 2 = 0$

2 These lines cut the y-axis at $(0, c)$. Work out the value of c in each case.

a $y = -x + 4$ **b** $y = 2x - 5$ **c** $y = \frac{1}{2}x - \frac{2}{3}$

d $y = -3x$ **e** $y = \frac{6}{7}x + \frac{7}{5}$ **f** $y = 2 - 7x$

g $3x - 4y + 8 = 0$ **h** $4x - 5y - 10 = 0$ **i** $-2x + y - 9 = 0$

j $7x + 4y + 12 = 0$ **k** $7x - 2y + 3 = 0$ **l** $-5x + 4y + 2 = 0$

3 Write these lines in the form $ax + by + c = 0$.

a $y = 4x + 3$ **b** $y = 3x - 2$ **c** $y = -6x + 7$

d $y = \frac{4}{5}x - 6$ **e** $y = \frac{5}{3}x + 2$ **f** $y = \frac{7}{3}x$

g $y = 2x - \frac{4}{7}$ **h** $y = -3x + \frac{2}{9}$ **i** $y = -6x - \frac{2}{3}$

j $y = -\frac{1}{3}x + \frac{1}{2}$ **k** $y = \frac{2}{3}x + \frac{5}{6}$ **l** $y = \frac{3}{5}x + \frac{1}{2}$

4 The line $y = 6x - 18$ meets the x-axis at the point P. Work out the coordinates of P.

5 The line $3x + 2y = 0$ meets the x-axis at the point R. Work out the coordinates of R.

6 The line $5x - 4y + 20 = 0$ meets the y-axis at the point A and the x-axis at the point B. Work out the coordinates of A and B.

7 A line l passes through the points with coordinates (0, 5) and (6, 7).

a Find the gradient of the line.

b Find an equation of the line in the form $ax + by + c = 0$.

(E) 8 A line l cuts the x-axis at (5, 0) and the y-axis at (0, 2).

a Find the gradient of the line. **(1 mark)**

b Find an equation of the line in the form $ax + by + c = 0$. **(2 marks)**

(P) 9 Show that the line with equation $ax + by + c = 0$ has gradient $-\frac{a}{b}$ and cuts the y-axis at $-\frac{c}{b}$

Problem-solving

Try solving a similar problem with numbers first:

Find the gradient and y-intercept of the straight line with equation $3x + 7y + 2 = 0$.

(E/P) 10 The line l with gradient 3 and y-intercept (0, 5) has the equation $ax - 2y + c = 0$. Find the values of a and c. **(2 marks)**

(E/P) 11 The straight line l passes through (0, 6) and has gradient −2. It intersects the line with equation $5x - 8y - 15 = 0$ at point P. Find the coordinates of P. **(4 marks)**

(E/P) 12 The straight line l_1 with equation $y = 3x - 7$ intersects the straight line l_2 with equation $ax + 4y - 17 = 0$ at the point $P(-3, b)$.

a Find the value of b. **(1 mark)**

b Find the value of a. **(2 marks)**

Challenge

Show that the equation of a straight line through $(0, a)$ and $(b, 0)$ is $ax + by - ab = 0$.

5.2 Equations of straight lines

You can define a straight line by giving:

- one point on the line and the gradient
- two different points on the line

You can find an equation of the line from either of these conditions.

- **The equation of a line with gradient m that passes through the point with coordinates (x_1, y_1) can be written as $y - y_1 = m(x - x_1)$.**

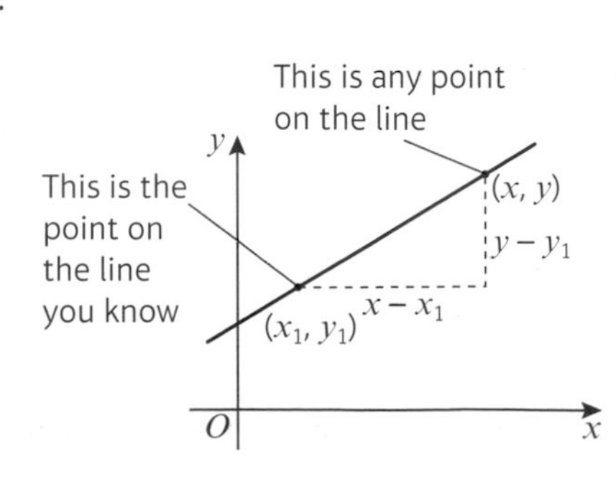

Example 6

Find the equation of the line with gradient 5 that passes through the point (3, 2).

$y - 2 = 5(x - 3)$

$y - 2 = 5x - 15$

$y = 5x - 13$

Online Explore lines of a given gradient passing through a given point using GeoGebra.

This is in the form $y - y_1 = m(x - x_1)$. Here $m = 5$ and $(x_1, y_1) = (3, 2)$.

Example 7

Find the equation of the line that passes through the points (5, 7) and (3, −1).

$m = \frac{y_2 - y_1}{x_2 - x_1} = \frac{7 - (-1)}{5 - 3} = \frac{8}{2} = 4$

So $y - y_1 = m(x - x_1)$

$y + 1 = 4(x - 3)$

$y + 1 = 4x - 12$

$y = 4x - 13$

First find the slope of the line.
Here $(x_1, y_1) = (3, -1)$ and $(x_2, y_2) = (5, 7)$.
(x_1, y_1) and (x_2, y_2) have been chosen to make the denominators positive.

You know the gradient and a point on the line, so use $y - y_1 = m(x - x_1)$.
Use $m = 4$, $x_1 = 3$ and $y_1 = -1$.

Exercise 5C

1 Find the equation of the line with gradient m that passes through the point (x_1, y_1) when:

a $m = 2$ and $(x_1, y_1) = (2, 5)$
b $m = 3$ and $(x_1, y_1) = (-2, 1)$
c $m = -1$ and $(x_1, y_1) = (3, -6)$
d $m = -4$ and $(x_1, y_1) = (-2, -3)$
e $m = \frac{1}{2}$ and $(x_1, y_1) = (-4, 10)$
f $m = -\frac{2}{3}$ and $(x_1, y_1) = (-6, -1)$
g $m = 2$ and $(x_1, y_1) = (a, 2a)$
h $m = -\frac{1}{2}$ and $(x_1, y_1) = (-2b, 3b)$

2 Find the equations of the lines that pass through these pairs of points:

a (2, 4) and (3, 8)
b (0, 2) and (3, 5)
c (−2, 0) and (2, 8)
d (5, −3) and (7, 5)
e (3, −1) and (7, 3)
f (−4, −1) and (6, 4)
g (−1, −5) and (−3, 3)
h (−4, −1) and (−3, −9)
i $(\frac{1}{3}, \frac{2}{5})$ and $(\frac{2}{3}, \frac{4}{5})$
j $(-\frac{3}{4}, \frac{1}{7})$ and $(\frac{1}{4}, \frac{3}{7})$

Hint In each case find the gradient m then use $y - y_1 = m(x - x_1)$.

(E) **3** Find the equation of the line l which passes through the points $A(7, 2)$ and $B(9, -8)$.
Give your answer in the form $ax + by + c = 0$. **(3 marks)**

4 The vertices of the triangle ABC have coordinates $A(3, 5)$, $B(-2, 0)$ and $C(4, -1)$.
Find the equations of the sides of the triangle.

E/P **5** The straight line l passes through $(a, 4)$ and $(3a, 3)$. An equation of l is $x + 6y + c = 0$.
Find the value of a and the value of c. **(3 marks)**

Problem-solving

It is often easier to find unknown values in the order they are given in the question. Find the value of a first then find the value of c.

E/P **6** The straight line l passes through $(7a, 5)$ and $(3a, 3)$.
An equation of l is $x + by - 12 = 0$.
Find the value of a and the value of b. **(3 marks)**

Challenge

Consider the line passing through points (x_1, y_1) and (x_2, y_2).

a Write down the formula for the gradient, m, of the line.

b Show that the general equation of the line can be written in the form $\frac{y - y_1}{y_2 - y_1} = \frac{x - x_1}{x_2 - x_1}$

c Use the equation from part **b** to find the equation of the line passing through the points (–8, 4) and (–1, 7).

Example 8

The line $y = 3x - 9$ meets the x-axis at the point A. Find the equation of the line with gradient $\frac{2}{3}$ that passes through point A. Write your answer in the form $ax + by + c = 0$, where a, b and c are integers.

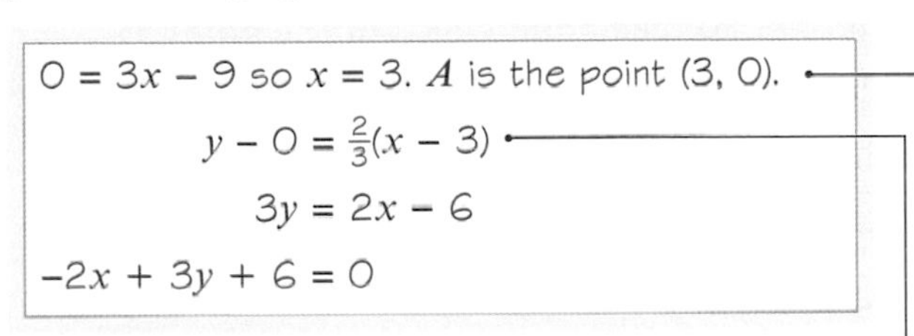

$0 = 3x - 9$ so $x = 3$. A is the point $(3, 0)$.

$y - 0 = \frac{2}{3}(x - 3)$

$3y = 2x - 6$

$-2x + 3y + 6 = 0$

Online Plot the solution on a graph using technology.

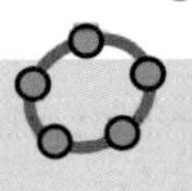

The line meets the x-axis when $y = 0$, so substitute $y = 0$ into $y = 3x - 9$.

Use $y - y_1 = m(x - x_1)$. Here $m = \frac{2}{3}$ and $(x_1, y_1) = (3, 0)$.

Rearrange the equation into the form $ax + by + c = 0$.

Example 9

The lines $y = 4x - 7$ and $2x + 3y - 21 = 0$ intersect at the point A. The point B has coordinates $(-2, 8)$. Find the equation of the line that passes through the points A and B. Write your answer in the form $ax + by + c = 0$, where a, b and c are integers.

$2x + 3(4x - 7) - 21 = 0$

$2x + 12x - 21 - 21 = 0$

$14x = 42$

$x = 3$

$y = 4(3) - 7 = 5$ so A is the point $(3, 5)$.

$m = \frac{y_2 - y_1}{x_2 - x_1} = \frac{8 - 5}{-2 - 3} = \frac{3}{-5} = -\frac{3}{5}$

$y - 5 = -\frac{3}{5}(x - 3)$

$5y - 25 = -3x + 9$

$3x + 5y - 34 = 0$

Online Check solutions to simultaneous equations using your calculator.

Solve the equations simultaneously to find point A. Substitute $y = 4x - 7$ into $2x + 3y - 21 = 0$.

Find the slope of the line connecting A and B.

Use $y - y_1 = m(x - x_1)$ with $m = -\frac{3}{5}$ and $(x_1, y_1) = (3, 5)$.

Exercise 5D

1 The line $y = 4x - 8$ meets the x-axis at the point A. Find the equation of the line with gradient 3 that passes through the point A.

2 The line $y = -2x + 8$ meets the y-axis at the point B. Find the equation of the line with gradient 2 that passes through the point B.

3 The line $y = \frac{1}{2}x + 6$ meets the x-axis at the point C. Find the equation of the line with gradient $\frac{2}{3}$ that passes through the point C. Write your answer in the form $ax + by + c = 0$, where a, b and c are integers.

(P) 4 The line $y = \frac{1}{4}x + 2$ meets the y-axis at the point B. The point C has coordinates $(-5, 3)$. Find the gradient of the line joining the points B and C.

(P) 5 The line that passes through the points $(2, -5)$ and $(-7, 4)$ meets the x-axis at the point P. Work out the coordinates of the point P.

Problem-solving

A sketch can help you check whether your answer looks right.

(P) 6 The line that passes through the points $(-3, -5)$ and $(4, 9)$ meets the y-axis at the point G. Work out the coordinates of the point G.

(P) 7 The line that passes through the points $(3, 2\frac{1}{2})$ and $(-1\frac{1}{2}, 4)$ meets the y-axis at the point J. Work out the coordinates of the point J.

(P) 8 The lines $y = x$ and $y = 2x - 5$ intersect at the point A. Find the equation of the line with gradient $\frac{2}{5}$ that passes through the point A.

(P) 9 The lines $y = 4x - 10$ and $y = x - 1$ intersect at the point T. Find the equation of the line with gradient $-\frac{2}{3}$ that passes through the point T. Write your answer in the form $ax + by + c = 0$, where a, b and c are integers.

(P) 10 The line p has gradient $\frac{2}{3}$ and passes through the point $(6, -12)$. The line q has gradient -1 and passes through the point $(5, 5)$. The line p meets the y-axis at A and the line q meets the x-axis at B. Work out the gradient of the line joining the points A and B.

(P) 11 The line $y = -2x + 6$ meets the x-axis at the point P. The line $y = \frac{3}{2}x - 4$ meets the y-axis at the point Q. Find the equation of the line joining the points P and Q.

(P) 12 The line $y = 3x - 5$ meets the x-axis at the point M. The line $y = -\frac{2}{3}x + \frac{2}{3}$ meets the y-axis at the point N. Find the equation of the line joining the points M and N. Write your answer in the form $ax + by + c = 0$, where a, b and c are integers.

(P) 13 The line $y = 2x - 10$ meets the x-axis at the point A. The line $y = -2x + 4$ meets the y-axis at the point B. Find the equation of the line joining the points A and B.

(P) 14 The line $y = 4x + 5$ meets the y-axis at the point C. The line $y = -3x - 15$ meets the x-axis at the point D. Find the equation of the line joining the points C and D. Write your answer in the form $ax + by + c = 0$, where a, b and c are integers.

(P) 15 The lines $y = x - 5$ and $y = 3x - 13$ intersect at the point S. The point T has coordinates $(-4, 2)$. Find the equation of the line that passes through the points S and T.

(P) 16 The lines $y = -2x + 1$ and $y = x + 7$ intersect at the point L. The point M has coordinates $(-3, 1)$. Find the equation of the line that passes through the points L and M.

5.3 Parallel and perpendicular lines

- **Parallel lines have the same gradient.**

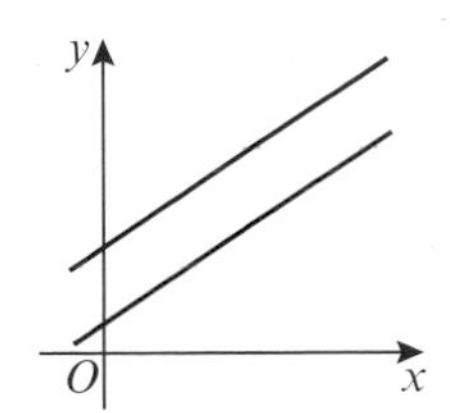

Example 10

A line is parallel to the line $6x + 3y - 2 = 0$ and it passes through the point (0, 3).
Work out the equation of the line.

$6x + 3y - 2 = 0$

$3y - 2 = -6x$

$3y = -6x + 2$

$y = -2x + \frac{2}{3}$

The gradient of this line is -2.

The equation of the line is $y = -2x + 3$.

Rearrange the equation into the form $y = mx + c$ to find m.

Compare $y = -2x + \frac{2}{3}$ with $y = mx + c$, so $m = -2$.
Parallel lines have the same gradient, so the gradient of the required line = -2.

(0, 3) is the intercept on the y-axis, so $c = 3$.

Exercise 5E

1 Work out whether each pair of lines is parallel.

a $y = 5x - 2$
$15x - 3y + 9 = 0$

b $7x + 14y - 1 = 0$
$y = \frac{1}{2}x + 9$

c $4x - 3y - 8 = 0$
$3x - 4y - 8 = 0$

(P) **2** The line r passes through the points (1, 4) and (6, 8) and the line s passes through the points (5, −3) and (20, 9). Show that the lines r and s are parallel.

(P) **3** The coordinates of a quadrilateral $ABCD$ are $A(-6, 2)$, $B(4, 8)$, $C(6, 1)$ and $D(-9, -8)$. Show that the quadrilateral is a trapezium.

Hint A trapezium has exactly one pair of parallel sides.

4 A line is parallel to the line $y = 5x + 8$ and its y-intercept is (0, 3). Write down the equation of the line.

Hint The line will have gradient 5.

5 A line is parallel to the line $y = -\frac{2}{5}x + 1$ and its y-intercept is (0, −4). Work out the equation of the line. Write your answer in the form $ax + by + c = 0$, where a, b and c are integers.

(P) **6** A line is parallel to the line $3x + 6y + 11 = 0$ and its intercept on the y-axis is (0, 7). Write down the equation of the line.

(P) **7** A line is parallel to the line $2x - 3y - 1 = 0$ and it passes through the point (0, 0). Write down the equation of the line.

8 Find an equation of the line that passes through the point (−2, 7) and is parallel to the line $y = 4x + 1$. Write your answer in the form $ax + by + c = 0$.

Perpendicular lines are at right angles to each other. If you know the gradient of one line, you can find the gradient of the other.

- **If a line has a gradient of m, a line perpendicular to it has a gradient of $-\frac{1}{m}$**
- **If two lines are perpendicular, the product of their gradients is −1.**

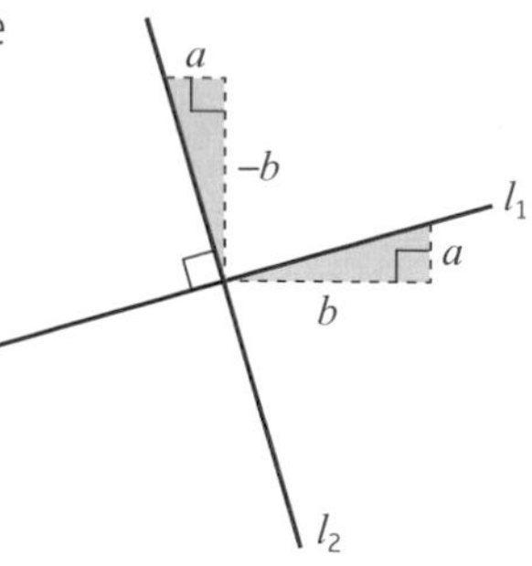

The shaded triangles are congruent.

Line l_1 has gradient $\frac{a}{b} = m$

Line l_2 has gradient $\frac{-b}{a} = -\frac{1}{m}$

Example 11

Work out whether these pairs of lines are parallel, perpendicular or neither:

a $3x - y - 2 = 0$
$x + 3y - 6 = 0$

b $y = \frac{1}{2}x$
$2x - y + 4 = 0$

a $3x - y - 2 = 0$ — Rearrange the equations into the form $y = mx + c$.

$3x - 2 = y$

So $y = 3x - 2$

The gradient of this line is 3.

$x + 3y - 6 = 0$

$3y - 6 = -x$

$3y = -x + 6$

$y = -\frac{1}{3}x + 2$

The gradient of this line is $-\frac{1}{3}$ — Compare $y = -\frac{1}{3}x + 2$ with $y = mx + c$, so $m = -\frac{1}{3}$

So the lines are perpendicular as $3 \times (-\frac{1}{3}) = -1$.

b $y = \frac{1}{2}x$ — Compare $y = \frac{1}{2}x$ with $y = mx + c$, so $m = \frac{1}{2}$.

The gradient of this line is $\frac{1}{2}$

$2x - y + 4 = 0$ — Rearrange the equation into the form $y = mx + c$ to find m.

$2x + 4 = y$

So $y = 2x + 4$

The gradient of this line is 2. — Compare $y = 2x + 4$ with $y = mx + c$, so $m = 2$.

The lines are not parallel as they have different gradients.

The lines are not perpendicular as $\frac{1}{2} \times 2 \neq -1$.

Online Explore this solution using technology.

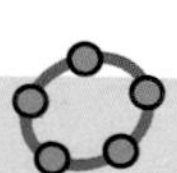

Example 12

A line is perpendicular to the line $2y - x - 8 = 0$ and passes through the point (5, −7). Find the equation of the line.

Rearranging, $y = \frac{1}{2}x + 4$

Gradient of $y = \frac{1}{2}x + 4$ is $\frac{1}{2}$

So the gradient of the perpendicular line is −2.

$y - y_1 = m(x - x_1)$

$y + 7 = -2(x - 5)$

$y + 7 = -2x + 10$

$y = -2x + 3$

Problem-solving

You need to fill in the steps of this problem yourself:

- Rearrange the equation into the form $y = mx + c$ to find the gradient.
- Use $-\frac{1}{m}$ to find the gradient of a perpendicular line.
- Use $y - y_1 = m(x - x_1)$ to find the equation of the line.

Exercise 5F

1 Work out whether these pairs of lines are parallel, perpendicular or neither:

a $y = 4x + 2$
$y = -\frac{1}{4}x - 7$

b $y = \frac{2}{3}x - 1$
$y = \frac{2}{3}x - 11$

c $y = \frac{1}{5}x + 9$
$y = 5x + 9$

d $y = -3x + 2$
$y = \frac{1}{3}x - 7$

e $y = \frac{3}{5}x + 4$
$y = -\frac{5}{3}x - 1$

f $y = \frac{5}{7}x$
$y = \frac{5}{7}x - 3$

g $y = 5x - 3$
$5x - y + 4 = 0$

h $5x - y - 1 = 0$
$y = -\frac{1}{5}x$

i $y = -\frac{3}{2}x + 8$
$2x - 3y - 9 = 0$

j $4x - 5y + 1 = 0$
$8x - 10y - 2 = 0$

k $3x + 2y - 12 = 0$
$2x + 3y - 6 = 0$

l $5x - y + 2 = 0$
$2x + 10y - 4 = 0$

2 A line is perpendicular to the line $y = 6x - 9$ and passes through the point (0, 1). Find an equation of the line.

(P) 3 A line is perpendicular to the line $3x + 8y - 11 = 0$ and passes through the point (0, −8). Find an equation of the line.

4 Find an equation of the line that passes through the point (6, −2) and is perpendicular to the line $y = 3x + 5$.

5 Find an equation of the line that passes through the point (−2, 5) and is perpendicular to the line $y = 3x + 6$.

(P) 6 Find an equation of the line that passes through the point (3, 4) and is perpendicular to the line $4x - 6y + 7 = 0$.

7 Find an equation of the line that passes through the point (5, −5) and is perpendicular to the line $y = \frac{2}{3}x + 5$. Write your answer in the form $ax + by + c = 0$, where a, b and c are integers.

8 Find an equation of the line that passes through the point (−2, −3) and is perpendicular to the line $y = -\frac{4}{7}x + 5$. Write your answer in the form $ax + by + c = 0$, where a, b and c are integers.

(P) 9 The line l passes through the points (−3, 0) and (3, −2) and the line n passes through the points (1, 8) and (−1, 2). Show that the lines l and n are perpendicular.

Problem-solving

Don't do more work than you need to. You only need to find the gradients of both lines, not their equations.

(P) 10 The vertices of a quadrilateral $ABCD$ have coordinates $A(-1, 5)$, $B(7, 1)$, $C(5, -3)$ and $D(-3, 1)$. Show that the quadrilateral is a rectangle.

Hint The sides of a rectangle are perpendicular.

(E/P) 11 A line l_1 has equation $5x + 11y - 7 = 0$ and crosses the x-axis at A. The line l_2 is perpendicular to l_1 and passes through A.

a Find the coordinates of the point A. **(1 mark)**

b Find the equation of the line l_2. Write your answer in the form $ax + by + c = 0$. **(3 marks)**

(E/P) 12 The points A and C lie on the y-axis and the point B lies on the x-axis as shown in the diagram.

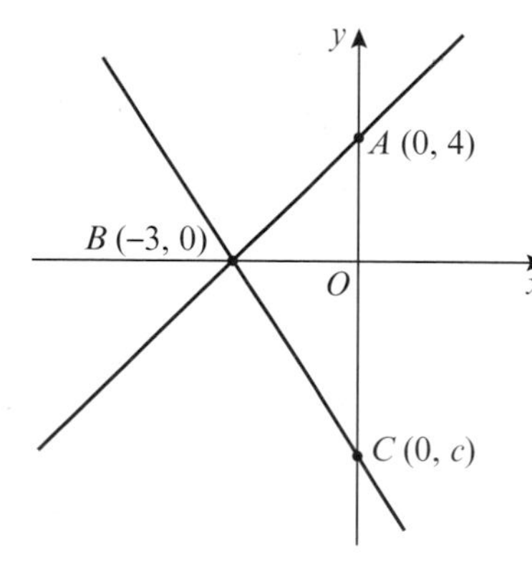

Problem-solving

Sketch graphs in coordinate geometry problems are not accurate, but you can use the graph to make sure that your answer makes sense. In this question c must be negative.

The line through points A and B is perpendicular to the line through points B and C. Find the value of c. **(6 marks)**

5.4 Length and area

You can find the distance between two points A and B by considering a right-angled triangle with hypotenuse AB.

- **You can find the distance d between (x_1, y_1) and (x_2, y_2) by using the formula**

$$d = \sqrt{(x_2 - x_1)^2 + (y_2 - y_1)^2}$$

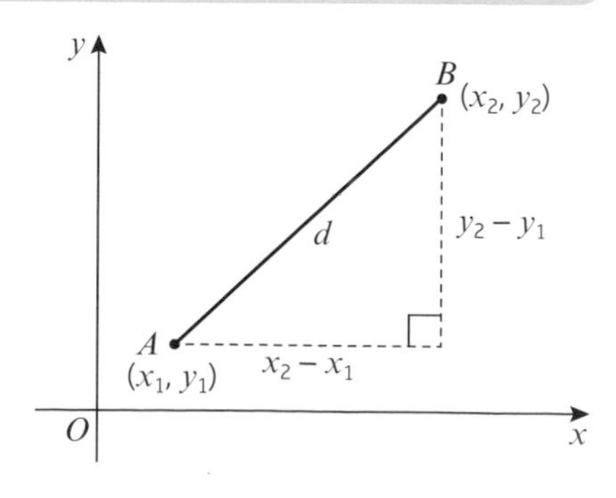

Example 13

Find the distance between (2, 3) and (5, 7).

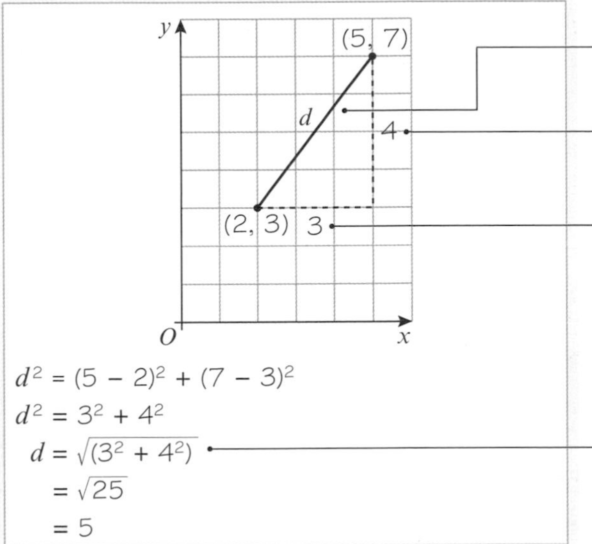

$d^2 = (5 - 2)^2 + (7 - 3)^2$

$d^2 = 3^2 + 4^2$

$d = \sqrt{(3^2 + 4^2)}$

$= \sqrt{25}$

$= 5$

Draw a sketch.
Let the distance between the points be d.
The difference in the y-coordinates is $7 - 3 = 4$.
The difference in the x-coordinates is $5 - 2 = 3$.

$d = \sqrt{(x_2 - x_1)^2 + (y_2 - y_1)^2}$ with $(x_1, y_1) = (2, 3)$ and $(x_2, y_2) = (5, 7)$.

Example 14

Online Draw both lines and the triangle AOB on a graph using technology.

The straight line l_1 with equation $4x - y = 0$ and the straight line l_2 with equation $2x + 3y - 21 = 0$ intersect at point A.

a Work out the coordinates of A.

b Work out the area of triangle AOB where B is the point where l_2 meets the x-axis.

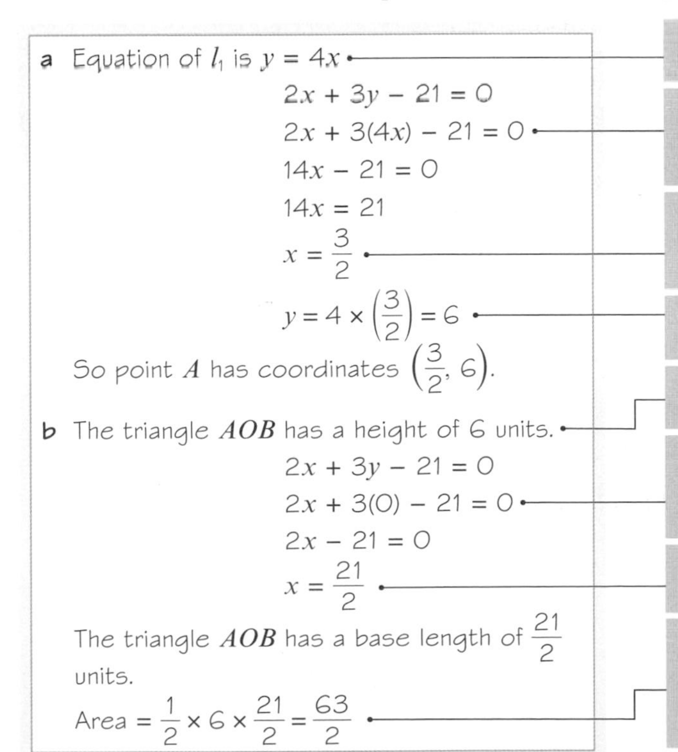

a Equation of l_1 is $y = 4x$

$2x + 3y - 21 = 0$

$2x + 3(4x) - 21 = 0$

$14x - 21 = 0$

$14x = 21$

$x = \frac{3}{2}$

$y = 4 \times \left(\frac{3}{2}\right) = 6$

So point A has coordinates $\left(\frac{3}{2}, 6\right)$.

b The triangle AOB has a height of 6 units.

$2x + 3y - 21 = 0$

$2x + 3(0) - 21 = 0$

$2x - 21 = 0$

$x = \frac{21}{2}$

The triangle AOB has a base length of $\frac{21}{2}$ units.

Area $= \frac{1}{2} \times 6 \times \frac{21}{2} = \frac{63}{2}$

Rewrite the equation of l_1 in the form $y = mx + c$.

Substitute $y = 4x$ into the equation for l_2 to find the point of intersection.

Solve the equation to find the x-coordinate of point A.

Substitute to find the y-coordinate of point A.

The height is the y-coordinate of point A.

B is the point where the line l_2 intersects the x-axis. At B, the y-coordinate is zero.

Solve the equation to find the x-coordinate of point B.

Area $= \frac{1}{2} \times$ base $\times$ height
You don't need to give units for length and area problems on coordinate grids.

Exercise 5G

1 Find the distance between these pairs of points:

a (0, 1), (6, 9) b (4, −6), (9, 6) c (3, 1), (−1, 4)
d (3, 5), (4, 7) e (0, −4), (5, 5) f (−2, −7), (5, 1)

2 Consider the points $A(-3, 5)$, $B(-2, -2)$ and $C(3, -7)$. Determine whether the line joining the points A and B is congruent to the line joining the points B and C.

Hint Two line segments are congruent if they are the same length.

3 Consider the points $P(11, -8)$, $Q(4, -3)$ and $R(7, 5)$. Show that the line segment joining the points P and Q is not congruent to the line joining the points Q and R.

(P) 4 The distance between the points $(-1, 13)$ and $(x, 9)$ is $\sqrt{65}$. Find two possible values of x.

Problem-solving Use the distance formula to formulate a quadratic equation in x.

(P) 5 The distance between the points $(2, y)$ and $(5, 7)$ is $3\sqrt{10}$. Find two possible values of y.

(P) 6 a Show that the straight line l_1 with equation $y = 2x + 4$ is parallel to the straight line l_2 with equation $6x - 3y - 9 = 0$.

b Find the equation of the straight line l_3 that is perpendicular to l_1 and passes through the point (3, 10).

c Find the point of intersection of the lines l_2 and l_3.

d Find the shortest distance between lines l_1 and l_2.

Problem-solving The shortest distance between two parallel lines is the perpendicular distance between them.

(E/P) 7 A point P lies on the line with equation $y = 4 - 3x$. The point P is a distance $\sqrt{34}$ from the origin. Find the two possible positions of point P. **(5 marks)**

(P) 8 The vertices of a triangle are $A(2, 7)$, $B(5, -6)$ and $C(8, -6)$.

a Show that the triangle is a scalene triangle.

b Find the area of the triangle ABC.

Notation Scalene triangles have three sides of different lengths.

Problem-solving Draw a sketch and label the points A, B and C. Find the length of the base and the height of the triangle.

9 The straight line l_1 has equation $y = 7x - 3$. The straight line l_2 has equation $4x + 3y - 41 = 0$. The lines intersect at the point A.

a Work out the coordinates of A.

The straight line l_2 crosses the x-axis at the point B.

b Work out the coordinates of B.

c Work out the area of triangle AOB.

10 The straight line l_1 has equation $4x - 5y - 10 = 0$ and intersects the x-axis at point A.
The straight line l_2 has equation $4x - 2y + 20 = 0$ and intersects the x-axis at the point B.

a Work out the coordinates of A.

b Work out the coordinates of B.

The straight lines l_1 and l_2 intersect at the point C.

c Work out the coordinates of C.

d Work out the area of triangle ABC.

(E) **11** The points $R(5, -2)$ and $S(9, 0)$ lie on the straight line l_1 as shown.

a Work out an equation for straight line l_1. **(2 marks)**

The straight line l_2 is perpendicular to l_1 and passes through the point R.

b Work out an equation for straight line l_2. **(2 marks)**

c Write down the coordinates of T. **(1 mark)**

d Work out the lengths of RS and TR leaving your answer in the form $k\sqrt{5}$. **(2 marks)**

e Work out the area of $\triangle RST$. **(2 marks)**

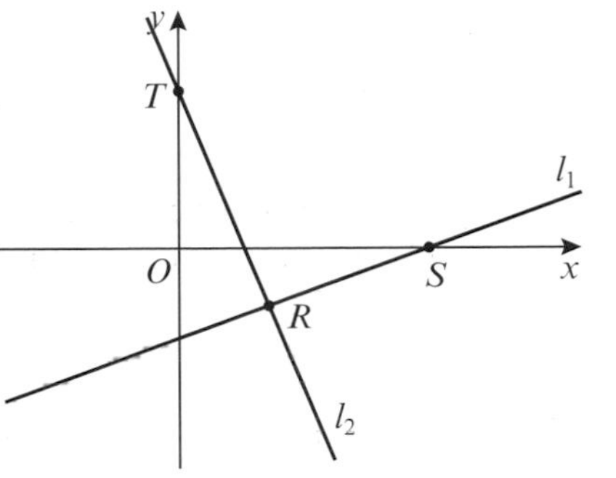

(E/P) **12** The straight line l_1 passes through the point $(-4, 14)$ and has gradient $-\frac{1}{4}$

a Find an equation for l_1 in the form $ax + by + c = 0$, where a, b and c are integers. **(3 marks)**

b Write down the coordinates of A, the point where straight line l_1 crosses the y-axis. **(1 mark)**

The straight line l_2 passes through the origin and has gradient 3. The lines l_1 and l_2 intersect at the point B.

c Calculate the coordinates of B. **(2 marks)**

d Calculate the exact area of $\triangle OAB$. **(2 marks)**

5.5 Modelling with straight lines

- **Two quantities are in direct proportion when they increase at the same rate. The graph of these quantities is a straight line through the origin.**

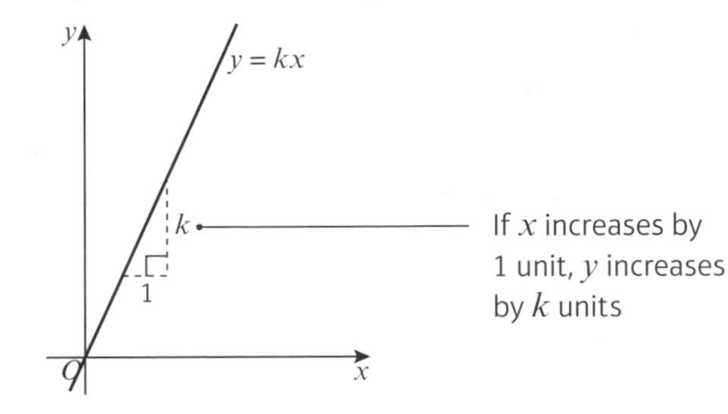

Notation These mean the same thing:
y is proportional to x
$y \propto x$
$y = kx$ for some real constant k.

Example 15

The graph shows the extension, E, of a spring when different masses, m, are attached to the end of the spring.

a Calculate the gradient, k, of the line.

b Write an equation linking E and m.

c Explain what the value of k represents in this situation.

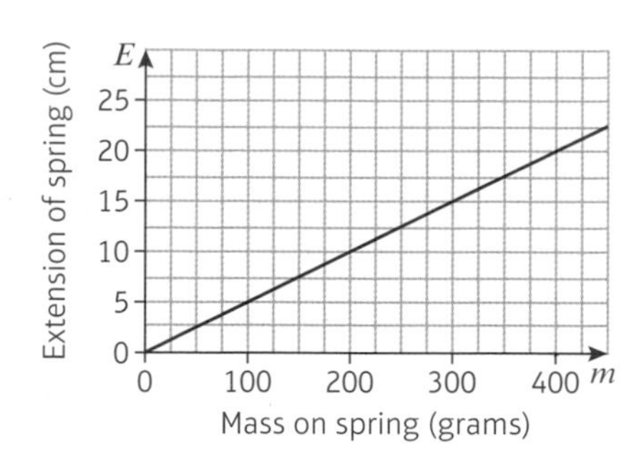

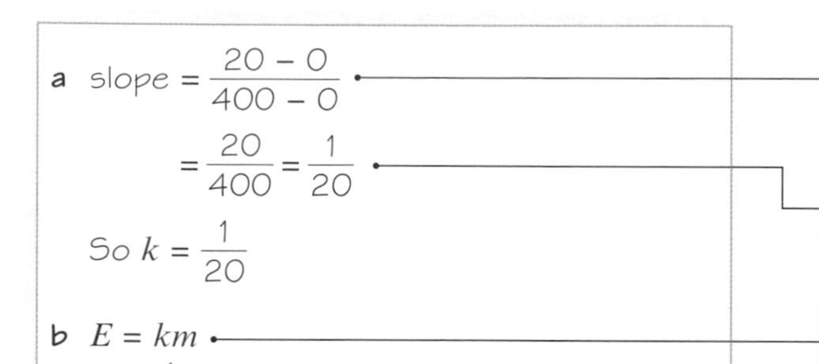

c k represents the increase in extension in cms when the mass increases by 1 gram.

Use any two points on the line to calculate the gradient. Here (0, 0) and (400, 20) are used.

Simplify the answer.

'$y = kx$' is the general form of a direct proportion equation. Here the variables are E and m.

k is the gradient. When the m-value increases by 1, the E-value increases by k.

You can sometimes use a **linear model** to show the relationship between two variables, x and y. The graph of a linear model is a straight line, and the variables are related by an equation of the form $y = ax + b$.

A linear model can still be appropriate even if all the points do not lie directly on the line. In this case, the points should be **close** to the line. The further the points are from the line, the less appropriate a linear model is for the data.

- **A mathematical model is an attempt to represent a real-life situation using mathematical concepts. It is often necessary to make assumptions about the real-life problem in order to create a model.**

Example 16

A container was filled with water. A hole was made in the bottom of the container. The depth of water remaining was recorded at certain time intervals. The table shows the results.

Time, t seconds	0	10	30	60	100	120
Depth of water, d cm	19.1	17.8	15.2	11.3	6.1	3.5

a Determine whether a linear model is appropriate by drawing a graph.

b Deduce an equation in the form $d = at + b$.

c Interpret the meaning of the coefficients a and b.

d Use the model to find the time when the container will be empty.

a

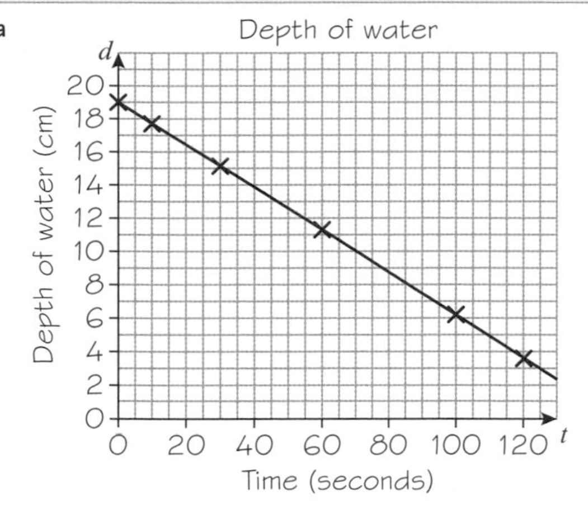

The points form a straight line, therefore a linear model is appropriate.

b $m = \frac{6.1 - 19.1}{100 - 0}$

$= -\frac{13}{100} = -0.13$

The d-intercept is 19.1. So $b = 19.1$

$d = at + b$

$d = -0.13t + 19.1$

c a is the change in depth of water in the container every second.

b is the depth of water in the container at the beginning of the experiment.

d $d = -0.13t + 19.1$

$0 = -0.13t + 19.1$

$0.13t = 19.1$

$t = 146.9$ seconds.

Problem-solving

You need to give your answer in the context of the question. Make sure you refer to the extension in the spring and the mass.

Pick any two points from the table. Here (0, 19.1) and (100, 6.1) are used.

The d-intercept is the d-value when $t = 0$.

State the linear equation using the variables in the question.

Substitute $a = -0.13$ and $b = 19.1$.

a represents the rate of change. Look at the problem and determine what is changing every second.

b is the value of d when $t = 0$. It represents the starting, or initial, value in the model.

State the linear equation using the variables in the question.

Substitute $d = 0$, as we want to know the time when the depth of water is zero.

Solve the equation to find t.

In 1991 there were 18 500 people living in Bradley Stoke. Planners projected that the number of people living in Bradley Stoke would increase by 350 each year.

a Write a linear model for the population p of Bradley Stoke t years after 1991.

b Write down one reason why this might not be a realistic model.

a 1991 is the first year, so $t = 0$.

When $t = 0$, the population is 18 500.

18 500 is the p-intercept.

The population is expected to increase by 350 each year.

350 represents the gradient of the line.

$p = at + b$

$p = 350t + 18\,500$

b The number of people living in Bradley Stoke would probably not increase by exactly the same amount each year.

The p-intercept is the population when $t = 0$.

The gradient is the yearly change in population.

State the linear equation using the variables in the question.

Substitute $a = 350$ and $b = 18\,500$.

Problem-solving

Look at the question carefully. Which points did you accept without knowing them to be true? These are your **assumptions**.

Exercise 5H

1 For each graph

i calculate the gradient, k, of the line

ii write a direct proportion equation connecting the two variables.

a

Distance, d (m)
0, 100, 200, 300, 400, 500, 600
0, 2, 4, 6, 8, 10, 12
Time, t (s)

b

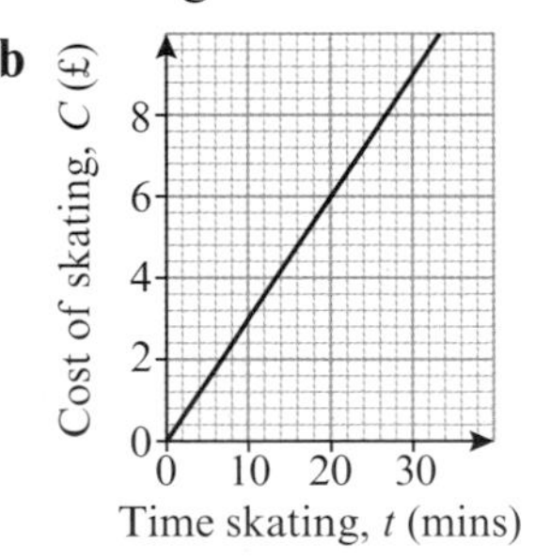

c

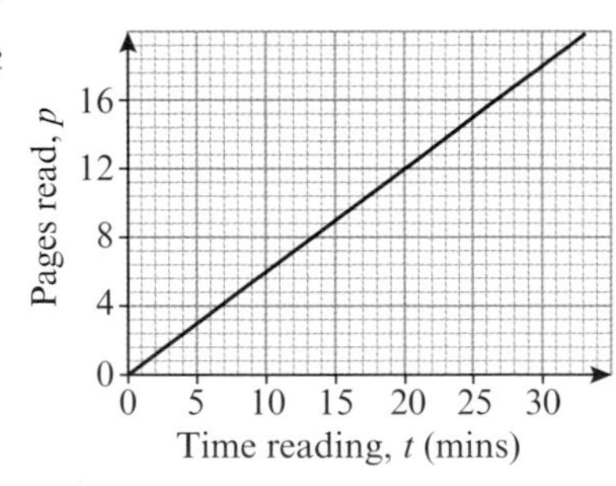

2 Draw a graph to determine whether a linear model would be appropriate for each set of data.

a

v	p
0	0
15	2
25	6
40	12
60	25
80	50

b

x	y
0	70
5	82.5
10	95
15	107.5
25	132.5
40	170

c

w	l
3.1	45
3.4	47
3.6	50
3.9	51
4.5	51
4.7	53

Hint A linear model can be appropriate even if all the points do not lie exactly in a straight line. In these cases, the points should lie close to a straight line.

E/P **3** The cost of electricity, E, in pounds and the number of kilowatt hours, h, are shown in the table.

kilowatt hours, h	0	15	40	60	80	110
cost of electricity, E	45	46.8	49.8	52.2	54.6	58.2

a Draw a graph of the data. **(3 marks)**
b Explain how you know a linear model would be appropriate. **(1 mark)**
c Deduce an equation in the form $E = ah + b$. **(2 marks)**
d Interpret the meaning of the coefficients a and b. **(2 marks)**
e Use the model to find the cost of 65 kilowatt hours. **(1 mark)**

(P) **4** A racing car accelerates from rest to 90 m/s in 10 seconds. The table shows the total distance travelled by the racing car in each of the first 10 seconds.

time, *t* seconds	0	1	2	3	4	5	6	7	8	9	10
distance, *d* m	0	4.5	18	40.5	72	112.5	162	220.5	288	364.5	450

a Draw a graph of the data.
b Explain how you know a linear model would not be appropriate.

(E/P) **5** A website designer charges a flat fee and then a daily rate in order to design new websites for companies.
Company A's new website takes 6 days and they are charged £7100.
Company B's new website take 13 days and they are charged £9550.

Hint Let $(d_1, C_1) = (6, 7100)$ and $(d_2, C_2) = (13, 9550)$.

a Write an equation linking days, d and website cost, C in the form $C = ad + b$. **(3 marks)**
b Interpret the values of a and b. **(2 marks)**
c The web designer charges a third company £13 400. Calculate the number of days the designer spent working on the website. **(1 mark)**

(E/P) **6** The average August temperature in Exeter is 20 °C or 68 °F. The average January temperature in the same place is 9 °C or 48.2 °F.
a Write an equation linking Fahrenheit F and Celsius C in the form $F = aC + b$. **(3 marks)**
b Interpret the values of a and b. **(2 marks)**
c The highest temperature recorded in the UK was 101.3 °F. Calculate this temperature in Celsius. **(1 mark)**
d For what value is the temperature in Fahrenheit the same as the temperature in Celsius? **(3 marks)**

(P) **7** In 2004, in a city, there were 17 500 homes with internet connections. A service provider predicts that each year an additional 750 homes will get internet connections.
a Write a linear model for the number of homes n with internet connections t years after 2004.
b Write down one assumption made by this model.

(E/P) **8** The scatter graph shows the height h and foot length f of 8 students. A line of best fit is drawn on the scatter graph.
a Explain why the data can be approximated to a linear model. **(1 mark)**
b Use points A and B on the scatter graph to write a linear equation in the form $h = af + b$. **(3 mark)**
c Calculate the expected height of a person with a foot length of 26.5 cm. **(1 mark)**

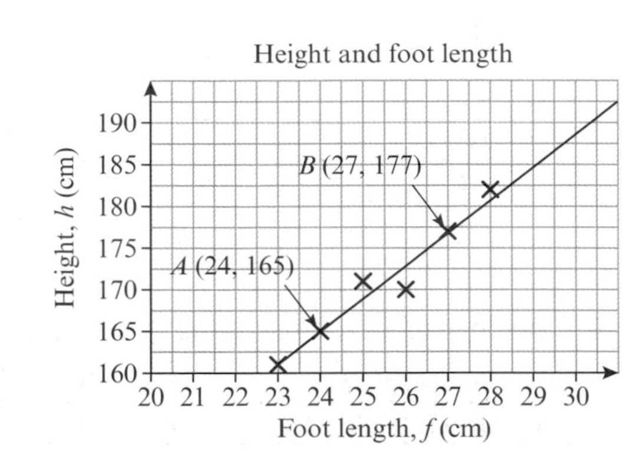

9 The price P of a good and the quantity Q of a good are linked.
The demand for a new pair of trainers can be modelled using the equation $P = -\frac{3}{4}Q + 35$.
The supply of the trainers can be modelled using the equation $P = \frac{2}{3}Q + 1$.
 a Draw a sketch showing the demand and supply lines on the same pair of axes.
 The equilibrium point is the point where the supply and demand lines meet.
 b Find the values of P and Q at the equilibrium point.

Mixed exercise 5

E/P 1 The straight line passing through the point $P(2, 1)$ and the point $Q(k, 11)$ has gradient $-\frac{5}{12}$
 a Find the equation of the line in terms of x and y only. **(2 marks)**
 b Determine the value of k. **(2 marks)**

E/P 2 The points A and B have coordinates $(k, 1)$ and $(8, 2k - 1)$ respectively, where k is a constant.
Given that the gradient of AB is $\frac{1}{3}$
 a show that $k = 2$ **(2 marks)**
 b find an equation for the line through A and B. **(3 marks)**

E 3 The line L_1 has gradient $\frac{1}{7}$ and passes through the point $A(2, 2)$. The line L_2 has gradient -1 and passes through the point $B(4, 8)$. The lines L_1 and L_2 intersect at the point C.
 a Find an equation for L_1 and an equation for L_2. **(4 marks)**
 b Determine the coordinates of C. **(2 marks)**

E 4 a Find an equation of the line l which passes through the points $A(1, 0)$ and $B(5, 6)$. **(2 marks)**
 The line m with equation $2x + 3y = 15$ meets l at the point C.
 b Determine the coordinates of C. **(2 marks)**

E 5 The line L passes through the points $A(1, 3)$ and $B(-19, -19)$.
Find an equation of L in the form $ax + by + c = 0$. where a, b and c are integers. **(3 marks)**

E 6 The straight line l_1 passes through the points A and B with coordinats $(2, 2)$ and $(6, 0)$ respectively.
 a Find an equation of l_1. **(3 marks)**
 The straight line l_2 passes through the point C with coordinate $(-9, 0)$ and has gradient $\frac{1}{4}$.
 b Find an equation of l_2. **(2 marks)**

E/P 7 The straight line l passes through $A(1, 3\sqrt{3})$ and $B(2 + \sqrt{3}, 3 + 4\sqrt{3})$.
Show that l meets the x-axis at the point $C(-2, 0)$. **(5 marks)**

E 8 The points A and B have coordinates $(-4, 6)$ and $(2, 8)$ respectively. A line p is drawn through B perpendicular to AB to meet the y-axis at the point C.
 a Find an equation of the line p. **(3 marks)**
 b Determine the coordinates of C. **(1 mark)**

(E/P) **9** The line l has equation $2x - y - 1 = 0$.

The line m passes through the point $A(0, 4)$ and is perpendicular to the line l.

a Find an equation of m. **(2 marks)**

b Show that the lines l and m intersect at the point $P(2, 3)$. **(2 marks)**

The line n passes through the point $B(3, 0)$ and is parallel to the line m.

c Find the coordinates of the point of intersection of the lines l and n. **(3 marks)**

(E/P) **10** The line l_1 passes through the points A and B with coordinates $(0, -2)$ and $(6, 7)$ respectively.
The line l_2 has equation $x + y = 8$ and cuts the y-axis at the point C.
The line l_1 and l_2 intersect at D.
Find the area of triangle ACD. **(6 marks)**

(E) **11** The points A and B have coordinates $(2, 16)$ and $(12, -4)$ respectively.
A straight line l_1 passes through A and B.

a Find an equation for l_1 in the form $ax + by = c$. **(2 marks)**

The line l_2 passes through the point C with coordinates $(-1, 1)$ and has gradient $\frac{1}{3}$

b Find an equation for l_2. **(2 marks)**

(E/P) **12** The points $A(-1, -2)$, $B(7, 2)$ and $C(k, 4)$, where k is a constant, are the vertices of $\triangle ABC$.
Angle ABC is a right angle.

a Find the gradient of AB. **(1 mark)**

b Calculate the value of k. **(2 marks)**

c Find an equation of the straight line passing through B and C. Give your answer in the form $ax + by + c = 0$, where a, b and c are integers **(2 marks)**

d Calculate the area of $\triangle ABC$. **(2 marks)**

(E/P) **13** **a** Find an equation of the straight line passing through the points with coordinates $(-1, 5)$ and $(4, -2)$, giving your answer in the form $ax + by + c = 0$, where a, b and c are integers. **(3 marks)**

The line crosses the x-axis at the point A and the y-axis at the point B, and O is the origin.

b Find the area of $\triangle AOB$. **(3 marks)**

(E) **14** The straight line l_1 has equation $4y + x = 0$.

The straight line l_2 has equation $y = 2x - 3$.

a On the same axes, sketch the graphs of l_1 and l_2. Show clearly the coordinates of all points at which the graphs meet the coordinate axes. **(2 marks)**

The lines l_1 and l_2 intersect at the point A.

b Calculate, as exact fractions, the coordinates of A. **(2 marks)**

c Find an equation of the line through A which is perpendicular to l_1.
Give your answer in the form $ax + by + c = 0$, where a, b and c are integers. **(2 marks)**

(E) **15** The points A and B have coordinates (4, 6) and (12, 2) respectively.
The straight line l_1 passes through A and B.

a Find an equation for l_1 in the form $ax + by + c = 0$, where a, b and c are integers. **(3 marks)**

The straight line l_2 passes through the origin and has gradient $-\frac{2}{3}$

b Write down an equation for l_2. **(1 mark)**

The lines l_1 and l_2 intersect at the point C.

c Find the coordinates of C. **(2 marks)**

d Show that the lines OA and OC are perpendicular, where O is the origin. **(2 marks)**

e Work out the lengths of OA and OC. Write your answers in the form $k\sqrt{13}$. **(2 marks)**

f Hence calculate the area of $\triangle OAC$. **(2 marks)**

16 **a** Use the distance formula to find the distance between $(4a, a)$ and $(-3a, 2a)$.

Hence find the distance between the following pairs of points:

b (4, 1) and (−3, 2) **c** (12, 3) and (−9, 6) **d** (−20, −5) and (15, −10)

(E/P) **17** A is the point (−1, 5). Let (x, y) be any point on the line $y = 3x$.

a Write an equation in terms of x for the distance between (x, y) and $A(-1, 5)$. **(3 marks)**

b Find the coordinates of the two points, B and C, on the line $y = 3x$ which are a distance of $\sqrt{74}$ from (−1, 5). **(3 marks)**

c Find the equation of the line l_1 that is perpendicular to $y = 3x$ and goes through the point (−1, 5). **(2 marks)**

d Find the coordinates of the point of intersection between l_1 and $y = 3x$. **(2 marks)**

e Find the area of triangle ABC. **(2 marks)**

(E/P) **18** The scatter graph shows the oil production P and carbon dioxide emissions C for various years since 1970. A line of best fit has been added to the scatter graph.

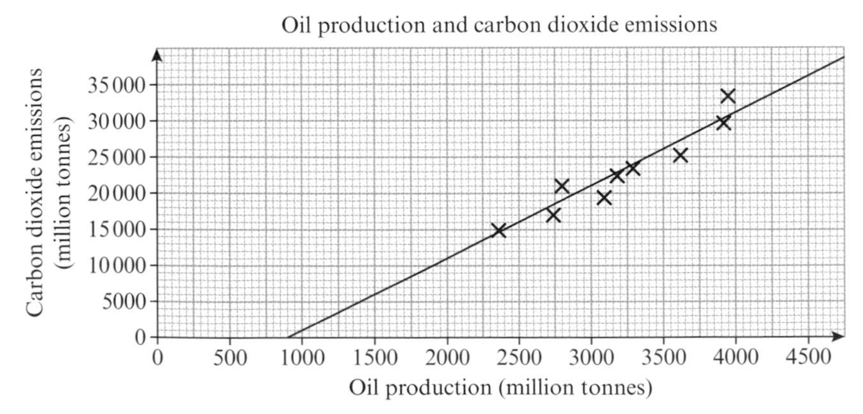

a Use two points on the line to calculate its gradient. **(1 mark)**

b Formulate a linear model linking oil production P and carbon dioxide emissions C, giving the relationship in the form $C = aP + b$. **(2 marks)**

c Interpret the value of a in your model. **(1 mark)**

d With reference to your value of b, comment on the validity of the model for small values of P. **(1 mark)**

Challenge

1 Find the area of the triangle with vertices $A(-2, -2)$, $B(13, 8)$ and $C(-4, 14)$.

2 A triangle has vertices $A(3, 8)$, $B(9, 9)$ and $C(5, 2)$ as shown in the diagram.
The line l_1 is perpendicular to AB and passes through C.
The line l_2 is perpendicular to BC and passes through A.
The line l_3 is perpendicular to AC and passes through B.
Show that the lines l_1, l_2 and l_3 meet at a point and find the coordinates of that point.

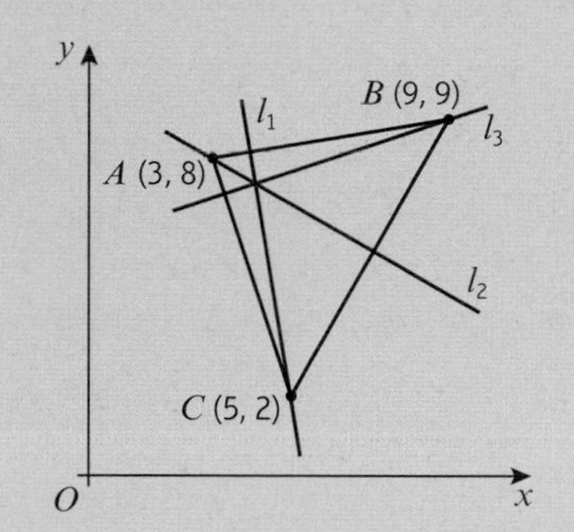

3 A triangle has vertices $A(0, 0)$, $B(a, b)$ and $C(c, 0)$ as shown in the diagram.
The line l_1 is perpendicular to AB and passes through C.
The line l_2 is perpendicular to BC and passes through A.
The line l_3 is perpendicular to AC and passes through B.
Find the coordinates of the point of intersection of l_1, l_2 and l_3.

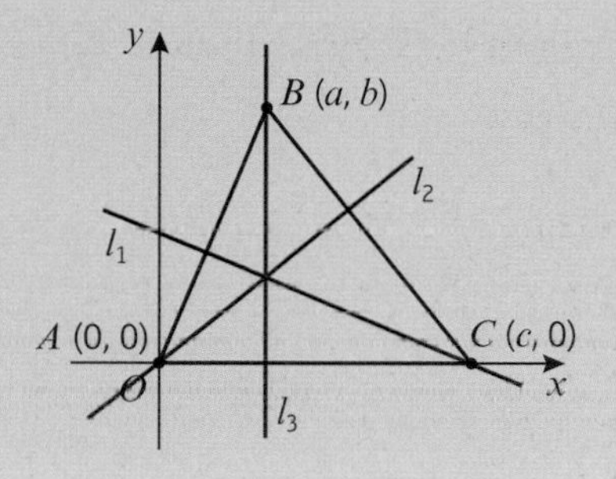

Summary of key points

1 The gradient m of the line joining the point with coordinates (x_1, y_1) to the point with coordinates (x_2, y_2) can be calculated using the formula

$$m = \frac{y_2 - y_1}{x_2 - x_1}$$

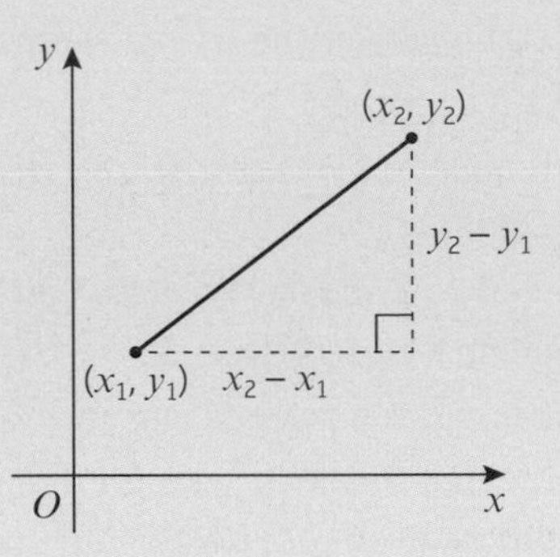

2
- The equation of a straight line can be written in the form

  $y = mx + c,$

  where m is the gradient and $(0, c)$ is the y-intercept.
- The equation of a straight line can also be written in the form

  $ax + by + c = 0,$

  where a, b and c are integers.

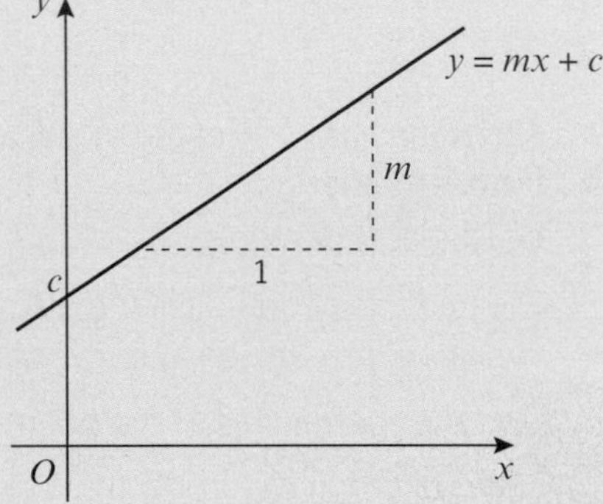

3 The equation of a line with gradient m that passes through the point with coordinates (x_1, y_1) can be written as $y - y_1 = m(x - x_1)$.

4 Parallel lines have the same gradient.

5 If a line has a gradient m, a line perpendicular to it has a gradient of $-\frac{1}{m}$

6 If two lines are perpendicular, the product of their gradients is -1.

7 You can find the distance d between (x_1, y_1) and (x_2, y_2) by using the formula $d = \sqrt{(x_2 - x_1)^2 + (y_2 - y_1)^2}$.

8 The point of intersection of two lines can be found using simultaneous equations.

9 Two quantities are in direct proportion when they increase at the same rate. The graph of these quantities is a straight line through the origin.

10 A mathematical model is an attempt to represent a real-life situation using mathematical concepts. It is often necessary to make assumptions about the real-life problems in order to create a model.

Circles

6

Objectives

After completing this unit you should be able to:

- Find the mid point of a line segment → pages 114 – 115
- Find the equation of the perpendicular bisector to a line segment → pages 116 – 117
- Know how to find the equation of a circle → pages 117 – 120
- Solve geometric problems involving straight lines and circles → pages 121 – 122
- Use circle properties to solve problems on coordinate grids → pages 123 – 128
- Find the angle in a semicircle and solve other problems involving circles and triangles → pages 128 – 132

Geostationary orbits are circular orbits around the Earth. Meteorologists use geostationary satellites to provide information about the Earth's surface and atmosphere.

Prior knowledge check

1 Write each of the following in the form $(x + p)^2 + q$:

a $x^2 + 10x + 28$ **b** $x^2 - 6x + 1$

c $x^2 - 12x$ **d** $x^2 + 7x$ ← Section 2.2

2 Find the equation of the line passing through each of the following pairs of points:

a $A(0, -6)$ and $B(4, 3)$

b $P(7, -5)$ and $Q(-9, 3)$

c $R(-4, -2)$ and $T(5, 10)$ ← Section 5.2

3 Use the discriminant to determine whether the following have two real solutions, one real solution or no real solutions.

a $x^2 - 7x + 14 = 0$

b $x^2 + 11x + 8 = 0$

c $4x^2 + 12x + 9 = 0$ ← Section 2.5

4 Find the equation of the line that passes through the point $(3, -4)$ and is perpendicular to the line with equation $6x - 5y - 1 = 0$ ← Section 5.3

6.1 Midpoints and perpendicular bisectors

You can find the **midpoint** of a line segment by averaging the x- and y-coordinates of its endpoints.

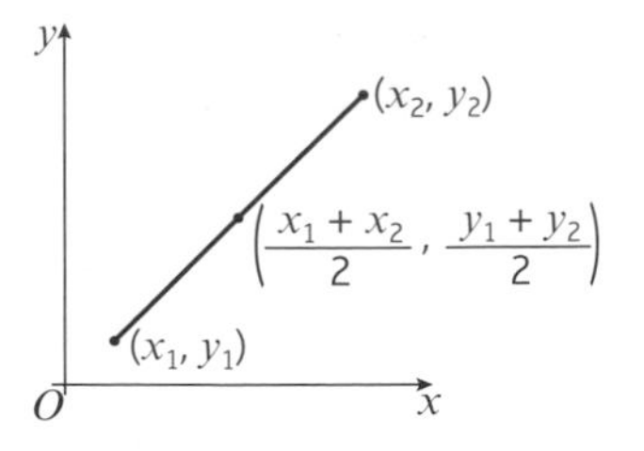

- **The midpoint of a line segment with endpoints (x_1, y_1) and (x_2, y_2) is $\left(\frac{x_1 + x_2}{2}, \frac{y_1 + y_2}{2}\right)$.**

Notation A **line segment** is a finite part of a straight line with two distinct endpoints.

Example 1

The line segment AB is a diameter of a circle, where A and B are $(-3, 8)$ and $(5, 4)$ respectively. Find the coordinates of the centre of the circle.

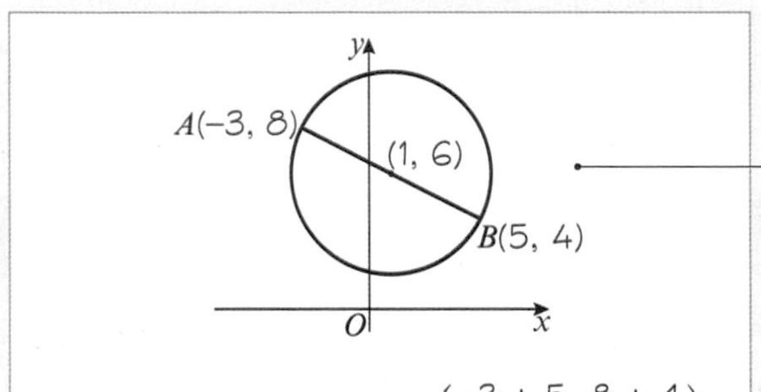

Draw a sketch.

Remember the centre of a circle is the midpoint of a diameter.

The centre of the circle is $\left(\frac{-3 + 5}{2}, \frac{8 + 4}{2}\right)$

$= \left(\frac{2}{2}, \frac{12}{2}\right) = (1, 6)$

Use $\left(\frac{x_1 + x_2}{2}, \frac{y_1 + y_2}{2}\right)$.

Here $(x_1, y_1) = (-3, 8)$ and $(x_2, y_2) = (5, 4)$.

Example 2

The line segment PQ is a diameter of the circle centre $(2, -2)$. Given that P is $(8, -5)$, find the coordinates of Q.

Problem-solving

In coordinate geometry problems, it is often helpful to draw a sketch showing the information given in the question.

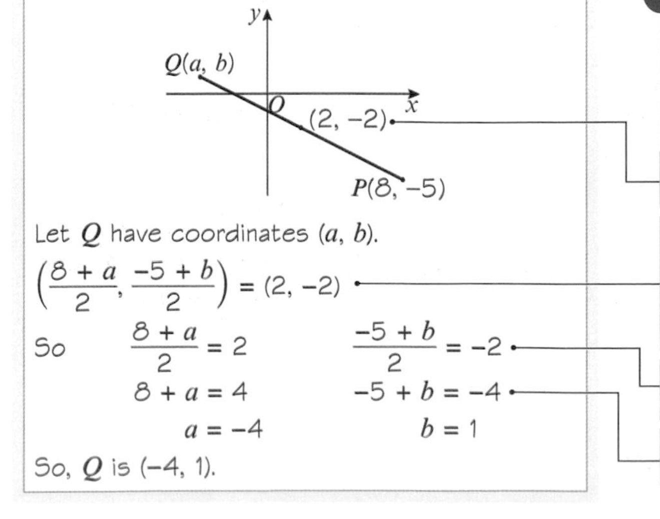

$(2, -2)$ is the mid-point of (a, b) and $(8, -5)$.

Let Q have coordinates (a, b).

$\left(\frac{8 + a}{2}, \frac{-5 + b}{2}\right) = (2, -2)$

Use $\left(\frac{x_1 + x_2}{2}, \frac{y_1 + y_2}{2}\right)$.

Here $(x_1, y_1) = (8, -5)$ and $(x_2, y_2) = (a, b)$.

So $\frac{8 + a}{2} = 2$ $\frac{-5 + b}{2} = -2$

Compare the x- and y-coordinates separately.

$8 + a = 4$ $-5 + b = -4$

Rearrange the equations to find a and b.

$a = -4$ $b = 1$

So, Q is $(-4, 1)$.

Exercise 6A

1 Find the midpoint of the line segment joining each pair of points:

a $(4, 2), (6, 8)$ **b** $(0, 6), (12, 2)$ **c** $(2, 2), (-4, 6)$

d $(-6, 4), (6, -4)$ **e** $(7, -4), (-3, 6)$ **f** $(-5, -5), (-11, 8)$

g $(6a, 4b), (2a, -4b)$ **h** $(-4u, 0), (3u, -2v)$ **i** $(a + b, 2a - b), (3a - b, -b)$

j $(4\sqrt{2}, 1)\ (2\sqrt{2}, 7)$ **k** $(\sqrt{2} - \sqrt{3}, 3\sqrt{2} + 4\sqrt{3}), (3\sqrt{2} + \sqrt{3}, -\sqrt{2} + 2\sqrt{3})$

(P) 2 The line segment AB has endpoints $A(-2, 5)$ and $B(a, b)$. The midpoint of AB is $M(4, 3)$. Find the values of a and b.

3 The line segment PQ is a diameter of a circle, where P and Q are $(-4, 6)$ and $(7, 8)$ respectively. Find the coordinates of the centre of the circle.

(P) 4 The line segment RS is a diameter of a circle, where R and S are $\left(\frac{4a}{5}, -\frac{3b}{4}\right)$ and $\left(\frac{2a}{5}, \frac{5b}{4}\right)$ respectively. Find the coordinates of the centre of the circle.

Problem-solving

Your answer will be in terms of a and b.

5 The line segment AB is a diameter of a circle, where A and B are $(-3, -4)$ and $(6, 10)$ respectively.

a Find the coordinates of the centre of the circle.

b Show the centre of the circle lies on the line $y = 2x$.

(P) 6 The line segment JK is a diameter of a circle, where J and K are $\left(\frac{3}{4}, \frac{4}{3}\right)$ and $\left(-\frac{1}{2}, 2\right)$ respectively. The centre of the circle lies on the line segment with equation $y = 8x + b$. Find the value of b.

(P) 7 The line segment AB is a diameter of a circle, where A and B are $(0, -2)$ and $(6, -5)$ respectively. Show that the centre of the circle lies on the line $x - 2y - 10 = 0$.

(P) 8 The line segment FG is a diameter of the circle centre $(6, 1)$. Given F is $(2, -3)$, find the coordinates of G.

(P) 9 The line segment CD is a diameter of the circle centre $(-2a, 5a)$. Given D has coordinates $(3a, -7a)$, find the coordinates of C.

(P) 10 The points $M(3, p)$ and $N(q, 4)$ lie on the circle centre $(5, 6)$. The line segment MN is a diameter of the circle. Find the values of p and q.

Problem-solving

Use the formula for finding the midpoint: $\left(\frac{3+q}{2}, \frac{p+4}{2}\right) = (5, 6)$

(P) 11 The points $V(-4, 2a)$ and $W(3b, -4)$ lie on the circle centre $(b, 2a)$. The line segment VW is a diameter of the circle. Find the values of a and b.

Challenge

A triangle has vertices at $A(3, 5)$, $B(7, 11)$ and $C(p, q)$. The midpoint of side BC is $M(8, 5)$.

a Find the values of p and q.

b Find the equation of the straight line joining the midpoint of AB to the point M.

c Show that the line in part **b** is parallel to the line AC.

Links You can also prove results like this using vectors.

→ Section 11.5

- **The perpendicular bisector of a line segment AB is the straight line that is perpendicular to AB and passes through the midpoint of AB.**

If the gradient of AB is m then the gradient of its perpendicular bisector, l, will be $-\frac{1}{m}$

Example 3

The line segment AB is a diameter of the circle centre C, where A and B are $(-1, 4)$ and $(5, 2)$ respectively. The line l passes through C and is perpendicular to AB. Find the equation of l.

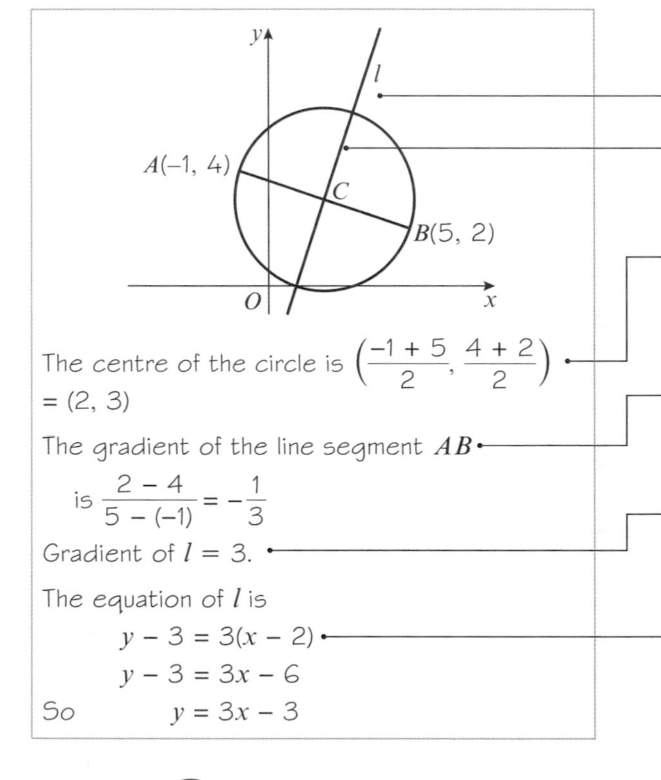

Draw a sketch.

l is the perpendicular bisector of AB.

The centre of the circle is $\left(\frac{-1+5}{2}, \frac{4+2}{2}\right)$
$= (2, 3)$

Use $\left(\frac{x_1+x_2}{2}, \frac{y_1+y_2}{2}\right)$.
Here $(x_1, y_1) = (-1, 4)$ and $(x_2, y_2) = (5, 2)$.

The gradient of the line segment AB is $\frac{2-4}{5-(-1)} = -\frac{1}{3}$

Use $m = \frac{y_2 - y_1}{x_2 - x_1}$. Here $(x_1, y_1) = (-1, 4)$ and $(x_2, y_2) = (5, 2)$.

Gradient of $l = 3$.

Remember the product of the gradients of two perpendicular lines is $= -1$, so $-\frac{1}{3} \times 3 = -1$.

The equation of l is

$y - 3 = 3(x - 2)$
$y - 3 = 3x - 6$
So $y = 3x - 3$

The perpendicular line l passes through the point $(2, 3)$ and has gradient 3, so use $y - y_1 = m(x - x_1)$ with $m = 3$ and $(x_1, y_1) = (2, 3)$.
Rearrange the equation into the form $y = mx + c$.

Exercise 6B

1 Find the perpendicular bisector of the line segment joining each pair of points:

a $A(-5, 8)$ and $B(7, 2)$ **b** $C(-4, 7)$ and $D(2, 25)$ **c** $E(3, -3)$ and $F(13, -7)$

d $P(-4, 7)$ and $Q(-4, -1)$ **e** $S(4, 11)$ and $T(-5, -1)$ **f** $X(13, 11)$ and $Y(5, 11)$

E/P **2** The line FG is a diameter of the circle centre C, where F and G are $(-2, 5)$ and $(2, 9)$ respectively. The line l passes through C and is perpendicular to FG. Find the equation of l. **(7 marks)**

P **3** The line JK is a diameter of the circle centre P, where J and K are $(0, -3)$ and $(4, -5)$ respectively. The line l passes through P and is perpendicular to JK. Find the equation of l. Write your answer in the form $ax + by + c = 0$, where a, b and c are integers.

4 Points A, B, C and D have coordinates $A(-4, -9)$, $B(6, -3)$, $C(11, 5)$ and $D(-1, 9)$.

a Find the equation of the perpendicular bisector of line segment AB.

b Find the equation of the perpendicular bisector of line segment CD.

c Find the coordinates of the point of intersection of the two perpendicular bisectors.

(P) **5** Point X has coordinates $(7, -2)$ and point Y has coordinates $(4, q)$. The perpendicular bisector of XY has equation $y = 4x + b$. Find the value of q and the value of b.

Problem-solving

It is often easier to find unknown values in the order they are given in the question. Find q first, then find b.

Challenge

Triangle PQR has vertices at $P(6, 9)$, $Q(3, -3)$ and $R(-9, 3)$.

a Find the perpendicular bisectors of each side of the triangle.

b Show that all three perpendicular bisectors meet at a single point, and find the coordinates of that point.

Links This point of intersection is called the **circumcentre** of the triangle. → Section 6.5

6.2 Equation of a circle

A circle is the set of points that are equidistant from a fixed point. You can use Pythagoras' theorem to derive equations of circles on a coordinate grid.

For any point (x, y) on the circumference of a circle, you can use Pythagoras' theorem to show the relationship between x, y and the radius r.

- **The equation of a circle with centre (0, 0) and radius r is $x^2 + y^2 = r^2$.**

When a circle has a centre (a, b) and radius r, you can use the following general form of the equation of a circle.

- **The equation of the circle with centre (a, b) and radius r is $(x - a)^2 + (y - b)^2 = r^2$.**

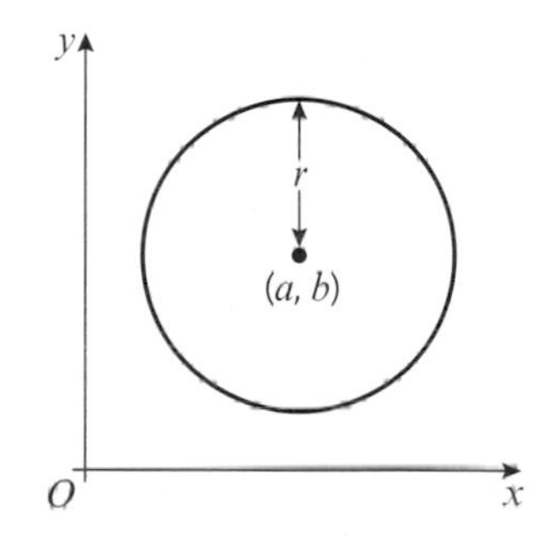

Links This circle is a translation of the circle $x^2 + y^2 = r^2$ by the vector $\begin{pmatrix} a \\ b \end{pmatrix}$. ← Section 4.5

Example 4

Write down the equation of the circle with centre (5, 7) and radius 4.

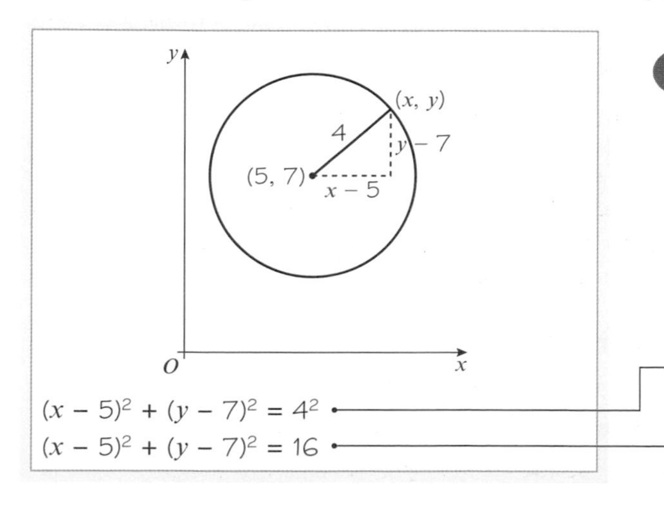

Online Explore the general form of the equation of a circle using technology.

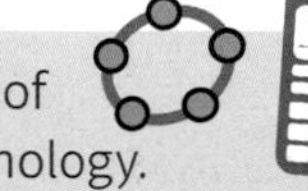

$(x - 5)^2 + (y - 7)^2 = 4^2$ — Substitute $a = 5$, $b = 7$ and $r = 4$ into the equation.

$(x - 5)^2 + (y - 7)^2 = 16$ — Simplify by calculating $4^2 = 16$.

Example 5

A circle has equation $(x - 3)^2 + (y + 4)^2 = 20$.

a Write down the centre and radius of the circle.

b Show that the circle passes through $(5, -8)$.

a Centre $(3, -4)$, radius $\sqrt{20} = 2\sqrt{5}$

$r^2 = 20$ so $r = \sqrt{20}$

b $(x - 3)^2 + (y + 4)^2 = 20$

Substitute $(5, -8)$

Substitute $x = 5$ and $y = -8$ into the equation of the circle.

$$(5 - 3)^2 + (-8 + 4)^2 = 2^2 + (-4)^2$$
$$= 4 + 16$$
$$= 20$$

$(5, -8)$ satisfies the equation of the circle.

So the circle passes through the point $(5, -8)$.

Example 6

The line segment AB is a diameter of a circle, where A and B are $(4, 7)$ and $(-8, 3)$ respectively. Find the equation of the circle.

Length of $AB = \sqrt{(4 - (-8))^2 + (7 - 3)^2}$

Use $d = \sqrt{(x_2 - x_1)^2 + (y_2 - y_1)^2}$
Here $(x_1, y_1) = (-8, 3)$ and $(x_2, y_2) = (4, 7)$

$$= \sqrt{12^2 + 4^2}$$
$$= \sqrt{160}$$
$$= \sqrt{16} \times \sqrt{10}$$
$$= 4\sqrt{10}$$

So the radius is $2\sqrt{10}$.

The centre is $\left(\frac{4 + (-8)}{2}, \frac{7 + 3}{2}\right) = (-2, 5)$.

Remember the centre of a circle is at the midpoint of a diameter. Use $\left(\frac{x_1 + x_2}{2}, \frac{y_1 + y_2}{2}\right)$.

The equation of the circle is

$$(x + 2)^2 + (y - 5)^2 = (2\sqrt{10})^2$$

Or $(x + 2)^2 + (y - 5)^2 = 40$.

Problem-solving

You need to work out the steps of this problem yourself:

- Find the radius of the circle by finding the length of the diameter and dividing by 2.
- Find the centre of the circle by finding the midpoint of AB.
- Write down the equation of the circle.

You can multiply out the brackets in the equation of a circle to find it in an alternate form:

$$(x - a)^2 + (y - b)^2 = r^2$$
$$x^2 - 2ax + a^2 + y^2 - 2by + b^2 = r^2$$
$$x^2 + y^2 - 2ax - 2by + b^2 + a^2 - r^2 = 0$$

Compare the constant terms with the equation given in the key point: $b^2 + a^2 - r^2 = c$ so $r = \sqrt{f^2 + g^2 - c}$

- **The equation of a circle can be given in the form:**
  $\mathbf{x^2 + y^2 + 2fx + 2gy + c = 0}$
- **This circle has centre $(-f, -g)$ and radius $\sqrt{f^2 + g^2 - c}$**

If you need to find the centre and radius of a circle with an equation given in expanded form it is usually safest to **complete the square** for the x and y terms.

Example 7

Find the centre and the radius of the circle with the equation $x^2 + y^2 - 14x + 16y - 12 = 0$.

Rearrange into the form $(x - a)^2 + (y - b)^2 = r^2$.

$x^2 + y^2 - 14x + 16y - 12 = 0$

$x^2 - 14x + y^2 + 16y - 12 = 0$ (1)

Group the x terms and y terms together.

Completing the square for x terms and y terms.

$x^2 - 14x = (x - 7)^2 - 49$

$y^2 + 16y = (y + 8)^2 - 64$

Substituting back into (1)

$(x - 7)^2 - 49 + (y + 8)^2 - 64 - 12 = 0$

$(x - 7)^2 + (y + 8)^2 = 125$

Move the number terms to the right-hand side of the equation.

$(x - 7)^2 + (y + 8)^2 = (\sqrt{125})^2$

Write the equation in the form $(x - a)^2 + (y - b)^2 = r^2$.

$\sqrt{125} = \sqrt{25} \times \sqrt{5} = 5\sqrt{5}$

Simplify $\sqrt{125}$.

The equation of the circle is

$(x - 7)^2 + (y + 8)^2 = (5\sqrt{5})^2$

The circle has centre $(7, -8)$ and radius $= 5\sqrt{5}$.

You could also compare the original equation with:

$x^2 + y^2 + 2fx + 2gy + c = 0$

$f = -7$, $g = 8$ and $c = -12$ so the circle has centre $(7, -8)$ and radius $\sqrt{(-7)^2 + 8^2 - (-12)} = 5\sqrt{5}$.

Links You need to **complete the square** for the terms in x and for the terms in y. ← Section 2.2

Exercise 6C

1 Write down the equation of each circle:

a Centre $(3, 2)$, radius 4
b Centre $(-4, 5)$, radius 6
c Centre $(5, -6)$, radius $2\sqrt{3}$
d Centre $(2a, 7a)$, radius $5a$
e Centre $(-2\sqrt{2}, -3\sqrt{2})$, radius 1

2 Write down the coordinates of the centre and the radius of each circle:

a $(x + 5)^2 + (y - 4)^2 = 9^2$
b $(x - 7)^2 + (y - 1)^2 = 16$
c $(x + 4)^2 + y^2 = 25$
d $(x + 4a)^2 + (y + a)^2 = 144a^2$
e $(x - 3\sqrt{5})^2 + (y + \sqrt{5})^2 = 27$

3 In each case, show that the circle passes through the given point:

a $(x - 2)^2 + (y - 5)^2 = 13$, point $(4, 8)$
b $(x + 7)^2 + (y - 2)^2 = 65$, point $(0, -2)$
c $x^2 + y^2 = 25^2$, point $(7, -24)$
d $(x - 2a)^2 + (y + 5a)^2 = 20a^2$, point $(6a, -3a)$
e $(x - 3\sqrt{5})^2 + (y - \sqrt{5})^2 = (2\sqrt{10})^2$ point, $(\sqrt{5}, -\sqrt{5})$

(P) **4** The point $(4, -2)$ lies on the circle centre $(8, 1)$. Find the equation of the circle.

Hint First find the radius of the circle.

E/P **5** The line PQ is the diameter of the circle, where P and Q are (5, 6) and (−2, 2) respectively. Find the equation of the circle. **(5 marks)**

E/P **6** The point (1, −3) lies on the circle $(x - 3)^2 + (y + 4)^2 = r^2$. Find the value of r. **(3 marks)**

E/P **7** The points $P(2, 2)$, $Q(2 + \sqrt{3}, 5)$ and $R(2 - \sqrt{3}, 5)$ lie on the circle $(x - 2)^2 + (y - 4)^2 = r^2$.

a Find the value of r. **(2 marks)**

b Show that $\triangle PQR$ is equilateral. **(3 marks)**

E/P **8** **a** Show that $x^2 + y^2 - 4x - 11 = 0$ can be written in the form $(x - a)^2 + y^2 = r^2$, where a and r are numbers to be found. **(2 marks)**

b Hence write down the centre and radius of the circle with equation $x^2 + y^2 - 4x - 11 = 0$ **(2 marks)**

Problem-solving

Start by writing $(x^2 - 4x)$ in the form $(x - a)^2 - b$.

E/P **9** **a** Show that $x^2 + y^2 - 10x + 4y - 20 = 0$ can be written in the form $(x - a)^2 + (y - b)^2 = r^2$, where a, b and r are numbers to be found. **(2 marks)**

b Hence write down the centre and radius of the circle with equation $x^2 + y^2 - 10x + 4y - 20 = 0$. **(2 marks)**

10 Find the centre and radius of the circle with each of the following equations.

a $x^2 + y^2 - 2x + 8y - 8 = 0$

b $x^2 + y^2 + 12x - 4y = 9$

c $x^2 + y^2 - 6y = 22x - 40$

d $x^2 + y^2 + 5x - y + 4 = 2y + 8$

e $2x^2 + 2y^2 - 6x + 5y = 2x - 3y - 3$

Hint Start by writing the equation in one of the following forms:
$(x - a)^2 + (y - b)^2 = r^2$
$x^2 + y^2 + 2fx + 2gy + c = 0$

E/P **11** A circle C has equation $x^2 + y^2 + 12x + 2y = k$, where k is a constant.

a Find the coordinates of the centre of C. **(2 marks)**

b State the range of possible values of k. **(2 marks)**

Problem-solving

A circle must have a positive radius.

E/P **12** The point $P(7, -14)$ lies on the circle with equation $x^2 + y^2 + 6x - 14y = 483$. The point Q also lies on the circle such that PQ is a diameter. Find the coordinates of point Q. **(4 marks)**

E/P **13** The circle with equation $(x - k)^2 + y^2 = 41$ passes through the point (3, 4). Find the two possible values of k. **(5 marks)**

Challenge

1 A circle with equation $(x - k)^2 + (y - 2)^2 = 50$ passes through the point (4, −5). Find the possible values of k and the equation of each circle.

2 By completing the square for x and y, show that the equation $x^2 + y^2 + 2fx + 2gy + c = 0$ describes a circle with centre $(-f, -g)$ and radius $\sqrt{f^2 + g^2 - c}$.

6.3 Intersections of straight lines and circles

You can use algebra to find the coordinates of intersection of a straight line and a circle.

- **A straight line can intersect a circle once, by just touching the circle, or twice. Not all straight lines will intersect a given circle.**

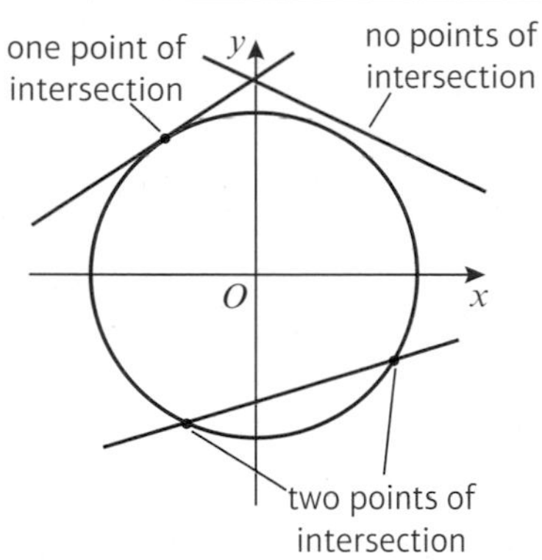

Example 8

Find the coordinates of the points where the line $y = x + 5$ meets the circle $x^2 + (y - 2)^2 = 29$.

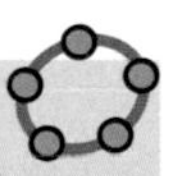

Online Explore intersections of straight lines and circles using GeoGebra.

$x^2 + (y - 2)^2 = 29$

$x^2 + (x + 5 - 2)^2 = 29$

$x^2 + (x + 3)^2 = 29$

$x^2 + x^2 + 6x + 9 = 29$

$2x^2 + 6x - 20 = 0$

$x^2 + 3x - 10 = 0$

$(x + 5)(x - 2) = 0$

So $x = -5$ and $x = 2$.

$x = -5$: $y = -5 + 5 = 0$

$x = 2$: $y = 2 + 5 = 7$

The line meets the circle at $(-5, 0)$ and $(2, 7)$.

Solve the equations simultaneously, so substitute $y = x + 5$ into the equation of the circle. ← Section 3.2

Simplify the equation to form a quadratic equation.

The resulting quadratic equation has two distinct solutions, so the line intersects the circle at two distinct points.

Now find the y-coordinates, so substitute the values of x into the equation of the line.

Remember to write the answers as coordinates.

Example 9

Show that the line $y = x - 7$ does not meet the circle $(x + 2)^2 + y^2 = 33$.

$(x + 2)^2 + y^2 = 33$

$(x + 2)^2 + (x - 7)^2 = 33$

$x^2 + 4x + 4 + x^2 - 14x + 49 = 33$

$2x^2 - 10x + 20 = 0$

$x^2 - 5x + 10 = 0$

Now $b^2 - 4ac = (-5)^2 - 4 \times 1 \times 10$

$= 25 - 40$

$= -15$

$b^2 - 4ac < 0$, so the line does not meet the circle.

Try to solve the equations simultaneously, so substitute $y = x - 7$ into the equation of the circle.

Use the discriminant $b^2 - 4ac$ to test for roots of the quadratic equation. ← Section 2.5

Problem-solving

If $b^2 - 4ac > 0$ there are two distinct roots.
If $b^2 - 4ac = 0$ there is a repeated root.
If $b^2 - 4ac < 0$ there are no real roots.

Exercise 6D

1 Find the coordinates of the points where the circle $(x - 1)^2 + (y - 3)^2 = 45$ meets the x-axis.

Hint Substitute $y = 0$ into the equation.

2 Find the coordinates of the points where the circle $(x - 2)^2 + (y + 3)^2 = 29$ meets the y-axis.

3 The line $y = x + 4$ meets the circle $(x - 3)^2 + (y - 5)^2 = 34$ at A and B. Find the coordinates of A and B.

4 Find the coordinates of the points where the line $x + y + 5 = 0$ meets the circle $x^2 + 6x + y^2 + 10y - 31 = 0$.

(P) 5 Show that the line $x - y - 10 = 0$ does not meet the circle $x^2 - 4x + y^2 = 21$.

Problem-solving

Attempt to solve the equations simultaneously. Use the discriminant to show that the resulting quadratic equation has no solutions.

(E/P) 6 **a** Show that the line $x + y = 11$ meets the circle with equation $x^2 + (y - 3)^2 = 32$ at only one point. **(4 marks)**

b Find the coordinates of the point of intersection. **(1 mark)**

(E/P) 7 The line $y = 2x - 2$ meets the circle $(x - 2)^2 + (y - 2)^2 = 20$ at A and B.

a Find the coordinates of A and B. **(5 marks)**

b Show that AB is a diameter of the circle. **(2 marks)**

(E/P) 8 The line $x + y = a$ meets the circle $(x - p)^2 + (y - 6)^2 = 20$ at $(3, 10)$, where a and p are constants.

a Work out the value of a. **(1 mark)**

b Work out the two possible values of p. **(5 marks)**

(E/P) 9 The circle with equation $(x - 4)^2 + (y + 7)^2 = 50$ meets the straight line with equation $x - y - 5 = 0$ at points A and B.

a Find the coordinates of the points A and B. **(5 marks)**

b Find the equation of the perpendicular bisector of line segment AB. **(3 marks)**

c Show that the perpendicular bisector of AB passes through the centre of the circle. **(1 mark)**

d Find the area of triangle OAB. **(2 marks)**

(E/P) 10 The line with equation $y = kx$ intersects the circle with equation $x^2 - 10x + y^2 - 12y + 57 = 0$ at two distinct points.

a Show that $21k^2 - 60k + 32 < 0$. **(5 marks)**

b Determine the range of possible values for k. Round your answer to 2 decimal places. **(3 marks)**

(E/P) 11 The line with equation $y = 4x - 1$ does not intersect the circle with equation $x^2 + 2x + y^2 = k$. Find the range of possible values of k.

(E/P) 12 The line with equation $y = 2x + 5$ meets the circle with equation $x^2 + kx + y^2 = 4$ at exactly one point. Find two possible values of k. **(7 marks)**

Problem-solving

If you are solving a problem where there are 0, 1 or 2 solutions (or points of intersection), you might be able to use the discriminant.

6.4 Use tangent and chord properties

You can use the properties of tangents and chords within circles to solve geometric problems. A tangent to a circle is a straight line that intersects the circle at only one point.

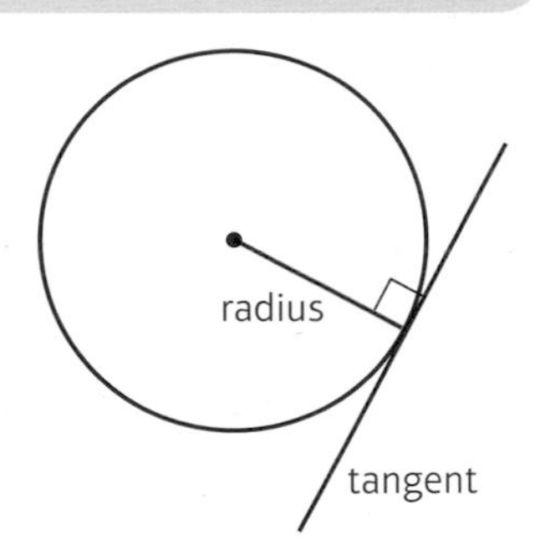

- **A tangent to a circle is perpendicular to the radius of the circle at the point of intersection.**

A chord is a line segment that joins two points on the circumference of a circle.

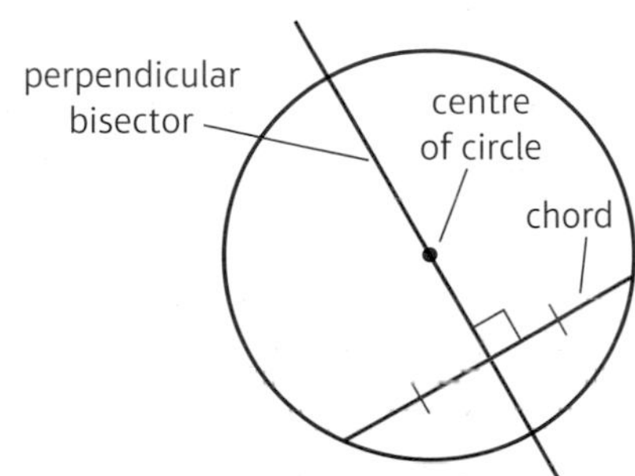

- **The perpendicular bisector of a chord will go through the centre of a circle.**

Online Explore the circle theorems using GeoGebra.

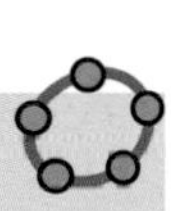

Example 10

The circle C has equation $(x - 2)^2 + (y - 6)^2 = 100$.

a Verify that the point $P(10, 0)$ lies on C.

b Find an equation of the tangent to C at the point $(10, 0)$, giving your answer in the form $ax + by + c = 0$.

a $(x - 2)^2 + (y - 6)^2 = (10 - 2)^2 + (0 - 6)^2$
$= 8^2 + (-6)^2$ — Substitute $(x, y) = (10, 0)$ into the equation for the circle.
$= 64 + 36$
$= 100$ ✓ — The point $P(10, 0)$ satisfies the equation, so P lies on C.

b The centre of circle C is (2, 6). Find the gradient of the line between (2, 6) and P. — A circle with equation $(x - a)^2 + (y - b)^2 = r^2$ has centre (a, b).

$m = \frac{y_2 - y_1}{x_2 - x_1} = \frac{6 - 0}{2 - 10} = \frac{6}{-8} = -\frac{3}{4}$ — Use the gradient formula with $(x_1, y_1) = (10, 0)$ and $(x_2, y_2) = (2, 6)$

The gradient of the tangent is $\frac{4}{3}$ — The tangent is perpendicular to the radius at that point. If the gradient of the radius is m then the gradient of the tangent will be $-\frac{1}{m}$

$y - y_1 = m(x - x_1)$

$y - 0 = \frac{4}{3}(x - 10)$ — Substitute $(x_1, y_1) = (10, 0)$ and $m = \frac{4}{3}$ into the equation for a straight line.

$3y = 4x - 40$ — Simplify.

$4x - 3y - 40 = 0$ — Leave the answer in the correct form.

Example 11

A circle C has equation $(x - 5)^2 + (y + 3)^2 = 10$.

The line l is a tangent to the circle and has gradient -3.

Find two possible equations for l, giving your answers in the form $y = mx + c$.

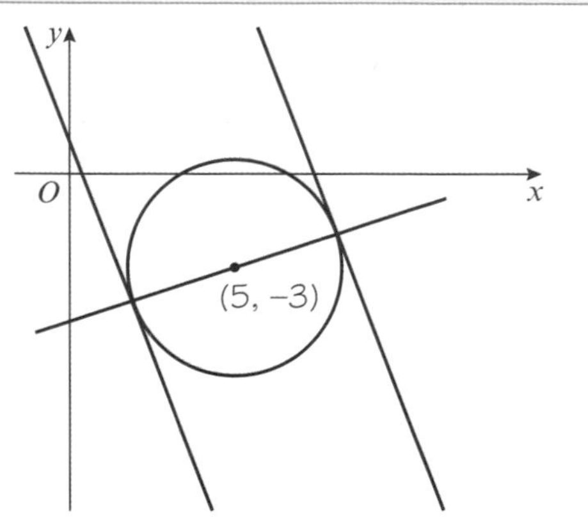

Problem-solving

Draw a sketch showing the circle and the two possible tangents with gradient −3. If you are solving a problem involving tangents and circles there is a good chance you will need to use the radius at the point of intersection, so draw this on your sketch.

Find a line that passes through the centre of the circle that is perpendicular to the tangents.

This line will intersect the circle at the same points where the tangent intersects the circle.

The gradient of this line is $\frac{1}{3}$

The gradient of the tangents is −3, so the gradient of a perpendicular line will be $\frac{-1}{-3} = \frac{1}{3}$

The coordinates of the centre of circle are (5, −3)

A circle with equation $(x - a)^2 + (y - b)^2 = r^2$ has centre (a, b).

$$y - y_1 = m(x - x_1)$$

$$y + 3 = \frac{1}{3}(x - 5)$$

$$y + 3 = \frac{1}{3}x - \frac{5}{3}$$

Substitute $(x_1, y_1) = (5, -3)$ and $m = \frac{1}{3}$ into the equation for a straight line.

$$y = \frac{1}{3}x - \frac{14}{3}$$

This is the equation of the line passing through the circle.

$$(x - 5)^2 + (y + 3)^2 = 10$$

$$(x - 5)^2 + \left(\frac{1}{3}x - \frac{14}{3} + 3\right)^2 = 10$$

Substitute $y = \frac{1}{3}x - \frac{14}{3}$ into the equation for a circle to find the points of intersection.

$$(x - 5)^2 + \left(\frac{1}{3}x - \frac{5}{3}\right)^2 = 10$$

$$x^2 - 10x + 25 + \frac{1}{9}x^2 - \frac{10}{9}x + \frac{25}{9} = 10$$

Simplify the expression.

$$\frac{10}{9}x^2 - \frac{100}{9}x + \frac{250}{9} = 10$$

$$10x^2 - 100x + 250 = 90$$

$$10x^2 - 100x + 160 = 0$$

$$x^2 - 10x + 16 = 0$$

$$(x - 8)(x - 2) = 0$$

Factorise to find the values of x.

$$x = 8 \text{ or } x = 2$$

$$y = -\frac{6}{3} = -2 \text{ or } y = -4$$

Substitute $x = 8$ and $x = 2$ into $y = \frac{1}{3}x - \frac{14}{3}$.

So the tangents will intersect the circle at (8, −2) and (2, −4)

$y - y_1 = m(x - x_1)$

$y + 2 = -3(x - 8)$ — Substitute $(x_1, y_1) = (8, -2)$ and $m = -3$ into the equation for a straight line.

$y = -3x + 22$ — This is one possible equation for the tangent.

$y - y_1 = m(x - x_1)$

$y + 4 = -3(x - 2)$ — Substitute $(x_1, y_1) = (2, -4)$ and $m = -3$ into the equation for a straight line.

$y = -3x + 2$ — This is the other possible equation for the tangent.

Example 12

The points P and Q lie on a circle with centre C, as shown in the diagram.
The point P has coordinates (−7, −1) and the point Q has coordinates (3, −5).
M is the midpoint of the line segment PQ.
The line l passes through the points M and C.

a Find an equation for l.

Given that the y-coordinate of C is −8,

b show that the x-coordinate of C is −4

c find an equation of the circle.

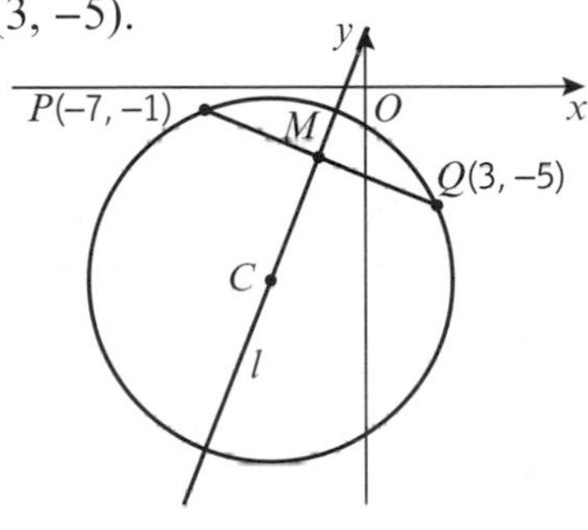

a The midpoint M of line segment PQ is:

$$\left(\frac{x_1 + x_2}{2}, \frac{y_1 + y_2}{2}\right) = \left(\frac{-7 + 3}{2}, \frac{-1 + (-5)}{2}\right)$$

Use the midpoint formula with $(x_1, y_1) = (-7, -1)$ and $(x_1, y_1) = (3, -5)$.

$$= (-2, -3)$$

$$\text{Gradient of } PQ = \frac{y_2 - y_1}{x_2 - x_1} = \frac{-5 - (-1)}{3 - (-7)}$$

Use the gradient formula with $(x_1, y_1) = (-7, -1)$ and $(x_1, y_1) = (3, -5)$.

$$= \frac{-4}{10} = -\frac{2}{5}$$

The gradient of a line perpendicular to PQ is $\frac{5}{2}$

Problem-solving

If a gradient is given as a fraction, you can find the perpendicular gradient quickly by turning the fraction upside down and changing the sign.

$y - y_1 = m(x - x_1)$

$y + 3 = \frac{5}{2}(x + 2)$ — Substitute $(x_1, y_1) = (-2, -3)$ and $m = \frac{5}{2}$ into the equation of a straight line.

$y + 3 = \frac{5}{2}x + 5$

$y = \frac{5}{2}x + 2$ — Simplify and leave in the form $y = mx + c$.

b $y = \frac{5}{2}x + 2$

$-8 = \frac{5}{2}x + 2$ — The perpendicular bisector of any chord passes through the centre of the circle. Substitute $y = -8$ into the equation of the straight line to find the corresponding x-coordinate.

$\frac{5}{2}x = -10$

$x = -4$ — Solve the equation to find x.

c The centre of the circle is $(-4, -8)$.
To find the radius of the circle:

$CQ = \sqrt{(x_2 - x_1)^2 + (y_2 - y_1)^2}$

$= \sqrt{(3 - (-4))^2 + (-5 - (-8))^2}$

$= \sqrt{49 + 9} = \sqrt{58}$

So the circle has a radius of $\sqrt{58}$.

The equation of the circle is:

$(x - a)^2 + (y - b)^2 = r^2$

$(x + 4)^2 + (y + 8)^2 = 58$

The radius is the length of the line segment CP or CQ.

Substitute $(x_1, y_1) = (-4, -8)$ and $(x_2, y_2) = (3, -5)$.

Substitute $(a, b) = (-4, -8)$ and $r = \sqrt{58}$ into the equation of a circle.

Exercise 6E

1 The line $x + 3y - 11 = 0$ touches the circle $(x + 1)^2 + (y + 6)^2 = r^2$ at $(2, 3)$.

a Find the radius of the circle.

b Show that the radius at $(2, 3)$ is perpendicular to the line.

2 The point $P(1, -2)$ lies on the circle centre $(4, 6)$.

a Find the equation of the circle.

b Find the equation of the tangent to the circle at P.

3 The points A and B with coordinates $(-1, -9)$ and $(7, -5)$ lie on the circle C with equation $(x - 1)^2 + (y + 3)^2 = 40$.

a Find the equation of the perpendicular bisector of the line segment AB.

b Show that the perpendicular bisector of AB passes through the centre of the circle C.

(P) **4** The points P and Q with coordinates $(3, 1)$ and $(5, -3)$ lie on the circle C with equation $x^2 - 4x + y^2 + 4y = 2$.

a Find the equation of the perpendicular bisector of the line segment PQ.

b Show that the perpendicular bisector of PQ passes through the centre of the circle C.

(E) **5** The circle C has equation $x^2 + 18x + y^2 - 2y + 29 = 0$.

a Verify the point $P(-7, -6)$ lies on C. **(2 marks)**

b Find an equation for the tangent to C at the point P, giving your answer in the form $y = mx + b$. **(4 marks)**

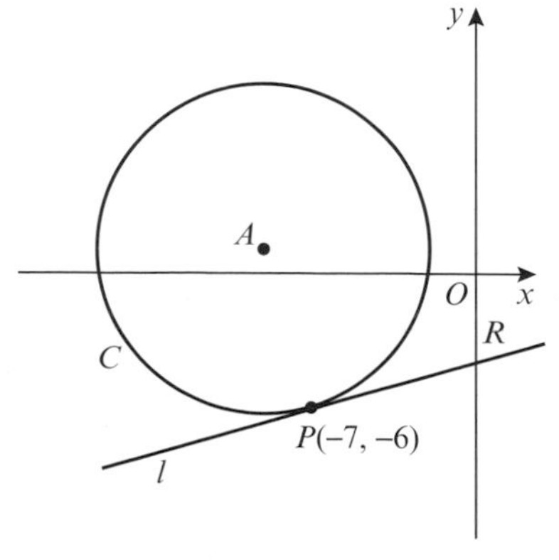

c Find the coordinates of R, the point of intersection of the tangent and the y-axis. **(2 marks)**

d Find the area of the triangle APR. **(2 marks)**

E/P **6** The tangent to the circle $(x + 4)^2 + (y - 1)^2 = 242$ at $(7, -10)$ meets the y-axis at S and the x-axis at T.

a Find the coordinates of S and T. **(5 marks)**

b Hence, find the area of $\triangle OST$, where O is the origin. **(3 marks)**

E/P **7** The circle C has equation $(x + 5)^2 + (y + 3)^2 = 80$.

The line l is a tangent to the circle and has gradient 2.

Find two possible equations for l giving your answers in the form $y = mx + c$. **(8 marks)**

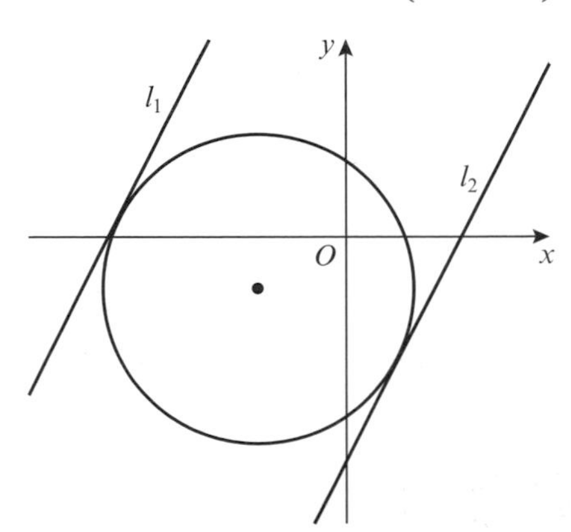

8 The line with equation $2x + y - 5 = 0$ is a tangent to the circle with equation

$$(x - 3)^2 + (y - p)^2 = 5$$

a Find the two possible values of p. **(8 marks)**

b Write down the coordinates of the centre of the circle in each case. **(2 marks)**

Problem-solving

The line is a tangent to the circle so it must intersect at exactly one point. You can use the discriminant to determine the values of p for which this occurs.

9 The circle C has centre $P(11, -5)$ and passes through the point $Q(5, 3)$.

a Find an equation for C. **(3 marks)**

The line l_1 is a tangent to C at the point Q.

b Find an equation for l_1. **(4 marks)**

The line l_2 is parallel to l_1 and passes through the midpoint of PQ. Given that l_2 intersects C at A and B

c find the coordinates of points A and B **(4 marks)**

d find the length of the line segment AB, leaving your answer in its simplest surd form. **(3 marks)**

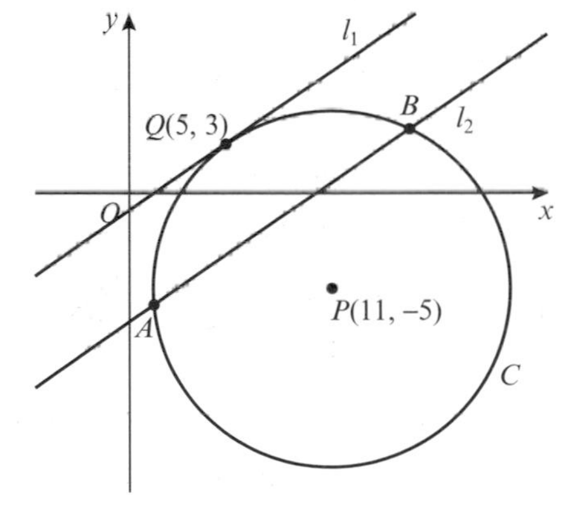

10 The points R and S lie on a circle with centre $C(a, -2)$, as shown in the diagram.

The point R has coordinates $(2, 3)$ and the point S has coordinates $(10, 1)$.

M is the midpoint of the line segment RS.

The line l passes through M and C.

a Find an equation for l. **(4 marks)**

b Find the value of a. **(2 marks)**

c Find the equation of the circle. **(3 marks)**

d Find the points of intersection, A and B, of the line l and the circle. **(5 marks)**

E/P **11** The circle C has equation $x^2 - 4x + y^2 - 6y = 7$.

The line l with equation $x - 3y + 17 = 0$ intersects the circle at the points P and Q.

a Find the coordinates of the point P and the point Q. **(4 marks)**

b Find the equation of the tangent at the point P and the point Q. **(4 marks)**

c Find the equation of the perpendicular bisector of the chord PQ. **(3 marks)**

d Show that the two tangents and the perpendicular bisector intersect at a single point and find the coordinates of the point of intersection. **(2 marks)**

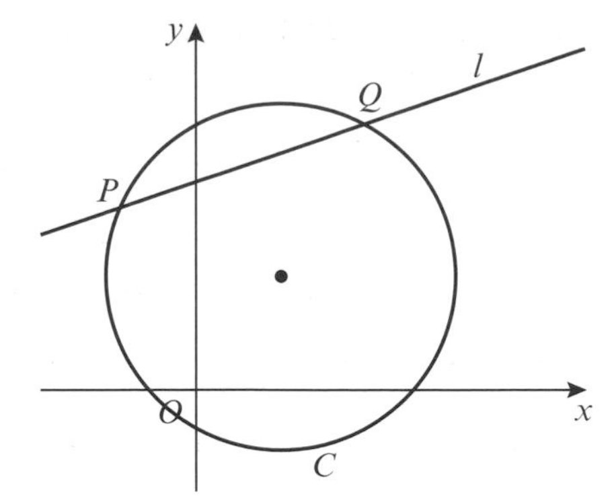

Challenge

1 The circle C has equation $(x - 7)^2 + (y + 1)^2 = 5$.

The line l with positive gradient passes through $(0, -2)$ and is a tangent to the circle.

Find an equation of l, giving your answer in the form $y = mx + c$.

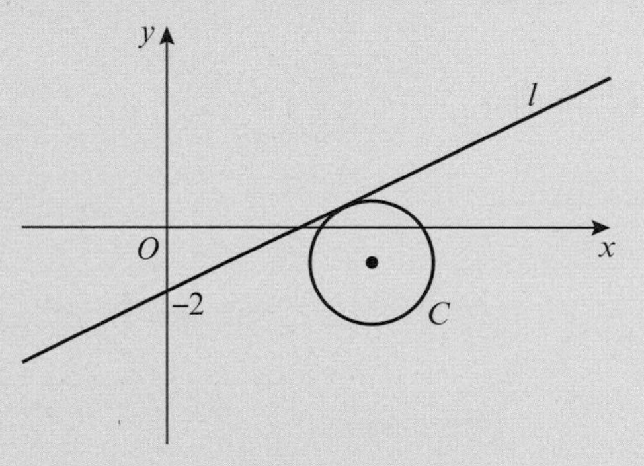

2 The circle with centre C has equation $(x - 2)^2 + (y - 1)^2 = 10$.

The tangents to the circle at points P and Q meet at the point R with coordinates $(6, -1)$.

a Show that $CPRQ$ is a square.

b Hence find the equations of both tangents.

Problem-solving

Use the point $(0, -2)$ to write an equation for the tangent in terms of m. Substitute this equation into the equation for the circle.

6.5 Circles and triangles

A triangle consists of three points, called vertices.
It is always possible to draw a unique circle through the three vertices of any triangle. This circle is called the **circumcircle** of the triangle. The centre of the circle is called the **circumcentre** of the triangle and is the point where the perpendicular bisectors of each side intersect.

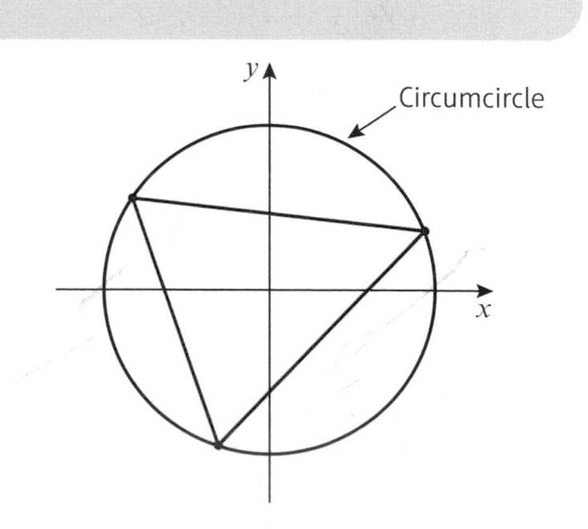

For a right-angled triangle, the hypotenuse of the triangle is a diameter of the circumcircle.

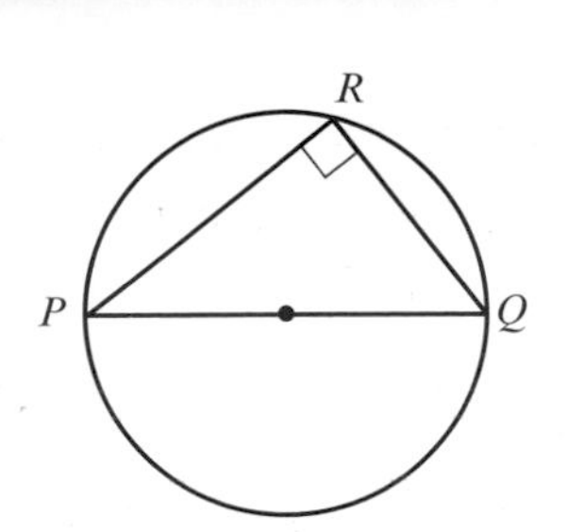

You can state this result in two other ways:

- **If $\angle PRQ = 90°$ then R lies on the circle with diameter PQ.**
- **The angle in a semicircle is always a right angle.**

To find the centre of a circle given any three points on the circumference:

- **Find the equations of the perpendicular bisectors of two different chords.**
- **Find the coordinates of the point of intersection of the perpendicular bisectors.**

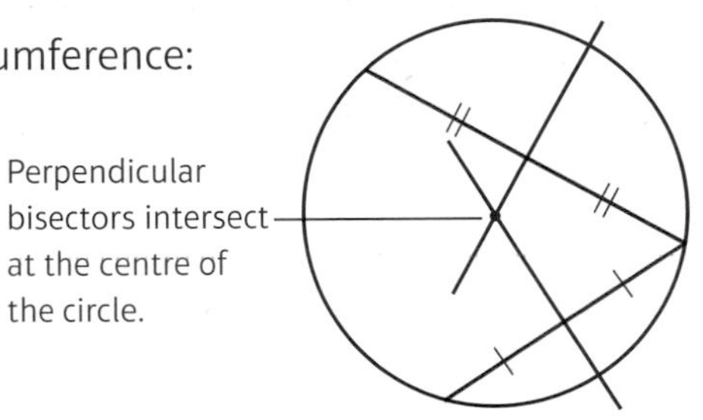

Example 13

The points $A(-8, 1)$, $B(4, 5)$ and $C(-4, 9)$ lie on the circle, as shown in the diagram.

a Show that AB is a diameter of the circle.

b Find an equation of the circle.

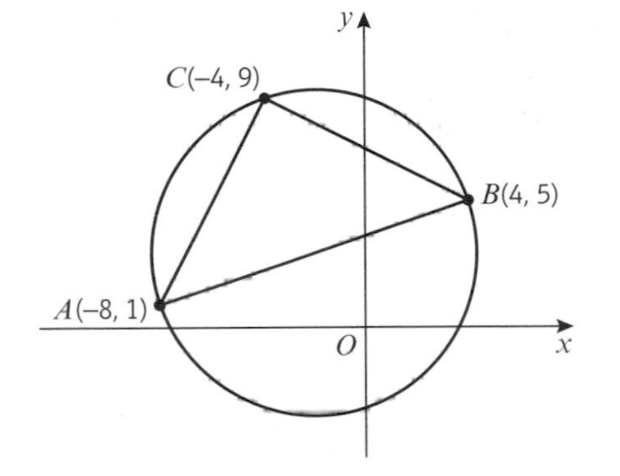

a Test triangle ABC to see if it is a right-angled triangle.

$AB^2 = (4-(-8))^2 + (5 - 1)^2$
$= 12^2 + 4^2 = 160$

$AC^2 = (-4-(-8))^2 + (9 - 1)^2$
$= 4^2 + 8^2 = 80$

$BC^2 = (-4 - 4)^2 + (9 - 5)^2$
$= (-8)^2 + 4^2 = 80$

Use $d^2 = (x_2 - x_1)^2 + (y_2 - y_1)^2$ to determine the length of each side of the triangle ABC.

Now, $80 + 80 = 160$ so $AC^2 + BC^2 = AB^2$

Use Pythagoras' theorem to test if triangle ABC is a right-angled triangle.

So triangle ABC is a right-angled triangle and AB is the diameter of the circle.

If ABC is a right-angled triangle, its longest side must be a diameter of the circle that passes through all three points.

b Find the midpoint M of AB.

The centre of the circle is the midpoint of AB.

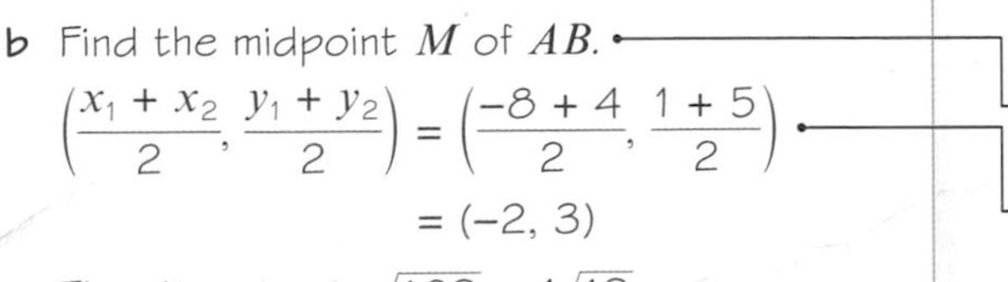

$\left(\frac{x_1 + x_2}{2}, \frac{y_1 + y_2}{2}\right) = \left(\frac{-8 + 4}{2}, \frac{1 + 5}{2}\right)$
$= (-2, 3)$

Substitute $(x_1, y_1) = (-8, 1)$ and $(x_2, y_2) = (4, 5)$.

The diameter is $\sqrt{160} = 4\sqrt{10}$

From part **a**, $AB^2 = 160$.

The radius is $2\sqrt{10}$

The radius is half the diameter.

$(x - a)^2 + (y - b)^2 = r^2$

$(x + 2)^2 + (y - 3)^2 = (2\sqrt{10})^2$

$(x + 2)^2 + (y - 3)^2 = 40$

Substitute $(a, b) = (-2, 3)$ and $r = 2\sqrt{10}$ into the equation for a circle.

Example 14

The points $P(3, 16)$, $Q(11, 12)$ and $R(-7, 6)$ lie on the circumference of a circle. The equation of the perpendicular bisector of PQ is $y = 2x$.

a Find the equation of the perpendicular bisector of PR.

b Find the centre of the circle.

c Work out the equation of the circle.

Online Explore triangles and their circumcircles using GeoGebra.

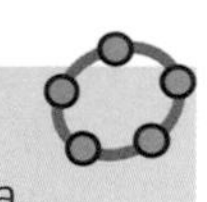

a The midpoint of PR is $\left(\frac{x_1 + x_2}{2}, \frac{y_1 + y_2}{2}\right)$

$= \left(\frac{3 + (-7)}{2}, \frac{16 + 6}{2}\right) = (-2, 11)$

The perpendicular bisector of PR passes through the midpoint of PR. Substitute $(x_1, y_1) = (3, 16)$ and $(x_2, y_2) = (-7, 6)$ into the midpoint formula.

The gradient of PR is $\frac{y_2 - y_1}{x_2 - x_1} = \frac{6 - 16}{-7 - 3}$

$= \frac{-10}{-10} = 1$

Substitute $(x_1, y_1) = (3, 16)$ and $(x_2, y_2) = (-7, 6)$ into the gradient formula.

The gradient of a line perpendicular to PR is -1.

$y - y_1 = m(x - x_1)$

$y - 11 = -1(x - (-2))$

Substitute $m = -1$ and $(x_1, y_1) = (-2, 11)$ into the equation for a straight line.

$y - 11 = -x - 2$

$y = -x + 9$

Simplify and leave in the form $y = mx + c$.

b Equation of perpendicular bisector to PQ: $y = 2x$

Equation of perpendicular bisector to PR: $y = -x + 9$

Solve these two equations simultaneously to find the point of intersection. The two perpendicular bisectors intersect at the centre of the circle.

$2x = -x + 9$

$3x = 9$

$x = 3$

This is the x-coordinate of the centre of the circle.

$y = 2x$

$y = 2(3) = 6$

Substitute $x = 3$ to find the y-coordinate of the centre of the circle.

The centre of the circle is at $(3, 6)$.

c Find the distance between $(3, 6)$ and $Q(11, 12)$.

The radius of the circle is the distance from the centre to a point on the circumference of the circle.

$d = \sqrt{(x_2 - x_1)^2 + (y_2 - y_1)^2}$

$d = \sqrt{(11 - 3)^2 + (12 - 6)^2}$

Substitute $(x_1, y_1) = (3, 6)$ and $(x_2, y_2) = (11, 12)$ into the distance formula.

$d = \sqrt{64 + 36}$

$d = \sqrt{100} = 10$

Simplify to find the radius of the circle.

The circle through the points P, Q and R has a radius of 10.

The centre of the circle is $(3, 6)$.

The equation for the circle is $(x - 3)^2 + (y - 6)^2 = 100$

Substitute $(a, b) = (3, 6)$ and $r = 10$ into $(x - a)^2 + (y - b)^2 = r^2$

Exercise 6F

1 The points $U(-2, 8)$, $V(7, 7)$ and $W(-3, -1)$ lie on a circle.

a Show that triangle UVW has a right angle.

b Find the coordinates of the centre of the circle.

c Write down an equation for the circle.

2 The points $A(2, 6)$, $B(5, 7)$ and $C(8, -2)$ lie on a circle.

a Show that AC is a diameter of the circle.

b Write down an equation for the circle.

c Find the area of the triangle ABC.

3 The points $A(-3, 19)$, $B(9, 11)$ and $C(-15, 1)$ lie on the circumference of a circle.

a Find the equation of the perpendicular bisector of
i AB **ii** AC

b Find the coordinates of the centre of the circle.

c Write down an equation for the circle.

4 The points $P(-11, 8)$, $Q(-6, -7)$ and $R(4, -7)$ lie on the circumference of a circle.

a Find the equation of the perpendicular bisector of
i PQ **ii** QR

b Find an equation for the circle.

(P) **5** The points $R(-2, 1)$, $S(4, 3)$ and $T(10, -5)$ lie on the circumference of a circle C. Find an equation for the circle.

Problem-solving

Use headings in your working to keep track of what you are working out at each stage.

(P) **6** Consider the points $A(3, 15)$, $B(-13, 3)$, $C(-7, -5)$ and $D(8, 0)$.

a Show that ABC is a right-angled triangle.

b Find the equation of the circumcircle of triangle ABC.

c Hence show that A, B, C and D all lie on the circumference of this circle.

(P) **7** The points $A(-1, 9)$, $B(6, 10)$, $C(7, 3)$ and $D(0, 2)$ lie on a circle.

a Show that $ABCD$ is a square.

b Find the area of $ABCD$.

c Find the centre of the circle.

(E/P) **8** The points $D(-12, -3)$, $E(-10, b)$ and $F(2, -5)$ lie on the circle C as shown in the diagram.

Given that $\angle DEF = 90°$ and $b > 0$

a show that $b = 1$ **(5 marks)**

b find an equation for C. **(4 marks)**

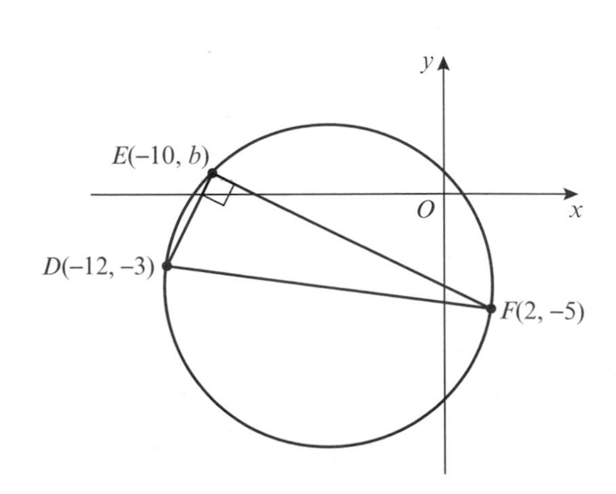

(E/P) **9** A circle has equation $x^2 + 2x + y^2 - 24y - 24 = 0$

a Find the centre and radius of the circle. **(3 marks)**

b The points $A(-13, 17)$ and $B(11, 7)$ both lie on the circumference of the circle. Show that AB is a diameter of the circle. **(3 marks)**

c The point C lies on the negative x-axis and the angle $ACB = 90°$. Find the coordinates of C. **(3 marks)**

Mixed exercise 6

(P) **1** The line segment QR is a diameter of the circle centre C, where Q and R have coordinates $(11, 12)$ and $(-5, 0)$ respectively. The point P has coordinates $(13, 6)$.

a Find the coordinates of C.

b Find the radius of the circle.

c Write down the equation of the circle.

d Show that P lies on the circle.

(P) **2** Show that $(0, 0)$ lies inside the circle $(x - 5)^2 + (y + 2)^2 = 30$.

(E/P) **3** The circle C has equation $x^2 + 3x + y^2 + 6y = 3x - 2y - 7$.

a Find the centre and radius of the circle. **(4 marks)**

b Find the points of intersection of the circle and the y-axis. **(3 marks)**

c Show that the circle does not intersect the x-axis. **(2 marks)**

4 The centres of the circles $(x - 8)^2 + (y - 8)^2 = 117$ and $(x + 1)^2 + (y - 3)^2 = 106$ are P and Q respectively.

a Show that P lies on $(x + 1)^2 + (y - 3)^2 = 106$.

b Find the length of PQ.

(P) **5** The points $A(-1, 0)$, $B\left(\frac{1}{2}, \frac{\sqrt{3}}{2}\right)$ and $C\left(\frac{1}{2}, -\frac{\sqrt{3}}{2}\right)$ are the vertices of a triangle.

a Show that the circle $x^2 + y^2 = 1$ passes through the vertices of the triangle.

b Show that $\triangle ABC$ is equilateral.

(E/P) **6** A circle with equation $(x - k)^2 + (y - 3k)^2 = 13$ passes through the point $(3, 0)$.

a Find two possible values of k. **(6 marks)**

b Given that $k > 0$, write down the equation of the circle. **(1 mark)**

(E/P) **7** The line with $3x - y - 9 = 0$ does not intersect the circle with equation $x^2 + px + y^2 + 4y = 20$. Show that $42 - 10\sqrt{10} < p < 42 + 10\sqrt{10}$. **(6 marks)**

(P) **8** The line $y = 2x - 8$ meets the coordinate axes at A and B. The line segment AB is a diameter of the circle. Find the equation of the circle.

(P) **9** The circle centre $(8, 10)$ meets the x-axis at $(4, 0)$ and $(a, 0)$.

a Find the radius of the circle.

b Find the value of a.

10 The circle $(x - 5)^2 + y^2 = 36$ meets the x-axis at P and Q. Find the coordinates of P and Q.

11 The circle $(x + 4)^2 + (y - 7)^2 = 121$ meets the y-axis at $(0, m)$ and $(0, n)$. Find the values of m and n.

(E) **12** The circle C with equation $(x + 5)^2 + (y + 2)^2 = 125$ meets the positive coordinate axes at $A(a, 0)$ and $B(0, b)$.

a Find the values of a and b. **(2 marks)**

b Find the equation of the line AB. **(2 marks)**

c Find the area of the triangle OAB, where O is the origin. **(2 marks)**

(P) **13** The circle, centre (p, q) radius 25, meets the x-axis at $(-7, 0)$ and $(7, 0)$, where $q > 0$.

a Find the values of p and q.

b Find the coordinates of the points where the circle meets the y-axis.

(P) **14** The point $A(-3, -7)$ lies on the circle centre $(5, 1)$. Find the equation of the tangent to the circle at A.

(P) **15** The line segment AB is a chord of a circle centre $(2, -1)$, where A and B are $(3, 7)$ and $(-5, 3)$ respectively. AC is a diameter of the circle. Find the area of $\triangle ABC$.

(E/P) **16** The circle C has equation $(x - 6)^2 + (y - 5)^2 = 17$.

The lines l_1 and l_2 are each a tangent to the circle and intersect at the point $(0, 12)$.

Find the equations of l_1 and l_2, giving your answers in the form $y = mx + c$. **(8 marks)**

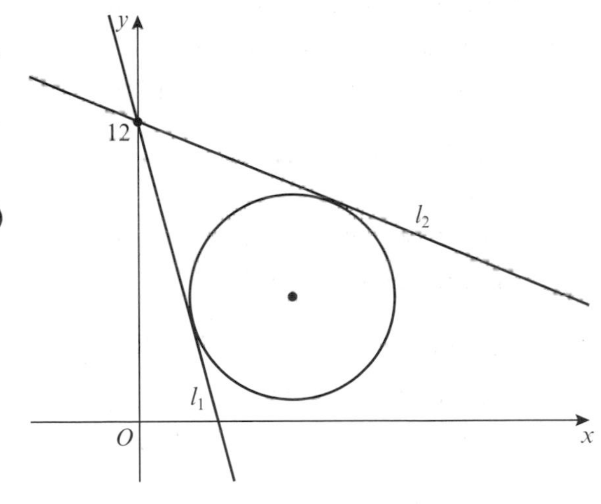

(E/P) **17** The points A and B lie on a circle with centre C, as shown in the diagram.

The point A has coordinates $(3, 7)$ and the point B has coordinates $(5, 1)$.

M is the midpoint of the line segment AB.

The line l passes through the points M and C.

a Find an equation for l. **(4 marks)**

Given that the x-coordinate of C is -2:

b find an equation of the circle **(4 marks)**

c find the area of the triangle ABC. **(3 marks)**

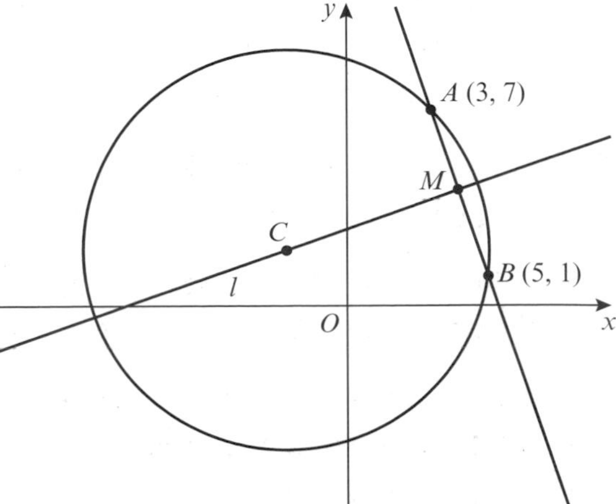

E/P **18** The circle C has equation $(x - 3)^2 + (y + 3)^2 = 52$.

The baselines l_1 and l_2 are tangents to the circle and have gradient $\frac{3}{2}$

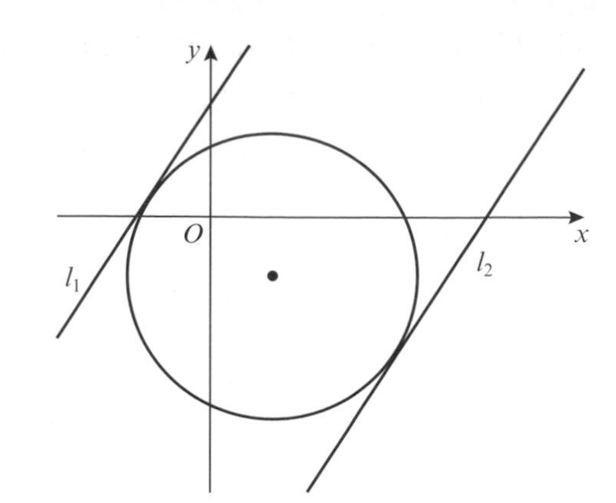

a Find the points of intersection, P and Q, of the tangents and the circle. **(6 marks)**

b Find the equations of lines l_1 and l_2, giving your answers in the form $ax + by + c = 0$. **(2 marks)**

E/P **19** The circle C has equation $x^2 + 6x + y^2 - 2y = 7$.

The lines l_1 and l_2 are tangents to the circle.

They intersect at the point $R(0, 6)$.

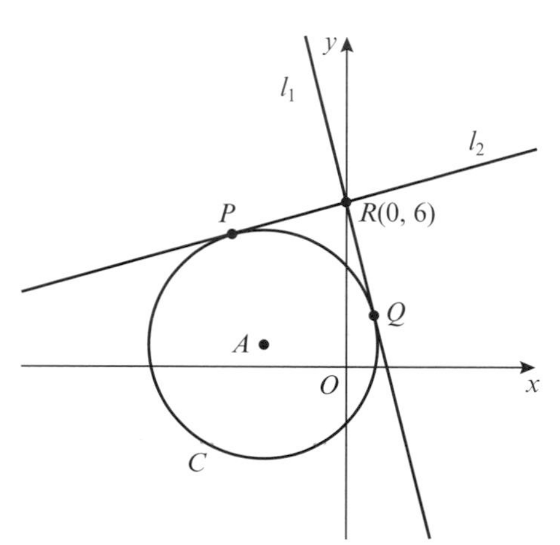

a Find the equations of lines l_1 and l_2, giving your answers in the form $y = mx + b$. **(6 marks)**

b Find the points of intersection, P and Q, of the tangents and the circle. **(4 marks)**

c Find the area of quadrilateral $APRQ$. **(2 marks)**

E/P **20** The circle C has a centre at (6, 9) and a radius of $\sqrt{50}$.

The line l_1 with equation $x + y - 21 = 0$ intersects the circle at the points P and Q.

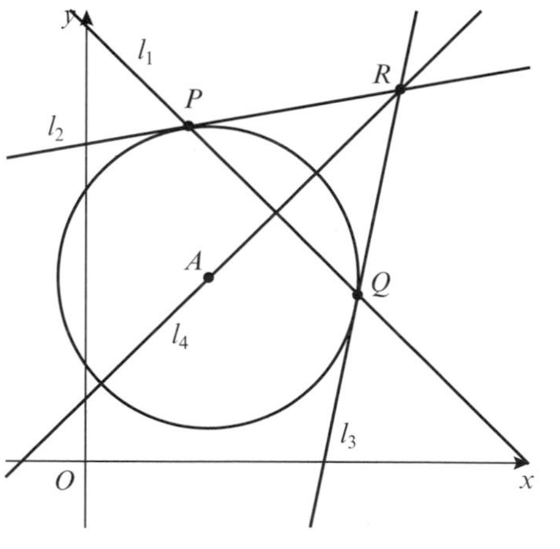

a Find the coordinates of the point P and the point Q. **(5 marks)**

b Find the equations of l_2 and l_3, the tangents at the points P and Q respectively. **(4 marks)**

c Find the equation of l_4, the perpendicular bisector of the chord PQ. **(4 marks)**

d Show that the two tangents and the perpendicular bisector intersect and find the coordinates of R, the point of intersection. **(2 marks)**

e Calculate the area of the kite $APRQ$. **(3 marks)**

P **21** The line $y = -3x + 12$ meets the coordinate axes at A and B.

a Find the coordinates of A and B.

b Find the coordinates of the midpoint of AB.

c Find the equation of the circle that passes through A, B and O, where O is the origin.

E/P **22** The points $A(-3, -2)$, $B(-6, 0)$ and $C(1, q)$ lie on the circumference of a circle such that $\angle BAC = 90°$.

a Find the value of q. **(4 marks)**

b Find the equation of the circle. **(4 marks)**

E/P **23** The points $R(-4, 3)$, $S(7, 4)$ and $T(8, -7)$ lie on the circumference of a circle.

a Show that RT is the diameter of the circle. **(4 marks)**

b Find the equation of the circle. **(4 marks)**

P **24** The points $A(-4, 0)$, $B(4, 8)$ and $C(6, 0)$ lie on the circumference of circle C.
Find the equation of the circle.

P **25** The points $A(-7, 7)$, $B(1, 9)$, $C(3, 1)$ and $D(-7, 1)$ lie on a circle.

a Find the equation of the perpendicular bisector of:

i AB **ii** CD

b Find the equation of the circle.

Challenge

The circle with equation $(x - 5)^2 + (y - 3)^2 = 20$ with centre A intersects the circle with equation $(x - 10)^2 + (y - 8)^2 = 10$ with centre B at the points P and Q.

a Find the equation of the line containing the points P and Q in the form $ax + by + c = 0$.

b Find the coordinates of the points P and Q.

c Find the area of the kite $APBQ$.

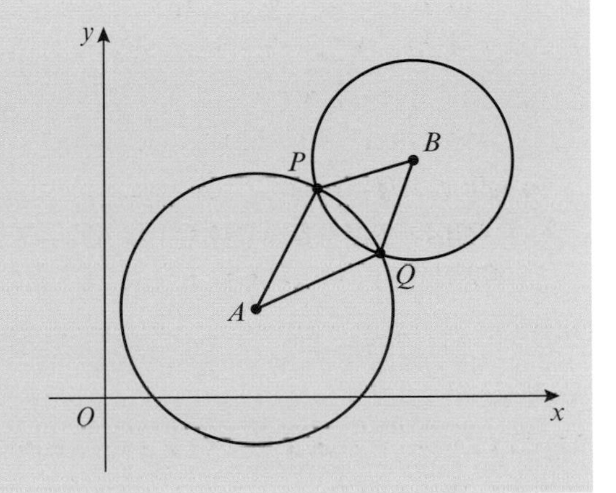

Summary of key points

1 The midpoint of a line segment with endpoints (x_1, y_1) and (x_2, y_2) is $\left(\frac{x_1 + x_2}{2}, \frac{y_1 + y_2}{2}\right)$.

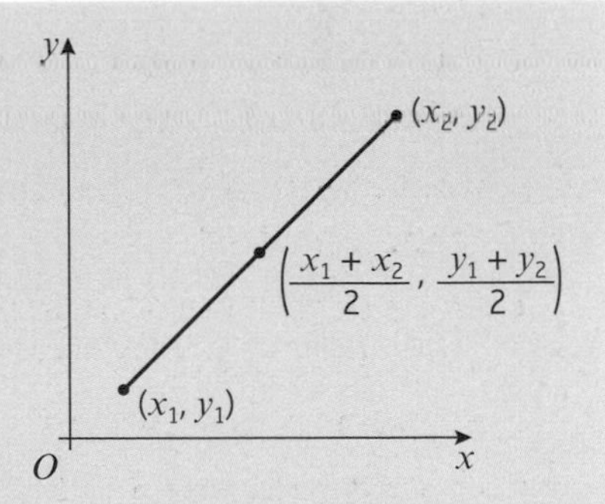

2 The perpendicular bisector of a line segment AB is the straight line that is perpendicular to AB and passes through the midpoint of AB.

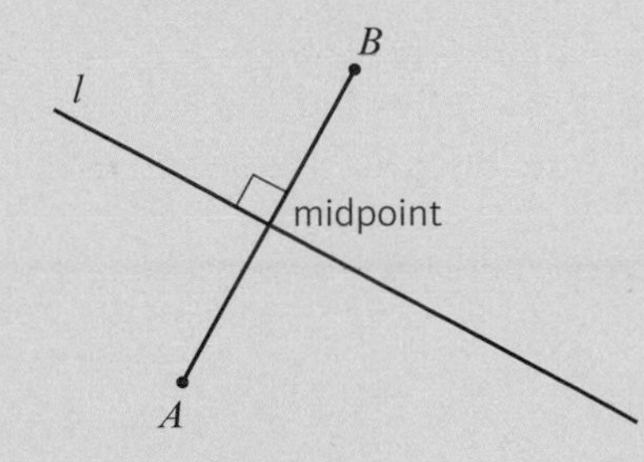

If the gradient of AB is m then the gradient of its perpendicular bisector, l, will be $-\frac{1}{m}$

3 The equation of a circle with centre (0, 0) and radius r is $x^2 + y^2 = r^2$.

4 The equation of the circle with centre (a, b) and radius r is $(x - a)^2 + (y - b)^2 = r^2$.

5 The equation of a circle can be given in the form: $x^2 + y^2 + 2fx + 2gy + c = 0$
This circle has centre $(-f, -g)$ and radius $\sqrt{f^2 + g^2 - c}$

6 A straight line can intersect a circle once, by just touching the circle, or twice. Not all straight lines will intersect a given circle.

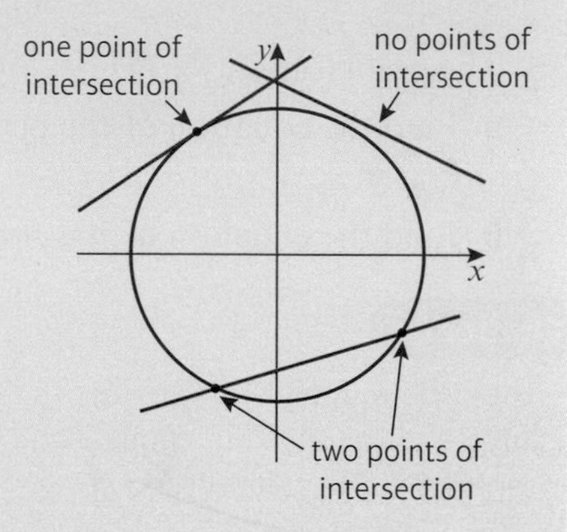

7 A tangent to a circle is perpendicular to the radius of the circle at the point of intersection.

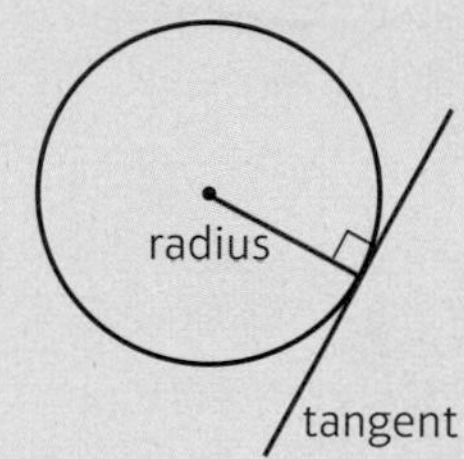

8 The perpendicular bisector of a chord will go through the centre of a circle.

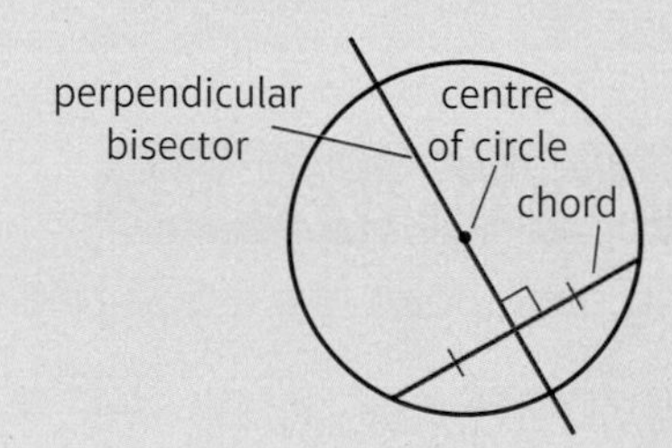

9 • If $\angle PRQ = 90°$ then R lies on the circle with diameter PQ.
• The angle in a semicircle is always a right angle.

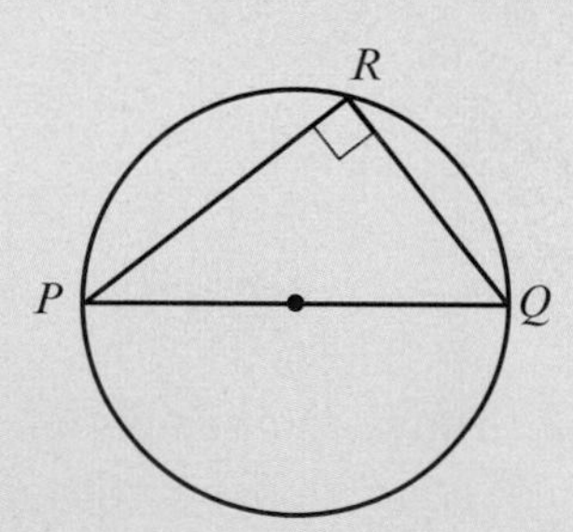

10 To find the centre of a circle given any three points:
- Find the equations of the perpendicular bisectors of two different chords.
- Find the coordinates of intersection of the perpendicular bisectors.

Perpendicular bisectors intersect at the centre of the circle

Algebraic methods

7

Objectives

After completing this unit you should be able to:

Proof is the cornerstone of mathematics. Mathematicians need to prove theorems (such as Pythagoras' theorem) before they can use them to solve problems. Pythagoras' theorem can be use to find approximate values for π.

→ Mixed exercise challenge Q1

Prior knowledge check

1 Simplify:

a $3x^2 \times 5x^5$ **b** $\dfrac{5x^3y^2}{15x^2y^3}$ ← Section 1.1

2 Factorise:

a $x^2 - 2x - 24$ **b** $3x^2 - 17x + 20$

← Section 1.3

3 Use long division to calculate:

a 197 041 ÷ 23 **b** 56 168 ÷ 34

← GCSE Mathematics

4 Find the equations of the lines that pass through these pairs of points:

a $(-1, 4)$ and $(5, -14)$

b $(2, -6)$ and $(8, -3)$ ← GCSE Mathematics

5 Complete the square for the expressions:

a $x^2 - 2x - 20$ **b** $2x^2 + 4x + 15$

← Section 2.2

7.1 Algebraic fractions

You can simplify algebraic fractions using division.

- **When simplifying an algebraic fraction, where possible factorise the numerator and denominator and then cancel common factors.**

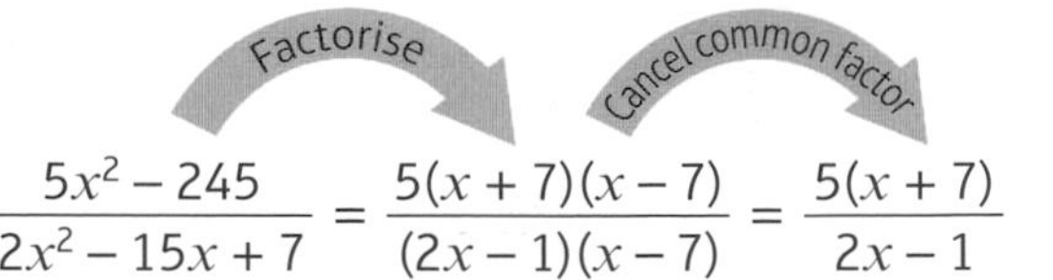

$$\frac{5x^2 - 245}{2x^2 - 15x + 7} = \frac{5(x+7)(x-7)}{(2x-1)(x-7)} = \frac{5(x+7)}{2x-1}$$

Example 1

Simplify these fractions:

a $\frac{7x^4 - 2x^3 + 6x}{x}$ **b** $\frac{(x+7)(2x-1)}{(2x-1)}$ **c** $\frac{x^2 + 7x + 12}{(x+3)}$ **d** $\frac{x^2 + 6x + 5}{x^2 + 3x - 10}$ **e** $\frac{2x^2 + 11x + 12}{(x+3)(x+4)}$

a $\frac{7x^4 - 2x^3 + 6x}{x}$

$= \frac{7x^4}{x} - \frac{2x^3}{x} + \frac{6x}{x}$ — Divide each part of the numerator by x.

$= 7x^3 - 2x^2 + 6$

b $\frac{(x+7)(2x-1)}{(2x-1)} = x + 7$ — Simplify by cancelling the common factor of $(2x - 1)$.

c $\frac{x^2 + 7x + 12}{(x+3)} = \frac{(x+3)(x+4)}{(x+3)}$ — Factorise: $x^2 + 7x + 12 = (x + 3)(x + 4)$.

$= x + 4$ — Cancel the common factor of $(x + 3)$.

d $\frac{x^2 + 6x + 5}{x^2 + 3x - 10} = \frac{(x+5)(x+1)}{(x+5)(x-2)}$ — Factorise: $x^2 + 6x + 5 = (x + 5)(x + 1)$ and $x^2 + 3x - 10 = (x + 5)(x - 2)$.

$= \frac{x+1}{x-2}$ — Cancel the common factor of $(x + 5)$.

e $2x^2 + 11x + 12 = 2x^2 + 3x + 8x + 12$

$= x(2x + 3) + 4(2x + 3)$

$= (2x + 3)(x + 4)$ — Factorise $2x^2 + 11x + 12$

So $\frac{2x^2 + 11x + 12}{(x+3)(x+4)}$

$= \frac{(2x+3)(x+4)}{(x+3)(x+4)}$

$= \frac{2x+3}{x+3}$ — Cancel the common factor of $(x + 4)$.

Exercise 7A

1 Simplify these fractions:

a $\frac{4x^4 + 5x^2 - 7x}{x}$ **b** $\frac{7x^5 - 5x^5 + 9x^3 + x^2}{x}$ **c** $\frac{-x^4 + 4x^2 + 6}{x}$

d $\frac{7x^5 - x^3 - 4}{x}$ **e** $\frac{8x^4 - 4x^3 + 6x}{2x}$ **f** $\frac{9x^2 - 12x^3 - 3x}{3x}$

g $\frac{7x^3 - x^4 - 2}{5x}$ **h** $\frac{-4x^2 + 6x^4 - 2x}{-2x}$ **i** $\frac{-x^8 + 9x^4 - 4x^3 + 6}{-2x}$

j $\frac{-9x^9 - 6x^6 + 4x^4 - 2}{-3x}$

2 Simplify these fractions as far as possible:

a $\frac{(x + 3)(x - 2)}{(x - 2)}$ **b** $\frac{(x + 4)(3x - 1)}{(3x - 1)}$ **c** $\frac{(x + 3)^2}{(x + 3)}$

d $\frac{x^2 + 10x + 21}{(x + 3)}$ **e** $\frac{x^2 + 9x + 20}{(x + 4)}$ **f** $\frac{x^2 + x - 12}{(x - 3)}$

g $\frac{x^2 + x - 20}{x^2 + 2x - 15}$ **h** $\frac{x^2 + 3x + 2}{x^2 + 5x + 4}$ **i** $\frac{x^2 + x - 12}{x^2 - 9x + 18}$

j $\frac{2x^2 + 7x + 6}{(x - 5)(x + 2)}$ **k** $\frac{2x^2 + 9x - 18}{(x + 6)(x + 1)}$ **l** $\frac{3x^2 - 7x + 2}{(3x - 1)(x + 2)}$

m $\frac{2x^2 + 3x + 1}{x^2 - x - 2}$ **n** $\frac{x^2 + 6x + 8}{3x^2 + 7x + 2}$ **o** $\frac{2x^2 - 5x - 3}{2x^2 - 9x + 9}$

3 $\frac{6x^3 + 3x^2 - 84x}{6x^2 - 33x + 42} = \frac{ax(x + b)}{x + c}$, where a, b and c are constants.

Work out the values of a, b and c. **(4 marks)**

7.2 Dividing polynomials

A **polynomial** is a finite expression with positive whole number indices.

- **You can use long division to divide a polynomial by $(x \pm p)$, where p is a constant.**

Polynomials	Not polynomials
$2x + 4$	$\sqrt{x}$
$4xy^2 + 3x - 9$	$6x^{-2}$
8	$\frac{4}{x}$

Example 2

Divide $x^3 + 2x^2 - 17x + 6$ by $(x - 3)$.

①

$$\begin{array}{r} x^2 \\ x - 3 \overline{)x^3 + 2x^2 - 17x + 6} \\ \underline{x^3 - 3x^2} \\ 5x^2 - 17x \end{array}$$

Start by dividing the first term of the polynomial by x, so that $x^3 \div x = x^2$.

Next multiply $(x - 3)$ by x^2, so that $x^2 \times (x - 3) = x^3 - 3x^2$.

Now subtract, so that $(x^3 + 2x^2) - (x^3 - 3x^2) = 5x^2$.

Copy $-17x$.

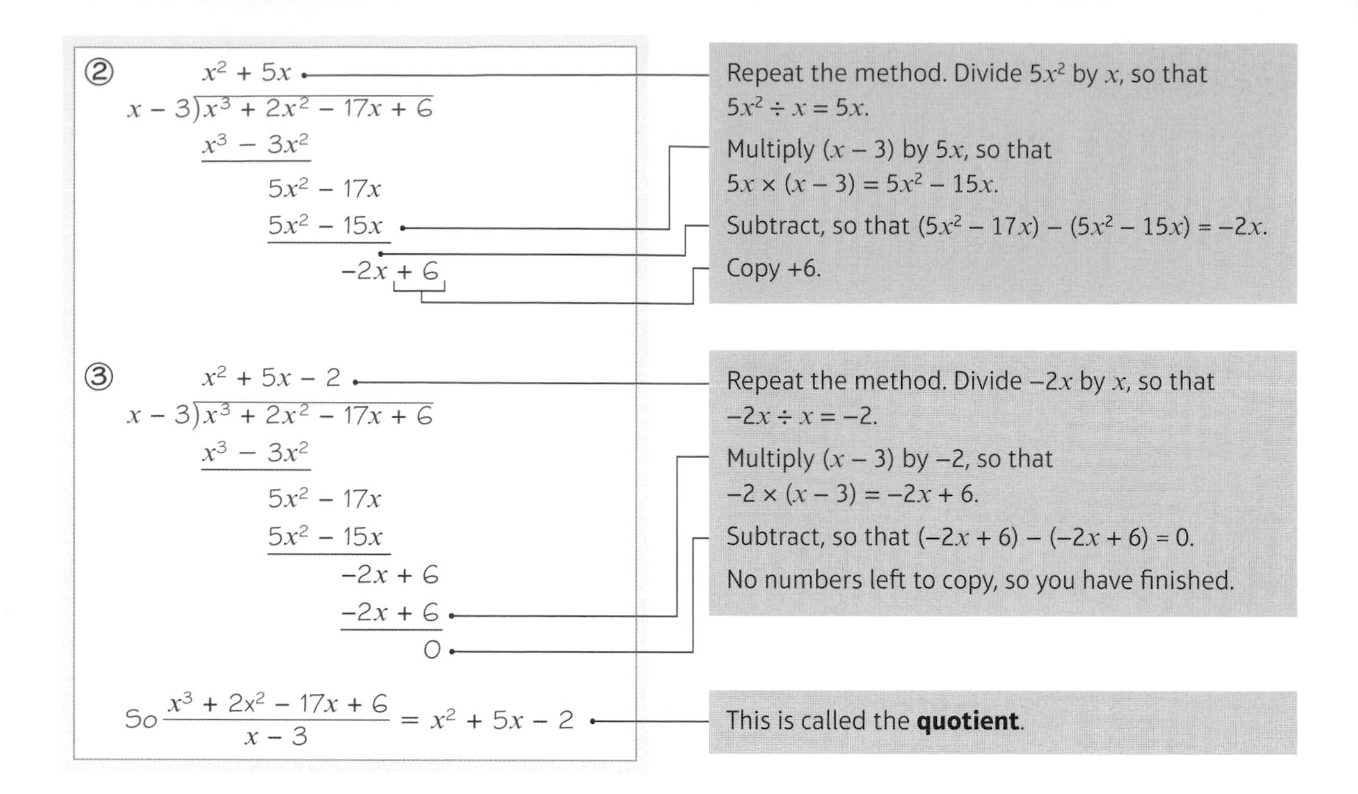

Example 3

$f(x) = 4x^4 - 17x^2 + 4$

Divide $f(x)$ by $(2x + 1)$, giving your answer in the form $f(x) = (2x + 1)(ax^3 + bx^2 + cx + d)$.

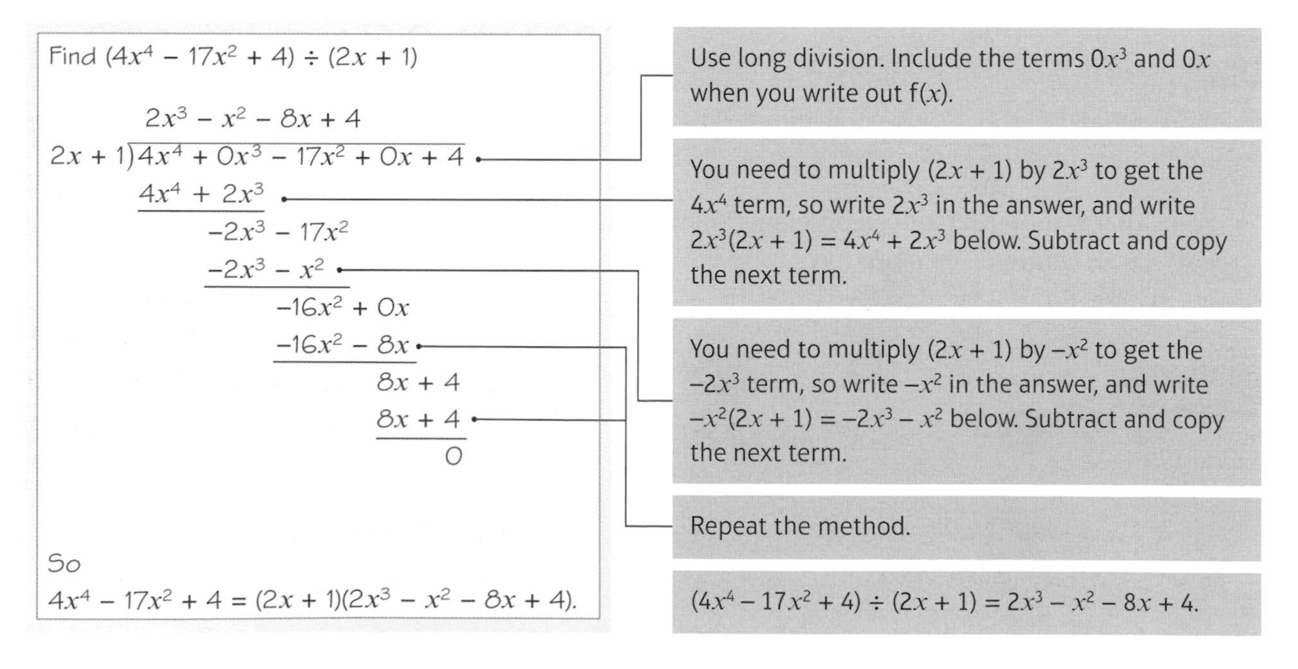

Example 4

Find the remainder when $2x^3 - 5x^2 - 16x + 10$ is divided by $(x - 4)$.

$$\begin{array}{r} 2x^2 + 3x - 4 \\ x - 4 \overline{)2x^3 - 5x^2 - 16x + 10} \\ \underline{2x^3 - 8x^2 } \\ 3x^2 - 16x \\ \underline{3x^2 - 12x } \\ -4x + 10 \\ \underline{-4x + 16} \\ -6 \end{array}$$

So the remainder is −6.

$(x - 4)$ is not a factor of $2x^3 - 5x^2 - 16x + 10$ as the remainder $\neq 0$.

This means you cannot write the expression in the form $(x - 4)(ax^2 + bx + c)$.

Exercise 7B

1 Write each polynomial in the form $(x \pm p)(ax^2 + bx + c)$ by dividing:

a $x^3 + 6x^2 + 8x + 3$ by $(x + 1)$ **b** $x^3 + 10x^2 + 25x + 4$ by $(x + 4)$

c $x^3 - x^2 + x + 14$ by $(x + 2)$ **d** $x^3 + x^2 - 7x - 15$ by $(x - 3)$

e $x^3 - 8x^2 + 13x + 10$ by $(x - 5)$ **f** $x^3 - 5x^2 - 6x - 56$ by $(x - 7)$

2 Write each polynomial in the form $(x \pm p)(ax^2 + bx + c)$ by dividing:

a $6x^3 + 27x^2 + 14x + 8$ by $(x + 4)$ **b** $4x^3 + 9x^2 - 3x - 10$ by $(x + 2)$

c $2x^3 + 4x^2 - 9x - 9$ by $(x + 3)$ **d** $2x^3 - 15x^2 + 14x + 24$ by $(x - 6)$

e $-5x^3 - 27x^2 + 23x + 30$ by $(x + 6)$ **f** $-4x^3 + 9x^2 - 3x + 2$ by $(x - 2)$

3 Divide:

a $x^4 + 5x^3 + 2x^2 - 7x + 2$ by $(x + 2)$ **b** $4x^4 + 14x^3 + 3x^2 - 14x - 15$ by $(x + 3)$

c $-3x^4 + 9x^3 - 10x^2 + x + 14$ by $(x - 2)$ **d** $-5x^5 + 7x^4 + 2x^3 - 7x^2 + 10x - 7$ by $(x - 1)$

4 Divide:

a $3x^4 + 8x^3 - 11x^2 + 2x + 8$ by $(3x + 2)$ **b** $4x^4 - 3x^3 + 11x^2 - x - 1$ by $(4x + 1)$

c $4x^4 - 6x^3 + 10x^2 - 11x - 6$ by $(2x - 3)$ **d** $6x^5 + 13x^4 - 4x^3 - 9x^2 + 21x + 18$ by $(2x + 3)$

e $6x^5 - 8x^4 + 11x^3 + 9x^2 - 25x + 7$ by $(3x - 1)$ **f** $8x^5 - 26x^4 + 11x^3 + 22x^2 - 40x + 25$ by $(2x - 5)$

g $25x^4 + 75x^3 + 6x^2 - 28x - 6$ by $(5x + 3)$ **h** $21x^5 + 29x^4 - 10x^3 + 42x - 12$ by $(7x - 2)$

5 Divide:

a $x^3 + x + 10$ by $(x + 2)$ **b** $2x^3 - 17x + 3$ by $(x + 3)$

c $-3x^3 + 50x - 8$ by $(x - 4)$

Hint Include $0x^2$ when you write out $f(x)$.

6 Divide:

a $x^3 + x^2 - 36$ by $(x - 3)$ **b** $2x^3 + 9x^2 + 25$ by $(x + 5)$

c $-3x^3 + 11x^2 - 20$ by $(x - 2)$

7 Show that $x^3 + 2x^2 - 5x - 10 = (x + 2)(x^2 - 5)$

8 Find the remainder when:

a $x^3 + 4x^2 - 3x + 2$ is divided by $(x + 5)$

b $3x^3 - 20x^2 + 10x + 5$ is divided by $(x - 6)$

c $-2x^3 + 3x^2 + 12x + 20$ is divided by $(x - 4)$

9 Show that when $3x^3 - 2x^2 + 4$ is divided by $(x - 1)$ the remainder is 5.

10 Show that when $3x^4 - 8x^3 + 10x^2 - 3x - 25$ is divided by $(x + 1)$ the remainder is -1.

11 Show that $(x + 4)$ is a factor of $5x^3 - 73x + 28$.

12 Simplify $\frac{3x^3 - 8x - 8}{x - 2}$

Hint Divide $3x^3 - 8x - 8$ by $(x - 2)$.

13 Divide $x^3 - 1$ by $(x - 1)$.

Hint Write $x^3 - 1$ as $x^3 + 0x^2 + 0x - 1$.

14 Divide $x^4 - 16$ by $(x + 2)$.

(E) **15** $f(x) = 10x^3 + 43x^2 - 2x - 10$

Find the remainder when $f(x)$ is divided by $(5x + 4)$. **(2 marks)**

(E/P) **16** $f(x) = 3x^3 - 14x^2 - 47x - 14$

a Find the remainder when $f(x)$ is divided by $(x - 3)$. **(2 marks)**

b Given that $(x + 2)$ is a factor of $f(x)$, factorise $f(x)$ completely. **(4 marks)**

Problem-solving

Write $f(x)$ in the form $(x + 2)(ax^2 + bx + c)$ then factorise the quadratic factor.

(E/P) **17** **a** Find the remainder when $x^3 + 6x^2 + 5x - 12$ is divided by

i $x - 2$,

ii $x + 3$. **(3 marks)**

b Hence, or otherwise, find all the solutions to the equation $x^3 + 6x^2 + 5x - 12 = 0$. **(4 marks)**

(E/P) **18** $f(x) = 2x^3 + 3x^2 - 8x + 3$

a Show that $f(x) = (2x - 1)(ax^2 + bx + c)$ where a, b and c are constants to be found. **(2 marks)**

b Hence factorise $f(x)$ completely. **(4 marks)**

c Write down all the real roots of the equation $f(x) = 0$. **(2 marks)**

(E/P) **19** $f(x) = 12x^3 + 5x^2 + 2x - 1$

a Show that $(4x - 1)$ is a factor of $f(x)$ and write $f(x)$ in the form $(4x - 1)(ax^2 + bx + c)$. **(6 marks)**

b Hence, show that the equation $12x^3 + 5x^2 + 2x - 1 = 0$ has exactly 1 real solution. **(2 marks)**

7.3 The factor theorem

The factor theorem is a quick way of finding simple linear factors of a polynomial.

- **The factor theorem states that if f(x) is a polynomial then:**
  - **If f(p) = 0, then ($x - p$) is a factor of f(x).**
  - **If ($x - p$) is a factor of f(x), then f(p) = 0.**

Watch out

These two statements are not the same. Here are two similar statements, only one of which is true:

If $x = -2$ then $x^2 = 4$ ✓

If $x^2 = 4$ then $x = -2$ ✗

You can use the factor theorem to quickly factorise a cubic function, g(x):

1 Substitute values into the function until you find a value p such that g(p) = 0.

2 Divide the function by ($x - p$). — The remainder will be 0 because ($x - p$) is a factor of g(x).

3 Write $g(x) = (x - p)(ax^2 + bx + c)$ — The other factor will be quadratic.

4 Factorise the quadratic factor, if possible, to write g(x) as a product of three linear factors.

Example 5

Show that $(x - 2)$ is a factor of $x^3 + x^2 - 4x - 4$ by:

a algebraic division **b** the factor theorem

a

$$\begin{array}{r} x^2 + 3x + 2 \\ x - 2\,\overline{)\,x^3 + x^2 - 4x - 4} \\ \underline{x^3 - 2x^2} \qquad\qquad \\ 3x^2 - 4x \qquad \\ \underline{3x^2 - 6x} \qquad \\ 2x - 4 \\ \underline{2x - 4} \\ 0 \end{array}$$

Divide $x^3 + x^2 - 4x - 4$ by $(x - 2)$.

The remainder is 0, so $(x - 2)$ is a factor of $x^3 + x^2 - 4x - 4$.

So $(x - 2)$ is a factor of $x^3 + x^2 - 4x - 4$.

b $f(x) = x^3 + x^2 - 4x - 4$ — Write the polynomial as a function.

$f(2) = (2)^3 + (2)^2 - 4(2) - 4$ — Substitute $x = 2$ into the polynomial.

$= 8 + 4 - 8 - 4$

$= 0$

So $(x - 2)$ is a factor of $x^3 + x^2 - 4x - 4$.

Use the factor theorem:
If f(p) = 0, then ($x - p$) is a factor of f(x).
Here $p = 2$, so $(x - 2)$ is a factor of $x^3 + x^2 - 4x - 4$.

Example 6

a Fully factorise $2x^3 + x^2 - 18x - 9$

b Hence sketch the graph of $y = 2x^3 + x^2 - 18x - 9$

a $f(x) = 2x^3 + x^2 - 18x - 9$

Write the polynomial as a function.

$f(-1) = 2(-1)^3 + (-1)^2 - 18(-1) - 9 = 8$
$f(1) = 2(1)^3 + (1)^2 - 18(1) - 9 = -24$
$f(2) = 2(2)^3 + (2)^2 - 18(2) - 9 = -25$
$f(3) = 2(3)^3 + (3)^2 - 18(3) - 9 = 0$

Try values of x, e.g. −1, 1, 2, 3, … until you find $f(p) = 0$.

$f(p) = 0$.

So $(x - 3)$ is a factor of $2x^3 + x^2 - 18x - 9$.

Use statement 1 from the factor theorem: If $f(p) = 0$, then $(x - p)$ is a factor of $f(x)$. Here $p = 3$.

$$\begin{array}{r} 2x^2 + 7x + 3 \\ x - 3 \overline{)\, 2x^3 + x^2 - 18x - 9} \\ \underline{2x^3 - 6x^2} \qquad\qquad \\ 7x^2 - 18x \qquad \\ \underline{7x^2 - 21x} \qquad \\ 3x - 9 \\ \underline{3x - 9} \\ 0 \end{array}$$

Use long division to find the quotient when dividing by $(x - 3)$.

You can check your division here: $(x - 3)$ is a factor of $2x^3 + x^2 - 18x - 9$, so the remainder must be 0.

$2x^3 + x^2 - 18x - 9 = (x - 3)(2x^2 + 7x + 3)$
$= (x - 3)(2x + 1)(x + 3)$

$2x^2 + 7x + 3$ can also be factorised.

b $0 = (x - 3)(2x + 1)(x + 3)$

Set $y = 0$ to find the points where the curve crosses the x-axis.

So the curve crosses the x-axis at $(3, 0)$, $(-\frac{1}{2}, 0)$ and $(-3, 0)$.

When $x = 0$, $y = (-3)(1)(3) = -9$

Set $x = 0$ to find the y-intercept.

The curve crosses the y-axis at $(0, -9)$.

$x \to \infty, y \to \infty$
$x \to -\infty, y \to -\infty$

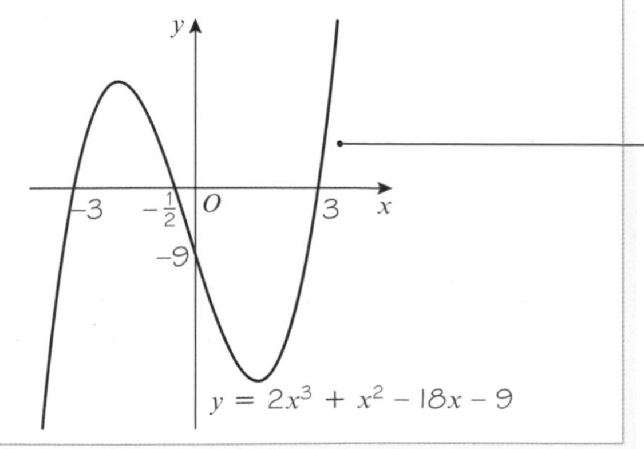

This is a cubic graph with a positive coefficient of x^3 and three distinct roots. You should be familiar with its general shape. ← Section 4.1

Example 7

Given that $(x + 1)$ is a factor of $4x^4 - 3x^2 + a$, find the value of a.

$f(x) = 4x^4 - 3x^2 + a$ — Write the polynomial as a function.

$f(-1) = 0$ — Use statement 2 from the factor theorem. $(x - p)$ is a factor of $f(x)$, so $f(p) = 0$. Here $p = -1$.

$4(-1)^4 - 3(-1)^2 + a = 0$

$4 - 3 + a = 0$

$a = -1$

Substitute $x = -1$ and solve the equation for a. Remember $(-1)^4 = 1$.

Exercise 7C

1 Use the factor theorem to show that:

a $(x - 1)$ is a factor of $4x^3 - 3x^2 - 1$

b $(x + 3)$ is a factor of $5x^4 - 45x^2 - 6x - 18$

c $(x - 4)$ is a factor of $-3x^3 + 13x^2 - 6x + 8$.

2 Show that $(x - 1)$ is a factor of $x^3 + 6x^2 + 5x - 12$ and hence factorise the expression completely.

3 Show that $(x + 1)$ is a factor of $x^3 + 3x^2 - 33x - 35$ and hence factorise the expression completely.

4 Show that $(x - 5)$ is a factor of $x^3 - 7x^2 + 2x + 40$ and hence factorise the expression completely.

5 Show that $(x - 2)$ is a factor of $2x^3 + 3x^2 - 18x + 8$ and hence factorise the expression completely.

6 Each of these expressions has a factor $(x \pm p)$. Find a value of p and hence factorise the expression completely.

a $x^3 - 10x^2 + 19x + 30$

b $x^3 + x^2 - 4x - 4$

c $x^3 - 4x^2 - 11x + 30$

7 i Fully factorise the right-hand side of each equation.

ii Sketch the graph of each equation.

a $y = 2x^3 + 5x^2 - 4x - 3$

b $y = 2x^3 - 17x^2 + 38x - 15$

c $y = 3x^3 + 8x^2 + 3x - 2$

d $y = 6x^3 + 11x^2 - 3x - 2$

e $y = 4x^3 - 12x^2 - 7x + 30$

(P) 8 Given that $(x - 1)$ is a factor of $5x^3 - 9x^2 + 2x + a$, find the value of a.

(P) 9 Given that $(x + 3)$ is a factor of $6x^3 - bx^2 + 18$, find the value of b.

(P) 10 Given that $(x - 1)$ and $(x + 1)$ are factors of $px^3 + qx^2 - 3x - 7$, find the values of p and q.

(P) 11 Given that $(x + 1)$ and $(x - 2)$ are factors of $cx^3 + dx^2 - 9x - 10$, find the values of c and d.

Problem-solving

Use the factor theorem to form simultaneous equations.

(P) 12 Given that $(x + 2)$ and $(x - 3)$ are factors of $gx^3 + hx^2 - 14x + 24$, find the values of g and h.

(E) **13** $f(x) = 3x^3 - 12x^2 + 6x - 24$

a Use the factor theorem to show that $(x - 4)$ is a factor of $f(x)$. **(2 marks)**

b Hence, show that 4 is the only real root of the equation $f(x) = 0$. **(4 marks)**

(E) **14** $f(x) = 4x^3 + 4x^2 - 11x - 6$

a Use the factor theorem to show that $(x + 2)$ is a factor of $f(x)$. **(2 marks)**

b Factorise $f(x)$ completely. **(4 marks)**

c Write down all the solutions of the equation $4x^3 + 4x^2 - 11x - 6 = 0$. **(1 mark)**

(E) **15** **a** Show that $(x - 2)$ is a factor of $9x^4 - 18x^3 - x^2 + 2x$. **(2 marks)**

b Hence, find four real solutions to the equation $9x^4 - 18x^3 - x^2 + 2x = 0$. **(5 marks)**

Challenge

$f(x) = 2x^4 - 5x^3 - 42x^2 - 9x + 54$

a Show that $f(1) = 0$ and $f(-3) = 0$.

b Hence, solve $f(x) = 0$.

7.4 Mathematical proof

A proof is a logical and structured argument to show that a mathematical statement (or **conjecture**) is always true.
A mathematical proof usually starts with previously established mathematical facts (or **theorems**) and then works through a series of logical steps. The final step in a proof is a **statement** of what has been proven.

Notation A statement that has been proven is called a **theorem**.

A statement that has yet to be proven is called a **conjecture**.

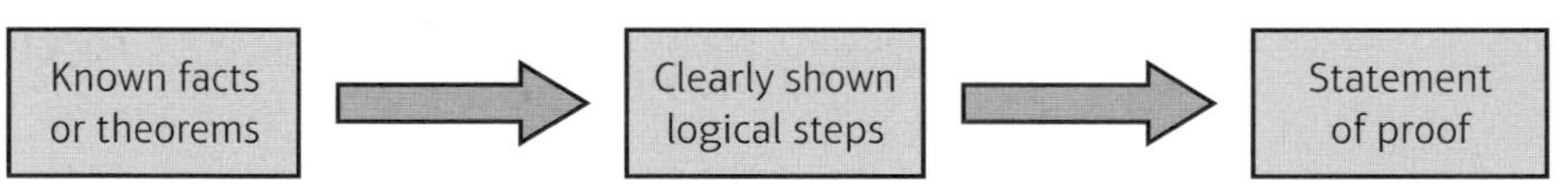

A mathematical proof needs to show that something is true in every case.

- **You can prove a mathematical statement is true by deduction. This means starting from known facts or definitions, then using logical steps to reach the desired conclusion.**

Here is an example of proof by deduction:

Statement: The product of two odd numbers is odd.

Demonstration: $5 \times 7 = 35$, which is odd — This is demonstration but it is not a proof. You have only shown one case.

Proof: p and q are integers, so $2p + 1$ and $2q + 1$ are odd numbers. — You can use $2p + 1$ and $2q + 1$ to represent any odd numbers. If you can show that $(2p + 1) \times (2q + 1)$ is always an odd number then you have proved the statement for all cases.

$$(2p + 1) \times (2q + 1) = 4pq + 2p + 2q + 1$$
$$= 2(2pq + p + q) + 1$$

Since p and q are integers, $2pq + p + q$ is also an integer.

So $2(2pq + p + q) + 1$ is one more than an even number.

So the product of two odd numbers is an odd number. — This is the statement of proof.

- **In a mathematical proof you must**
    - **State any information or assumptions you are using**
    - **Show every step of your proof clearly**
    - **Make sure that every step follows logically from the previous step**
    - **Make sure you have covered all possible cases**
    - **Write a statement of proof at the end of your working**

You need to be able to prove results involving identities, such as $(a + b)(a - b) \equiv a^2 - b^2$

- **To prove an identity you should**
    - **Start with the expression on one side of the identity**
    - **Manipulate that expression algebraically until it matches the other side**
    - **Show every step of your algebraic working**

Notation The symbol $\equiv$ means 'is always equal to'. It shows that two expressions are mathematically **identical**.

Watch out Don't try to 'solve' an identity like an equation. Start from one side and manipulate the expression to match the other side.

Example 8

Prove that $(3x + 2)(x - 5)(x + 7) \equiv 3x^3 + 8x^2 - 101x - 70$

$(3x + 2)(x - 5)(x + 7)$

$= (3x + 2)(x^2 + 2x - 35)$

$= 3x^3 + 6x^2 - 105x + 2x^2 + 4x - 70$

$= 3x^3 + 8x^2 - 101x - 70$

So

$(3x + 2)(x - 5)(x + 7) \equiv 3x^3 + 8x^2 - 101x - 70$

Start with the left-hand side and expand the brackets.

In proof questions you need to show all your working.

Left-hand side = right-hand side.

Example 9

Prove that if $(x - p)$ is a factor of $f(x)$ then $f(p) = 0$.

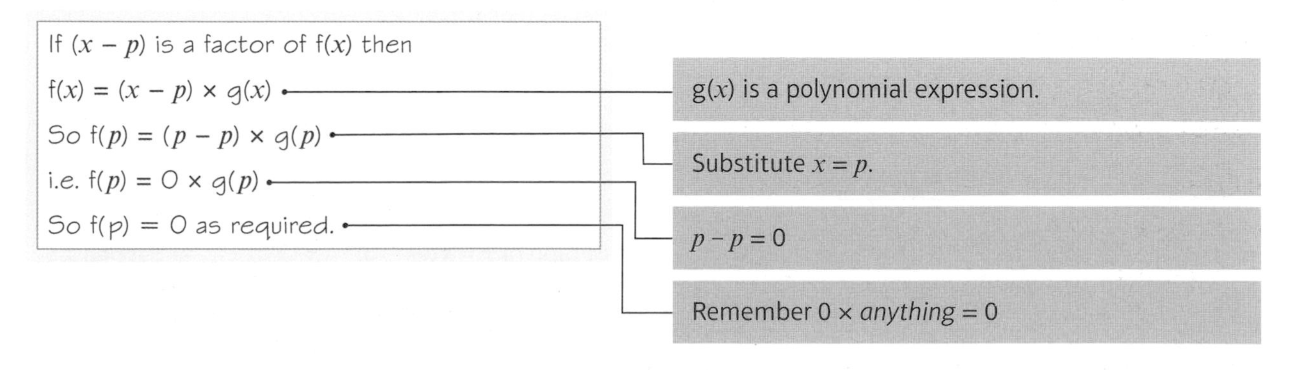

Example 10

Prove that $A(1, 1)$, $B(3, 3)$ and $C(4, 2)$ are the vertices of a right-angled triangle.

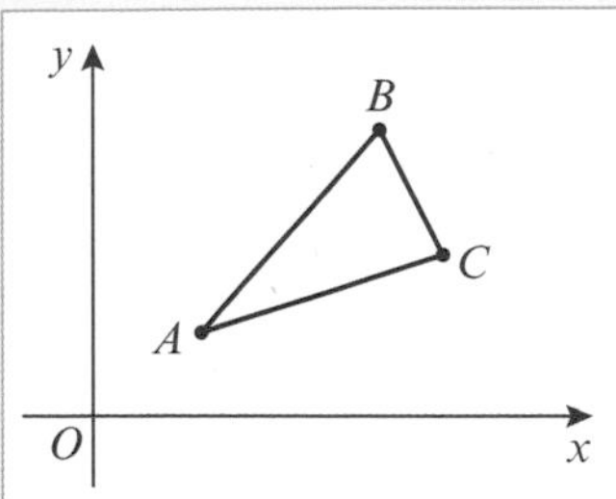

Problem-solving

If you need to prove a geometrical result, it can sometimes help to sketch a diagram as part of your working.

The gradient of line $AB = \frac{3-1}{3-1} = \frac{2}{2} = 1$

The gradient of a line $= \frac{y_2 - y_1}{x_2 - x_1}$

The gradient of line $BC = \frac{2-3}{4-3} = \frac{-1}{1} = -1$

The gradient of line $AC = \frac{2-1}{4-1} = \frac{1}{3}$

The gradients are different so the three points are not collinear.

Hence ABC is a triangle.

Gradient of AB × gradient of $BC = 1 \times (-1)$
$= -1$

If the product of two gradients is –1 then the two lines are perpendicular.
Gradient of line AB × gradient of line $BC = -1$

So AB is perpendicular to BC,
and the triangle is a right-angled triangle.

Remember to state what you have proved.

Example 11

The equation $kx^2 + 3kx + 2 = 0$, where k is a constant, has no real roots.

Prove that k satisfies the inequality $0 \leqslant k < \frac{8}{9}$

$kx^2 + 3kx + 2 = 0$ has no real roots,
so $b^2 - 4ac < 0$

State which assumption or information you are using at each stage of your proof.

$(3k)^2 - 4k(2) < 0$
$9k^2 - 8k < 0$

Use the discriminant. ← Section 2.5

$k(9k - 8) < 0$

Solve this quadratic inequality by sketching the graph of $y = k(9k - 8)$ ← Section 3.5

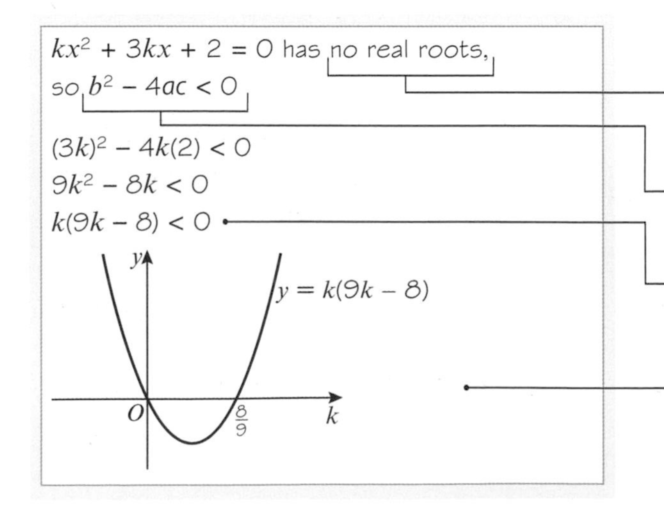

The graph shows that when $k(9k - 8) < 0$,
$0 < k < \frac{8}{9}$

$0 < k < \frac{8}{9}$

When $k = 0$:

$(0)x^2 + 3(0)x + 2 = 0$

$2 = 0$

Which is impossible, so no real roots

So combining these:

$0 \leqslant k < \frac{8}{9}$ as required

Be really careful to consider all the possible situations. You can't use the discriminant if $k = 0$ so look at this case separately.

Write out all of your conclusions clearly.

$0 < k < \frac{8}{9}$ together with $k = 0$, gives $0 \leqslant k < \frac{8}{9}$

Exercise 7D

Hint The proofs in this exercise are all proofs by deduction.

(P) **1** Prove that $n^2 - n$ is an even number for all values of n.

(P) **2** Prove that $\frac{x}{1 + \sqrt{2}} \equiv x\sqrt{2} - x$.

(P) **3** Prove that $(x + \sqrt{y})(x - \sqrt{y}) \equiv x^2 - y$.

(P) **4** Prove that $(2x - 1)(x + 6)(x - 5) \equiv 2x^3 + x^2 - 61x + 30$.

(P) **5** Prove that $x^2 + bx \equiv \left(x + \frac{b}{2}\right)^2 - \left(\frac{b}{2}\right)^2$

(P) **6** Prove that the solutions of $x^2 + 2bx + c = 0$ are $x = -b \pm \sqrt{b^2 - c}$.

(P) **7** Prove that $\left(x - \frac{2}{x}\right)^3 \equiv x^3 - 6x + \frac{12}{x} - \frac{8}{x^3}$

(P) **8** Prove that $\left(x^3 - \frac{1}{x}\right)\left(x^{\frac{3}{2}} + x^{-\frac{5}{2}}\right) \equiv x^{\frac{1}{2}}\left(x^4 - \frac{1}{x^4}\right)$

(P) **9** Use completing the square to prove that $3n^2 - 4n + 10$ is positive for all values of n.

(P) **10** Use completing the square to prove that $-n^2 - 2n - 3$ is negative for all values of n.

Problem-solving

Any expression that is squared must be $\geqslant 0$.

(E/P) **11** Prove that $x^2 + 8x + 20 \geqslant 4$ for all values of x. **(3 marks)**

(E/P) **12** The equation $kx^2 + 5kx + 3 = 0$, where k is a constant, has no real roots. Prove that k satisfies the inequality $0 \leqslant k < \frac{12}{25}$ **(4 marks)**

E/P **13** The equation $px^2 - 5x - 6 = 0$, where p is a constant, has two distinct real roots. Prove that p satisfies the inequality $p > -\frac{25}{24}$ **(4 marks)**

P **14** Prove that $A(3, 1)$, $B(1, 2)$ and $C(2, 4)$ are the vertices of a right-angled triangle.

P **15** Prove that quadrilateral $A(1, 1)$, $B(2, 4)$, $C(6, 5)$ and $D(5, 2)$ is a parallelogram.

P **16** Prove that quadrilateral $A(2, 1)$, $B(5, 2)$, $C(4, -1)$ and $D(1, -2)$ is a rhombus.

P **17** Prove that $A(-5, 2)$, $B(-3, -4)$ and $C(3, -2)$ are the vertices of an isosceles right-angled triangle.

E/P **18** A circle has equation $(x - 1)^2 + y^2 = k$, where $k > 0$.
The straight line L with equation $y = ax$ cuts the circle at two distinct points.
Prove that $k > \frac{a^2}{1 + a^2}$ **(6 marks)**

E/P **19** Prove that the line $4y - 3x + 26 = 0$ is a tangent to the circle $(x + 4)^2 + (y - 3)^2 = 100$. **(5 marks)**

P **20** The diagram shows a square and four congruent right-angled triangles.
Use the diagram to prove that $a^2 + b^2 = c^2$.

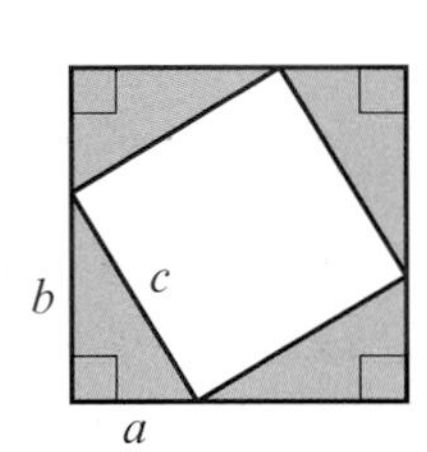

Problem-solving

Find an expression for the area of the large square in terms of a and b.

Challenge

1 Prove that $A(7, 8)$, $B(-1, 8)$, $C(6, 1)$ and $D(0, 9)$ are points on the same circle.

2 Prove that any odd number can be written as the difference of two squares.

7.5 Methods of proof

A mathematical statement can be proved by **exhaustion**. For example, you can prove that the sum of two consecutive square numbers between 100 and 200 is an odd number. The square numbers between 100 and 200 are 121, 144, 169, 196.

$121 + 144 = 265$ which is odd $\quad$ $144 + 169 = 313$ which is odd $\quad$ $169 + 196 = 365$ which is odd

So the sum of two consecutive square numbers between 100 and 200 is an odd number.

- **You can prove a mathematical statement is true by exhaustion. This means breaking the statement into smaller cases and proving each case separately.**

This method is better suited to a small number of results. You cannot use one example to prove a statement is true, as one example is only one case.

Example 12

Prove that all square numbers are either a multiple of 4 or 1 more than a multiple of 4.

For odd numbers:

$(2n + 1)^2 = 4n^2 + 4n + 1 = 4n(n + 1) + 1$

$4n(n + 1)$ is a multiple of 4, so $4n(n + 1) + 1$ is 1 more than a multiple of 4.

For even numbers:

$(2n)^2 = 4n^2$

$4n^2$ is a multiple of 4.

All integers are either odd or even, so all square numbers are either a multiple of 4 or 1 more than a multiple of 4.

Problem-solving

Consider the two cases, odd and even numbers, separately.

You can write any odd number in the form $2n + 1$ where n is a positive integer.

You can write any even number in the form $2n$ where n is a positive integer.

A mathematical statement can be disproved using a **counter-example**. For example, to prove that the statement '$3n + 3$ is a multiple of 6 for all values of n' is not true you can use the counter-example when $n = 2$, as $3 \times 2 + 3 = 9$ and 9 is not a multiple of 6.

- **You can prove a mathematical statement is not true by a counter-example. A counter-example is one example that does not work for the statement. You do not need to give more than one example, as one is sufficient to disprove a statement.**

Example 13

Prove that the following statement is **not** true:

'The sum of two consecutive prime numbers is always even.'

2 and 3 are both prime

$2 + 3 = 5$

5 is odd

So the statement is not true.

You only need one counter-example to show that the statement is false.

Example 14

a Prove that for all positive values of x and y:

$$\frac{x}{y} + \frac{y}{x} \geqslant 2$$

b Use a counter-example to show that this is not true when x and y are not both positive.

Watch out You must always start a proof from **known facts**. Never start your proof with the statement you are trying to prove.

a Jottings:

$\frac{x}{y} + \frac{y}{x} \geqslant 2$

$\frac{x^2 + y^2}{xy} \geqslant 2$

$x^2 + y^2 - 2xy \geqslant 0$

$(x - y)^2 \geqslant 0$

Proof:

Consider $(x - y)^2$

$(x - y)^2 \geqslant 0$

$x^2 + y^2 - 2xy \geqslant 0$

$\frac{x^2 + y^2 - 2xy}{xy} \geqslant 0$

This step is valid because x and y are both positive so $xy > 0$.

$\frac{x}{y} + \frac{y}{x} - 2 \geqslant 0$

$\frac{x}{y} + \frac{y}{x} \geqslant 2$

b Try $x = -3$ and $y = 6$

$\frac{-3}{6} + \frac{6}{-3} = -\frac{1}{2} - 2 = -\frac{5}{2}$

This is not $\geqslant 2$ so the statement is not true.

Problem-solving

Use jottings to get some ideas for a good starting point. These don't form part of your proof, but can give you a clue as to what expression you can consider to begin your proof.

Now you are ready to start your proof. You know that any expression squared is $\geqslant 0$. This is a **known fact** so this is a valid way to begin your proof.

State how you have used the fact that x and y are positive in your proof. If $xy = 0$ you couldn't divide the LHS by xy, and if $xy < 0$, then the direction of the inequality would be reversed.

This was what you wanted to prove so you have finished.

Your working for part **a** tells you that the proof fails when $xy < 0$, so try one positive and one negative value.

Exercise 7E

(P) **1** Prove that when n is an integer and $1 \leqslant n \leqslant 6$, then $m = n + 2$ is not divisible by 10.

Hint You can try each integer for $1 \leqslant n \leqslant 6$.

(P) **2** Prove that every odd integer between 2 and 26 is either prime or the product of two primes.

(P) **3** Prove that the sum of two consecutive square numbers from 1^2 to 8^2 is an odd number.

(E/P) **4** Prove that all cube numbers are either a multiple of 9 or 1 more or 1 less than a multiple of 9. **(4 marks)**

(P) **5** Find a counter-example to disprove each of the following statements:

a If n is a positive integer then $n^4 - n$ is divisible by 4.

b Integers always have an even number of factors.

c $2n^2 - 6n + 1$ is positive for all values of n.

d $2n^2 - 2n - 4$ is a multiple of 3 for all integer values of n.

(E/P) **6** A student is trying to prove that $x^3 + y^3 < (x + y)^3$.

The student writes:

> $(x + y)^3 = x^3 + 3x^2y + 3xy^2 + y^3$
> which is less than $x^3 + y^3$ since
> $3x^2y + 3xy^2 > 0$

Problem-solving For part **b** you need to write down suitable values of x and y and show that they do not satisfy the inequality.

a Identify the error made in the proof. **(1 mark)**

b Provide a counter-example to show that the statement is not true. **(2 marks)**

(E/P) **7** Prove that for all real values of x

$$(x + 6)^2 \geqslant 2x + 11$$

(3 marks)

(E/P) **8** Given that a is a positive real number, prove that:

$$a + \frac{1}{a} \geqslant 2$$

Watch out Remember to state how you use the condition that a is positive.

(2 marks)

(E/P) **9 a** Prove that for any positive numbers p and q:

$$p + q \geqslant \sqrt{4pq}$$

(3 marks)

b Show, by means of a counter-example, that this inequality does not hold when p and q are both negative. **(2 marks)**

Problem-solving Use jottings and work backwards to work out what expression to consider.

(E/P) **10** It is claimed that the following inequality is true for all negative numbers x and y:

$$x + y \geqslant \sqrt{x^2 + y^2}$$

The following proof is offered by a student:

> $x + y \geqslant \sqrt{x^2 + y^2}$
> $(x + y)^2 \geqslant x^2 + y^2$
> $x^2 + y^2 + 2xy \geqslant x^2 + y^2$
> $2xy > 0$ which is true because x and y are both negative, so xy is positive.

a Explain the error made by the student. **(2 marks)**

b By use of a counter-example, verify that the inequality is not satisfied if both x and y are negative. **(1 mark)**

c Prove that this inequality is true if x and y are both positive. **(2 marks)**

Mixed exercise 7

1 Simplify these fractions as far as possible:

a $\frac{3x^4 - 21x}{3x}$

b $\frac{x^2 - 2x - 24}{x^2 - 7x + 6}$

c $\frac{2x^2 + 7x - 4}{2x^2 + 9x + 4}$

2 Divide $3x^3 + 12x^2 + 5x + 20$ by $(x + 4)$.

3 Simplify $\frac{2x^3 + 3x + 5}{x + 1}$

(E) 4 a Show that $(x - 3)$ is a factor of $2x^3 - 2x^2 - 17x + 15$. **(2 marks)**

b Hence express $2x^3 - 2x^2 - 17x + 15$ in the form $(x - 3)(Ax^2 + Bx + C)$, where the values A, B and C are to be found. **(3 marks)**

(E) 5 a Show that $(x - 2)$ is a factor of $x^3 + 4x^2 - 3x - 18$. **(2 marks)**

b Hence express $x^3 + 4x^2 - 3x - 18$ in the form $(x - 2)(px + q)^2$, where the values p and q are to be found. **(4 marks)**

(E) 6 Factorise completely $2x^3 + 3x^2 - 18x + 8$. **(6 marks)**

(E/P) 7 Find the value of k if $(x - 2)$ is a factor of $x^3 - 3x^2 + kx - 10$. **(4 marks)**

(E/P) 8 $f(x) = 2x^2 + px + q$. Given that $f(-3) = 0$, and $f(4) = 21$:

a find the value of p and q **(6 marks)**

b factorise $f(x)$. **(3 marks)**

(E/P) 9 $h(x) = x^3 + 4x^2 + rx + s$. Given $h(-1) = 0$, and $h(2) = 30$:

a find the values of r and s **(6 marks)**

b factorise $h(x)$. **(3 marks)**

(E) 10 $g(x) = 2x^3 + 9x^2 - 6x - 5$.

a Factorise $g(x)$. **(6 marks)**

b Solve $g(x) = 0$. **(2 marks)**

(E) **11** **a** Show that $(x - 2)$ is a factor of $f(x) = x^3 + x^2 - 5x - 2$. **(2 marks)**

b Hence, or otherwise, find the exact solutions of the equation $f(x) = 0$. **(4 marks)**

(E) **12** Given that -1 is a root of the equation $2x^3 - 5x^2 - 4x + 3$, find the two positive roots. **(4 marks)**

(E) **13** $f(x) = x^3 - 2x^2 - 19x + 20$

a Show that $(x + 4)$ is a factor of $f(x)$. **(3 marks)**

b Hence, or otherwise, find all the solutions to the equation $x^3 - 2x^2 - 19x + 20 = 0$. **(4 marks)**

(E) **14** $f(x) = 6x^3 + 17x^2 - 5x - 6$

a Show that $f(x) = (3x - 2)(ax^2 + bx + c)$, where a, b and c are constants to be found. **(2 marks)**

b Hence factorise $f(x)$ completely. **(4 marks)**

c Write down all the real roots of the equation $f(x) = 0$. **(2 marks)**

15 Prove that $\dfrac{x - y}{\sqrt{x} - \sqrt{y}} \equiv \sqrt{x} + \sqrt{y}$.

(P) **16** Use completing the square to prove that $n^2 - 8n + 20$ is positive for all values of n.

(P) **17** Prove that the quadrilateral $A(1, 1)$, $B(3, 2)$, $C(4, 0)$ and $D(2, -1)$ is a square.

(P) **18** Prove that the sum of two consecutive positive odd numbers less than ten gives an even number.

(P) **19** Prove that the statement '$n^2 - n + 3$ is a prime number for all values of n' is untrue.

(P) **20** Prove that $\left(x - \dfrac{1}{x}\right)\left(x^{\frac{4}{3}} + x^{-\frac{2}{3}}\right) \equiv x^{\frac{1}{3}}\left(x^2 - \dfrac{1}{x^2}\right)$.

(P) **21** Prove that $2x^3 + x^2 - 43x - 60 \equiv (x + 4)(x - 5)(2x + 3)$.

(E) **22** The equation $x^2 - kx + k = 0$, where k is a positive constant, has two equal roots. Prove that $k = 4$. **(3 marks)**

(P) **23** Prove that the distance between opposite edges of a regular hexagon of side length $\sqrt{3}$ is a rational value.

(P) **24** **a** Prove that the difference of the squares of two consecutive even numbers is always divisible by 4.

b Is this statement true for odd numbers? Give a reason for your answer.

(E) **25** A student is trying to prove that $1 + x^2 < (1 + x)^2$.

The student writes:

$(1 + x)^2 = 1 + 2x + x^2$.
So $1 + x^2 < 1 + 2x + x^2$.

a Identify the error made in the proof. **(1 mark)**

b Provide a counter-example to show that the statement is not true. **(2 marks)**

Challenge

1 The diagram shows two squares and a circle.

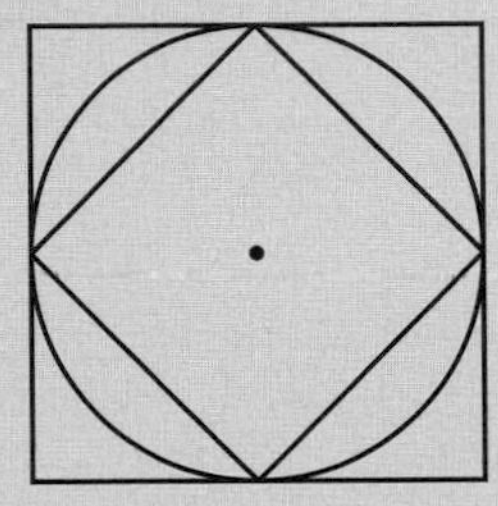

a Given that π is defined as the circumference of a circle of diameter 1 unit, prove that $2\sqrt{2} < \pi < 4$.

b By similarly constructing regular hexagons inside and outside a circle, prove that $3 < \pi < 2\sqrt{3}$.

2 Prove that if $f(x) = ax^3 + bx^2 + cx + d$ and $f(p) = 0$, then $(x - p)$ is a factor of $f(x)$.

Summary of key points

1. When simplifying an algebraic fraction, factorise the numerator and denominator where possible and then cancel common factors.

2. You can use long division to divide a polynomial by $(x \pm p)$, where p is a constant.

3. The **factor theorem** states that if $f(x)$ is a polynomial then:
   - If $f(p) = 0$, then $(x - p)$ is a factor of $f(x)$
   - If $(x - p)$ is a factor of $f(x)$, then $f(p) = 0$

4. You can prove a mathematical statement is true by **deduction**. This means starting from known facts or definitions, then using logical steps to reach the desired conclusion.

5. In a mathematical proof you must
   - State any information or assumptions you are using
   - Show every step of your proof clearly
   - Make sure that every step follows logically from the previous step
   - Make sure you have covered all possible cases
   - Write a statement of proof at the end of your working

6. To prove an identity you should
   - Start with the expression on one side of the identity
   - Manipulate that expression algebraically until it matches the other side
   - Show every step of your algebraic working

7. You can prove a mathematical statement is true by **exhaustion**. This means breaking the statement into smaller cases and proving each case separately.

8. You can prove a mathematical statement is not true by a **counter-example**. A counter-example is one example that does not work for the statement. You do not need to give more than one example, as one is sufficient to disprove a statement.

8 The binomial expansion

Objectives

After completing this chapter you should be able to:

- Use Pascal's triangle to identify binomial coefficients and use them to expand simple binomial expressions → pages 159–161
- Use combinations and factorial notation → pages 161–163
- Use the binomial expansion to expand brackets → pages 163–165
- Find individual coefficients in a binomial expansion → pages 165–167
- Make approximations using the binomial expansion → pages 167–169

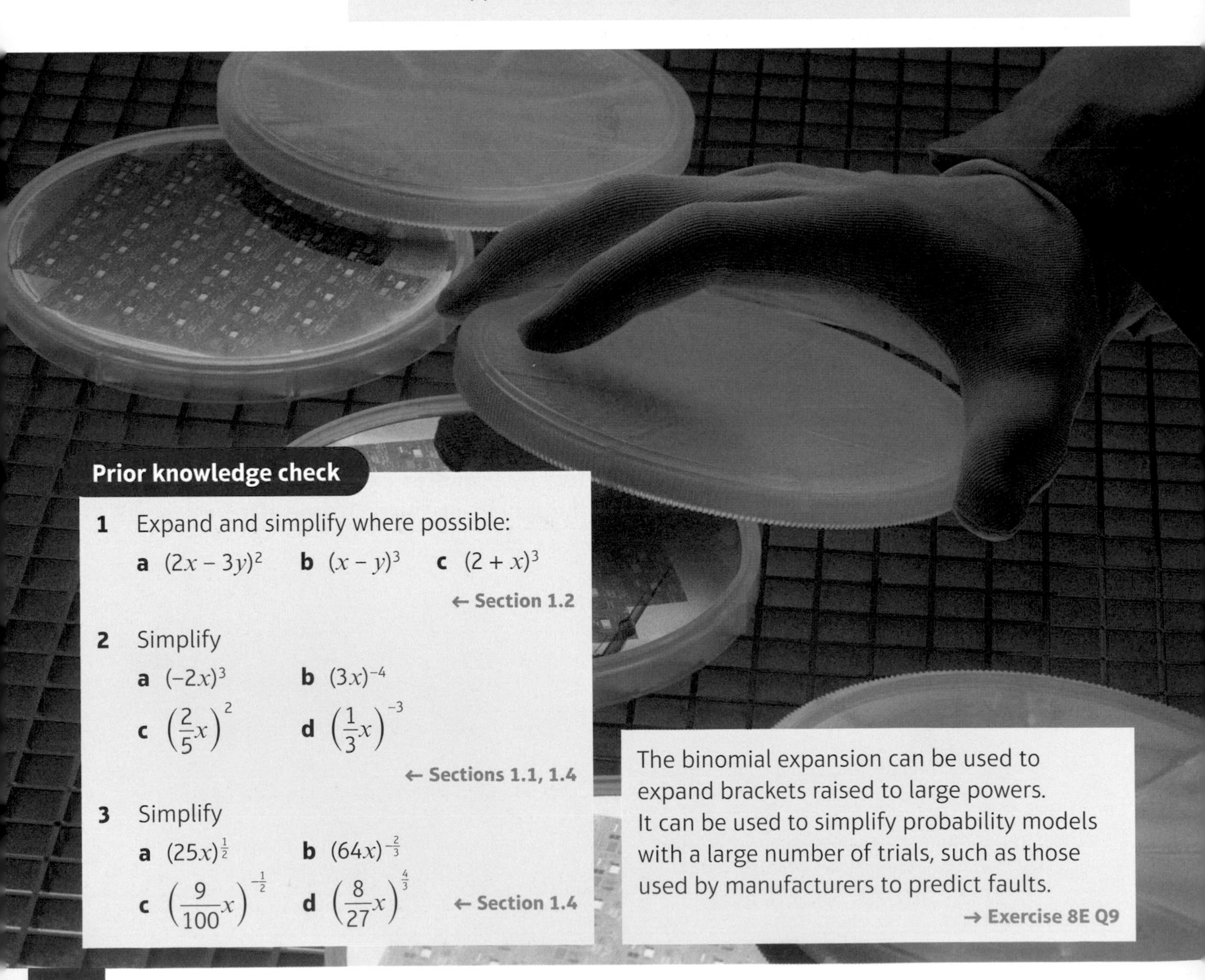

Prior knowledge check

1 Expand and simplify where possible:

a $(2x - 3y)^2$ **b** $(x - y)^3$ **c** $(2 + x)^3$

← Section 1.2

2 Simplify

a $(-2x)^3$ **b** $(3x)^{-4}$

c $\left(\frac{2}{5}x\right)^2$ **d** $\left(\frac{1}{3}x\right)^{-3}$

← Sections 1.1, 1.4

3 Simplify

a $(25x)^{\frac{1}{2}}$ **b** $(64x)^{-\frac{2}{3}}$

c $\left(\frac{9}{100}x\right)^{-\frac{1}{2}}$ **d** $\left(\frac{8}{27}x\right)^{\frac{4}{3}}$

← Section 1.4

The binomial expansion can be used to expand brackets raised to large powers. It can be used to simplify probability models with a large number of trials, such as those used by manufacturers to predict faults.

→ Exercise 8E Q9

8.1 Pascal's triangle

You can use **Pascal's triangle** to quickly expand expressions such as $(x + 2y)^3$.

Consider the expansions of $(a + b)^n$ for $n = 0, 1, 2, 3$ and 4:

$$(a + b)^0 = 1$$
$$(a + b)^1 = 1a + 1b$$
$$(a + b)^2 = 1a^2 + 2ab + 1b^2$$
$$(a + b)^3 = 1a^3 + 3a^2b + 3ab^2 + 1b^3$$
$$(a + b)^4 = 1a^4 + 4a^3b + 6a^2b^2 + 4ab^3 + 1b^4$$

Each coefficient is the sum of the two coefficients immediately above it.

Every term in the expansion of $(a + b)^n$ has total index n:
In the $6a^2b^2$ term the total index is $2 + 2 = 4$.
In the $4ab^3$ term the total index is $1 + 3 = 4$.

The coefficients in the expansions form a pattern that is known as Pascal's triangle.

- **Pascal's triangle is formed by adding adjacent pairs of numbers to find the numbers on the next row.**

Here are the first 7 rows of Pascal's triangle:

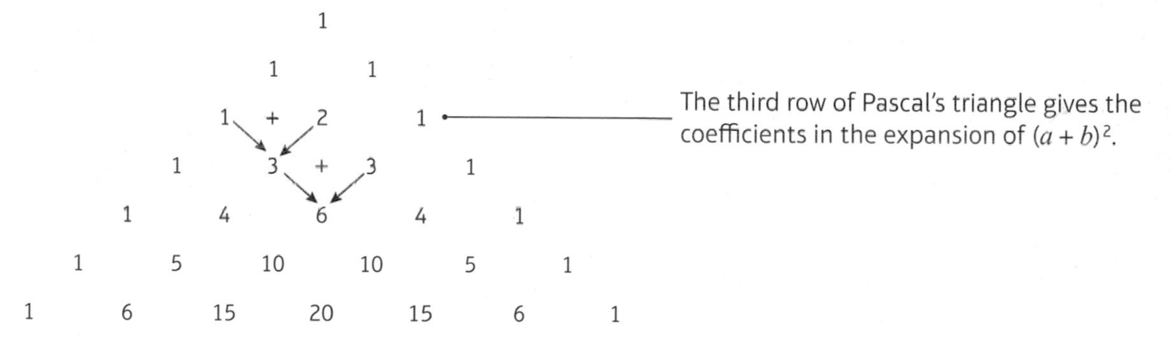

- **The $(n + 1)$th row of Pascal's triangle gives the coefficients in the expansion of $(a + b)^n$.**

Example 1

Use Pascal's triangle to find the expansions of:

a $(x + 2y)^3$ **b** $(2x - 5)^4$

a $(x + 2y)^3$

The coefficients are 1, 3, 3, 1 so:

$(x + 2y)^3 = 1x^3 + 3x^2(2y) + 3x(2y)^2 + 1(2y)^3$
$= x^3 + 6x^2y + 12xy^2 + 8y^3$

Index = 3 so look at the 4th row of Pascal's triangle to find the coefficients.

This is the expansion of $(a + b)^3$ with $a = x$ and $b = 2y$. Use brackets to make sure you don't make a mistake. $(2y)^2 = 4y^2$.

b $(2x - 5)^4$

The coefficients are 1, 4, 6, 4, 1 so:

$(2x - 5)^4 = 1(2x)^4 + 4(2x)^3(-5)^1 + 6(2x)^2(-5)^2 + 4(2x)^1(-5)^3 + 1(-5)^4$

$= 16x^4 - 160x^3 + 600x^2 - 1000x + 625$

Index = 4 so look at the 5th row of Pascal's triangle.

This is the expansion of $(a + b)^4$ with $a = 2x$ and $b = -5$.

Be careful with the negative numbers.

Example 2

The coefficient of x^2 in the expansion of $(2 - cx)^3$ is 294. Find the possible value(s) of the constant c.

Problem-solving

If there is an unknown in the original expression, you might be able to form an equation involving that unknown.

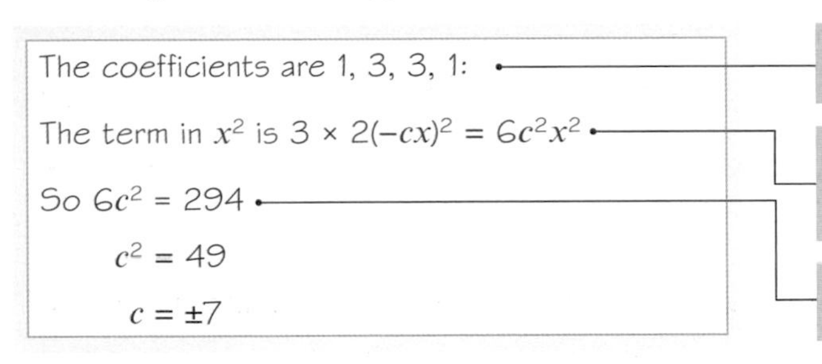

The coefficients are 1, 3, 3, 1:

The term in x^2 is $3 \times 2(-cx)^2 = 6c^2x^2$

So $6c^2 = 294$

$c^2 = 49$

$c = \pm 7$

Index = 3 so use the 4th row of Pascal's triangle.

From the expansion of $(a + b)^3$ the x^2 term is $3ab^2$ where $a = 2$ and $b = -cx$.

Form and solve an equation in c.

Exercise 8A

1 State the row of Pascal's triangle that would give the coefficients of each expansion:

a $(x + y)^3$ **b** $(3x - 7)^{15}$ **c** $\left(2x + \frac{1}{2}\right)^n$ **d** $(y - 2x)^{n+4}$

2 Write down the expansion of:

a $(x + y)^4$ **b** $(p + q)^5$ **c** $(a - b)^3$ **d** $(x + 4)^3$

e $(2x - 3)^4$ **f** $(a + 2)^5$ **g** $(3x - 4)^4$ **h** $(2x - 3y)^4$

3 Find the coefficient of x^3 in the expansion of:

a $(4 + x)^4$ **b** $(1 - x)^5$ **c** $(3 + 2x)^3$ **d** $(4 + 2x)^5$

e $(2 + x)^6$ **f** $\left(4 - \frac{1}{2}x\right)^4$ **g** $(x + 2)^5$ **h** $(3 - 2x)^4$

(P) **4** Fully expand the expression $(1 + 3x)(1 + 2x)^3$.

Problem-solving

Expand $(1 + 2x)^3$, then multiply each term by 1 and by $3x$.

(P) **5** Expand $(2 + y)^3$. Hence or otherwise, write down the expansion of $(2 + x - x^2)^3$ in ascending powers of x.

(P) **6** The coefficient of x^2 in the expansion of $(2 + ax)^3$ is 54. Find the possible values of the constant a.

(P) **7** The coefficient of x^2 in the expansion of $(2 - x)(3 + bx)^3$ is 45. Find possible values of the constant b.

(P) **8** Work out the coefficient of x^2 in the expansion of $(p - 2x)^3$. Give your answer in terms of p.

(P) **9** After 5 years, the value of an investment of £500 at an interest rate of X% per annum is given by:

$$500\left(1 + \frac{X}{100}\right)^5$$

Find an approximation for this expression in the form $A + BX + CX^2$, where A, B and C are constants to be found. You can ignore higher powers of X.

Challenge

Find the constant term in the expansion of $\left(x^2 - \frac{1}{2x}\right)^3$.

8.2 Factorial notation

You can use combinations and factorial notation to help you expand binomial expressions. For larger indices, it is quicker than using Pascal's triangle.

Using **factorial notation** $3 \times 2 \times 1 = 3!$

Notation You say 'n factorial'. By definition, $0! = 1$.

- **You can use factorial notation and your calculator to find entries in Pascal's triangle quickly.**
  - **The number of ways of choosing r items from a group of n items is written as ^nC_r or $\binom{n}{r}$:**
    $$^nC_r = \binom{n}{r} = \frac{n!}{r!(n-r)!}$$
  - **The rth entry in the nth row of Pascal's triangle is given by $^{n-1}C_{r-1} = \binom{n-1}{r-1}$**

Notation You can say 'n choose r' for ^nC_r. It is sometimes written without superscripts and subscripts as nCr.

Example 3

Calculate:

a $5!$ **b** $^5C_2$ **c** the 6th entry in the 10th row of Pascal's triangle

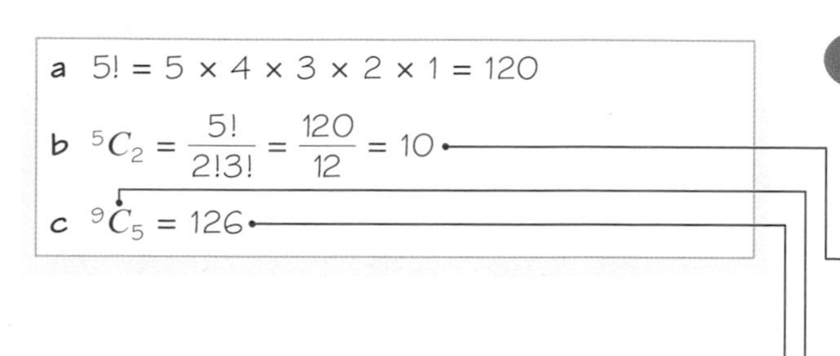

a $5! = 5 \times 4 \times 3 \times 2 \times 1 = 120$

b $^5C_2 = \frac{5!}{2!3!} = \frac{120}{12} = 10$

c $^9C_5 = 126$

Online Use the ^nC_r and ! functions on your calculator to answer this question.

You can calculate $^5C_2$ by using the ^nC_r function on your calculator.
$^nC_r = \frac{n!}{r!(n-r)!} = \frac{5!}{2!(5-2)!}$

The rth entry in the nth row is $^{n-1}C_{r-1}$.

In the expansion of $(a + b)^9$ this would give the term $126a^4b^5$.

Exercise 8B

1 Work out:

a 4! **b** 9! **c** $\frac{10!}{7!}$ **d** $\frac{15!}{13!}$

2 Without using a calculator, work out:

a $\binom{4}{2}$ **b** $\binom{6}{4}$ **c** ${}^6C_3$ **d** $\binom{5}{4}$ **e** ${}^{10}C_8$ **f** $\binom{9}{5}$

3 Use a calculator to work out:

a $\binom{15}{6}$ **b** ${}^{10}C_7$ **c** $\binom{20}{10}$ **d** $\binom{20}{17}$ **e** ${}^{14}C_9$ **f** ${}^{18}C_5$

4 Write each value a to d from Pascal's triangle using ${}^nC_r$ notation:

1
1 1
1 2 1
1 3 3 1
1 a 6 4 1
1 5 b 10 5 1
1 6 c d 15 6 1

5 Work out the 5th number on the 12th row from Pascal's triangle.

6 The 11th row of Pascal's triangle is shown below.

1 10 45 … …

a Find the next two values in the row.

b Hence find the coefficient of x^3 in the expansion of $(1 + 2x)^{10}$.

7 The 14th row of Pascal's triangle is shown below.

1 13 78 … …

a Find the next two values in the row.

b Hence find the coefficient of x^4 in the expansion of $(1 + 3x)^{13}$.

8 The probability of throwing exactly 10 heads when a fair coin is tossed 20 times is given by $\binom{20}{10}0.5^{20}$. Calculate the probability and describe the likelihood of this occurring.

(P) **9** Show that:

a ${}^nC_1 = n$ **b** ${}^nC_2 = \frac{n(n-1)}{2}$

(E) **10** Given that $\binom{50}{13} = \frac{50!}{13!a!}$, write down the value of a. **(1 mark)**

(E) **11** Given that $\binom{35}{p} = \frac{35!}{p!18!}$, write down the value of p. **(1 mark)**

Challenge

a Work out ${}^{10}C_3$ and ${}^{10}C_7$

b Work out ${}^{14}C_5$ and ${}^{14}C_9$

c What do you notice about your answers to parts **a** and **b**?

d Prove that ${}^nC_r = {}^nC_{n-r}$

8.3 The binomial expansion

A binomial expression has two terms. The binomial expansion allows you to expand powers of binomial expressions. For example, in the expansion of $(a + b)^5 = (a + b)(a + b)(a + b)(a + b)(a + b)$ the term a^2b^3 occurs $\binom{5}{3}$ times. This is because you need to choose b 3 times from the 5 brackets. You can do this in $\binom{5}{3}$ ways so when the expansion is simplified, the term in a^2b^3 is $\binom{5}{3}a^2b^3$.

- **The binomial expansion is:**

$$(a+b)^n = a^n + \binom{n}{1}a^{n-1}b + \binom{n}{2}a^{n-2}b^2 + \ldots + \binom{n}{r}a^{n-r}b^r + \ldots + b^n \quad (n \in \mathbb{N})$$

where $\binom{n}{r} = {}^nC_r = \dfrac{n!}{r!(n-r)!}$

Notation $n \in \mathbb{N}$ means that n must be a member of the **natural numbers**. This is all the **positive integers**.

Example 4

Use the binomial theorem to find the expansion of $(3 - 2x)^5$.

$$(3-2x)^5 = 3^5 + \binom{5}{1}3^4(-2x) + \binom{5}{2}3^3(-2x)^2 + \binom{5}{3}3^2(-2x)^3 + \binom{5}{4}3^1(-2x)^4 + (-2x)^5$$

$$= 243 - 810x + 1080x^2 - 720x^3 + 240x^4 - 32x^5$$

There will be 6 terms.
Each term has a total index of 5.
Use $(a + b)^n$ where $a = 3$, $b = -2x$ and $n = 5$.

There are $\binom{5}{2}$ ways of choosing two '$-2x$' terms from five brackets.

Online Work out each coefficient quickly using the ${}^nC_r$ and power functions on your calculator.

Example 5

Find the first four terms in the binomial expansion of:

a $(1 + 2x)^{10}$ **b** $\left(10 - \frac{1}{2}x\right)^6$

a $(1 + 2x)^{10}$

$= 1^{10} + \binom{10}{1}1^9(2x) + \binom{10}{2}1^8(2x)^2 + \binom{10}{3}1^7(2x)^3 + \ldots$

$= 1 + 20x + 180x^2 + 960x^3 + \ldots$

b $\left(10 - \frac{1}{2}x\right)^6$

$= 10^6 + \binom{6}{1}10^5\left(-\frac{1}{2}x\right) + \binom{6}{2}10^4\left(-\frac{1}{2}x\right)^2 + \binom{6}{3}10^3\left(-\frac{1}{2}x\right)^3 + \ldots$

$= 1000000 - 300000x + 37500x^2 - 2500x^3 + \ldots$

Notation This is sometimes called the expansion in **ascending powers of** x.

Write each coefficient in its simplest form.

Exercise 8C

1 Write down the expansion of the following:

a $(1 + x)^4$ **b** $(3 + x)^4$ **c** $(4 - x)^4$ **d** $(x + 2)^6$ **e** $(1 + 2x)^4$ **f** $\left(1 - \frac{1}{2}x\right)^4$

2 Use the binomial theorem to find the first four terms in the expansion of:

a $(1 + x)^{10}$ **b** $(1 - 2x)^5$ **c** $(1 + 3x)^6$ **d** $(2 - x)^8$ **e** $\left(2 - \frac{1}{2}x\right)^{10}$ **f** $(3 - x)^7$

3 Use the binomial theorem to find the first four terms in the expansion of:

a $(2x + y)^6$ **b** $(2x + 3y)^5$ **c** $(p - q)^8$ **d** $(3x - y)^6$ **e** $(x + 2y)^8$ **f** $(2x - 3y)^9$

4 Use the binomial expansion to find the first four terms, in ascending powers of x, of:

a $(1 + x)^8$ **b** $(1 - 2x)^6$ **c** $\left(1 + \frac{x}{2}\right)^{10}$

d $(1 - 3x)^5$ **e** $(2 + x)^7$ **f** $(3 - 2x)^3$

g $(2 - 3x)^6$ **h** $(4 + x)^4$ **i** $(2 + 5x)^7$

Hint Your answers should be in the form $a + bx + cx^2 + dx^3$ where a, b, c and d are numbers.

(E) 5 Find the first 3 terms, in ascending powers of x, of the binomial expansion of $(2 - x)^6$ and simplify each term. **(4 marks)**

(E) 6 Find the first 3 terms, in ascending powers of x, of the binomial expansion of $(3 - 2x)^5$ giving each term in its simplest form. **(4 marks)**

(E/P) 7 Find the binomial expansion of $\left(x + \frac{1}{x}\right)^5$ giving each term in its simplest form. **(4 marks)**

Challenge

a Show that $(a + b)^4 - (a - b)^4 = 8ab(a^2 + b^2)$.

b Given that $82\,896 = 17^4 - 5^4$, write 82 896 as a product of its prime factors.

8.4 Solving binomial problems

You can use the general term of the binomial expansion to find individual coefficients in a binomial expansion.

- **In the expansion of $(a + b)^n$ the general term is given by $\binom{n}{r}a^{n-r}b^r$.**

Example 6

a Find the coefficient of x^4 in the binomial expansion of $(2 + 3x)^{10}$.

b Find the coefficient of x^3 in the binomial expansion of $(2 + x)(3 - 2x)^7$.

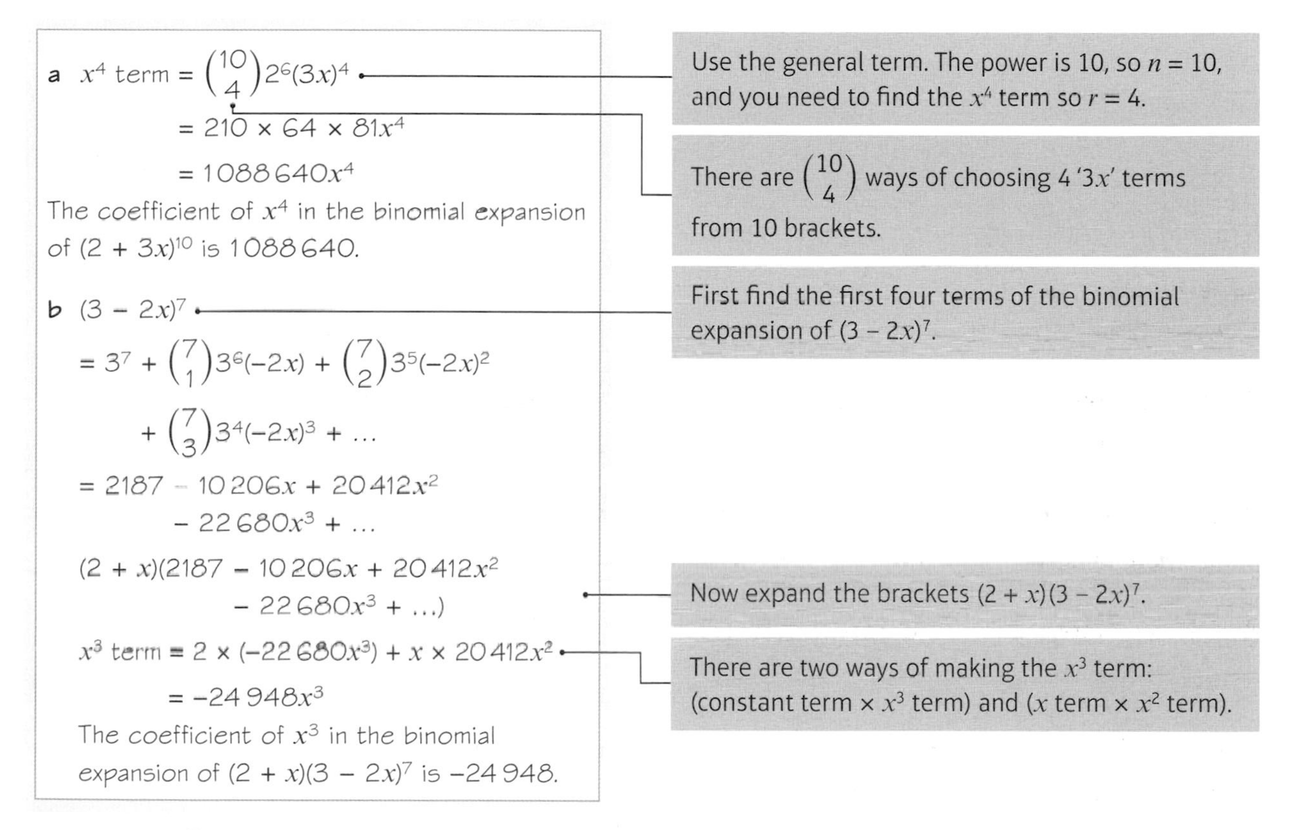

a x^4 term $= \binom{10}{4}2^6(3x)^4$

$= 210 \times 64 \times 81x^4$

$= 1088640x^4$

The coefficient of x^4 in the binomial expansion of $(2 + 3x)^{10}$ is 1088640.

b $(3 - 2x)^7$

$= 3^7 + \binom{7}{1}3^6(-2x) + \binom{7}{2}3^5(-2x)^2 + \binom{7}{3}3^4(-2x)^3 + \ldots$

$= 2187 - 10206x + 20412x^2 - 22680x^3 + \ldots$

$(2 + x)(2187 - 10206x + 20412x^2 - 22680x^3 + \ldots)$

x^3 term $= 2 \times (-22680x^3) + x \times 20412x^2$

$= -24948x^3$

The coefficient of x^3 in the binomial expansion of $(2 + x)(3 - 2x)^7$ is -24948.

Use the general term. The power is 10, so $n = 10$, and you need to find the x^4 term so $r = 4$.

There are $\binom{10}{4}$ ways of choosing 4 '$3x$' terms from 10 brackets.

First find the first four terms of the binomial expansion of $(3 - 2x)^7$.

Now expand the brackets $(2 + x)(3 - 2x)^7$.

There are two ways of making the x^3 term: (constant term × x^3 term) and (x term × x^2 term).

Example 7

$g(x) = (1 + kx)^{10}$, where k is a constant.

Given that the coefficient of x^3 in the binomial expansion of $g(x)$ is 15, find the value of k.

x^3 term $= \binom{10}{3}1^7(kx)^3 = 15x^3$

$120k^3x^3 = 15x^3$

$k = \frac{1}{2}$

$a = 1$, $b = kx$, $n = 10$ and $r = 3$.

$k^3x^3 = \frac{15}{120}x^3$

$k^3x^3 = \frac{1}{8}x^3$

$k^3 = \frac{1}{8}$, $k = \sqrt[3]{\frac{1}{8}}$

Example 8

a Write down the first three terms, in ascending powers of x, of the binomial expansion of $(1 + qx)^8$, where q is a non-zero constant.

b Given that, in the expansion of $(1 + qx)^8$, the coefficient of x is $-r$ and the coefficient of x^2 is $7r$, find the value of q and the value of r.

a $(1 + qx)^8$

$= 1^8 + \binom{8}{1}1^7(qx)^1 + \binom{8}{2}1^6(qx)^2 + \ldots$

$= 1 + 8qx + 28q^2x^2 + \ldots$

b $8q = -r$ and $28q^2 = 7r$

$8q = -4q^2$ — Using $28q^2 = 7r$, $r = 4q^2$ and $-r = -4q^2$.

$4q^2 + 8q = 0$

$4q(q + 2) = 0$

$q = -2$, $r = 16$ — q is non-zero so $q = -2$.

Problem-solving

There are two unknowns in this expression. Your expansion will be in terms of q and x.

Exercise 8D

1 Find the coefficient of x^3 in the binomial expansion of:

a $(3 + x)^5$ **b** $(1 + 2x)^5$ **c** $(1 - x)^6$ **d** $(3x + 2)^5$

e $(1 + x)^{10}$ **f** $(3 - 2x)^6$ **g** $(1 + x)^{20}$ **h** $(4 - 3x)^7$

i $\left(1 - \frac{1}{2}x\right)^6$ **j** $\left(3 + \frac{1}{2}x\right)^7$ **k** $\left(2 - \frac{1}{2}x\right)^8$ **l** $\left(5 + \frac{1}{4}x\right)^5$

(P) **2** The coefficient of x^2 in the expansion of $(2 + ax)^6$ is 60. Find two possible values of the constant a.

Problem-solving

$a = 2$, $b = ax$, $n = 6$. Use brackets when you substitute ax.

(P) **3** The coefficient of x^3 in the expansion of $(3 + bx)^5$ is -720. Find the value of the constant b.

(P) **4** The coefficient of x^3 in the expansion of $(2 + x)(3 - ax)^4$ is 30. Find the three possible values of the constant a.

(E/P) **5** When $(1 - 2x)^p$ is expanded, the coefficient of x^2 is 40. Given that $p > 0$, use this information to find:

a the value of the constant p **(6 marks)**

b the coefficient of x **(1 mark)**

c the coefficient of x^3 **(2 marks)**

Problem-solving

You will need to use the definition of $\binom{n}{r}$ to find an expression for $\binom{p}{2}$.

(E/P) **6** **a** Find the first three terms, in ascending powers of x, of the binomial expansion of $(5 + px)^{30}$, where p is a non-zero constant. **(2 marks)**

b Given that in this expansion the coefficient of x^2 is 29 times the coefficient of x work out the value of p. **(4 marks)**

(E/P) **7** **a** Find the first four terms, in ascending powers of x, of the binomial expansion of $(1 + qx)^{10}$, where q is a non-zero constant. **(2 marks)**

b Given that in the expansion of $(1 + qx)^{10}$ the coefficient of x^3 is 108 times the coefficient of x, work out the value of q. **(4 marks)**

(E/P) **8** **a** Find the first three terms, in ascending powers of x of the binomial expansion of $(1 + px)^{11}$, where p is a constant. **(2 marks)**

b The first 3 terms in the same expansion are 1, $77x$ and qx^2, where q is a constant. Find the value of p and the value of q. **(4 marks)**

(E/P) **9** **a** Write down the first three terms, in ascending powers of x, of the binomial expansion of $(1 + px)^{15}$, where p is a non-zero constant. **(2 marks)**

b Given that, in the expansion of $(1 + px)^{15}$, the coefficient of x is $(-q)$ and the coefficient of x^2 is $5q$, find the value of p and the value of q. **(4 marks)**

(E/P) **10** In the binomial expansion of $(1 + x)^{30}$, the coefficients of x^9 and x^{10} are p and q respectively. Find the value of $\frac{q}{p}$. **(4 marks)**

Challenge

Find the coefficient of x^4 in the binomial expansion of: **a** $(3 - 2x^2)^9$ **b** $\left(\frac{5}{x} + x^2\right)^8$

8.5 Binomial estimation

In engineering and science, it is often useful to find simple **approximations** for complicated functions. If the value of x is less than 1, then x^n gets smaller as n gets larger. If x is small you can sometimes **ignore large powers** of x to approximate a function or estimate a value.

Example 9

a Find the first four terms of the binomial expansion, in ascending powers of x, of $\left(1 - \frac{x}{4}\right)^{10}$.

b Use your expansion to estimate the value of 0.975^{10}, giving your answer to 4 decimal places.

a $\left(1 - \frac{x}{4}\right)^{10}$

$= 1^{10} + \binom{10}{1}1^9\left(-\frac{x}{4}\right) + \binom{10}{2}1^8\left(-\frac{x}{4}\right)^2 + \binom{10}{3}1^7\left(-\frac{x}{4}\right)^3 + \ldots$

$= 1 - 2.5x + 2.8125x^2 - 1.875x^3 + \ldots$

b We want $\left(1 - \frac{x}{4}\right) = 0.975$

$\frac{x}{4} = 0.025$

$x = 0.1$

Calculate the value of x.

Substitute $x = 0.1$ into the expansion for $\left(1 - \frac{x}{4}\right)^{10}$ from part **a**:

$0.975^{10} \approx 1 - 0.25 + 0.028125 - 0.001875$

Substitute $x = 0.1$ into your expansion.

$= 0.77625$

$0.975^{10} \approx 0.7763$ to 4 d.p.

Using a calculator, $0.975^{10} = 0.776\,329\,62$.
So approximation is correct to 4 decimal places.

Online Use technology to find the values of x for which the first four terms of this expansion give a good approximation to the value of the function.

Exercise 8E

1 **a** Find the first four terms of the binomial expansion, in ascending powers of x, of $\left(1 - \frac{x}{10}\right)^6$.

b By substituting an appropriate value for x, find an approximate value for 0.99^6.

2 **a** Write down the first four terms of the binomial expansion of $\left(2 + \frac{x}{5}\right)^{10}$.

b By substituting an appropriate value for x, find an approximate value for 2.1^{10}.

(P) **3** If x is so small that terms of x^3 and higher can be ignored, show that:

$$(2 + x)(1 - 3x)^5 \approx 2 - 29x + 165x^2$$

Hint Start by using the binomial expansion to expand $(1 - 3x)^5$. You can ignore terms of x^3 and higher so you only need to expand up to and including the x^2 term.

(P) **4** If x is so small that terms of x^3 and higher can be ignored, and

$$(2 - x)(3 + x)^4 \approx a + bx + cx^2$$

find the values of the constants a, b and c.

Problem-solving

Find the first 3 terms in the expansion of $(2 - x)(3 + x)^4$, compare with $a + bx + cx^2$ and write down the values of a, b and c.

5 **a** Write down the first four terms in the expansion of $(1 + 2x)^8$.

b By substituting an appropriate value of x (which should be stated), find an approximate value of 1.02^8.

6 $f(x) = (1 - 5x)^{30}$

a Find the first four terms, in ascending powers of x, in the binomial expansion of $f(x)$.

b Use your answer to part **a** to estimate the value of $(0.995)^{30}$, giving your answer to 6 decimal places.

c Use your calculator to evaluate 0.995^{30} and calculate the percentage error in your answer to part **b**.

(E/P) **7** **a** Find the first 3 terms, in ascending powers of x, of the binomial expansion of $\left(3 - \frac{x}{5}\right)^{10}$, giving each term in its simplest form. **(4 marks)**

b Explain how you would use your expansion to give an estimate for the value of 2.98^{10}. **(1 mark)**

(E) **8** **a** Find the first 4 terms, in ascending powers of x, of the binomial expansion of $(1 - 3x)^5$. Give each term in its simplest form. **(4 marks)**

b If x is small, so that x^2 and higher powers can be ignored, show that $(1 + x)(1 - 3x)^5 \approx 1 - 14x$. **(2 marks)**

(E/P) **9** A microchip company models the probability of having no faulty chips on a single production run as:

$$\text{P(no fault)} = (1 - p)^n,\ p < 0.001$$

where p is the probability of a single chip being faulty, and n being the total number of chips produced.

a State why the model is restricted to small values of p. **(1 mark)**

b Given that $n = 200$, find an approximate expression for P(no fault) in the form $a + bp + cp^2$. **(2 marks)**

c The company wants to achieve a 92% likelihood of having no faulty chips on a production run of 200 chips. Use your answer to part **b** to suggest a maximum value of p for this to be the case. **(4 marks)**

Mixed exercise 8

(P) **1** The 16th row of Pascal's triangle is shown below.

1 15 105 … …

a Find the next two values in the row.

b Hence find the coefficient of x^3 in the expansion of $(1 + 2x)^{15}$.

(E) **2** Given that $\binom{45}{17} = \frac{45!}{17!a!}$, write down the value of a. **(1 mark)**

3 20 people play a game at a school fete.
The probability that exactly n people win a prize is modelled as $\binom{20}{n}p^n(1 - p)^{20-n}$, where p is the probability of any one person winning.
Calculate the probability of:

a 5 people winning when $p = \frac{1}{2}$

b nobody winning when $p = 0.7$

c 13 people winning when $p = 0.6$

Give your answers to 3 significant figures.

(E/P) **4** When $\left(1 - \frac{3}{2}x\right)^p$ is expanded in ascending powers of x, the coefficient of x is -24.

a Find the value of p. **(2 marks)**

b Find the coefficient of x^2 in the expansion. **(3 marks)**

c Find the coefficient of x^3 in the expansion. **(1 mark)**

(E/P) **5** Given that:

$$(2 - x)^{13} \equiv A + Bx + Cx^2 + \dots$$

find the values of the integers A, B and C. **(4 marks)**

(E) **6** **a** Expand $(1 - 2x)^{10}$ in ascending powers of x up to and including the term in x^3, simplifying each coefficient in the expansion. **(4 marks)**

b Use your expansion to find an approximation of 0.98^{10}, stating clearly the substitution which you have used for x. **(3 marks)**

(E) **7** **a** Use the binomial series to expand $(2 - 3x)^{10}$ in ascending powers of x up to and including the term in x^3, giving each coefficient as an integer. **(4 marks)**

b Use your series expansion, with a suitable value for x, to obtain an estimate for 1.97^{10}, giving your answer to 2 decimal places. **(3 marks)**

(E/P) **8** **a** Expand $(3 + 2x)^4$ in ascending powers of x, giving each coefficient as an integer. **(4 marks)**

b Hence, or otherwise, write down the expansion of $(3 - 2x)^4$ in ascending powers of x. **(2 marks)**

c Hence by choosing a suitable value for x show that $(3 + 2\sqrt{2})^4 + (3 - 2\sqrt{2})^4$ is an integer and state its value. **(2 marks)**

(E/P) **9** The coefficient of x^2 in the binomial expansion of $\left(1 + \frac{x}{2}\right)^n$, where n is a positive integer, is 7.

a Find the value of n. **(2 marks)**

b Using the value of n found in part **a**, find the coefficient of x^4. **(4 marks)**

(E) **10** **a** Use the binomial theorem to expand $(3 + 10x)^4$ giving each coefficient as an integer. **(4 marks)**

b Use your expansion, with an appropriate value for x, to find the exact value of 1003^4. State the value of x which you have used. **(3 marks)**

(E) **11** **a** Expand $(1 + 2x)^{12}$ in ascending powers of x up to and including the term in x^3, simplifying each coefficient. **(4 marks)**

b By substituting a suitable value for x, which must be stated, into your answer to part **a**, calculate an approximate value of 1.02^{12}. **(3 marks)**

c Use your calculator, writing down all the digits in your display, to find a more exact value of 1.02^{12}. **(1 mark)**

d Calculate, to 3 significant figures, the percentage error of the approximation found in part **b**. **(1 mark)**

(E/P) **12** Expand $\left(x - \frac{1}{x}\right)^5$, simplifying the coefficients. **(4 marks)**

(E/P) **13** In the binomial expansion of $(2k + x)^n$, where k is a constant and n is a positive integer, the coefficient of x^2 is equal to the coefficient of x^3.

a Prove that $n = 6k + 2$. **(3 marks)**

b Given also that $k = \frac{2}{3}$, expand $(2k + x)^n$ in ascending powers of x up to and including the term in x^3, giving each coefficient as an exact fraction in its simplest form. **(4 marks)**

(E/P) **14** **a** Expand $(2 + x)^6$ as a binomial series in ascending powers of x, giving each coefficient as an integer. **(4 marks)**

b By making suitable substitutions for x in your answer to part **a**, show that $(2 + \sqrt{3})^6 - (2 - \sqrt{3})^6$ can be simplified to the form $k\sqrt{3}$, stating the value of the integer k. **(3 marks)**

(E/P) **15** The coefficient of x^2 in the binomial expansion of $(2 + kx)^8$, where k is a positive constant, is 2800.

a Use algebra to calculate the value of k. **(2 marks)**

b Use your value of k to find the coefficient of x^3 in the expansion. **(4 marks)**

(E/P) **16** **a** Given that

$$(2 + x)^5 + (2 - x)^5 \equiv A + Bx^2 + Cx^4,$$

find the value of the constants A, B and C. **(4 marks)**

b Using the substitution $y = x^2$ and your answers to part **a**, solve

$$(2 + x)^5 + (2 - x)^5 = 349.$$ **(3 marks)**

(E/P) **17** In the binomial expansion of $(2 + px)^5$, where p is a constant, the coefficient of x^3 is 135. Calculate:

a the value of p, **(4 marks)**

b the value of the coefficient of x^4 in the expansion. **(2 marks)**

(P) **18** Find the constant term in the expansion of $\left(\frac{x^2}{2} - \frac{2}{x}\right)^9$.

(E/P) **19** **a** Find the first three terms, in ascending powers of x of the binomial expansion of $(2 + px)^7$, where p is a constant. **(2 marks)**

The first 3 terms are 128, $2240x$ and qx^2, where q is a constant.

b Find the value of p and the value of q. **(4 marks)**

(E/P) **20** **a** Write down the first three terms, in ascending powers of x, of the binomial expansion of $(1 - px)^{12}$, where p is a non-zero constant. **(2 marks)**

b Given that, in the expansion of $(1 - px)^{12}$, the coefficient of x is q and the coefficient of x^2 is $6q$, find the value of p and the value of q. **(4 marks)**

(E/P) **21** **a** Find the first 3 terms, in ascending powers of x, of the binomial expansion of $\left(2 + \frac{x}{2}\right)^7$, giving each term in its simplest form. **(4 marks)**

b Explain how you would use your expansion to give an estimate for the value of 2.05^7. **(1 mark)**

(E/P) **22** $g(x) = (4 + kx)^5$, where k is a constant.

Given that the coefficient of x^3 in the binomial expansion of $g(x)$ is 20, find the value of k. **(3 marks)**

Challenge

1 $f(x) = (2 - px)(3 + x)^5$ where p is a constant.
There is no x^2 term in the expansion of $f(x)$.
Show that $p = \frac{4}{3}$

2 Find the coefficient of x^2 in the expansion of $(1 + 2x)^8(2 - 5x)^7$.

Summary of key points

1 Pascal's triangle is formed by adding adjacent pairs of numbers to find the numbers on the next row.

2 The $(n + 1)$th row of Pascal's triangle gives the coefficients in the expansion of $(a + b)^n$.

3 $n! = n \times (n - 1) \times (n - 2) \times \ldots \times 3 \times 2 \times 1$.

4 You can use factorial notation and your calculator to find entries in Pascal's triangle quickly.

- The number of ways of choosing r items from a group of n items is written as ${}^nC_r$ or $\binom{n}{r}$: ${}^nC_r = \binom{n}{r} = \frac{n!}{r!(n-r)!}$
- The rth entry in the nth row of Pascal's triangle is given by ${}^{n-1}C_{r-1} = \binom{n-1}{r-1}$.

5 The binomial expansion is:
$(a + b)^n = a^n + \binom{n}{1}a^{n-1}b + \binom{n}{2}a^{n-2}b^2 + \ldots + \binom{n}{r}a^{n-r}b^r + \ldots + b^n \ (n \in \mathbb{N})$
where $\binom{n}{r} = {}^nC_r = \frac{n!}{r!(n-r)!}$

6 In the expansion of $(a + b)^n$ the general term is given by $\binom{n}{r}a^{n-r}b^r$.

7 If x is small, the first few terms in the binomial expansion can be used to find an approximate value for a complicated expression.

Trigonometric ratios 9

Objectives

After completing this unit you should be able to:

- Use the cosine rule to find a missing side or angle → pages 174–179
- Use the sine rule to find a missing side or angle → pages 179–185
- Find the area of a triangle using an appropriate formula → pages 185–187
- Solve problems involving triangles → pages 187–192
- Sketch the graphs of the sine, cosine and tangent functions → pages 192–194
- Sketch simple transformations of these graphs → pages 194–198

Trigonometry in both two and three dimensions is used by surveyors to work out distances and areas when planning building projects. You will also use trigonometry when working with vector quantities in mechanics.
→ Exercise 9B Q12 and Mixed exercise Q10, Q11

Prior knowledge check

1 Use trigonometry to find the lengths of the marked sides.

a 23°, 7.3 cm, x

b

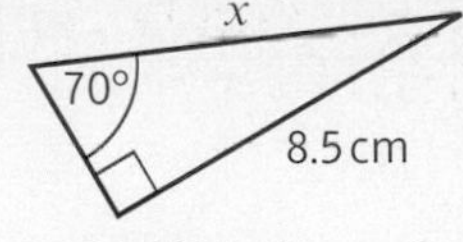

← GCSE Mathematics

2 Find the sizes of the angles marked.

a θ, 6.2 cm, 2.7 cm

b

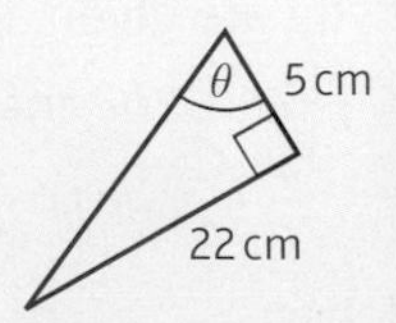

← GCSE Mathematics

3 $f(x) = x^2 + 3x$. Sketch the graphs of

a $y = f(x)$ **b** $y = f(x + 2)$

c $y = f(x) - 3$ **d** $y = f(\frac{1}{2}x)$

← Sections 4.5, 4.6

9.1 The cosine rule

The cosine rule can be used to work out missing sides or angles in triangles.

- **This version of the cosine rule is used to find a missing side if you know two sides and the angle between them:**

$$a^2 = b^2 + c^2 - 2bc \cos A$$

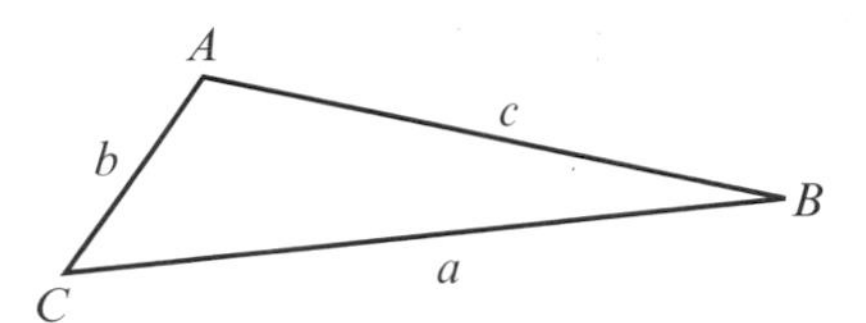

Watch out You can exchange the letters depending on which side you want to find, as long as each side has the same letter as the **opposite** angle.

You can use the standard trigonometric ratios for right-angled triangles to prove the cosine rule:

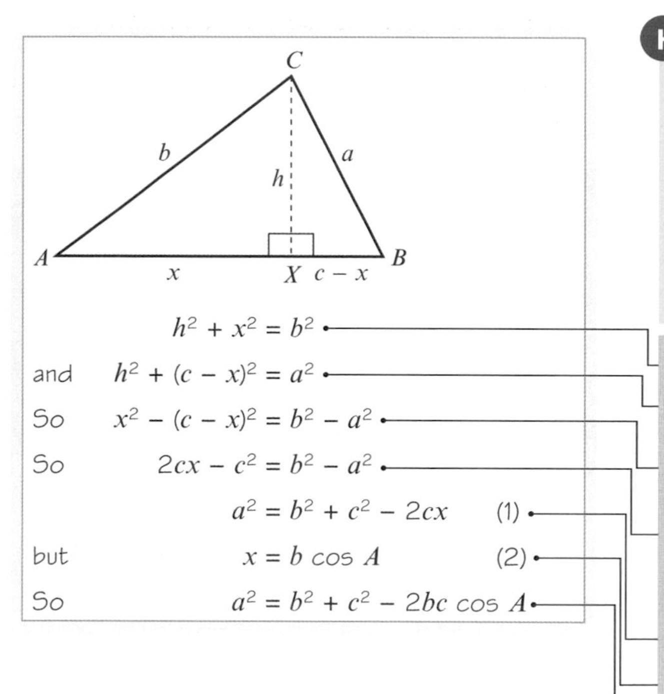

$h^2 + x^2 = b^2$ — Use Pythagoras' theorem in $\triangle CAX$.

and $h^2 + (c - x)^2 = a^2$ — Use Pythagoras' theorem in $\triangle CBX$.

So $x^2 - (c - x)^2 = b^2 - a^2$ — Subtract the two equations.

So $2cx - c^2 = b^2 - a^2$ — $(c - x)^2 = c^2 - 2cx + x^2$. So $x^2 - (c - x)^2 = x^2 - c^2 + 2cx - x^2$.

$a^2 = b^2 + c^2 - 2cx \quad (1)$ — Rearrange.

but $x = b \cos A \quad (2)$ — Use the cosine ratio $\cos A = \frac{x}{b}$ in $\triangle CAX$.

So $a^2 = b^2 + c^2 - 2bc \cos A$ — Combine (1) and (2). This is the cosine rule.

Hint For a right-angled triangle

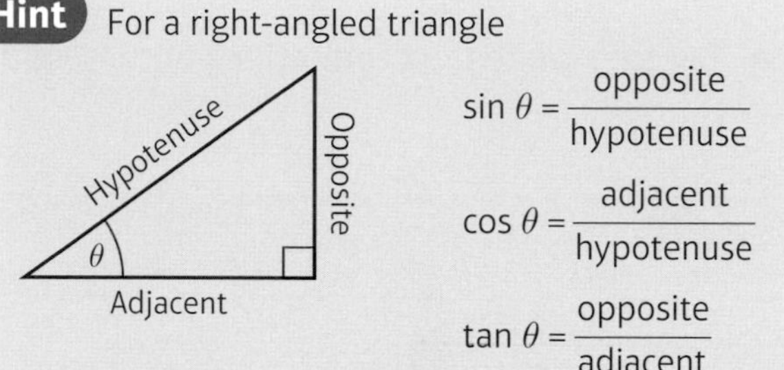

$\sin \theta = \frac{\text{opposite}}{\text{hypotenuse}}$

$\cos \theta = \frac{\text{adjacent}}{\text{hypotenuse}}$

$\tan \theta = \frac{\text{opposite}}{\text{adjacent}}$

If you are given all three sides and asked to find an angle, the cosine rule can be rearranged.

$$a^2 + 2bc \cos A = b^2 + c^2$$

$$2bc \cos A = b^2 + c^2 - a^2$$

Hence $\cos A = \frac{b^2 + c^2 - a^2}{2bc}$

You can exchange the letters depending on which angle you want to find.

- **This version of the cosine rule is used to find an angle if you know all three sides:**

$$\cos A = \frac{b^2 + c^2 - a^2}{2bc}$$

Online Explore the cosine rule using technology.

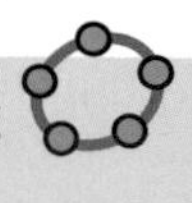

Example 1

Calculate the length of the side AB of the triangle ABC in which $AC = 6.5$ cm, $BC = 8.7$ cm and $\angle ACB = 100°$.

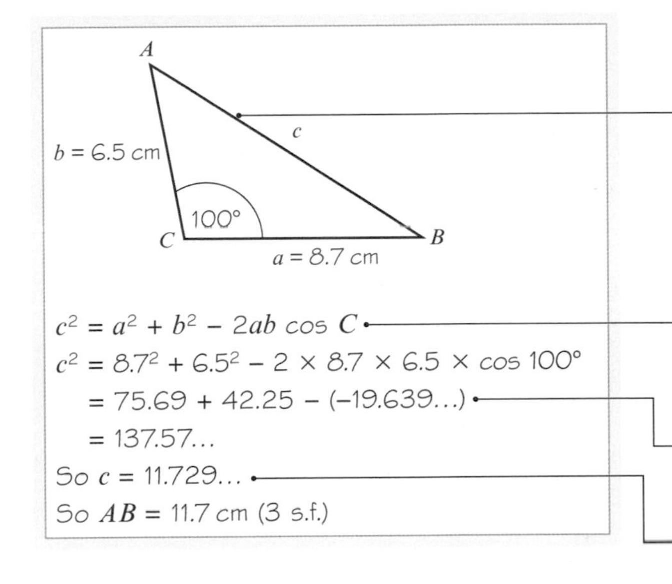

Label the sides of the triangle with small letters a, b and c opposite the angles marked.

Write out the formula you are using as the first line of working, then substitute in the values given.

Don't round any values until the end of your working. You can write your final answer to 3 significant figures.

Find the square root.

Example 2

Find the size of the smallest angle in a triangle whose sides have lengths 3 cm, 5 cm and 6 cm.

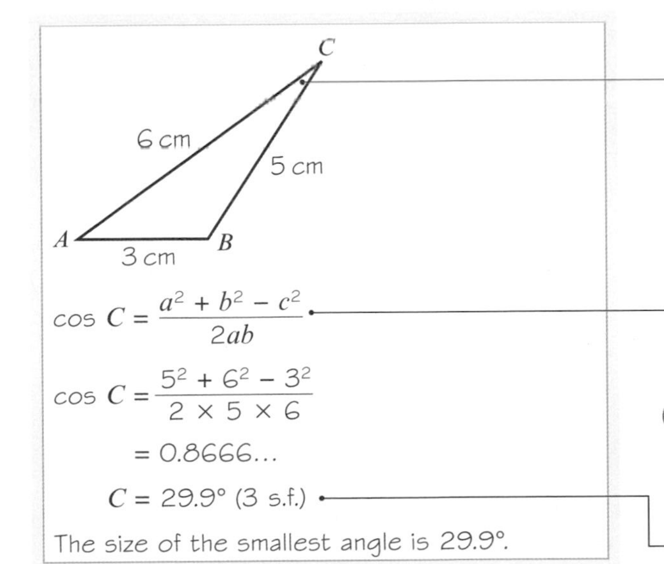

Label the triangle ABC.
The smallest angle is opposite the smallest side so angle C is required.

Use the cosine rule $\cos C = \frac{a^2 + b^2 - c^2}{2ab}$

Online Use your calculator to work this out efficiently.

$C = \cos^{-1}(0.8666...)$

Example 3

Coastguard station B is 8 km, on a bearing of 060°, from coastguard station A. A ship C is 4.8 km, on a bearing of 018°, away from A. Calculate how far C is from B.

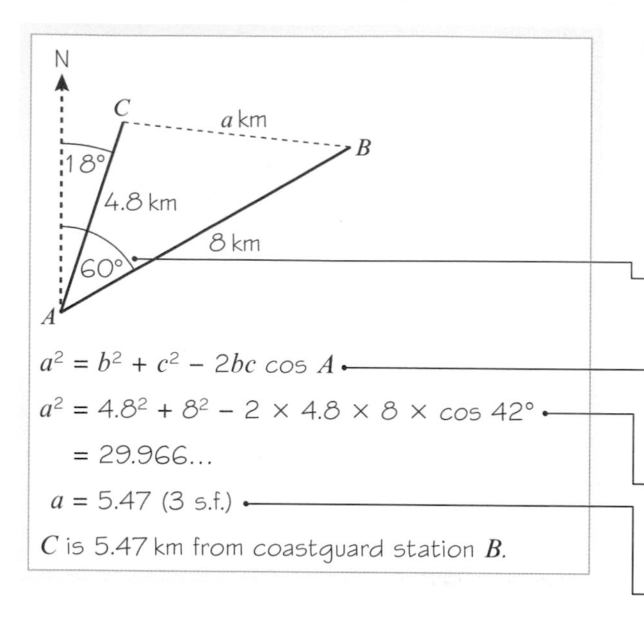

$a^2 = b^2 + c^2 - 2bc \cos A$

$a^2 = 4.8^2 + 8^2 - 2 \times 4.8 \times 8 \times \cos 42°$

$= 29.966...$

$a = 5.47$ (3 s.f.)

C is 5.47 km from coastguard station B.

Problem-solving

If no diagram is given with a question you should draw one carefully. Double-check that the information given in the question matches your sketch.

In $\triangle ABC$, $\angle CAB = 60° - 18° = 42°$.

You now have $b = 4.8$ km, $c = 8$ km and $A = 42°$. Use the cosine rule $a^2 = b^2 + c^2 - 2bc \cos A$.

If possible, work this out in one go using your calculator.

Take the square root of 29.966... and round your final answer to 3 significant figures.

Example 4

In $\triangle ABC$, $AB = x$ cm, $BC = (x + 2)$ cm, $AC = 5$ cm and $\angle ABC = 60°$. Find the value of x.

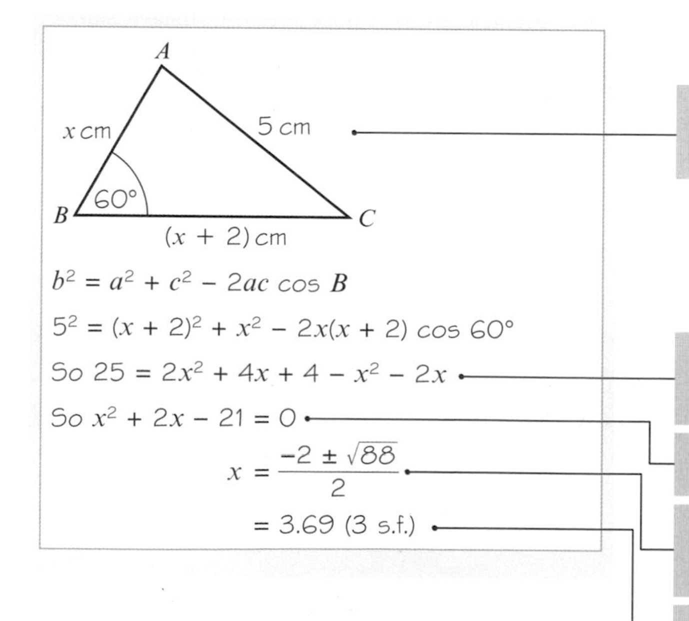

$b^2 = a^2 + c^2 - 2ac \cos B$

$5^2 = (x + 2)^2 + x^2 - 2x(x + 2) \cos 60°$

So $25 = 2x^2 + 4x + 4 - x^2 - 2x$

So $x^2 + 2x - 21 = 0$

$x = \dfrac{-2 \pm \sqrt{88}}{2}$

$= 3.69$ (3 s.f.)

Use the information given in the question to draw a sketch.

Carefully expand and simplify the right-hand side. Note that $\cos 60° = \frac{1}{2}$

Rearrange to the form $ax^2 + bx + c = 0$.

Solve the quadratic equation using the quadratic formula. ← Section 2.1

x represents a length so it cannot be negative.

Exercise 9A

Give answers to 3 significant figures, where appropriate.

1 In each of the following triangles calculate the length of the missing side.

a

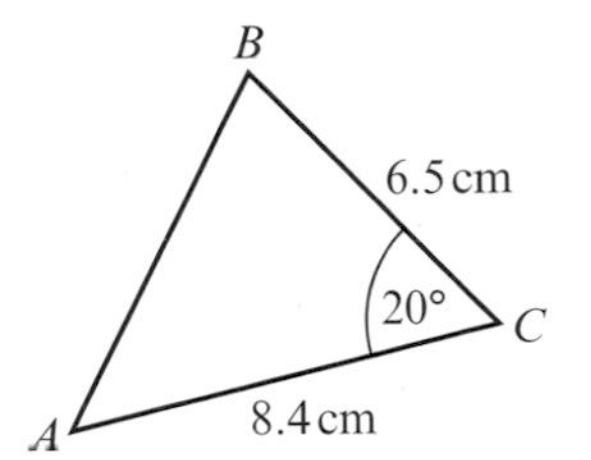

b

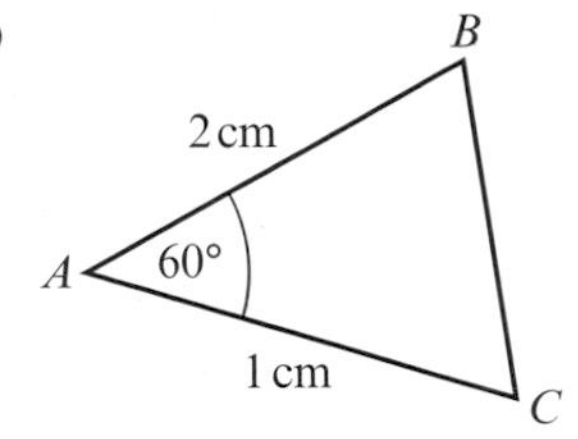

c

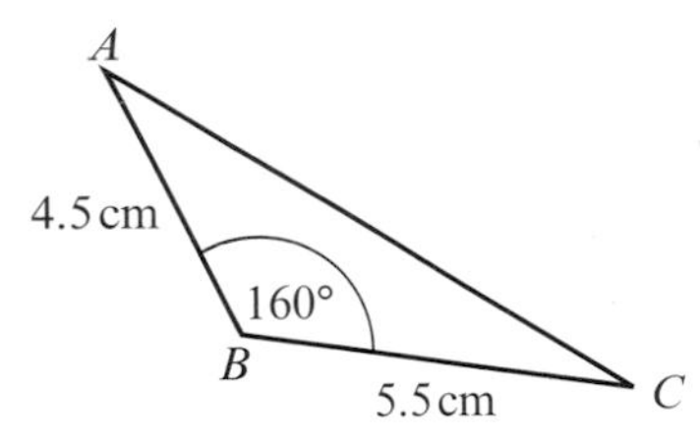

d

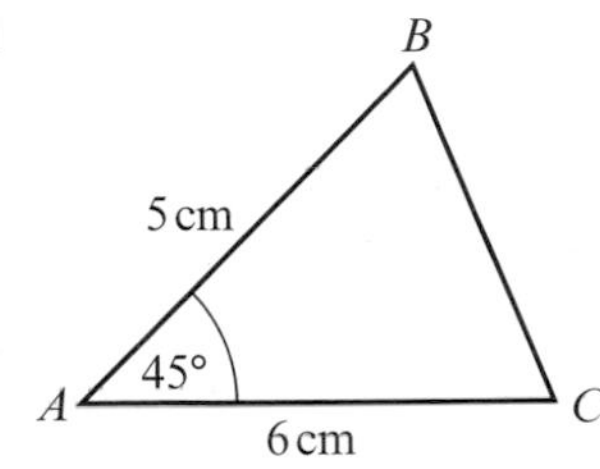

e

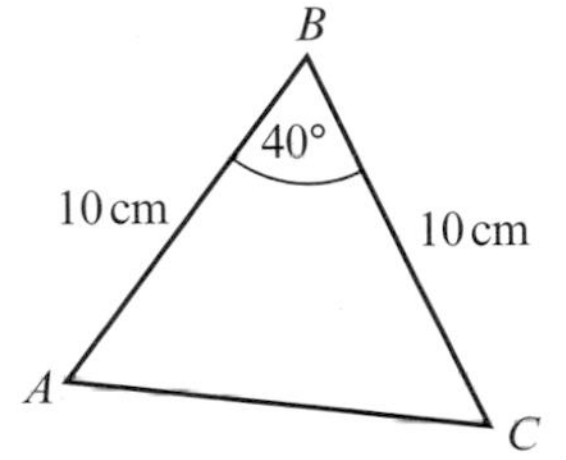

f

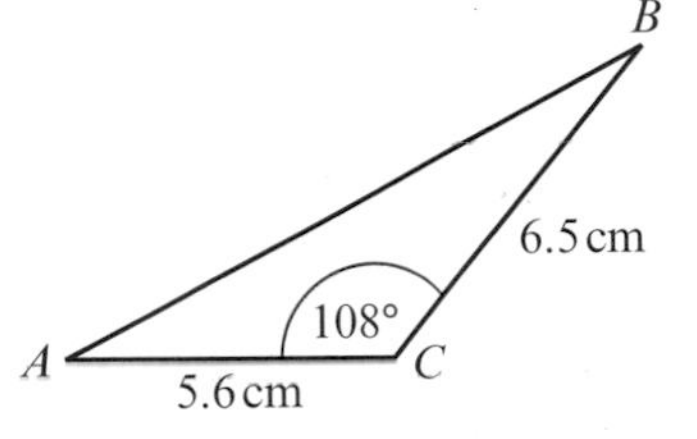

2 In the following triangles calculate the size of the angle marked x:

a

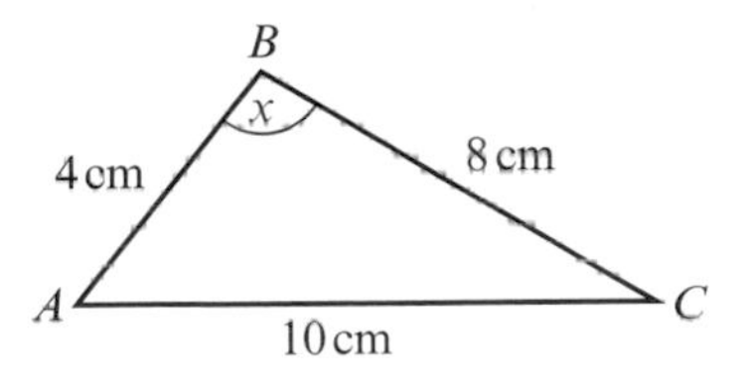

b

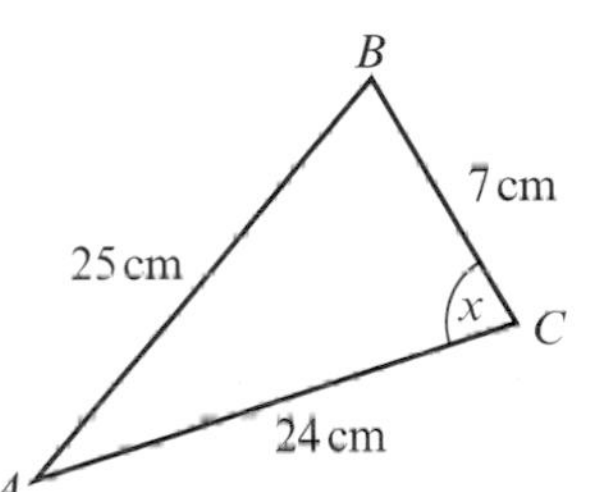

c

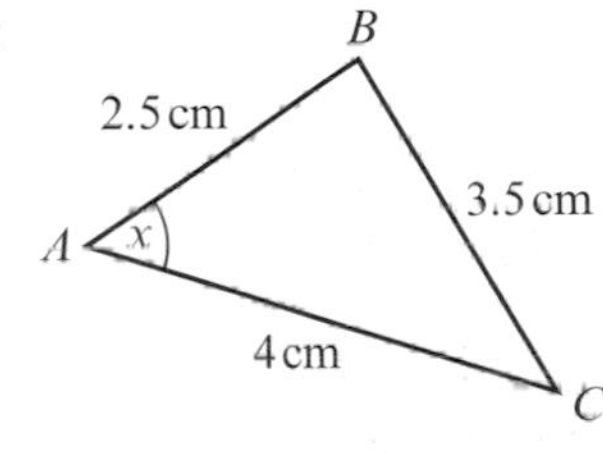

d

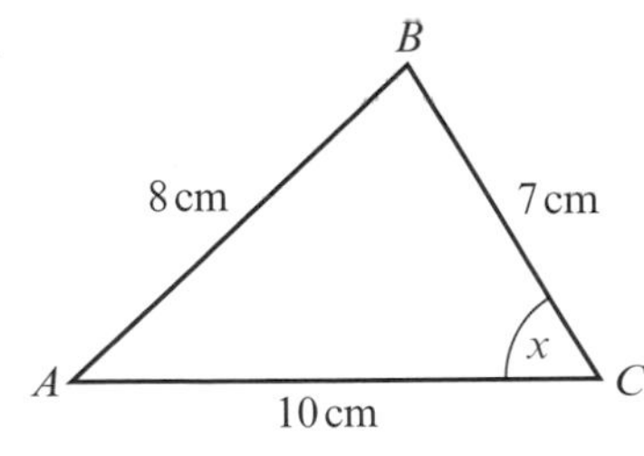

e

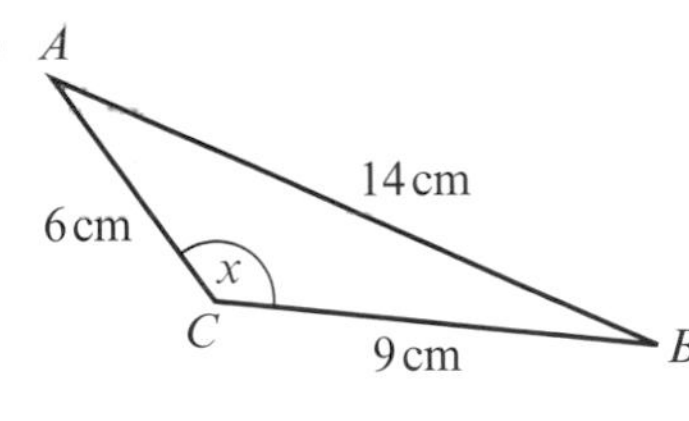

f

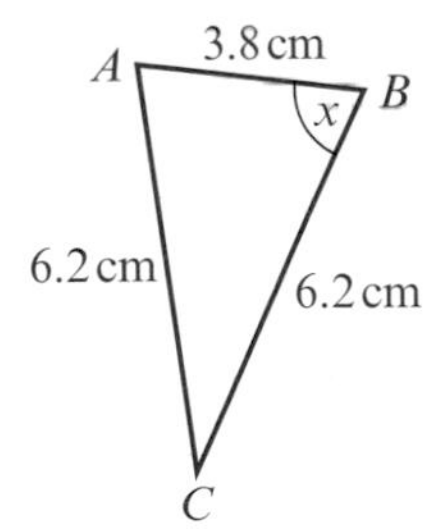

3 A plane flies from airport A on a bearing of 040° for 120 km and then on a bearing of 130° for 150 km. Calculate the distance of the plane from the airport.

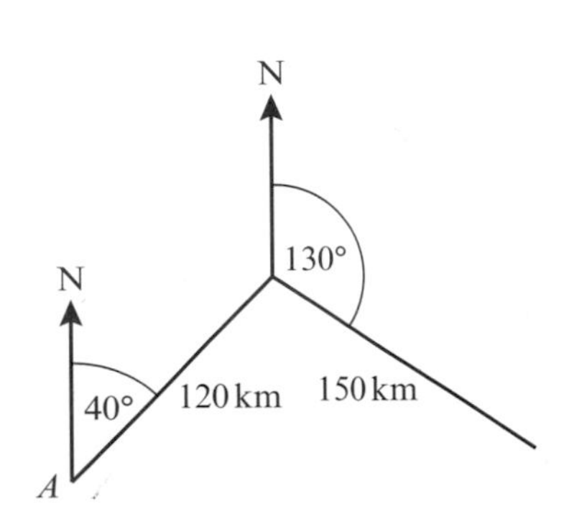

4 From a point A a boat sails due north for 7 km to B. The boat leaves B and moves on a bearing of 100° for 10 km until it reaches C. Calculate the distance of C from A.

5 A helicopter flies on a bearing of 080° from A to B, where $AB = 50$ km.
It then flies for 60 km to a point C.
Given that C is 80 km from A, calculate the bearing of C from A.

6 The distance from the tee, T, to the flag, F, on a particular hole on a golf course is 494 yards.
A golfer's tee shot travels 220 yards and lands at the point S, where $\angle STF = 22°$.
Calculate how far the ball is from the flag.

(P) **7** Show that $\cos A = \frac{1}{8}$

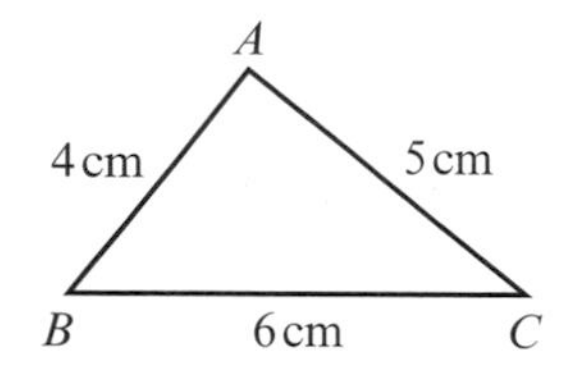

(P) **8** Show that $\cos P = -\frac{1}{4}$

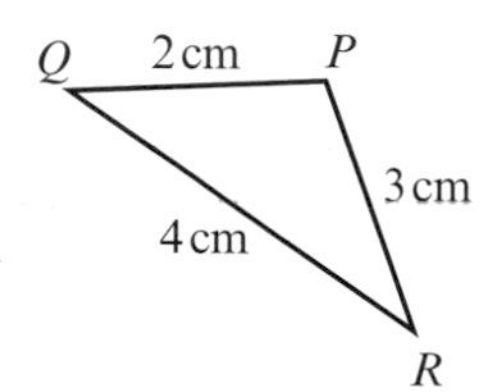

9 In $\triangle ABC$, $AB = 5$ cm, $BC = 6$ cm and $AC = 10$ cm.
Calculate the size of the smallest angle.

10 In $\triangle ABC$, $AB = 9.3$ cm, $BC = 6.2$ cm and $AC = 12.7$ cm.
Calculate the size of the largest angle.

(P) **11** The lengths of the sides of a triangle are in the ratio $2:3:4$.
Calculate the size of the largest angle.

12 In $\triangle ABC$, $AB = (x - 3)$ cm, $BC = (x + 3)$ cm, $AC = 8$ cm and $\angle BAC = 60°$.
Use the cosine rule to find the value of x.

(P) **13** In $\triangle ABC$, $AB = x$ cm, $BC = (x - 4)$ cm, $AC = 10$ cm and $\angle BAC = 60°$.
Calculate the value of x.

(P) **14** In $\triangle ABC$, $AB = (5 - x)$ cm, $BC = (4 + x)$ cm, $\angle ABC = 120°$ and $AC = y$ cm.

a Show that $y^2 = x^2 - x + 61$.

b Use the method of completing the square to find the minimum value of y^2, and give the value of x for which this occurs.

P **15** In $\triangle ABC$, $AB = x$ cm, $BC = 5$ cm, $AC = (10 - x)$ cm.

a Show that $\cos \angle ABC = \dfrac{4x - 15}{2x}$

b Given that $\cos \angle ABC = -\frac{1}{7}$, work out the value of x.

P **16** A farmer has a field in the shape of a quadrilateral as shown.

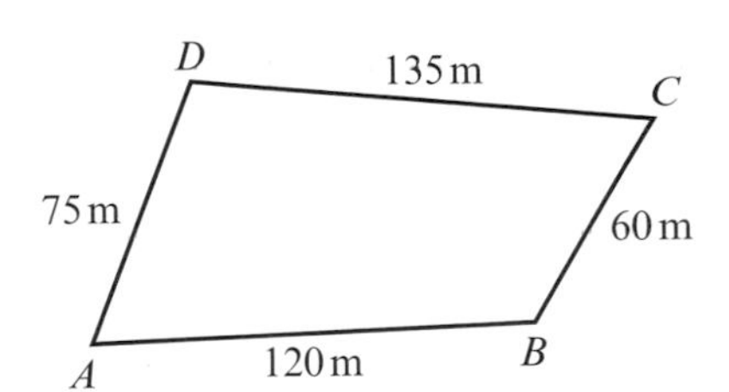

Problem-solving

You will have to use the cosine rule twice. Copy the diagram and write any angles or lengths you work out on your copy.

The angle between fences AB and AD is 74°. Find the angle between fences BC and CD.

E/P **17** The diagram shows three cargo ships, A, B and C, which are in the same horizontal plane. Ship B is 50 km due north of ship A and ship C is 70 km from ship A. The bearing of C from A is 020°.

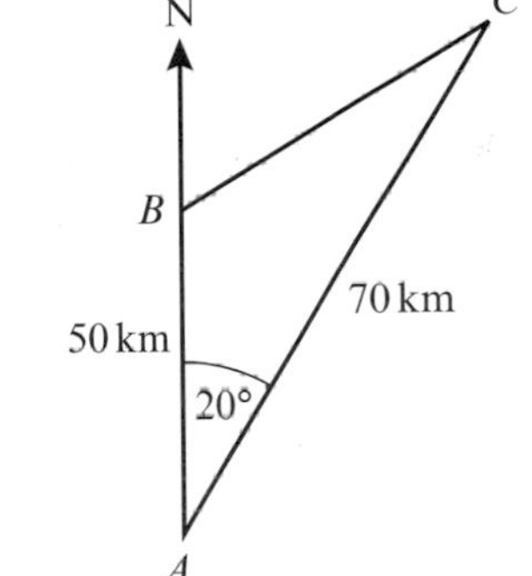

a Calculate the distance between ships B and C, in kilometres to 3 s.f. **(3 marks)**

b Calculate the bearing of ship C from ship B. **(4 marks)**

9.2 The sine rule

The sine rule can be used to work out missing sides or angles in triangles.

- **This version of the sine rule is used to find the length of a missing side:**

$$\frac{a}{\sin A} = \frac{b}{\sin B} = \frac{c}{\sin C}$$

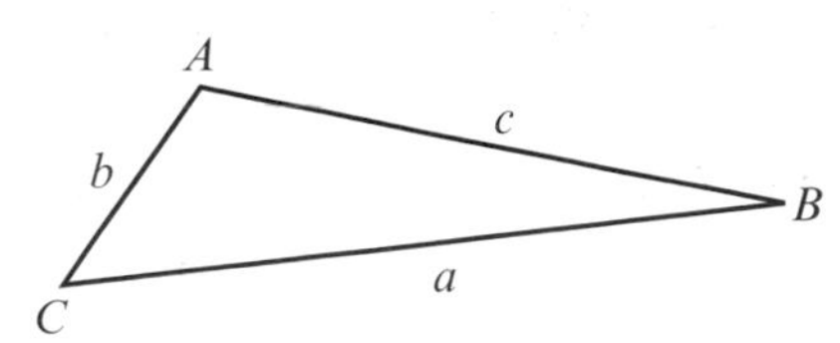

You can use the standard trigonometric ratios for right-angled triangles to prove the sine rule:

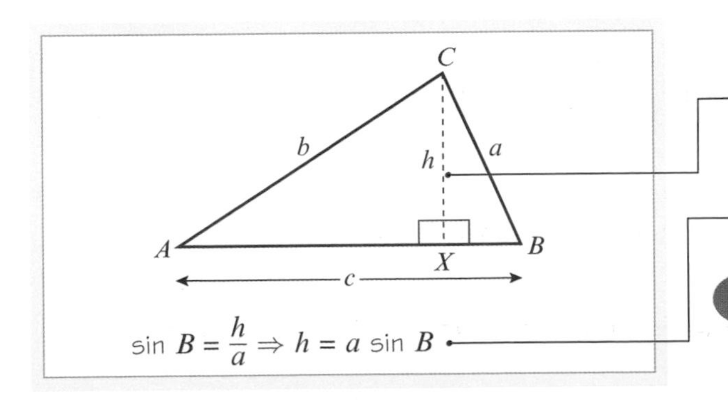

$\sin B = \frac{h}{a} \Rightarrow h = a \sin B$

In a general triangle ABC, draw the perpendicular from C to AB. It meets AB at X.

The length of CX is h.

Use the sine ratio in triangle CBX.

Online Explore the sine rule using technology.

and $\sin A = \frac{h}{b} \Rightarrow h = b \sin A$ — Use the sine ratio in triangle CAX.

So $a \sin B = b \sin A$

So $\frac{a}{\sin A} = \frac{b}{\sin B}$ — Divide throughout by $\sin A \sin B$.

In a similar way, by drawing the perpendicular from B to the side AC, you can show that:

$\frac{a}{\sin A} = \frac{c}{\sin C}$

So $\frac{a}{\sin A} = \frac{b}{\sin B} = \frac{c}{\sin C}$ — This is the sine rule and is true for all triangles.

- **This version of the sine rule is used to find a missing angle:**

$$\frac{\sin A}{a} = \frac{\sin B}{b} = \frac{\sin C}{c}$$

Example 5

In $\triangle ABC$, $AB = 8$ cm, $\angle BAC = 30°$ and $\angle BCA = 40°$. Find BC.

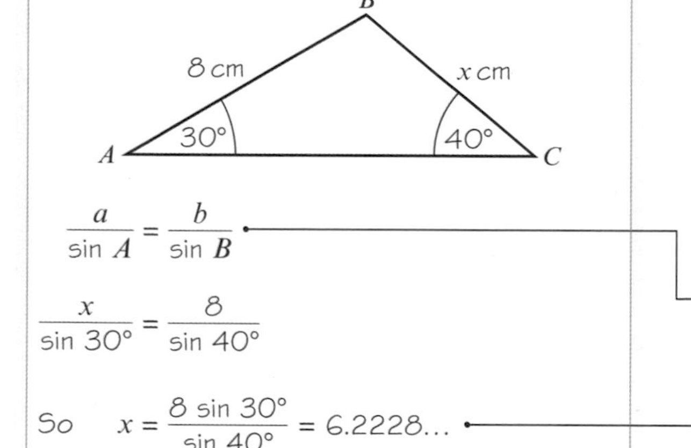

Always draw a diagram and carefully add the data. Here $c = 8$ (cm), $C = 40°$, $A = 30°$, $a = x$ (cm).

In a triangle, the larger a side is, the larger the opposite angle is. Here, as $C > A$, then $c > a$, so you know that $8 > x$.

$\frac{a}{\sin A} = \frac{b}{\sin B}$ — Use this version of the sine rule to find a missing side. Write the formula you are going to use as the first line of working.

$\frac{x}{\sin 30°} = \frac{8}{\sin 40°}$

So $x = \frac{8 \sin 30°}{\sin 40°} = 6.2228...$ — Multiply throughout by sin 30°.

$= 6.22$ cm (3 s.f.) — Give your answer to 3 significant figures.

Example 6

In $\triangle ABC$, $AB = 3.8$ cm, $BC = 5.2$ cm and $\angle BAC = 35°$. Find $\angle ABC$.

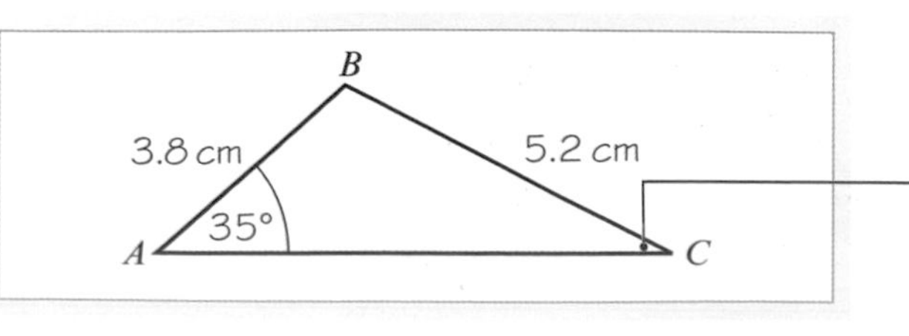

Here $a = 5.2$ cm, $c = 3.8$ cm and $A = 35°$. You first need to find angle C.

$\frac{\sin C}{c} = \frac{\sin A}{a}$

$\frac{\sin C}{3.8} = \frac{\sin 35°}{5.2}$

So $\sin C = \frac{3.8 \sin 35°}{5.2}$

$C = 24.781...$

So $B = 120°$ (3 s.f.)

Use $\frac{\sin C}{c} = \frac{\sin A}{a}$

Write the formula you are going to use as the first line of working.

Use your calculator to find the value of C in a single step. Don't round your answer at this point.

$B = 180° - (24.781...° + 35°) = 120.21...$ which rounds to 120° to 3 s.f.

Exercise 9B

Give answers to 3 significant figures, where appropriate.

1 In each of parts **a** to **d**, the given values refer to the general triangle.

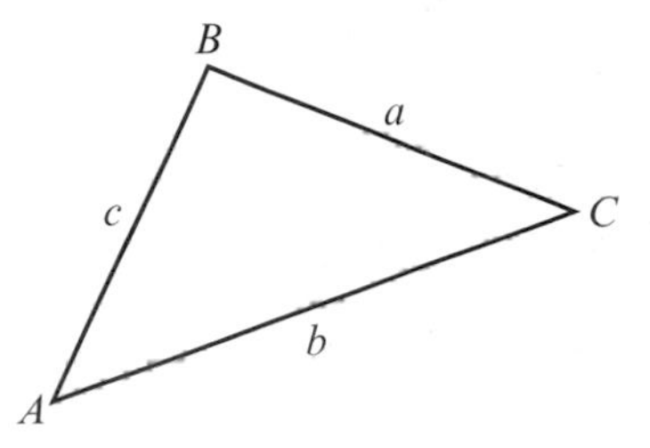

a Given that $a = 8$ cm, $A = 30°$, $B = 72°$, find b.

b Given that $a = 24$ cm, $A = 110°$, $C = 22°$, find c.

c Given that $b = 14.7$ cm, $A = 30°$, $C = 95°$, find a.

d Given that $c = 9.8$ cm, $B = 68.4°$, $C = 83.7°$, find a.

2 In each of the following triangles calculate the values of x and y.

a

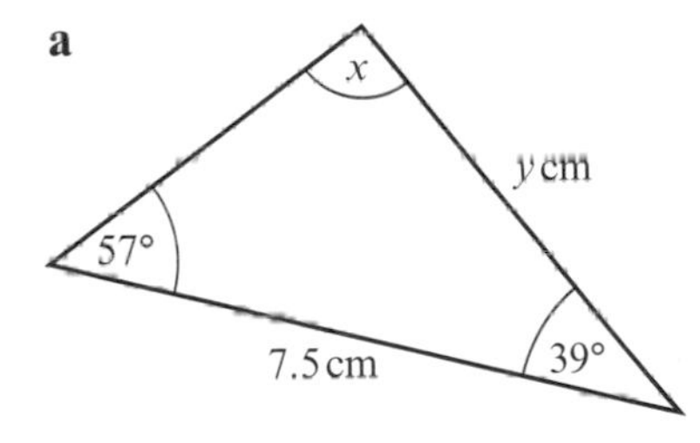

b

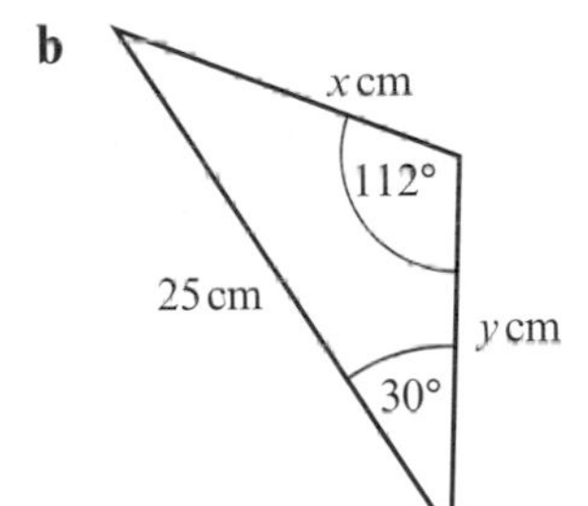

c

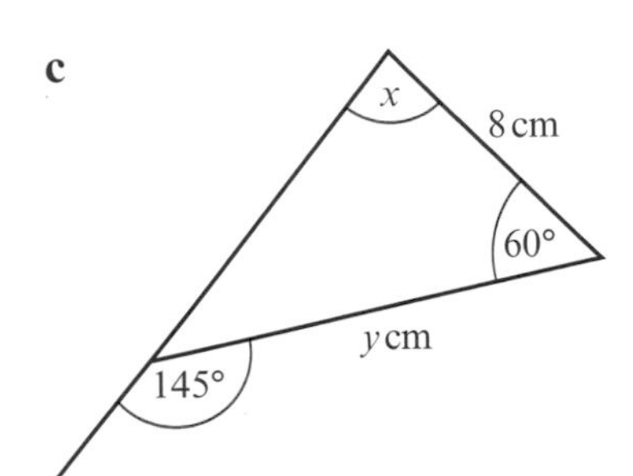

d

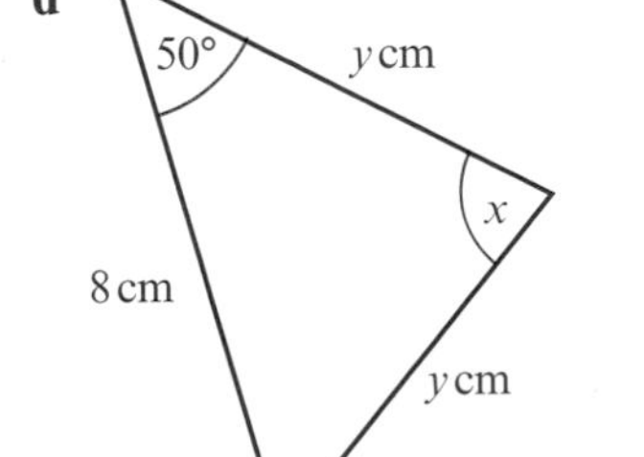

Hint In parts **c** and **d**, start by finding the size of the third angle.

e

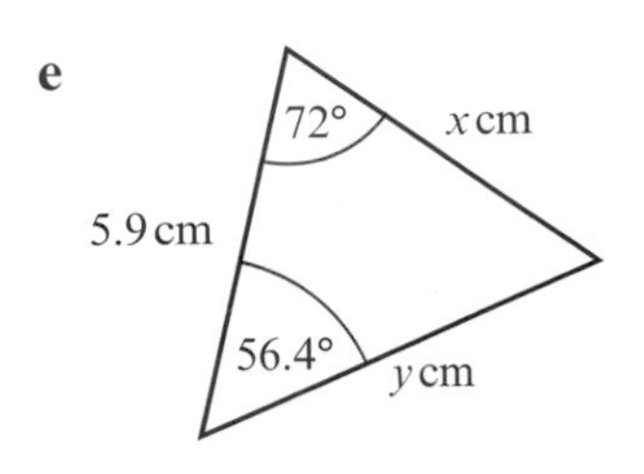

f

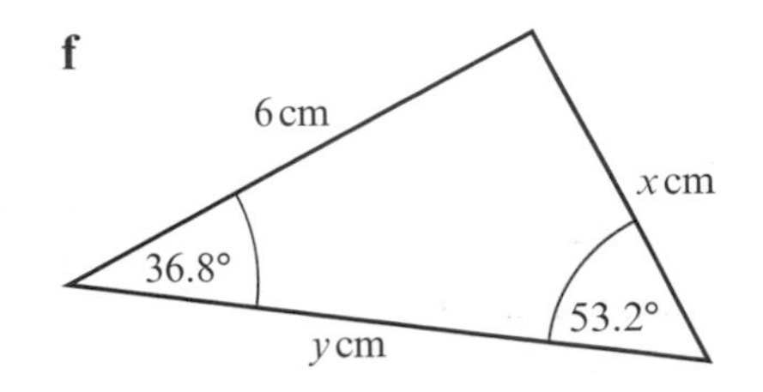

3 In each of the following sets of data for a triangle ABC, find the value of x.

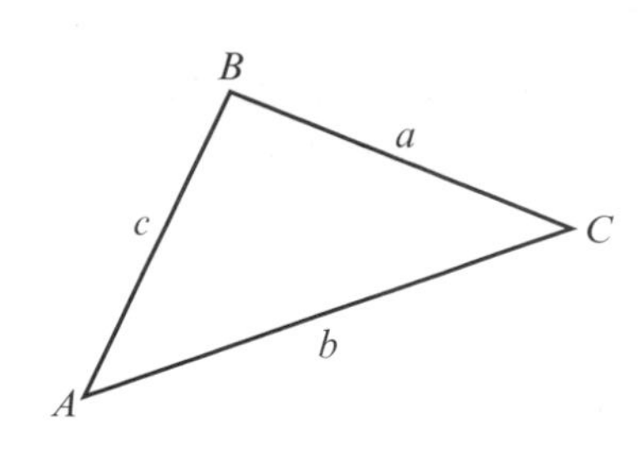

a $AB = 6$ cm, $BC = 9$ cm, $\angle BAC = 117°$, $\angle ACB = x$

b $AC = 11$ cm, $BC = 10$ cm, $\angle ABC = 40°$, $\angle CAB = x$

c $AB = 6$ cm, $BC = 8$ cm, $\angle BAC = 60°$, $\angle ACB = x$

d $AB = 8.7$ cm, $AC = 10.8$ cm, $\angle ABC = 28°$, $\angle BAC = x$

4 In each of the diagrams shown below, work out the size of angle x.

a

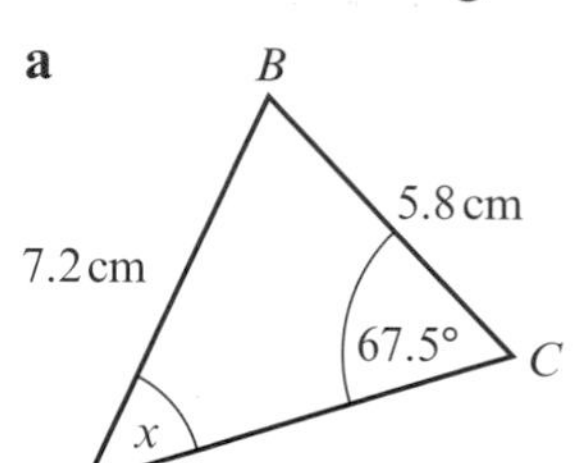

b

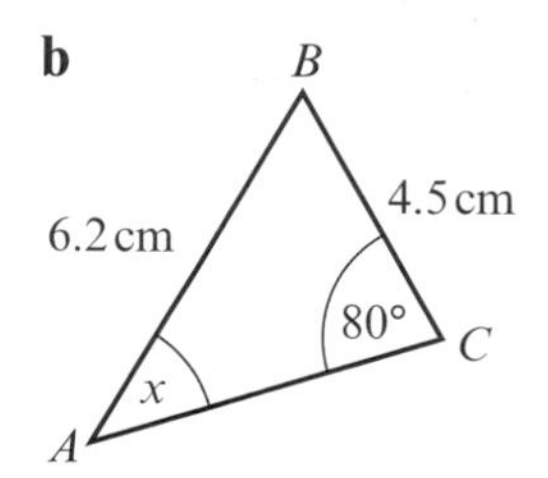

c

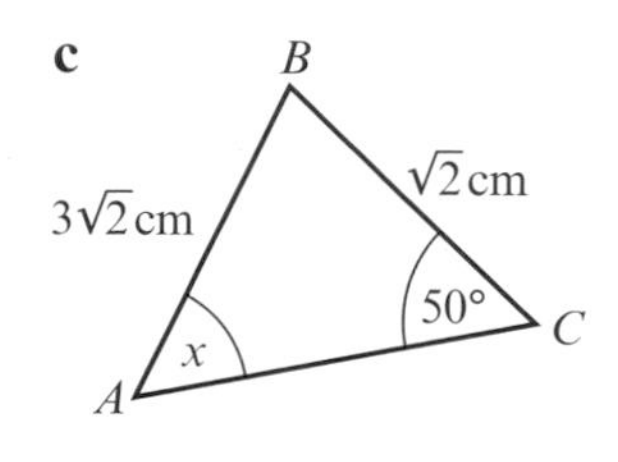

d

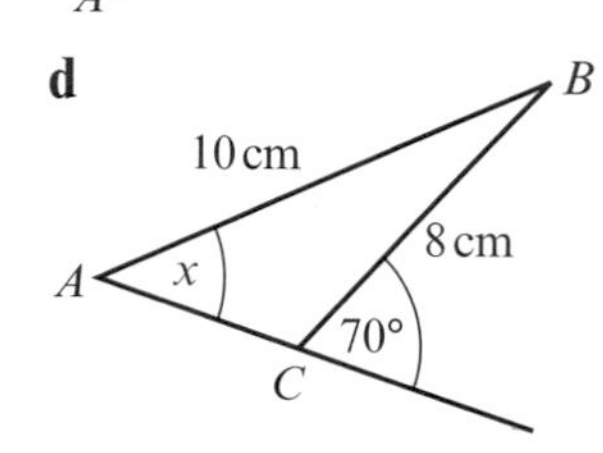

e

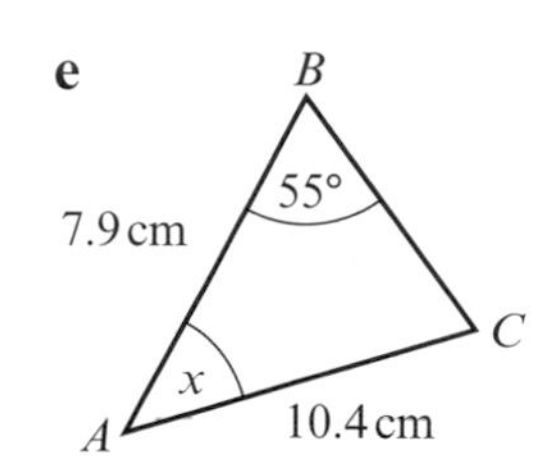

f

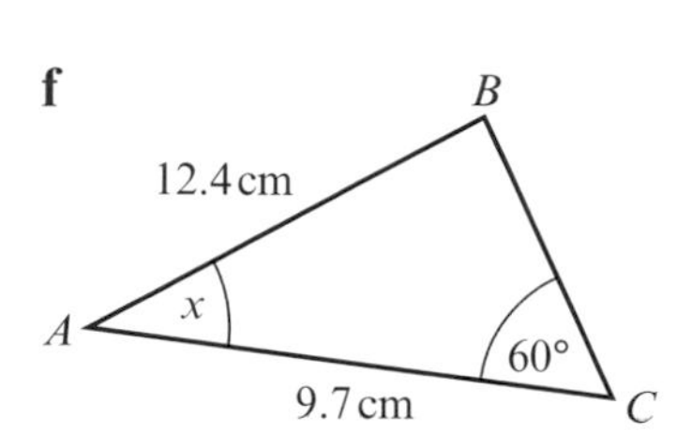

5 In $\triangle PQR$, $QR = \sqrt{3}$ cm, $\angle PQR = 45°$ and $\angle QPR = 60°$. Find **a** PR and **b** PQ.

6 In $\triangle PQR$, $PQ = 15$ cm, $QR = 12$ cm and $\angle PRQ = 75°$. Find the two remaining angles.

7 In each of the following diagrams work out the values of x and y.

a

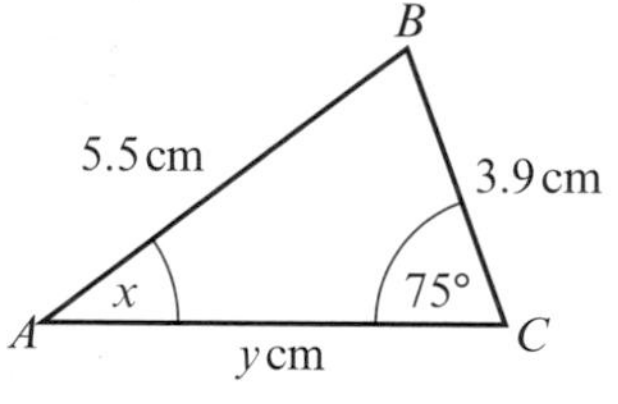

b

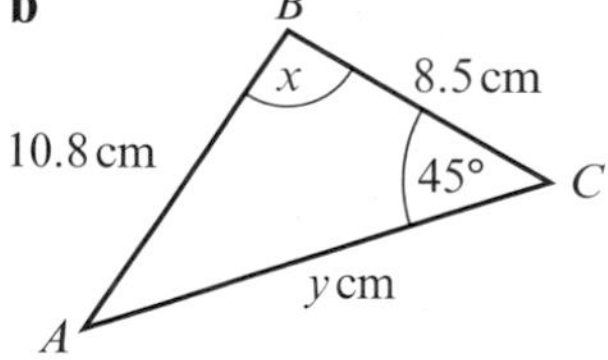

c

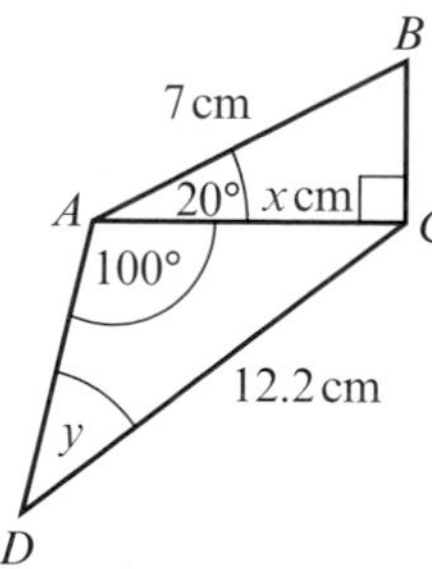

d

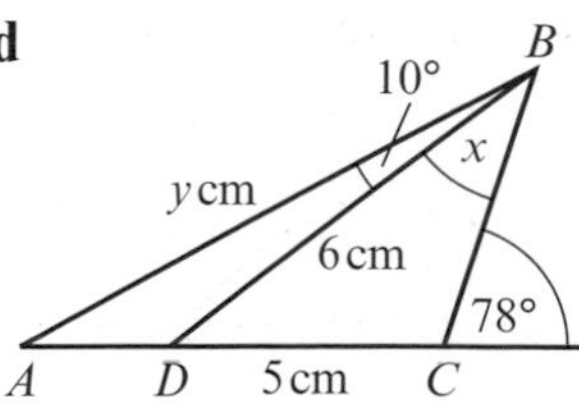

e

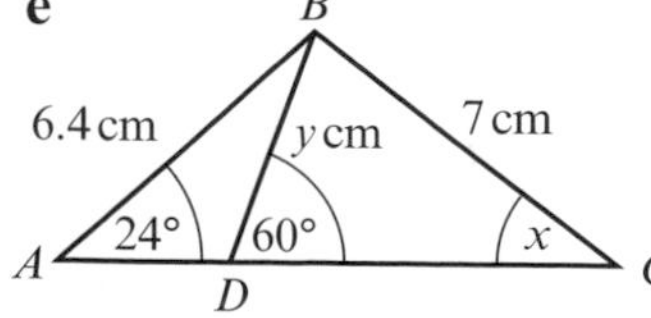

f

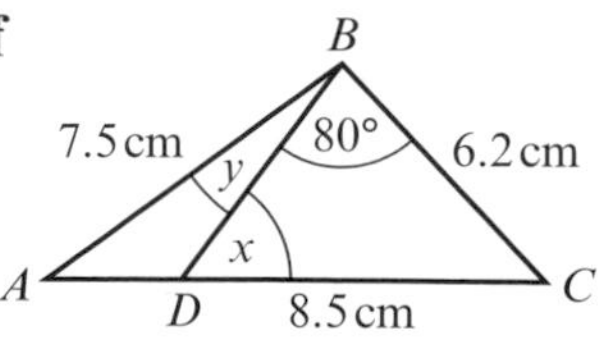

P **8** Town B is 6 km, on a bearing of 020°, from town A. Town C is located on a bearing of 055° from town A and on a bearing of 120° from town B. Work out the distance of town C from:

a town A **b** town B

Problem-solving

Draw a sketch to show the information.

9 In the diagram $AD = DB = 5$ cm, $\angle ABC = 43°$ and $\angle ACB = 72°$.
Calculate:
a AB
b CD

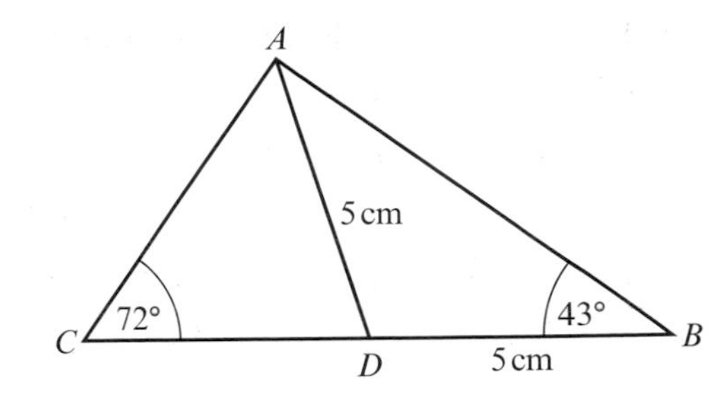

10 A zookeeper is building an enclosure for some llamas. The enclosure is in the shape of a quadrilateral as shown.
If the length of the diagonal BD is 136 m
a find the angle between the fences AB and BC
b find the length of fence AB

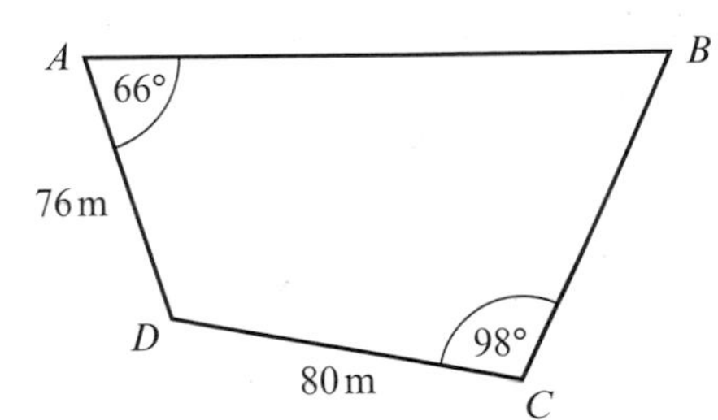

E/P 11 In $\triangle ABC$, $AB = x$ cm, $BC = (4 - x)$ cm, $\angle BAC = y$ and $\angle BCA = 30°$.
Given that $\sin y = \frac{1}{\sqrt{2}}$, show that
$x = 4(\sqrt{2} - 1)$ **(5 marks)**

Problem-solving
You can use the value of sin y directly in your calculation. You don't need to work out the value of y.

E/P 12 A surveyor wants to determine the height of a building. She measures the angle of elevation of the top of the building at two points 15 m apart on the ground.
a Use this information to determine the height of the building. **(4 marks)**
b State one assumption made by the surveyor in using this mathematical model. **(1 mark)**

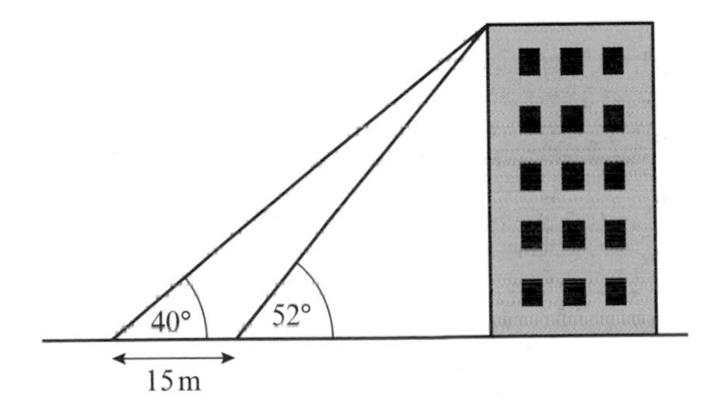

For given side lengths b and c and given angle B, you can draw the triangle in two different ways.

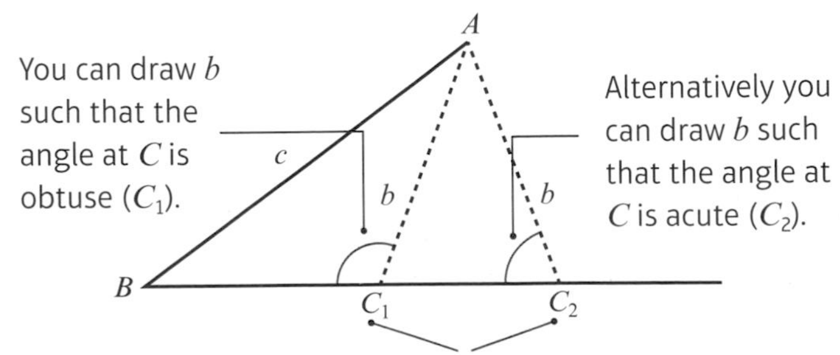

Since AC_1C_2 is an isosceles triangle, it follows that the angles AC_1B and AC_2B add together to make 180°.

Links You can confirm this relationship by considering the graph of $y = \sin x$.

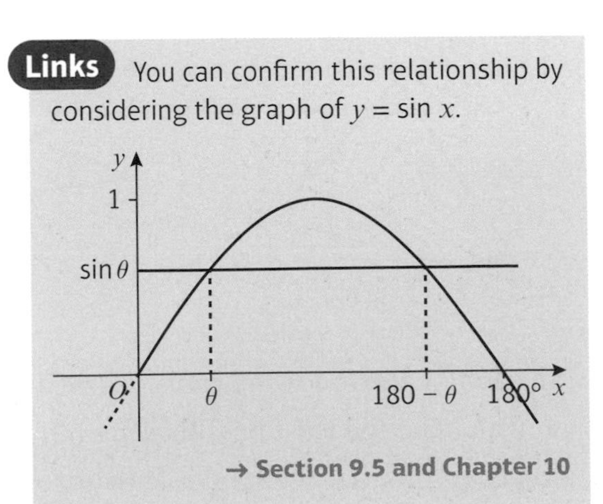

→ Section 9.5 and Chapter 10

- **The sine rule sometimes produces two possible solutions for a missing angle:**
  - **$\sin\theta = \sin(180° - \theta)$**

Example 7

In $\triangle ABC$, $AB = 4$ cm, $AC = 3$ cm and $\angle ABC = 44°$. Work out the two possible values of $\angle ACB$.

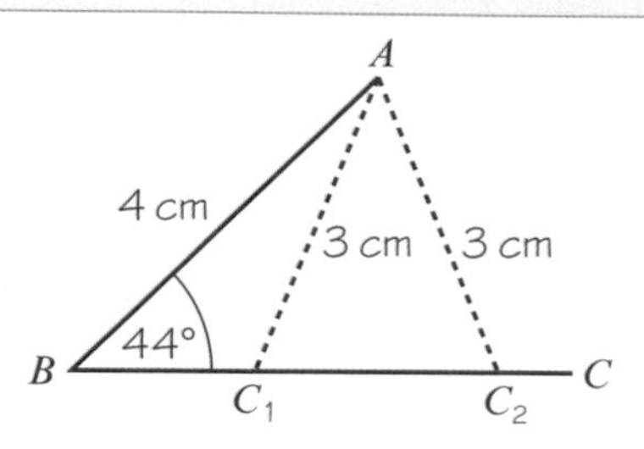

Problem-solving

Think about which lengths and angles are fixed, and which ones can vary. The length AC is fixed. If you drew a circle with radius 3 cm and centre A it would intersect the horizontal side of the triangle at two points, C_1 and C_2.

$$\frac{\sin C}{c} = \frac{\sin B}{b}$$

Use $\frac{\sin C}{c} = \frac{\sin B}{b}$, where $b = 3$, $c = 4$, $B = 44°$.

$$\frac{\sin C}{4} = \frac{\sin 44°}{3}$$

$$\sin C = \frac{4 \sin 44°}{3}$$

So $C = 67.851... = 67.9°$ (3 s.f.)

or $C = 180 - 67.851... = 112.14...$

$= 112°$ (3 s.f.)

As $\sin (180 - \theta) = \sin \theta$.

Exercise 9C

Give answers to 3 significant figures, where appropriate.

1 In $\triangle ABC$, $BC = 6$ cm, $AC = 4.5$ cm and $\angle ABC = 45°$.

a Calculate the two possible values of $\angle BAC$.

b Draw a diagram to illustrate your answers.

2 In each of the diagrams shown below, calculate the possible values of x and the corresponding values of y.

a

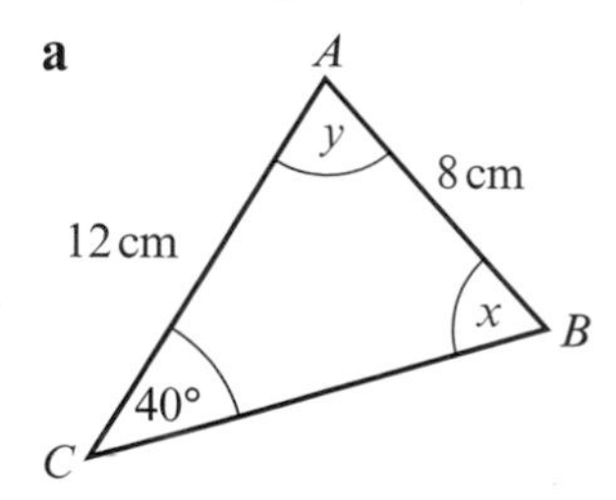

b

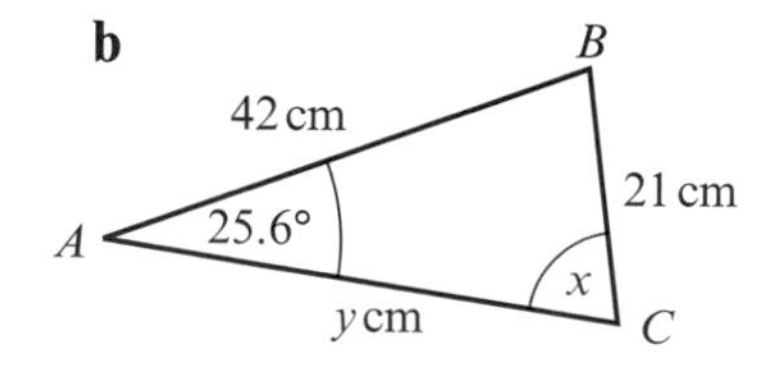

c

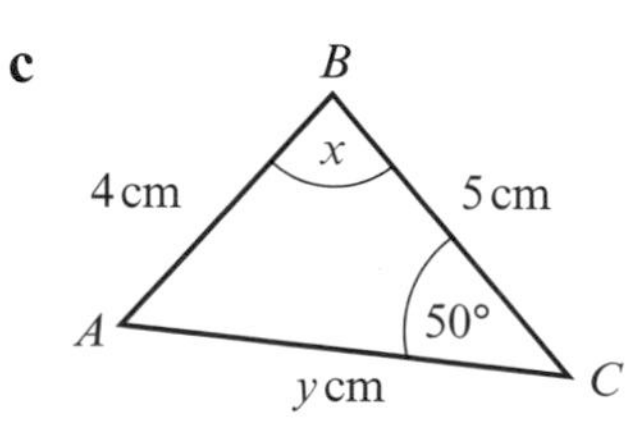

P **3** In each of the following cases $\triangle ABC$ has $\angle ABC = 30°$ and $AB = 10$ cm.

a Calculate the least possible length that AC could be.

b Given that $AC = 12$ cm, calculate $\angle ACB$.

c Given instead that $AC = 7$ cm, calculate the two possible values of $\angle ACB$.

(P) **4** Triangle ABC is such that $AB = 4$ cm, $BC = 6$ cm and $\angle ACB = 36°$. Show that one of the possible values of $\angle ABC$ is 25.8° (to 3 s.f.). Using this value, calculate the length of AC.

(P) **5** Two triangles ABC are such that $AB = 4.5$ cm, $BC = 6.8$ cm and $\angle ACB = 30°$. Work out the value of the largest angle in each of the triangles.

(E/P) **6** **a** A crane arm AB of length 80 m is anchored at point B at an angle of 40° to the horizontal. A wrecking ball is suspended on a cable of length 60 m from A. Find the angle x through which the wrecking ball rotates as it passes the two points level with the base of the crane arm at B. **(6 marks)**

b Write down one modelling assumption you have made. **(1 mark)**

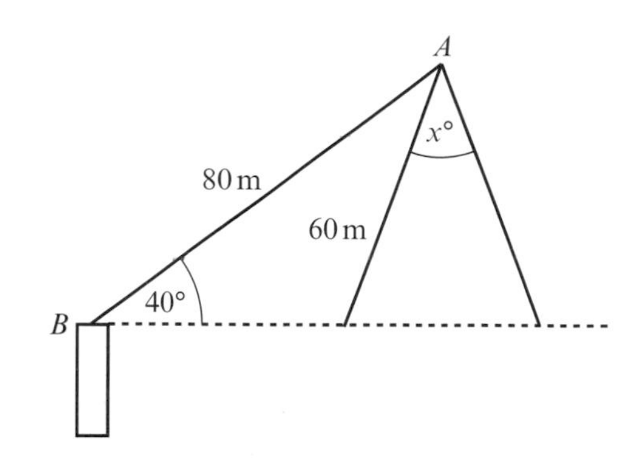

9.3 Areas of triangles

You need to be able to use the formula for finding the area of any triangle when you know two sides and the angle between them.

- **Area = $\frac{1}{2}ab \sin C$**

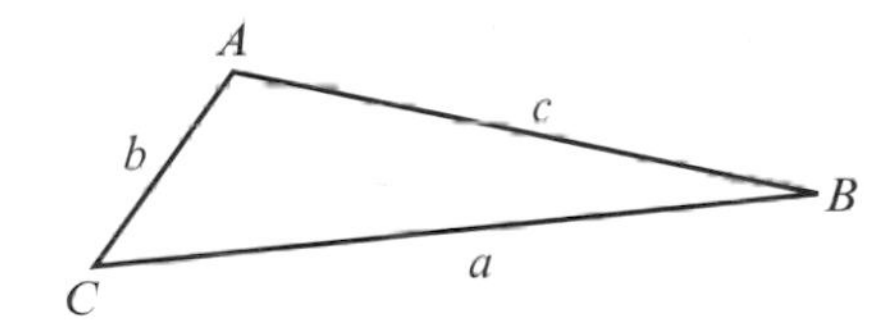

Hint As with the cosine rule, the letters are interchangeable. For example, if you know angle B and sides a and c, the formula becomes Area = $\frac{1}{2}ac \sin B$.

A proof of the formula:

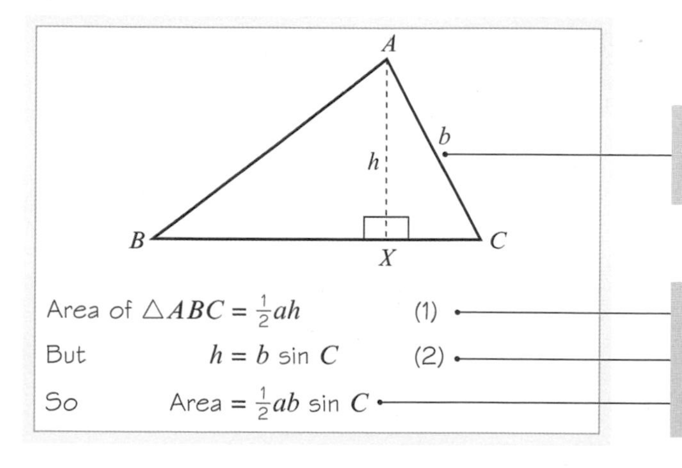

Area of $\triangle ABC = \frac{1}{2}ah$ (1)

But $h = b \sin C$ (2)

So Area $= \frac{1}{2}ab \sin C$

The perpendicular from A to BC is drawn and it meets BC at X. The length of $AX = h$.

Area of triangle = $\frac{1}{2}$ base × height.

Use the sine ratio $\sin C = \frac{h}{b}$ in $\triangle AXC$.

Substitute (2) into (1).

Example 8

Work out the area of the triangle shown below.

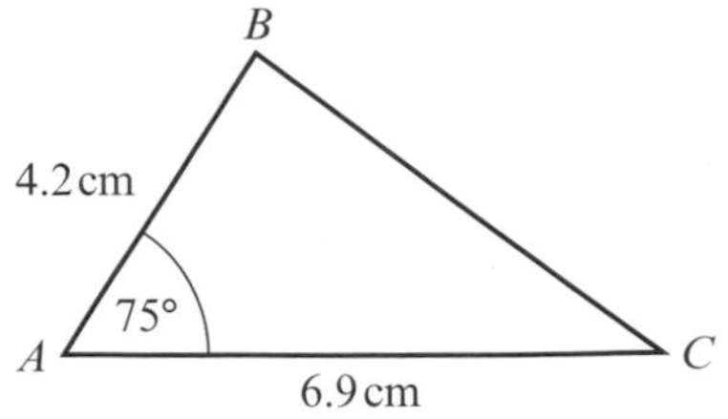

Online Explore the area of a triangle using technology.

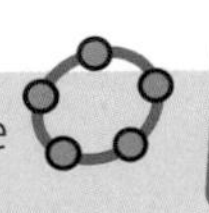

Area $= \frac{1}{2}bc \sin A$

Area of $\triangle ABC = \frac{1}{2} \times 6.9 \times 4.2 \times \sin 75° \text{cm}^2$

$= 14.0 \text{cm}^2$ (3 s.f.)

Here $b = 6.9$ cm, $c = 4.2$ cm and angle $A = 75°$, so use:
Area $= \frac{1}{2}bc \sin A$.

Example 9

In $\triangle ABC$, $AB = 5$ cm, $BC = 6$ cm and $\angle ABC = x$. Given that the area of $\triangle ABC$ is 12cm^2 and that AC is the longest side, find the value of x.

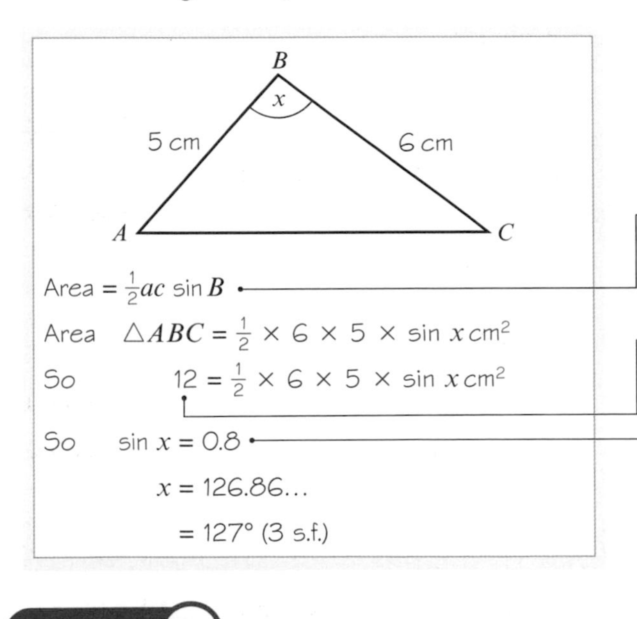

Area $= \frac{1}{2}ac \sin B$

Area $\triangle ABC = \frac{1}{2} \times 6 \times 5 \times \sin x \text{cm}^2$

So $12 = \frac{1}{2} \times 6 \times 5 \times \sin x \text{cm}^2$

So $\sin x = 0.8$

$x = 126.86\ldots$

$= 127°$ (3 s.f.)

Here $a = 6$ cm, $c = 5$ cm and angle $B = x$, so use:
Area $= \frac{1}{2}ac \sin B$.

Area of $\triangle ABC$ is 12cm^2.

$\sin x = \frac{12}{15}$

Problem-solving

There are two values of x for which $\sin x = 0.8$, $53.13\ldots°$ and $126.86\ldots°$, but here you know B is the largest angle because AC is the largest side.

Exercise 9D

1 Calculate the area of each triangle.

a

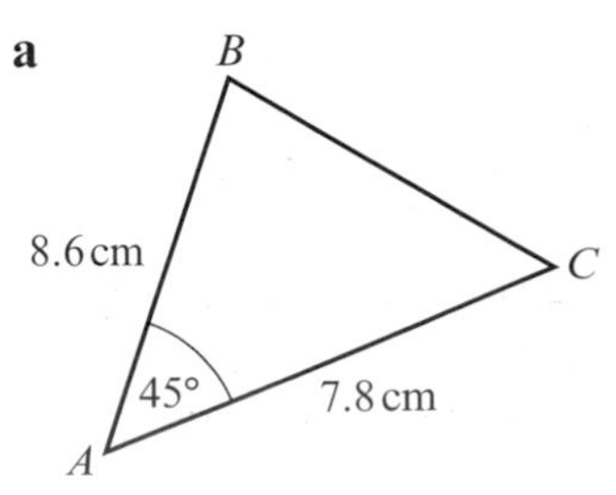

b

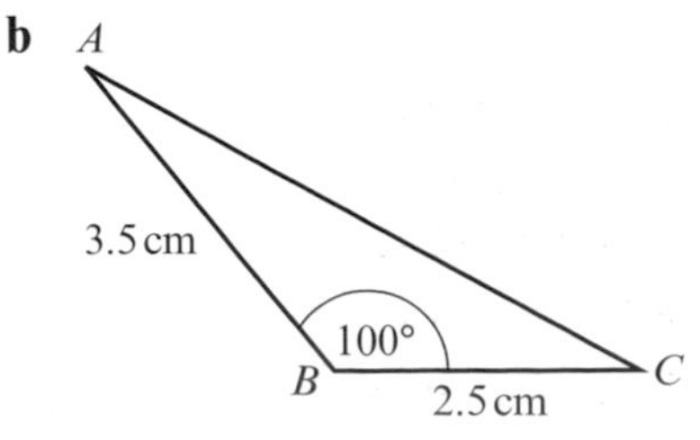

c

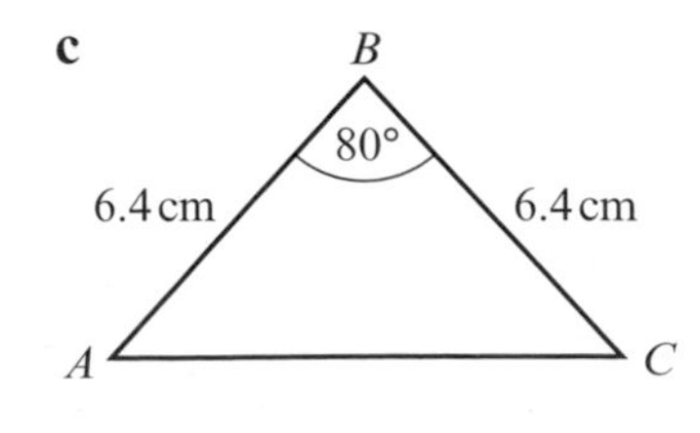

2 Work out the possible sizes of x in the following triangles.

a

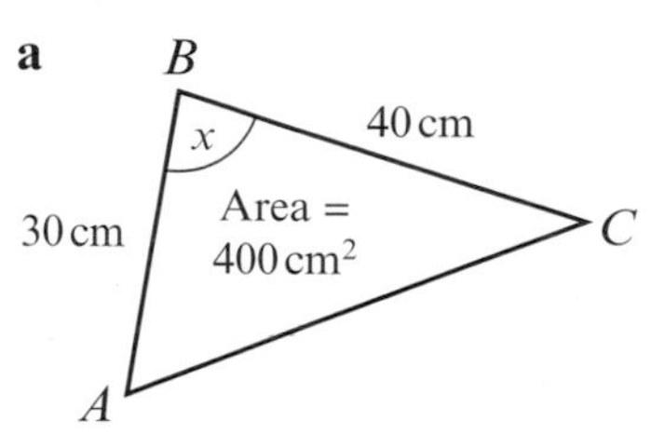

b

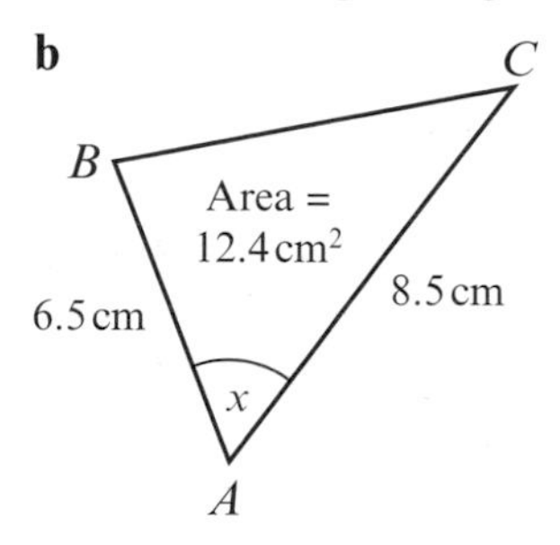

c

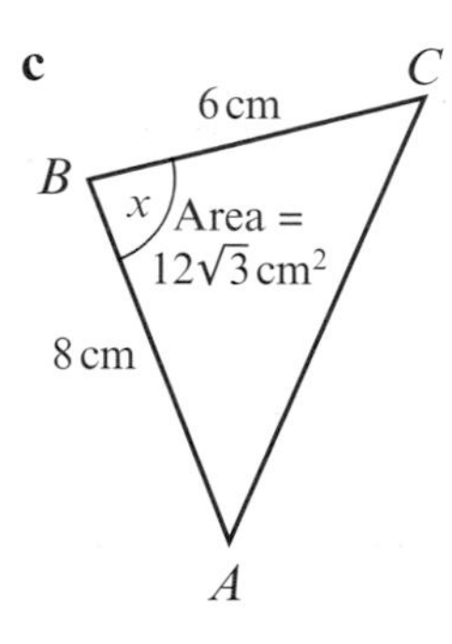

3 A fenced triangular plot of ground has area 1200 m^2. The fences along the two smaller sides are 60 m and 80 m respectively and the angle between them is θ. Show that $\theta = 150°$, and work out the total length of fencing.

(P) **4** In triangle ABC, $BC = (x + 2)$ cm, $AC = x$ cm and $\angle BCA = 150°$. Given that the area of the triangle is 5 cm^2, work out the value of x, giving your answer to 3 significant figures.

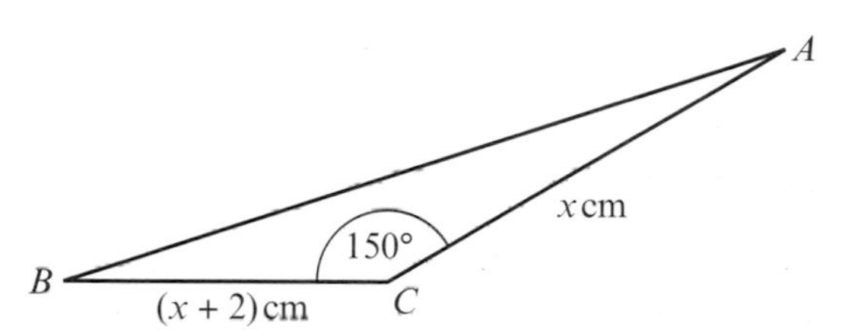

(E/P) **5** In $\triangle PQR$, $PQ = (x + 2)$ cm, $PR = (5 - x)$ cm and $\angle QPR = 30°$. The area of the triangle is $A\text{ cm}^2$.

a Show that $A = \frac{1}{4}(10 + 3x - x^2)$. **(3 marks)**

b Use the method of completing the square, or otherwise, to find the maximum value of A, and give the corresponding value of x. **(4 marks)**

(E/P) **6** In $\triangle ABC$, $AB = x$ cm, $AC = (5 + x)$ cm and $\angle BAC = 150°$. Given that the area of the triangle is $3\frac{3}{4}\text{ cm}^2$

Problem-solving

x represents a length so it must be positive.

a Show that x satisfies the equation $x^2 + 5x - 15 = 0$. **(3 marks)**

b Calculate the value of x, giving your answer to 3 significant figures. **(3 marks)**

9.4 Solving triangle problems

You can solve problems involving triangles by using the sine and cosine rules along with Pythagoras' theorem and standard right-angled triangle trigonometry.

If some of the triangles are right-angled, try to use basic trigonometry and Pythagoras' theorem first to work out other information.

If you encounter a triangle which is not right-angled, you will need to decide whether to use the sine rule or the cosine rule. Generally, use the sine rule when you are considering two angles and two sides and the cosine rule when you are considering three sides and one angle.

Watch out The sine rule is often easier to use than the cosine rule. If you know one side and an opposite angle in a triangle, try to use the sine rule to find other missing sides and angles.

For questions involving area, check first whether you can use Area = $\frac{1}{2}$ × base × height, before using the formula involving sine.

- to find an unknown angle given two sides and one opposite angle, use the sine rule
- to find an unknown side given two angles and one opposite side, use the sine rule
- to find an unknown angle given all three sides, use the cosine rule
- to find an unknown side given two sides and the angle between them, use the cosine rule
- to find the area given two sides and the angle between them, use Area = $\frac{1}{2}ab\sin C$

Example 10

The diagram shows the locations of four mobile phone masts in a field. $BC = 75$ m, $CD = 80$ m, angle $BCD = 55°$ and angle $ADC = 140°$.

In order that the masts do not interfere with each other, they must be at least 70 m apart.
Given that A is the minimum distance from D, find:

a the distance A is from B

b the angle BAD

c the area enclosed by the four masts

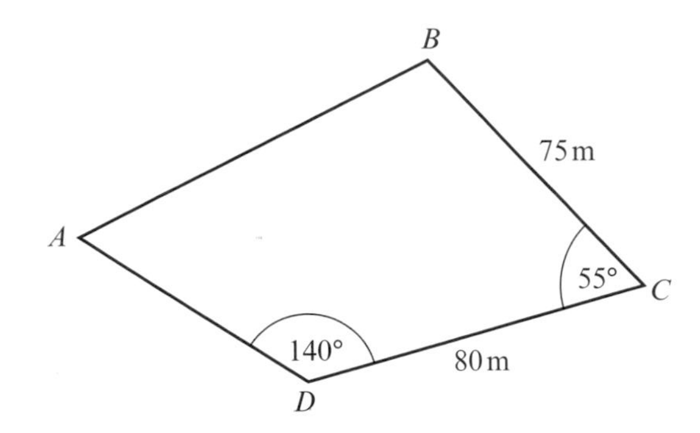

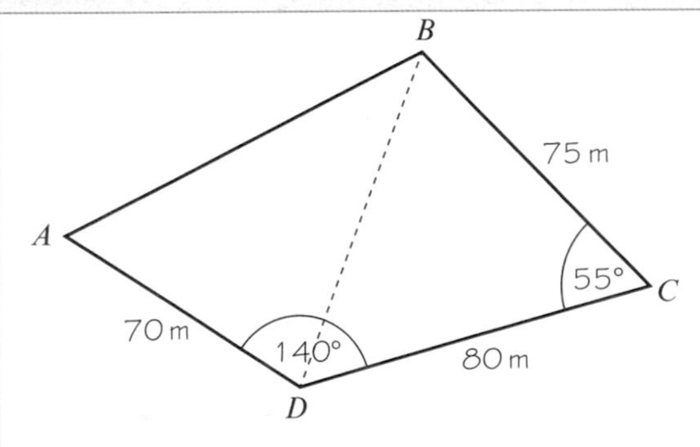

Problem-solving

Split the diagram into two triangles. Use the information in triangle BCD to work out the length BD. You are using three sides and one angle so use the **cosine rule**.

a $BD^2 = BC^2 + CD^2 - 2(BC)(CD)\cos(\angle BCD)$ — Find BD first using the cosine rule.

$BD^2 = 75^2 + 80^2 - 2(75)(80)\cos 55°$

$BD^2 = 5142.08...$

$BD = \sqrt{5142.08...} = 71.708...$ — Store this value in your calculator, or write down all the digits from your calculator display.

$\frac{\sin(\angle BDC)}{BC} = \frac{\sin(\angle BCD)}{BD}$ — You know a side and its opposite angle (BD and $\angle BCD$), so use the sine rule to calculate angle BDC.

$\sin(\angle BDC) = \frac{\sin(55°) \times 75}{71.708} = 0.85675...$

$\angle BDC = 58.954...$

$\angle BDA = 140 - 58.954... = 81.045...$ — Find BDA and store this value, or write down all the digits from your calculator display.

$AB^2 = AD^2 + BD^2 - 2(AD)(BD)\cos(\angle BDA)$

You can now use the cosine rule in triangle ABD to find AB.

$AB^2 = 70^2 + 71.708...^2$
$- 2(70)(71.708...)\cos(81.045...)$

$AB^2 = 8479.55...$

$AB = \sqrt{8479.55...} = 92.084...$
$= 92.1$ m (3 s.f.)

AB is a length, so you are only interested in the positive solution.

b $\dfrac{\sin(\angle BAD)}{BD} = \dfrac{\sin(\angle BDA)}{AB}$

Use the sine rule to calculate angle BAD.

$\sin(\angle BAD) = \dfrac{\sin(81.045...) \times 71.708...}{92.084...}$

$= 0.769...$

$\angle BAD = 50.28... = 50.3°$ (3 s.f.)

Alternatively you could have used the cosine rule with sides AB, BD and AD.

c Area $ABCD$ = area BCD + area BDA

Area $ABCD = \frac{1}{2}(BC)(CD)\sin(\angle BCD)$
$+ \frac{1}{2}(AB)(AD)\sin(\angle BAD)$

Use the area formula twice.

Area $ABCD = \frac{1}{2}(75)(80)\sin(55°)$
$+ \frac{1}{2}(92.084...)(70)\sin(50.28...°)$

Area $ABCD = 2457.4... + 2479.2...$

Area $ABCD = 4936.6... = 4940$ m^2 (3 s.f.)

Online Explore the solution step-by-step using technology.

Exercise 9E

Try to use the most efficient method, and give answers to 3 significant figures.

1 In each triangle below find the values of x, y and z.

a

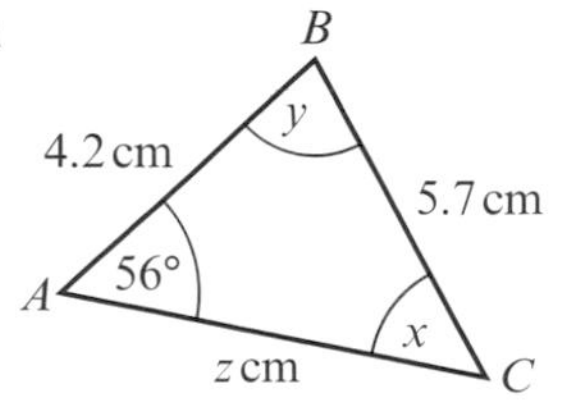

b

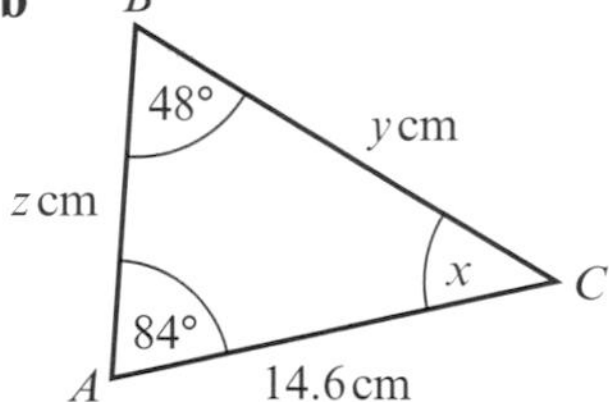

c

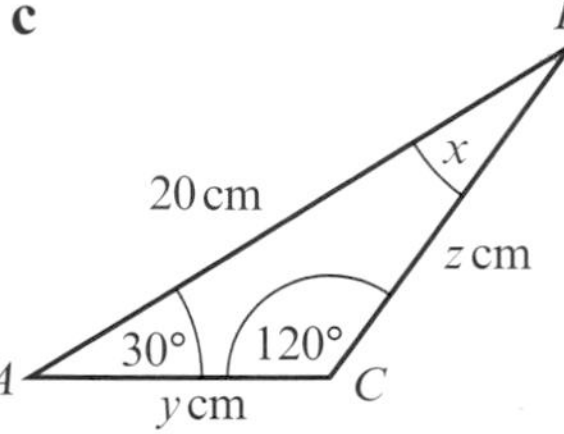

d

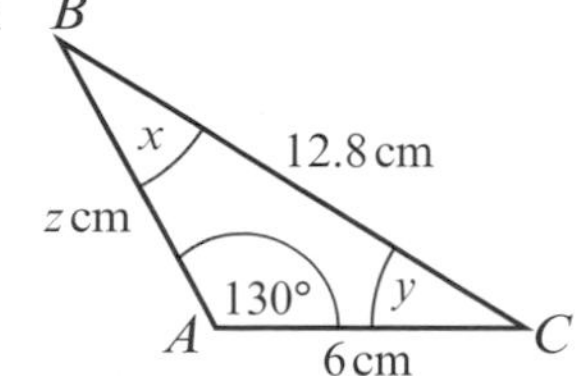

e

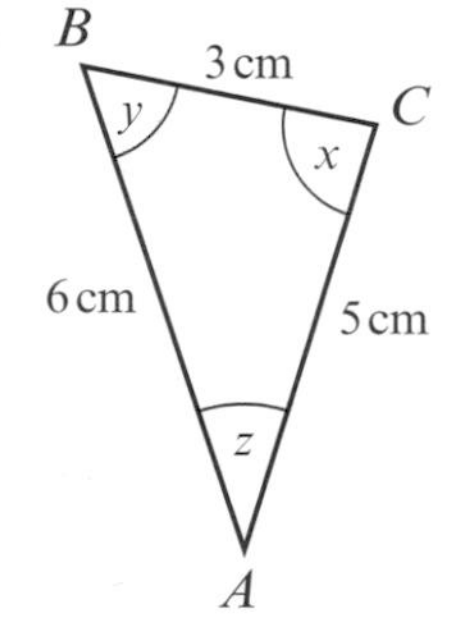

f

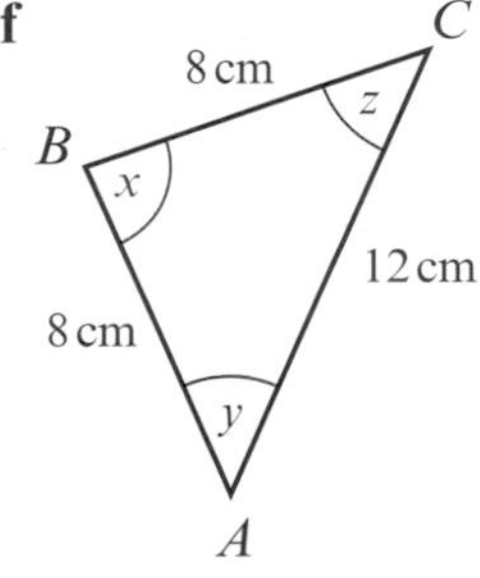

g

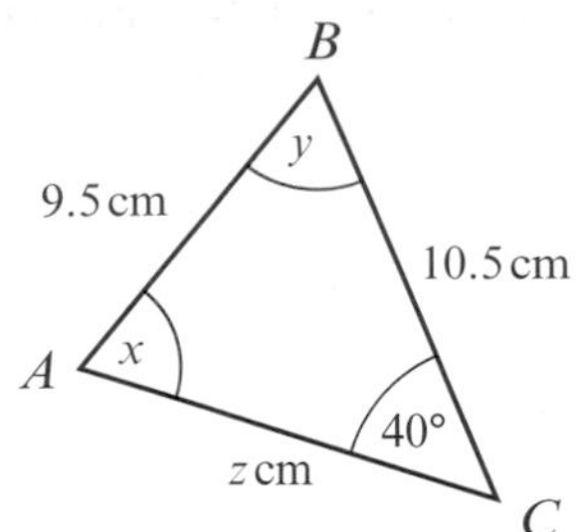

h

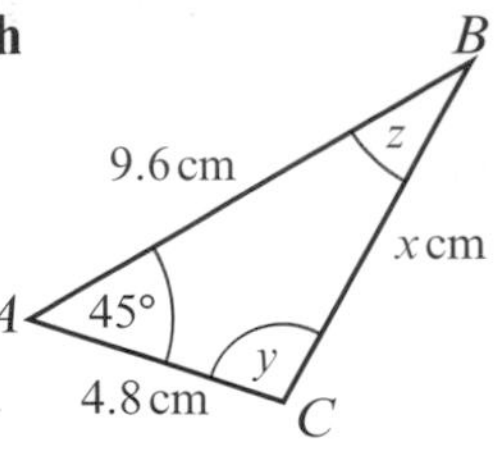

i

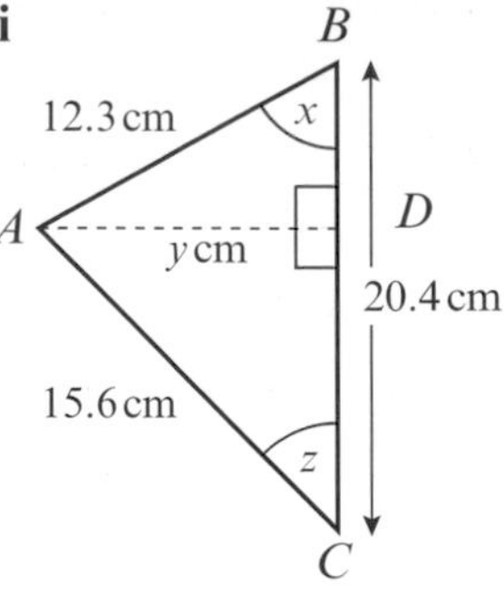

2 In $\triangle ABC$, calculate the size of the remaining angles, the lengths of the third side and the area of the triangle given that

a $\triangle BAC = 40°$, $AB = 8.5$ cm and $BC = 10.2$ cm

b $\triangle ACB = 110°$, $AC = 4.9$ cm and $BC = 6.8$ cm

3 A hiker walks due north from A and after 8 km reaches B. She then walks a further 8 km on a bearing of 120° to C. Work out **a** the distance from A to C and **b** the bearing of C from A.

(P) **4** A helicopter flies on a bearing of 200° from A to B, where $AB = 70$ km. It then flies on a bearing of 150° from B to C, where C is due south of A. Work out the distance of C from A.

(P) **5** Two radar stations A and B are 16 km apart and A is due north of B. A ship is known to be on a bearing of 150° from A and 10 km from B. Show that this information gives two positions for the ship, and calculate the distance between these two positions.

(P) **6** Find x in each of the following diagrams:

a

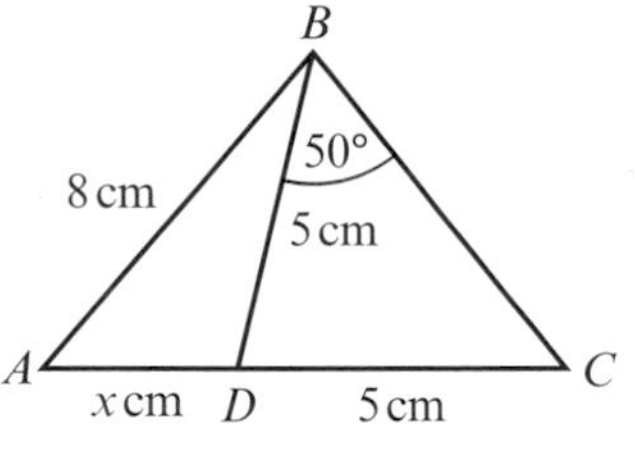

b

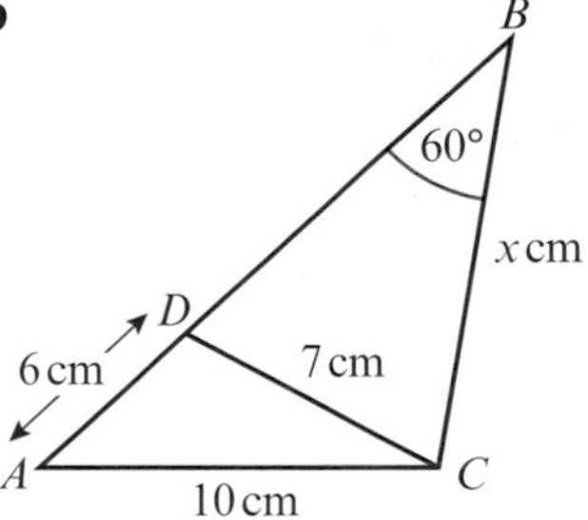

c

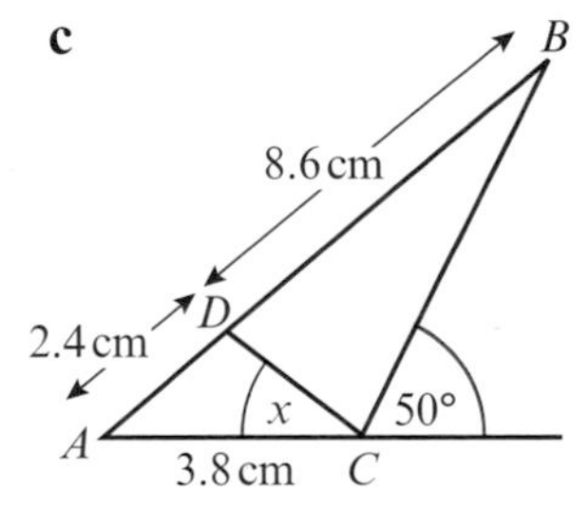

(P) **7** In $\triangle ABC$, $AB = 4$ cm, $BC = (x + 2)$ cm and $AC = 7$ cm.

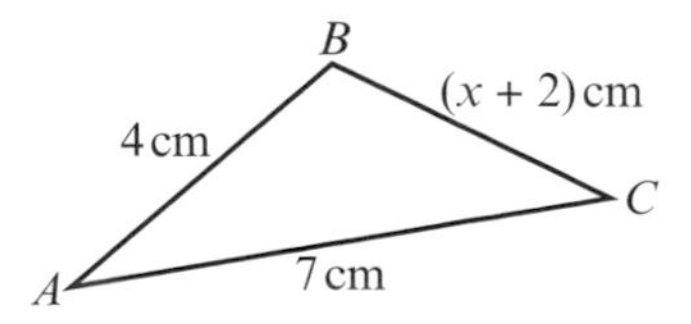

a Explain how you know that $1 < x < 9$.

b Work out the value of x and the area of the triangle for the cases when

i $\angle ABC = 60°$ and

ii $\angle ABC = 45°$, giving your answers to 3 significant figures.

(P) **8** In the triangle, $\cos \angle ABC = \frac{5}{8}$

a Calculate the value of x.

b Find the area of triangle ABC.

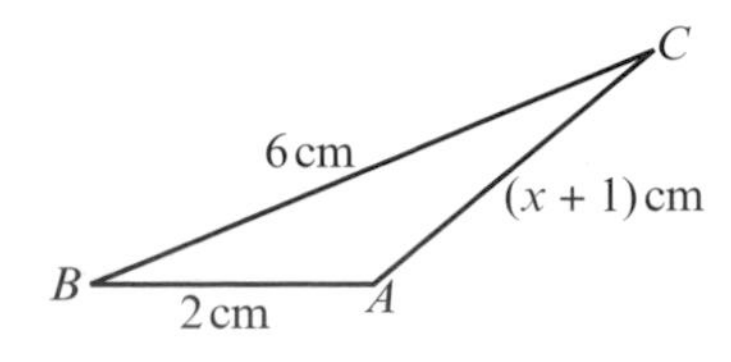

(P) **9** In $\triangle ABC$, $AB = \sqrt{2}$ cm, $BC = \sqrt{3}$ cm and $\angle BAC = 60°$. Show that $\angle ACB = 45°$ and find AC.

(P) **10** In $\triangle ABC$, $AB = (2 - x)$ cm, $BC = (x + 1)$ cm and $\angle ABC = 120°$.

a Show that $AC^2 = x^2 - x + 7$.

b Find the value of x for which AC has a minimum value.

Problem-solving

Complete the square for the expression $x^2 - x + 7$ to find the minimum value of AC^2 and the value of x where it occurs.

11 Triangle ABC is such that $BC = 5\sqrt{2}$ cm, $\angle ABC = 30°$ and $\angle BAC = \theta$, where $\sin\theta = \frac{\sqrt{5}}{8}$

Work out the length of AC, giving your answer in the form $a\sqrt{b}$, where a and b are integers.

(P) **12** The perimeter of $\triangle ABC = 15$ cm. Given that $AB = 7$ cm and $\angle BAC = 60°$, find the lengths of AC and BC and the area of the triangle.

(E) **13** In the triangle ABC, $AB = 14$ cm, $BC = 12$ cm and $CA = 15$ cm.

a Find the size of angle C, giving your answer to 3 s.f. **(3 marks)**

b Find the area of triangle ABC, giving your answer in cm^2 to 3 s.f. **(3 marks)**

(E/P) **14** A flower bed is in the shape of a quadrilateral as shown in the diagram.

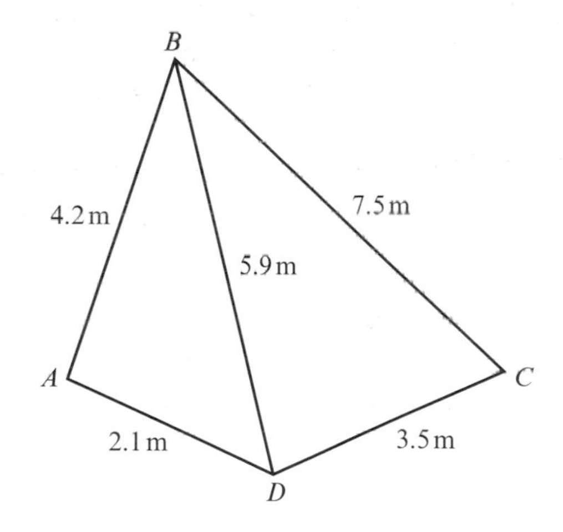

a Find the sizes of angles DAB and BCD. **(4 marks)**

b Find the total area of the flower bed. **(3 marks)**

c Find the length of the diagonal AC. **(4 marks)**

(E/P) **15** $ABCD$ is a square. Angle CED is obtuse.

Find the area of the shaded triangle. **(7 marks)**

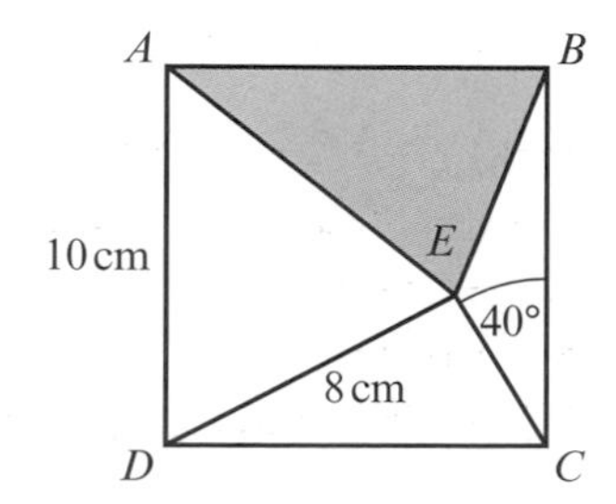

9.5 Graphs of sine, cosine and tangent

- **The graphs of sine, cosine and tangent are periodic. They repeat themselves after a certain interval.**

You need to be able to draw the graphs for a given range of angles.

- **The graph of $y = \sin\theta$:**
  - **repeats every 360° and crosses the x-axis at ..., −180°, 0, 180°, 360°, ...**
  - **has a maximum value of 1 and a minimum value of −1.**

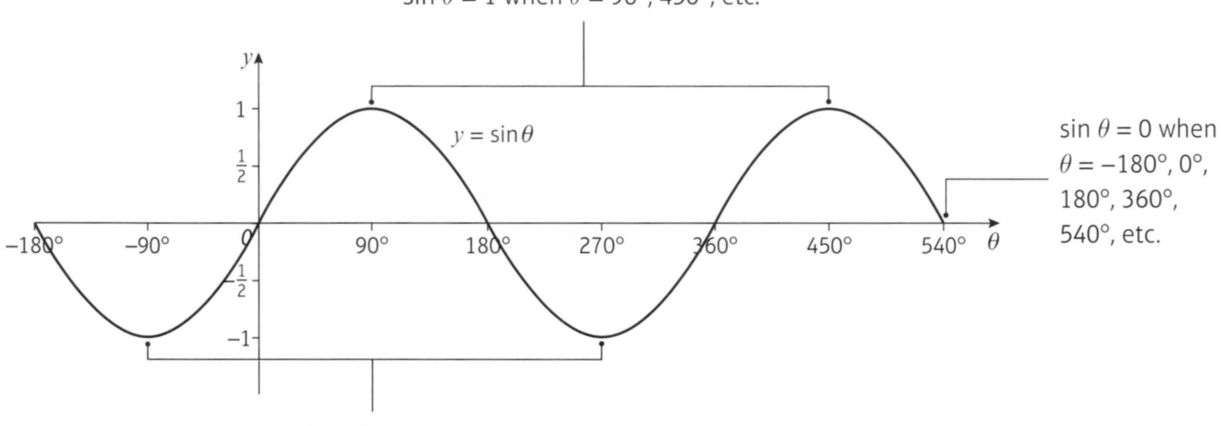

- **The graph of $y = \cos\theta$:**
  - **repeats every 360° and crosses the x-axis at ..., −90°, 90°, 270°, 450°, ...**
  - **has a maximum value of 1 and a minimum value of −1.**

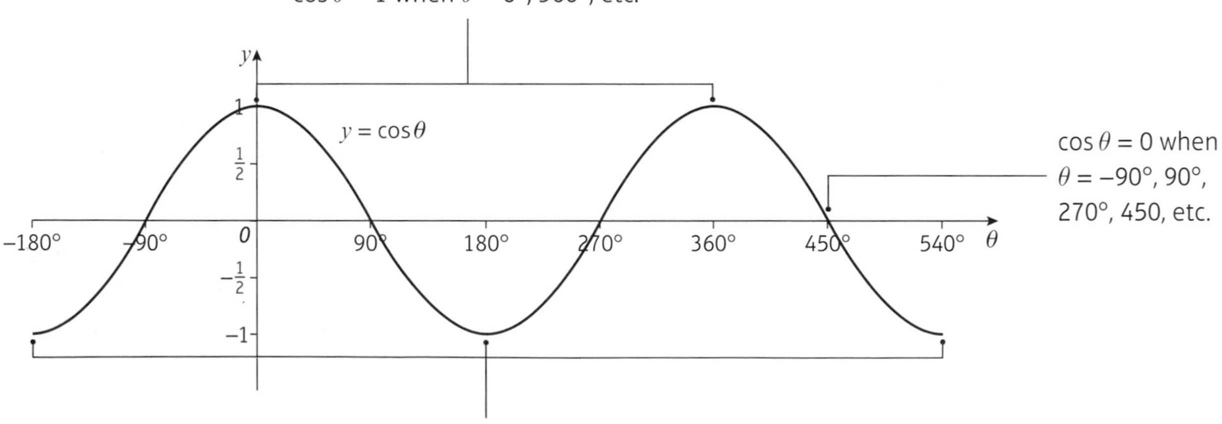

- **The graph of $y = \tan\theta$:**
  - **repeats every 180° and crosses the x-axis at ... −180°, 0°, 180°, 360°, ...**
  - **has no maximum or minimum value**
  - **has vertical asymptotes at $x = -90°$, $x = 90°$, $x = 270°$, ...**

tan θ does **not** have maximum and minimum points but approaches negative or positive infinity as the curve approaches the **asymptotes** at −90°, 90°, 270°, etc. tan θ is **undefined** for these values of θ.

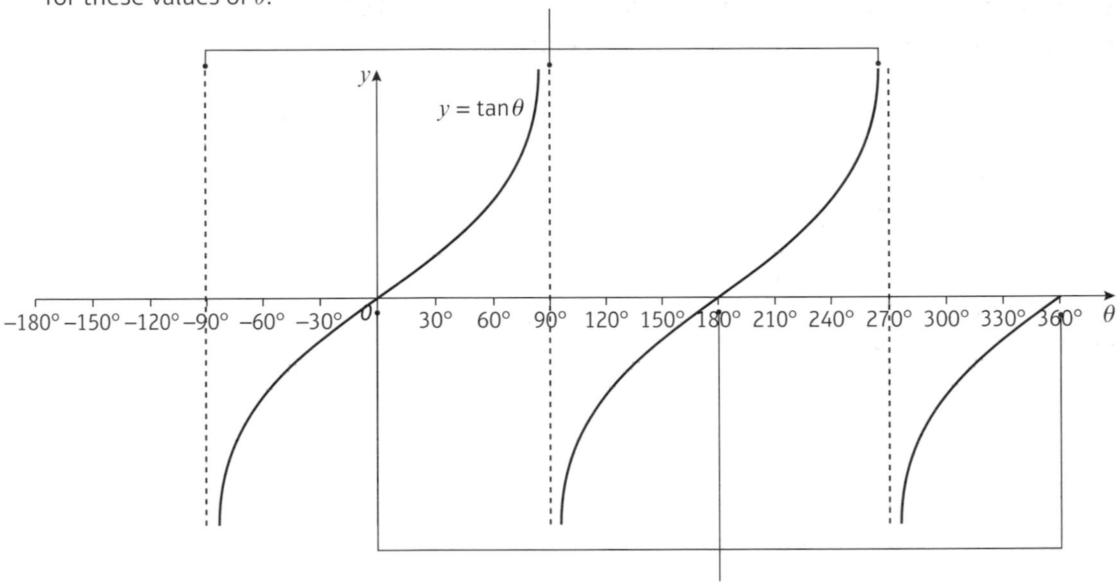

tan $\theta = 0$ when $\theta = 0°$, 180°, 360°, etc.

Example 11

a Sketch the graph of $y = \cos\theta$ in the interval $-360° \leqslant \theta \leqslant 360°$.

b **i** Sketch the graph of $y = \sin x$ in the interval $-180° \leqslant x \leqslant 270°$

ii $\sin(-30°) = -0.5$. Use your graph to determine two further values of x for which $\sin x = -0.5$.

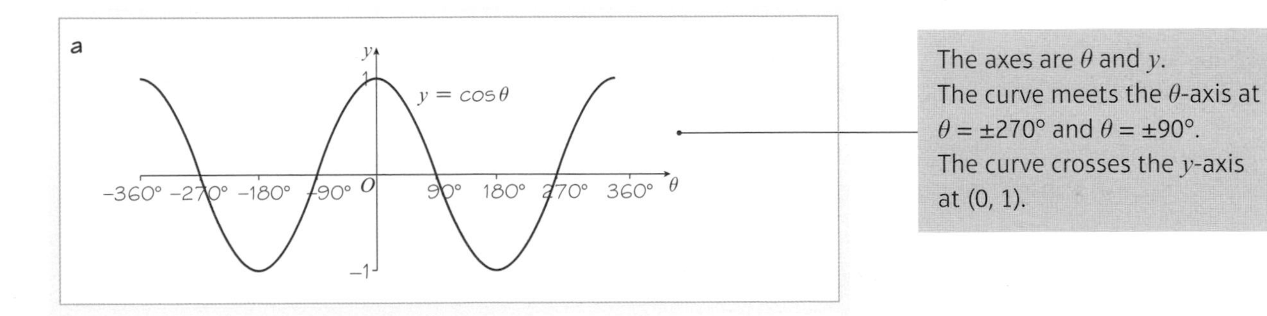

The axes are θ and y.
The curve meets the θ-axis at $\theta = \pm 270°$ and $\theta = \pm 90°$.
The curve crosses the y-axis at (0, 1).

b i

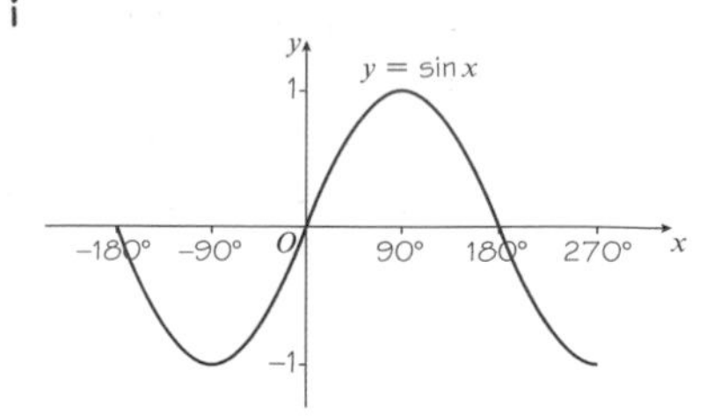

ii Using the symmetry of the graph:

$\sin(-150°) = -0.5$

$\sin 210° = -0.5$

$x = -150°$ or $210°$

The line $x = -90°$ is a line of symmetry.

The line $x = 90°$ is a line of symmetry.
You could also find this value by working out sin (180° – (–30°)).

Exercise 9F

1 Sketch the graph of $y = \cos\theta$ in the interval $-180° \leqslant \theta \leqslant 180°$.

2 Sketch the graph of $y = \tan\theta$ in the interval $-180° \leqslant \theta \leqslant 180°$.

3 Sketch the graph of $y = \sin\theta$ in the interval $-180° \leqslant \theta \leqslant 180°$.

4 **a** $\cos 30° = \frac{\sqrt{3}}{2}$ Use your graph in question **1** to find another value of θ for which $\cos\theta = \frac{\sqrt{3}}{2}$

b $\tan 60° = \sqrt{3}$. Use your graph in question **2** to find other values of θ for which:

i $\tan\theta = \sqrt{3}$ **ii** $\tan\theta = -\sqrt{3}$

c $\sin 45° = \frac{1}{\sqrt{2}}$ Use your graph in question **3** to find other values of θ for which:

i $\sin\theta = \frac{1}{\sqrt{2}}$ **ii** $\sin\theta = -\frac{1}{\sqrt{2}}$

9.6 Transforming trigonometric graphs

You can use your knowledge of transforming graphs to transform the graphs of trigonometric functions.

Links You need to be able to apply translations and stretches to graphs of trigonometric functions.
← Chapter 4

Example 12

Sketch on separate sets of axes the graphs of:

a $y = 3\sin x, 0 \leqslant x \leqslant 360°$

b $y = -\tan\theta, -180° \leqslant \theta \leqslant 180°$

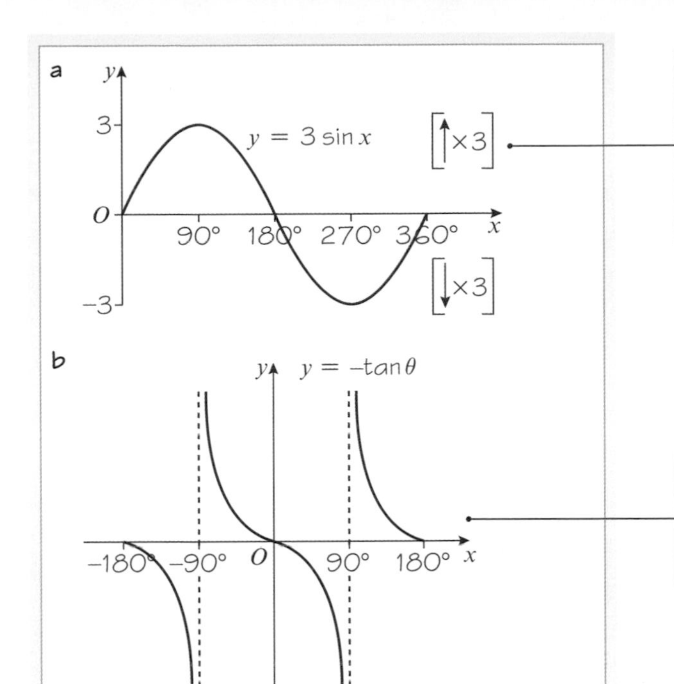

$y = 3f(x)$ is a vertical stretch of the graph $y = f(x)$ with scale factor 3. The intercepts on the x-axis remain unchanged, and the graph has a maximum point at (90°, 3) and a minimum point at (270°, −3).

$y = -f(x)$ is a reflection of the graph $y = f(x)$ in the x-axis. So this graph is a reflection of the graph $y = \tan x$ in the x-axis.

Example 13

Sketch on separate sets of axes the graphs of:

a $y = -1 + \sin x$, $0 \leqslant x \leqslant 360°$ **b** $y = \frac{1}{2} + \cos x$, $0 \leqslant x \leqslant 360°$

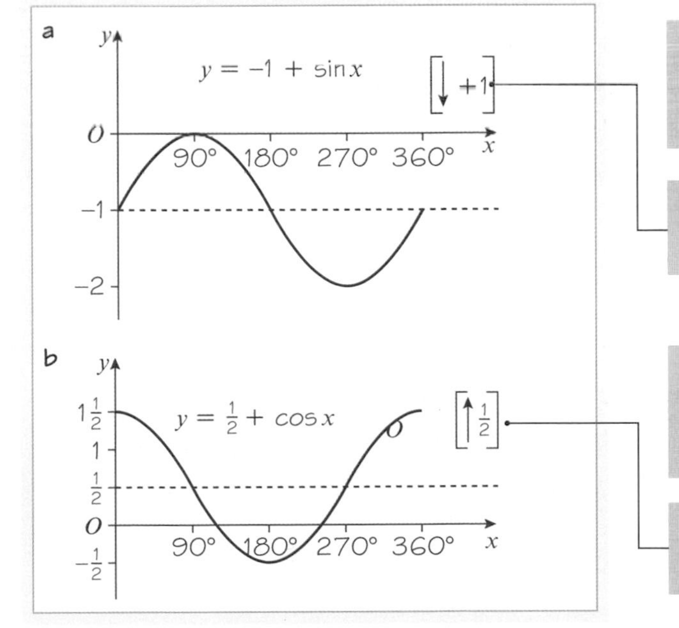

$y = f(x) - 1$ is a translation of the graph $y = f(x)$ by vector $\begin{pmatrix} 0 \\ -1 \end{pmatrix}$.

The graph of $y = \sin x$ is translated by 1 unit in the negative y-direction.

$y = f(x) + \frac{1}{2}$ is a translation of the graph $y = f(x)$ by vector $\begin{pmatrix} 0 \\ \frac{1}{2} \end{pmatrix}$.

The graph of $y = \cos x$ is translated by $\frac{1}{2}$ unit in the positive y-direction.

Example 14

Sketch on separate sets of axes the graphs of:

a $y = \tan(\theta + 45°)$, $0 \leqslant \theta \leqslant 360°$ **b** $y = \cos(\theta - 90°)$, $-360° \leqslant \theta \leqslant 360°$

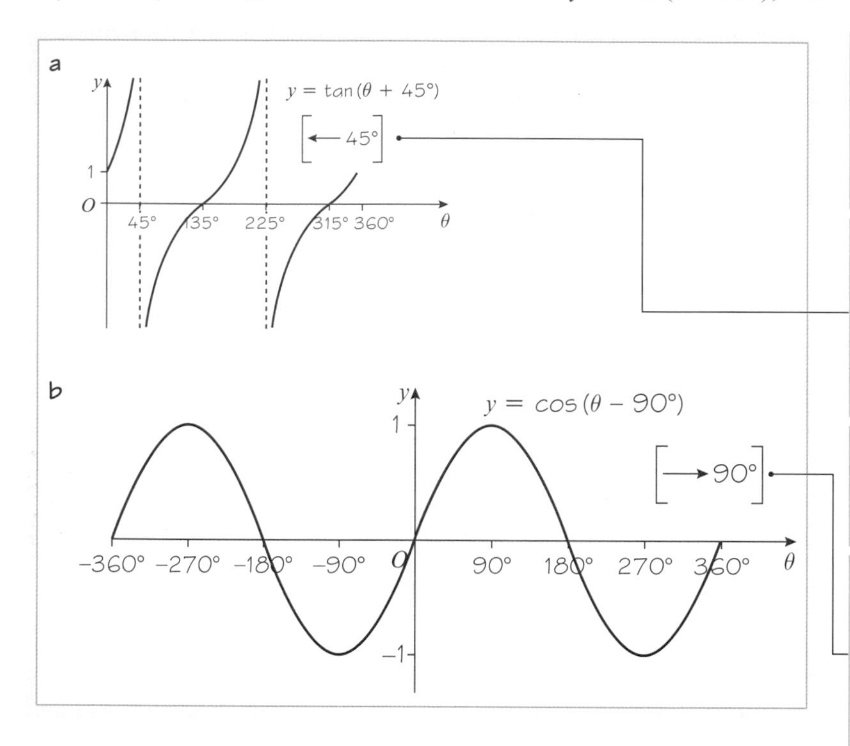

$y = f(\theta + 45°)$ is a translation of the graph $y = f(\theta)$ by vector $\begin{pmatrix} -45° \\ 0 \end{pmatrix}$. Remember to translate any asymptotes as well.

The graph of $y = \tan\theta$ is translated by 45° to the left. The asymptotes are now at $\theta = 45°$ and $\theta = 225°$. The curve meets the y-axis where $\theta = 0$ so $y = 1$.

$y = f(\theta - 90°)$ is a translation of the graph $y = f(\theta)$ by vector $\begin{pmatrix} 90° \\ 0 \end{pmatrix}$.

The graph of $y = \cos\theta$ is translated by 90° to the right. Note that this is exactly the same curve as $y = \sin\theta$, so another property is that $\cos(\theta - 90°) = \sin\theta$.

Example 15

Sketch on separate sets of axes the graphs of:

a $y = \sin 2x$, $0 \leqslant x \leqslant 360°$ **b** $y = \cos\frac{\theta}{3}$, $-540° \leqslant \theta \leqslant 540°$ **c** $y = \tan(-x)$, $-360° \leqslant x \leqslant 360°$

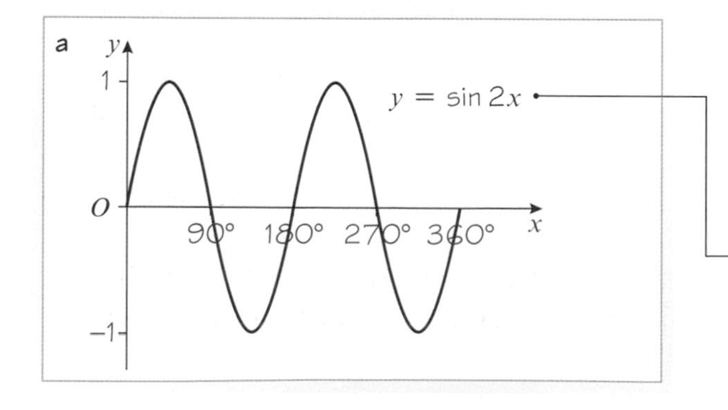

$y = f(2x)$ is a horizontal stretch of the graph $y = f(x)$ with scale factor $\frac{1}{2}$.

The graph of $y = \sin x$ is stretched horizontally with scale factor $\frac{1}{2}$

The period is now 180° and two complete 'waves' are seen in the interval $0 \leqslant x \leqslant 360°$.

b

$y = f(\frac{1}{3}\theta)$ is a horizontal stretch of the graph $y = f(\theta)$ with scale factor 3.

The graph of $y = \cos\theta$ is stretched horizontally with scale factor 3. The period of $\cos\frac{\theta}{3}$ is 1080° and only one complete wave is seen while $-540 \leqslant \theta \leqslant 540°$. The curve crosses the θ-axis at $\theta = \pm 270°$.

c

$y = f(-x)$ is a reflection of the graph $y = f(x)$ in the y-axis.

The graph of $y = \tan(-x)$ is reflected in the y-axis. In this case the asymptotes are all vertical so they remain unchanged.

Online Plot transformations of trigonometric graphs using technology.

Exercise 9G

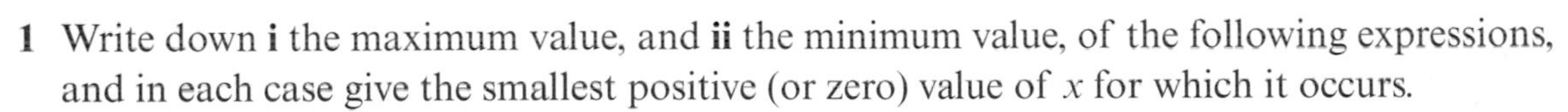

1 Write down **i** the maximum value, and **ii** the minimum value, of the following expressions, and in each case give the smallest positive (or zero) value of x for which it occurs.

a $\cos x$ **b** $4\sin x$ **c** $\cos(-x)$

d $3 + \sin x$ **e** $-\sin x$ **f** $\sin 3x$

2 Sketch, on the same set of axes, in the interval $0 \leqslant \theta \leqslant 360°$, the graphs of $\cos\theta$ and $\cos 3\theta$.

3 Sketch, on separate sets of axes, the graphs of the following, in the interval $0 \leqslant \theta \leqslant 360°$. Give the coordinates of points of intersection with the axes, and of maximum and minimum points where appropriate.

a $y = -\cos\theta$ **b** $y = \frac{1}{3}\sin\theta$ **c** $y = \sin\frac{1}{3}\theta$ **d** $y = \tan(\theta - 45°)$

4 Sketch, on separate sets of axes, the graphs of the following, in the interval $-180° \leqslant \theta \leqslant 180°$. Give the coordinates of points of intersection with the axes, and of maximum and minimum points where appropriate.

a $y = -2\sin\theta$ **b** $y = \tan(\theta + 180°)$ **c** $y = \cos 4\theta$ **d** $y = \sin(-\theta)$

5 Sketch, on separate sets of axes, the graphs of the following in the interval $-360° \leqslant \theta \leqslant 360°$. In each case give the periodicity of the function.

a $y = \sin\frac{1}{2}\theta$ **b** $y = -\frac{1}{2}\cos\theta$ **c** $y = \tan(\theta - 90°)$ **d** $y = \tan 2\theta$

(P) **6 a** By considering the graphs of the functions, or otherwise, verify that:

i $\cos\theta = \cos(-\theta)$

ii $\sin\theta = -\sin(-\theta)$

iii $\sin(\theta - 90°) = -\cos\theta$.

b Use the results in **a ii** and **iii** to show that $\sin(90° - \theta) = \cos\theta$.

c In Example 14 you saw that $\cos(\theta - 90°) = \sin\theta$.
Use this result with part **a i** to show that $\cos(90° - \theta) = \sin\theta$.

(E) **7** The graph shows the curve
$y = \cos(x + 30°)$, $-360° \leqslant x \leqslant 360°$.

a Write down the coordinates of the points where the curve crosses the x-axis. **(2 marks)**

b Find the coordinates of the point where the curve crosses the y-axis. **(1 mark)**

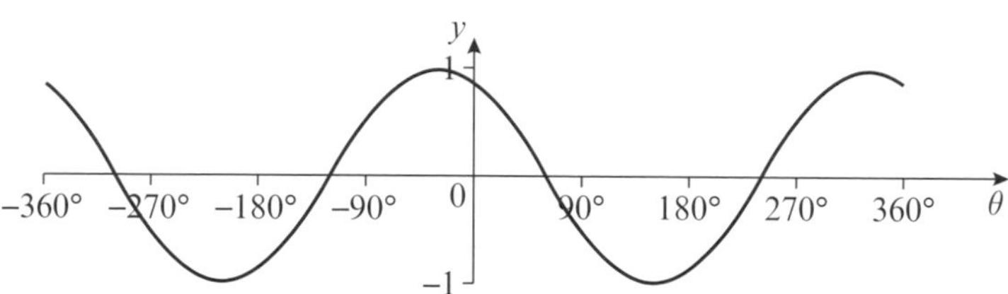

(E/P) **8** The graph shows the curve with equation
$y = \sin(x + k)$, $-360° \leqslant x \leqslant 360°$,
where k is a constant.

a Find one possible value for k. **(2 marks)**

b Is there more than one possible answer to part **a**? Give a reason for your answer. **(2 marks)**

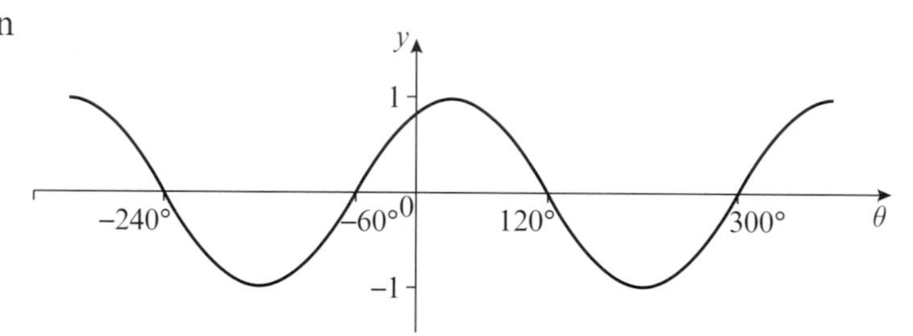

(E/P) **9** The variation in the depth of water in a rock pool can be modelled using the function $y = \sin(30t)°$, where t is the time in hours and $0 \leqslant t \leqslant 6$.

a Sketch the function for the given interval. **(2 marks)**

b If $t = 0$ represents midday, during what times will the rock pool be at least half full? **(3 marks)**

Mixed exercise 9

Give non-exact answers to 3 significant figures.

1 Triangle ABC has area 10 cm^2. $AB = 6$ cm, $BC = 8$ cm and $\angle ABC$ is obtuse. Find:

a the size of $\angle ABC$

b the length of AC

2 In each triangle below, find the size of x and the area of the triangle.

a

2.4 cm, 1.2 cm, 3 cm, x

b

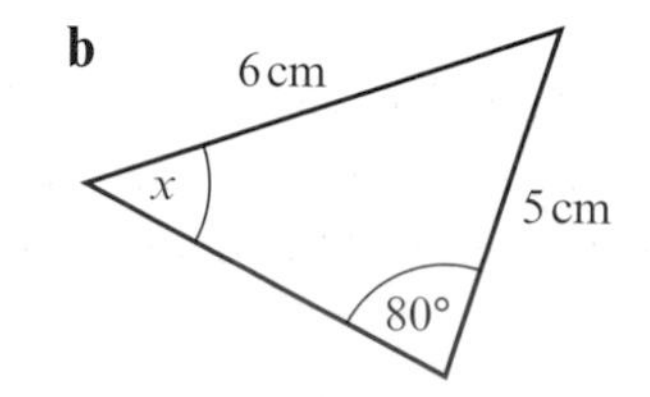

c

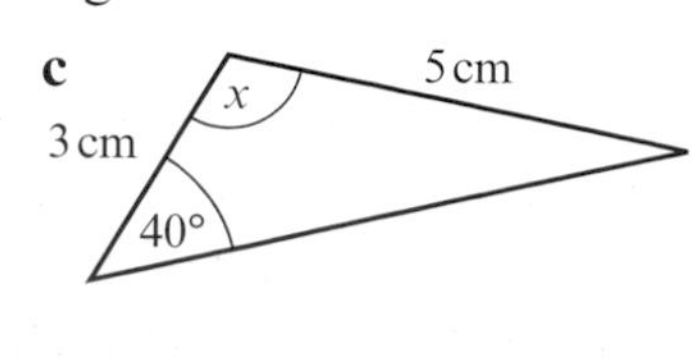

3 The sides of a triangle are 3 cm, 5 cm and 7 cm respectively. Show that the largest angle is 120°, and find the area of the triangle.

(P) **4** In each of the figures below calculate the total area.

a

10.4 cm
B
100°
8.2 cm
30.6°
A
D
C

b

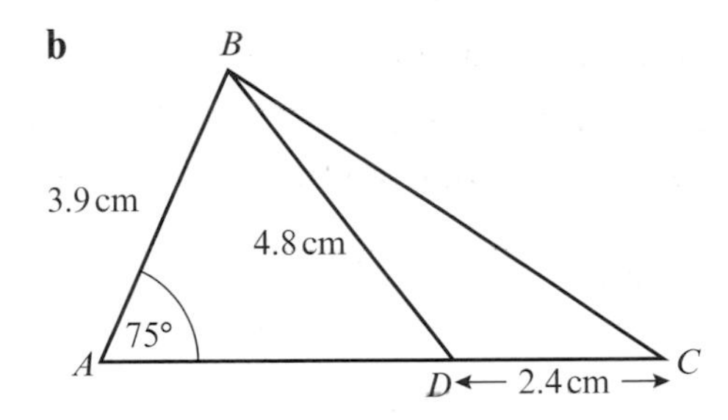

5 In $\triangle ABC$, $AB = 10$ cm, $BC = a\sqrt{3}$ cm, $AC = 5\sqrt{13}$ cm and $\angle ABC = 150°$. Calculate:

a the value of a

b the exact area of $\triangle ABC$.

(P) **6** In a triangle, the largest side has length 2 cm and one of the other sides has length $\sqrt{2}$ cm. Given that the area of the triangle is 1 cm^2, show that the triangle is right-angled and isosceles.

(E/P) **7** The three points A, B and C, with coordinates $A(0, 1)$, $B(3, 4)$ and $C(1, 3)$ respectively, are joined to form a triangle.

a Show that $\cos \angle ACB = -\frac{4}{5}$ **(5 marks)**

b Calculate the area of $\triangle ABC$. **(2 marks)**

(E/P) **8** The longest side of a triangle has length $(2x - 1)$ cm. The other sides have lengths $(x - 1)$ cm and $(x + 1)$ cm. Given that the largest angle is 120°, work out

a the value of x **(5 marks)**

b the area of the triangle. **(3 marks)**

(E/P) **9** A park keeper walks 1.2 km due north from his hut at point A to point B. He then walks 1.4 km on a bearing of 110° from point B to point C.

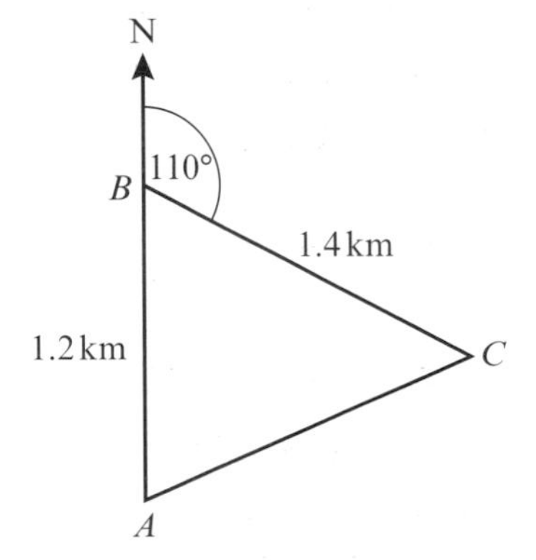

a Find how far he is from his hut when at point C. Give your answer in km to 3 s.f. **(3 marks)**

b Work out the bearing of the hut from point C. Give your answer to the nearest degree. **(3 marks)**

c Work out the area enclosed by his walk. **(3 marks)**

(E/P) **10** A windmill has four identical triangular sails made from wood. If each triangle has sides of length 12 m, 15 m and 20 m, work out the total area of wood needed. **(5 marks)**

(E/P) **11** Two points, A and B are on level ground. A church tower at point C has an angle of elevation from A of 15° and an angle of elevation from B of 32°. A and B are both on the same side of C, and A, B and C lie on the same straight line. The distance $AB = 75$ m.
Find the height of the church tower. **(4 marks)**

12 Describe geometrically the transformations which map:

a the graph of $y = \tan x$ onto the graph of $\tan \frac{1}{2}x$

b the graph of $y = \tan \frac{1}{2}x$ onto the graph of $3 + \tan \frac{1}{2}x$

c the graph of $y = \cos x$ onto the graph of $-\cos x$

d the graph of $y = \sin (x - 10)$ onto the graph of $\sin (x + 10)$.

13 a Sketch on the same set of axes, in the interval $0 \leqslant x \leqslant 180°$, the graphs of $y = \tan (x - 45°)$ and $y = -2\cos x$, showing the coordinates of points of intersection with the axes. **(6 marks)**

b Deduce the number of solutions of the equation $\tan (x - 45°) + 2\cos x = 0$, in the interval $0 \leqslant x \leqslant 180°$. **(2 marks)**

E **14** The diagram shows part of the graph of $y = \text{f}(x)$.
It crosses the x-axis at $A(120°, 0)$ and $B(p, 0)$.
It crosses the y-axis at $C(0, q)$ and has a maximum value at D, as shown.

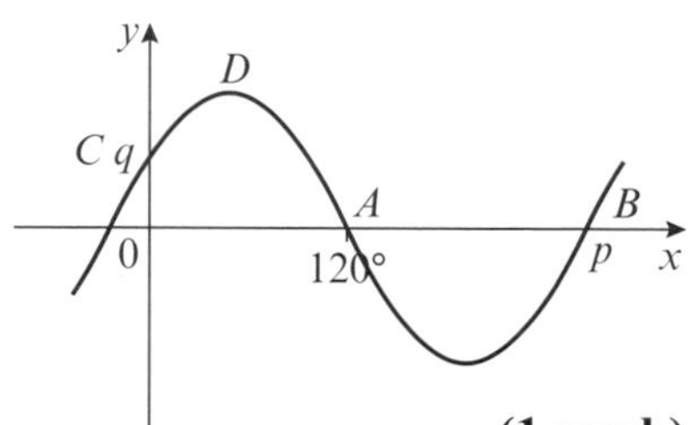

Given that $\text{f}(x) = \sin (x + k)$, where $k > 0$, write down

a the value of p **(1 mark)**

b the coordinates of D **(1 mark)**

c the smallest value of k **(1 mark)**

d the value of q. **(1 mark)**

15 Consider the function $\text{f}(x) = \sin px, p \in \mathbb{R}, 0 \leqslant x \leqslant 360°$.
The closest point to the origin that the graph of $\text{f}(x)$ crosses the x-axis has x-coordinate $36°$.

a Determine the value of p and sketch the graph of $y = \text{f}(x)$. **(5 marks)**

b Write down the period of $\text{f}(x)$. **(1 mark)**

16 The graph shows $y = \sin \theta, 0 \leqslant \theta \leqslant 360°$, with one value of $\theta\,(\theta = \alpha)$ marked on the axis.

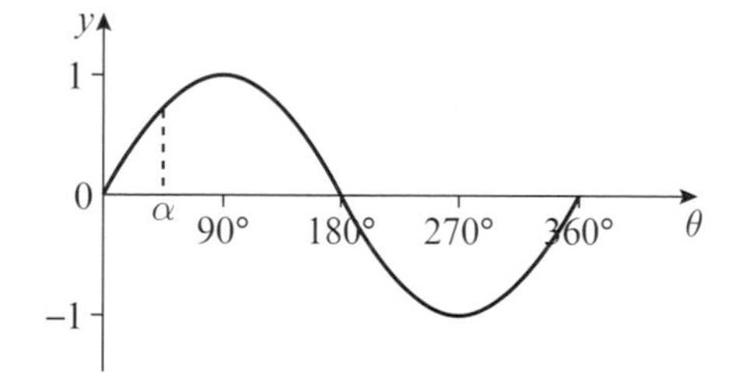

a Copy the graph and mark on the θ-axis the positions of $180° - \alpha$, $180° + \alpha$, and $360° - \alpha$.

b Verify that:
$\sin \alpha = \sin (180° - \alpha) = -\sin (180° + \alpha) = -\sin (360° - \alpha)$.

17 a Sketch on separate sets of axes the graphs of $y = \cos \theta\,(0 \leqslant \theta \leqslant 360°)$ and $y = \tan \theta\,(0 \leqslant \theta \leqslant 360°)$, and on each θ-axis mark the point $(\alpha, 0)$ as in question **16**.

b Verify that:

i $\cos \alpha = -\cos (180° - \alpha) = -\cos (180° + \alpha) = \cos (360° - \alpha)$

ii $\tan \alpha = -\tan (180° - \alpha) = \tan (180° + \alpha) = -\tan (360° - \alpha)$

18 A series of sand dunes has a cross-section which can be modelled using a sine curve of the form $y = \sin (60x)°$ where x is the length of the series of dunes in metres.

a Draw the graph of $y = \sin (60x)°$ for $0 \leqslant x \leqslant 24°$. **(3 marks)**

b Write down the number of sand dunes in this model. **(1 mark)**

c Give one reason why this may not be a realistic model. **(1 mark)**

Challenge

In this diagram $AB = BC = CD = DE = 1$ m.

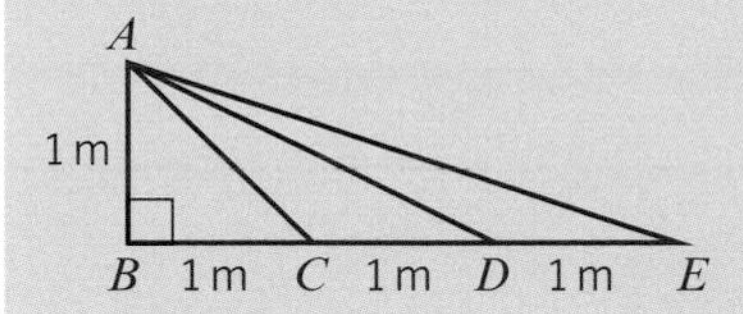

Prove that $\angle AEB + \angle ADB = \angle ACB$.

Hint Try drawing triangles ADB and AEB back to back.

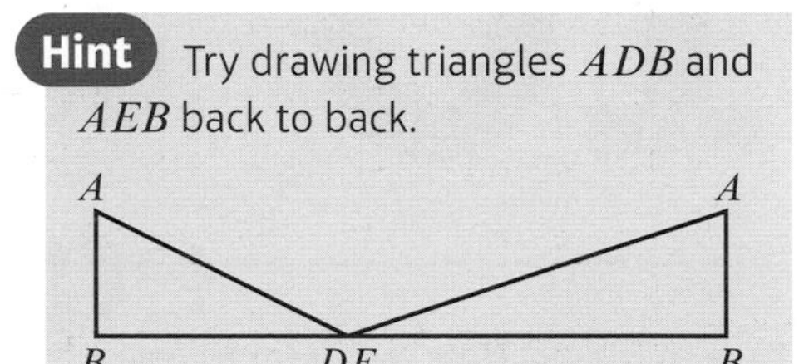

Summary of key points

1 This version of the cosine rule is used to find a missing side if you know two sides and the angle between them:

$$a^2 = b^2 + c^2 - 2bc \cos A$$

2 This version of the cosine rule is used to find an angle if you know all three sides:

$$\cos A = \frac{b^2 + c^2 - a^2}{2bc}$$

3 This version of the sine rule is used to find the length of a missing side:

$$\frac{a}{\sin A} = \frac{b}{\sin B} = \frac{c}{\sin C}$$

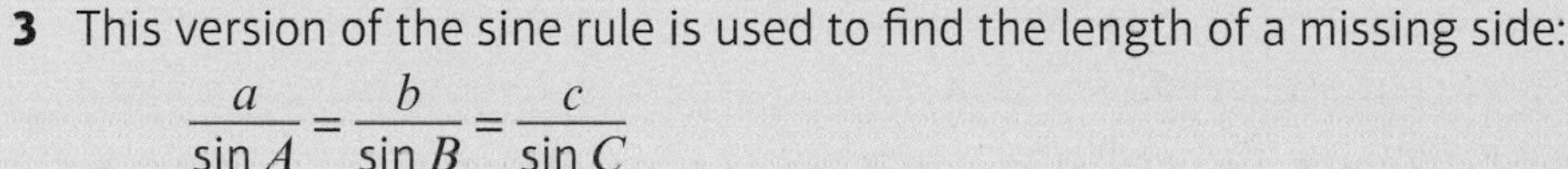

4 This version of the sine rule is used to find a missing angle:

$$\frac{\sin A}{a} = \frac{\sin B}{b} = \frac{\sin C}{c}$$

5 The sine rule sometimes produces two possible solutions for a missing angle:

$$\sin \theta = \sin (180° - \theta)$$

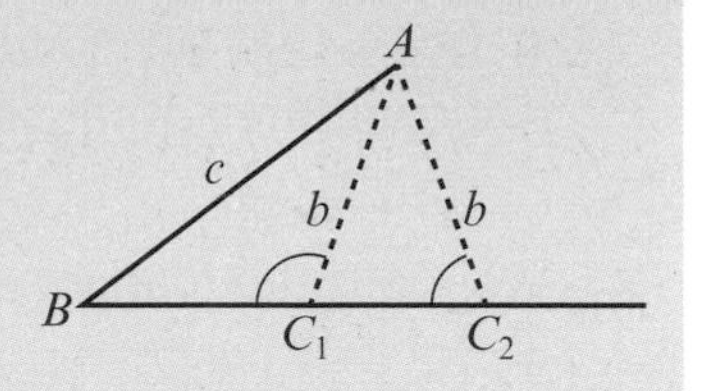

6 Area of a triangle $= \frac{1}{2}ab \sin C$.

7 The graphs of sine, cosine and tangent are **periodic**. They repeat themselves after a certain interval.

- The graph of $y = \sin \theta$: repeats every 360° and crosses the x-axis at …, −180°, 0, 180°, 360°, …
- has a maximum value of 1 and a minimum value of −1.
- The graph of $y = \cos \theta$: repeats every 360° and crosses the x-axis at …, −90°, 90°, 270°, 450°, …
- has a maximum value of 1 and a minimum value of −1
- The graph of $y = \tan \theta$: repeats every 180° and crosses the x-axis at … −180°, 0°, 180°, 360°, …
- has no maximum or minimum value
- has vertical asymptotes at $x = -90°$, $x = 90°$, $x = 270°$, …

10 Trigonometric identities and equations

Objectives

After completing this chapter you should be able to:

- Calculate the sine, cosine and tangent of any angle → pages 203–208
- Know the exact trigonometric ratios for 30°, 45° and 60° → pages 208–209
- Know and use the relationships $\tan\theta \equiv \dfrac{\sin\theta}{\cos\theta}$ and $\sin^2\theta + \cos^2\theta \equiv 1$ → pages 209–213
- Solve simple trigonometric equations of the forms $\sin\theta = k$, $\cos\theta = k$ and $\tan\theta = k$ → pages 213–217
- Solve more complicated trigonometric equations of the forms $\sin n\theta = k$ and $\sin(\theta \pm \alpha) = k$ and equivalent equations involving cos and tan → pages 217–219
- Solve trigonometric equations that produce quadratics → pages 219–222

Prior knowledge check

1 **a** Sketch the graph of $y = \sin x$ for $0 \leqslant x \leqslant 540°$.
b How many solutions are there to the equation $\sin x = 0.6$ in the range $0 \leqslant x \leqslant 540°$?
c Given that $\sin^{-1}(0.6) = 36.9°$ (to 3 s.f.), write down three other solutions to the equation $\sin x = 0.6$. ← Section 9.5

2 Work out the marked angles in these triangles.

a

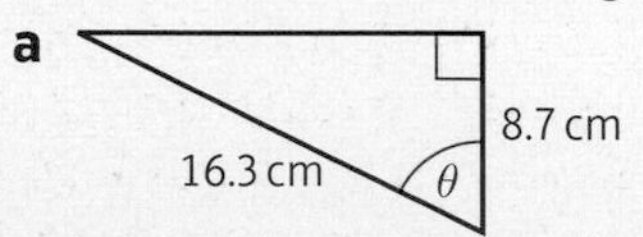

b

θ
6.1 cm
20 cm

← GCSE Mathematics

3 Solve the following equations.
a $2x - 7 = 15$ **b** $3x + 5 = 7x - 4$
c $\sin x = -0.7$ ← GCSE Mathematics

4 Solve the following equations.
a $x^2 - 4x + 3 = 0$ **b** $x^2 + 8x - 9 = 0$
c $2x^2 - 3x - 7 = 0$ ← Section 2.1

Trigonometric equations can be used to model many real-life situations such as the rise and fall of the tides or the angle of elevation of the sun at different times of the day.

10.1 Angles in all four quadrants

You can use a unit circle with its centre at the origin to help you understand the trigonometric ratios.

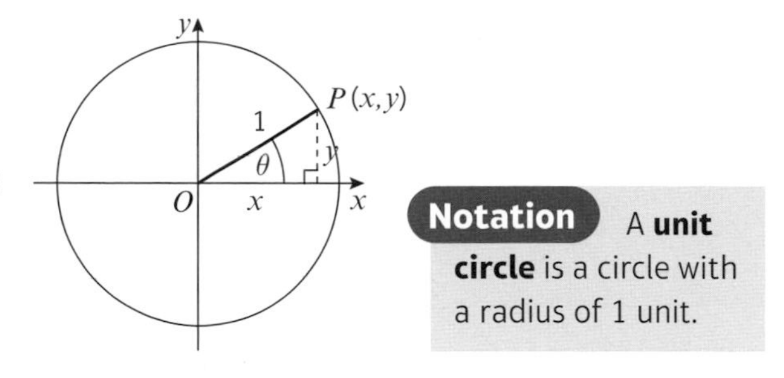

- **For a point $P(x, y)$ on a unit circle such that OP makes an angle θ with the positive x-axis:**
  - **$\cos\theta = x$ = x-coordinate of P**
  - **$\sin\theta = y$ = y-coordinate of P**
  - **$\tan\theta = \frac{y}{x}$ = gradient of OP**

Notation A **unit circle** is a circle with a radius of 1 unit.

You can use these definitions to find the values of sine, cosine and tangent for any angle θ. You always measure positive angles θ **anticlockwise** from the **positive x-axis**.

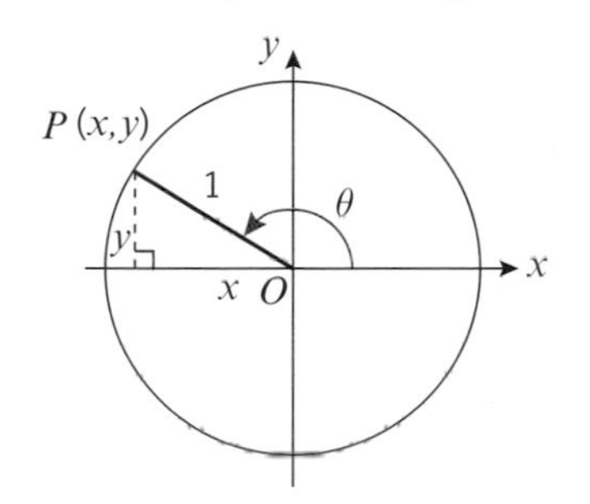

When θ is **obtuse**, $\cos\theta$ is negative because the x-coordinate of P is negative.

Online Use GeoGebra to explore the values of $\sin\theta$, $\cos\theta$ and $\tan\theta$ for any angle θ in a unit circle.

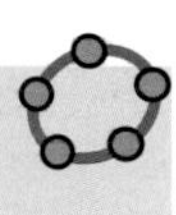

You can also use these definitions to generate the graphs of $y = \sin\theta$ and $y = \cos\theta$.

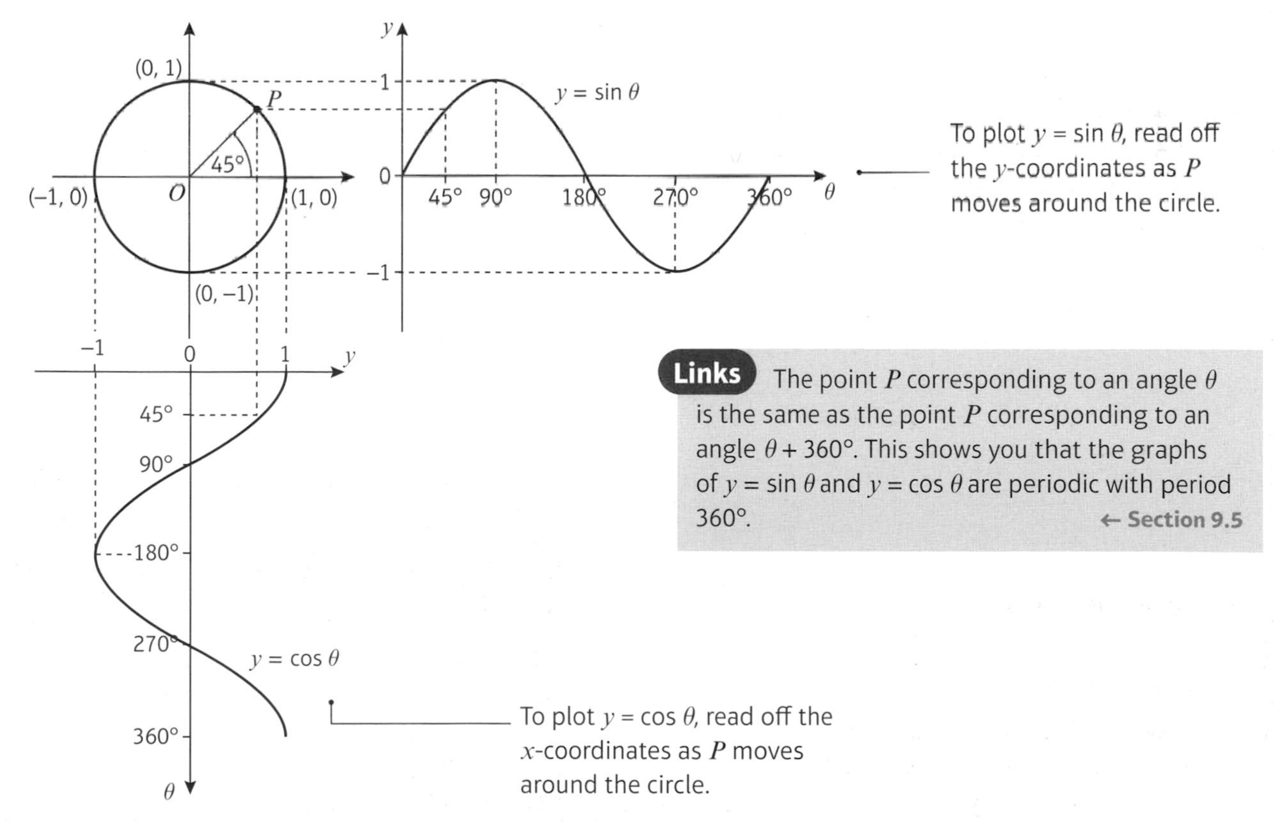

To plot $y = \sin\theta$, read off the y-coordinates as P moves around the circle.

Links The point P corresponding to an angle θ is the same as the point P corresponding to an angle $\theta + 360°$. This shows you that the graphs of $y = \sin\theta$ and $y = \cos\theta$ are periodic with period 360°. ← Section 9.5

To plot $y = \cos\theta$, read off the x-coordinates as P moves around the circle.

Example 1

Write down the values of:

a $\sin 90°$ **b** $\sin 180°$ **c** $\sin 270°$
d $\cos 180°$ **e** $\cos(-90)°$ **f** $\cos 450°$

a $\sin 90° = 1$

b $\sin 180° = 0$

c $\sin 270° = -1$

d $\cos 180° = -1$

e $\cos(-90°) = 0$

f $\cos 450° = 0$

The y-coordinate is 1 when $\theta = 90°$.

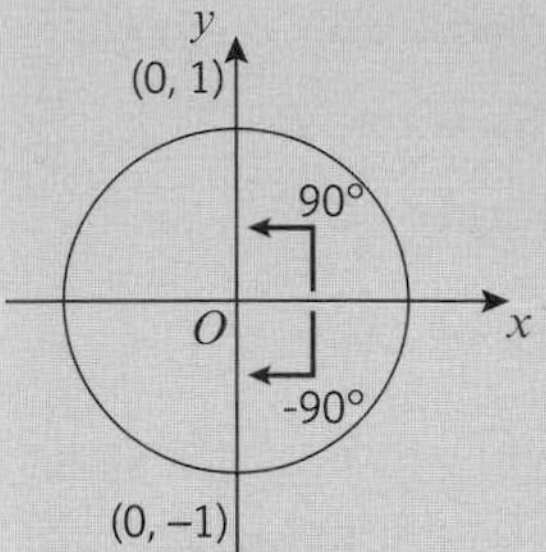

If θ is negative, then measure **clockwise** from the positive x-axis.

An angle of $-90°$ is equivalent to a positive angle of 270°. The x-coordinate is 0 when $\theta = -90°$ or 270°.

Example 2

Write down the values of:

a $\tan 45°$ **b** $\tan 135°$ **c** $\tan 225°$
d $\tan(-45°)$ **e** $\tan 180°$ **f** $\tan 90°$

a $\tan 45° = 1$

b $\tan 135° = -1$

c $\tan 225° = 1$

d $\tan(-45°) = \tan 315° = -1$

e $\tan 180° = 0$

f $\tan 90° =$ undefined

When $\theta = 45°$, the coordinates of OP are $\left(\frac{1}{\sqrt{2}}, \frac{1}{\sqrt{2}}\right)$ so the gradient of OP is 1.

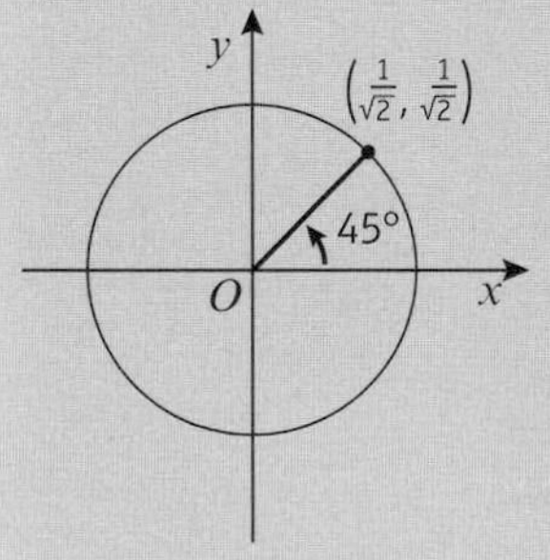

When $\theta = -45°$ the gradient of OP is -1.

When $\theta = 180°$, P has coordinates $(-1, 0)$ so the gradient of $OP = \frac{0}{1} = 0$.

When $\theta = 90°$, P has coordinates $(0, 1)$ so the gradient of $OP = \frac{1}{0}$. This is undefined since you cannot divide by zero.

Links $\tan\theta$ is undefined when $\theta = 270°$ or any other odd multiple of 90°. These values of θ correspond to the asymptotes on the graph of $y = \tan\theta$. ←Section 9.5

The x-y plane is divided into **quadrants**:

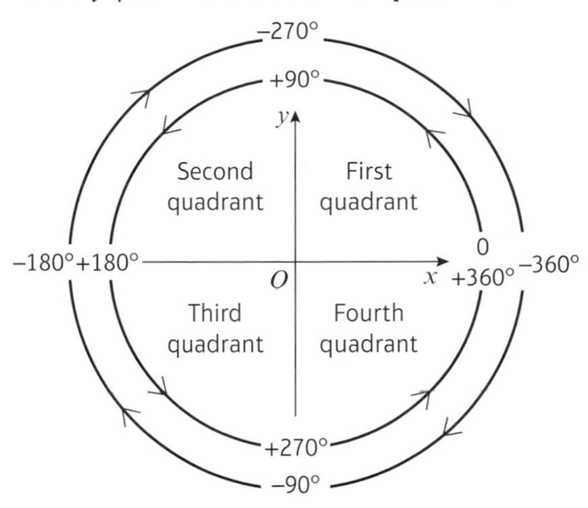

Angles may lie outside the range 0–360°, but they will always lie in one of the four quadrants.
For example, an angle of 600° would be equivalent to 600° − 360° = 240°, so it would lie in the third quadrant.

Example 3

Find the signs of $\sin\theta$, $\cos\theta$ and $\tan\theta$ in the second quadrant.

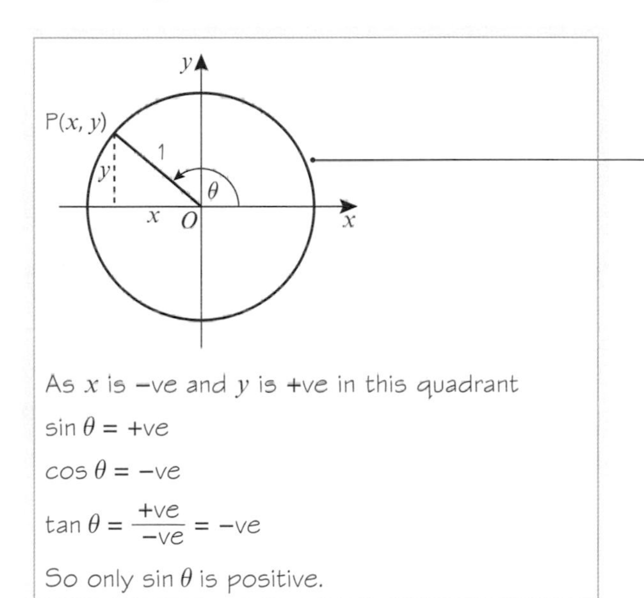

As x is −ve and y is +ve in this quadrant

$\sin\theta = +\text{ve}$

$\cos\theta = -\text{ve}$

$\tan\theta = \frac{+\text{ve}}{-\text{ve}} = -\text{ve}$

So only $\sin\theta$ is positive.

In the second quadrant, θ is obtuse, or $90° < \theta < 180°$.

Draw a circle, centre O and radius 1, with $P(x, y)$ on the circle in the second quadrant.

- **You can use the quadrants to determine whether each of the trigonometric ratios is positive or negative.**

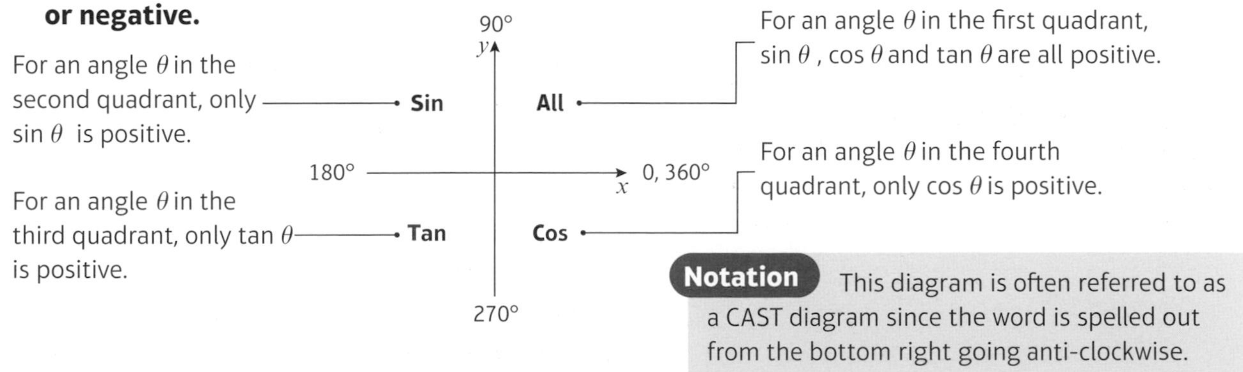

Notation This diagram is often referred to as a CAST diagram since the word is spelled out from the bottom right going anti-clockwise.

- **You can use these rules to find sin, cos or tan of any positive or negative angle using the corresponding acute angle made with the x-axis, θ.**

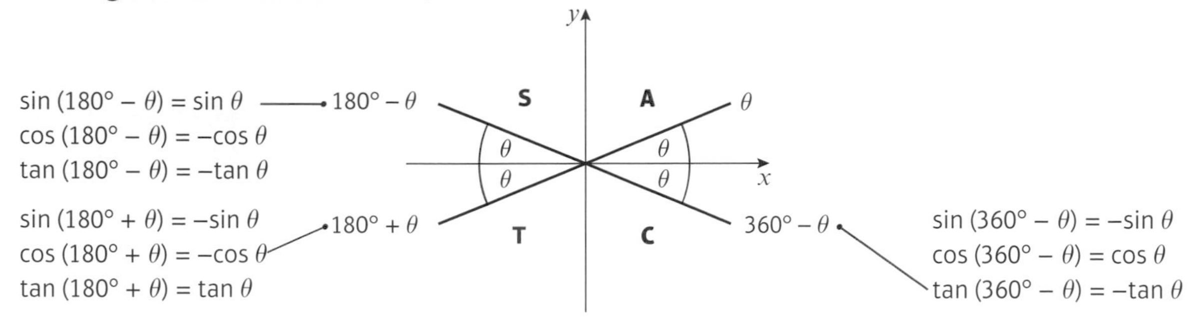

Example 4

Express in terms of trigonometric ratios of acute angles:

a $\sin(-100°)$ **b** $\cos 330°$ **c** $\tan 500°$

a

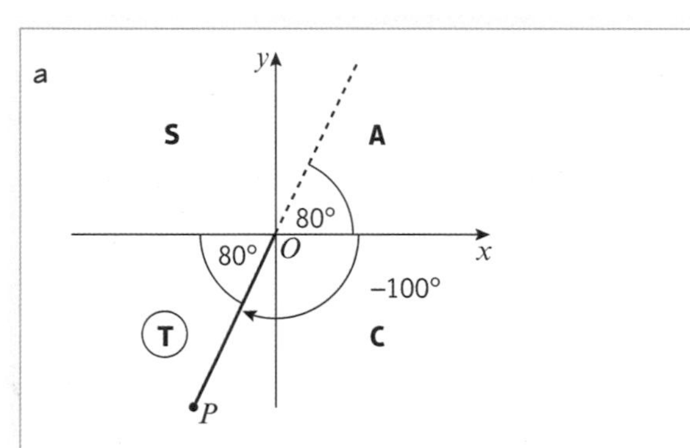

The acute angle made with the x-axis is 80°.

In the third quadrant only tan is +ve, so sin is –ve.

So $\sin(-100)° = -\sin 80°$

b

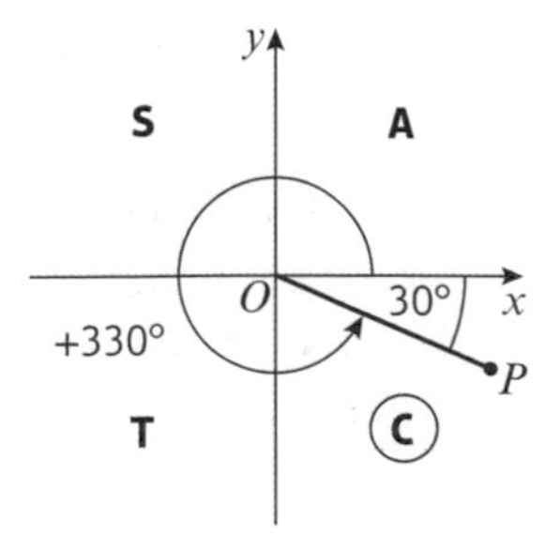

The acute angle made with the x-axis is 30°.

In the fourth quadrant only cos is +ve.

So $\cos 330° = +\cos 30°$

For each part, draw diagrams showing the position of OP for the given angle and insert the acute angle that OP makes with the x-axis.

c

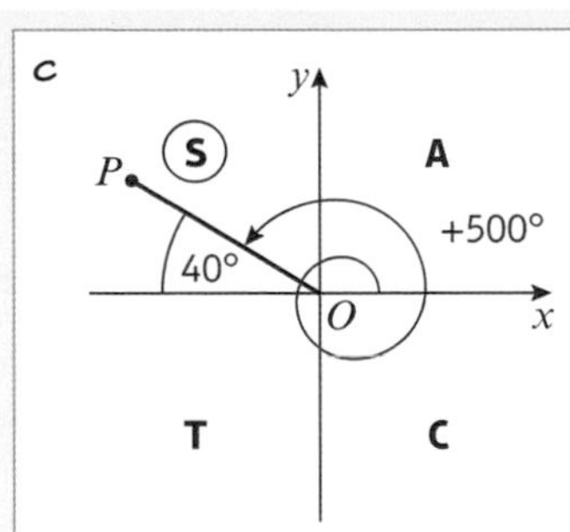

The acute angle made with the x-axis is 40°.

In the second quadrant only sin is +ve.

So $\tan 500° = -\tan 40°$

Exercise 10A

1 Draw diagrams to show the following angles. Mark in the acute angle that OP makes with the x-axis.

a $-80°$ **b** $100°$ **c** $200°$ **d** $165°$ **e** $-145°$
f $225°$ **g** $280°$ **h** $330°$ **i** $-160°$ **j** $-280°$

2 State the quadrant that OP lies in when the angle that OP makes with the positive x-axis is:

a $400°$ **b** $115°$ **c** $-210°$ **d** $255°$ **e** $-100°$

3 Without using a calculator, write down the values of:

a $\sin(-90°)$ **b** $\sin 450°$ **c** $\sin 540°$ **d** $\sin(-450°)$ **e** $\cos(-180°)$
f $\cos(-270°)$ **g** $\cos 270°$ **h** $\cos 810°$ **i** $\tan 360°$ **j** $\tan(-180°)$

4 Express the following in terms of trigonometric ratios of acute angles:

a $\sin 240°$ **b** $\sin(-80°)$ **c** $\sin(-200°)$ **d** $\sin 300°$ **e** $\sin 460°$
f $\cos 110°$ **g** $\cos 260°$ **h** $\cos(-50°)$ **i** $\cos(-200°)$ **j** $\cos 545°$
k $\tan 100°$ **l** $\tan 325°$ **m** $\tan(-30°)$ **n** $\tan(-175°)$ **o** $\tan 600°$

5 Given that θ is an acute angle, express in terms of $\sin\theta$:

a $\sin(-\theta)$ **b** $\sin(180° + \theta)$ **c** $\sin(360° - \theta)$
d $\sin(-(180° + \theta))$ **e** $\sin(-180° + \theta)$ **f** $\sin(-360° + \theta)$
g $\sin(540° + \theta)$ **h** $\sin(720° - \theta)$ **i** $\sin(\theta + 720°)$

Hint The results obtained in questions **5** and **6** are true for all values of θ.

6 Given that θ is an acute angle, express in terms of $\cos\theta$ or $\tan\theta$:

a $\cos(180° - \theta)$ **b** $\cos(180° + \theta)$ **c** $\cos(-\theta)$ **d** $\cos(-(180° - \theta))$
e $\cos(\theta - 360°)$ **f** $\cos(\theta - 540°)$ **g** $\tan(-\theta)$ **h** $\tan(180° - \theta)$
i $\tan(180° + \theta)$ **j** $\tan(-180° + \theta)$ **k** $\tan(540° - \theta)$ **l** $\tan(\theta - 360°)$

Challenge

a Prove that $\sin(180° - \theta) = \sin\theta$

b Prove that $\cos(-\theta) = \cos\theta$

c Prove that $\tan(180° - \theta) = -\tan\theta$

Problem-solving

Draw a diagram showing the positions of θ and $180° - \theta$ on the unit circle.

10.2 Exact values of trigonometrical ratios

You can find sin, cos and tan of 30°, 45° and 60° exactly using triangles.

Consider an **equilateral** triangle ABC of side 2 units.

Draw a perpendicular from A to meet BC at D.

Apply the trigonometric ratios in the right-angled triangle ABD.

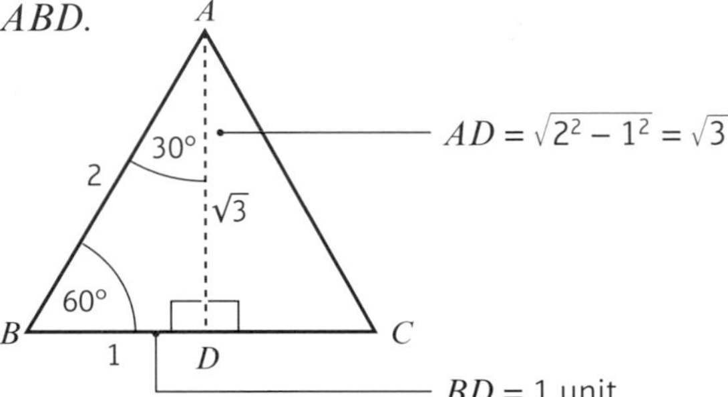

- $\sin 30° = \frac{1}{2}$ $\quad\cos 30° = \frac{\sqrt{3}}{2}$ $\quad\tan 30° = \frac{1}{\sqrt{3}} = \frac{\sqrt{3}}{3}$

  $\sin 60° = \frac{\sqrt{3}}{2}$ $\quad\cos 60° = \frac{1}{2}$ $\quad\tan 60° = \sqrt{3}$

Consider an **isosceles right-angled** triangle PQR with $PQ = RQ = 1$ unit.

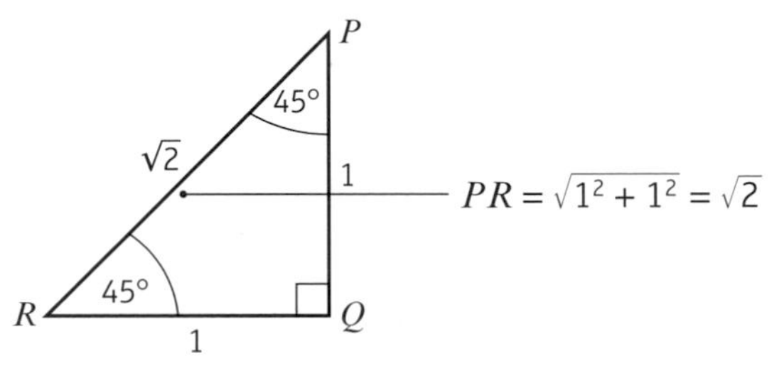

- $\sin 45° = \frac{1}{\sqrt{2}} = \frac{\sqrt{2}}{2}$ $\quad\cos 45° = \frac{1}{\sqrt{2}} = \frac{\sqrt{2}}{2}$ $\quad\tan 45° = 1$

Example 5

Find the exact value of $\sin(-210°)$.

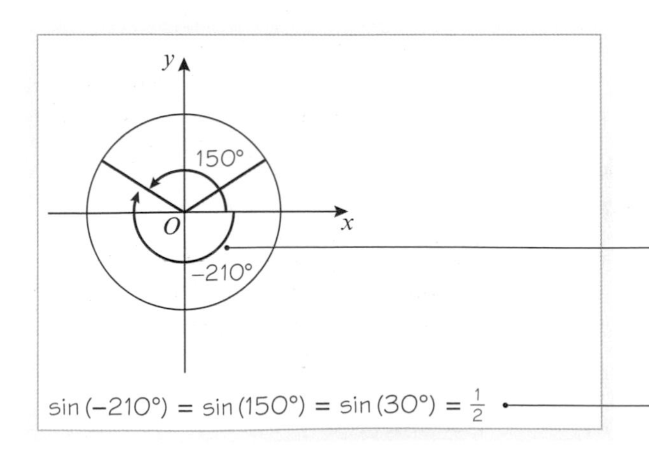

$\sin(-210°) = \sin(150°)$

Use $\sin(180° - \theta) = \sin\theta$

Exercise 10B

1 Express the following as trigonometric ratios of either 30°, 45° or 60°, and hence find their exact values.

a $\sin 135°$ **b** $\sin(-60°)$ **c** $\sin 330°$ **d** $\sin 420°$ **e** $\sin(-300°)$

f $\cos 120°$ **g** $\cos 300°$ **h** $\cos 225°$ **i** $\cos(-210°)$ **j** $\cos 495°$

k $\tan 135°$ **l** $\tan(-225°)$ **m** $\tan 210°$ **n** $\tan 300°$ **o** $\tan(-120°)$

Challenge

The diagram shows an isosceles right-angled triangle ABC. $AE = DE = 1$ unit. Angle $ACD = 30°$.

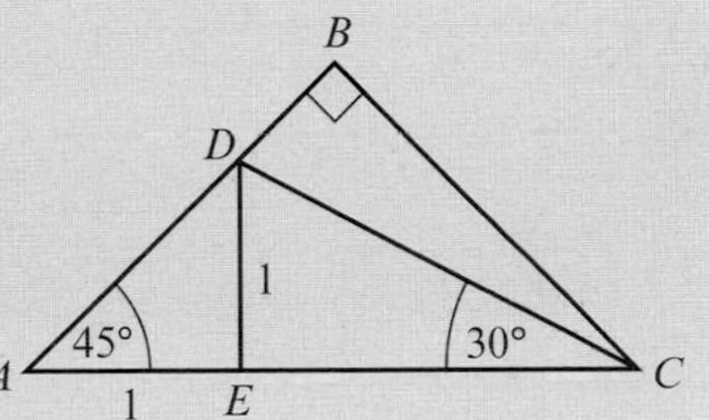

a Calculate the exact lengths of

i CE **ii** DC **iii** BC **iv** DB

b State the size of angle BCD.

c Hence find exact values for

i $\sin 15°$ **ii** $\cos 15°$

10.3 Trigonometric identities

You can use the definitions of sin, cos and tan, together with Pythagoras' theorem, to find two useful identities.

The unit circle has equation $x^2 + y^2 = 1$.

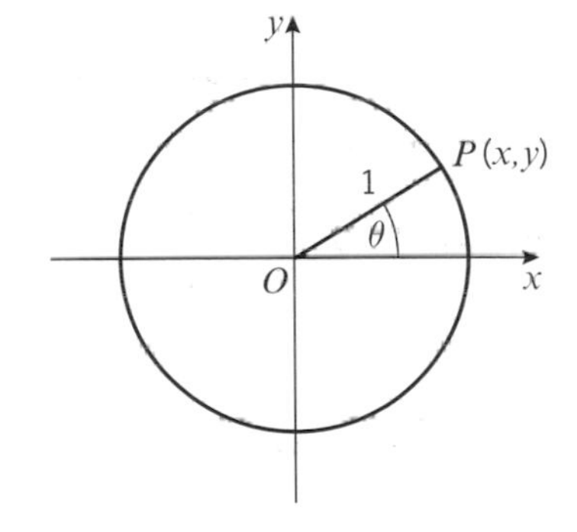

Links The equation of a circle with radius r and centre at the origin is $x^2 + y^2 = r^2$. ← Section 6.2

Since $\cos\theta = x$ and $\sin\theta = y$, it follows that $\cos^2\theta + \sin^2\theta = 1$.

- **For all values of θ, $\sin^2\theta + \cos^2\theta \equiv 1$.**

Since $\tan\theta = \frac{y}{x}$ it follows that $\tan\theta = \frac{\sin\theta}{\cos\theta}$

- **For all values of θ such that $\cos\theta \neq 0$, $\tan\theta \equiv \frac{\sin\theta}{\cos\theta}$**

You can use these two **identities** to simplify trigonometrical expressions and complete proofs.

Notation These results are called trigonometric identities. You use the $\equiv$ symbol instead of $=$ to show that they are always true for all values of θ (subject to any conditions given).

Watch out $\tan\theta$ is undefined when the denominator $= 0$. This occurs when $\cos\theta = 0$, so when $\theta = \ldots -90°, 90°, 270°, 450°, \ldots$

Example 6

Simplify the following expressions:

a $\sin^2 3\theta + \cos^2 3\theta$ **b** $5 - 5\sin^2\theta$ **c** $\dfrac{\sin 2\theta}{\sqrt{1 - \sin^2 2\theta}}$

a $\sin^2 3\theta + \cos^2 3\theta = 1$

$\sin^2\theta + \cos^2\theta = 1$, with θ replaced by 3θ.

b $5 - 5\sin^2\theta = 5(1 - \sin^2\theta)$

$= 5\cos^2\theta$

Always look for factors.
$\sin^2\theta + \cos^2\theta = 1$, so $1 - \sin^2\theta = \cos^2\theta$.

c $\dfrac{\sin 2\theta}{\sqrt{1 - \sin^2 2\theta}} = \dfrac{\sin 2\theta}{\sqrt{\cos^2 2\theta}}$

$\sin^2 2\theta + \cos^2 2\theta = 1$, so $1 - \sin^2 2\theta = \cos^2 2\theta$.

$= \dfrac{\sin 2\theta}{\cos 2\theta}$

$= \tan 2\theta$

$\tan\theta = \dfrac{\sin\theta}{\cos\theta}$, so $\dfrac{\sin 2\theta}{\cos 2\theta} = \tan 2\theta$.

Example 7

Prove that $\dfrac{\cos^4\theta - \sin^4\theta}{\cos^2\theta} \equiv 1 - \tan^2\theta$

Problem-solving

When you have to prove an identity like this you may quote the basic identities like '$\sin^2 + \cos^2 \equiv 1$'.

$\text{LHS} \equiv \dfrac{\cos^4\theta - \sin^4\theta}{\cos^2\theta}$

To prove an identity, start from the left-hand side, and manipulate the expression until it matches the right-hand side. ← Sections 7.4, 7.5

$\equiv \dfrac{(\cos^2\theta + \sin^2\theta)(\cos^2\theta - \sin^2\theta)}{\cos^2\theta}$

The numerator can be factorised as the 'difference of two squares'.

$\sin^2\theta + \cos^2\theta \equiv 1$.

$\equiv \dfrac{(\cos^2\theta - \sin^2\theta)}{\cos^2\theta}$

$\equiv \dfrac{\cos^2\theta}{\cos^2\theta} - \dfrac{\sin^2\theta}{\cos^2\theta}$

$\equiv 1 - \tan^2\theta = \text{RHS}$

Divide through by $\cos^2\theta$ and note that $\dfrac{\sin^2\theta}{\cos^2\theta} \equiv \left(\dfrac{\sin\theta}{\cos\theta}\right)^2 \equiv \tan^2\theta$.

Example 8

a Given that $\cos\theta = -\dfrac{3}{5}$ and that θ is reflex, find the value of $\sin\theta$.

b Given that $\sin\alpha = \dfrac{2}{5}$ and that α is obtuse, find the exact value of $\cos\alpha$.

a Since $\sin^2\theta + \cos^2\theta \equiv 1$,

$$\sin^2\theta = 1 - \left(-\frac{3}{5}\right)^2$$

$$= 1 - \frac{9}{25}$$

$$= \frac{16}{25}$$

So $\sin\theta = -\frac{4}{5}$

b Using $\sin^2\alpha + \cos^2\alpha \equiv 1$,

$$\cos^2\alpha = 1 - \frac{4}{25} = \frac{21}{25}$$

As α is obtuse, $\cos\alpha$ is negative

so $\cos\alpha = -\frac{\sqrt{21}}{5}$

Watch out If you use your calculator to find $\cos^{-1}\left(-\frac{3}{5}\right)$, then the sine of the result, you will get an incorrect answer. This is because the $\cos^{-1}$ function on your calculator gives results between 0 and 180°.

'θ is reflex' means θ is in the 3rd or 4th quadrants, but as $\cos\theta$ is negative, θ must be in the 3rd quadrant. $\sin\theta = \pm\frac{4}{5}$ but in the third quadrant $\sin\theta$ is negative.

Obtuse angles lie in the second quadrant, and have a negative cosine.

The question asks for the exact value so leave your answer as a surd.

Example 9

Given that $p = 3\cos\theta$, and that $q = 2\sin\theta$, show that $4p^2 + 9q^2 = 36$.

As $p = 3\cos\theta$, and $q = 2\sin\theta$,

$$\cos\theta = \frac{p}{3} \text{ and } \sin\theta = \frac{q}{2}$$

Using $\sin^2\theta + \cos^2\theta \equiv 1$,

$$\left(\frac{q}{2}\right)^2 + \left(\frac{p}{3}\right)^2 = 1$$

so $$\frac{q^2}{4} + \frac{p^2}{9} = 1$$

$\therefore$ $$4p^2 + 9q^2 = 36$$

Problem-solving

You need to eliminate θ from the equations. As you can find $\sin\theta$ and $\cos\theta$ in terms of p and q, use the identity $\sin^2\theta + \cos^2\theta \equiv 1$.

Multiply both sides by 36.

Exercise 10C

1 Simplify each of the following expressions:

a $1 - \cos^2\frac{1}{2}\theta$

b $5\sin^2 3\theta + 5\cos^2 3\theta$

c $\sin^2 A - 1$

d $\dfrac{\sin\theta}{\tan\theta}$

e $\dfrac{\sqrt{1 - \cos^2 x}}{\cos x}$

f $\dfrac{\sqrt{1 - \cos^2 3A}}{\sqrt{1 - \sin^2 3A}}$

g $(1 + \sin x)^2 + (1 - \sin x)^2 + 2\cos^2 x$

h $\sin^4\theta + \sin^2\theta\cos^2\theta$

i $\sin^4\theta + 2\sin^2\theta\cos^2\theta + \cos^4\theta$

2 Given that $2\sin\theta = 3\cos\theta$, find the value of $\tan\theta$.

3 Given that $\sin x\cos y = 3\cos x\sin y$, express $\tan x$ in terms of $\tan y$.

4 Express in terms of $\sin\theta$ only:

a $\cos^2\theta$ **b** $\tan^2\theta$ **c** $\cos\theta\tan\theta$

d $\dfrac{\cos\theta}{\tan\theta}$ **e** $(\cos\theta - \sin\theta)(\cos\theta + \sin\theta)$

(P) **5** Using the identities $\sin^2 A + \cos^2 A \equiv 1$ and/or $\tan A = \dfrac{\sin A}{\cos A}$ $(\cos A \neq 0)$, prove that:

a $(\sin\theta + \cos\theta)^2 \equiv 1 + 2\sin\theta\cos\theta$ **b** $\dfrac{1}{\cos\theta} - \cos\theta \equiv \sin\theta\tan\theta$

c $\tan x + \dfrac{1}{\tan x} \equiv \dfrac{1}{\sin x\cos x}$ **d** $\cos^2 A - \sin^2 A \equiv 1 - 2\sin^2 A$

e $(2\sin\theta - \cos\theta)^2 + (\sin\theta + 2\cos\theta)^2 \equiv 5$ **f** $2 - (\sin\theta - \cos\theta)^2 \equiv (\sin\theta + \cos\theta)^2$

g $\sin^2 x\cos^2 y - \cos^2 x\sin^2 y \equiv \sin^2 x - \sin^2 y$

6 Find, without using your calculator, the values of:

a $\sin\theta$ and $\cos\theta$, given that $\tan\theta = \frac{5}{12}$ and θ is acute.

b $\sin\theta$ and $\tan\theta$, given that $\cos\theta = -\frac{3}{5}$ and θ is obtuse.

c $\cos\theta$ and $\tan\theta$, given that $\sin\theta = -\frac{7}{25}$ and $270° < \theta < 360°$.

7 Given that $\sin\theta = \frac{2}{3}$ and that θ is obtuse, find the exact value of: **a** $\cos\theta$ **b** $\tan\theta$

8 Given that $\tan\theta = -\sqrt{3}$ and that θ is reflex, find the exact value of: **a** $\sin\theta$ **b** $\cos\theta$

9 Given that $\cos\theta = \frac{3}{4}$ and that θ is reflex, find the exact value of: **a** $\sin\theta$ **b** $\tan\theta$

(P) **10** In each of the following, eliminate θ to give an equation relating x and y:

a $x = \sin\theta,\ y = \cos\theta$ **b** $x = \sin\theta,\ y = 2\cos\theta$

c $x = \sin\theta,\ y = \cos^2\theta$ **d** $x = \sin\theta,\ y = \tan\theta$

e $x = \sin\theta + \cos\theta,\ y = \cos\theta - \sin\theta$

Problem-solving

In part **e** find expressions for $x + y$ and $x - y$.

(E/P) **11** The diagram shows the triangle ABC with $AB = 12$ cm, $BC = 8$ cm and $AC = 10$ cm.

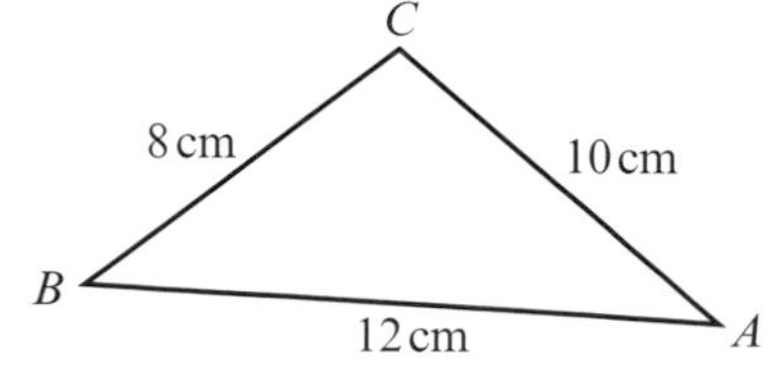

a Show that $\cos B = \dfrac{9}{16}$ **(3 marks)**

b Hence find the exact value of $\sin B$. **(2 marks)**

Hint Use the cosine rule: $a^2 = b^2 + c^2 - 2bc\cos A$ ← Section 9.1

(E/P) **12** The diagram shows triangle PQR with $PR = 8$ cm, $QR = 6$ cm and angle $QPR = 30°$.

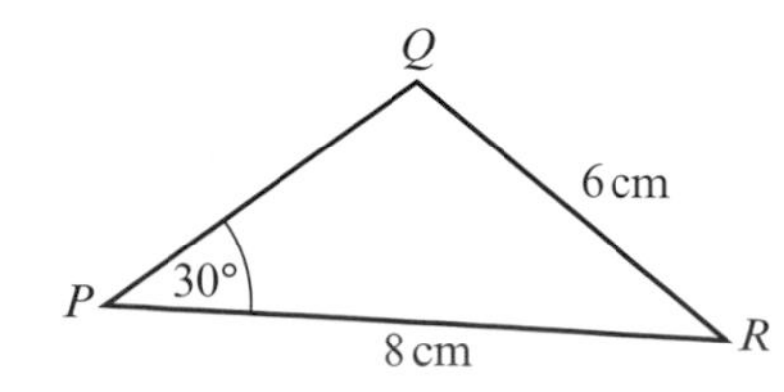

a Show that $\sin Q = \dfrac{2}{3}$ **(3 marks)**

b Given that Q is obtuse, find the exact value of $\cos Q$ **(2 marks)**

10.4 Simple trigonometric equations

You need to be able to solve simple trigonometric equations of the form $\sin\theta = k$ and $\cos\theta = k$ (where $-1 \leqslant k \leqslant 1$) and $\tan\theta = p$ (where $p \in \mathbb{R}$) for given intervals of θ.

- **Solutions to $\sin\theta = k$ and $\cos\theta = k$ only exist when $-1 \leqslant k \leqslant 1$.**
- **Solutions to $\tan\theta = p$ exist for all values of p.**

Links The graphs of $y = \sin\theta$ and $y = \cos\theta$ have a maximum value of 1 and a minimum value of –1.
The graph of $y = \tan\theta$ has no maximum or minimum value. ← Section 9.5

Example 10

Find the solutions of the equation $\sin\theta = \frac{1}{2}$ in the interval $0 \leqslant \theta \leqslant 360°$.

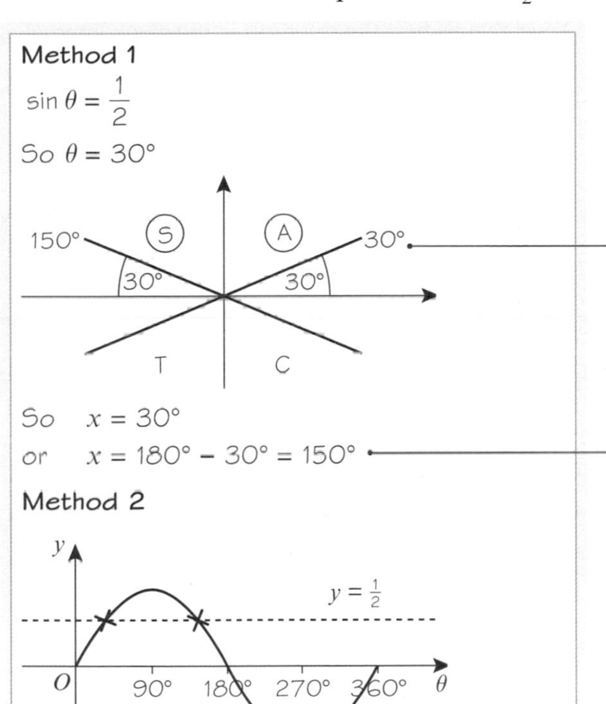

Putting 30° in the four positions shown gives the angles 30°, 150°, 210° and 330° but sine is only positive in the 1st and 2nd quadrants.

You can check this by putting sin 150° in your calculator.

Draw the graph of $y = \sin\theta$ for the given interval.

$\sin\theta = \frac{1}{2}$ where the line $y = \frac{1}{2}$ cuts the curve.
Hence $\theta = 30°$ or $150°$

Use the symmetry properties of the $y = \sin\theta$ graph. ← Sections 9.5

- **When you use the inverse trigonometric functions on your calculator, the angle you get is called the principal value.**

Your calculator will give principal values in the following ranges:

$\cos^{-1}$ in the range $0 \leqslant \theta \leqslant 180°$
$\sin^{-1}$ in the range $-90° \leqslant \theta \leqslant 90°$
$\tan^{-1}$ in the range $-90° \leqslant \theta \leqslant 90°$

Notation The inverse trigonometric functions are also called **arccos**, **arcsin** and **arctan**.

Example 11

Solve, in the interval $0 \leqslant x \leqslant 360°$, $5 \sin x = -2$.

Method 1

$5 \sin x = -2$

$\sin x = -0.4$

First rewrite in the form $\sin x = \ldots$

Principal value is $x = -23.6°$ (3 s.f.)

Watch out The principal value will not always be a solution to the equation.

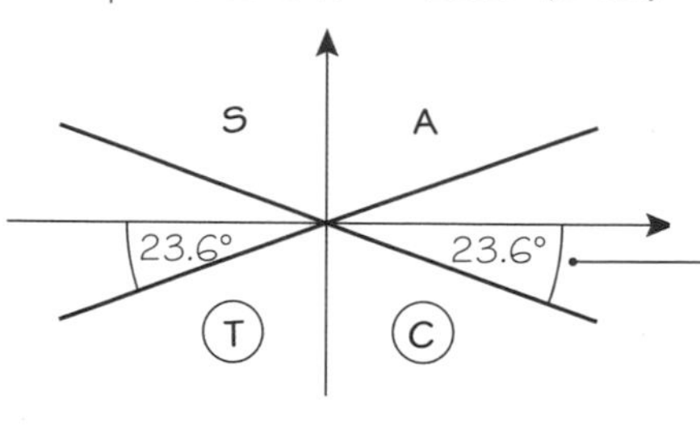

Sine is negative so you need to look in the 3rd and 4th quadrants for your solutions.

$x = 203.6°$ (204° to 3 s.f.)

or $x = 336.4°$ (336° to 3 s.f.)

You can now find the solutions in the given interval. Note that in this case, if $\alpha = \sin^{-1}(-0.4)$, the solutions are $180 - \alpha$ and $360 + \alpha$.

Method 2

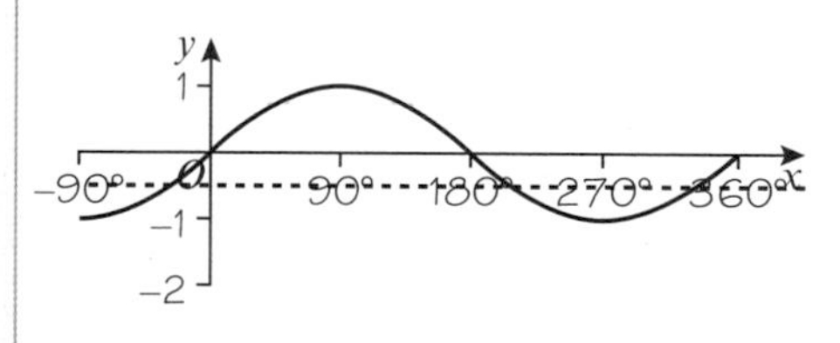

Draw the graph of $y = \sin x$ starting from $-90°$ since the principal solution given by $\sin^{-1}(-0.4)$ is negative.

$\sin^{-1}(-0.4) = -23.578\ldots°$

$x = 203.578\ldots°$ (204° to 3 s.f.)

or $x = 336.421\ldots°$ (336° to 3 s.f.)

Use the symmetry properties of the $y = \sin \theta$ graph.

Example 12

Solve, in the interval $0 < x \leqslant 360°$, $\cos x = \frac{\sqrt{3}}{2}$

A student writes down the following working:

$\cos^{-1}\left(\frac{\sqrt{3}}{2}\right) = 30°$

So $x = 30°$ or $x = 180° - 30° = 150°$

a Identify the error made by the student.

b Write down the correct answer.

Problem-solving

In your exam you might have to analyse student working and identify errors. One strategy is to solve the problem yourself, then compare your working with the incorrect working that has been given.

a The principal solution is correct but the student has found a second solution in the second quadrant where cos is negative.

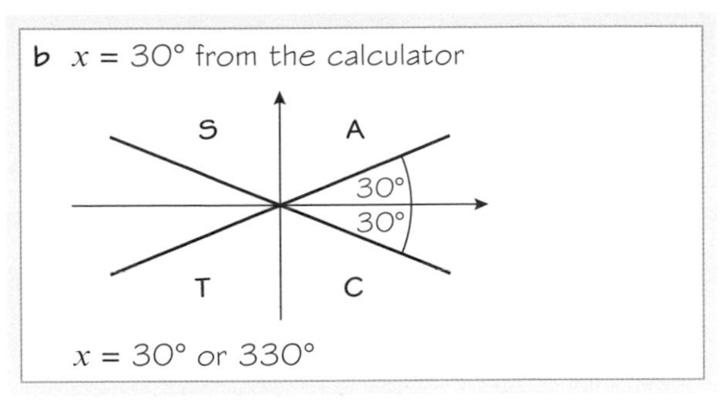

$\cos x$ is positive so you need to look in the 1st and 4th quadrants.

Find the solutions, in $0 < x \leqslant 360°$, from your diagram.

Note that these results are α and $360° - \alpha$ where $\alpha = \cos^{-1}\left(\frac{\sqrt{3}}{2}\right)$.

You can use the identity $\tan\theta \equiv \frac{\sin\theta}{\cos\theta}$ to solve equations.

Example 13

Find the values of θ in the interval $0 < \theta \leqslant 360°$ that satisfy the equation $\sin\theta = \sqrt{3}\cos\theta$.

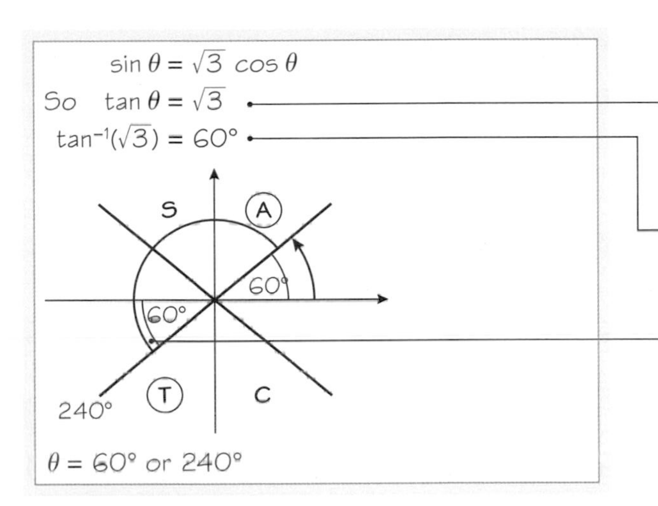

Since $\cos\theta = 0$ does not satisfy the equation, divide both sides by $\cos\theta$ and use the identity $\tan\theta \equiv \frac{\sin\theta}{\cos\theta}$

This is the principal solution.

Tangent is positive in the 1st and 3rd quadrants, so insert the angle in the correct positions.

Exercise 10D

1 The diagram shows a sketch of $y = \tan x$.

a Use your calculator to find the principal solution to the equation $\tan x = -2$.

Hint The principal solution is marked A on the diagram.

b Use the graph and your answer to part **a** to find solutions to the equation $\tan x = -2$ in the range $0 \leqslant x \leqslant 360°$.

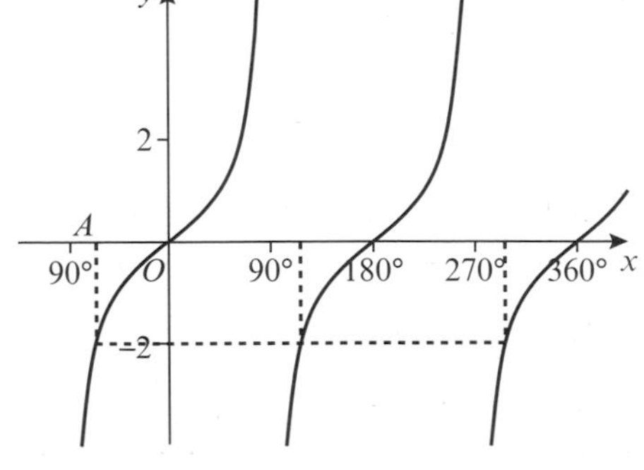

2 The diagram shows a sketch of $y = \cos x$.

a Use your calculator to find the principal solution to the equation $\cos x = 0.4$.

b Use the graph and your answer to part **a** to find solutions to the equation $\cos x = \pm 0.4$ in the range $0 \leqslant x \leqslant 360°$.

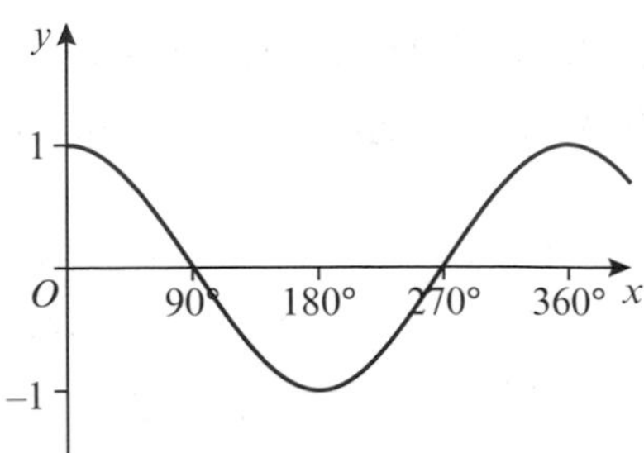

3 Solve the following equations for θ, in the interval $0 < \theta \leqslant 360°$:

a $\sin\theta = -1$ **b** $\tan\theta = \sqrt{3}$ **c** $\cos\theta = \frac{1}{2}$

d $\sin\theta = \sin 15°$ **e** $\cos\theta = -\cos 40°$ **f** $\tan\theta = -1$

g $\cos\theta = 0$ **h** $\sin\theta = -0.766$

Hint Give your answers exactly where possible, or round to 3 significant figures.

4 Solve the following equations for θ, in the interval $0 < \theta \leqslant 360°$:

a $7\sin\theta = 5$ **b** $2\cos\theta = -\sqrt{2}$ **c** $3\cos\theta = -2$ **d** $4\sin\theta = -3$

e $7\tan\theta = 1$ **f** $8\tan\theta = 15$ **g** $3\tan\theta = -11$ **h** $3\cos\theta = \sqrt{5}$

5 Solve the following equations for θ, in the interval $0 < \theta \leqslant 360°$:

a $\sqrt{3}\sin\theta = \cos\theta$ **b** $\sin\theta + \cos\theta = 0$ **c** $3\sin\theta = 4\cos\theta$

d $2\sin\theta - 3\cos\theta = 0$ **e** $\sqrt{2}\sin\theta = 2\cos\theta$ **f** $\sqrt{5}\sin\theta + \sqrt{2}\cos\theta = 0$

6 Solve the following equations for x, giving your answers to 3 significant figures where appropriate, in the intervals indicated:

a $\sin x = -\frac{\sqrt{3}}{2}$, $-180° \leqslant x \leqslant 540°$ **b** $2\sin x = -0.3$, $-180° \leqslant x \leqslant 180°$

c $\cos x = -0.809$, $-180° \leqslant x \leqslant 180°$ **d** $\cos x = 0.84$, $-360° < x < 0°$

e $\tan x = -\frac{\sqrt{3}}{3}$, $0 \leqslant x \leqslant 720°$ **f** $\tan x = 2.90$, $80° \leqslant x \leqslant 440°$

E/P **7** A teacher asks two students to solve the equation $2\cos x = 3\sin x$ for $-180° \leqslant x \leqslant 180°$. The attempts are shown:

Student A:
$\tan x = \frac{3}{2}$
$x = 56.3°$ or $x = -123.7°$

Student B:
$4\cos^2 x = 9\sin^2 x$
$4(1 - \sin^2 x) = 9\sin^2 x$
$4 = 13\sin^2 x$
$\sin x = \pm\sqrt{\frac{4}{13}}$, $x = \pm 33.7°$ or $x = \pm 146.3°$

a Identify the mistake made by Student A. **(1 mark)**

b Identify the mistake made by Student B and explain the effect it has on their solution. **(2 marks)**

c Write down the correct answers to the question. **(1 mark)**

8 **a** Sketch the graphs of $y = 2\sin x$ and $y = \cos x$ on the same set of axes ($0 \leqslant x \leqslant 360°$).

b Write down how many solutions there are in the given range for the equation $2\sin x = \cos x$.

c Solve the equation $2\sin x = \cos x$ algebraically, giving your answers to 1 d.p.

E/P **9** Find all the values of θ, to 1 decimal place, in the interval $0 < \theta < 360°$ for which $\tan^2\theta = 9$. **(5 marks)**

Problem-solving
When you take square roots of both sides of an equation you need to consider both the positive and the negative square roots.

E/P **10** **a** Show that $4\sin^2 x - 3\cos^2 x = 2$ can be written as $7\sin^2 x = 5$. **(2 marks)**

b Hence solve, for $0 \leqslant x \leqslant 360°$, the equation $4\sin^2 x - 3\cos^2 x = 2$. Give your answers to 1 decimal place. **(7 marks)**

E/P **11** **a** Show that the equation $2\sin^2 x + 5\cos^2 x = 1$ can be written as $3\sin^2 x = 4$. **(2 marks)**

b Use your result in part **a** to explain why the equation $2\sin^2 x + 5\cos^2 x = 1$ has no solutions. **(1 marks)**

10.5 Harder trigonometric equations

You need to be able to solve equations of the form $\sin n\theta = k$, $\cos n\theta = k$ and $\tan n\theta = p$.

Example 14

a Solve the equation $\cos 3\theta = 0.766$, in the interval $0 \leqslant \theta \leqslant 360°$.

b Solve the equation $2\sin 2\theta = \cos 2\theta$, in the interval $0 \leqslant \theta \leqslant 360°$.

a Let $X = 3\theta$

So $\cos X° = 0.766$

As $X = 3\theta$,
then as $0 \leqslant \theta \leqslant 360°$

So $3 \times 0 \leqslant X \leqslant 3 \times 360°$

So the interval for X is
$0 \leqslant X \leqslant 1080°$

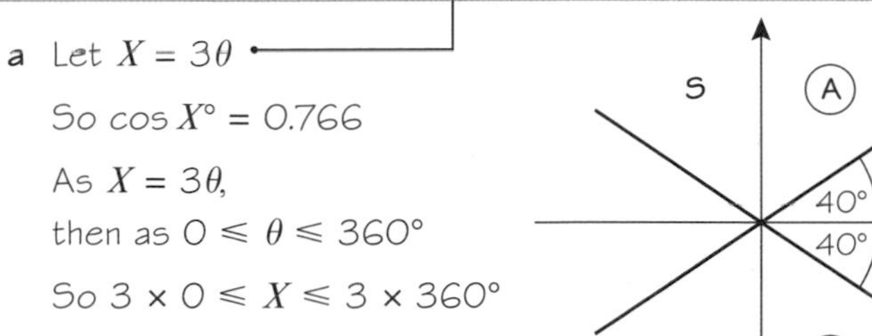

$X = 40.0°, 320°, 400°, 680°, 760°, 1040°$

i.e. $3\theta = 40.0°, 320°, 400°, 680°, 760°, 1040°$

So $\theta = 13.3°, 107°, 133°, 227°, 253°, 347°$

Replace 3θ by X and solve.

Watch out If the range of values for θ is $0 \leqslant \theta \leqslant 360°$, then the range of values for 3θ is $0 \leqslant 3\theta \leqslant 1080°$.

The value of X from your calculator is 40.0. You need to list all values in the 1st and 4th quadrants for three complete revolutions.

Remember $X = 3\theta$.

b $\frac{\sin 2\theta}{\cos 2\theta} = \frac{1}{2}$, so $\tan 2\theta = \frac{1}{2}$

Let $X = 2\theta$

So $\tan X = \frac{1}{2}$

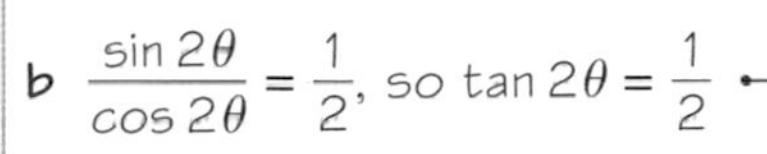

As $X = 2\theta$, then as $0 \leqslant \theta \leqslant 360°$

The interval for X is $0 \leqslant X \leqslant 720°$

Use the identity for tan to rearrange the equation.

Let $X = 2\theta$, and double both values to find the interval for X.

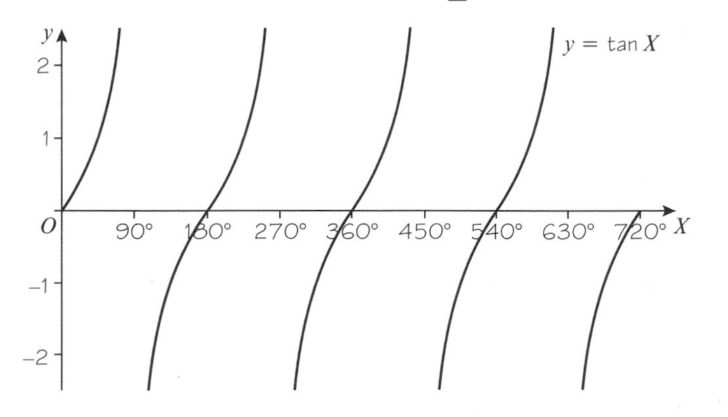

Draw a graph of $\tan X$ for this interval.
Alternatively, you could use a CAST diagram as in part **a**.

The principal solution for X is $26.565...°$

Add multiples of 180°:

$X = 26.565...°, 206.565...°, 386.565...°, 566.565...°$

$\theta = 13.3°, 103°, 193°, 283°$

Convert your values of X back into values of θ.
Round each answer to a sensible degree of accuracy at the end.

You need to be able to solve equations of the form $\sin(\theta + \alpha) = k$, $\cos(\theta + \alpha) = k$ and $\tan(\theta + \alpha) = p$.

Example 15

Solve the equation $\sin(x + 60°) = 0.3$ in the interval $0 \leqslant x \leqslant 360°$.

Let $X = x + 60°$

So $\sin X = 0.3$

The interval for X is

$0° + 60° \leqslant X \leqslant 360° + 60°$

So $60° \leqslant X \leqslant 420°$

Adjust the interval by adding 60° to both values.

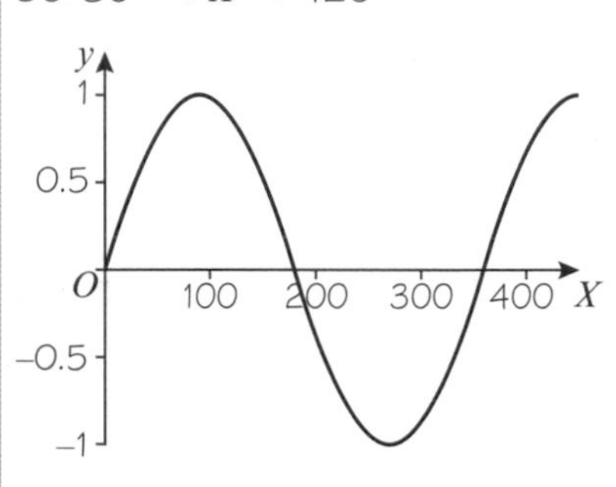

Draw a sketch of the sin graph for the given interval.

The principal value for X is 17.45...°

X = 162.54...°, 377.45...°

This is not in the given interval so it does not correspond to a solution of the equation. Use the symmetry of the sin graph to find other solutions.

Subtract 60° from each value:

x = 102.54...°, 317.45...°

Hence x = 102.5° or 317.5°

You could also use a CAST diagram to solve this problem.

Exercise 10E

1 Find the values of θ, in the interval $0 \leqslant \theta \leqslant 360°$, for which:

a $\sin 4\theta = 0$ **b** $\cos 3\theta = -1$ **c** $\tan 2\theta = 1$

d $\cos 2\theta = \frac{1}{2}$ **e** $\tan \frac{1}{2}\theta = -\frac{1}{\sqrt{3}}$ **f** $\sin(-\theta) = \frac{1}{\sqrt{2}}$

2 Solve the following equations in the interval given:

a $\tan(45° - \theta) = -1$, $0 \leqslant \theta \leqslant 360°$ **b** $2\sin(\theta - 20°) = 1$, $0 \leqslant \theta \leqslant 360°$

c $\tan(\theta + 75°) = \sqrt{3}$, $0 \leqslant \theta \leqslant 360°$ **d** $\sin(\theta - 10°) = -\frac{\sqrt{3}}{2}$, $0 \leqslant \theta \leqslant 360°$

e $\cos(70° - x) = 0.6$, $0 \leqslant \theta \leqslant 180°$

3 Solve the following equations in the interval given:

a $3\sin 3\theta = 2\cos 3\theta$, $0 \leqslant \theta \leqslant 180°$

b $4\sin(\theta + 45°) = 5\cos(\theta + 45°)$, $0 \leqslant \theta \leqslant 450°$

c $2\sin 2x - 7\cos 2x = 0$, $0 \leqslant x \leqslant 180°$

d $\sqrt{3}\sin(x - 60°) + \cos(x - 60°) = 0$, $-180° \leqslant x \leqslant 180°$

(E) **4** Solve for $0 \leqslant x \leqslant 180°$ the equations:

a $\sin(x + 20°) = \frac{1}{2}$ **(4 marks)**

b $\cos 2x = -0.8$, giving your answers to 1 decimal place. **(4 marks)**

(E) **5** **a** Sketch for $0 \leqslant x \leqslant 360°$ the graph of $y = \sin(x + 60°)$ **(2 marks)**

b Write down the exact coordinates of the points where the graph meets the coordinate axes. **(3 marks)**

c Solve, for $0 \leqslant x \leqslant 360°$, the equation $\sin(x + 60°) = 0.55$, giving your answers to 1 decimal place. **(5 marks)**

(E) **6** **a** Given that $4 \sin x = 3 \cos x$, write down the value of $\tan x$. **(1 mark)**

b Solve, for $0 \leqslant \theta \leqslant 360°$, $4 \sin 2\theta = 3 \cos 2\theta$ giving your answers to 1 decimal place. **(5 marks)**

(E/P) **7** The equation $\tan kx = -\frac{1}{\sqrt{3}}$, where k is a constant and $k > 0$, has a solution at $x = 60°$

a Find a possible value of k. **(3 marks)**

b State, with justification, whether this is the only such possible value of k. **(1 mark)**

Challenge

Solve the equation $\sin(3x - 45°) = \frac{1}{2}$ in the interval $0 \leqslant x \leqslant 180°$.

10.6 Equations and identities

You need to be able to solve quadratic equations in $\sin \theta$, $\cos \theta$ or $\tan \theta$. This may give rise to two sets of solutions.

$5 \sin^2 x + 3 \sin x - 2 = 0$ — This is a quadratic equation in the form $5A^2 + 3A - 2 = 0$ where $A = \sin x$.

$(5 \sin x - 2)(\sin x + 1) = 0$ — Factorise

$5 \sin x - 2 = 0$ $\quad$ $\sin x + 1 = 0$ — Setting each factor equal to zero produces two linear equations in $\sin x$.

Example 16

Solve for θ, in the interval $0 \leqslant x \leqslant 360°$, the equations

a $2 \cos^2 \theta - \cos \theta - 1 = 0$ $\quad$ **b** $\sin^2(\theta - 30°) = \frac{1}{2}$

a $2 \cos^2 \theta - \cos \theta - 1 = 0$

So $(2 \cos \theta + 1)(\cos \theta - 1) = 0$ — Compare with $2x^2 - x - 1 = (2x + 1)(x - 1)$

So $\cos \theta = -\frac{1}{2}$ or $\cos \theta = 1$

Set each factor equal to 0 to find two sets of solutions.

$\cos\theta = -\frac{1}{2}$ so $\theta = 120°$

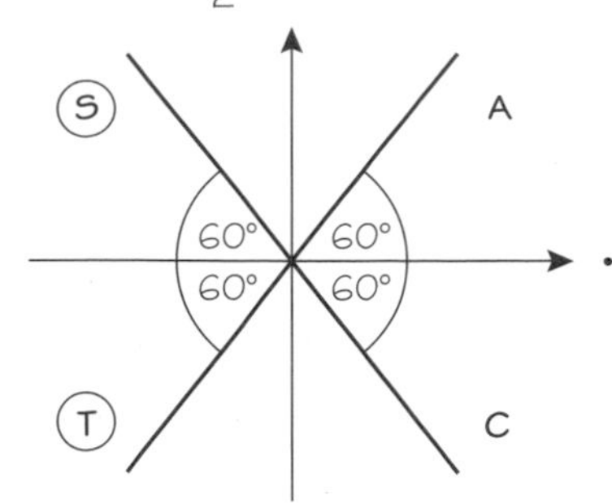

120° makes an angle of 60° with the horizontal. But cosine is negative in the 2nd and 3rd quadrants so $\theta = 120°$ or $\theta = 240°$.

$\theta = 120°$ or $\theta = 240°$

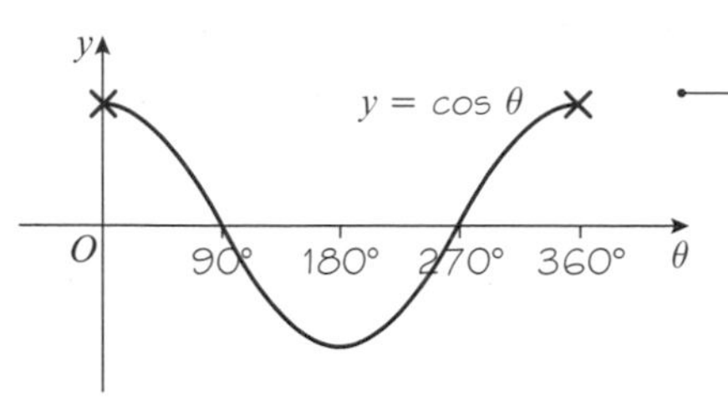

Sketch the graph of $y = \cos\theta$.

Or $\cos\theta = 1$ so $\theta = 0$ or $360°$

So the solutions are

$\theta = 0°, 120°, 240°, 360°$

There are four solutions within the given interval.

b $\sin^2(\theta - 30°) = \frac{1}{2}$

$\sin(\theta - 30°) = \frac{1}{\sqrt{2}}$

or $\sin(\theta - 30°) = -\frac{1}{\sqrt{2}}$

The solutions of $x^2 = k$ are $x = \pm\sqrt{k}$.

So $\theta - 30° = 45°$ or $\theta - 30° = -45°$

Use your calculator to find one solution for each equation.

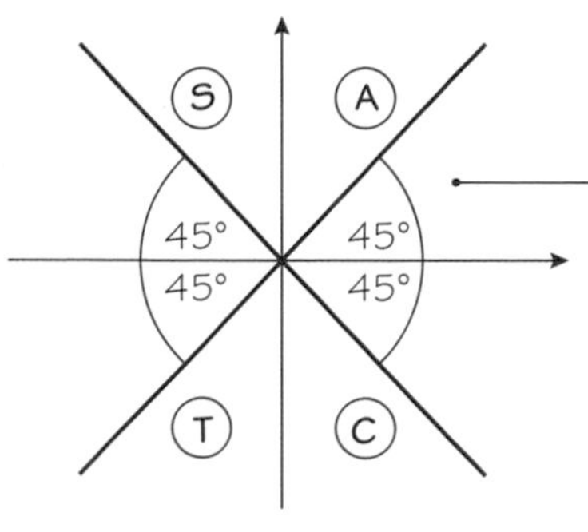

Draw a diagram to find the quadrants where sine is positive and the quadrants where sine is negative.

So from $\sin(\theta - 30°) = \frac{1}{\sqrt{2}}$

$\theta - 30° = 45°, 135°$

and from $\sin(\theta - 30°) = -\frac{1}{\sqrt{2}}$

$\theta - 30° = 225°, 315°$

So the solutions are: $\theta = 75°, 165°, 255°, 345°$

In some equations you may need to use the identity $\sin^2 \theta + \cos^2 \theta \equiv 1$.

Example 17

Find the values of x, in the interval $-180° \leqslant x \leqslant 180°$, satisfying the equation $2\cos^2 x + 9\sin x = 3\sin^2 x$.

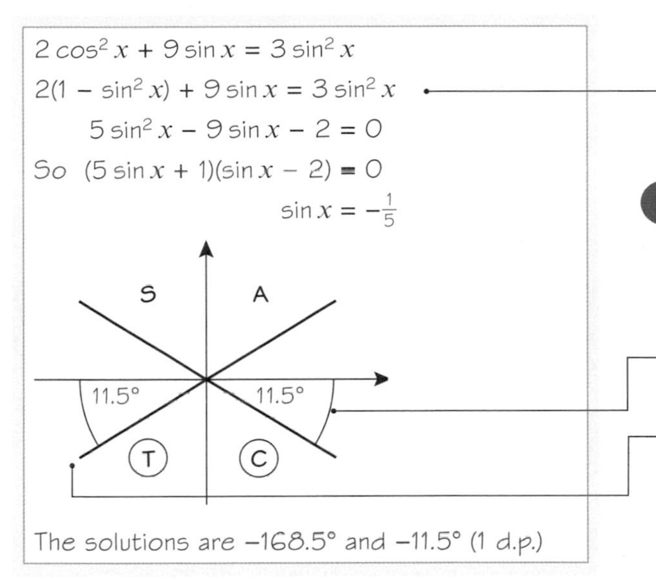

As $\sin^2 x + \cos^2 x \equiv 1$, you are able to rewrite $\cos^2 x$ as $(1 - \sin^2 x)$, and so form a quadratic equation in $\sin x$.

Watch out The factor $(\sin x - 2)$ does not produce any solutions, because $\sin x = 2$ has no solutions.

Your calculator value of x is $x = -11.5°$ (1 d.p.). Insert into the CAST diagram.

The smallest angle in the interval, in the 3rd quadrant, is $(-180 + 11.5) = -168.5°$; there are no values between 0 and 180°.

Exercise 10F

1 Solve for θ, in the interval $0 \leqslant \theta \leqslant 360°$, the following equations.
Give your answers to 3 significant figures where they are not exact.

a $4\cos^2 \theta = 1$ **b** $2\sin^2 \theta - 1 = 0$ **c** $3\sin^2 \theta + \sin \theta = 0$

d $\tan^2 \theta - 2\tan \theta - 10 = 0$ **e** $2\cos^2 \theta - 5\cos \theta + 2 = 0$ **f** $\sin^2 \theta - 2\sin \theta - 1 = 0$

g $\tan^2 2\theta = 3$

Hint In part **e**, only one factor leads to valid solutions.

2 Solve for θ, in the interval $-180° \leqslant \theta \leqslant 180°$, the following equations.
Give your answers to 3 significant figures where they are not exact.

a $\sin^2 2\theta = 1$ **b** $\tan^2 \theta = 2\tan \theta$

c $\cos \theta(\cos \theta - 2) = 1$ **d** $4\sin \theta = \tan \theta$

3 Solve for θ, in the interval $0 \leqslant \theta \leqslant 180°$, the following equations.
Give your answers to 3 significant figures where they are not exact.

a $4(\sin^2 \theta - \cos \theta) = 3 - 2\cos \theta$ **b** $2\sin^2 \theta = 3(1 - \cos \theta)$ **c** $4\cos^2 \theta - 5\sin \theta - 5 = 0$

4 Solve for θ, in the interval $-180° \leqslant \theta \leqslant 180°$, the following equations.
Give your answers to 3 significant figures where they are not exact.

a $5\sin^2 \theta = 4\cos^2 \theta$ **b** $\tan \theta = \cos \theta$

(E) **5** Find all the solutions, in the interval $0 \leqslant x \leqslant 360°$, to the equation $8\sin^2 x + 6\cos x - 9 = 0$ giving each solution to one decimal place. **(6 marks)**

(E) **6** Find, for $0 \leqslant x \leqslant 360°$, all the solutions of $\sin^2 x + 1 = \frac{7}{2}\cos^2 x$ giving each solution to one decimal place. **(6 marks)**

(E/P) **7** Show that the equation $2\cos^2 x + \cos x - 6 = 0$ has no solutions. **(3 marks)**

(E/P) **8** **a** Show that the equation $\cos^2 x = 2 - \sin x$ can be written as $\sin^2 x - \sin x + 1 = 0$. **(2 marks)**

b Hence show that the equation $\cos^2 x = 2 - \sin x$ has no solutions. **(3 marks)**

Problem-solving

If you have to answer a question involving the number of solutions to a quadratic equation, see if you can make use of the discriminant.

(E/P) **9** $\tan^2 x - 2\tan x - 4 = 0$

a Show that $\tan x = p \pm \sqrt{q}$ where p and q are numbers to be found. **(3 marks)**

b Hence solve the equation $\tan^2 x - 2\tan x - 4 = 0$ in the interval $0 \leqslant x \leqslant 540°$. **(5 marks)**

Challenge

1 Solve the equation $\cos^2 3\theta - \cos 3\theta = 2$ in the interval $-180° \leqslant \theta \leqslant 180°$.

2 Solve the equation $\tan^2(\theta - 45°) = 1$ in the interval $0 \leqslant \theta \leqslant 360°$.

Mixed exercise 10

1 Write each of the following as a trigonometric ratio of an acute angle:

a $\cos 237°$ **b** $\sin 312°$ **c** $\tan 190°$

2 Without using your calculator, work out the values of:

a $\cos 270°$ **b** $\sin 225°$ **c** $\cos 180°$ **d** $\tan 240°$ **e** $\tan 135°$

(P) **3** Given that angle A is obtuse and $\cos A = -\sqrt{\frac{7}{11}}$, show that $\tan A = \frac{-2\sqrt{7}}{7}$

(P) **4** Given that angle B is reflex and $\tan B = +\frac{\sqrt{21}}{2}$, find the exact value of: **a** $\sin B$ **b** $\cos B$

5 Simplify the following expressions:

a $\cos^4\theta - \sin^4\theta$ **b** $\sin^2 3\theta - \sin^2 3\theta\cos^2 3\theta$

c $\cos^4\theta + 2\sin^2\theta\cos^2\theta + \sin^4\theta$

6 **a** Given that $2(\sin x + 2\cos x) = \sin x + 5\cos x$, find the exact value of $\tan x$.

b Given that $\sin x\cos y + 3\cos x\sin y = 2\sin x\sin y - 4\cos x\cos y$, express $\tan y$ in terms of $\tan x$.

(P) **7** Prove that, for all values of θ:

a $(1 + \sin\theta)^2 + \cos^2\theta \equiv 2(1 + \sin\theta)$ **b** $\cos^4\theta + \sin^2\theta \equiv \sin^4\theta + \cos^2\theta$

(P) **8** Without attempting to solve them, state how many solutions the following equations have in the interval $0 \leqslant \theta \leqslant 360°$. Give a brief reason for your answer.

a $2\sin\theta = 3$

b $\sin\theta = -\cos\theta$

c $2\sin\theta + 3\cos\theta + 6 = 0$

d $\tan\theta + \dfrac{1}{\tan\theta} = 0$

(E) **9** **a** Factorise $4xy - y^2 + 4x - y$. **(2 marks)**

b Solve the equation $4\sin\theta\cos\theta - \cos^2\theta + 4\sin\theta - \cos\theta = 0$, in the interval $0 \leqslant \theta \leqslant 360°$. **(5 marks)**

(E) **10** **a** Express $4\cos 3\theta - \sin(90° - 3\theta)$ as a single trigonometric function. **(1 mark)**

b Hence solve $4\cos 3\theta - \sin(90° - 3\theta) = 2$ in the interval $0 \leqslant \theta \leqslant 360°$. Give your answers to 3 significant figures. **(3 marks)**

(E/P) **11** Given that $2\sin 2\theta = \cos 2\theta$:

a Show that $\tan 2\theta = 0.5$. **(1 mark)**

b Hence find the values of θ, to one decimal place, in the interval $0 \leqslant \theta \leqslant 360°$ for which $2\sin 2\theta = \cos 2\theta$. **(4 marks)**

12 Find all the values of θ in the interval $0 \leqslant \theta \leqslant 360°$ for which:

a $\cos(\theta + 75°) = 0.5$,

b $\sin 2\theta = 0.7$, giving your answers to one decimal place.

(E) **13** Find the values of x in the interval $0 < x < 270°$ which satisfy the equation

$$\frac{\cos 2x + 0.5}{1 - \cos 2x} = 2$$ **(6 marks)**

(E) **14** Find, in degrees, the values of θ in the interval $0 \leqslant \theta \leqslant 360°$ for which $2\cos^2\theta - \cos\theta - 1 = \sin^2\theta$

Give your answers to 1 decimal place, where appropriate. **(6 marks)**

(E/P) **15** A teacher asks one of his students to solve the equation $2\sin 3x = 1$ for $-360° \leqslant x \leqslant 360°$. The attempt is shown below:

$\sin 3x = \frac{1}{2}$
$3x = 30°$
$x = 10°$
Additional solution at $180° - 10° = 170°$

a Identify two mistakes made by the student. **(2 marks)**

b Solve the equation. **(2 marks)**

16 **a** Sketch the graphs of $y = 3\sin x$ and $y = 2\cos x$ on the same set of axes $(0 \leqslant x \leqslant 360°)$.

b Write down how many solutions there are in the given range for the equation $3\sin x = 2\cos x$.

c Solve the equation $3\sin x = 2\cos x$ algebraically, giving your answers to one decimal place.

(E) **17** The diagram shows the triangle ABC with $AB = 11$ cm, $BC = 6$ cm and $AC = 7$ cm.

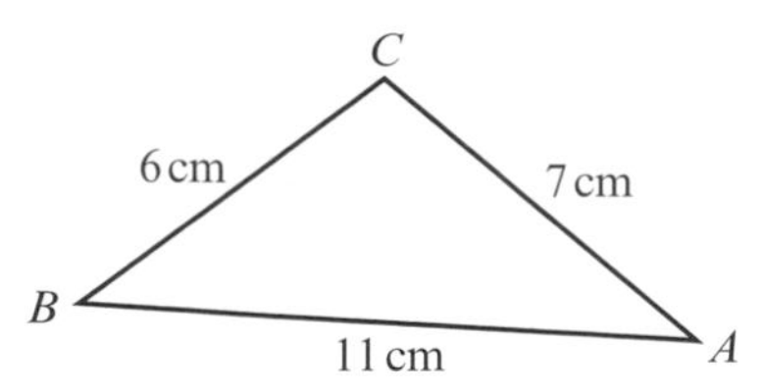

a Find the exact value of $\cos B$, giving your answer in simplest form. **(3 marks)**

b Hence find the exact value of $\sin B$. **(2 marks)**

(E/P) **18** The diagram shows triangle PQR with $PR = 6$ cm, $QR = 5$ cm and angle $QPR = 45°$.

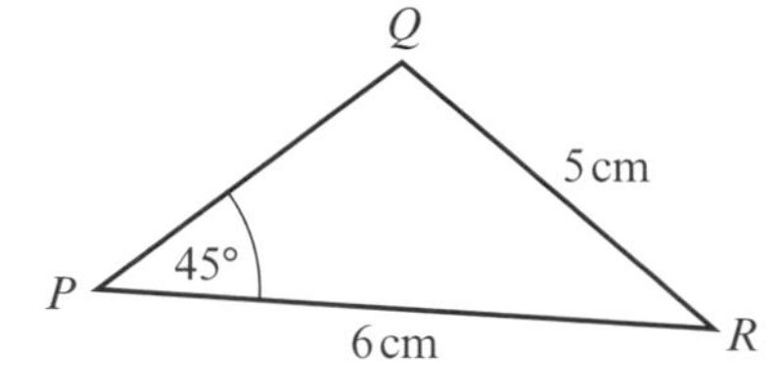

a Show that $\sin Q = \frac{3\sqrt{2}}{5}$ **(3 marks)**

b Given that Q is obtuse, find the exact value of $\cos Q$. **(2 marks)**

(E/P) **19** **a** Show that the equation $3\sin^2 x - \cos^2 x = 2$ can be written as $4\sin^2 x = 3$. **(2 marks)**

b Hence solve the equation $3\sin^2 x - \cos^2 x = 2$ in the interval $-180° \leqslant x \leqslant 180°$, giving your answers to 1 decimal place. **(7 marks)**

(E) **20** Find all the solutions to the equation $3\cos^2 x + 1 = 4\sin x$ in the interval $-360° \leqslant x \leqslant 360°$, giving your answers to 1 decimal place. **(6 marks)**

Challenge

Solve the equation $\tan^4 x - 3\tan^2 x + 2 = 0$ in the interval $0 \leqslant x \leqslant 360°$.

Summary of key points

1 For a point $P(x, y)$ on a unit circle such that OP makes an angle θ with the positive x-axis:

- $\cos\theta = x = x$-coordinate of P
- $\sin\theta = y = y$-coordinate of P
- $\tan\theta = \frac{y}{x} =$ gradient of OP

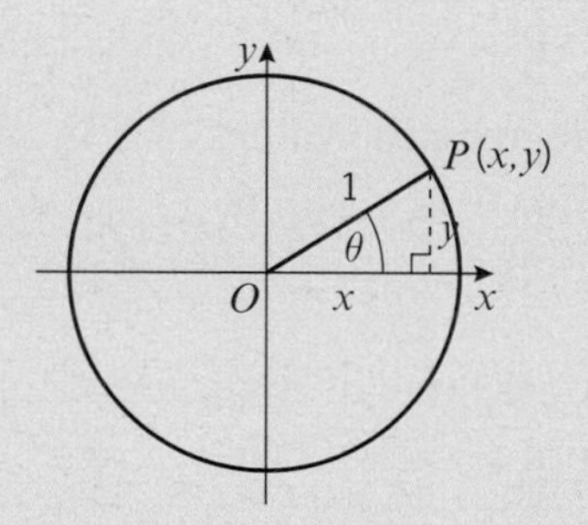

2 You can use the quadrants to determine whether each of the trigonometric ratios is positive or negative.

90°
y
For an angle θ in the second quadrant, only $\sin\theta$ is positive. → **Sin**
All ← For an angle θ in the first quadrant, $\sin\theta$, $\cos\theta$ and $\tan\theta$ are all positive.
180° → 0, 360°
x
For an angle θ in the third quadrant, only $\tan\theta$ is positive. → **Tan**
Cos ← For an angle θ in the fourth quadrant, only $\cos\theta$ is positive.
270°

3 You can use these rules to find sin, cos or tan of any positive or negative angle using the corresponding **acute** angle made with the x-axis, θ.

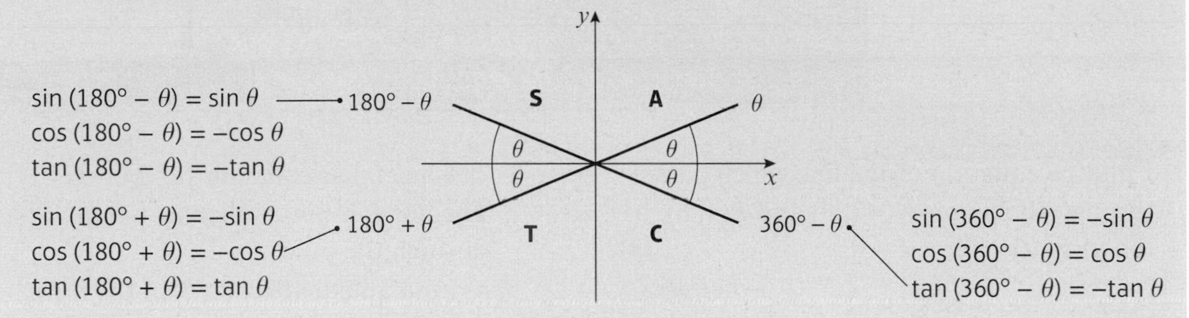

4 The trigonometric ratios of 30°, 45° and 60° have exact forms, given below:

$\sin 30° = \frac{1}{2}$ $\quad$ $\cos 30° = \frac{\sqrt{3}}{2}$ $\quad$ $\tan 30° = \frac{1}{\sqrt{3}} = \frac{\sqrt{3}}{3}$

$\sin 45° = \frac{1}{\sqrt{2}} = \frac{\sqrt{2}}{2}$ $\quad$ $\cos 45° = \frac{1}{\sqrt{2}} = \frac{\sqrt{2}}{2}$ $\quad$ $\tan 45° = 1$

$\sin 60° = \frac{\sqrt{3}}{2}$ $\quad$ $\cos 60° = \frac{1}{2}$ $\quad$ $\tan 60° = \sqrt{3}$

5 For all values of θ, $\sin^2\theta + \cos^2\theta \equiv 1$

6 For all values of θ such that $\cos\theta \neq 0$, $\tan\theta \equiv \frac{\sin\theta}{\cos\theta}$

7
- Solutions to $\sin\theta = k$ and $\cos\theta = k$ only exist when $-1 \leqslant k \leqslant 1$
- Solutions to $\tan\theta = p$ exist for all values of p.

8 When you use the inverse trigonometric functions on your calculator, the angle you get is called the **principal value**.

9 Your calculator will give principal values in the following ranges:
- $\cos^{-1}$ in the range $0 \leqslant \theta \leqslant 180°$
- $\sin^{-1}$ in the range $-90° \leqslant \theta \leqslant 90°$
- $\tan^{-1}$ in the range $-90° \leqslant \theta \leqslant 90°$

2 Review exercise

(E) **1** Find the equation of the line which passes through the points $A(-2, 8)$ and $B(4, 6)$, in the form $ax + by + c = 0$. **(3)**

← Section 5.2

(E) **2** The line l passes through the point $(9, -4)$ and has gradient $\frac{1}{3}$. Find an equation for l, in the form $ax + by + c = 0$, where a, b and c are integers. **(3)**

← Section 5.2

(E/P) **3** The points $A(0, 3)$, $B(k, 5)$ and $C(10, 2k)$, where k is a constant, lie on the same straight line. Find the two possible values of k. **(5)**

← Section 5.1

(E/P) **4** The scatter graph shows the height, h cm, and inseam leg measurement, l cm, of six adults. A line of best fit has been added to the scatter graph.

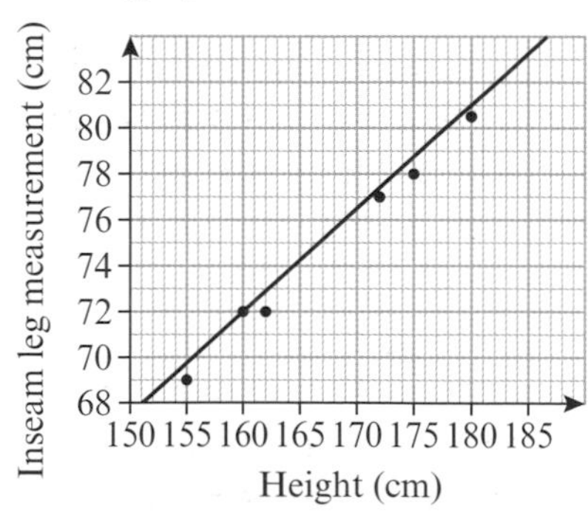

a Use two points on the scatter graph to calculate the gradient of the line. **(2)**

b Use your answer to part **a** to write a linear model relating height and inseam in the form $l = kh$, where k is a constant to be found. **(1)**

c Comment on the validity of your model for small values of h. **(1)**

← Section 5.5

(E) **5** The line l_1 has equation $y = 3x - 6$.
The line l_2 is perpendicular to l_1 and passes through the point $(6, 2)$.

a Find an equation for l_2 in the form $y = mx + c$, where m and c are constants **(3)**

The lines l_1 and l_2 intersect at the point C.

b Use algebra to find the coordinates of C. **(2)**

The lines l_1 and l_2 cross the x-axis at the points A and B respectively.

c Calculate the exact area of triangle ABC. **(4)**

← Sections 5.3, 5.4

(E) **6** The lines $y = 2x$ and $5y + x - 33 = 0$ intersect at the point P. Find the distance of the point from the origin O, giving your answer as a surd in its simplest form. **(4)**

← Sections 5.2, 5.4

(E/P) **7** The perpendicular bisector of the line segment joining $(5, 8)$ and $(7, -4)$ crosses the x-axis at the point Q. Find the coordinates of Q. **(4)**

← Section 6.1

(E) **8** The circle C has centre $(-3, 8)$ and passes through the point $(0, 9)$. Find an equation for C. **(4)**

← Section 6.2

(E/P) **9** **a** Show that $x^2 + y^2 - 6x + 2y - 10 = 0$ can be written in the form $(x - a)^2 + (y - b)^2 = r^2$, where a, b and r are numbers to be found. **(2)**

b Hence write down the centre and radius of the circle with equation $x^2 + y^2 - 6x + 2y - 10 = 0$. **(2)**

← Section 6.2

E/P **10** The line $3x + y = 14$ intersects the circle $(x - 2)^2 + (y - 3)^2 = 5$ at the points A and B.

a Find the coordinates of A and B. **(4)**

b Determine the length of the chord AB. **(2)**

← Section 6.3

E/P **11** The line with equation $y = 3x - 2$ does not intersect the circle with centre $(0, 0)$ and radius r. Find the range of possible values of r. **(8)**

← Section 6.3

E/P **12** The circle C has centre $(1, 5)$ and passes through the point $P(4, -2)$. Find:

a an equation for the circle C. **(4)**

b an equation for the tangent to the circle at P. **(3)**

← Section 6.4

E/P **13** The points $A(2, 1)$, $B(6, 5)$ and $C(8, 3)$ lie on a circle.

a Show that $\angle ABC = 90°$. **(2)**

b Deduce a geometrical property of the line segment AC. **(1)**

c Hence find the equation of the circle. **(4)**

← Section 6.5

E/P **14** $\dfrac{2x^2 + 20x + 42}{224x + 4x^2 - 4x^3} = \dfrac{x + a}{bx(x + c)}$

where a, b and c are constants. Work out the values of a, b and c. **(4)**

← Section 7.1

E **15** **a** Show that $(2x - 1)$ is a factor of $2x^3 - 7x^2 - 17x + 10$. **(2)**

b Factorise $2x^3 - 7x^2 - 17x + 10$ completely. **(4)**

c Hence, or otherwise, sketch the graph of $y = 2x^3 - 7x^2 - 17x + 10$, labelling any intersections with the coordinate axes clearly. **(2)**

← Section 7.3

E/P **16** $f(x) = 3x^3 + x^2 - 38x + c$

Given that $f(3) = 0$,

a find the value of c, **(2)**

b factorise $f(x)$ completely, **(4)**

← Section 7.3

E **17** $g(x) = x^3 - 13x + 12$

a Use the factor theorem to show that $(x - 3)$ is a factor of $g(x)$. **(2)**

b Factorise $g(x)$ completely. **(4)**

← Section 7.3

E/P **18** **a** It is claimed that the following inequality is true for all real numbers a and b. Use a counter-example to show that the claim is false:

$$a^2 + b^2 < (a + b)^2$$ **(2)**

b Specify conditions on a and b that make this inequality true. Prove your result. **(4)**

← Section 7.5

E/P **19** **a** Use proof by exhaustion to prove that for all prime numbers p, $3 < p < 20$, p^2 is one greater than a multiple of 24. **(2)**

b Find a counterexample that disproves the statement 'All numbers which are one greater than a multiple of 24 are the squares of prime numbers.' **(2)**

← Sections 7.5

E/P **20** **a** Show that $x^2 + y^2 - 10x - 8y + 32 = 0$ can be written in the form $(x - a)^2 + (y - b)^2 = r^2$, where a, b and r are numbers to be found. **(2)**

b Circle C has equation $x^2 + y^2 - 10x - 8y + 32 = 0$ and circle D has equation $x^2 + y^2 = 9$. Calculate the distance between the centre of circle C and the centre of circle D. **(3)**

c Using your answer to part **b**, or otherwise, prove that circles C and D do not touch. **(2)**

← Sections 6.4, 7.5

(E) **21** **a** Expand $(1 - 2x)^{10}$ in ascending powers of x up to and including the term in x^3. **(3)**

b Use your answer to part **a** to evaluate $(0.98)^{10}$ correct to 3 decimal places. **(1)**

← Section 8.5

(E/P) **22** If x is so small that terms of x^3 and higher can be ignored,
$(2 - x)(1 + 2x)^5 \approx a + bx + cx^2$.
Find the values of the constants a, b and c. **(5)**

← Section 8.4

(E/P) **23** The coefficient of x in the binomial expansion of $(2 - 4x)^q$, where q is a positive integer, is $-32q$. Find the value of q. **(4)**

← Section 8.4

(E) **24** The diagram shows triangle ABC, with $AB = \sqrt{5}$ cm, $\angle ABC = 45°$ and $\angle BCA = 30°$. Find the exact length of AC. **(3)**

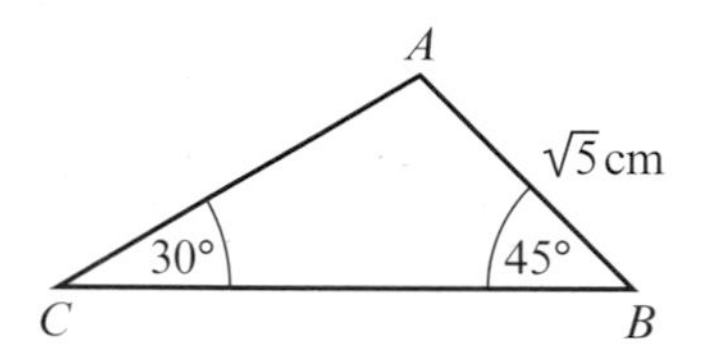

Not to scale

← Section 9.2

(E/P) **25** The diagram shows triangle ABC, with $AB = 5$ cm, $BC = (2x - 3)$ cm, $CA = (x + 1)$ cm and $\angle ABC = 60°$.

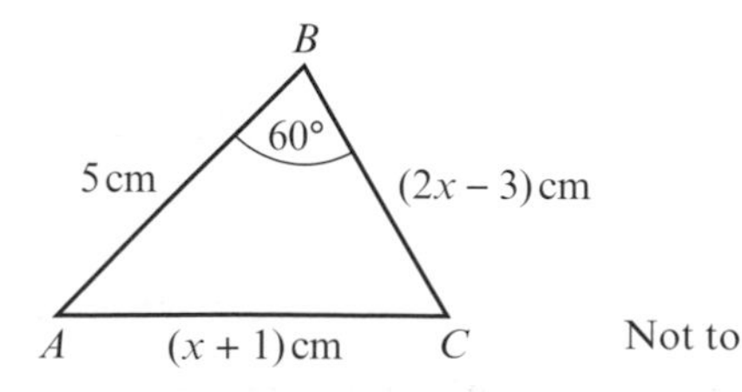

Not to scale

a Show that x satisfies the equation $x^2 - 8x + 16 = 0$. **(3)**

b Find the value of x. **(1)**

c Calculate the area of the triangle, giving your answer to 3 significant figures. **(2)**

← Section 9.4

(E/P) **26** Ship B is 8 km, on a bearing of 030°, from ship A.
Ship C is 12 km, on a bearing of 140°, from ship B.

a Calculate the distance of ship C from ship A. **(4)**

b Calculate the bearing of ship C from ship A. **(3)**

← Section 9.4

(E/P) **27** The triangle ABC has vertices $A(-2, 4)$, $B(6, 10)$ and $C(16, 10)$.

a Prove that ABC is an isosceles triangle. **(2)**

b Calculate the size of $\angle ABC$. **(3)**

← Sections 5.4, 9.4

(E/P) **28** The diagram shows ΔABC.
Calculate the area of ΔABC. **(6)**

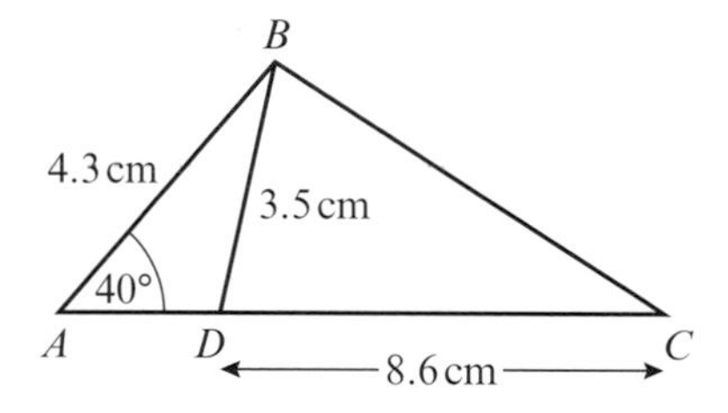

← Section 9.4

(E/P) **29** The circle C has centre (5, 2) and radius 5. The points $X(1, -1)$, $Y(10, 2)$ and $Z(8, k)$ lie on the circle, where k is a positive integer.

a Write down the equation of the circle. **(2)**

b Calculate the value of k. **(1)**

c Show that $\cos \angle XYZ = \frac{\sqrt{2}}{10}$ **(5)**

← Sections 6.2, 9.4

(E) **30** **a** On the same set of axes, in the interval $0 \leqslant x \leqslant 360°$, sketch the graphs of $y = \tan(x - 90°)$ and $y = \sin x$. Label clearly any points at which the graphs cross the coordinate axes. **(5)**

b Hence write down the number of solutions of the equation $\tan(x - 90°) = \sin x$ in the interval $0 \leqslant x \leqslant 360°$. **(1)**

← Section 9.6

E **31** The graph shows the curve $y = \sin(x + 45°)$, $-360° \leqslant x \leqslant 360°$.

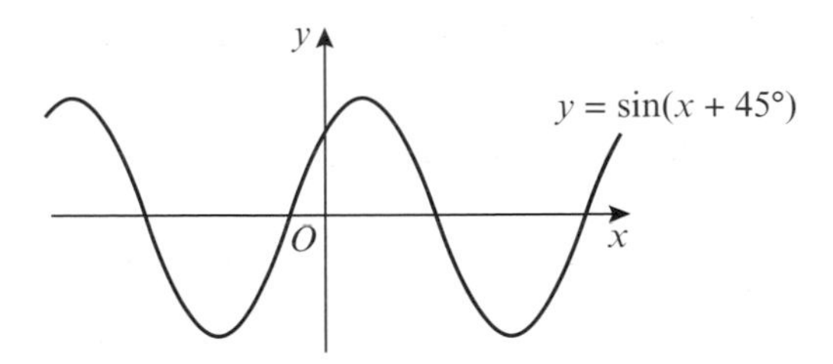

a Write down the coordinates of each point where the curve crosses the x-axis. **(2)**

b Write down the coordinates of the point where the curve crosses the y-axis. **(1)**

← Section 9.6

E/P **32** A pyramid has four triangular faces and a square base. All the edges of the pyramid are the same length, s cm. Show that the total surface area of the pyramid is $(\sqrt{3} + 1)s^2$ cm². **(3)**

← Sections 9.4, 10.2

E **33 a** Given that $\sin\theta = \cos\theta$, find the value of $\tan\theta$. **(1)**

b Find the values of θ in the interval $0 \leqslant \theta < 360°$ for which $\sin\theta = \cos\theta$. **(2)**

← Sections 10.3, 10.4

E **34** Find all the values of x in the interval $0 \leqslant x < 360°$ for which $3\tan^2 x = 1$. **(4)**

← Section 10.4

E **35** Find all the values of θ in the interval $0 \leqslant \theta < 360°$ for which $2\sin(\theta - 30°) = \sqrt{3}$. **(4)**

← Section 10.5

E **36 a** Show that the equation $2\cos^2 x = 4 - 5\sin x$ may be written as $2\sin^2 x - 5\sin x + 2 = 0$. **(2)**

b Hence solve, for $0 \leqslant x < 360°$, the equation $2\cos^2 x = 4 - 5\sin x$. **(4)**

E **37** Find all of the solutions in the interval $0 \leqslant x < 360°$ of $2\tan^2 x - 4 = 5\tan x$ giving each solution, in degrees, to one decimal place. **(6)**

← Section 10.6

E **38** Find all of the solutions in the interval $0 \leqslant x < 360°$ of $5\sin^2 x = 6(1 - \cos x)$ giving each solution, in degrees, to one decimal place. **(7)**

← Section 10.6

E/P **39** Prove that $\cos^2 x(\tan^2 x + 1) = 1$ for all values of x where $\cos x$ and $\tan x$ are defined. **(4)**

← Sections 7.4, 10.3

Challenge

1 The diagram shows a square $ABCD$ on a set of coordinate axes. The square intersects the x-axis at the points B and S, and the equation of the line which passes through B and C is $y = 3x - 12$.

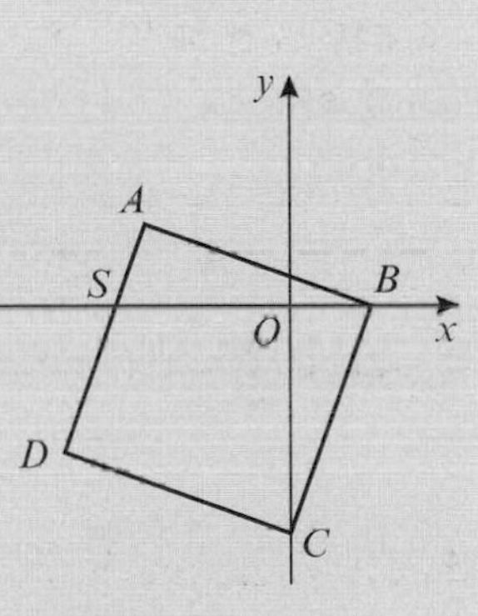

a Calculate the area of the square.

b Find the coordinates of S.

← Sections 5.2, 5.4

2 Prove that the circle $(x + 4)^2 + (y - 5)^2 = 8^2$ lies completely inside the circle $x^2 + y^2 + 8x - 10y = 59$.

← Sections 1.5, 6.2

3 Prove that for all positive integers n and k, $\binom{n}{k} + \binom{n}{k+1} = \binom{n+1}{k+1}$.

← Sections 7.4, 8.2

4 Solve for $0° \leqslant x \leqslant 360°$ the equation $2\sin^3 x - \sin x + 1 = \cos^2 x$.

← Section 10.6

11 Vectors

Objectives

After completing this chapter you should be able to:

- Use vectors in two dimensions → pages 231–235
- Use column vectors and carry out arithmetic operations on vectors → pages 235–238
- Calculate the magnitude and direction of a vector → pages 239–242
- Understand and use position vectors → pages 242–244
- Use vectors to solve geometric problems → pages 244–247
- Understand vector magnitude and use vectors in speed and distance calculations → pages 248–251
- Use vectors to solve problems in context → pages 248–251

Prior knowledge check

1 Write the column vector for the translation of shape

a A to B

b A to C

c A to D

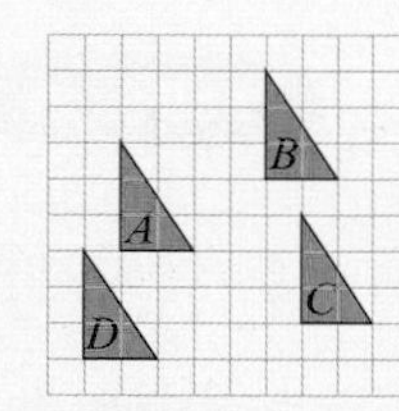

← GCSE Mathematics

2 P divides the line AB in the ratio $AP : PB = 7 : 2$.

Find:

a $\frac{AP}{AB}$ **b** $\frac{PB}{AB}$ **c** $\frac{AP}{PB}$

← GCSE Mathematics

3 Find x to one decimal place.

a

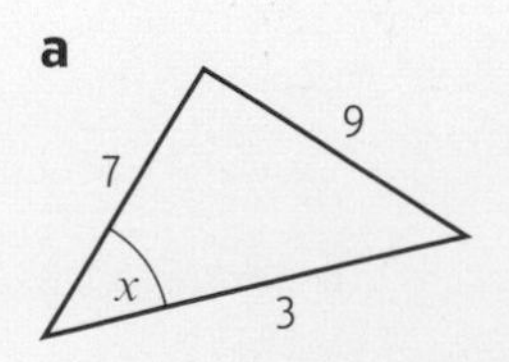

b

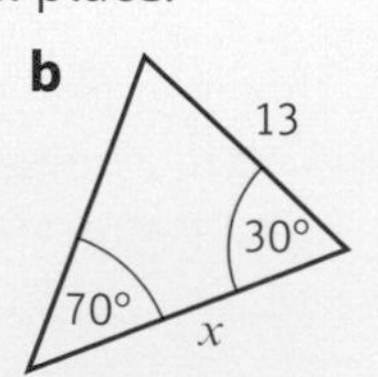

c

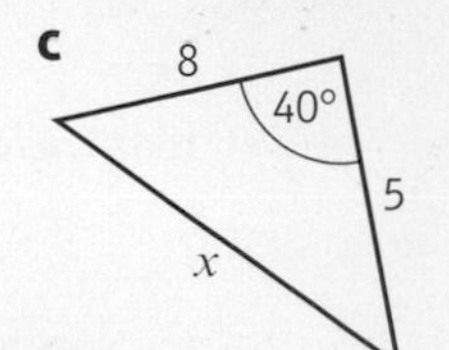

d

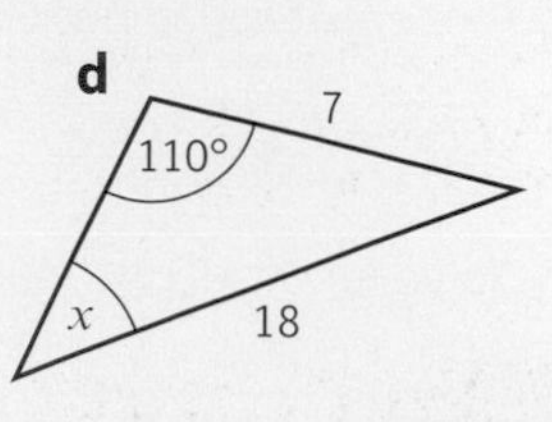

← Sections 9.1, 9.2

Pilots use **vector addition** to work out the resultant vector for their speed and heading when a plane encounters a strong cross-wind. Engineers also use vectors to work out the resultant forces acting on structures in construction.

11.1 Vectors

A **vector** has both **magnitude** and **direction**.

You can represent a vector using a **directed line segment**.

This is vector $\overrightarrow{PQ}$. It starts at P and finishes at Q.

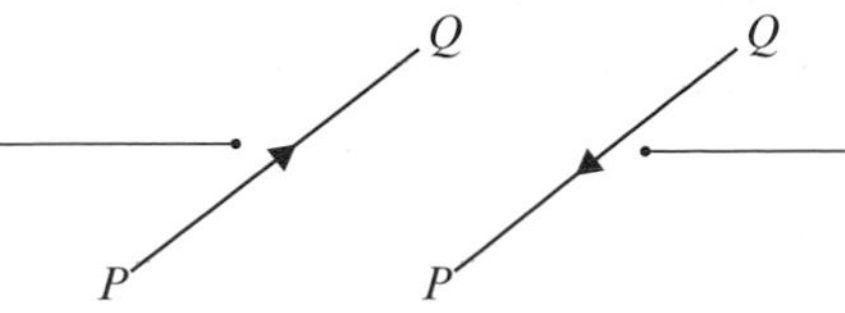

This is vector $\overrightarrow{QP}$. It starts at Q and finishes at P.

The direction of the arrow shows the direction of the vector. Small (lower case) letters are also used to represent vectors. In print, the small letter will be in bold type. In writing, you should underline the small letter to show it is a vector: $\underline{a}$ or $\underset{\sim}{a}$

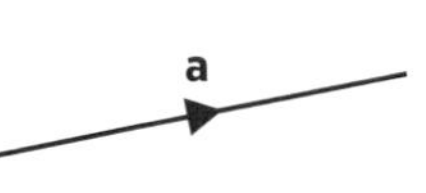

- **If $\overrightarrow{PQ} = \overrightarrow{RS}$ then the line segments PQ and RS are equal in length and are parallel.**

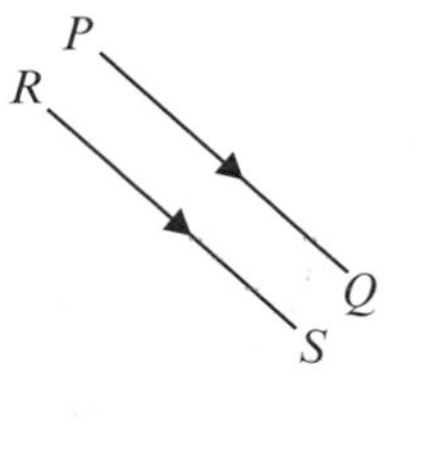

- **$\overrightarrow{AB} = -\overrightarrow{BA}$ as the line segment AB is equal in length, parallel and in the opposite direction to BA.**

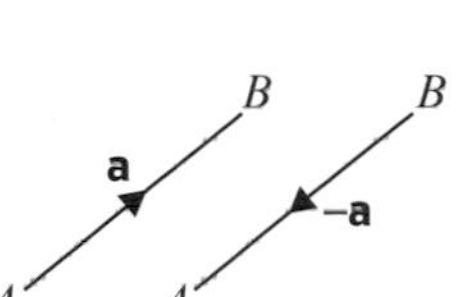

You can add two vectors together using the **triangle law** for vector addition.

- **Triangle law for vector addition:**
  $\overrightarrow{AB} + \overrightarrow{BC} = \overrightarrow{AC}$

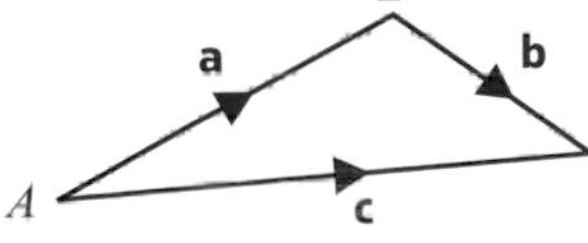

If $\overrightarrow{AB}$ = a, $\overrightarrow{BC}$ = b and $\overrightarrow{AC}$ = c, then a + b = c

Notation The **resultant** is the **vector sum** of two or more vectors.

$\overrightarrow{AB} + \overrightarrow{BC} + \overrightarrow{CD} = \overrightarrow{AD}$

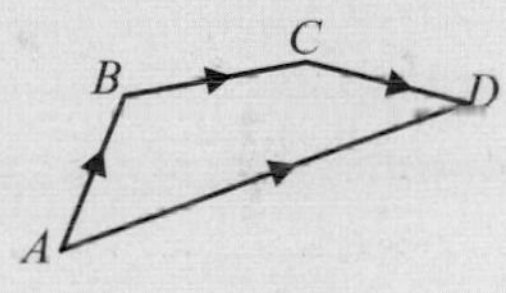

Example 1

The diagram shows vectors **a**, **b** and **c**.

Draw a diagram to illustrate the vector addition **a** + **b** + **c**.

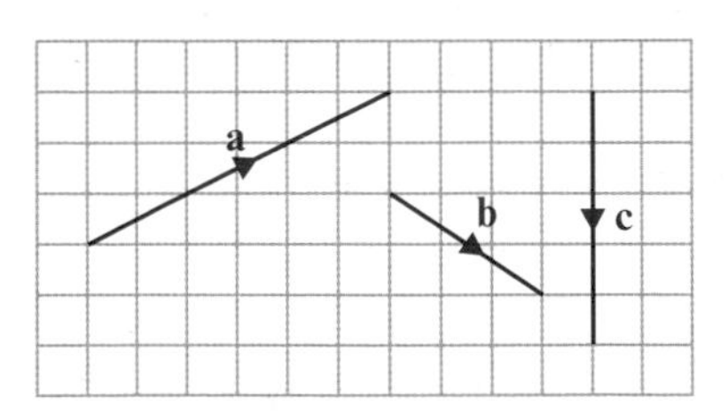

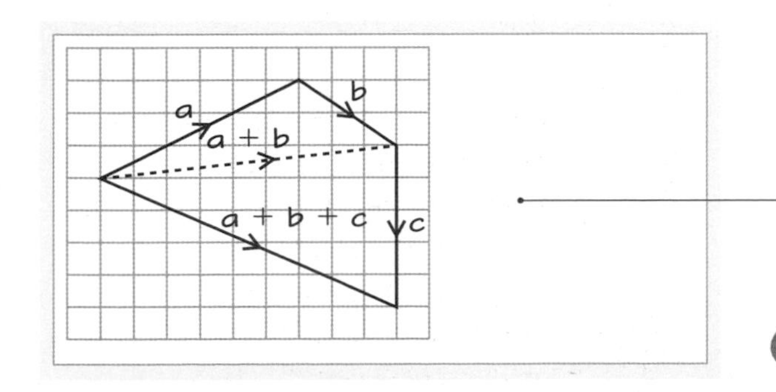

First use the triangle law for **a** + **b**, then use it again for (**a** + **b**) + **c**.
The resultant goes from the start of **a** to the end of **c**.

Online Explore vector addition using GeoGebra.

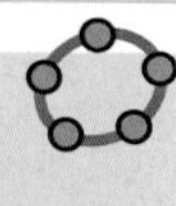

- **Subtracting a vector is equivalent to 'adding a negative vector':**
  a − b = a + (−b)

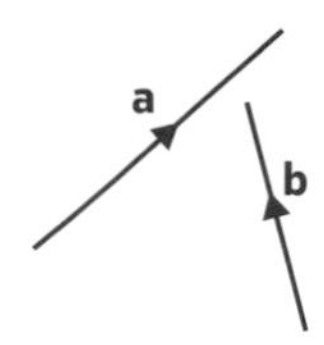

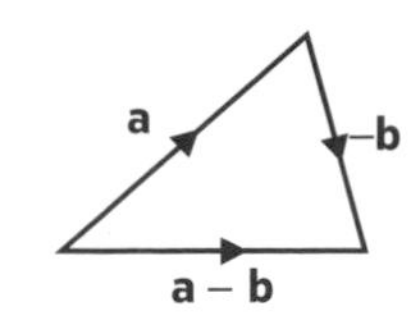

Hint To subtract **b**, you reverse the direction of **b** then add.

If you travel from P to Q, then back from Q to P, you are back where you started, so your displacement is zero.

- **Adding the vectors $\overrightarrow{PQ}$ and $\overrightarrow{QP}$ gives the zero vector 0: $\overrightarrow{PQ} + \overrightarrow{QP} = \mathbf{0}$**

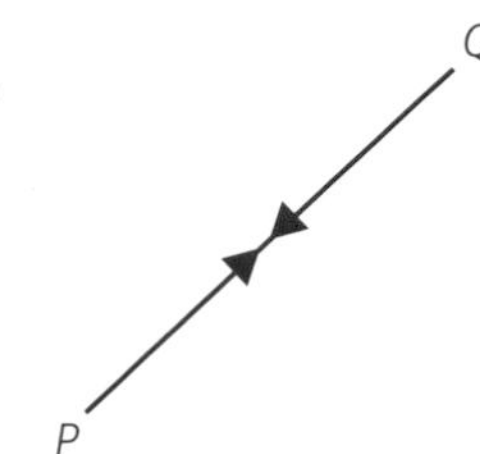

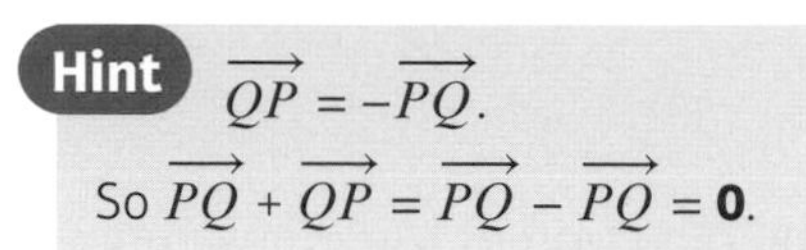

Hint $\overrightarrow{QP} = -\overrightarrow{PQ}$.
So $\overrightarrow{PQ} + \overrightarrow{QP} = \overrightarrow{PQ} - \overrightarrow{PQ} = \mathbf{0}$.

You can multiply a vector by a scalar (or number).

If the number is positive (≠ 1) the new vector has a different length but the **same** direction.

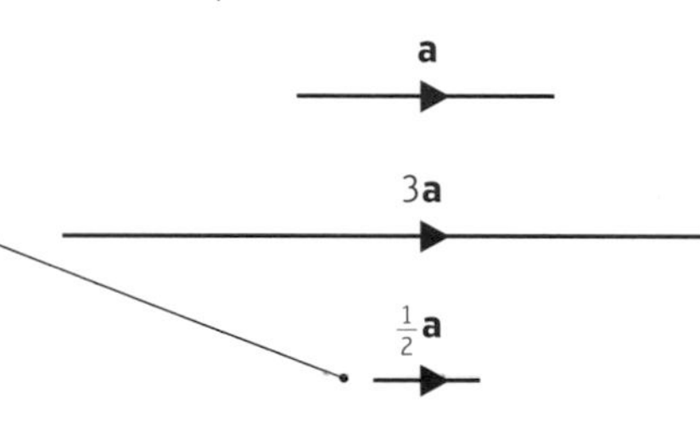

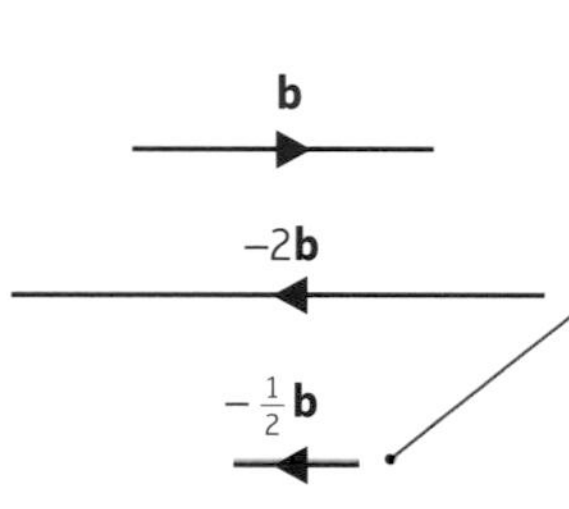

If the number is negative (≠ −1) the new vector has a different length and the **opposite** direction.

- **Any vector parallel to the vector a may be written as λa, where λ is a non-zero scalar.**

Notation Real numbers are examples of **scalars**. They have magnitude but no direction.

Example 2

In the diagram, $\overrightarrow{QP} = \mathbf{a}$, $\overrightarrow{QR} = \mathbf{b}$, $\overrightarrow{QS} = \mathbf{c}$ and $\overrightarrow{RT} = \mathbf{d}$.
Find in terms of **a**, **b**, **c** and **d**:

a $\overrightarrow{PS}$ **b** $\overrightarrow{RP}$
c $\overrightarrow{PT}$ **d** $\overrightarrow{TS}$

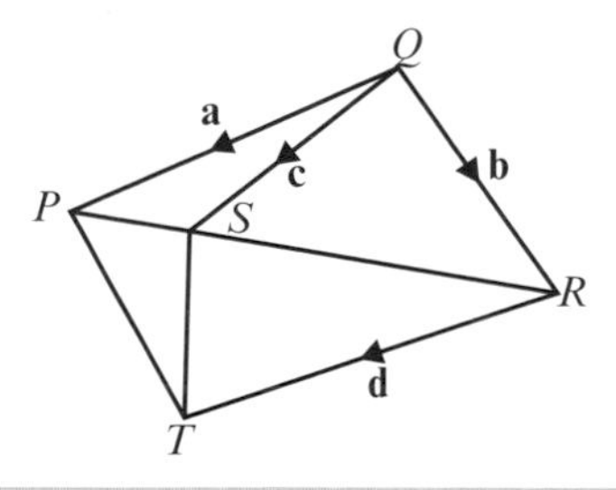

a $\overrightarrow{PS} = \overrightarrow{PQ} + \overrightarrow{QS} = -\mathbf{a} + \mathbf{c}$
$= \mathbf{c} - \mathbf{a}$

Add vectors using $\triangle PQS$.

b $\overrightarrow{RP} = \overrightarrow{RQ} + \overrightarrow{QP} = -\mathbf{b} + \mathbf{a}$
$= \mathbf{a} - \mathbf{b}$

Add vectors using $\triangle RQP$.

c $\overrightarrow{PT} = \overrightarrow{PR} + \overrightarrow{RT} = (\mathbf{b} - \mathbf{a}) + \mathbf{d}$
$= \mathbf{b} + \mathbf{d} - \mathbf{a}$

Add vectors using $\triangle PRT$.
Use $\overrightarrow{PR} = -\overrightarrow{RP} = -(\mathbf{a} - \mathbf{b}) = \mathbf{b} - \mathbf{a}$.

d $\overrightarrow{TS} = \overrightarrow{TR} + \overrightarrow{RS} = -\mathbf{d} + (\overrightarrow{RQ} + \overrightarrow{QS})$
$= -\mathbf{d} + (-\mathbf{b} + \mathbf{c})$
$= \mathbf{c} - \mathbf{b} - \mathbf{d}$

Add vectors using $\triangle TRS$ and $\triangle RQS$.

Example 3

$ABCD$ is a parallelogram. $\overrightarrow{AB} = \mathbf{a}$, $\overrightarrow{AD} = \mathbf{b}$. Find $\overrightarrow{AC}$.

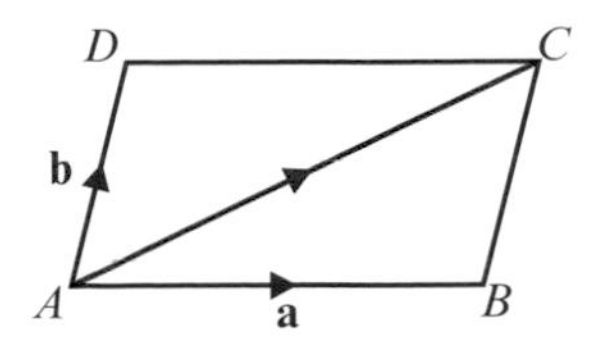

Notation This is called the **parallelogram law** for vector addition.

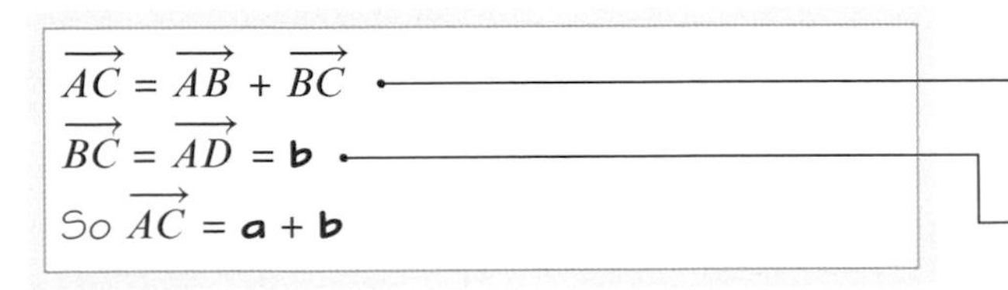

$\overrightarrow{AC} = \overrightarrow{AB} + \overrightarrow{BC}$

$\overrightarrow{BC} = \overrightarrow{AD} = \mathbf{b}$

So $\overrightarrow{AC} = \mathbf{a} + \mathbf{b}$

Using the triangle law for addition of vectors.

AD and BC are opposite sides of a parallelogram so they are parallel and equal in magnitude.

Example 4

Show that the vectors $6\mathbf{a} + 8\mathbf{b}$ and $9\mathbf{a} + 12\mathbf{b}$ are parallel.

$9\mathbf{a} + 12\mathbf{b} = \frac{3}{2}(6\mathbf{a} + 8\mathbf{b})$

$\therefore$ the vectors are parallel.

Here $\lambda = \frac{3}{2}$

Example 5

In triangle ABC, $\overrightarrow{AB} = \mathbf{a}$ and $\overrightarrow{AC} = \mathbf{b}$.

P is the midpoint of AB.

Q divides AC in the ratio $3:2$.

Write in terms of $\mathbf{a}$ and $\mathbf{b}$:

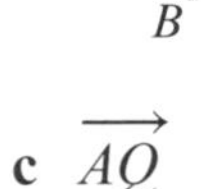

a $\overrightarrow{BC}$ **b** $\overrightarrow{AP}$ **c** $\overrightarrow{AQ}$ **d** $\overrightarrow{PQ}$

a $\overrightarrow{BC} = \overrightarrow{BA} + \overrightarrow{AC}$

$= -\overrightarrow{AB} + \overrightarrow{AC}$

$\overrightarrow{BC} = \mathbf{b} - \mathbf{a}$

b $\overrightarrow{AP} = \frac{1}{2}\overrightarrow{AB} = \frac{1}{2}\mathbf{a}$

c $\overrightarrow{AQ} = \frac{3}{5}\overrightarrow{AC} = \frac{3}{5}\mathbf{b}$

d $\overrightarrow{PQ} = \overrightarrow{PA} + \overrightarrow{AQ}$

$= -\overrightarrow{AP} + \overrightarrow{AQ}$

$= \frac{3}{5}\mathbf{b} - \frac{1}{2}\mathbf{a}$

$\overrightarrow{BA} = -\overrightarrow{AB}$

$AP = \frac{1}{2}AB$ so $\overrightarrow{AP} = \frac{1}{2}\mathbf{a}$

Watch out AP is the line segment between A and P, whereas $\overrightarrow{AP}$ is the vector from A to P.

Q divides AC in the ratio $3:2$ so $AQ = \frac{3}{5}AC$.

Going from P to Q is the same as going from P to A, then from A to Q.

Exercise 11A

1 The diagram shows the vectors **a**, **b**, **c** and **d**.

Draw a diagram to illustrate these vectors:

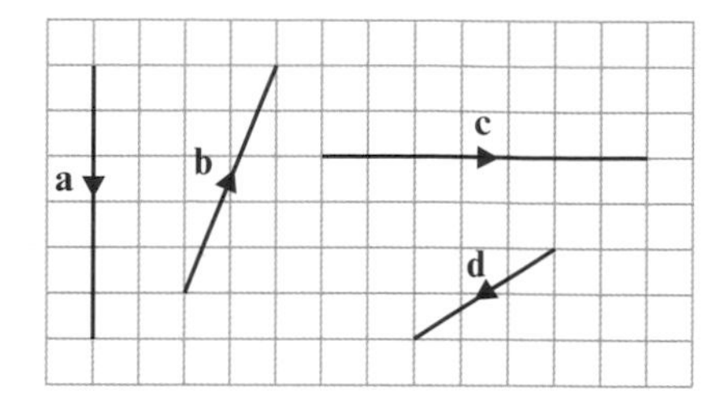

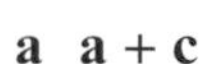

a $\mathbf{a} + \mathbf{c}$ **b** $-\mathbf{b}$

c $\mathbf{c} - \mathbf{d}$ **d** $\mathbf{b} + \mathbf{c} + \mathbf{d}$

e $2\mathbf{c} + 3\mathbf{d}$ **f** $\mathbf{a} - 2\mathbf{b}$

g $\mathbf{a} + \mathbf{b} + \mathbf{c} + \mathbf{d}$

2 $ACGI$ is a square, B is the midpoint of AC, F is the midpoint of CG, H is the midpoint of GI, D is the midpoint of AI.

$\overrightarrow{AB} = \mathbf{b}$ and $\overrightarrow{AD} = \mathbf{d}$. Find, in terms of **b** and **d**:

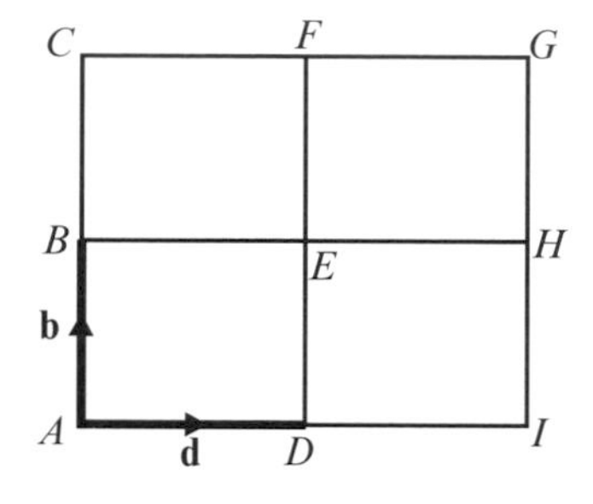

a $\overrightarrow{AC}$ **b** $\overrightarrow{BE}$ **c** $\overrightarrow{HG}$ **d** $\overrightarrow{DF}$

e $\overrightarrow{AE}$ **f** $\overrightarrow{DH}$ **g** $\overrightarrow{HB}$ **h** $\overrightarrow{FE}$

i $\overrightarrow{AH}$ **j** $\overrightarrow{BI}$ **k** $\overrightarrow{EI}$ **l** $\overrightarrow{FB}$

3 $OACB$ is a parallelogram. M, Q, N and P are the midpoints of OA, AC, BC and OB respectively.

Vectors **p** and **m** are equal to $\overrightarrow{OP}$ and $\overrightarrow{OM}$ respectively. Express in terms of **p** and **m**.

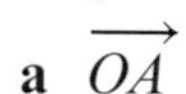

a $\overrightarrow{OA}$ **b** $\overrightarrow{OB}$ **c** $\overrightarrow{BN}$ **d** $\overrightarrow{DQ}$

e $\overrightarrow{OD}$ **f** $\overrightarrow{MQ}$ **g** $\overrightarrow{OQ}$ **h** $\overrightarrow{AD}$

i $\overrightarrow{CD}$ **j** $\overrightarrow{AP}$ **k** $\overrightarrow{BM}$ **l** $\overrightarrow{NO}$

4 In the diagram, $\overrightarrow{PQ} = \mathbf{a}$, $\overrightarrow{QS} = \mathbf{b}$, $\overrightarrow{SR} = \mathbf{c}$ and $\overrightarrow{PT} = \mathbf{d}$.
Find in terms of **a**, **b**, **c** and **d**:

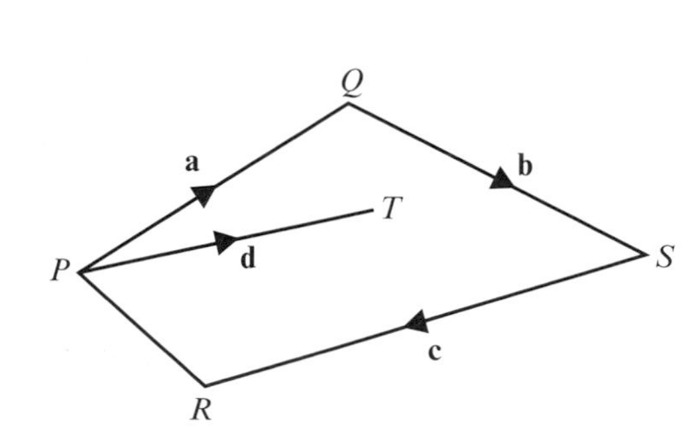

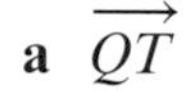

a $\overrightarrow{QT}$ **b** $\overrightarrow{PR}$

c $\overrightarrow{TS}$ **d** $\overrightarrow{TR}$

5 In the triangle PQR, $PQ = 2\mathbf{a}$ and $QR = 2\mathbf{b}$.
The midpoint of PR is M. Find, in terms of **a** and **b**:

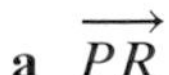

a $\overrightarrow{PR}$ **b** $\overrightarrow{PM}$ **c** $\overrightarrow{QM}$

P **6** $ABCD$ is a trapezium with AB parallel to DC and $DC = 3AB$. M divides DC such that $DM : MC = 2 : 1$. $\overrightarrow{AB} = \mathbf{a}$ and $\overrightarrow{BC} = \mathbf{b}$.

Find, in terms of **a** and **b**:

a $\overrightarrow{AM}$ **b** $\overrightarrow{BD}$ **c** $\overrightarrow{MB}$ **d** $\overrightarrow{DA}$

Problem-solving

Draw a sketch to show the information given in the question.

7 *OABC* is a parallelogram. $\overrightarrow{OA} = \mathbf{a}$ and $\overrightarrow{OC} = \mathbf{b}$.

The point *P* divides *OB* in the ratio 5:3.

Find, in terms of **a** and **b**:

a $\overrightarrow{OB}$ **b** $\overrightarrow{OP}$ **c** $\overrightarrow{AP}$

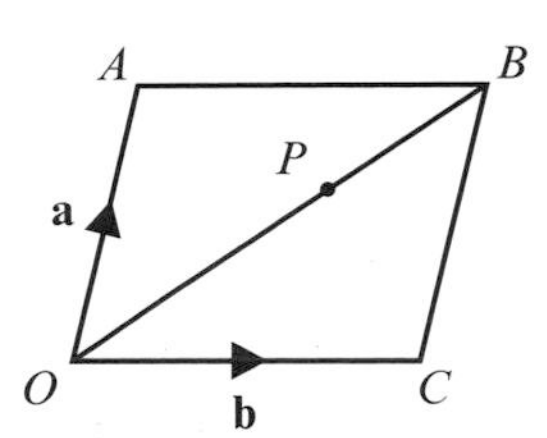

8 State with a reason whether each of these vectors is parallel to the vector $\mathbf{a} - 3\mathbf{b}$:

a $2\mathbf{a} - 6\mathbf{b}$ **b** $4\mathbf{a} - 12\mathbf{b}$ **c** $\mathbf{a} + 3\mathbf{b}$ **d** $3\mathbf{b} - \mathbf{a}$ **e** $9\mathbf{b} - 3\mathbf{a}$ **f** $\frac{1}{2}\mathbf{a} - \frac{2}{3}\mathbf{b}$

(P) **9** In triangle *ABC*, $\overrightarrow{AB} = \mathbf{a}$ and $\overrightarrow{AC} = \mathbf{b}$.

P is the midpoint of *AB* and *Q* is the midpoint of *AC*.

a Write in terms of **a** and **b**:

i $\overrightarrow{BC}$ **ii** $\overrightarrow{AP}$ **iii** $\overrightarrow{AQ}$ **iv** $\overrightarrow{PQ}$

b Show that *PQ* is parallel to *BC*.

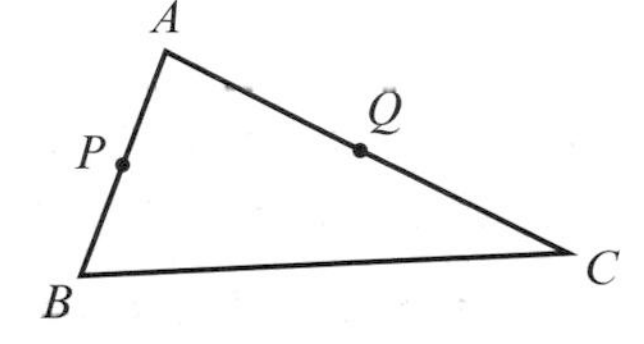

(P) **10** *OABC* is a quadrilateral. $\overrightarrow{OA} = \mathbf{a}$, $\overrightarrow{OC} = 3\mathbf{b}$ and $\overrightarrow{OB} = \mathbf{a} + 2\mathbf{b}$.

a Find, in terms of **a** and **b**:

i $\overrightarrow{AB}$ **ii** $\overrightarrow{CB}$

b Show that *AB* is parallel to *OC*.

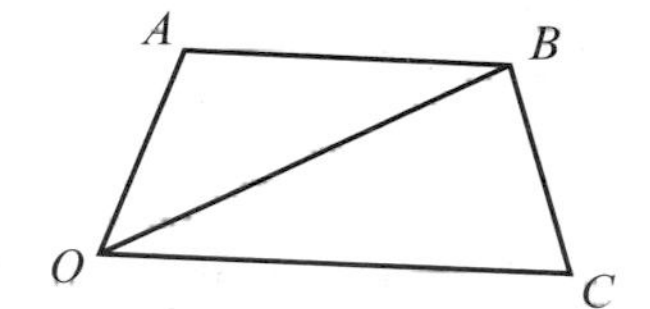

(P) **11** The vectors $2\mathbf{a} + k\mathbf{b}$ and $5\mathbf{a} + 3\mathbf{b}$ are parallel. Find the value of k.

11.2 Representing vectors

A vector can be described by its change in position or **displacement** relative to the x- and y-axes.

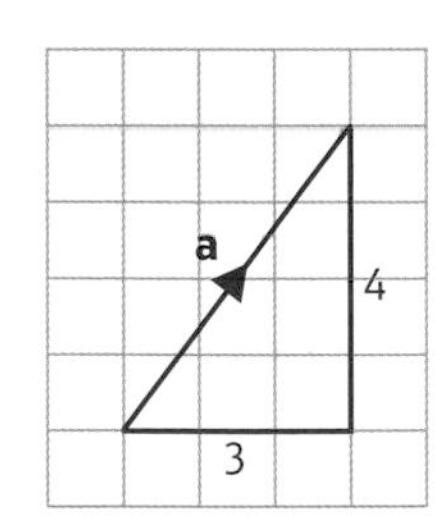

$\mathbf{a} = \begin{pmatrix} 3 \\ 4 \end{pmatrix}$ where 3 is the change in the x-direction and 4 is the change in the y-direction.

This is called **column vector** form.

Notation The top number is the x-component and the bottom number is the y-component.

- **To multiply a column vector by a scalar, multiply each component by the scalar:** $\lambda\begin{pmatrix} p \\ q \end{pmatrix} = \begin{pmatrix} \lambda p \\ \lambda q \end{pmatrix}$
- **To add two column vectors, add the x-components and the y-components:** $\begin{pmatrix} p \\ q \end{pmatrix} + \begin{pmatrix} r \\ s \end{pmatrix} = \begin{pmatrix} p + r \\ q + s \end{pmatrix}$

Example 6

$\mathbf{a} = \begin{pmatrix} 2 \\ 6 \end{pmatrix}$ and $\mathbf{b} = \begin{pmatrix} 3 \\ -1 \end{pmatrix}$

Find **a** $\frac{1}{3}\mathbf{a}$ **b** $\mathbf{a} + \mathbf{b}$ **c** $2\mathbf{a} - 3\mathbf{b}$

a $\frac{1}{3}\mathbf{a} = \begin{pmatrix} \frac{2}{3} \\ 2 \end{pmatrix}$

Both of the components are divided by 3.

b $\mathbf{a} + \mathbf{b} = \begin{pmatrix} 2 \\ 6 \end{pmatrix} + \begin{pmatrix} 3 \\ -1 \end{pmatrix} = \begin{pmatrix} 5 \\ 5 \end{pmatrix}$

Add the x-components and the y-components.

c $2\mathbf{a} - 3\mathbf{b} = 2\begin{pmatrix} 2 \\ 6 \end{pmatrix} - 3\begin{pmatrix} 3 \\ -1 \end{pmatrix}$

$= \begin{pmatrix} 4 \\ 12 \end{pmatrix} - \begin{pmatrix} 9 \\ -3 \end{pmatrix} = \begin{pmatrix} 4-9 \\ 12+3 \end{pmatrix} = \begin{pmatrix} -5 \\ 15 \end{pmatrix}$

Multiply each of the vectors by the scalars then subtract the x- and y-components.

You can use **unit vectors** to represent vectors in two dimensions.

- **A unit vector is a vector of length 1. The unit vectors along the x- and y-axes are usually denoted by i and j respectively.**
  - $\mathbf{i} = \begin{pmatrix} 1 \\ 0 \end{pmatrix}$ $\mathbf{j} = \begin{pmatrix} 0 \\ 1 \end{pmatrix}$

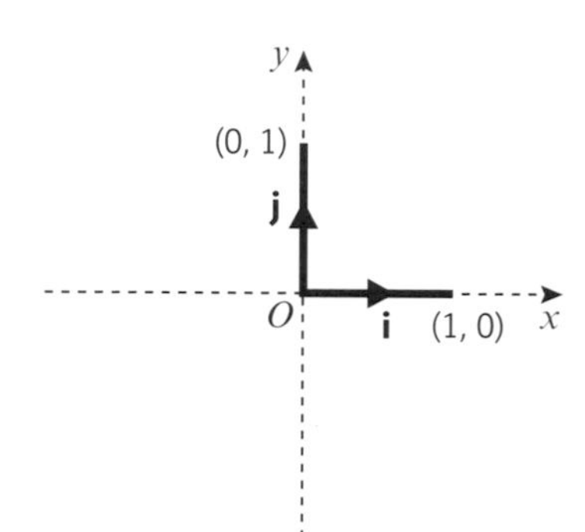

- **You can write any two-dimensional vector in the form $p\mathbf{i} + q\mathbf{j}$.**

By the triangle law of addition:

$\overrightarrow{AC} = \overrightarrow{AB} + \overrightarrow{BC}$

$= 5\mathbf{i} + 2\mathbf{j}$

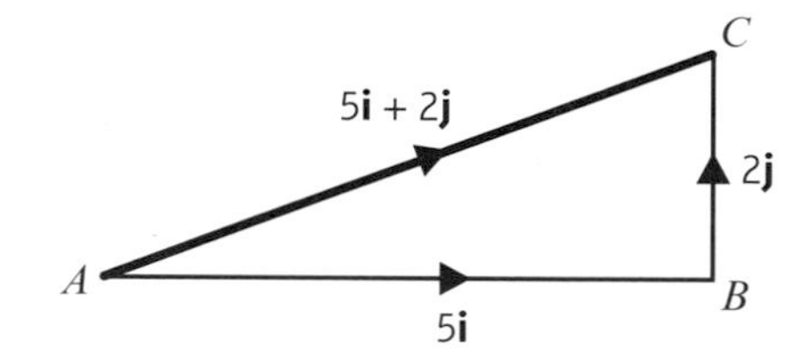

You can also write this as a column vector: $5\mathbf{i} + 2\mathbf{j} = \begin{pmatrix} 5 \\ 2 \end{pmatrix}$

- **For any two-dimensional vector: $\begin{pmatrix} p \\ q \end{pmatrix} = p\mathbf{i} + q\mathbf{j}$**

Example 7

$\mathbf{a} = 3\mathbf{i} - 4\mathbf{j}$, $\mathbf{b} = 2\mathbf{i} + 7\mathbf{j}$

Find **a** $\frac{1}{2}\mathbf{a}$ **b** $\mathbf{a} + \mathbf{b}$ **c** $3\mathbf{a} - 2\mathbf{b}$

a $\frac{1}{2}\mathbf{a} = \frac{1}{2}(3\mathbf{i} - 4\mathbf{j}) = 1.5\mathbf{i} - 2\mathbf{j}$

Divide the **i** component and the **j** component by 2.

b $\mathbf{a} + \mathbf{b} = 3\mathbf{i} - 4\mathbf{j} + 2\mathbf{i} + 7\mathbf{j}$

$= (3 + 2)\mathbf{i} + (-4 + 7)\mathbf{j} = 5\mathbf{i} + 3\mathbf{j}$

Add the **i** components and the **j** components.

c $3\mathbf{a} - 2\mathbf{b} = 3(3\mathbf{i} - 4\mathbf{j}) - 2(2\mathbf{i} + 7\mathbf{j})$

$= 9\mathbf{i} - 12\mathbf{j} - (4\mathbf{i} + 14\mathbf{j})$

$= (9 - 4)\mathbf{i} + (-12 - 14)\mathbf{j}$

$= 5\mathbf{i} - 26\mathbf{j}$

Multiply each of the vectors by the scalar then subtract the **i** and **j** components.

Example 8

a Draw a diagram to represent the vector $-3\mathbf{i} + \mathbf{j}$

b Write this as a column vector.

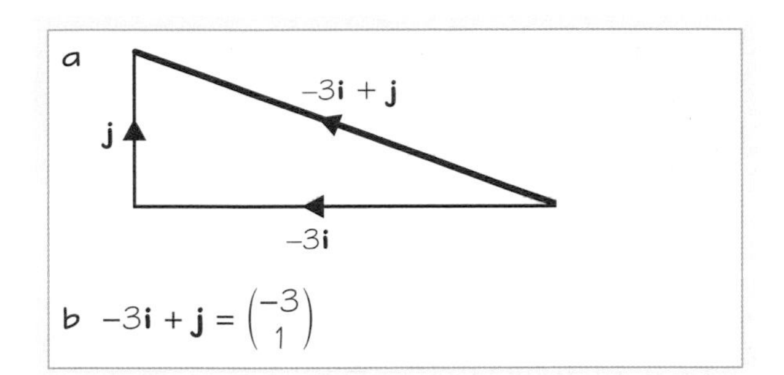

b $-3\mathbf{i} + \mathbf{j} = \begin{pmatrix} -3 \\ 1 \end{pmatrix}$

3 units in the direction of the unit vector $-\mathbf{i}$ and 1 unit in the direction of the unit vector $\mathbf{j}$.

Example 9

Given that $\mathbf{a} = 2\mathbf{i} + 5\mathbf{j}$, $\mathbf{b} = 12\mathbf{i} - 10\mathbf{j}$ and $\mathbf{c} = -3\mathbf{i} + 9\mathbf{j}$, find $\mathbf{a} + \mathbf{b} + \mathbf{c}$, using column vector notation in your working.

$\mathbf{a} + \mathbf{b} + \mathbf{c} = \begin{pmatrix} 2 \\ 5 \end{pmatrix} + \begin{pmatrix} 12 \\ -10 \end{pmatrix} + \begin{pmatrix} -3 \\ 9 \end{pmatrix} = \begin{pmatrix} 11 \\ 4 \end{pmatrix}$

Add the numbers in the top line to get 11 (the x-component), and the bottom line to get 4 (the y-component). This is $11\mathbf{i} + 4\mathbf{j}$.

Example 10

Given $\mathbf{a} = 5\mathbf{i} + 2\mathbf{j}$ and $\mathbf{b} = 3\mathbf{i} - 4\mathbf{j}$, find $2\mathbf{a} - \mathbf{b}$ in terms of $\mathbf{i}$ and $\mathbf{j}$.

Online Explore this solution as a vector diagram on a coordinate grid using GeoGebra.

$2\mathbf{a} = 2\begin{pmatrix} 5 \\ 2 \end{pmatrix} = \begin{pmatrix} 10 \\ 4 \end{pmatrix}$

To find the column vector for vector $2\mathbf{a}$ multiply the $\mathbf{i}$ and $\mathbf{j}$ components of vector $\mathbf{a}$ by 2.

$2\mathbf{a} - \mathbf{b} = \begin{pmatrix} 10 \\ 4 \end{pmatrix} - \begin{pmatrix} 3 \\ -4 \end{pmatrix} = \begin{pmatrix} 10 - 3 \\ 4 - (-4) \end{pmatrix} = \begin{pmatrix} 7 \\ 8 \end{pmatrix}$

To find the column vector for $2\mathbf{a} - \mathbf{b}$ subtract the components of vector $\mathbf{b}$ from those of vector $2\mathbf{a}$.

$2\mathbf{a} - \mathbf{b} = 7\mathbf{i} + 8\mathbf{j}$

Remember to give your answer in terms of $\mathbf{i}$ and $\mathbf{j}$.

Exercise 11B

1 These vectors are drawn on a grid of unit squares. Express the vectors $\mathbf{v}_1$, $\mathbf{v}_2$, $\mathbf{v}_3$, $\mathbf{v}_4$, $\mathbf{v}_5$ and $\mathbf{v}_6$ in $\mathbf{i}$, $\mathbf{j}$ notation and column vector form.

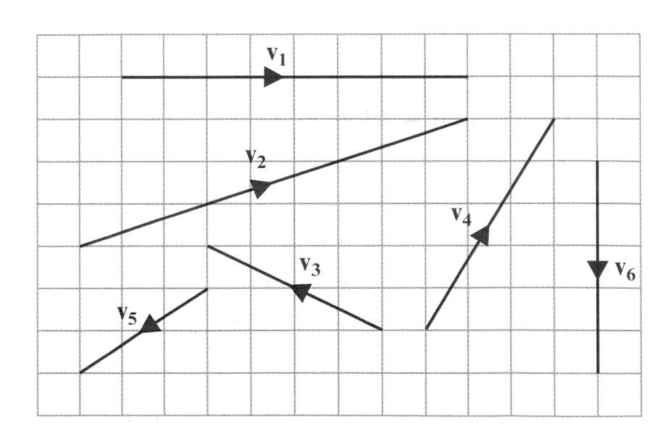

2 Given that $\mathbf{a} = 2\mathbf{i} + 3\mathbf{j}$ and $\mathbf{b} = 4\mathbf{i} - \mathbf{j}$, find these vectors in terms of $\mathbf{i}$ and $\mathbf{j}$.

a $4\mathbf{a}$ **b** $\frac{1}{2}\mathbf{a}$ **c** $-\mathbf{b}$ **d** $2\mathbf{b} + \mathbf{a}$

e $3\mathbf{a} - 2\mathbf{b}$ **f** $\mathbf{b} - 3\mathbf{a}$ **g** $4\mathbf{b} - \mathbf{a}$ **h** $2\mathbf{a} - 3\mathbf{b}$

3 Given that $\mathbf{a} = \begin{pmatrix} 9 \\ 7 \end{pmatrix}$, $\mathbf{b} = \begin{pmatrix} 11 \\ -3 \end{pmatrix}$ and $\mathbf{c} = \begin{pmatrix} -8 \\ -1 \end{pmatrix}$ find:

a $5\mathbf{a}$ **b** $-\frac{1}{2}\mathbf{c}$ **c** $\mathbf{a} + \mathbf{b} + \mathbf{c}$ **d** $2\mathbf{a} - \mathbf{b} + \mathbf{c}$

e $2\mathbf{b} + 2\mathbf{c} - 3\mathbf{a}$ **f** $\frac{1}{2}\mathbf{a} + \frac{1}{2}\mathbf{b}$

(P) **4** Given that $\mathbf{a} = 2\mathbf{i} + 5\mathbf{j}$ and $\mathbf{b} = 3\mathbf{i} - \mathbf{j}$, find:

a λ if $\mathbf{a} + \lambda\mathbf{b}$ is parallel to the vector $\mathbf{i}$ **b** μ if $\mu\mathbf{a} + \mathbf{b}$ is parallel to the vector $\mathbf{j}$

(P) **5** Given that $\mathbf{c} = 3\mathbf{i} + 4\mathbf{j}$ and $\mathbf{d} = \mathbf{i} - 2\mathbf{j}$, find:

a λ if $\mathbf{c} + \lambda\mathbf{d}$ is parallel to $\mathbf{i} + \mathbf{j}$ **b** μ if $\mu\mathbf{c} + \mathbf{d}$ is parallel to $\mathbf{i} + 3\mathbf{j}$

c s if $\mathbf{c} - s\mathbf{d}$ is parallel to $2\mathbf{i} + \mathbf{j}$ **d** t if $\mathbf{d} - t\mathbf{c}$ is parallel to $-2\mathbf{i} + 3\mathbf{j}$

(E) **6** In triangle ABC, $\overrightarrow{AB} = 4\mathbf{i} + 3\mathbf{j}$ and $\overrightarrow{AC} = 5\mathbf{i} + 2\mathbf{j}$.
Find $\overrightarrow{BC}$.

(2 marks)

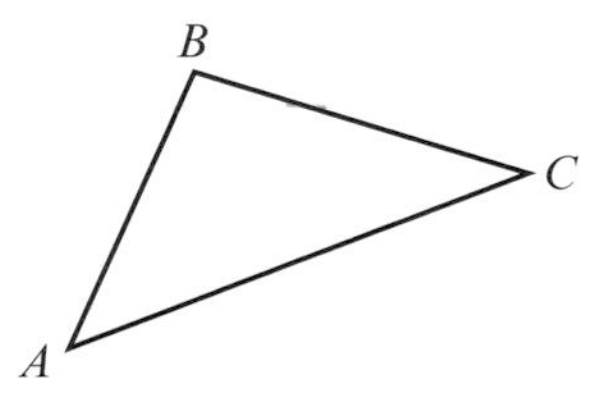

(P) **7** $OABC$ is a parallelogram.
P divides AC in the ratio $3:2$. $\overrightarrow{OA} = 2\mathbf{i} + 4\mathbf{j}$, $\overrightarrow{OC} = 7\mathbf{i}$.
Find in $\mathbf{i}$, $\mathbf{j}$ format and column vector format:

a $\overrightarrow{AC}$ **b** $\overrightarrow{AP}$ **c** $\overrightarrow{OP}$

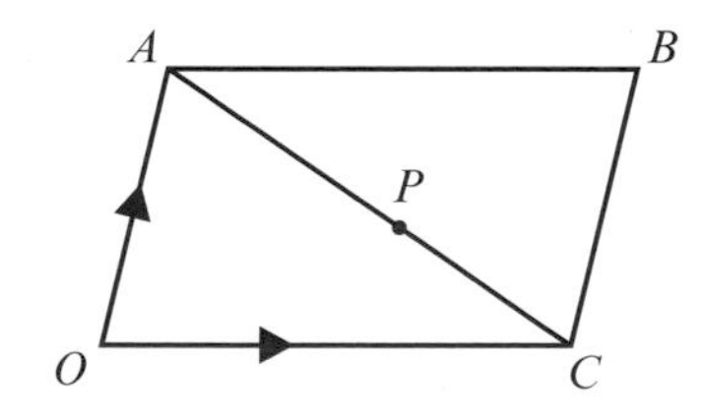

(E/P) **8** $\mathbf{a} = \begin{pmatrix} j \\ 3 \end{pmatrix}$, $\mathbf{b} = \begin{pmatrix} 10 \\ k \end{pmatrix}$, $\mathbf{c} = \begin{pmatrix} 2 \\ 5 \end{pmatrix}$

Given that $\mathbf{b} - 2\mathbf{a} = \mathbf{c}$, find the values of j and k.

(2 marks)

Problem-solving

You can consider $\mathbf{b} - 2\mathbf{a} = \mathbf{c}$ as two linear equations. One for the x-components and one for the y-components.

(E/P) **9** $\mathbf{a} = \begin{pmatrix} p \\ -q \end{pmatrix}$, $\mathbf{b} = \begin{pmatrix} q \\ p \end{pmatrix}$, $\mathbf{c} = \begin{pmatrix} 7 \\ 4 \end{pmatrix}$

Given that $\mathbf{a} + 2\mathbf{b} = \mathbf{c}$, find the values of p and q. **(2 marks)**

(E/P) **10** The resultant of the vectors $\mathbf{a} = 3\mathbf{i} - 2\mathbf{j}$ and $\mathbf{b} = p\mathbf{i} - 2p\mathbf{j}$ is parallel to the vector $\mathbf{c} = 2\mathbf{i} - 3\mathbf{j}$.
Find:

a the value of p **(4 marks)**

b the resultant of vectors $\mathbf{a}$ and $\mathbf{b}$. **(1 mark)**

11.3 Magnitude and direction

You can use Pythagoras' theorem to calculate the **magnitude** of a vector.

- **For the vector $\mathbf{a} = x\mathbf{i} + y\mathbf{j} = \begin{pmatrix} x \\ y \end{pmatrix}$, the magnitude of the vector is given by:**

$$|\mathbf{a}| = \sqrt{x^2 + y^2}$$

Notation You use straight lines on either side of the vector:

$|\mathbf{a}| = |x\mathbf{i} + y\mathbf{j}| = \left|\begin{pmatrix} x \\ y \end{pmatrix}\right|$

You need to be able to find a **unit vector** in the direction of a given vector.

- **A unit vector in the direction of a is $\dfrac{\mathbf{a}}{|\mathbf{a}|}$**

If $|\mathbf{a}| = 5$ then a unit vector in the direction of $\mathbf{a}$ is $\frac{\mathbf{a}}{5}$.

Notation A unit vector is any vector with magnitude 1.
A unit vector in the direction of $\mathbf{a}$ is sometimes written as $\hat{\mathbf{a}}$.

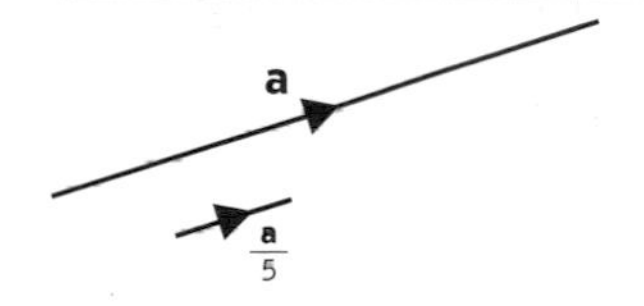

Example 11

Given that $\mathbf{a} = 3\mathbf{i} + 4\mathbf{j}$ and $\mathbf{b} = -2\mathbf{i} - 4\mathbf{j}$:

a find $|\mathbf{a}|$

b find a unit vector in the direction of $\mathbf{a}$

c find the exact value of $|2\mathbf{a} + \mathbf{b}|$

Online Explore the magnitude of a vector using GeoGebra.

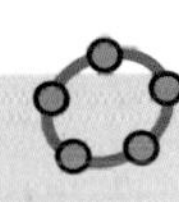

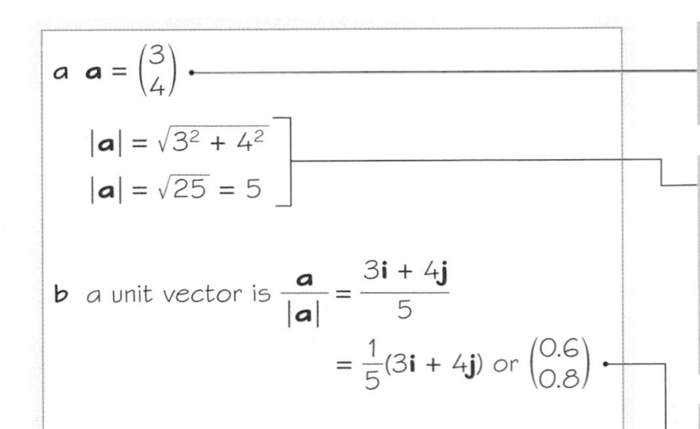

a $\mathbf{a} = \begin{pmatrix} 3 \\ 4 \end{pmatrix}$

$|\mathbf{a}| = \sqrt{3^2 + 4^2}$

$|\mathbf{a}| = \sqrt{25} = 5$

b a unit vector is $\dfrac{\mathbf{a}}{|\mathbf{a}|} = \dfrac{3\mathbf{i} + 4\mathbf{j}}{5}$

$= \frac{1}{5}(3\mathbf{i} + 4\mathbf{j})$ or $\begin{pmatrix} 0.6 \\ 0.8 \end{pmatrix}$

c $2\mathbf{a} + \mathbf{b} = 2\begin{pmatrix} 3 \\ 4 \end{pmatrix} + \begin{pmatrix} -2 \\ -4 \end{pmatrix} = \begin{pmatrix} 6 - 2 \\ 8 - 4 \end{pmatrix} = \begin{pmatrix} 4 \\ 4 \end{pmatrix}$

$|2\mathbf{a} + \mathbf{b}| = \sqrt{4^2 + 4^2} = \sqrt{32} = 4\sqrt{2}$

It is often quicker and easier to convert from **i**, **j** form to column vector form for calculations.

Using Pythagoras.

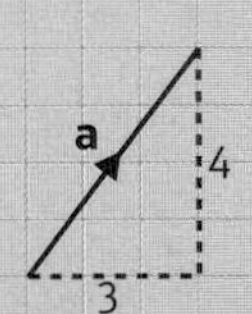

Unless specified in the question it is acceptable to give your answer in **i**, **j** form or column vector form.

You need to give an exact answer, so leave your answer in surd form:
$\sqrt{32} = \sqrt{16 \times 2} = 4\sqrt{2}$ ← Section 1.5

You can define a vector by giving its magnitude, and the angle between the vector and one of the coordinate axes. This is called **magnitude-direction form**.

Example 12

Find the angle between the vector $4\mathbf{i} + 5\mathbf{j}$ and the positive x-axis.

This might be referred to as the angle between the vector and **i**.

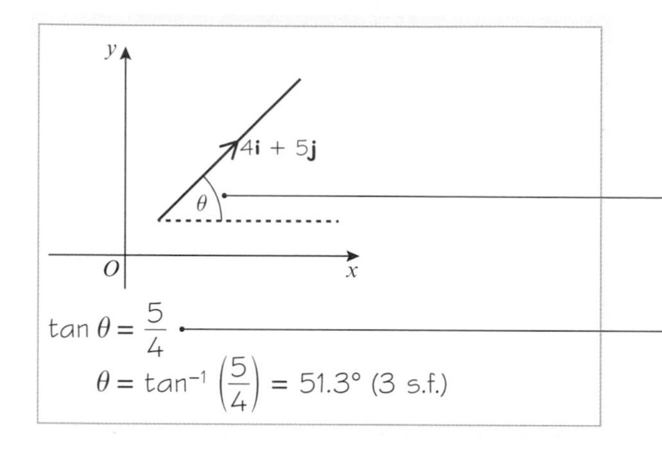

Identify the angle that you need to find. A diagram always helps.

$\tan\theta = \frac{5}{4}$

You have a right-angled triangle with base 4 units and height 5 units, so use trigonometry.

$\theta = \tan^{-1}\left(\frac{5}{4}\right) = 51.3°$ (3 s.f.)

Example 13

Vector **a** has magnitude 10 and makes an angle of 30° with **j**.
Find **a** in **i**, **j** and column vector format.

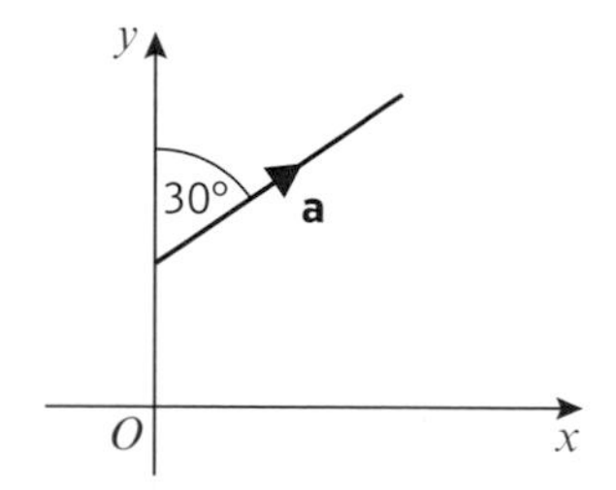

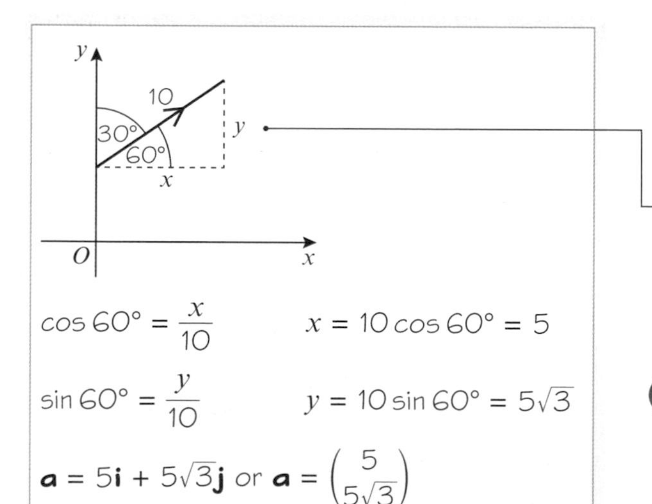

Use trigonometry to find the lengths of the x- and y-components for vector **a**.

$\cos 60° = \frac{x}{10}$ $\quad x = 10\cos 60° = 5$

$\sin 60° = \frac{y}{10}$ $\quad y = 10\sin 60° = 5\sqrt{3}$

$\mathbf{a} = 5\mathbf{i} + 5\sqrt{3}\mathbf{j}$ or $\mathbf{a} = \begin{pmatrix} 5 \\ 5\sqrt{3} \end{pmatrix}$

Watch out The direction of a vector can be given relative to either the positive x-axis (the **i** direction) or the positive y-axis (or the **j** direction).

Exercise 11C

1 Find the magnitude of each of these vectors.

a $3\mathbf{i} + 4\mathbf{j}$ **b** $6\mathbf{i} - 8\mathbf{j}$ **c** $5\mathbf{i} + 12\mathbf{j}$ **d** $2\mathbf{i} + 4\mathbf{j}$

e $3\mathbf{i} - 5\mathbf{j}$ **f** $4\mathbf{i} + 7\mathbf{j}$ **g** $-3\mathbf{i} + 5\mathbf{j}$ **h** $-4\mathbf{i} - \mathbf{j}$

2 $\mathbf{a} = 2\mathbf{i} + 3\mathbf{j}$, $\mathbf{b} = 3\mathbf{i} - 4\mathbf{j}$ and $\mathbf{c} = 5\mathbf{i} - \mathbf{j}$. Find the exact value of the magnitude of:

a $\mathbf{a} + \mathbf{b}$ **b** $2\mathbf{a} - \mathbf{c}$ **c** $3\mathbf{b} - 2\mathbf{c}$

3 For each of the following vectors, find the unit vector in the same direction.

a $\mathbf{a} = 4\mathbf{i} + 3\mathbf{j}$ **b** $\mathbf{b} = 5\mathbf{i} - 12\mathbf{j}$ **c** $\mathbf{c} = -7\mathbf{i} + 24\mathbf{j}$ **d** $\mathbf{d} = \mathbf{i} - 3\mathbf{j}$

4 Find the angle that each of these vectors makes with the positive x-axis.

a $3\mathbf{i} + 4\mathbf{j}$ **b** $6\mathbf{i} - 8\mathbf{j}$ **c** $5\mathbf{i} + 12\mathbf{j}$ **d** $2\mathbf{i} + 4\mathbf{j}$

5 Find the angle that each of these vectors makes with $\mathbf{j}$.

a $3\mathbf{i} - 5\mathbf{j}$ **b** $4\mathbf{i} + 7\mathbf{j}$ **c** $-3\mathbf{i} + 5\mathbf{j}$ **d** $-4\mathbf{i} - \mathbf{j}$

6 Write these vectors in **i**, **j** and column vector form.

a

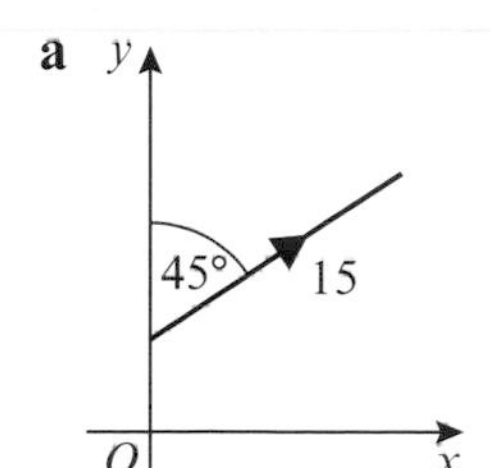

b

y
20°
8
O
x

c

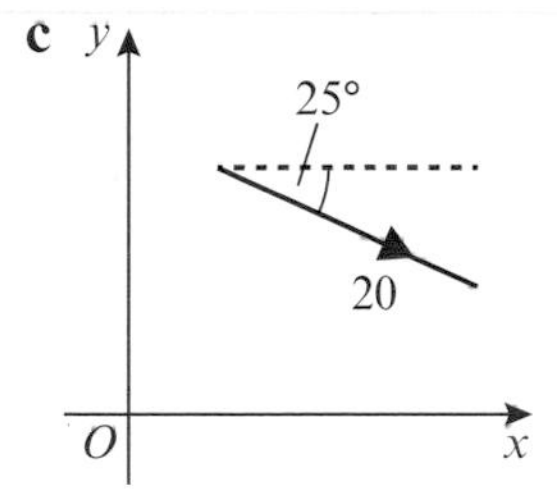

d

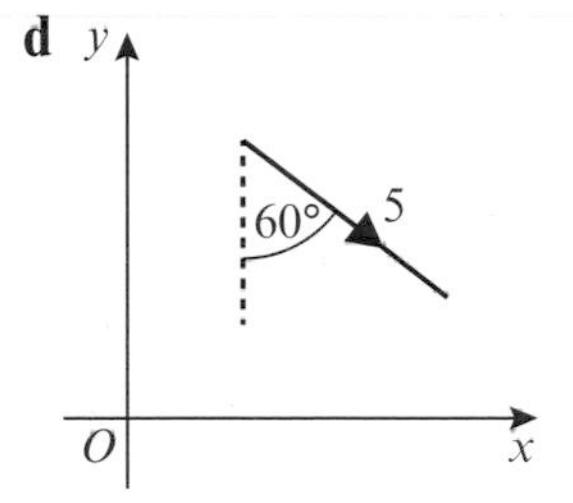

7 Draw a sketch for each vector and work out the exact value of its magnitude and the angle it makes with the positive x-axis to one decimal place.

a $3\mathbf{i} + 4\mathbf{j}$ **b** $2\mathbf{i} - \mathbf{j}$ **c** $-5\mathbf{i} + 2\mathbf{j}$

E/P **8** Given that $|2\mathbf{i} - k\mathbf{j}| = 2\sqrt{10}$, find the exact value of k. **(3 marks)**

E/P **9** Vector $\mathbf{a} = p\mathbf{i} + q\mathbf{j}$ has magnitude 10 and makes an angle θ with the positive x-axis where $\sin\theta = \frac{3}{5}$. Find the possible values of p and q.

(4 marks)

Problem-solving
Make sure you consider all the possible cases.

10 In triangle ABC, $\overrightarrow{AB} = 4\mathbf{i} + 3\mathbf{j}$, $\overrightarrow{AC} = 6\mathbf{i} - 4\mathbf{j}$.

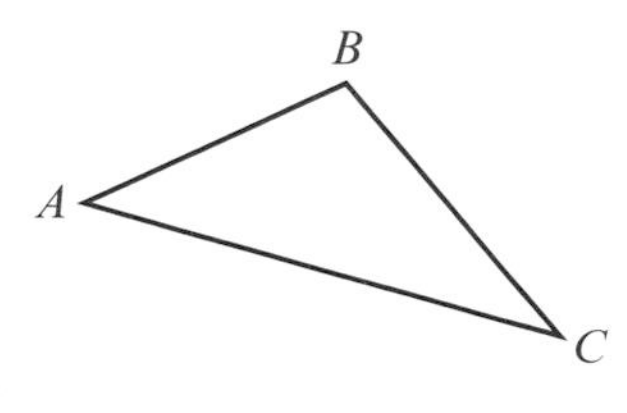

a Find the angle between $\overrightarrow{AB}$ and **i**.

b Find the angle between $\overrightarrow{AC}$ and **i**.

c Hence find the size of $\angle BAC$, in degrees, to one decimal place.

E/P **11** In triangle PQR, $\overrightarrow{PQ} = 4\mathbf{i} + \mathbf{j}$, $\overrightarrow{PR} = 6\mathbf{i} - 8\mathbf{j}$.

a Find the size of $\angle QPR$, in degrees, to one decimal place. **(5 marks)**

b Find the area of triangle PQR. **(2 marks)**

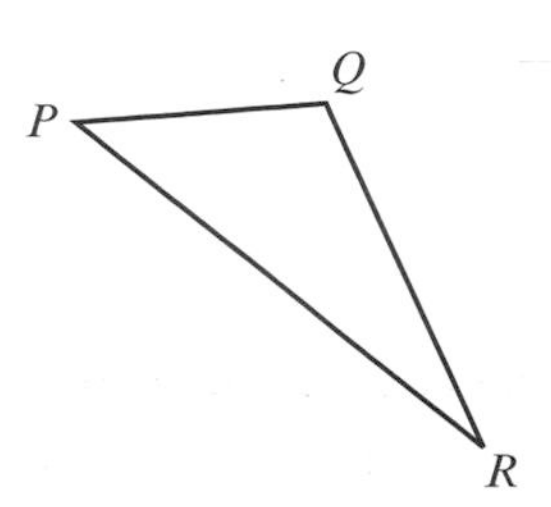

Hint The area of a triangle is $\frac{1}{2}ab\sin\theta$.

← Section 9.3

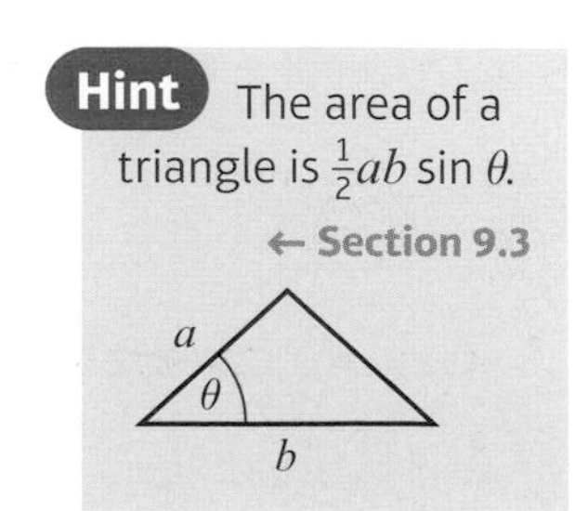

Challenge

In the diagram $\overrightarrow{AB} = p\mathbf{i} + q\mathbf{j}$ and $\overrightarrow{AD} = r\mathbf{i} + s\mathbf{j}$.
$ABCD$ is a parallelogram.
Prove that the area of $ABCD$ is $ps - qr$.

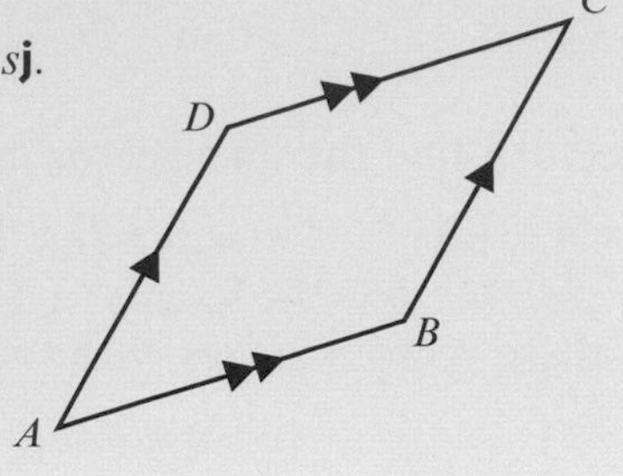

Problem-solving

Draw the parallelogram on a coordinate grid, and choose a position for the origin that will simplify your calculations.

11.4 Position vectors

You need to be able to use vectors to describe the position of a point in two dimensions.

Position vectors are vectors giving the position of a point, relative to a fixed origin.

The position vector of a point A is the vector $\overrightarrow{OA}$, where O is the origin.

If $\overrightarrow{OA} = a\mathbf{i} + b\mathbf{j}$ then the position vector of A is $\begin{pmatrix} a \\ b \end{pmatrix}$.

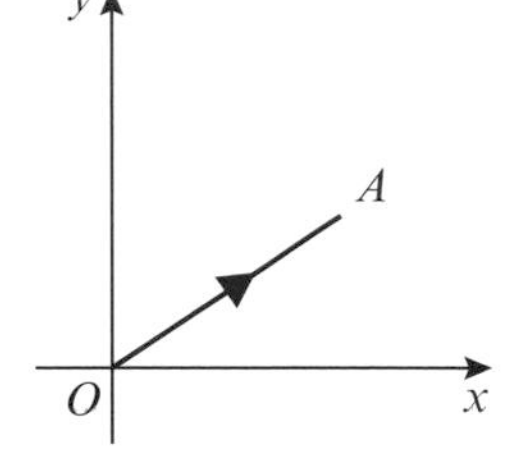

- **In general, a point P with coordinates (p, q) has a position vector $\overrightarrow{OP} = p\mathbf{i} + q\mathbf{j} = \begin{pmatrix} p \\ q \end{pmatrix}$.**

- **$\overrightarrow{AB} = \overrightarrow{OB} - \overrightarrow{OA}$, where $\overrightarrow{OA}$ and $\overrightarrow{OB}$ are the position vectors of A and B respectively.**

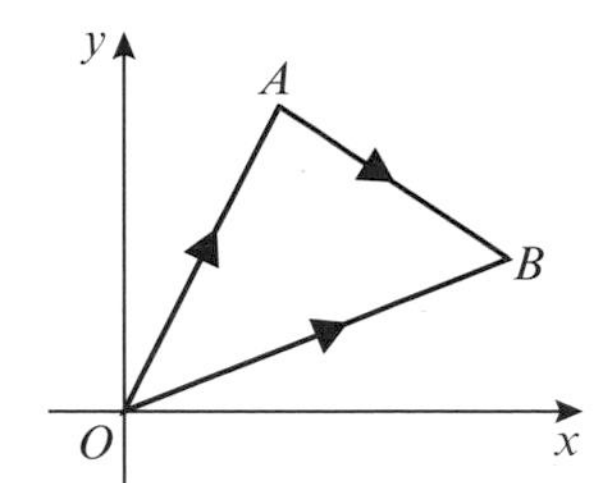

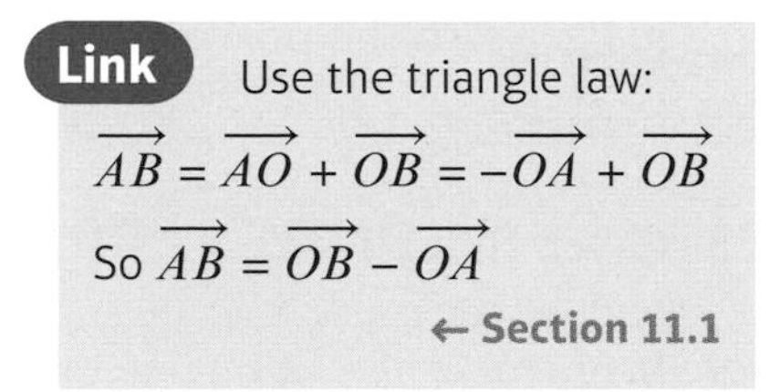
Link Use the triangle law:
$\overrightarrow{AB} = \overrightarrow{AO} + \overrightarrow{OB} = -\overrightarrow{OA} + \overrightarrow{OB}$
So $\overrightarrow{AB} = \overrightarrow{OB} - \overrightarrow{OA}$

← Section 11.1

Example 14

The points A and B in the diagram have coordinates (3, 4) and (11, 2) respectively.
Find, in terms of **i** and **j**:

a the position vector of A **b** the position vector of B

c the vector $\overrightarrow{AB}$

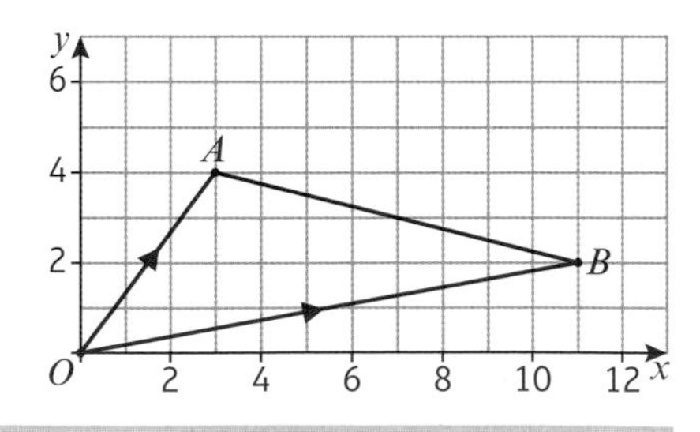

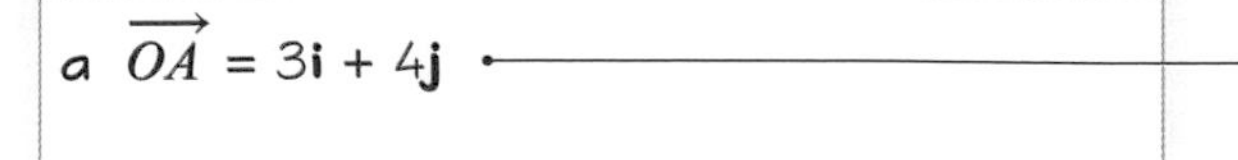
a $\overrightarrow{OA} = 3\mathbf{i} + 4\mathbf{j}$

In column vector form this is $\begin{pmatrix} 3 \\ 4 \end{pmatrix}$.

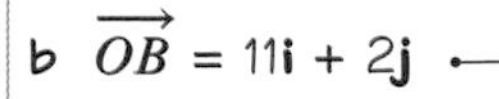
b $\overrightarrow{OB} = 11\mathbf{i} + 2\mathbf{j}$

In column vector form this is $\begin{pmatrix} 11 \\ 2 \end{pmatrix}$.

c $\overrightarrow{AB} = \overrightarrow{OB} - \overrightarrow{OA}$
$= (11\mathbf{i} + 2\mathbf{j}) - (3\mathbf{i} + 4\mathbf{j}) = 8\mathbf{i} - 2\mathbf{j}$

In column vector form this is $\begin{pmatrix} 8 \\ -2 \end{pmatrix}$.

Example 15

$\overrightarrow{OA} = 5\mathbf{i} - 2\mathbf{j}$ and $\overrightarrow{AB} = 3\mathbf{i} + 4\mathbf{j}$. Find:

a the position vector of B

b the exact value of $|\overrightarrow{OB}|$ in simplified surd form.

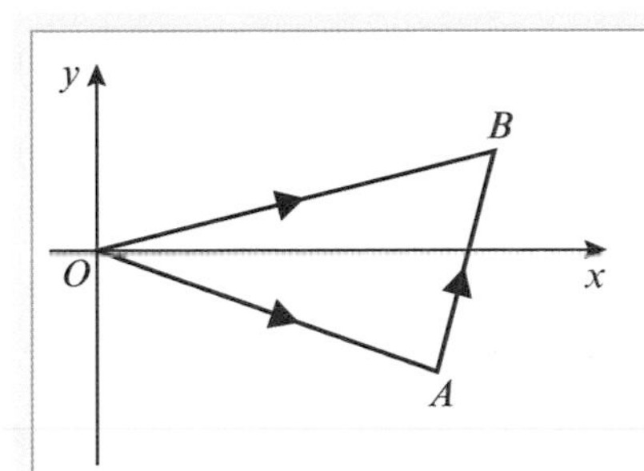

a $\overrightarrow{OA} = \begin{pmatrix} 5 \\ -2 \end{pmatrix}$ and $\overrightarrow{AB} = \begin{pmatrix} 3 \\ 4 \end{pmatrix}$

It is usually quicker to use column vector form for calculations.

$\overrightarrow{OB} = \overrightarrow{OA} + \overrightarrow{AB} = \begin{pmatrix} 5 \\ -2 \end{pmatrix} + \begin{pmatrix} 3 \\ 4 \end{pmatrix} = \begin{pmatrix} 8 \\ 2 \end{pmatrix}$

In **i**, **j** form the answer is $8\mathbf{i} + 2\mathbf{j}$.

b $|\overrightarrow{OB}| = \sqrt{8^2 + 2^2} = \sqrt{64 + 4} = \sqrt{68} = 2\sqrt{17}$

$\sqrt{68} = \sqrt{4 \times 17} = 2\sqrt{17}$ in simplified surd form.

Exercise 11D

1 The points A, B and C have coordinates $(3, -1)$, $(4, 5)$ and $(-2, 6)$ respectively, and O is the origin.
Find, in terms of **i** and **j**:

a **i** the position vectors of A, B and C **ii** $\overrightarrow{AB}$ **iii** $\overrightarrow{AC}$

b Find, in surd form: **i** $|\overrightarrow{OC}|$ **ii** $|\overrightarrow{AB}|$ **iii** $|\overrightarrow{AC}|$

2 $\overrightarrow{OP} = 4\mathbf{i} - 3\mathbf{j}$, $\overrightarrow{OQ} = 3\mathbf{i} + 2\mathbf{j}$

a Find $\overrightarrow{PQ}$

b Find, in surd form: **i** $|\overrightarrow{OP}|$ **ii** $|\overrightarrow{OQ}|$ **iii** $|\overrightarrow{PQ}|$

3 $\overrightarrow{OQ} = 4\mathbf{i} - 3\mathbf{j}$, $\overrightarrow{PQ} = 5\mathbf{i} + 6\mathbf{j}$

a Find $\overrightarrow{OP}$

b Find, in surd form: **i** $|\overrightarrow{OP}|$ **ii** $|\overrightarrow{OQ}|$ **iii** $|\overrightarrow{PQ}|$

(P) **4** $OABCDE$ is a regular hexagon. The points A and B have position vectors **a** and **b** respectively, where O is the origin.
Find, in terms of **a** and **b**, the position vectors of

a C **b** D **c** E.

(P) **5** The position vectors of 3 vertices of a parallelogram are $\begin{pmatrix}4\\2\end{pmatrix}$, $\begin{pmatrix}3\\5\end{pmatrix}$ and $\begin{pmatrix}8\\6\end{pmatrix}$.
Find the possible position vectors of the fourth vertex.

Problem-solving

Use a sketch to check that you have considered all the possible positions for the fourth vertex.

(E) **6** Given that the point A has position vector $4\mathbf{i} - 5\mathbf{j}$ and the point B has position vector $6\mathbf{i} + 3\mathbf{j}$,

a find the vector $\overrightarrow{AB}$. **(2 marks)**

b find $|\overrightarrow{AB}|$ giving your answer as a simplified surd. **(2 marks)**

(E/P) **7** The point A lies on the circle with equation $x^2 + y^2 = 9$. Given that $\overrightarrow{OA} = 2k\mathbf{i} + k\mathbf{j}$, find the exact value of k. **(3 marks)**

Challenge

The point B lies on the line with equation $2y = 12 - 3x$. Given that $|OB| = \sqrt{13}$, find possible expressions for $\overrightarrow{OB}$ in the form $p\mathbf{i} + q\mathbf{j}$.

11.5 Solving geometric problems

You need to be able to use vectors to solve geometric problems and to find the position vector of a point that divides a line segment in a given ratio.

- **If the point P divides the line segment AB in the ratio $\lambda : \mu$, then**

$$\overrightarrow{OP} = \overrightarrow{OA} + \frac{\lambda}{\lambda + \mu}\overrightarrow{AB}$$

$$= \overrightarrow{OA} + \frac{\lambda}{\lambda + \mu}(\overrightarrow{OB} - \overrightarrow{OA})$$

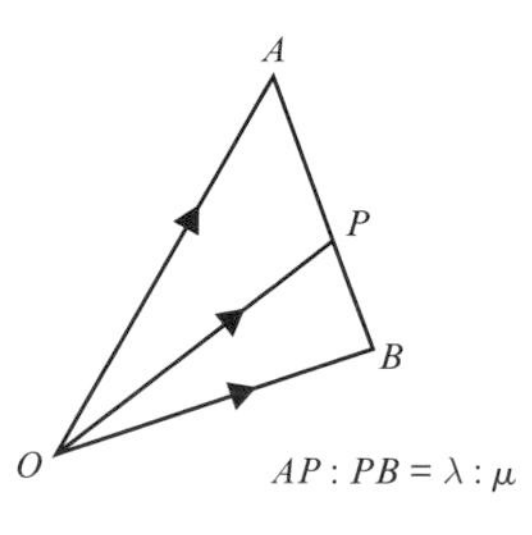

Example 16

In the diagram the points A and B have position vectors $\mathbf{a}$ and $\mathbf{b}$ respectively (referred to the origin O). The point P divides AB in the ratio $1:2$.

Find the position vector of P.

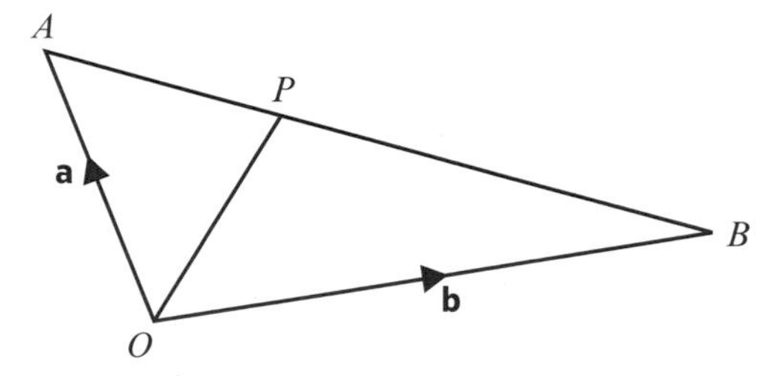

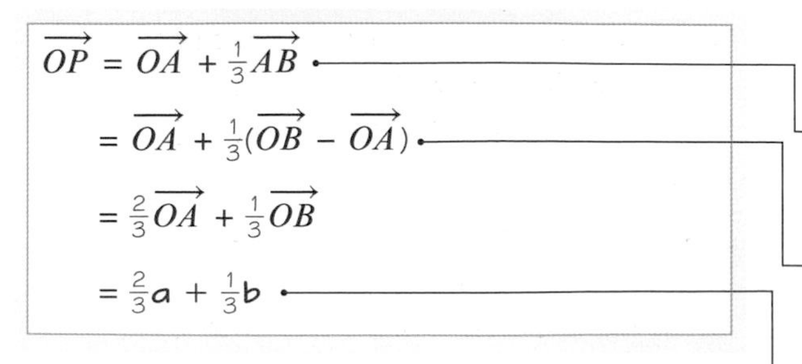

$\overrightarrow{OP} = \overrightarrow{OA} + \frac{1}{3}\overrightarrow{AB}$

$= \overrightarrow{OA} + \frac{1}{3}(\overrightarrow{OB} - \overrightarrow{OA})$

$= \frac{2}{3}\overrightarrow{OA} + \frac{1}{3}\overrightarrow{OB}$

$= \frac{2}{3}\mathbf{a} + \frac{1}{3}\mathbf{b}$

There are 3 parts in the ratio in total, so P is $\frac{1}{3}$ of the way along the line segment AB.

Rewrite $\overrightarrow{AB}$ in terms of the position vectors for A and B.

Give your final answer in terms of **a** and **b**.

You can solve geometric problems by comparing coefficients on both sides of an equation:

- **If a and b are two non-parallel vectors and $p\mathbf{a} + q\mathbf{b} = r\mathbf{a} + s\mathbf{b}$ then $p = r$ and $q = s$.**

Example 17

$OABC$ is a parallelogram. P is the point where the diagonals OB and AC intersect.

The vectors **a** and **c** are equal to $\overrightarrow{OA}$ and $\overrightarrow{OC}$ respectively.

Prove that the diagonals bisect each other.

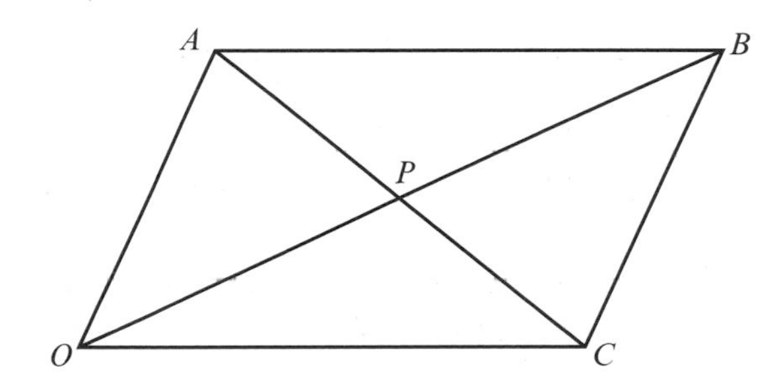

If the diagonals bisect each other, then P must be the midpoint of OB and the midpoint of AC.

From the diagram,

$\overrightarrow{OB} = \overrightarrow{OC} + \overrightarrow{CB} = \mathbf{c} + \mathbf{a}$

and $\overrightarrow{AC} = \overrightarrow{AO} + \overrightarrow{OC}$.

$= -\overrightarrow{OA} + \overrightarrow{OC} = -\mathbf{a} + \mathbf{c}$

*Express $\overrightarrow{OB}$ and $\overrightarrow{AC}$ in terms of **a** and **c**.*

P lies on OB $\Rightarrow$ $\overrightarrow{OP} = \lambda(\mathbf{c} + \mathbf{a})$

P lies on AC $\Rightarrow$ $\overrightarrow{OP} = \overrightarrow{OA} + \overrightarrow{AP}$

$= \mathbf{a} + \mu(-\mathbf{a} + \mathbf{c})$

Use the fact that P lies on both diagonals to find two different routes from O to P, giving two different forms of $\overrightarrow{OP}$.

$\Rightarrow$ $\lambda(\mathbf{c} + \mathbf{a}) = \mathbf{a} + \mu(-\mathbf{a} + \mathbf{c})$

The two expressions for $\overrightarrow{OP}$ must be equal.

$\Rightarrow$ $\lambda = 1 - \mu$ and $\lambda = \mu$

*Form and solve a pair of simultaneous equations by equating the coefficients of **a** and **c**.*

$\Rightarrow$ $\lambda = \mu = \frac{1}{2}$, so P is the midpoint of both diagonals, so the diagonals bisect each other.

If P is halfway along the line segment then it must be the midpoint.

Online Use technology to show that diagonals of a parallelogram bisect each other.

Example 18

In triangle ABC, $\overrightarrow{AB} = 3\mathbf{i} - 2\mathbf{j}$ and $\overrightarrow{AC} = \mathbf{i} - 5\mathbf{j}$,

Find the exact size of $\angle BAC$ in degrees.

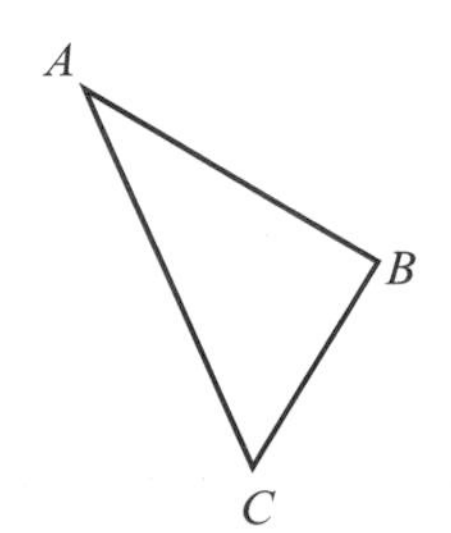

Problem-solving

Work out what information you would need to find the angle. You could:

- find the lengths of all three sides then use the cosine rule
- convert $\overrightarrow{AB}$ and $\overrightarrow{AC}$ to magnitude-direction form

The working here shows the first method.

$\overrightarrow{BC} = \overrightarrow{AC} - \overrightarrow{AB} = \begin{pmatrix} 1 \\ -5 \end{pmatrix} - \begin{pmatrix} 3 \\ -2 \end{pmatrix} = \begin{pmatrix} -2 \\ -3 \end{pmatrix}$

Use the triangle law to find $\overrightarrow{BC}$.

$|\overrightarrow{AB}| = \sqrt{3^2 + (-2)^2} = \sqrt{13}$

$|\overrightarrow{AC}| = \sqrt{1^2 + (-5)^2} = \sqrt{26}$

$|\overrightarrow{BC}| = \sqrt{(-2)^2 + (-3)^2} = \sqrt{13}$

Leave your answers in surd form.

$$\cos \angle BAC = \frac{|\overrightarrow{AB}|^2 + |\overrightarrow{AC}|^2 - |\overrightarrow{BC}|^2}{2 \times |\overrightarrow{AB}| \times |\overrightarrow{AC}|}$$

$$= \frac{13 + 26 - 13}{2 \times \sqrt{13} \times \sqrt{26}} = \frac{26}{26\sqrt{2}} = \frac{1}{\sqrt{2}}$$

$\cos A = \dfrac{b^2 + c^2 - a^2}{2bc}$ ← Section 9.1

$\angle BAC = \cos^{-1}\left(\frac{1}{\sqrt{2}}\right) = 45°$

Online Check your answer by entering the vectors directly into your calculator.

Exercise 11E

(P) **1** In the diagram, $\overrightarrow{WX} = \mathbf{a}$, $\overrightarrow{WY} = \mathbf{b}$ and $\overrightarrow{WZ} = \mathbf{c}$. It is given that $\overrightarrow{XY} = \overrightarrow{YZ}$.
Prove that $\mathbf{a} + \mathbf{c} = 2\mathbf{b}$.

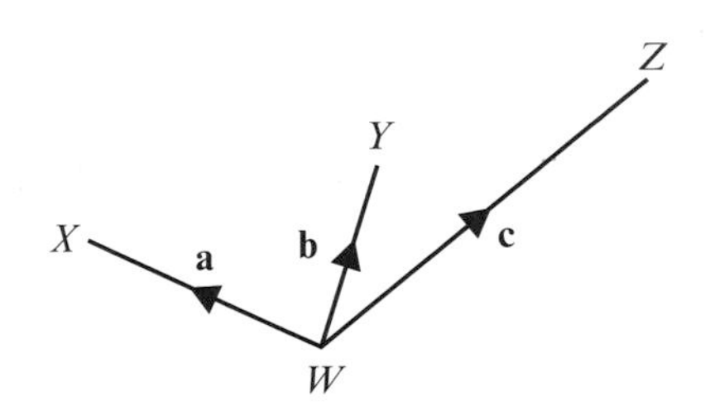

(P) **2** OAB is a triangle. P, Q and R are the midpoints of OA, AB and OB respectively.
OP and OR are equal to $\mathbf{p}$ and $\mathbf{r}$ respectively.

a Find **i** $\overrightarrow{OB}$ **ii** $\overrightarrow{PQ}$

b Hence prove that triangle PAQ is similar to triangle OAB.

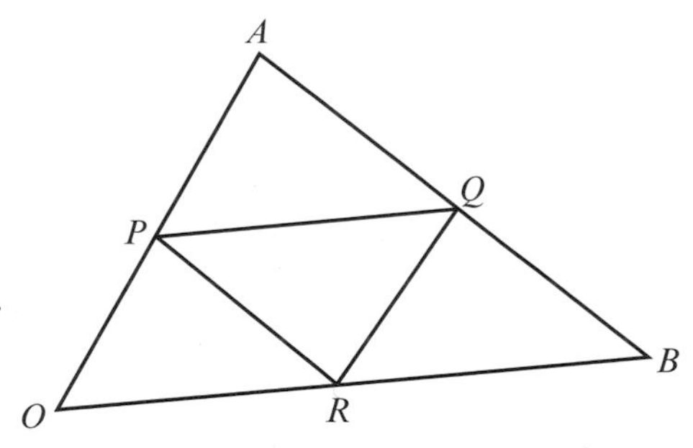

(P) **3** OAB is a triangle. $\overrightarrow{OA} = \mathbf{a}$ and $\overrightarrow{OB} = \mathbf{b}$.
The point M divides OA in the ratio 2 : 1.
MN is parallel to OB.

a Express the vector $\overrightarrow{ON}$ in terms of $\mathbf{a}$ and $\mathbf{b}$.

b Show that $AN : NB = 1 : 2$

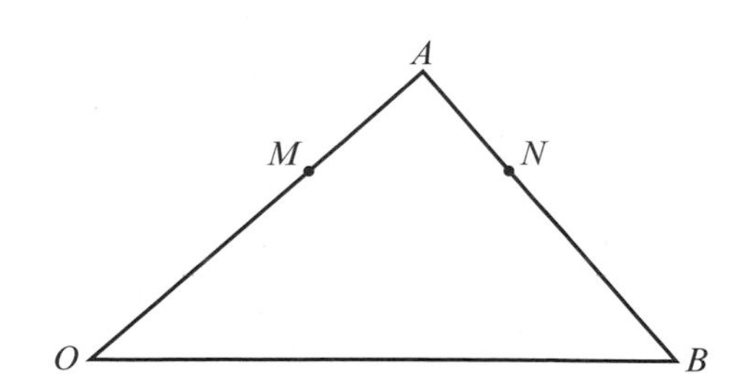

(P) **4** $OABC$ is a square. M is the midpoint of OA, and Q divides BC in the ratio $1:3$.
AC and MQ meet at P.

a If $\overrightarrow{OA} = \mathbf{a}$ and $\overrightarrow{OC} = \mathbf{c}$, express $\overrightarrow{OP}$ in terms of $\mathbf{a}$ and $\mathbf{c}$.

b Show that P divides AC in the ratio $2:3$.

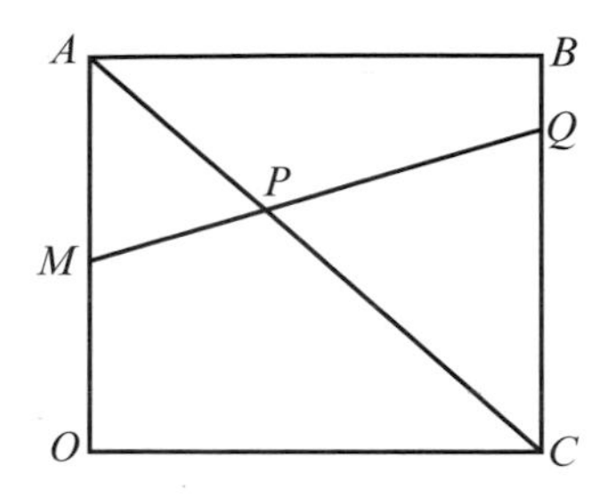

5 In triangle ABC the position vectors of the vertices A, B and C are $\begin{pmatrix}5\\8\end{pmatrix}$, $\begin{pmatrix}4\\3\end{pmatrix}$ and $\begin{pmatrix}7\\6\end{pmatrix}$. Find:

a $|\overrightarrow{AB}|$ **b** $|\overrightarrow{AC}|$ **c** $|\overrightarrow{BC}|$

d the size of $\angle BAC$, $\angle ABC$ and $\angle ACB$ to the nearest degree.

(P) **6** OPQ is a triangle.
$2\overrightarrow{PR} = \overrightarrow{RQ}$ and $3\overrightarrow{OR} = \overrightarrow{OS}$
$\overrightarrow{OP} = \mathbf{a}$ and $\overrightarrow{OQ} = \mathbf{b}$.

a Show that $\overrightarrow{OS} = 2\mathbf{a} + \mathbf{b}$.

b Point T is added to the diagram such that $\overrightarrow{OT} = -\mathbf{b}$.
Prove that points T, P and S lie on a straight line.

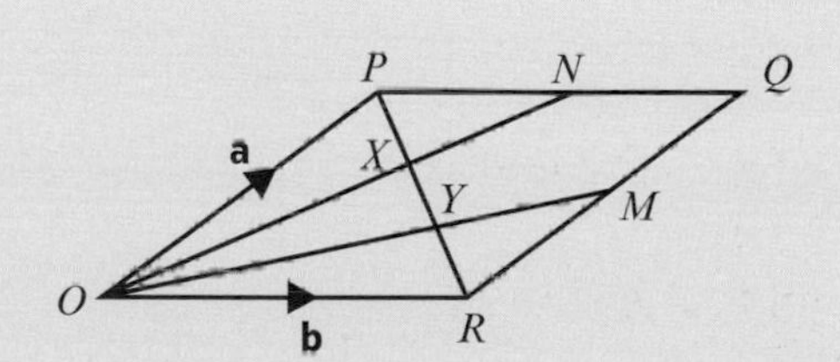

Problem-solving

To show that T, P and S lie on the same straight line you need to show that any **two** of the vectors $\overrightarrow{TP}$, $\overrightarrow{TS}$ or $\overrightarrow{PS}$ are parallel.

Challenge

$OPQR$ is a parallelogram.

N is the midpoint of PQ and M is the midpoint of QR.
$\overrightarrow{OP} = \mathbf{a}$ and $\overrightarrow{OR} = \mathbf{b}$. The lines ON and OM intersect the diagonal PR at points X and Y respectively.

a Explain why $\overrightarrow{PX} = -j\mathbf{a} + j\mathbf{b}$, where j is a constant.

b Show that $\overrightarrow{PX} = (k-1)\mathbf{a} + \frac{1}{2}k\mathbf{b}$, where k is a constant.

c Explain why the values of j and k must satisfy these simultaneous equations:
$k - 1 = -j$
$\frac{1}{2}k = j$

d Hence find the values of j and k.

e Deduce that the lines ON and OM divide the diagonal PR into 3 equal parts.

11.6 Modelling with vectors

You need to be able to use vectors to solve problems in context.

In mechanics, **vector quantities** have both magnitude and direction. Here are three examples:

- velocity
- displacement
- force

You can also refer to the **magnitude** of these vectors. The magnitude of a vector is a **scalar quantity** – it has size but no direction:

- **speed** is the magnitude of the velocity vector
- **distance** in a straight line between A and B is the magnitude of the displacement vector $\overrightarrow{AB}$

When modelling with vectors in mechanics, it is common to use the unit vector **j** to represent north and the unit vector **i** to represent east.

Example 19

A girl walks 2 km due east from a fixed point O to A, and then 3 km due south from A to B. Find:

a the total distance travelled

b the position vector of B relative to O

c $|\overrightarrow{OB}|$

d the bearing of B from O.

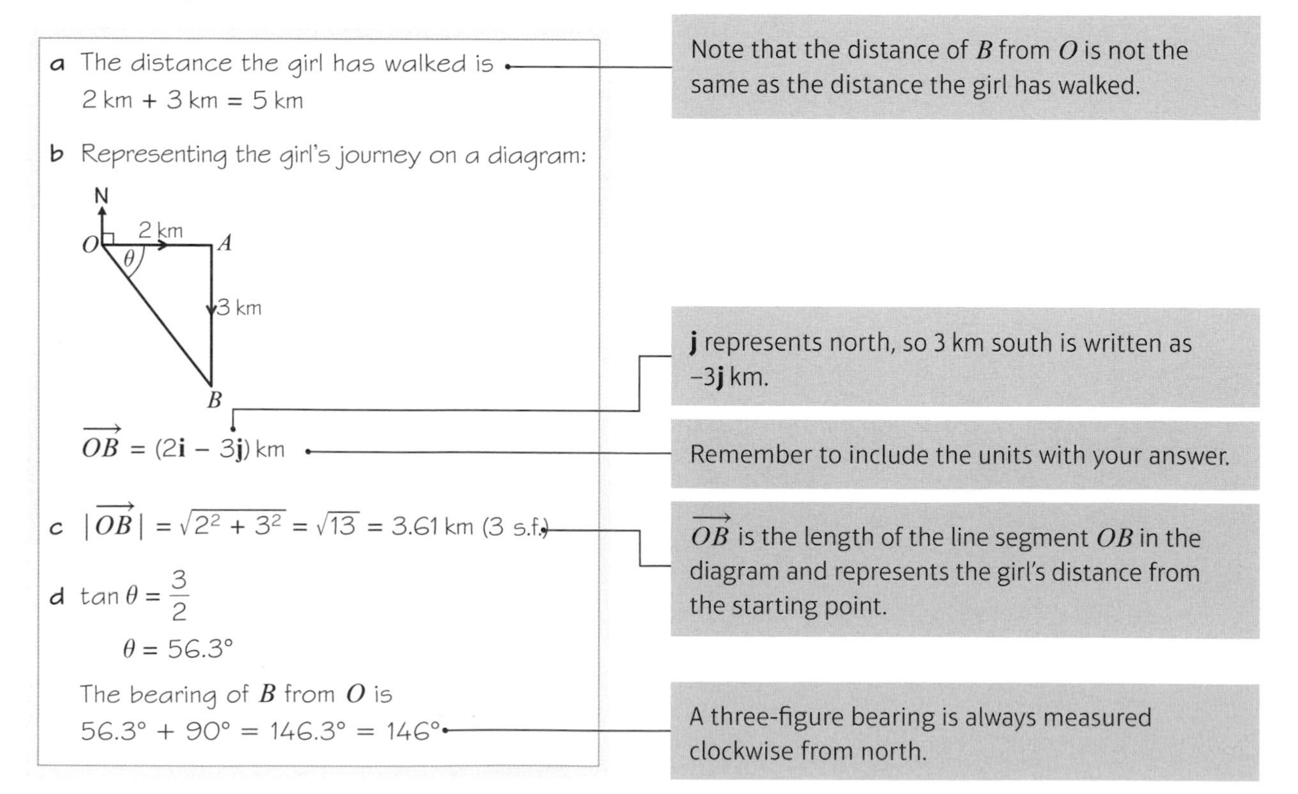

a The distance the girl has walked is $2\text{ km} + 3\text{ km} = 5\text{ km}$

Note that the distance of B from O is not the same as the distance the girl has walked.

b Representing the girl's journey on a diagram:

$\overrightarrow{OB} = (2\mathbf{i} - 3\mathbf{j})\text{ km}$

j represents north, so 3 km south is written as $-3\mathbf{j}$ km.

Remember to include the units with your answer.

c $|\overrightarrow{OB}| = \sqrt{2^2 + 3^2} = \sqrt{13} = 3.61\text{ km}$ (3 s.f.)

$\overrightarrow{OB}$ is the length of the line segment OB in the diagram and represents the girl's distance from the starting point.

d $\tan\theta = \frac{3}{2}$

$\theta = 56.3°$

The bearing of B from O is $56.3° + 90° = 146.3° = 146°$

A three-figure bearing is always measured clockwise from north.

Example 20

In an orienteering exercise, a cadet leaves the starting point O and walks 15 km on a bearing of 120° to reach A, the first checkpoint. From A he walks 9 km on a bearing of 240° to the second checkpoint, at B. From B he returns directly to O.

Find:

a the position vector of A relative to O

b $|\overrightarrow{OB}|$

c the bearing of B from O

d the position vector of B relative to O.

a

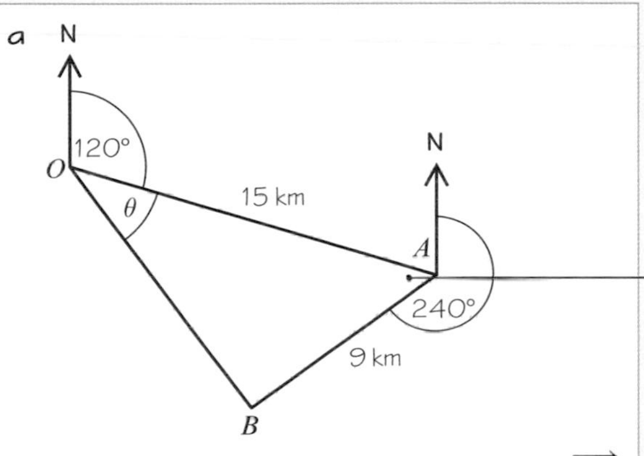

Start by drawing a diagram.

$\angle OAB = 360° - (240° + 60°) = 60°$

The position vector of A relative to O is $\overrightarrow{OA}$.

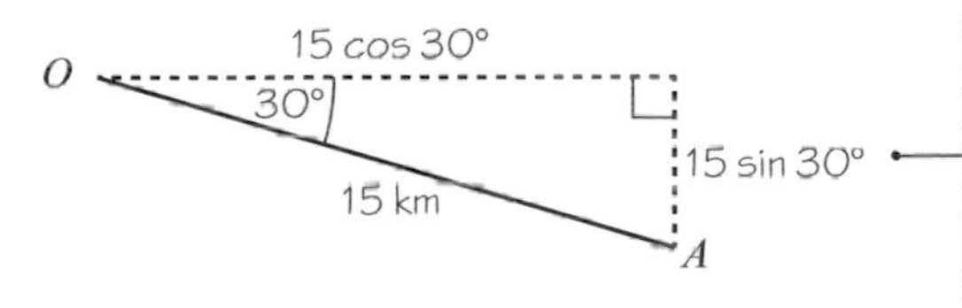

Draw a right angled triangle to work out the lengths of the **i** and **j** components for the position vector of A relative to O.

$\overrightarrow{OA} = (15 \cos 30°\mathbf{i} - 15 \sin 30°\mathbf{j})$ km

$= (13.0\mathbf{i} - 7.5\mathbf{j})$ km

b

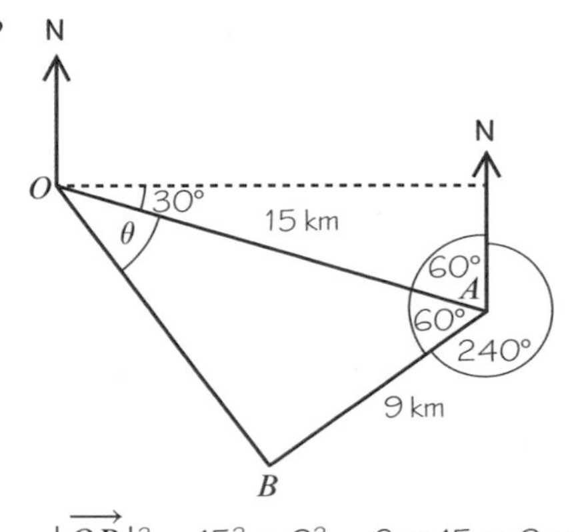

$|\overrightarrow{OB}|^2 = 15^2 + 9^2 - 2 \times 15 \times 9 \times \cos 60°$

$= 171$

$|\overrightarrow{OB}| = \sqrt{171} = 13.1$ km (3 s.f.)

$|\overrightarrow{OB}|$ is the length of OB in triangle OAB.
Use the cosine rule in triangle OAB.

c $\frac{\sin\theta}{9} = \frac{\sin 60°}{\sqrt{171}}$

$\sin\theta = \frac{9 \times \sin 60°}{\sqrt{171}} = 0.596...$

$\theta = 36.6...°$

The bearing of B from $O = 120 + 36.6...$
$= 157°$ (3 s.f.)

Use the sine rule to work out θ.

d

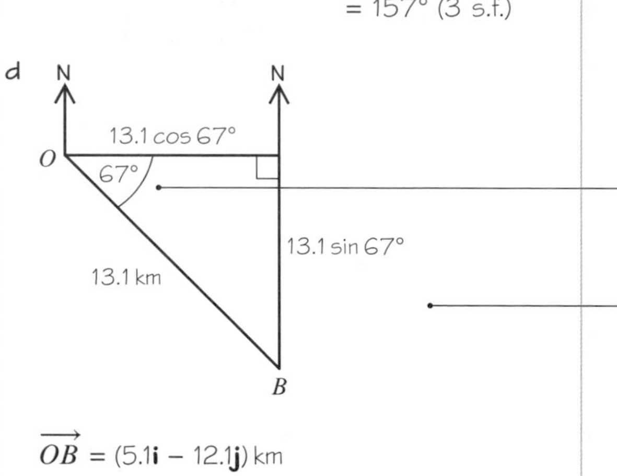

$\overrightarrow{OB} = (5.1\mathbf{i} - 12.1\mathbf{j})$ km

$157° - 90° = 67°$

Draw a right angled triangle to work out the lengths of the **i** and **j** components for the position vector of B relative to O.

Exercise 11F

1 Find the speed of a particle moving with these velocities:

a $(3\mathbf{i} + 4\mathbf{j})\,\text{m s}^{-1}$ **b** $(24\mathbf{i} - 7\mathbf{j})\,\text{km h}^{-1}$

c $(5\mathbf{i} + 2\mathbf{j})\,\text{m s}^{-1}$ **d** $(-7\mathbf{i} + 4\mathbf{j})\,\text{cm s}^{-1}$

Hint Speed is the magnitude of the velocity vector.

2 Find the distance moved by a particle which travels for:

a 5 hours at velocity $(8\mathbf{i} + 6\mathbf{j})\,\text{km h}^{-1}$

b 10 seconds at velocity $(5\mathbf{i} - \mathbf{j})\,\text{m s}^{-1}$

c 45 minutes at velocity $(6\mathbf{i} + 2\mathbf{j})\,\text{km h}^{-1}$

d 2 minutes at velocity $(-4\mathbf{i} - 7\mathbf{j})\,\text{cm s}^{-1}$.

Hint Find the speed in each case then use:
Distance travelled = speed × time

3 Find the speed and the distance travelled by a particle moving in a straight line with:

a velocity $(-3\mathbf{i} + 4\mathbf{j})\,\text{m s}^{-1}$ for 15 seconds **b** velocity $(2\mathbf{i} + 5\mathbf{j})\,\text{m s}^{-1}$ for 3 seconds

c velocity $(5\mathbf{i} - 2\mathbf{j})\,\text{km h}^{-1}$ for 3 hours **d** velocity $(12\mathbf{i} - 5\mathbf{j})\,\text{km h}^{-1}$ for 30 minutes.

4 A particle P is accelerating at a constant speed.
When $t = 0$, P has velocity $\mathbf{u} = (2\mathbf{i} + 3\mathbf{j})\,\text{m s}^{-1}$
and at time $t = 5$ s, P has velocity $\mathbf{v} = (16\mathbf{i} - 5\mathbf{j})\,\text{m s}^{-1}$.

Hint The units of acceleration will be m/s² or m s^{-2}.

The acceleration vector of the particle is given by the formula: $\mathbf{a} = \frac{\mathbf{v} - \mathbf{u}}{t}$

Find the acceleration of P in terms of **i** and **j**.

(E) **5** A particle P of mass $m = 0.3$ kg moves under the action of a single constant force $\mathbf{F}$ newtons. The acceleration of P is $\mathbf{a} = (5\mathbf{i} + 7\mathbf{j})$ m s^{-2}.

a Find the angle between the acceleration and $\mathbf{i}$. **(2 marks)**

Force, mass and acceleration are related by the formula $\mathbf{F} = m\mathbf{a}$.

b Find the magnitude of $\mathbf{F}$. **(3 marks)**

(E/P) **6** Two forces, $\mathbf{F}_1$ and $\mathbf{F}_2$, are given by the vectors $\mathbf{F}_1 = (3\mathbf{i} - 4\mathbf{j})$ N and $\mathbf{F}_2 = (p\mathbf{i} + q\mathbf{j})$ N. The resultant force, $\mathbf{R} = \mathbf{F}_1 + \mathbf{F}_2$ acts in a direction which is parallel to the vector $(2\mathbf{i} - \mathbf{j})$.

a Find the angle between $\mathbf{R}$ and the vector $\mathbf{i}$. **(2 marks)**

b Show that $p + 2q = 5$. **(3 marks)**

c Given that $p = 1$, find the magnitude of $\mathbf{R}$. **(3 marks)**

(E/P) **7** The diagram shows a sketch of a field in the shape of a triangle ABC. Given $\overrightarrow{AB} = 30\mathbf{i} + 40\mathbf{j}$ metres and $\overrightarrow{AC} = 40\mathbf{i} - 60\mathbf{j}$ metres,

a find $\overrightarrow{BC}$ **(2 marks)**

b find the size of $\angle BAC$, in degrees, to one decimal place **(4 marks)**

c find the area of the field in square metres. **(3 marks)**

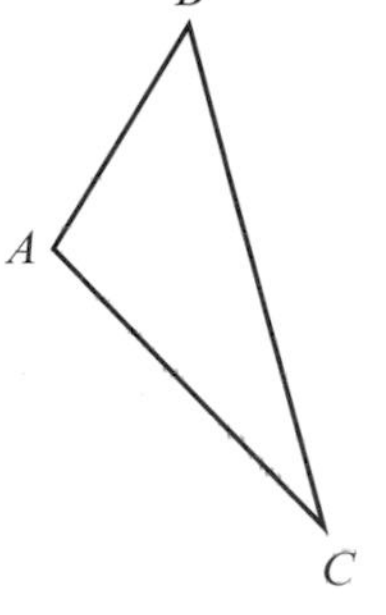

(P) **8** A boat A has a position vector of $(2\mathbf{i} + \mathbf{j})$ km and a buoy B has a position vector of $(6\mathbf{i} - 4\mathbf{j})$ km, relative to a fixed origin O.

a Find the distance of the boat from the buoy.

b Find the bearing of the boat from the buoy.

The boat travels with constant velocity $(8\mathbf{i} - 10\mathbf{j})$ km/h.

c Verify that the boat is travelling directly towards the buoy

d Find the speed of the boat.

e Work out how long it will take the boat to reach the buoy.

> **Problem-solving**
> Draw a sketch showing the initial positions of the boat, the buoy and the origin.

Mixed exercise 11

(E) **1** Two forces $\mathbf{F}_1$ and $\mathbf{F}_2$ act on a particle.

$\mathbf{F}_1 = -3\mathbf{i} + 7\mathbf{j}$ newtons

$\mathbf{F}_2 = \mathbf{i} - \mathbf{j}$ newtons

The resultant force $\mathbf{R}$ acting on the particle is given by $\mathbf{R} = \mathbf{F}_1 + \mathbf{F}_2$.

a Calculate the magnitude of $\mathbf{R}$ in newtons. **(3 marks)**

b Calculate, to the nearest degree, the angle between the line of action of $\mathbf{R}$ and the vector $\mathbf{j}$. **(2 marks)**

(P) **2** A small boat S, drifting in the sea, is modelled as a particle moving in a straight line at constant speed. When first sighted at 09:00, S is at a point with position vector $(-2\mathbf{i} - 4\mathbf{j})$ km relative to a fixed origin O, where $\mathbf{i}$ and $\mathbf{j}$ are unit vectors due east and due north respectively. At 09:40, S is at the point with position vector $(4\mathbf{i} - 6\mathbf{j})$ km.

a Calculate the bearing on which S is drifting.

b Find the speed of S.

(P) **3** A football player kicks a ball from point A on a flat football field. The motion of the ball is modelled as that of a particle travelling with constant velocity $(4\mathbf{i} + 9\mathbf{j})\,\text{m s}^{-1}$.

a Find the speed of the ball.

b Find the distance of the ball from A after 6 seconds.

c Comment on the validity of this model for large values of t.

(P) **4** $ABCD$ is a trapezium with AB parallel to DC and $DC = 4AB$.
M divides DC such that $DM : MC = 3 : 2$, $\overrightarrow{AB} = \mathbf{a}$ and $\overrightarrow{BC} = \mathbf{b}$.
Find, in terms of $\mathbf{a}$ and $\mathbf{b}$:

a $\overrightarrow{AM}$ **b** $\overrightarrow{BD}$ **c** $\overrightarrow{MB}$ **d** $\overrightarrow{DA}$

(E/P) **5** The vectors $5\mathbf{a} + k\mathbf{b}$ and $8\mathbf{a} + 2\mathbf{b}$ are parallel. Find the value of k. **(3 marks)**

6 Given that $\mathbf{a} = \begin{pmatrix} 7 \\ 4 \end{pmatrix}$, $\mathbf{b} = \begin{pmatrix} 10 \\ -2 \end{pmatrix}$ and $\mathbf{c} = \begin{pmatrix} -5 \\ -3 \end{pmatrix}$ find:

a $\mathbf{a} + \mathbf{b} + \mathbf{c}$ **b** $\mathbf{a} - 2\mathbf{b} + \mathbf{c}$ **c** $2\mathbf{a} + 2\mathbf{b} - 3\mathbf{c}$

(E) **7** In triangle ABC, $\overrightarrow{AB} = 3\mathbf{i} + 5\mathbf{j}$ and $\overrightarrow{AC} = 6\mathbf{i} + 3\mathbf{j}$, find:

a $\overrightarrow{BC}$ **(2 marks)**

b $\angle BAC$ **(4 marks)**

c the area of the triangle. **(2 marks)**

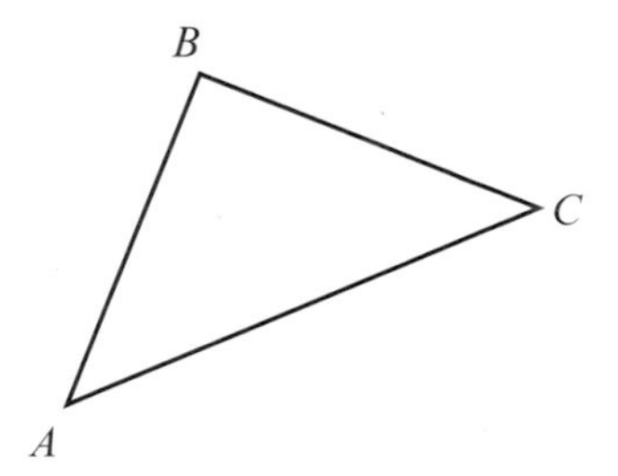

(E/P) **8** The resultant of the vectors $\mathbf{a} = 4\mathbf{i} - 3\mathbf{j}$ and $\mathbf{b} = 2p\mathbf{i} - p\mathbf{j}$ is parallel to the vector $\mathbf{c} = 2\mathbf{i} - 3\mathbf{j}$. Find:

a the value of p **(3 marks)**

b the resultant of vectors $\mathbf{a}$ and $\mathbf{b}$. **(1 mark)**

9 For each of the following vectors, find
i a unit vector in the same direction **ii** the angle the vector makes with $\mathbf{i}$

a $\mathbf{a} = 8\mathbf{i} + 15\mathbf{j}$ **b** $\mathbf{b} = 24\mathbf{i} - 7\mathbf{j}$ **c** $\mathbf{c} = -9\mathbf{i} + 40\mathbf{j}$ **d** $\mathbf{d} = 3\mathbf{i} - 2\mathbf{j}$

(P) **10** The vector $\mathbf{a} = p\mathbf{i} + q\mathbf{j}$, where p and q are positive constants, is such that $|\mathbf{a}| = 15$. Given that $\mathbf{a}$ makes an angle of 55° with $\mathbf{i}$, find the values of p and q.

(E/P) **11** Given that $|3\mathbf{i} - k\mathbf{j}| = 3\sqrt{5}$, find the value of k. **(3 marks)**

(E/P) **12** OAB is a triangle. $\overrightarrow{OA} = \mathbf{a}$ and $\overrightarrow{OB} = \mathbf{b}$. The point M divides OA in the ratio $3:2$. MN is parallel to OB.

a Express the vector $\overrightarrow{ON}$ in terms of $\mathbf{a}$ and $\mathbf{b}$. **(4 marks)**

b Find vector $\overrightarrow{MN}$. **(2 marks)**

c Show that $AN:NB = 2:3$. **(2 marks)**

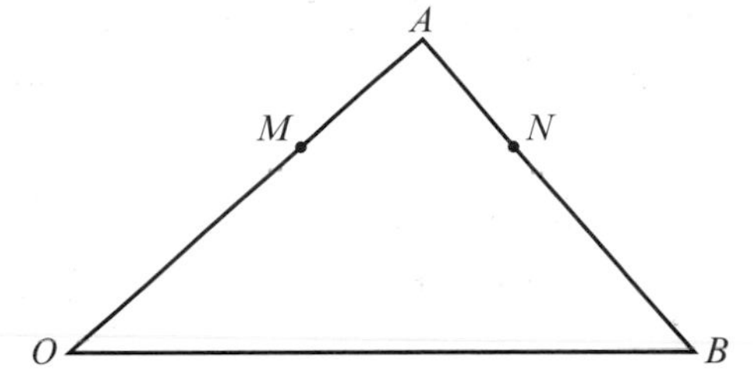

(E/P) **13** Two forces, $\mathbf{F}_1$ and $\mathbf{F}_2$, are given by the vectors $\mathbf{F}_1 = (4\mathbf{i} - 5\mathbf{j})$ N and $\mathbf{F}_2 = (p\mathbf{i} + q\mathbf{j})$ N. The resultant force, $\mathbf{R} = \mathbf{F}_1 + \mathbf{F}_2$ acts in a direction which is parallel to the vector $(3\mathbf{i} - \mathbf{j})$

a Find the angle between $\mathbf{R}$ and the vector $\mathbf{i}$. **(3 marks)**

b Show that $p + 3q = 11$. **(4 marks)**

c Given that $p = 2$, find the magnitude of $\mathbf{R}$. **(2 marks)**

(E) **14** A particle P is accelerating at a constant speed. When $t = 0$, P has velocity $\mathbf{u} = (3\mathbf{i} + 4\mathbf{j})\text{ m s}^{-1}$ and at time $t = 2$ s, P has velocity $\mathbf{v} = (15\mathbf{i} - 3\mathbf{j})\text{ m s}^{-1}$.

The acceleration vector of the particle is given by the formula: $\mathbf{a} = \dfrac{\mathbf{v} - \mathbf{u}}{t}$

Find the magnitude of the acceleration of P. **(3 marks)**

Challenge

The point B lies on the line with equation $3y = 15 - 5x$.

Given that $|\overrightarrow{OB}| = \dfrac{\sqrt{34}}{2}$, find two possible expressions for $\overrightarrow{OB}$ In the form $p\mathbf{i} + q\mathbf{j}$.

Summary of key points

1 If $\overrightarrow{PQ} = \overrightarrow{RS}$ then the line segments PQ and RS are equal in length and are parallel.

2 $\overrightarrow{AB} = -\overrightarrow{BA}$ as the line segment AB is equal in length, parallel and in the opposite direction to BA.

3 **Triangle law for vector addition:** $\overrightarrow{AB} + \overrightarrow{BC} = \overrightarrow{AC}$

If $\overrightarrow{AB} = \mathbf{a}$, $\overrightarrow{BC} = \mathbf{b}$ and $\overrightarrow{AC} = \mathbf{c}$, then $\mathbf{a} + \mathbf{b} = \mathbf{c}$

4 Subtracting a vector is equivalent to 'adding a negative vector': $\mathbf{a} - \mathbf{b} = \mathbf{a} + (-\mathbf{b})$

5 Adding the vectors $\overrightarrow{PQ}$ and $\overrightarrow{QP}$ gives the zero vector $\mathbf{0}$: $\overrightarrow{PQ} + \overrightarrow{QP} = \mathbf{0}$.

6 Any vector parallel to the vector $\mathbf{a}$ may be written as $\lambda\mathbf{a}$, where λ is a non-zero scalar.

7 To multiply a column vector by a scalar, multiply each component by the scalar: $\lambda\begin{pmatrix}p\\q\end{pmatrix} = \begin{pmatrix}\lambda p\\\lambda q\end{pmatrix}$

8 To add two column vectors, add the x-components and the y-components $\begin{pmatrix}p\\q\end{pmatrix} + \begin{pmatrix}r\\s\end{pmatrix} = \begin{pmatrix}p+r\\q+s\end{pmatrix}$

9 A unit vector is a vector of length 1. The unit vectors along the x- and y-axes are usually denoted by $\mathbf{i}$ and $\mathbf{j}$ respectively. $\mathbf{i} = \begin{pmatrix}1\\0\end{pmatrix}$ $\mathbf{j} = \begin{pmatrix}0\\1\end{pmatrix}$

10 For any two-dimensional vector: $\begin{pmatrix}p\\q\end{pmatrix} = p\mathbf{i} + q\mathbf{j}$

11 For the vector $\mathbf{a} = x\mathbf{i} + y\mathbf{j} = \begin{pmatrix}x\\y\end{pmatrix}$, the magnitude of the vector is given by: $|\mathbf{a}| = \sqrt{x^2 + y^2}$

12 A unit vector in the direction of $\mathbf{a}$ is $\dfrac{\mathbf{a}}{|\mathbf{a}|}$

13 In general, a point P with coordinates (p, q) has position vector:

$$\overrightarrow{OP} = p\mathbf{i} + q\mathbf{j} = \begin{pmatrix}p\\q\end{pmatrix}$$

14 $\overrightarrow{AB} = \overrightarrow{OB} - \overrightarrow{OA}$, where $\overrightarrow{OA}$ and $\overrightarrow{OB}$ are the position vectors of A and B respectively.

15 If the point P divides the line segment AB in the ratio $\lambda : \mu$, then

$$\overrightarrow{OP} = \overrightarrow{OA} + \frac{\lambda}{\lambda + \mu}\overrightarrow{AB}$$

$$= \overrightarrow{OA} + \frac{\lambda}{\lambda + \mu}(\overrightarrow{OB} - \overrightarrow{OA})$$

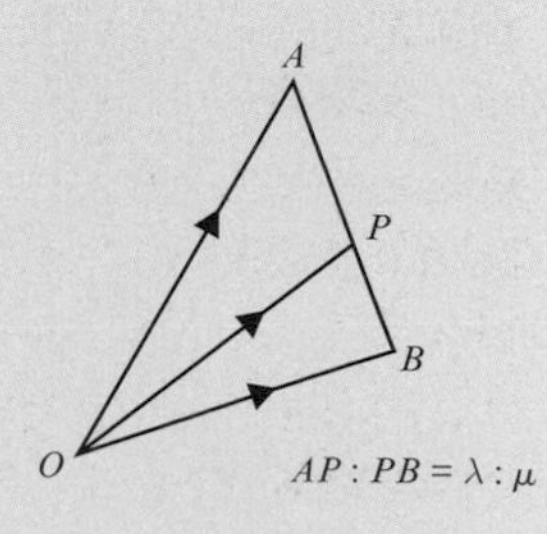

16 If $\mathbf{a}$ and $\mathbf{b}$ are two non-parallel vectors and $p\mathbf{a} + q\mathbf{b} = r\mathbf{a} + s\mathbf{b}$ then $p = r$ and $q = s$

Differentiation

12

Objectives

After completing this chapter you should be able to:

- Find the derivative, $f'(x)$ or $\frac{dy}{dx}$, of a simple function → pages 259–268
- Use the derivative to solve problems involving gradients, tangents and normals → pages 268–270
- Identify increasing and decreasing functions → pages 270–271
- Find the second order derivative, $f''(x)$ or $\frac{d^2y}{dx^2}$, of a simple function → pages 271–272
- Find stationary points of functions and determine their nature → pages 273–276
- Sketch the gradient function of a given function → pages 277–278
- Model real-life situations with differentiation → pages 279–281

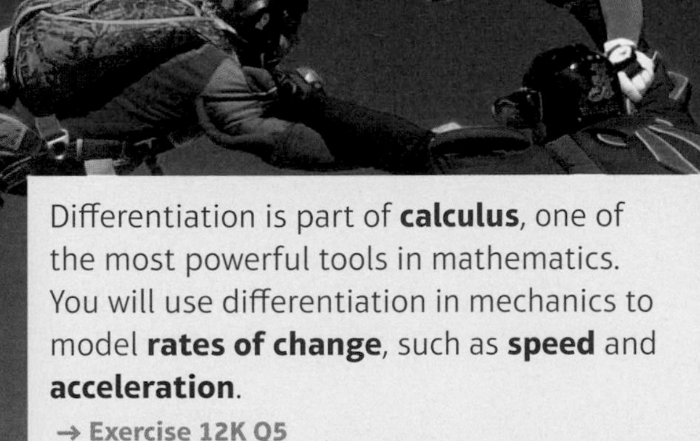

Differentiation is part of **calculus**, one of the most powerful tools in mathematics. You will use differentiation in mechanics to model **rates of change**, such as **speed** and **acceleration**.

→ Exercise 12K Q5

Prior knowledge check

1 Find the gradients of these lines.

a 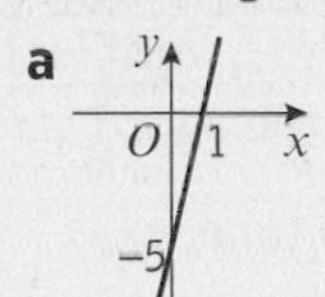

b 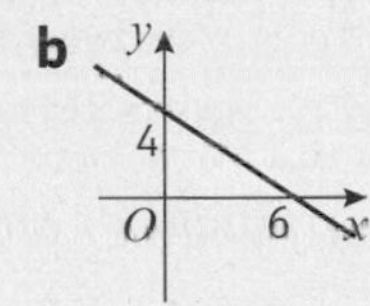

c 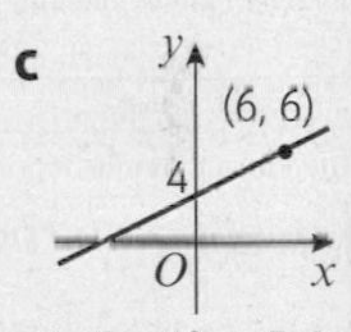

← Section 5.1

2 Write each of these expressions in the form x^n where n is a positive or negative real number.

a $x^3 \times x^7$ **b** $\sqrt[3]{x^2}$ **c** $\frac{x^2 \times x^3}{x^6}$

d $\sqrt{\frac{x^2}{\sqrt{x}}}$ ← Sections 1.1, 1.4

3 Find the equation of the straight line that passes through:

a (0, −2) and (6, 1) **b** (3, 7) and (9, 4)
c (10, 5) and (−2, 8) ← Section 5.2

4 Find the equation of the perpendicular to the line $y = 2x - 5$ at the point (2, −1). ← Section 5.3

12.1 Gradients of curves

The gradient of a curve is **constantly changing**. You can use a tangent to find the gradient of a curve at any point on the curve. The tangent to a curve at a point A is the straight line that just touches the curve at A.

- **The gradient of a curve at a given point is defined as the gradient of the tangent to the curve at that point.**

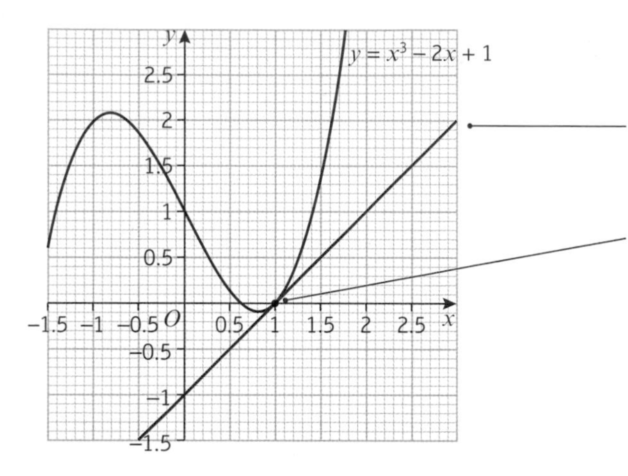

The tangent to the curve at (1, 0) has gradient 1, so the gradient of the curve at the point (1, 0) is equal to 1.

The tangent **just touches** the curve at (1, 0). It does not cut the curve at this point, although it may cut the curve at another point.

Example 1

The diagram shows the curve with equation $y = x^2$.
The tangent, T, to the curve at the point $A(1, 1)$ is shown.
Point A is joined to point P by the chord AP.

a Calculate the gradient of the tangent, T.

b Calculate the gradient of the chord AP when P has coordinates:

i (2, 4)

ii (1.5, 2.25)

iii (1.1, 1.21)

iv (1.01, 1.0201)

v $(1 + h, (1 + h)^2)$

c Comment on the relationship between your answers to parts **a** and **b**.

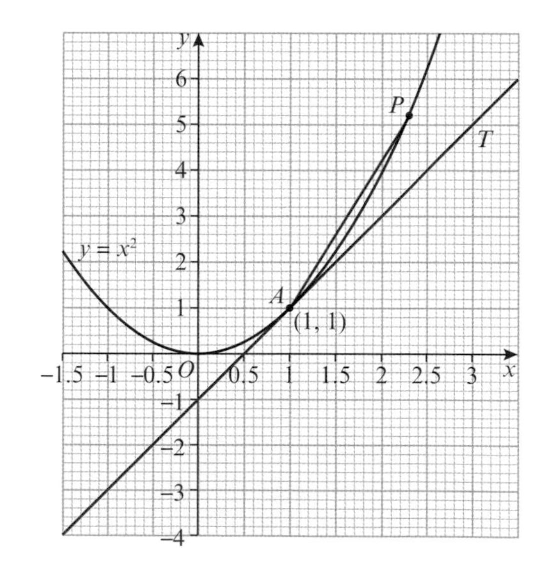

a Gradient of tangent $= \frac{y_2 - y_1}{x_2 - x_1}$

$= \frac{3 - 1}{2 - 1}$

$= 2$

Use the formula for the gradient of a straight line between points (x_1, y_1) and (x_2, y_2). ← Section 5.1

The points used are (1, 1) and (2, 3).

b **i** Gradient of chord joining (1, 1) to (2, 4)

$= \frac{4 - 1}{2 - 1}$

$= 3$

Online Explore the gradient of the chord AP using technology.

ii Gradient of the chord joining (1, 1) to (1.5, 2.25)

$= \frac{2.25 - 1}{1.5 - 1}$

$= \frac{1.25}{0.5}$

$= 2.5$

This time (x_1, y_1) is (1, 1) and (x_2, y_2) is (1.5, 2.25).

iii Gradient of the chord joining (1, 1) to (1.1, 1.21)

$= \frac{1.21 - 1}{1.1 - 1}$

$= \frac{0.21}{0.1}$

$= 2.1$

iv Gradient of the chord joining (1, 1) to (1.01, 1.0201)

$= \frac{1.0201 - 1}{1.01 - 1}$

$= \frac{0.0201}{0.01}$

$= 2.01$

This point is closer to (1, 1) than (1.1, 1.21) is.

This gradient is closer to 2.

v Gradient of the chord joining (1, 1) to $(1 + h, (1 + h)^2)$

$= \frac{(1 + h)^2 - 1}{(1 + h) - 1}$

$= \frac{1 + 2h + h^2 - 1}{1 + h - 1}$

$= \frac{2h + h^2}{h}$

$= 2 + h$

h is a constant.

$(1 + h)^2 = (1 + h)(1 + h) = 1 + 2h + h^2$

This becomes $\frac{h(2 + h)}{h}$

You can use this formula to confirm the answers to questions **i** to **iv**. For example, when $h = 0.5$, $(1 + h, (1 + h)^2) = (1.5, 2.25)$ and the gradient of the chord is $2 + 0.5 = 2.5$.

c As P gets closer to A, the gradient of the chord AP gets closer to the gradient of the tangent at A.

As h gets closer to zero, $2 + h$ gets closer to 2, so the gradient of the chord gets closer to the gradient of the tangent.

Exercise 12A

1 The diagram shows the curve with equation $y = x^2 - 2x$.

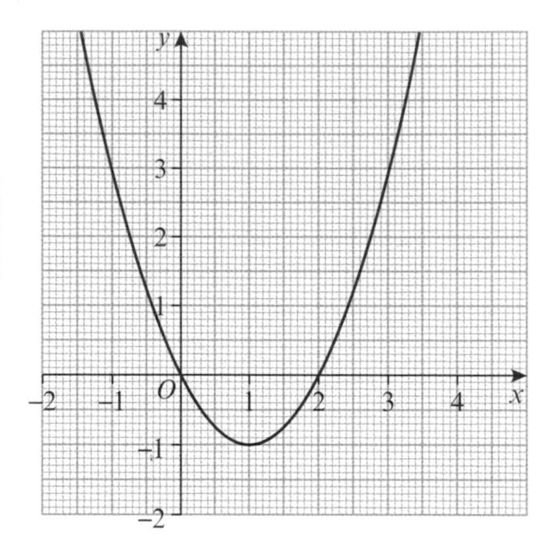

a Copy and complete this table showing estimates for the gradient of the curve.

x**-coordinate**	−1	0	1	2	3
Estimate for gradient of curve					

b Write a hypothesis about the gradient of the curve at the point where $x = p$.

c Test your hypothesis by estimating the gradient of the graph at the point (1.5, −0.75).

Hint Place a ruler on the graph to approximate each tangent.

2 The diagram shows the curve with equation $y = \sqrt{1 - x^2}$.
The point A has coordinates (0.6, 0.8).
The points B, C and D lie on the curve with x-coordinates 0.7, 0.8 and 0.9 respectively.

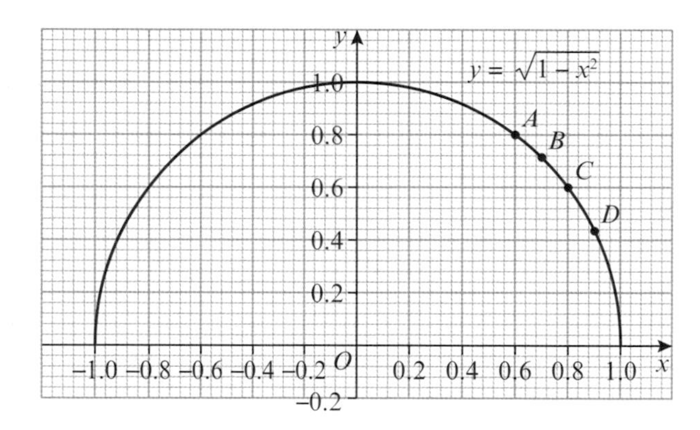

a Verify that point A lies on the curve.

b Use a ruler to estimate the gradient of the curve at point A.

c Find the gradient of the line segments:

i AD

ii AC

iii AB

Hint Use algebra for part **c**.

d Comment on the relationship between your answers to parts **b** and **c**.

3 F is the point with coordinates (3, 9) on the curve with equation $y = x^2$.

a Find the gradients of the chords joining the point F to the points with coordinates:

i (4, 16) **ii** (3.5, 12.25) **iii** (3.1, 9.61)

iv (3.01, 9.0601) **v** $(3 + h, (3 + h)^2)$

b What do you deduce about the gradient of the tangent at the point (3, 9)?

4 G is the point with coordinates (4, 16) on the curve with equation $y = x^2$.

a Find the gradients of the chords joining the point G to the points with coordinates:

i (5, 25) **ii** (4.5, 20.25) **iii** (4.1, 16.81)

iv (4.01, 16.0801) **v** $(4 + h, (4 + h)^2)$

b What do you deduce about the gradient of the tangent at the point (4, 16)?

12.2 Finding the derivative

You can use algebra to find the exact gradient of a curve at a given point. This diagram shows two points, A and B, that lie on the curve with equation $y = f(x)$.

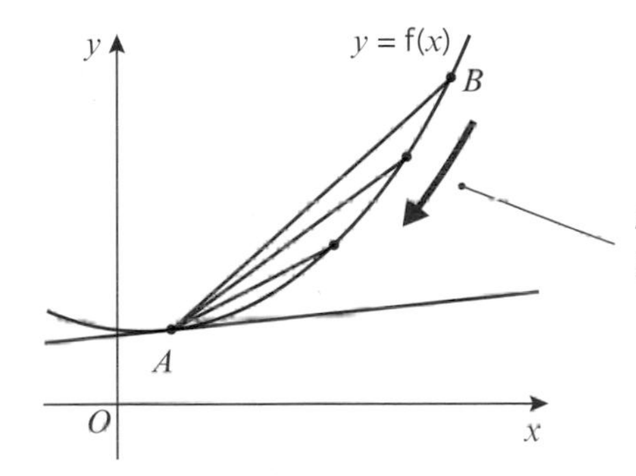

As point B moves closer to point A the gradient of chord AB gets closer to the gradient of the tangent to the curve at A.

You can formalise this approach by letting the x-coordinate of A be x_0 and the x-coordinate of B be $x_0 + h$. Consider what happens to the gradient of AB as h gets smaller.

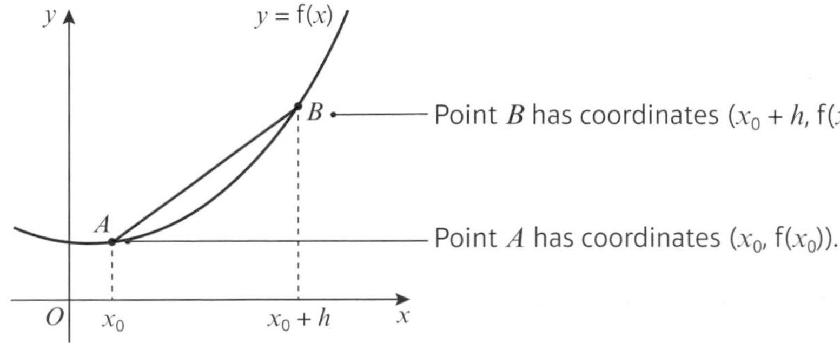

Notation h represents a **small change** in the value of x. You can also use δx to represent this small change. It is pronounced 'delta x'.

The vertical distance from A to B is $\mathrm{f}(x_0 + h) - \mathrm{f}(x_0)$.
The horizontal distance is $x_0 + h - x_0 = h$.

So the gradient of AB is $\dfrac{\mathrm{f}(x_0 + h) - \mathrm{f}(x_0)}{h}$

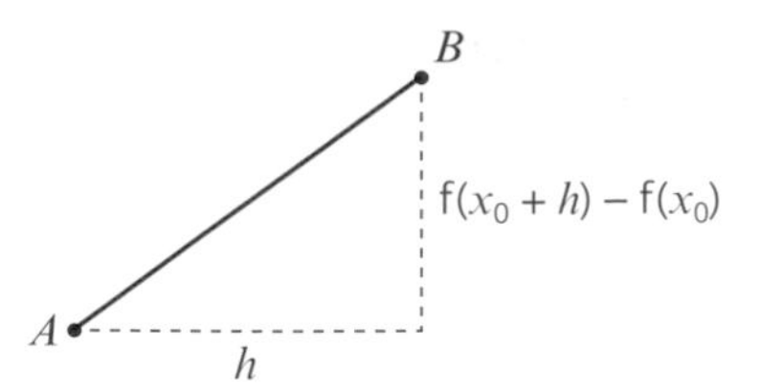

As h gets smaller, the gradient of AB gets closer to the gradient of the tangent to the curve at A. This means that the gradient of the **curve** at A is the **limit** of this expression as the value of h tends to 0.

You can use this to define the **gradient function**.

- **The gradient function, or derivative, of the curve $y = \mathrm{f}(x)$ is written as $\mathrm{f}'(x)$ or $\dfrac{\mathrm{d}y}{\mathrm{d}x}$.**

  $$\mathrm{f}'(x) = \lim_{h \to 0} \frac{\mathrm{f}(x + h) - \mathrm{f}(x)}{h}$$

  The gradient function can be used to find the gradient of the curve for any value of x.

Notation $\lim\limits_{h \to 0}$ means 'the limit as h tends to 0'. You can't evaluate the expression when $h = 0$, but as h gets smaller the expression gets closer to a fixed (or **limiting**) value.

Using this rule to find the derivative is called **differentiating from first principles**.

Example 2

The point A with coordinates (4, 16) lies on the curve with equation $y = x^2$.
At point A the curve has gradient g.

a Show that $g = \lim\limits_{h \to 0}(8 + h)$.

b Deduce the value of g.

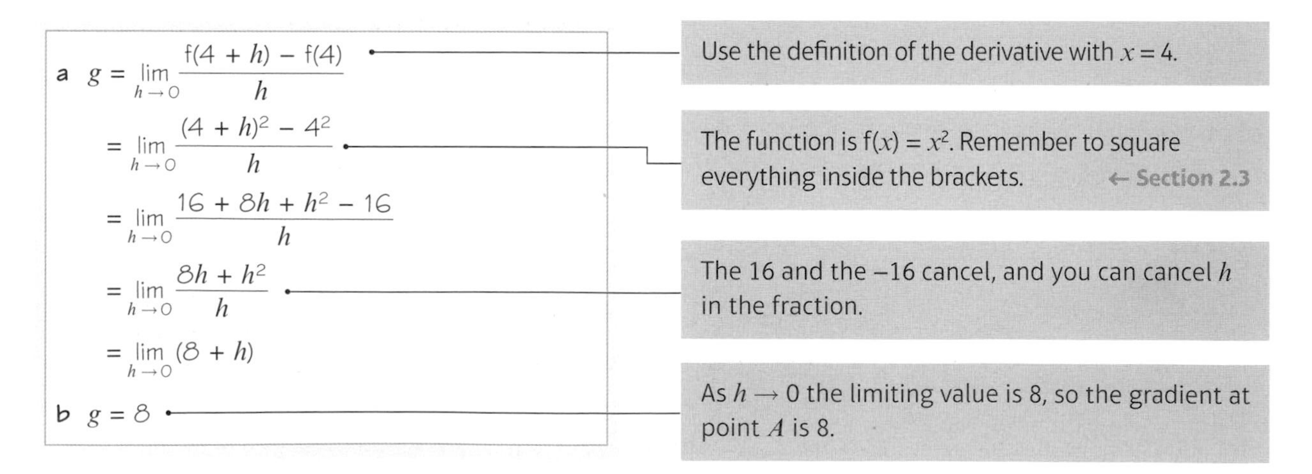

a $g = \lim\limits_{h \to 0} \dfrac{\mathrm{f}(4 + h) - \mathrm{f}(4)}{h}$ — Use the definition of the derivative with $x = 4$.

$= \lim\limits_{h \to 0} \dfrac{(4 + h)^2 - 4^2}{h}$ — The function is $\mathrm{f}(x) = x^2$. Remember to square everything inside the brackets. ← Section 2.3

$= \lim\limits_{h \to 0} \dfrac{16 + 8h + h^2 - 16}{h}$

$= \lim\limits_{h \to 0} \dfrac{8h + h^2}{h}$ — The 16 and the −16 cancel, and you can cancel h in the fraction.

$= \lim\limits_{h \to 0} (8 + h)$

b $g = 8$ — As $h \to 0$ the limiting value is 8, so the gradient at point A is 8.

Example 3

Prove, from first principles, that the derivative of x^3 is $3x^2$.

$f(x) = x^3$

$$f'(x) = \lim_{h \to 0} \frac{f(x+h) - f(x)}{h}$$

$$= \lim_{h \to 0} \frac{(x+h)^3 - (x)^3}{h}$$

$$= \lim_{h \to 0} \frac{x^3 + 3x^2h + 3xh^2 + h^3 - x^3}{h}$$

$(x + h)^3 = (x + h)(x + h)^2$
$= (x + h)(x^2 + 2hx + h^2)$
which expands to give $x^3 + 3x^2h + 3xh^2 + h^3$

$$= \lim_{h \to 0} \frac{3x^2h + 3xh^2 + h^3}{h}$$

$$= \lim_{h \to 0} \frac{h(3x^2 + 3xh + h^2)}{h}$$

Factorise the numerator.

$$= \lim_{h \to 0} (3x^2 + 3xh + h^2)$$

As $h \to 0$, $3xh \to 0$ and $h^2 \to 0$.

Any terms containing h, h^2, h^3, etc will have a limiting value of 0 as $h \to 0$.

So $f'(x) = 3x^2$

'From first principles' means that you have to use the definition of the derivative. You are starting your proof with a known definition, so this is an example of a proof by deduction.

Exercise 12B

1 For the function $f(x) = x^2$, use the definition of the derivative to show that:

a $f'(2) = 4$ **b** $f'(-3) = -6$ **c** $f'(0) = 0$ **d** $f'(50) = 100$

2 $f(x) = x^2$

a Show that $f'(x) = \lim_{h \to 0} (2x + h)$. **b** Hence deduce that $f'(x) = 2x$.

3 The point A with coordinates $(-2, -8)$ lies on the curve with equation $y = x^3$. At point A the curve has gradient g.

a Show that $g = \lim_{h \to 0} (12 - 6h + h^2)$. **b** Deduce the value of g.

(P) 4 The point A with coordinates $(-1, 4)$ lies on the curve with equation $y = x^3 - 5x$.
The point B also lies on the curve and has x-coordinate $(-1 + h)$.

a Show that the gradient of the line segment AB is given by $h^2 - 3h - 2$.

b Deduce the gradient of the curve at point A.

Problem-solving

Draw a sketch showing points A and B and the chord between them.

(E/P) 5 Prove, from first principles, that the derivative of $6x$ is 6. **(3 marks)**

(E/P) 6 Prove, from first principles, that the derivative of $4x^2$ is $8x$. **(4 marks)**

(E/P) 7 $f(x) = ax^2$, where a is a constant. Prove, from first principles, that $f'(x) = 2ax$. **(4 marks)**

Challenge

$f(x) = \frac{1}{x}$

a Given that $f'(x) = \lim_{h \to 0} \frac{f(x+h) - f(x)}{h}$, show that $f'(x) = \lim_{h \to 0} \frac{-1}{x^2 + xh}$

b Deduce that $f'(x) = -\frac{1}{x^2}$

12.3 Differentiating x^n

You can use the definition of the derivative to find an expression for the derivative of x^n where n is any number. This is called **differentiation**.

- **For all real values of n, and for a constant a:**
  - **If $f(x) = x^n$ then $f'(x) = nx^{n-1}$**

    If $y = x^n$ then $\frac{dy}{dx} = nx^{n-1}$
  - **If $f(x) = ax^n$ then $f'(x) = anx^{n-1}$**

    If $y = ax^n$ then $\frac{dy}{dx} = anx^{n-1}$

Notation

$f'(x)$ and $\frac{dy}{dx}$ both represent the derivative. You usually use $\frac{dy}{dx}$ when an expression is given in the form $y = \ldots$

Example 4

Find the derivative, $f'(x)$, when $f(x)$ equals:

a x^6 **b** $x^{\frac{1}{2}}$ **c** x^{-2} **d** $x^2 \times x^3$ **e** $\frac{x}{x^5}$

a $f(x) = x^6$

So $f'(x) = 6x^5$

Multiply by the power, then subtract 1 from the power:

$6 \times x^{6-1} = 6x^5$

b $f(x) = x^{\frac{1}{2}}$

So $f'(x) = \frac{1}{2}x^{-\frac{1}{2}}$

$= \frac{1}{2\sqrt{x}}$

The new power is $\frac{1}{2} - 1 = -\frac{1}{2}$

$x^{-\frac{1}{2}} = \frac{1}{\sqrt{x}}$

← Section 1.4

c $f(x) = x^{-2}$

So $f'(x) = -2x^{-3}$

$= -\frac{2}{x^3}$

You can leave your answer in this form or write it as a fraction.

d $f(x) = x^2 \times x^3$

$= x^5$

So $f'(x) = 5x^4$

You need to write the function in the form x^n before you can use the rule.

$x^2 \times x^3 = x^{2+3}$

$= x^5$

e $f(x) = x \div x^5$
$= x^{-4}$
So $f'(x) = -4x^{-5}$
$= -\frac{4}{x^5}$

Use the laws of indices to simplify the fraction:
$x^1 \div x^5 = x^{1-5} = x^{-4}$

Example 5

Find $\frac{dy}{dx}$ when y equals:

a $7x^3$ **b** $-4x^{\frac{1}{2}}$ **c** $3x^{-2}$ **d** $\frac{8x^7}{3x}$ **e** $\sqrt{36x^3}$

a $\frac{dy}{dx} = 7 \times 3x^{3-1} = 21x^2$

Use the rule for differentiating ax^n with $a = 7$ and $n = 3$. Multiply by 3 then subtract 1 from the power.

b $\frac{dy}{dx} = -4 \times \frac{1}{2}x^{-\frac{1}{2}} = -2x^{-\frac{1}{2}} = -\frac{2}{\sqrt{x}}$

This is the same as differentiating $x^{\frac{1}{2}}$ then multiplying the result by −4.

c $\frac{dy}{dx} = 3 \times -2x^{-3} = -6x^{-3} = -\frac{6}{x^3}$

d $y = \frac{8}{3}x^6$

Write the expression in the form ax^n. Remember a can be any number, including fractions.

$\frac{dy}{dx} = 6 \times \frac{8}{3}x^5 = 16x^5$

e $y = \sqrt{36} \times \sqrt{x^3} = 6 \times (x^3)^{\frac{1}{2}} = 6x^{\frac{3}{2}}$

$\frac{dy}{dx} = 6 \times \frac{3}{2}x^{\frac{1}{2}} = 9x^{\frac{1}{2}} = 9\sqrt{x}$

$\frac{3}{2} - 1 = \frac{1}{2}$

Simplify the number part as much as possible.

Exercise 12C

Hint Make sure that the functions are in the form x^n before you differentiate.

1 Find $f'(x)$ given that $f(x)$ equals:

a x^7 **b** x^8 **c** x^4 **d** $x^{\frac{1}{3}}$ **e** $x^{\frac{1}{4}}$ **f** $\sqrt[3]{x}$

g x^{-3} **h** x^{-4} **i** $\frac{1}{x^2}$ **j** $\frac{1}{x^5}$ **k** $\frac{1}{\sqrt{x}}$ **l** $\frac{1}{\sqrt[3]{x}}$

m $x^3 \times x^6$ **n** $x^2 \times x^3$ **o** $x \times x^2$ **p** $\frac{x^2}{x^4}$ **q** $\frac{x^3}{x^2}$ **r** $\frac{x^6}{x^3}$

2 Find $\frac{dy}{dx}$ given that y equals:

a $3x^2$ **b** $6x^9$ **c** $\frac{1}{2}x^4$ **d** $20x^{\frac{1}{4}}$ **e** $6x^{\frac{5}{4}}$

f $10x^{-1}$ **g** $\frac{4x^6}{2x^3}$ **h** $\frac{x}{8x^5}$ **i** $-\frac{2}{\sqrt{x}}$ **j** $\sqrt{\frac{5x^4 \times 10x}{2x^2}}$

3 Find the gradient of the curve with equation $y = 3\sqrt{x}$ at the point where:

a $x = 4$ b $x = 9$

c $x = \frac{1}{4}$ d $x = \frac{9}{16}$

E/P 4 Given that $2y^2 - x^3 = 0$ and $y > 0$, find $\frac{dy}{dx}$ **(2 marks)**

Problem-solving

Try rearranging unfamiliar equations into a form you recognise.

12.4 Differentiating quadratics

You can differentiate a function with more than one term by differentiating the terms **one-at-a-time**. The highest power of x in a **quadratic function** is x^2, so the highest power of x in its derivative will be x.

- **For the quadratic curve with equation $y = ax^2 + bx + c$, the derivative is given by**
  $$\frac{dy}{dx} = 2ax + b$$

Links The derivative is a **straight line** with gradient $2a$. It crosses the x-axis once, at the point where the quadratic curve has zero gradient. This is the **turning point** of the quadratic curve. ← Section 5.1

You can find this expression for $\frac{dy}{dx}$ by differentiating each of the terms one-at-a-time:

ax^2 Differentiate → $2ax^1 = 2ax$

The quadratic term tells you the slope of the gradient function.

$bx = bx^1$ Differentiate → $1bx^0 = b$

An x term differentiates to give a constant.

c Differentiate → 0

Constant terms disappear when you differentiate.

Example 6

Find $\frac{dy}{dx}$ given that y equals:

a $x^2 + 3x$ b $8x^2 - 7$ c $4x^2 - 3x + 5$

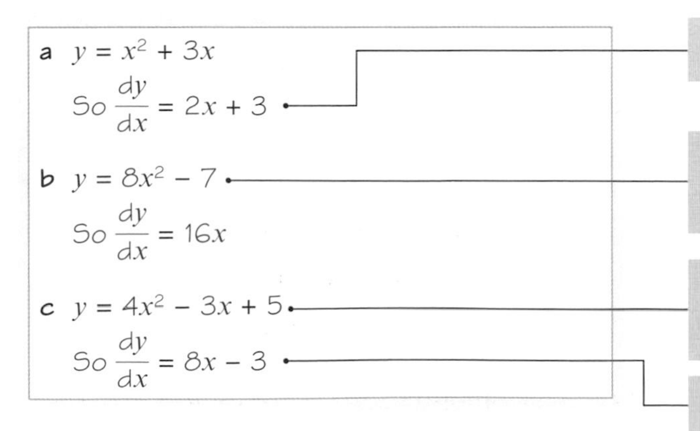

a $y = x^2 + 3x$

So $\frac{dy}{dx} = 2x + 3$

Differentiate the terms one-at-a-time.

b $y = 8x^2 - 7$

So $\frac{dy}{dx} = 16x$

The constant term disappears when you differentiate. The line $y = -7$ has **zero gradient**.

c $y = 4x^2 - 3x + 5$

So $\frac{dy}{dx} = 8x - 3$

$4x^2 - 3x + 5$ is a quadratic expression with $a = 4$, $b = -3$ and $c = 5$.

The derivative is $2ax + b = 2 \times 4x - 3 = 8x - 3$.

Example 7

Let $f(x) = 4x^2 - 8x + 3$.

a Find the gradient of $y = f(x)$ at the point $\left(\frac{1}{2}, 0\right)$.

b Find the coordinates of the point on the graph of $y = f(x)$ where the gradient is 8.

c Find the gradient of $y = f(x)$ at the points where the curve meets the line $y = 4x - 5$.

a As $y = 4x^2 - 8x + 3$

$\frac{dy}{dx} = f'(x) = 8x - 8 + 0$

So $f'\left(\frac{1}{2}\right) = -4$

Differentiate to find the gradient function. Then substitute the x-coordinate value to obtain the gradient.

b $\frac{dy}{dx} = f'(x) = 8x - 8 = 8$

So $x = 2$

So $y = f(2) = 3$

The point where the gradient is 8 is (2, 3).

Put the gradient function equal to 8. Then solve the equation you have obtained to give the value of x.

Substitute this value of x into $f(x)$ to give the value of y and interpret your answer in words.

c $4x^2 - 8x + 3 = 4x - 5$

$4x^2 - 12x + 8 = 0$

$x^2 - 3x + 2 = 0$

$(x - 2)(x - 1) = 0$

So $x = 1$ or $x = 2$

At $x = 1$, the gradient is 0.

At $x = 2$, the gradient is 8, as in part b.

To find the points of intersection, set the equation of the curve equal to the equation of the line. Solve the resulting quadratic equation to find the x-coordinates of the points of intersection. ← Section 4.4

Substitute the values of x into $f'(x) = 8x - 8$ to give the gradients at the specified points.

Online Use your calculator to check solutions to quadratic equations quickly.

Exercise 12D

1 Find $\frac{dy}{dx}$ when y equals:

a $2x^2 - 6x + 3$

b $\frac{1}{2}x^2 + 12x$

c $4x^2 - 6$

d $8x^2 + 7x + 12$

e $5 + 4x - 5x^2$

2 Find the gradient of the curve with equation:

a $y = 3x^2$ at the point (2, 12)

b $y = x^2 + 4x$ at the point (1, 5)

c $y = 2x^2 - x - 1$ at the point (2, 5)

d $y = \frac{1}{2}x^2 + \frac{3}{2}x$ at the point (1, 2)

e $y = 3 - x^2$ at the point (1, 2)

f $y = 4 - 2x^2$ at the point (−1, 2)

3 Find the y-coordinate and the value of the gradient at the point P with x-coordinate 1 on the curve with equation $y = 3 + 2x - x^2$.

4 Find the coordinates of the point on the curve with equation $y = x^2 + 5x - 4$ where the gradient is 3.

(P) **5** Find the gradients of the curve $y = x^2 - 5x + 10$ at the points A and B where the curve meets the line $y = 4$.

(P) **6** Find the gradients of the curve $y = 2x^2$ at the points C and D where the curve meets the line $y = x + 3$.

(P) **7** $f(x) = x^2 - 2x - 8$

a Sketch the graph of $y = f(x)$.

b On the same set of axes, sketch the graph of $y = f'(x)$.

c Explain why the x-coordinate of the turning point of $y = f(x)$ is the same as the x-coordinate of the point where the graph of $y = f'(x)$ crosses the x-axis.

12.5 Differentiating functions with two or more terms

You can use the rule for differentiating ax^n to differentiate functions with two or more terms. You need to be able to rearrange **each term** into the form ax^n, where a is a constant and n is a real number. Then you can differentiate the terms **one-at-a-time**.

- **If $y = f(x) \pm g(x)$, then $\frac{dy}{dx} = f'(x) \pm g'(x)$.**

Example 8

Find $\frac{dy}{dx}$ given that y equals:

a $4x^3 + 2x$ **b** $x^3 + x^2 - x^{\frac{1}{2}}$ **c** $\frac{1}{3}x^{\frac{1}{2}} + 4x^2$

a $y = 4x^3 + 2x$

So $\frac{dy}{dx} = 12x^2 + 2$

Differentiate the terms one-at-a-time.

b $y = x^3 + x^2 - x^{\frac{1}{2}}$

So $\frac{dy}{dx} = 3x^2 + 2x - \frac{1}{2}x^{-\frac{1}{2}}$

Be careful with the third term. You multiply the term by $\frac{1}{2}$ and then reduce the power by 1 to get $-\frac{1}{2}$

c $y = \frac{1}{3}x^{\frac{1}{2}} + 4x^2$

Check that each term is in the form ax^n before differentiating.

So $\frac{dy}{dx} = \frac{1}{3} \times \frac{1}{2}x^{-\frac{1}{2}} + 8x$

$= \frac{1}{6}x^{-\frac{1}{2}} + 8x$

Example 9

Differentiate:

a $\frac{1}{4\sqrt{x}}$ **b** $x^3(3x + 1)$ **c** $\frac{x - 2}{x^2}$

a Let $y = \frac{1}{4\sqrt{x}}$

$= \frac{1}{4}x^{-\frac{1}{2}}$

Therefore $\frac{dy}{dx} = -\frac{1}{8}x^{-\frac{3}{2}}$

Use the laws of indices to write the expression in the form ax^n.

$\frac{1}{4\sqrt{x}} = \frac{1}{4} \times \frac{1}{\sqrt{x}} = \frac{1}{4} \times \frac{1}{x^{\frac{1}{2}}} = \frac{1}{4}x^{-\frac{1}{2}}$

b Let $y = x^3(3x + 1)$

$= 3x^4 + x^3$

Therefore $\frac{dy}{dx} = 12x^3 + 3x^2$

$= 3x^2(4x + 1)$

Multiply out the brackets to give a polynomial function.

Differentiate each term.

c Let $y = \frac{x - 2}{x^2}$

$= \frac{1}{x} - \frac{2}{x^2}$

$= x^{-1} - 2x^{-2}$

Therefore $\frac{dy}{dx} = -x^{-2} + 4x^{-3}$

$= -\frac{1}{x^2} + \frac{4}{x^3}$

$= \frac{4 - x}{x^3}$

Express the single fraction as two separate fractions, and simplify: $\frac{x}{x^2} = \frac{1}{x}$

Write each term in the form ax^n then differentiate.

You can write the answer as a single fraction with denominator x^3.

Exercise 12E

1 Differentiate:

a $x^4 + x^{-1}$ **b** $2x^5 + 3x^{-2}$ **c** $6x^{\frac{3}{2}} + 2x^{-\frac{1}{2}} + 4$

2 Find the gradient of the curve with equation $y = f(x)$ at the point A where:

a $f(x) = x^3 - 3x + 2$ and A is at $(-1, 4)$ **b** $f(x) = 3x^2 + 2x^{-1}$ and A is at $(2, 13)$

3 Find the point or points on the curve with equation $y = f(x)$, where the gradient is zero:

a $f(x) = x^2 - 5x$ **b** $f(x) = x^3 - 9x^2 + 24x - 20$

c $f(x) = x^{\frac{3}{2}} - 6x + 1$ **d** $f(x) = x^{-1} + 4x$

4 Differentiate:

a $2\sqrt{x}$ **b** $\frac{3}{x^2}$ **c** $\frac{1}{3x^3}$ **d** $\frac{1}{3}x^3(x - 2)$

e $\frac{2}{x^3} + \sqrt{x}$ **f** $\sqrt[3]{x} + \frac{1}{2x}$ **g** $\frac{2x + 3}{x}$ **h** $\frac{3x^2 - 6}{x}$

i $\frac{2x^3 + 3x}{\sqrt{x}}$ **j** $x(x^2 - x + 2)$ **k** $3x^2(x^2 + 2x)$ **l** $(3x - 2)\left(4x + \frac{1}{x}\right)$

5 Find the gradient of the curve with equation $y = f(x)$ at the point A where:

a $f(x) = x(x + 1)$ and A is at $(0, 0)$

b $f(x) = \dfrac{2x - 6}{x^2}$ and A is at $(3, 0)$

c $f(x) = \dfrac{1}{\sqrt{x}}$ and A is at $\left(\frac{1}{4}, 2\right)$

d $f(x) = 3x - \dfrac{4}{x^2}$ and A is at $(2, 5)$

(E/P) **6** $f(x) = \dfrac{12}{p\sqrt{x}} + x$, where p is a real constant and $x > 0$.

Given that $f'(2) = 3$, find p, giving your answer in the form $a\sqrt{2}$ where a is a rational number. **(4 marks)**

(P) **7** $f(x) = (2 - x)^9$

a Find the first 3 terms, in ascending powers of x, of the binomial expansion of $f(x)$, giving each term in its simplest form.

b If x is small, so that x^2 and higher powers can be ignored, show that $f'(x) \approx 9216x - 2304$.

Hint Use the binomial expansion with $a = 2$, $b = -x$ and $n = 9$. ← Section 8.3

12.6 Gradients, tangents and normals

You can use the derivative to find the equation of the tangent to a curve at a given point. On the curve with equation $y = f(x)$, the gradient of the tangent at a point A with x-coordinate a will be $f'(a)$.

- **The tangent to the curve $y = f(x)$ at the point with coordinates $(a, f(a))$ has equation**

$$y - f(a) = f'(a)(x - a)$$

Links The equation of a straight line with gradient m that passes through the point (x_1, y_1) is $y - y_1 = m(x - x_1)$. ← Section 5.2

The **normal** to a curve at point A is the straight line through A which is perpendicular to the tangent to the curve at A. The gradient of the normal will be $-\dfrac{1}{f'(a)}$

- **The normal to the curve $y = f(x)$ at the point with coordinates $(a, f(a))$ has equation**

$$y - f(a) = -\frac{1}{f'(a)}(x - a)$$

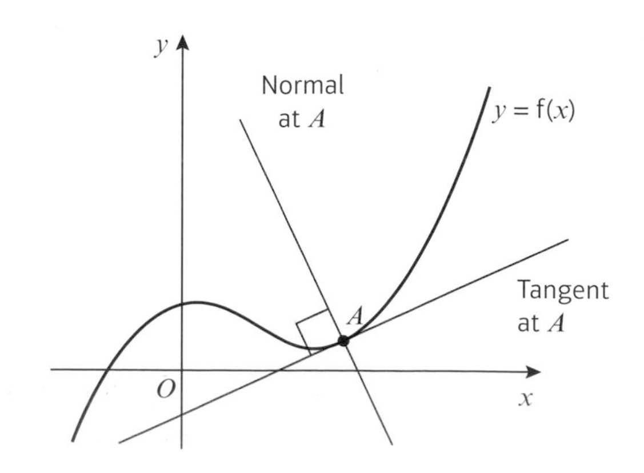

Example 10

Find the equation of the tangent to the curve $y = x^3 - 3x^2 + 2x - 1$ at the point (3, 5).

$y = x^3 - 3x^2 + 2x - 1$

$\frac{dy}{dx} = 3x^2 - 6x + 2$

When $x = 3$, the gradient is 11.

So the equation of the tangent at (3, 5) is

$y - 5 = 11(x - 3)$

$y = 11x - 28$

First differentiate to determine the gradient function.

Then substitute for x to calculate the value of the gradient of the curve and of the tangent when $x = 3$.

You can now use the line equation and simplify.

Example 11

Find the equation of the normal to the curve with equation $y = 8 - 3\sqrt{x}$ at the point where $x = 4$.

$y = 8 - 3\sqrt{x}$

$= 8 - 3x^{\frac{1}{2}}$

$\frac{dy}{dx} = -\frac{3}{2}x^{-\frac{1}{2}}$

When $x = 4$, $y = 2$ and gradient of curve and of tangent $= -\frac{3}{4}$

So gradient of normal is $\frac{4}{3}$.

Equation of normal is

$y - 2 = \frac{4}{3}(x - 4)$

$3y - 6 = 4x - 16$

$3y - 4x + 10 = 0$

Write each term in the form ax^n and differentiate to obtain the gradient function, which you can use to find the gradient at any point.

Find the y-coordinate when $x = 4$ by substituting into the equation of the curve and calculating $8 - 3\sqrt{4} = 8 - 6 = 2$.

Find the gradient of the curve, by calculating

$$\frac{dy}{dx} = -\frac{3}{2}(4)^{-\frac{1}{2}} = -\frac{3}{2} \times \frac{1}{2} = -\frac{3}{4}$$

$$\text{Gradient of normal} = -\frac{1}{\text{gradient of curve}} = -\frac{1}{\left(-\frac{3}{4}\right)} = \frac{4}{3}$$

Simplify by multiplying both sides by 3 and collecting terms.

Online Explore the tangent and normal to the curve using technology.

Exercise 12F

1 Find the equation of the tangent to the curve:

a $y = x^2 - 7x + 10$ at the point (2, 0)

b $y = x + \frac{1}{x}$ at the point $\left(2, 2\frac{1}{2}\right)$

c $y = 4\sqrt{x}$ at the point (9, 12)

d $y = \frac{2x - 1}{x}$ at the point (1, 1)

e $y = 2x^3 + 6x + 10$ at the point (−1, 2)

f $y = x^2 - \frac{7}{x^2}$ at the point (1, −6)

2 Find the equation of the normal to the curve:

a $y = x^2 - 5x$ at the point (6, 6)

b $y = x^2 - \frac{8}{\sqrt{x}}$ at the point (4, 12)

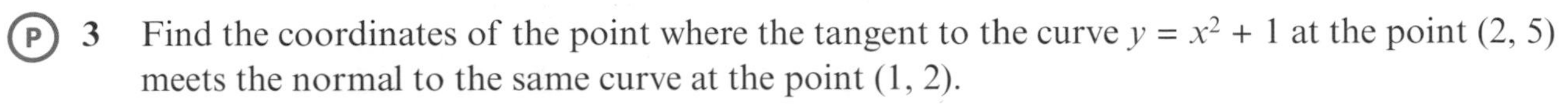

(P) **3** Find the coordinates of the point where the tangent to the curve $y = x^2 + 1$ at the point (2, 5) meets the normal to the same curve at the point (1, 2).

(P) 4 Find the equations of the normals to the curve $y = x + x^3$ at the points (0, 0) and (1, 2), and find the coordinates of the point where these normals meet.

(P) 5 For $f(x) = 12 - 4x + 2x^2$, find the equations of the tangent and the normal at the point where $x = -1$ on the curve with equation $y = f(x)$.

(E/P) 6 The point P with x-coordinate $\frac{1}{2}$ lies on the curve with equation $y = 2x^2$.
The normal to the curve at P intersects the curve at points P and Q.
Find the coordinates of Q. **(6 marks)**

Problem-solving

Draw a sketch showing the curve, the point P and the normal. This will help you check that your answer makes sense.

Challenge

The line L is a tangent to the curve with equation $y = 4x^2 + 1$. L cuts the y-axis at (0, −8) and has a positive gradient. Find the equation of L in the form $y = mx + c$.

Hint Use the discriminant to find the value of m when the line just touches the curve. ← Section 2.5

12.7 Increasing and decreasing functions

You can use the derivative to determine whether a function is **increasing** or **decreasing** on a given interval.

- **The function f(x) is increasing on the interval [a, b] if f′(x) ⩾ 0 for all values of x such that $a < x < b$.**
- **The function f(x) is decreasing on the interval [a, b] if f′(x) ⩽ 0 for all values of x such that $a < x < b$.**

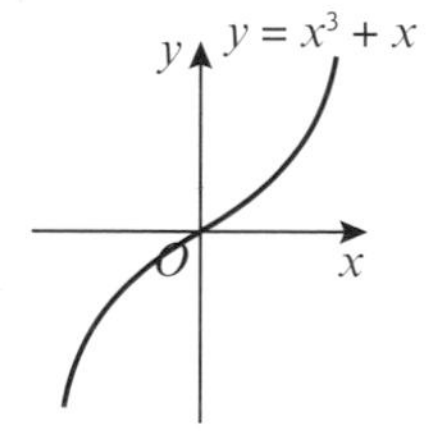

The function $f(x) = x^3 + x$ is increasing for all real values of x.

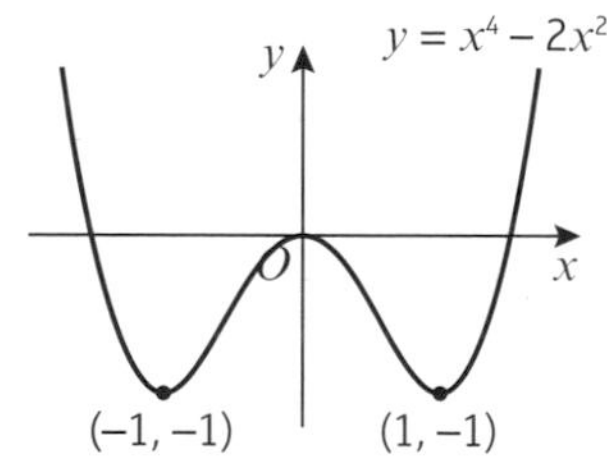

The function $f(x) = x^4 - 2x^2$ is increasing on the interval [−1, 0] and decreasing on the interval [0, 1].

Notation The **interval** [a, b] is the set of all real numbers, x, that satisfy $a \leqslant x \leqslant b$.

Example 12

Show that the function $f(x) = x^3 + 24x + 3$ is increasing for all real values of x.

$f(x) = x^3 + 24x + 3$
$f'(x) = 3x^2 + 24$
$x^2 \geqslant 0$ for all real values of x
So $3x^2 + 24 \geqslant 0$ for all real values of x.
So $f(x)$ is increasing for all real values of x.

First differentiate to obtain the gradient function.

State that the condition for an increasing function is met. In fact $f'(x) \geqslant 24$ for all real values of x.

Example 13

Find the interval on which the function $f(x) = x^3 + 3x^2 - 9x$ is decreasing.

$f(x) = x^3 + 3x^2 - 9x$
$f'(x) = 3x^2 + 6x - 9$
If $f'(x) \leqslant 0$ then $3x^2 + 6x - 9 \leqslant 0$

Find $f'(x)$ and put this expression $\leqslant 0$.

So $3(x^2 + 2x - 3) \leqslant 0$
$3(x + 3)(x - 1) \leqslant 0$
So $-3 \leqslant x \leqslant 1$

Solve the inequality by considering the three regions $x \leqslant -3$, $-3 \leqslant x \leqslant 1$ and $x \geqslant 1$, or by sketching the curve with equation
$y = 3(x + 3)(x - 1)$ ← Section 3.5

So $f(x)$ is decreasing on the interval $[-3, 1]$.

Write the answer clearly.

Online Explore increasing and decreasing functions using technology.

Exercise 12G

1 Find the values of x for which $f(x)$ is an increasing function, given that $f(x)$ equals:

a $3x^2 + 8x + 2$ **b** $4x - 3x^2$ **c** $5 - 8x - 2x^2$ **d** $2x^3 - 15x^2 + 36x$
e $3 + 3x - 3x^2 + x^3$ **f** $5x^3 + 12x$ **g** $x^4 + 2x^2$ **h** $x^4 - 8x^3$

2 Find the values of x for which $f(x)$ is a decreasing function, given that $f(x)$ equals:

a $x^2 - 9x$ **b** $5x - x^2$ **c** $4 - 2x - x^2$ **d** $2x^3 - 3x^2 - 12x$
e $1 - 27x + x^3$ **f** $x + \frac{25}{x}$ **g** $x^{\frac{1}{2}} + 9x^{-\frac{1}{2}}$ **h** $x^2(x + 3)$

E/P 3 Show that the function $f(x) = 4 - x(2x^2 + 3)$ is decreasing for all $x \in \mathbb{R}$. **(3 marks)**

E/P 4 **a** Given that the function $f(x) = x^2 + px$ is increasing on the interval $[-1, 1]$, find one possible value for p. **(2 marks)**

b State with justification whether this is the only possible value for p. **(1 mark)**

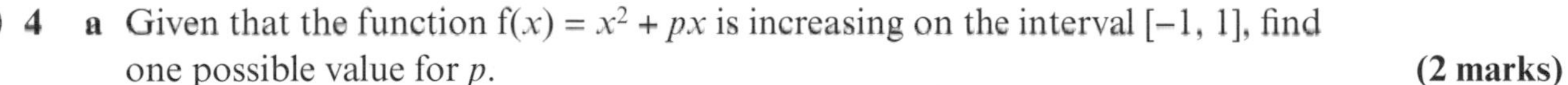

12.8 Second order derivatives

You can find the rate of change of the gradient function by differentiating a function twice.

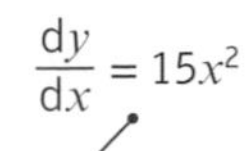

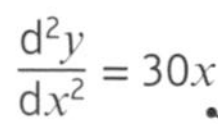

This is the gradient function. It describes the rate of change of the function with respect to x.

This is the **rate of change of the gradient function**. It is called the second order derivative. It can also be written as $f''(x)$.

- **Differentiating a function $y = f(x)$ twice gives you the second order derivative, $f''(x)$ or $\frac{d^2y}{dx^2}$**

Notation The derivative is also called the **first order derivative** or **first derivative**. The **second order derivative** is sometimes just called the **second derivative**.

Example 14

Given that $y = 3x^5 + \dfrac{4}{x^2}$ find:

a $\dfrac{dy}{dx}$ **b** $\dfrac{d^2y}{dx^2}$

a $y = 3x^5 + \dfrac{4}{x^2}$

$= 3x^5 + 4x^{-2}$ — Express the fraction as a negative power of x.

So $\dfrac{dy}{dx} = 15x^4 - 8x^{-3}$ — Differentiate once to get the first order derivative.

$= 15x^4 - \dfrac{8}{x^3}$

b $\dfrac{d^2y}{dx^2} = 60x^3 + 24x^{-4}$ — Differentiate a second time to get the second order derivative.

$= 60x^3 + \dfrac{24}{x^4}$

Example 15

Given that $f(x) = 3\sqrt{x} + \dfrac{1}{2\sqrt{x}}$, find:

a $f'(x)$ **b** $f''(x)$

a $f(x) = 3\sqrt{x} + \dfrac{1}{2\sqrt{x}}$

$= 3x^{\frac{1}{2}} + \frac{1}{2}x^{-\frac{1}{2}}$

$f'(x) = \frac{3}{2}x^{-\frac{1}{2}} - \frac{1}{4}x^{-\frac{3}{2}}$ — Don't rewrite your expression for $f'(x)$ as a fraction. It will be easier to differentiate again if you leave it in this form.

b $f''(x) = -\frac{3}{4}x^{-\frac{3}{2}} + \frac{3}{8}x^{-\frac{5}{2}}$ — The coefficient for the second term is $\left(-\frac{3}{2}\right) \times \left(-\frac{1}{4}\right) = +\frac{3}{8}$. The new power is $-\frac{3}{2} - 1 = -\frac{5}{2}$

Exercise 12H

1 Find $\dfrac{dy}{dx}$ and $\dfrac{d^2y}{dx^2}$ when y equals:

a $12x^2 + 3x + 8$ **b** $15x + 6 + \dfrac{3}{x}$ **c** $9\sqrt{x} - \dfrac{3}{x^2}$ **d** $(5x + 4)(3x - 2)$ **e** $\dfrac{3x + 8}{x^2}$

2 The displacement of a particle in metres at time t seconds is modelled by the function

$$f(t) = \frac{t^2 + 2}{\sqrt{t}}$$

The acceleration of the particle in m s^{-2} is the second derivative of this function. Find an expression for the acceleration of the particle at time t seconds.

Links The velocity of the particle will be $f'(t)$ and its acceleration will be $f''(t)$.
→ Statistics and Mechanics Year 2, Section 6.2

(P) 3 Given that $y = (2x - 3)^3$, find the value of x when $\dfrac{d^2y}{dx^2} = 0$.

(P) 4 $f(x) = px^3 - 3px^2 + x^2 - 4$
When $x = 2$, $f''(x) = -1$. Find the value of p.

Problem-solving When you differentiate with respect to x, you treat any other letters as constants.

12.9 Stationary points

A **stationary point** on a curve is any point where the curve has **gradient zero**. You can determine whether a stationary point is a **local maximum**, a **local minimum** or a **point of inflection** by looking at the gradient of the curve on either side.

- **Any point on the curve $y = f(x)$ where $f'(x) = 0$ is called a stationary point. For a small positive value h:**

Type of stationary point	$f'(x - h)$	$f'(x)$	$f'(x + h)$
Local maximum	**Positive**	**0**	**Negative**
Local minimum	**Negative**	**0**	**Positive**
Point of inflection	**Negative**	**0**	**Negative**
	Positive	**0**	**Positive**

Notation The plural of maximum is **maxima** and the plural of minimum is **minima**.

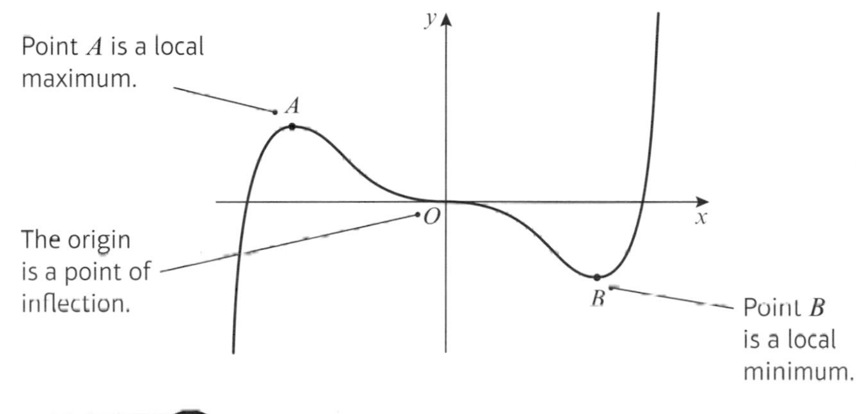

Notation Point A is called a **local** maximum because it is not the largest value the function can take. It is just the largest value in that immediate vicinity.

Example 16

a Find the coordinates of the stationary point on the curve with equation $y = x^4 - 32x$.

b By considering points on either side of the stationary point, determine whether it is a local maximum, a local minimum or a point of inflection.

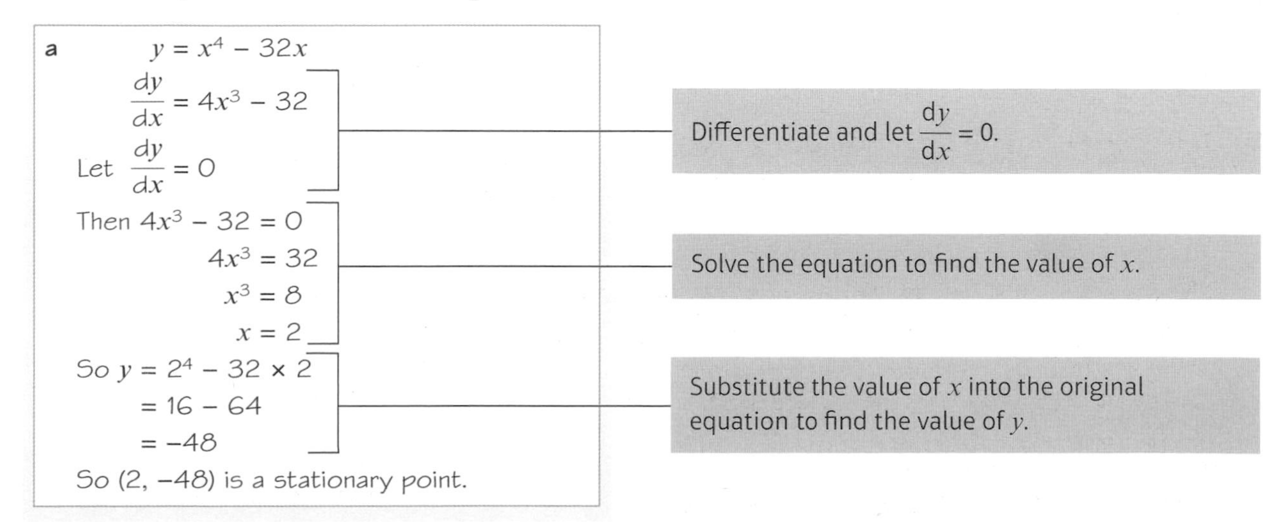

a $y = x^4 - 32x$

$\frac{dy}{dx} = 4x^3 - 32$

Let $\frac{dy}{dx} = 0$

Differentiate and let $\frac{dy}{dx} = 0$.

Then $4x^3 - 32 = 0$

$4x^3 = 32$

$x^3 = 8$

$x = 2$

Solve the equation to find the value of x.

So $y = 2^4 - 32 \times 2$

$= 16 - 64$

$= -48$

Substitute the value of x into the original equation to find the value of y.

So $(2, -48)$ is a stationary point.

b Now consider the gradient on either side of $(2, -48)$.

Value of x	$x = 1.9$	$x = 2$	$x = 2.1$
Gradient	-4.56 which is $-$ve	0	5.04 which is +ve
Shape of curve	\	—	/

From the shape of the curve, the point $(2, -48)$ is a local minimum point.

Make a table where you consider a value of x slightly less than 2 and a value of x slightly greater than 2.

Calculate the gradient for each of these values of x close to the stationary point.

Deduce the shape of the curve.

Online Explore the solution using technology.

In some cases you can use the **second derivative**, $f''(x)$, to determine the nature of a stationary point.

- **If a function $f(x)$ has a stationary point when $x = a$, then:**
  - **if $f''(a) > 0$, the point is a local minimum**
  - **if $f''(a) < 0$, the point is a local maximum**

  If $f''(a) = 0$, the point could be a local minimum, a local maximum or a point of inflection. You will need to look at points on either side to determine its nature.

Hint $f''(x)$ tells you the **rate of change** of the gradient function. When $f'(x) = 0$ and $f''(x) > 0$ the gradient is **increasing** from a negative value to a positive value, so the stationary point is a **minimum**.

Example 17

a Find the coordinates of the stationary points on the curve with equation

$$y = 2x^3 - 15x^2 + 24x + 6$$

b Find $\frac{d^2y}{dx^2}$ and use it to determine the nature of the stationary points.

a $y = 2x^3 - 15x^2 + 24x + 6$

$\frac{dy}{dx} = 6x^2 - 30x + 24$

Putting $6x^2 - 30x + 24 = 0$

$6(x - 4)(x - 1) = 0$

So $x = 4$ or $x = 1$

When $x = 1$,

$y = 2 - 15 + 24 + 6 = 17$

When $x = 4$,

$y = 2 \times 64 - 15 \times 16 + 24 \times 4 + 6$

$= -10$

So the stationary points are at $(1, 17)$ and $(4, -10)$.

Differentiate and put the derivative equal to zero.

Solve the equation to obtain the values of x for the stationary points.

Substitute $x = 1$ and $x = 4$ into the original equation of the curve to obtain the values of y which correspond to these values.

b $\frac{d^2y}{dx^2} = 12x - 30$

When $x = 1$, $\frac{d^2y}{dx^2} = -18$ which is < 0

So (1, 17) is a local maximum point.

When $x = 4$, $\frac{d^2y}{dx^2} = 18$ which is > 0

So (4, −10) is a local minimum point.

Differentiate again to obtain the second derivative.

Substitute $x = 1$ and $x = 4$ into the second derivative expression. If the second derivative is negative then the point is a local maximum point. If it is positive then the point is a local minimum point.

Example 18

a The curve with equation $y = \frac{1}{x} + 27x^3$ has stationary points at $x = \pm a$. Find the value of a.

b Sketch the graph of $y = \frac{1}{x} + 27x^3$.

a $y = x^{-1} + 27x^3$

$\frac{dy}{dx} = -x^{-2} + 81x^2 = -\frac{1}{x^2} + 81x^2$

When $\frac{dy}{dx} = 0$:

$-\frac{1}{x^2} + 81x^2 = 0$

$81x^2 = \frac{1}{x^2}$

$81x^4 = 1$

$x^4 = \frac{1}{81}$

$x = \pm\frac{1}{3}$

So $a = \frac{1}{3}$

Write $\frac{1}{x}$ as x^{-1} to differentiate.

Set $\frac{dy}{dx} = 0$ to determine the x-coordinates of the stationary points.

You need to consider the positive and negative roots: $\left(-\frac{1}{3}\right)^4 = \left(-\frac{1}{3}\right) \times \left(-\frac{1}{3}\right) \times \left(-\frac{1}{3}\right) \times \left(-\frac{1}{3}\right) = \frac{1}{81}$

b $\frac{d^2y}{dx^2} = 2x^{-3} + 162x = \frac{2}{x^3} + 162x$

When $x = -\frac{1}{3}$, $y = \frac{1}{\left(-\frac{1}{3}\right)} + 27\left(-\frac{1}{3}\right)^3 = -4$

and $\frac{d^2y}{dx^2} = \frac{2}{\left(-\frac{1}{3}\right)^3} + 162\left(-\frac{1}{3}\right) = -108$

which is negative.

So the curve has a local maximum at $\left(-\frac{1}{3}, -4\right)$.

When $x = \frac{1}{3}$,

$y = \frac{1}{\left(\frac{1}{3}\right)} + 27\left(\frac{1}{3}\right)^3 = 4$

and

$\frac{d^2y}{dx^2} = \frac{2}{\left(\frac{1}{3}\right)^3} + 162\left(\frac{1}{3}\right) = 108$

which is positive.

To sketch the curve, you need to find the coordinates of the stationary points and determine their natures. Differentiate your expression for $\frac{dy}{dx}$ to find $\frac{d^2y}{dx^2}$

Substitute $x = -\frac{1}{3}$ and $x = \frac{1}{3}$ into the equation of the curve to find the y-coordinates of the stationary points.

Online Check your solution using your calculator.

So the curve has a local minimum at $\left(\frac{1}{3}, 4\right)$.
The curve has an asymptote at $x = 0$.
As $x \to \infty$, $y \to \infty$.
As $x \to -\infty$, $y \to -\infty$.

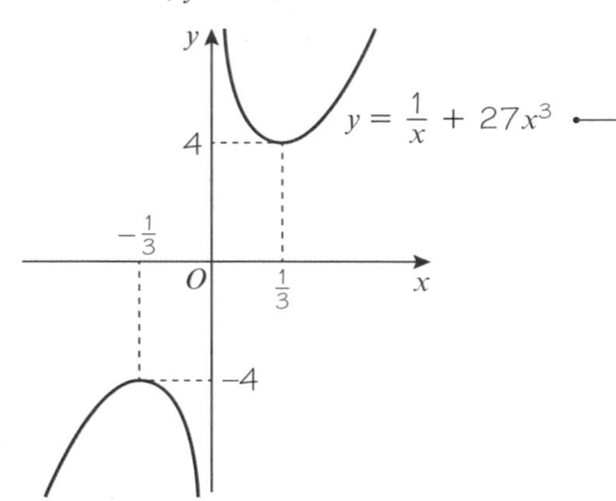

$\frac{1}{x} \to \pm\infty$ as $x \to 0$ so $x = 0$ is an asymptote of the curve.

Mark the coordinates of the stationary points on your sketch, and label the curve with its equation. You could check $\frac{dy}{dx}$ at specific points to help with your sketch:

- When $x = \frac{1}{4}$, $\frac{dy}{dx} = -10.9375$ which is negative.
- When $x = 1$, $\frac{dy}{dx} = 80$ which is positive.

Exercise 12I

Hint For each part of questions **1** and **2**:
- Find $f'(x)$.
- Set $f'(x) = 0$ and solve to find the value of x at the stationary point.
- Find the corresponding value of $f(x)$.

1 Find the least value of the following functions:

a $f(x) = x^2 - 12x + 8$ **b** $f(x) = x^2 - 8x - 1$ **c** $f(x) = 5x^2 + 2x$

2 Find the greatest value of the following functions:

a $f(x) = 10 - 5x^2$ **b** $f(x) = 3 + 2x - x^2$ **c** $f(x) = (6 + x)(1 - x)$

3 Find the coordinates of the points where the gradient is zero on the curves with the given equations. Establish whether these points are local maximum points, local minimum points or points of inflection in each case.

a $y = 4x^2 + 6x$ **b** $y = 9 + x - x^2$ **c** $y = x^3 - x^2 - x + 1$

d $y = x(x^2 - 4x - 3)$ **e** $y = x + \frac{1}{x}$ **f** $y = x^2 + \frac{54}{x}$

g $y = x - 3\sqrt{x}$ **h** $y = x^{\frac{1}{2}}(x - 6)$ **i** $y = x^4 - 12x^2$

4 Sketch the curves with equations given in question **3** parts **a**, **b**, **c** and **d**, labelling any stationary points with their coordinates.

(P) **5** By considering the gradient on either side of the stationary point on the curve $y = x^3 - 3x^2 + 3x$, show that this point is a point of inflection.
Sketch the curve $y = x^3 - 3x^2 + 3x$.

(P) **6** Find the maximum value and hence the range of values for the function $f(x) = 27 - 2x^4$.

(P) **7** $f(x) = x^4 + 3x^3 - 5x^2 - 3x + 1$

a Find the coordinates of the stationary points of $f(x)$, and determine the nature of each.

b Sketch the graph of $y = f(x)$.

Hint Use the **factor theorem** with small positive integer values of x to find one factor of $f'(x)$. ← Section 7.2

12.10 Sketching gradient functions

You can use the features of a given function to sketch the corresponding gradient function. This table shows you features of the graph of a function, $y = f(x)$, and the graph of its gradient function, $y = f'(x)$, at corresponding values of x.

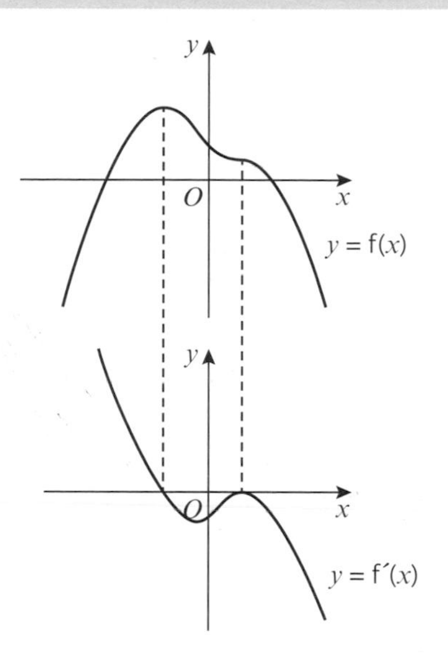

$y = f(x)$	$y = f'(x)$
Maximum or minimum	Cuts the x-axis
Point of inflection	Touches the x-axis
Positive gradient	Above the x-axis
Negative gradient	Below the x-axis
Vertical asymptote	Vertical asymptote
Horizontal asymptote	Horizontal asymptote at the x-axis

Example 19

The diagram shows the curve with equation $y = f(x)$. The curve has stationary points at $(-1, 4)$ and $(1, 0)$, and cuts the x-axis at $(-3, 0)$.

Sketch the gradient function, $y = f'(x)$, showing the coordinates of any points where the curve cuts or meets the x-axis.

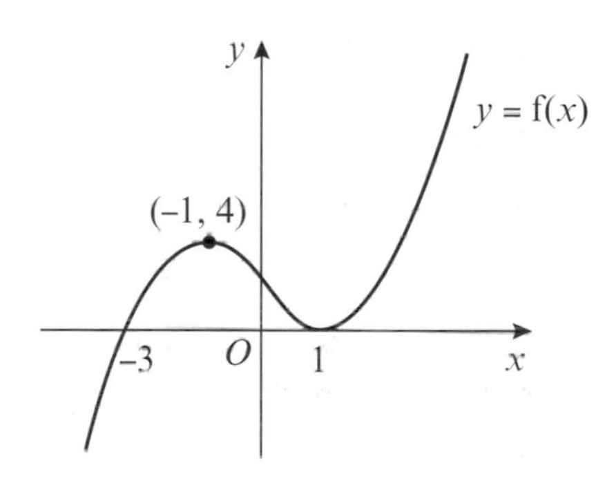

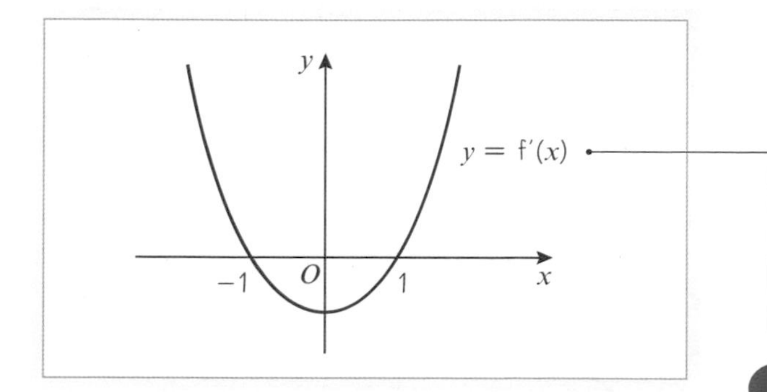

x	$y = f(x)$	$y = f'(x)$
$x < -1$	Positive gradient	Above x-axis
$x = -1$	Maximum	Cuts x-axis
$-1 < x < 1$	Negative gradient	Below x-axis
$x = 1$	Minimum	Cuts x-axis
$x > 1$	Positive gradient	Above x-axis

Watch out Ignore any points where the curve $y = f(x)$ cuts the x-axis. These will not tell you anything about the features of the graph of $y = f'(x)$.

Online Use technology to explore the key features linking $y = f(x)$ and $y = f'(x)$.

Example 20

The diagram shows the curve with equation $y = f(x)$. The curve has an asymptote at $y = -2$ and a turning point at $(-3, -8)$. It cuts the x-axis at $(-10, 0)$.

a Sketch the graph of $y = f'(x)$.

b State the equation of the asymptote of $y = f'(x)$.

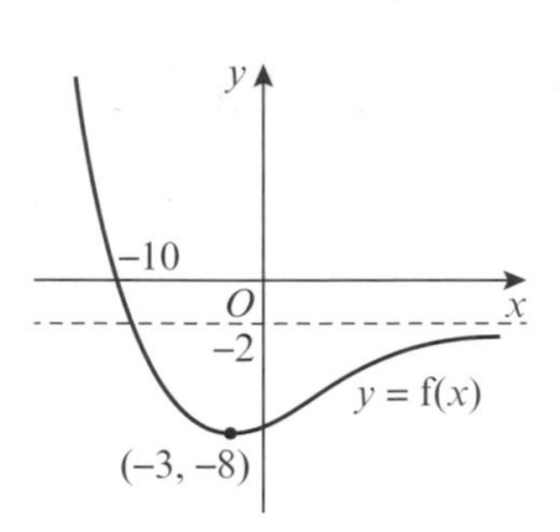

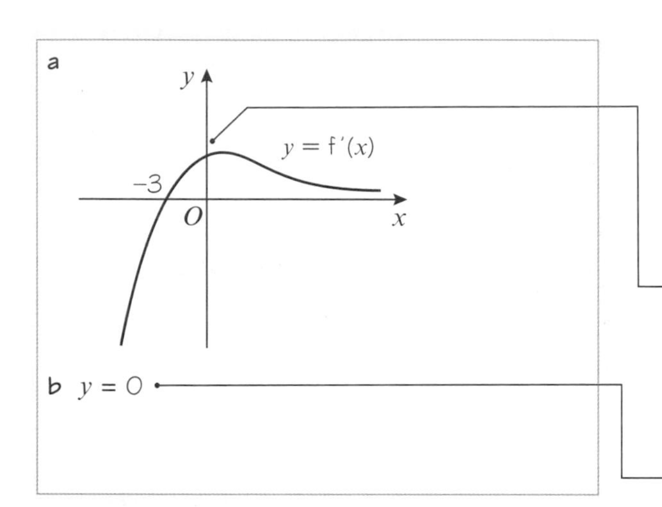

Draw your sketch on a separate set of axes. The graph of $y = f'(x)$ will have the same horizontal scale but will have a different vertical scale.

You don't have enough information to work out the coordinates of the y-intercept, or the local maximum, of the graph of the gradient function.

The graph of $y = f(x)$ is a smooth curve so the graph of $y = f'(x)$ will also be a smooth curve.

If $y = f(x)$ has any **horizontal asymptotes** then the graph of $y = f'(x)$ will have an asymptote at the x-axis.

Exercise 12J

1 For each graph given, sketch the graph of the corresponding gradient function on a separate set of axes. Show the coordinates of any points where the curve cuts or meets the x-axis, and give the equations of any asymptotes.

a

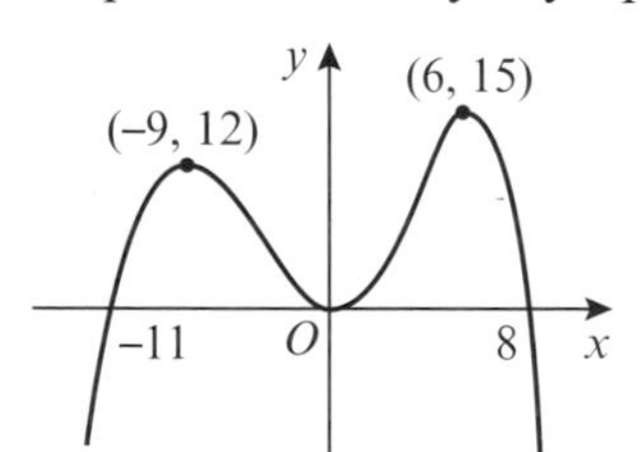

b

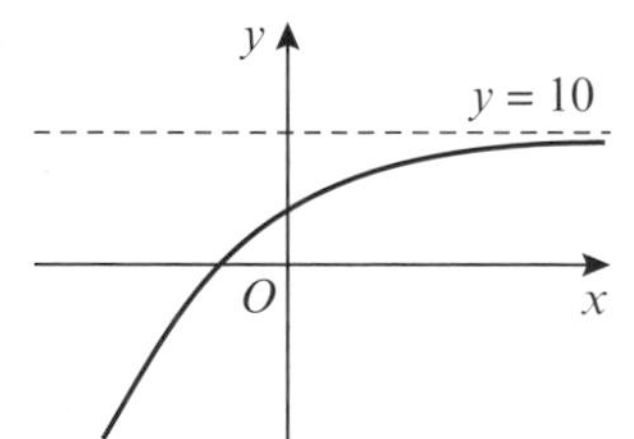

c

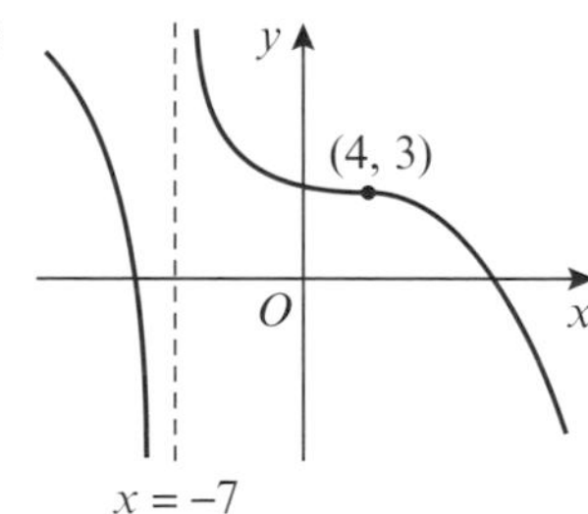

d

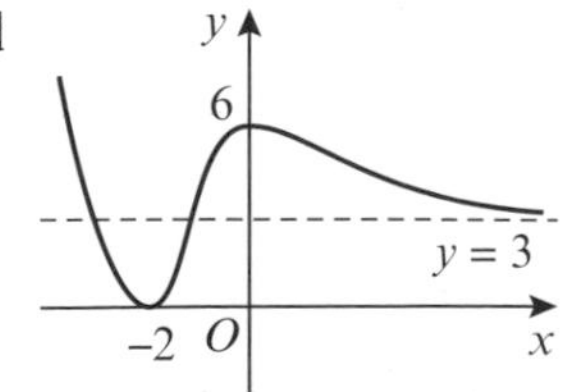

e

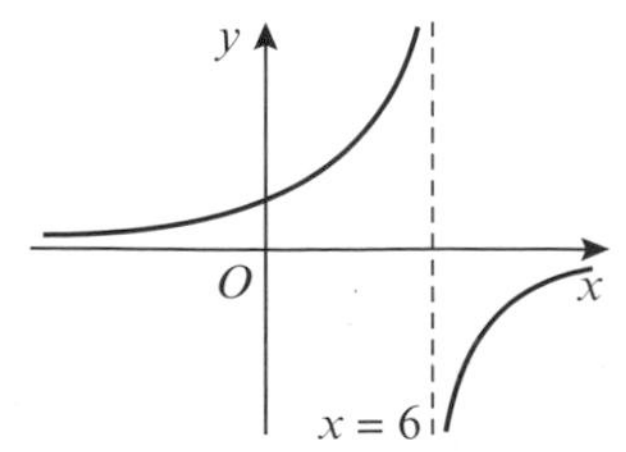

f

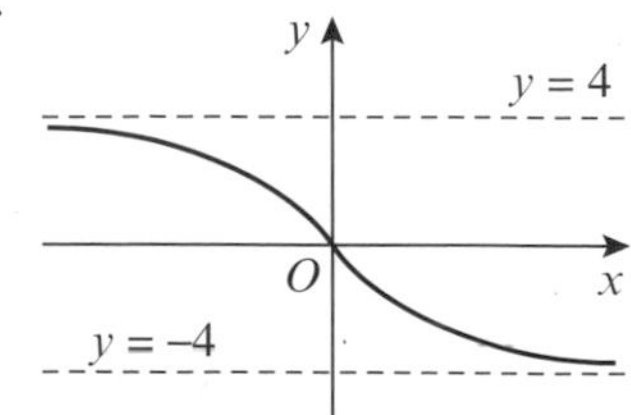

(P) **2** $f(x) = (x + 1)(x - 4)^2$

a Sketch the graph of $y = f(x)$.

b On a separate set of axes, sketch the graph of $y = f'(x)$.

c Show that $f'(x) = (x - 4)(3x - 2)$.

d Use the derivative to determine the exact coordinates of the points where the gradient function cuts the coordinate axes.

Hint This is an x^3 graph with a positive coefficient of x^3. ← Section 4.1

12.11 Modelling with differentiation

You can think of $\frac{dy}{dx}$ as $\frac{\text{small change in } y}{\text{small change in } x}$. It represents the **rate of change** of y with respect to x.

If you replace y and x with variables that represent real-life quantities, you can use the derivative to model lots of real-life situations involving rates of change.

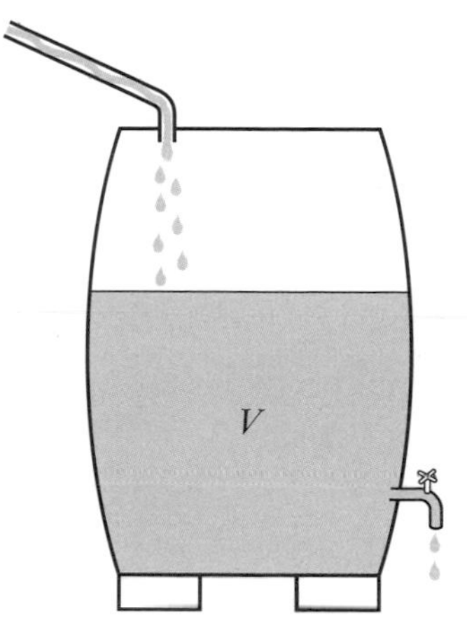

The volume of water in this water butt is constantly changing over time. If V represents the volume of water in the water butt in litres, and t represents the time in seconds, then you could model V as a function of t. If $V = f(t)$ then $\frac{dV}{dt} = f'(t)$ would represent the **rate of change** of volume with respect to time. The units of $\frac{dV}{dt}$ would be litres per second.

Example 21

Given that the volume, V cm³, of an expanding sphere is related to its radius, r cm, by the formula $V = \frac{4}{3}\pi r^3$, find the rate of change of volume with respect to radius at the instant when the radius is 5 cm.

$V = \frac{4}{3}\pi r^3$

$\frac{dV}{dr} = 4\pi r^2$ — Differentiate V with respect to r. Remember that π is a constant.

When $r = 5$, $\frac{dV}{dr} = 4\pi \times 5^2$ — Substitute $r = 5$.

$= 314$ (3 s.f.)

So the rate of change is 314 cm³ per cm. — Interpret the answer with units.

Example 22

A large tank in the shape of a cuboid is to be made from 54 m² of sheet metal. The tank has a horizontal base and no top. The height of the tank is x metres. Two opposite vertical faces are squares.

a Show that the volume, V m³, of the tank is given by $V = 18x - \frac{2}{3}x^3$

b Given that x can vary, use differentiation to find the maximum or minimum value of V.

c Justify that the value of V you have found is a maximum.

a Let the length of the tank be y metres.

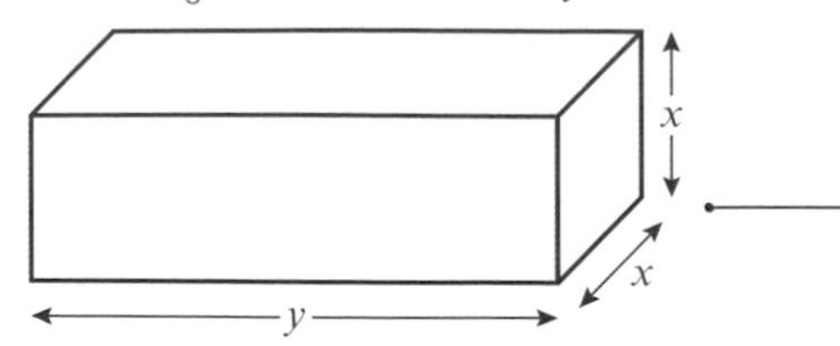

Total area, $A = 2x^2 + 3xy$

So $54 = 2x^2 + 3xy$

$$y = \frac{54 - 2x^2}{3x}$$

But $V = x^2y$

So $V = x^2\left(\frac{54 - 2x^2}{3x}\right)$

$$= \frac{x}{3}(54 - 2x^2)$$

So $V = 18x - \frac{2}{3}x^3$

b $\frac{dV}{dx} = 18 - 2x^2$

Put $\frac{dV}{dx} = 0$

$0 = 18 - 2x^2$

So $x^2 = 9$

$x = -3$ or 3

But x is a length so $x = 3$

When $x = 3$, $V = 18 \times 3 - \frac{2}{3} \times 3^3$

$= 54 - 18$

$= 36$

$V = 36$ is a maximum or minimum value of V.

c $\frac{d^2V}{dx^2} = -4x$

When $x = 3$, $\frac{d^2V}{dx^2} = -4 \times 3 = -12$

This is negative, so $V = 36$ is the maximum value of V.

Problem-solving

You don't know the length of the tank. Write it as y metres to simplify your working.
You could also draw a sketch to help you find the correct expressions for the surface area and volume of the tank.

Draw a sketch.

Rearrange to find y in terms of x.

Substitute the expression for y into the equation.

Simplify.

Differentiate V with respect to x and put $\frac{dV}{dx} = 0$.

Rearrange to find x.
x is a length so use the positive solution.

Substitute the value of x into the expression for V.

Find the second derivative of V.

$\frac{d^2V}{dx^2} < 0$ so $V = 36$ is a maximum.

Exercise 12K

1 Find $\frac{d\theta}{dt}$ where $\theta = t^2 - 3t$.

2 Find $\frac{dA}{dr}$ where $A = 2\pi r$.

3 Given that $r = \frac{12}{t}$, find the value of $\frac{dr}{dt}$ when $t = 3$.

4 The surface area, A cm², of an expanding sphere of radius r cm is given by $A = 4\pi r^2$. Find the rate of change of the area with respect to the radius at the instant when the radius is 6 cm.

5 The displacement, s metres, of a car from a fixed point at time t seconds is given by $s = t^2 + 8t$. Find the rate of change of the displacement with respect to time at the instant when $t = 5$.

(P) 6 A rectangular garden is fenced on three sides, and the house forms the fourth side of the rectangle.

a Given that the total length of the fence is 80 m, show that the area, A, of the garden is given by the formula $A = y(80 - 2y)$, where y is the distance from the house to the end of the garden.

b Given that the area is a maximum for this length of fence, find the dimensions of the enclosed garden, and the area which is enclosed.

(P) 7 A closed cylinder has total surface area equal to 600π.

a Show that the volume, V cm³, of this cylinder is given by the formula $V = 300\pi r - \pi r^3$, where r cm is the radius of the cylinder.

b Find the maximum volume of such a cylinder.

(P) 8 A sector of a circle has area 100 cm².

a Show that the perimeter of this sector is given by the formula

$$P = 2r + \frac{200}{r},\ r > \sqrt{\frac{100}{\pi}}$$

b Find the minimum value for the perimeter.

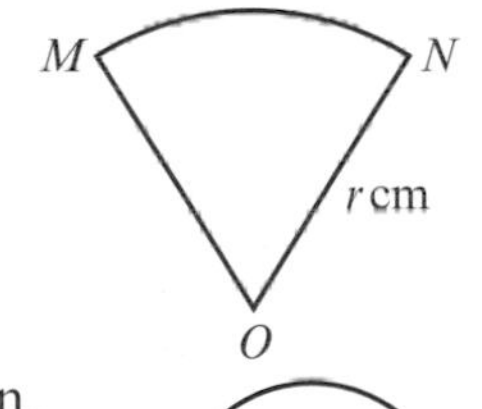

(E/P) 9 A shape consists of a rectangular base with a semicircular top, as shown.

a Given that the perimeter of the shape is 40 cm, show that its area, A cm², is given by the formula

$$A = 40r - 2r^2 - \frac{\pi r^2}{2}$$

where r cm is the radius of the semicircle. **(2 marks)**

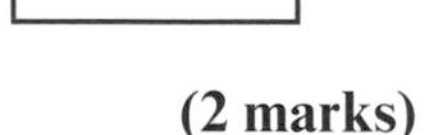

b Hence find the maximum value for the area of the shape. **(4 marks)**

(E/P) 10 The shape shown is a wire frame in the form of a large rectangle split by parallel lengths of wire into 12 smaller equal-sized rectangles.

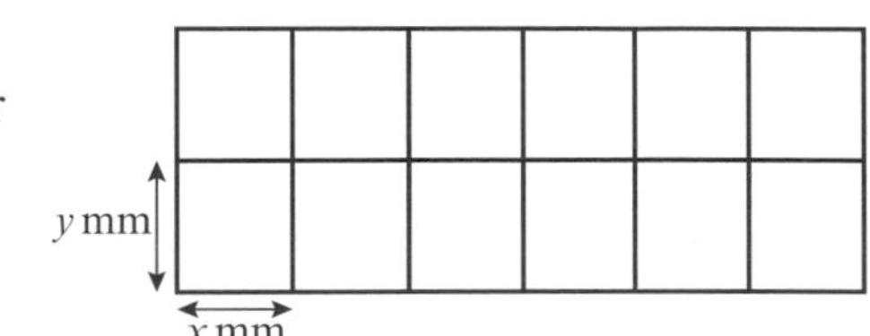

a Given that the total length of wire used to complete the whole frame is 1512 mm, show that the area of the whole shape, A mm², is given by the formula

$$A = 1296x - \frac{108x^2}{7}$$

where x mm is the width of one of the smaller rectangles. **(4 marks)**

b Hence find the maximum area which can be enclosed in this way. **(4 marks)**

Mixed exercise 12

E/P **1** Prove, from first principles, that the derivative of $10x^2$ is $20x$. **(4 marks)**

P **2** The point A with coordinates (1, 4) lies on the curve with equation $y = x^3 + 3x$.
The point B also lies on the curve and has x-coordinate $(1 + \delta x)$.

a Show that the gradient of the line segment AB is given by $(\delta x)^2 + 3\delta x + 6$.

b Deduce the gradient of the curve at point A.

3 A curve is given by the equation $y = 3x^2 + 3 + \frac{1}{x^2}$, where $x > 0$. At the points A, B and C on the curve, $x = 1$, 2 and 3 respectively. Find the gradient of the curve at A, B and C.

E **4** Calculate the x-coordinates of the points on the curve with equation $y = 7x^2 - x^3$ at which the gradient is equal to 16. **(4 marks)**

5 Find the x-coordinates of the two points on the curve with equation $y = x^3 - 11x + 1$ where the gradient is 1. Find the corresponding y-coordinates.

E **6** The function f is defined by $f(x) = x + \frac{9}{x}, x \in \mathbb{R}, x \neq 0$.

a Find $f'(x)$. **(2 marks)**

b Solve $f'(x) = 0$. **(2 marks)**

E **7** Given that

$$y = 3\sqrt{x} - \frac{4}{\sqrt{x}}, \quad x > 0,$$

find $\frac{dy}{dx}$ **(3 marks)**

E/P **8** A curve has equation $y = 12x^{\frac{1}{2}} - x^{\frac{3}{2}}$.

a Show that $\frac{dy}{dx} = \frac{3}{2}x^{-\frac{1}{2}}(4 - x)$. **(2 marks)**

b Find the coordinates of the point on the curve where the gradient is zero. **(2 marks)**

E **9** **a** Expand $(x^{\frac{3}{2}} - 1)(x^{-\frac{1}{2}} + 1)$. **(2 marks)**

b A curve has equation $y = (x^{\frac{3}{2}} - 1)(x^{-\frac{1}{2}} + 1), x > 0$. Find $\frac{dy}{dx}$ **(2 marks)**

c Use your answer to part **b** to calculate the gradient of the curve at the point where $x = 4$. **(1 mark)**

E **10** Differentiate with respect to x:

$$2x^3 + \sqrt{x} + \frac{x^2 + 2x}{x^2}$$ **(3 marks)**

E/P **11** The curve with equation $y = ax^2 + bx + c$ passes through the point (1, 2). The gradient of the curve is zero at the point (2, 1). Find the values of a, b and c. **(5 marks)**

E/P **12** A curve C has equation $y = x^3 - 5x^2 + 5x + 2$.

a Find $\frac{dy}{dx}$ in terms of x. **(2 marks)**

b The points P and Q lie on C. The gradient of C at both P and Q is 2. The x-coordinate of P is 3.

i Find the x-coordinate of Q. **(3 marks)**

ii Find an equation for the tangent to C at P, giving your answer in the form $y = mx + c$, where m and c are constants. **(3 marks)**

iii If this tangent intersects the coordinate axes at the points R and S, find the length of RS, giving your answer as a surd. **(3 marks)**

13 A curve has equation $y = \frac{8}{x} - x + 3x^2$, $x > 0$. Find the equations of the tangent and the normal to the curve at the point where $x = 2$.

E/P **14** The normals to the curve $2y = 3x^3 - 7x^2 + 4x$, at the points $O(0, 0)$ and $A(1, 0)$, meet at the point N.

a Find the coordinates of N. **(7 marks)**

b Calculate the area of triangle OAN. **(3 marks)**

E/P **15** A curve C has equation $y = x^3 - 2x^2 - 4x - 1$ and cuts the y-axis at a point P. The line L is a tangent to the curve at P, and cuts the curve at the point Q. Show that the distance PQ is $2\sqrt{17}$. **(7 marks)**

E **16** Given that $y = x^{\frac{3}{2}} + \frac{48}{x}$, $x > 0$

a find the value of x and the value of y when $\frac{dy}{dx} = 0$. **(5 marks)**

b show that the value of y which you found in part **a** is a minimum. **(2 marks)**

17 A curve has equation $y = x^3 - 5x^2 + 7x - 14$. Determine, by calculation, the coordinates of the stationary points of the curve.

E/P **18** The function f, defined for $x \in \mathbb{R}$, $x > 0$, is such that:

$$f'(x) = x^2 - 2 + \frac{1}{x^2}$$

a Find the value of $f''(x)$ at $x = 4$. **(4 marks)**

b Prove that f is an increasing function. **(3 marks)**

E **19** A curve has equation $y = x^3 - 6x^2 + 9x$. Find the coordinates of its local maximum. **(4 marks)**

20 $f(x) = 3x^4 - 8x^3 - 6x^2 + 24x + 20$

a Find the coordinates of the stationary points of $f(x)$, and determine the nature of each of them.

b Sketch the graph of $y = f(x)$.

(E) **21** The diagram shows part of the curve with equation $y = \mathrm{f}(x)$, where:

$$\mathrm{f}(x) = 200 - \frac{250}{x} - x^2, \ x > 0$$

The curve cuts the x-axis at the points A and C. The point B is the maximum point of the curve.

a Find $\mathrm{f}'(x)$. **(3 marks)**

b Use your answer to part **a** to calculate the coordinates of B. **(4 marks)**

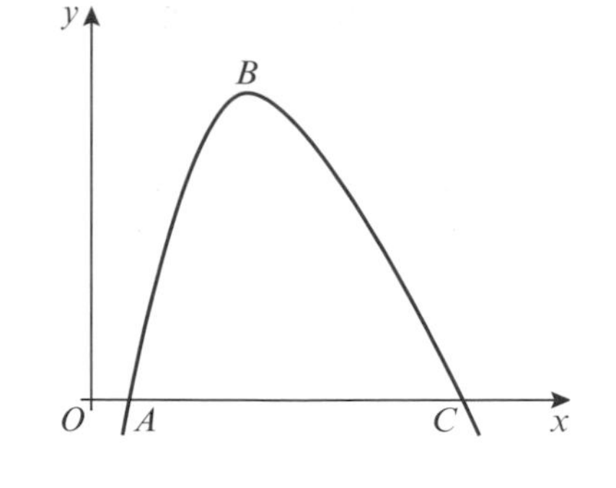

(E/P) **22** The diagram shows the part of the curve with equation $y = 5 - \frac{1}{2}x^2$ for which $y > 0$.
The point $P(x, y)$ lies on the curve and O is the origin.

a Show that $OP^2 = \frac{1}{4}x^4 - 4x^2 + 25$. **(3 marks)**

Taking $\mathrm{f}(x) = \frac{1}{4}x^4 - 4x^2 + 25$:

b Find the values of x for which $\mathrm{f}'(x) = 0$. **(4 marks)**

c Hence, or otherwise, find the minimum distance from O to the curve, showing that your answer is a minimum. **(4 marks)**

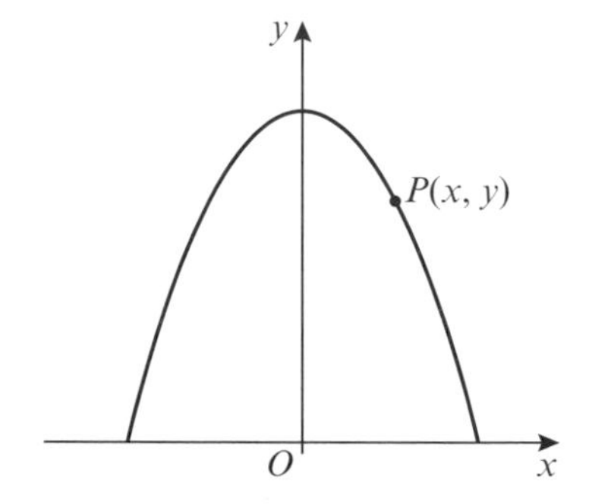

(E) **23** The diagram shows part of the curve with equation $y = 3 + 5x + x^2 - x^3$. The curve touches the x-axis at A and crosses the x-axis at C. The points A and B are stationary points on the curve.

a Show that C has coordinates $(3, 0)$. **(1 mark)**

b Using calculus and showing all your working, find the coordinates of A and B. **(5 marks)**

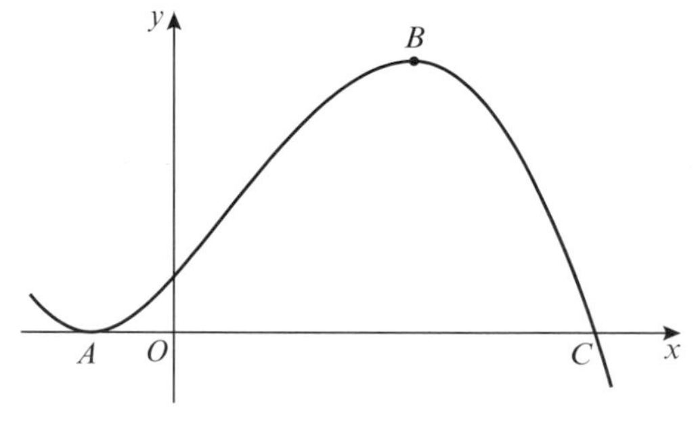

(P) **24** The motion of a damped spring is modelled using this graph.

On a separate graph, sketch the gradient function for this model. Choose suitable labels and units for each axis, and indicate the coordinates of any points where the gradient function crosses the horizontal axis.

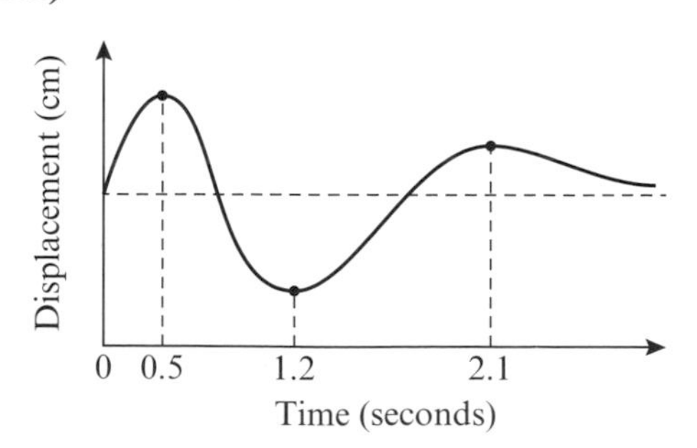

25 The volume, $V\,\mathrm{cm}^3$, of a tin of radius $r\,\mathrm{cm}$ is given by the formula $V = \pi(40r - r^2 - r^3)$.
Find the positive value of r for which $\frac{\mathrm{d}V}{\mathrm{d}r} = 0$, and find the value of V which corresponds to this value of r.

(P) **26** The total surface area, $A\,\mathrm{cm}^2$, of a cylinder with a fixed volume of $1000\,\mathrm{cm}^3$ is given by the formula $A = 2\pi x^2 + \frac{2000}{x}$, where $x\,\mathrm{cm}$ is the radius. Show that when the rate of change of the area with respect to the radius is zero, $x^3 = \frac{500}{\pi}$

(E/P) **27** A wire is bent into the plane shape *ABCDE* as shown. Shape *ABDE* is a rectangle and *BCD* is a semicircle with diameter *BD*. The area of the region enclosed by the wire is $R\,\text{m}^2$, $AE = x$ metres, and $AB = ED = y$ metres. The total length of the wire is 2 m.

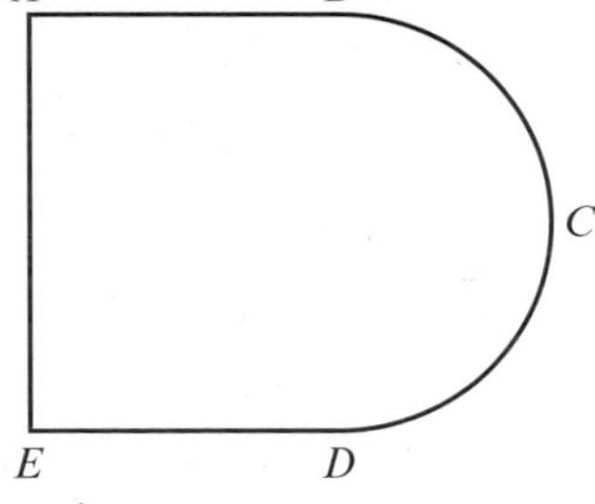

a Find an expression for y in terms of x. **(3 marks)**

b Prove that $R = \frac{x}{8}(8 - 4x - \pi x)$. **(4 marks)**

Given that x can vary, using calculus and showing your working:

c find the maximum value of R. (You do not have to prove that the value you obtain is a maximum.) **(5 marks)**

(E/P) **28** A cylindrical biscuit tin has a close-fitting lid which overlaps the tin by 1 cm, as shown. The radii of the tin and the lid are both x cm. The tin and the lid are made from a thin sheet of metal of area $80\pi\,\text{cm}^2$ and there is no wastage. The volume of the tin is $V\,\text{cm}^3$.

a Show that $V = \pi(40x - x^2 - x^3)$. **(5 marks)**

Given that x can vary:

b use differentiation to find the positive value of x for which V is stationary. **(3 marks)**

c Prove that this value of x gives a maximum value of V. **(2 marks)**

d Find this maximum value of V. **(1 mark)**

e Determine the percentage of the sheet metal used in the lid when V is a maximum. **(2 marks)**

29 The diagram shows an open tank for storing water, *ABCDEF*. The sides *ABFE* and *CDEF* are rectangles. The triangular ends *ADE* and *BCF* are isosceles, and $\angle AED = \angle BFC = 90°$. The ends *ADE* and *BCF* are vertical and *EF* is horizontal.

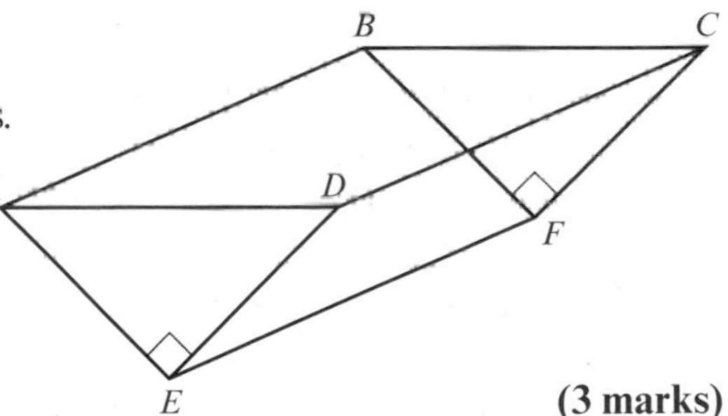

Given that $AD = x$ metres:

a show that the area of triangle *ADE* is $\frac{1}{4}x^2\,\text{m}^2$ **(3 marks)**

Given also that the capacity of the container is $4000\,\text{m}^3$ and that the total area of the two triangular and two rectangular sides of the container is $S\,\text{m}^2$:

b show that $S = \frac{x^2}{2} + \frac{16\,000\sqrt{2}}{x}$ **(4 marks)**

Given that x can vary:

c use calculus to find the minimum value of S. **(6 marks)**

d justify that the value of S you have found is a minimum. **(2 marks)**

Challenge

a Find the first four terms in the binomial expansion of $(x + h)^7$, in ascending powers of h.

b Hence prove, from first principles, that the derivative of x^7 is $7x^6$.

Summary of key points

1 The **gradient** of a **curve** at a given point is defined as the gradient of the **tangent** to the curve at that point.

2 The **gradient function**, or **derivative**, of the curve $y = \text{f}(x)$ is written as $\text{f}'(x)$ or $\frac{dy}{dx}$

$$\text{f}'(x) = \lim_{h \to 0} \frac{\text{f}(x+h) - \text{f}(x)}{h}$$

The gradient function can be used to find the gradient of the curve for any value of x.

3 For all real values of n, and for a constant a:

- If $\text{f}(x) = x^n$ then $\text{f}'(x) = nx^{n-1}$
- If $\text{f}(x) = ax^n$ then $\text{f}'(x) = anx^{n-1}$
- If $y = x^n$ then $\frac{dy}{dx} = nx^{n-1}$
- If $y = ax^n$ then $\frac{dy}{dx} = anx^{n-1}$

4 For the quadratic curve with equation $y = ax^2 + bx + c$, the derivative is given by

$$\frac{dy}{dx} = 2ax + b$$

5 If $y = \text{f}(x) \pm \text{g}(x)$, then $\frac{dy}{dx} = \text{f}'(x) \pm \text{g}'(x)$.

6 The tangent to the curve $y = \text{f}(x)$ at the point with coordinates $(a, \text{f}(a))$ has equation

$$y - \text{f}(a) = \text{f}'(a)(x - a)$$

7 The normal to the curve $y = \text{f}(x)$ at the point with coordinates $(a, \text{f}(a))$ has equation

$$y - \text{f}(a) = -\frac{1}{\text{f}'(a)}(x - a)$$

8
- The function $\text{f}(x)$ is **increasing** on the interval $[a, b]$ if $\text{f}'(x) \geqslant 0$ for all values of x such that $a < x < b$.
- The function $\text{f}(x)$ is **decreasing** on the interval $[a, b]$ if $\text{f}'(x) \leqslant 0$ for all values of x such that $a < x < b$.

9 Differentiating a function $y = \text{f}(x)$ twice gives you the second order derivative, $\text{f}''(x)$ or $\frac{d^2y}{dx^2}$

10 Any point on the curve $y = \text{f}(x)$ where $\text{f}'(x) = 0$ is called a **stationary point**. For a small positive value h:

Type of stationary point	$\text{f}'(x - h)$	$\text{f}'(x)$	$\text{f}'(x + h)$
Local maximum	Positive	0	Negative
Local minimum	Negative	0	Positive
Point of inflection	Negative	0	Negative
	Positive	0	Positive

11 If a function $\text{f}(x)$ has a stationary point when $x = a$, then:

- if $\text{f}''(a) > 0$, the point is a local minimum
- if $\text{f}''(a) < 0$, the point is a local maximum.

If $\text{f}''(a) = 0$, the point could be a local minimum, a local maximum or a point of inflection. You will need to look at points on either side to determine its nature.

Integration

13

Objectives

After completing this unit you should be able to:

- Find y given $\frac{dy}{dx}$ for x^n → pages 288–290
- Integrate polynomials → pages 290–293
- Find $f(x)$, given $f'(x)$ and a point on the curve → pages 293–295
- Evaluate a definite integral → pages 295–297
- Find the area bounded by a curve and the x-axis → pages 297–302
- Find areas bounded by curves and straight lines → pages 302–306

Integration is the opposite of differentiation. It is used to calculate areas of surfaces, volumes of irregular shapes and areas under curves. In mechanics, integration can be used to calculate the area under a velocity-time graph to find distance travelled.

→ Exercise 13D Q8

Prior knowledge check

1 Simplify these expressions

a $\frac{x^3}{\sqrt{x}}$ b $\frac{\sqrt{x} \times 2x^3}{x^2}$

c $\frac{x^3 - x}{\sqrt{x}}$ d $\frac{\sqrt{x} + 4x^3}{x^2}$

← Sections 1.1, 1.4

2 Find $\frac{dy}{dx}$ when y equals

a $2x^3 + 3x - 5$ b $\frac{1}{2}x^2 - x$

c $x^2(x + 1)$ d $\frac{x - x^5}{x^2}$ ← Section 12.5

3 Sketch the curves with the following equations:

a $y = (x + 1)(x - 3)$

b $y = (x + 1)^2(x + 5)$ ← Chapter 4

13.1 Integrating x^n

Integration is the reverse process of differentiation:

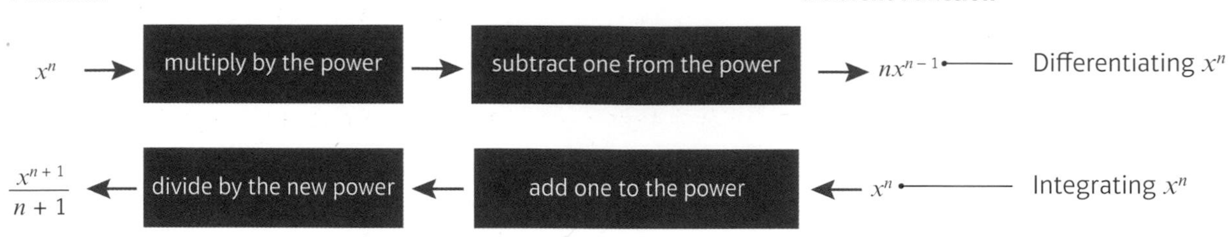

Constant terms disappear when you differentiate. This means that when you differentiate functions that only differ in the constant term, they will all differentiate to give the same function. To allow for this, you need to add a **constant of integration** at the end of a function when you integrate.

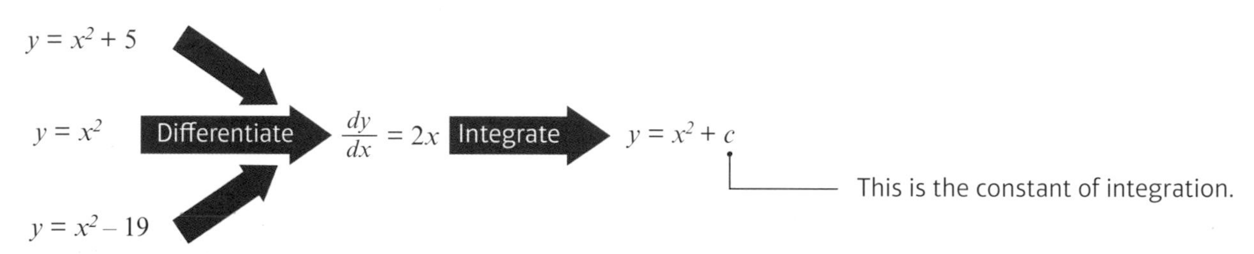

- **If $\frac{dy}{dx} = x^n$, then $y = \frac{1}{n+1}x^{n+1} + c, n \neq -1$.**
- **If $f'(x) = x^n$, then $f(x) = \frac{1}{n+1}x^{n+1} + c, n \neq -1$.**

Links You cannot use this rule if $n = -1$ because $\frac{1}{n+1} = \frac{1}{0}$ and so is not defined.
You will learn how to integrate the function x^{-1} in Year 2. → Year 2, Section 11.2

Example 1

Find y for the following:

a $\frac{dy}{dx} = x^4$ **b** $\frac{dy}{dx} = x^{-5}$

a $y = \frac{x^5}{5} + c$

Use $y = \frac{1}{n+1}x^{n+1} + c$ with $n = 4$.
Don't forget to add c.

b $y = \frac{x^{-4}}{-4} + c = -\frac{1}{4}x^{-4} + c$

Remember, adding 1 to the power gives $-5 + 1 = -4$.
Divide by the new power (-4) and add c.

Example 2

Find $f(x)$ for the following:

a $f'(x) = 3x^{\frac{1}{2}}$ **b** $f'(x) = 3$

a $f(x) = 3 \times \frac{x^{\frac{3}{2}}}{\frac{3}{2}} + c = 2x^{\frac{3}{2}} + c$

Remember $3 \div \frac{3}{2} = 3 \times \frac{2}{3} = 2$

Simplify your answer.

b $f'(x) = 3 = 3x^0$

So $f(x) = 3 \times \frac{x^1}{1} + c = 3x + c$

$x^0 = 1$, so 3 can be written as $3x^0$.

You can integrate a function in the form kx^n by integrating x^n and multiplying the integral by k.

- **If $\frac{dy}{dx} = kx^n$, then $y = \frac{k}{n+1}x^{n+1} + c, n \neq -1$.**
- **Using function notation, if $f'(x) = kx^n$, then $f(x) = \frac{k}{n+1}x^{n+1} + c, n \neq -1$.**
- **When integrating polynomials, apply the rule of integration separately to each term.**

Watch out You don't need to multiply the constant term (c) by k.

Example 3

Given $\frac{dy}{dx} = 6x + 2x^{-3} - 3x^{\frac{1}{2}}$, find y.

$y = \frac{6x^2}{2} + \frac{2}{-2}x^{-2} - \frac{3}{\frac{3}{2}}x^{\frac{3}{2}} + c$

$= 3x^2 - x^{-2} - 2x^{\frac{3}{2}} + c$

Apply the rule of integration to each term of the expression and add c.

Now simplify each term and remember to add c.

Exercise 13A

1 Find an expression for y when $\frac{dy}{dx}$ is the following:

a x^5 **b** $10x^4$ **c** $-x^{-2}$ **d** $-4x^{-3}$ **e** $x^{\frac{2}{3}}$ **f** $4x^{\frac{1}{2}}$

g $-2x^6$ **h** $x^{-\frac{1}{2}}$ **i** $5x^{-\frac{3}{2}}$ **j** $6x^{\frac{1}{3}}$ **k** $36x^{11}$ **l** $-14x^{-8}$

m $-3x^{-\frac{2}{3}}$ **n** -5 **o** $6x$ **p** $2x^{-0.4}$

2 Find y when $\frac{dy}{dx}$ is given by the following expressions. In each case simplify your answer.

a $x^3 - \frac{3}{2}x^{-\frac{1}{2}} - 6x^{-2}$ **b** $4x^3 + x^{-\frac{2}{3}} - x^{-2}$ **c** $4 - 12x^{-4} + 2x^{-\frac{1}{2}}$

d $5x^{\frac{2}{3}} - 10x^4 + x^{-3}$ **e** $-\frac{4}{3}x^{-\frac{4}{3}} - 3 + 8x$ **f** $5x^4 - x^{-\frac{3}{2}} - 12x^{-5}$

3 Find $f(x)$ when $f'(x)$ is given by the following expressions. In each case simplify your answer.

a $12x + \frac{3}{2}x^{-\frac{3}{2}} + 5$ **b** $6x^5 + 6x^{-7} - \frac{1}{6}x^{-\frac{7}{6}}$ **c** $\frac{1}{2}x^{-\frac{1}{2}} - \frac{1}{2}x^{-\frac{3}{2}}$

d $10x^4 + 8x^{-3}$ **e** $2x^{-\frac{1}{3}} + 4x^{-\frac{5}{3}}$ **f** $9x^2 + 4x^{-3} + \frac{1}{4}x^{-\frac{1}{2}}$

(E/P) **4** Find y given that $\frac{dy}{dx} = (2x + 3)^2$. **(4 marks)**

Problem-solving Start by expanding the brackets.

E **5** Find f(x) given that $f'(x) = 3x^{-2} + 6x^{\frac{1}{2}} + x - 4$. **(4 marks)**

Challenge

Find y when $\frac{dy}{dx} = (2\sqrt{x} - x^2)\left(\frac{3 + x}{x^5}\right)$

13.2 Indefinite integrals

You can use the symbol $\int$ to represent the process of **integration**.

- $\int f'(x)dx = f(x) + c$

You can write the process of integrating x^n as follows:

$$\int x^n dx = \frac{x^{n+1}}{n+1} + c, \quad n \neq -1$$

The elongated S means integrate.

The expression to be integrated.

The dx tells you to integrate with respect to x.

Notation This process is called **indefinite integration**. You will learn about **definite** integration later in this chapter.

When you are integrating a polynomial function, you can integrate the terms one at a time.

- $\int(f(x) + g(x))dx = \int f(x)dx + \int g(x)dx$

Example 4

Find:

a $\int(x^{\frac{1}{2}} + 2x^3)\,dx$ **b** $\int(x^{-\frac{3}{2}} + 2)\,dx$ **c** $\int(p^2x^{-2} + q)\,dx$ **d** $\int(4t^2 + 6)\,dt$

a $\int(x^{\frac{1}{2}} + 2x^3)dx = \frac{x^{\frac{3}{2}}}{\frac{3}{2}} + \frac{2x^4}{4} + c$

First apply the rule term by term.

$= \frac{2}{3}x^{\frac{3}{2}} + \frac{1}{2}x^4 + c$

Simplify each term.

b $\int(x^{-\frac{3}{2}} + 2)dx = \frac{x^{-\frac{1}{2}}}{-\frac{1}{2}} + 2x + c$

$= -2x^{-\frac{1}{2}} + 2x + c$

Remember $-\frac{3}{2} + 1 = -\frac{1}{2}$ and the integral of the constant 2 is $2x$.

c $\int(p^2x^{-2} + q)dx = \frac{p^2}{-1}x^{-1} + qx + c$

$= -p^2x^{-1} + qx + c$

The dx tells you to integrate with respect to the variable x, so any other letters must be treated as constants.

d $\int(4t^2 + 6)dt = \frac{4t^3}{3} + 6t + c$

The dt tells you that this time you must integrate with respect to t.

Use the rule for integrating x^n but replace x with t:

If $\frac{dy}{dt} = kt^n$, then $y = \frac{k}{n+1}t^{n+1} + c, n \neq -1$.

Before you integrate, you need to ensure that each term of the expression is in the form kx^n, where k and n are real numbers.

Example 5

Find:

a $\int\left(\frac{2}{x^3} - 3\sqrt{x}\right)dx$ **b** $\int x\left(x^2 + \frac{2}{x}\right)dx$ **c** $\int\left((2x)^2 + \frac{\sqrt{x}+5}{x^2}\right)dx$

a $\int\left(\frac{2}{x^3} - 3\sqrt{x}\right)dx$

$= \int(2x^{-3} - 3x^{\frac{1}{2}})dx$ — First write each term in the form x^n.

$= \frac{2}{-2}x^{-2} - \frac{3}{\frac{3}{2}}x^{\frac{3}{2}} + c$ — Apply the rule term by term.

$= -x^{-2} - 2x^{\frac{3}{2}} + c$ — Simplify each term.

$= -\frac{1}{x^2} - 2\sqrt{x^3} + c$ — Sometimes it is helpful to write the answer in the same form as the question.

b $\int x\left(x^2 + \frac{2}{x}\right)dx$

$= \int(x^3 + 2)dx$ — First multiply out the bracket.

$= \frac{x^4}{4} + 2x + c$ — Then apply the rule to each term.

c $\int\left((2x)^2 + \frac{\sqrt{x}+5}{x^2}\right)dx$

$= \int\left(4x^2 + \frac{x^{\frac{1}{2}}}{x^2} + \frac{5}{x^2}\right)dx$ — Simplify $(2x)^2$ and write $\sqrt{x}$ as $x^{\frac{1}{2}}$.

$= \int(4x^2 + x^{-\frac{3}{2}} + 5x^{-2})dx$ — Write each term in the form x^n.

$= \frac{4}{3}x^3 + \frac{x^{-\frac{1}{2}}}{-\frac{1}{2}} + \frac{5x^{-1}}{-1} + c$ — Apply the rule term by term.

$= \frac{4}{3}x^3 - 2x^{-\frac{1}{2}} - 5x^{-1} + c$

$= \frac{4}{3}x^3 - \frac{2}{\sqrt{x}} - \frac{5}{x} + c$ — Finally simplify the answer.

Exercise 13B

1 Find the following integrals:

a $\int x^3\,dx$ **b** $\int x^7\,dx$ **c** $\int 3x^{-4}\,dx$ **d** $\int 5x^2\,dx$

2 Find the following integrals:

a $\int(x^4 + 2x^3)dx$ **b** $\int(2x^3 - x^2 + 5x)dx$ **c** $\int(5x^{\frac{3}{2}} - 3x^2)dx$

3 Find the following integrals:

a $\int(4x^{-2} + 3x^{-\frac{1}{2}})dx$ **b** $\int(6x^{-2} - x^{\frac{1}{2}})dx$ **c** $\int(2x^{-\frac{3}{2}} + x^2 - x^{-\frac{1}{2}})dx$

4 Find the following integrals:

a $\int(4x^3 - 3x^{-4} + r)\mathrm{d}x$

b $\int(x + x^{-\frac{1}{2}} + x^{-\frac{3}{2}})\mathrm{d}x$

c $\int(px^4 + 2q + 3x^{-2})\mathrm{d}x$

Hint In **Q4** part **c** you are integrating with respect to x, so treat p and q as constants.

5 Find the following integrals:

a $\int(3t^2 - t^{-2})\mathrm{d}t$

b $\int(2t^2 - 3t^{-\frac{3}{2}} + 1)\mathrm{d}t$

c $\int(pt^3 + q^2 + pr^3)\mathrm{d}t$

6 Find the following integrals:

a $\int\frac{(2x^3 + 3)}{x^2}\mathrm{d}x$

b $\int(2x + 3)^2\,\mathrm{d}x$

c $\int(2x + 3)\sqrt{x}\,\mathrm{d}x$

7 Find $\int\mathrm{f}(x)\mathrm{d}x$ when $\mathrm{f}(x)$ is given by the following:

a $\left(x + \frac{1}{x}\right)^2$

b $(\sqrt{x} + 2)^2$

c $\left(\frac{1}{\sqrt{x}} + 2\sqrt{x}\right)$

8 Find the following integrals:

a $\int\left(x^{\frac{2}{3}} + \frac{4}{x^3}\right)\mathrm{d}x$

b $\int\left(\frac{2 + x}{x^3} + 3\right)\mathrm{d}x$

c $\int(x^2 + 3)(x - 1)\mathrm{d}x$

d $\int\frac{(2x + 1)^2}{\sqrt{x}}\mathrm{d}x$

e $\int\left(3 + \frac{\sqrt{x} + 6x^3}{x}\right)\mathrm{d}x$

f $\int\sqrt{x}(\sqrt{x} + 3)^2\,\mathrm{d}x$

9 Find the following integrals:

a $\int\left(\frac{A}{x^2} - 3\right)\mathrm{d}x$

b $\int\left(\sqrt{Px} + \frac{2}{x^3}\right)\mathrm{d}x$

c $\int\left(\frac{p}{x^2} + q\sqrt{x} + r\right)\mathrm{d}x$

(E) **10** Given that $\mathrm{f}(x) = \frac{6}{x^2} + 4\sqrt{x} - 3x + 2$, $x > 0$, find $\int\mathrm{f}(x)\mathrm{d}x$. **(5 marks)**

(E) **11** Find $\int\left(8x^3 + 6x - \frac{3}{\sqrt{x}}\right)\mathrm{d}x$, giving each term in its simplest form. **(4 marks)**

(E/P) **12** **a** Show that $(2 + 5\sqrt{x})^2$ can be written as $4 + k\sqrt{x} + 25x$, where k is a constant to be found. **(2 marks)**

b Hence find $\int(2 + 5\sqrt{x})^2\,\mathrm{d}x$. **(3 marks)**

(E) **13** Given that $y = 3x^5 - \frac{4}{\sqrt{x}}$, $x > 0$, find $\int y\,\mathrm{d}x$ in its simplest form. **(3 marks)**

(E/P) **14** $\int\left(\frac{p}{2x^2} + pq\right)\mathrm{d}x = \frac{2}{x} + 10x + c$ **(5 marks)**

Find the value of p and the value of q.

Problem-solving

Integrate the expression on the left-hand side, treating p and q as constants, then compare the result with the right-hand side.

15 $f(x) = (2 - x)^{10}$

Given that x is small, and so terms in x^3 and higher powers of x can be ignored:

a find an approximation for $f(x)$ in the form $A + Bx + Cx^2$ **(3 marks)**

b find an approximation for $\int f(x)dx$. **(3 marks)**

Hint Find the first three terms of the binomial expansion of $(2 - x)^{10}$. ← Section 8.3

13.3 Finding functions

You can find the constant of integration, c, when you are given (i) any point (x, y) that the curve of the function passes through or (ii) any value that the function takes. For example, if $\frac{dy}{dx} = 3x^2$ then $y = x^3 + c$. There are infinitely many curves with this equation, depending on the value of c.

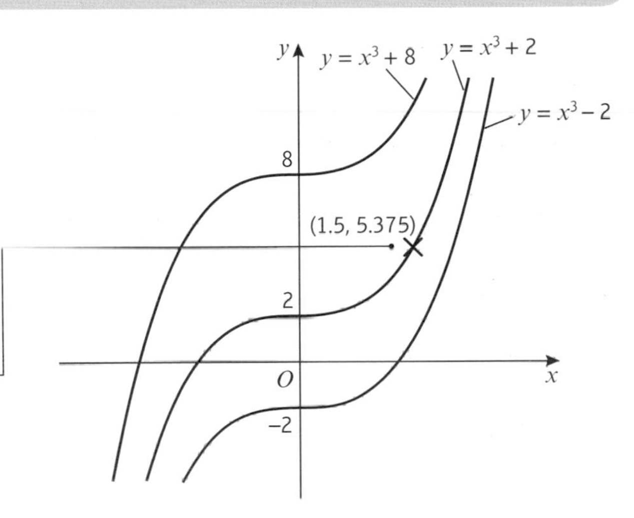

Only one of these curves passes through this point. Choosing a point on the curve determines the value of c.

- **To find the constant of integration, c**
  - **Integrate the function**
  - **Substitute the values (x, y) of a point on the curve, or the value of the function at a given point $f(x) = k$, into the integrated function**
  - **Solve the equation to find c**

Example 6

The curve C with equation $y = f(x)$ passes through the point (4, 5). Given that $f'(x) = \frac{x^2 - 2}{\sqrt{x}}$, find the equation of C.

$f'(x) = \frac{x^2 - 2}{\sqrt{x}} = x^{\frac{3}{2}} - 2x^{-\frac{1}{2}}$ — First write $f'(x)$ in a form suitable for integration.

So $f(x) = \frac{x^{\frac{5}{2}}}{\frac{5}{2}} - \frac{2x^{\frac{1}{2}}}{\frac{1}{2}} + c$ — Integrate as normal and don't forget the $+ c$.

$= \frac{2}{5}x^{\frac{5}{2}} - 4x^{\frac{1}{2}} + c$

But $f(4) = 5$ — Use the fact that the curve passes through (4, 5).

So $\quad 5 = \frac{2}{5} \times 2^5 - 4 \times 2 + c$

$5 = \frac{64}{5} - 8 + c$

$5 = \frac{24}{5} + c$

So $\quad c = \frac{1}{5}$

So $y = \frac{2}{5}x^{\frac{5}{2}} - 4x^{\frac{1}{2}} + \frac{1}{5}$

Remember $4^{\frac{5}{2}} = 2^5$.

Solve for c.

Finally write down the equation of the curve.

Online Explore the solution using technology.

Exercise 13C

1 Find the equation of the curve with the given derivative of y with respect to x that passes through the given point:

a $\frac{dy}{dx} = 3x^2 + 2x$; point (2, 10)

b $\frac{dy}{dx} = 4x^3 + \frac{2}{x^3} + 3$; point (1, 4)

c $\frac{dy}{dx} = \sqrt{x} + \frac{1}{4}x^2$; point (4, 11)

d $\frac{dy}{dx} = \frac{3}{\sqrt{x}} - x$; point (4, 0)

e $\frac{dy}{dx} = (x + 2)^2$; point (1, 7)

f $\frac{dy}{dx} = \frac{x^2 + 3}{\sqrt{x}}$; point (0, 1)

2 The curve C, with equation $y = f(x)$, passes through the point (1, 2) and $f'(x) = 2x^3 - \frac{1}{x^2}$. Find the equation of C in the form $y = f(x)$.

3 The gradient of a particular curve is given by $\frac{dy}{dx} = \frac{\sqrt{x} + 3}{x^2}$. Given that the curve passes through the point (9, 0), find an equation of the curve.

(E) **4** The curve with equation $y = f(x)$ passes through the point $(-1, 0)$. Given that $f'(x) = 9x^2 + 4x - 3$, find $f(x)$. **(5 marks)**

(E/P) **5** $\frac{dy}{dx} = 3x^{-\frac{1}{2}} - 2x\sqrt{x}$, $x > 0$.

Given that $y = 10$ at $x = 4$, find y in terms of x, giving each term in its simplest form. **(7 marks)**

(E/P) **6** Given that $\frac{6x + 5x^{\frac{3}{2}}}{\sqrt{x}}$ can be written in the form $6x^p + 5x^q$,

a write down the value of p and the value of q. **(2 marks)**

Given that $\frac{dy}{dx} = \frac{6x + 5x^{\frac{3}{2}}}{\sqrt{x}}$ and that $y = 100$ when $x = 9$,

b find y in terms of x, simplifying the coefficient of each term. **(5 marks)**

(P) 7 The displacement of a particle at time t is given by the function $f(t)$, where $f(0) = 0$.
Given that the velocity of the particle is given by $f'(t) = 10 - 5t$,

a find $f(t)$

b determine the displacement of the particle when $t = 3$.

Problem-solving

You don't need any specific knowledge of mechanics to answer this question. You are told that the displacement of the particle at time t is given by $f(t)$.

(P) 8 The height, in metres, of an arrow fired horizontally from the top of a castle is modelled by the function $f(t)$, where $f(0) = 35$. Given that $f'(t) = -9.8t$,

a find $f(t)$.

b determine the height of the arrow when $t = 1.5$.

c write down the height of the castle according to this model.

d estimate the time it will take the arrow to hit the ground.

e state one assumption used in your calculation.

Challenge

1 A set of curves, where each curve passes through the origin, has equations $y = f_1(x)$, $y = f_2(x)$, $y = f_3(x)$ ... where $f_n'(x) = f_{n-1}(x)$ and $f_1(x) = x^2$.

a Find $f_2(x)$, $f_3(x)$.

b Suggest an expression for $f_n(x)$.

2 A set of curves, with equations $y = f_1(x)$, $y = f_2(x)$, $y = f_3(x)$, ... all pass through the point (0, 1) and they are related by the property $f_n'(x) = f_{n-1}(x)$ and $f_1(x) = 1$. Find $f_2(x)$, $f_3(x)$, $f_4(x)$.

13.4 Definite integrals

You can calculate an integral between two **limits**. This is called a **definite integral**. A definite integral usually produces a **value** whereas an indefinite integral always produces a **function**.

Here are the steps for integrating the function $3x^2$ between the limits $x = 1$ and $x = 2$.

$$\int_1^2 3x^2\,dx = [x^3]_1^2$$
$$= (2^3) - (1^3)$$
$$= 8 - 1$$
$$= 7$$

The limits of the integral are from $x = 1$ to $x = 2$.

Write the integral in [ ] brackets.

Write this step in ( ) brackets.

Evaluate the integral at the upper limit.

Evaluate the integral at the lower limit.

There are three stages when you work out a definite integral:

Write the definite integral **statement** with its limits, a and b.	**Integrate**, and write the integral in square brackets	**Evaluate** the definite integral by working out $f(b) - f(a)$.
$\int_a^b \ldots dx$	$[\ldots]_a^b$	$(\ldots) - (\ldots)$

- **If $f'(x)$ is the derivative of $f(x)$ for all values of x in the interval $[a, b]$, then the definite integral is defined as $\int_a^b f'(x)dx = [f(x)]_a^b = f(b) - f(a)$.**

Problem-solving

The relationship between the derivative and the integral is called the **fundamental theorem of calculus**.

Example 

Evaluate

$$\int_0^1 (x^{\frac{1}{3}} - 1)^2 \, dx$$

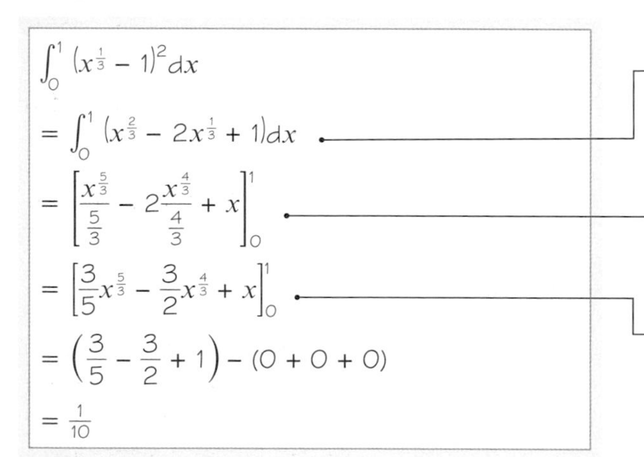

$$\int_0^1 (x^{\frac{1}{3}} - 1)^2 dx$$
$$= \int_0^1 (x^{\frac{2}{3}} - 2x^{\frac{1}{3}} + 1)dx$$
$$= \left[\frac{x^{\frac{5}{3}}}{\frac{5}{3}} - 2\frac{x^{\frac{4}{3}}}{\frac{4}{3}} + x\right]_0^1$$
$$= \left[\frac{3}{5}x^{\frac{5}{3}} - \frac{3}{2}x^{\frac{4}{3}} + x\right]_0^1$$
$$= \left(\frac{3}{5} - \frac{3}{2} + 1\right) - (0 + 0 + 0)$$
$$= \frac{1}{10}$$

First multiply out the bracket to put the expression in a form ready to be integrated.

For definite integrals you don't need to include $+c$ in your square brackets.

Simplify each term.

Example 8

Given that P is a constant and $\int_1^5 (2Px + 7)dx = 4P^2$, show that there are two possible values for P and find these values.

$$\int_1^5 (2Px + 7)dx = [Px^2 + 7x]_1^5$$
$$= (25P + 35) - (P + 7)$$
$$= 24P + 28$$
$$24P + 28 = 4P^2$$
$$4P^2 - 24P - 28 = 0$$
$$P^2 - 6P - 7 = 0$$
$$(P + 1)(P - 7) = 0$$
$$P = -1 \text{ or } 7$$

Problem-solving

You are integrating with respect to x so treat P as a constant. Find the definite integral in terms of P then set it equal to $4P^2$. The fact that the question asks for 'two possible values' gives you a clue that the resulting equation will be quadratic.

Divide every term by 4 to simplify.

Exercise 13D

1 Evaluate the following definite integrals:

a $\int_2^5 x^3\,dx$ b $\int_1^3 x^4\,dx$

c $\int_0^4 \sqrt{x}\,dx$ d $\int_1^3 \frac{3}{x^2}\,dx$

Watch out You must not use a calculator to work out definite integrals in your exam. You need to use calculus and show clear algebraic working.

2 Evaluate the following definite integrals:

a $\int_1^2 \left(\frac{2}{x^3} + 3x\right)dx$ b $\int_0^2 (2x^3 - 4x + 5)\,dx$ c $\int_4^9 \left(\sqrt{x} - \frac{6}{x^2}\right)dx$ d $\int_1^8 (x^{-\frac{1}{3}} + 2x - 1)\,dx$

3 Evaluate the following definite integrals:

a $\int_1^3 \frac{x^3 + 2x^2}{x}\,dx$ b $\int_3^6 \left(x - \frac{3}{x}\right)^2 dx$ c $\int_0^1 x^2\left(\sqrt{x} + \frac{1}{x}\right)dx$ d $\int_1^4 \frac{2 + \sqrt{x}}{x^2}\,dx$

E/P 4 Given that A is a constant and $\int_1^4 (6\sqrt{x} - A)\,dx = A^2$, show that there are two possible values for A and find these values. **(5 marks)**

E 5 Use calculus to find the value of $\int_1^9 (2x - 3\sqrt{x})\,dx$. **(5 marks)**

E 6 Evaluate $\int_4^{12} \frac{2}{\sqrt{x}}\,dx$, giving your answer in the form $a + b\sqrt{3}$, where a and b are integers. **(4 marks)**

E/P 7 Given that $\int_1^k \frac{1}{\sqrt{x}}\,dx = 3$, calculate the value of k. **(4 marks)**

Problem-solving You might encounter a definite integral with an unknown in the limits. Here, you can find an expression for the definite integral in terms of k then set that expression equal to 3.

8 The speed, v ms^{-1}, of a train at time t seconds is given by $v = 20 + 5t$, $0 \leqslant t \leqslant 10$.

The distance, s metres, travelled by the train in 10 seconds is given by $s = \int_0^{10} (20 + 5t)\,dt$. Find the value of s.

Challenge

Given that $\int_k^{3k} \frac{3x + 2}{8}\,dx = 7$ and $k > 0$, calculate the value of k.

13.5 Areas under curves

Definite integration can be used to find the area under a curve.

For any curve with equation $y = f(x)$, you can define the area under the curve to the left of x as a function of x called $A(x)$. As x increases, this area $A(x)$ also increases (since x moves further to the right).

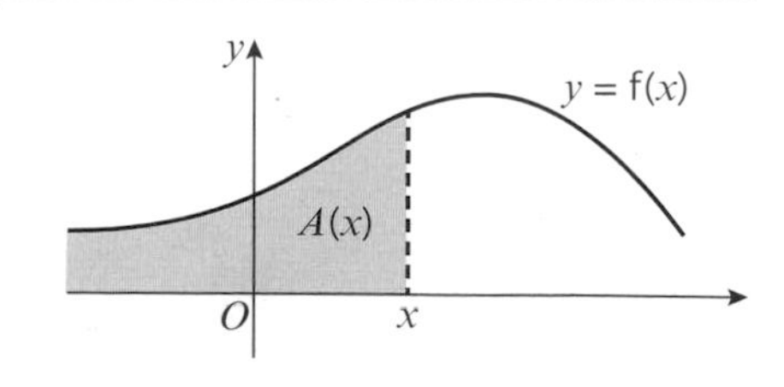

If you look at a small increase in x, say δx, then the area increases by an amount $\delta A = A(x + \delta x) - A(x)$.

This increase in the δA is approximately rectangular and of magnitude $y\delta x$. (As you make δx smaller any error between the actual area and this will be negligible.)

This vertical height will be $y = \mathrm{f}(x)$.

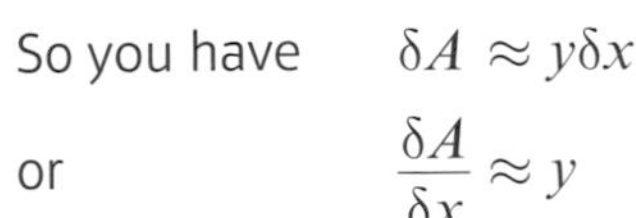

So you have $\quad \delta A \approx y\delta x$

or $\quad \dfrac{\delta A}{\delta x} \approx y$

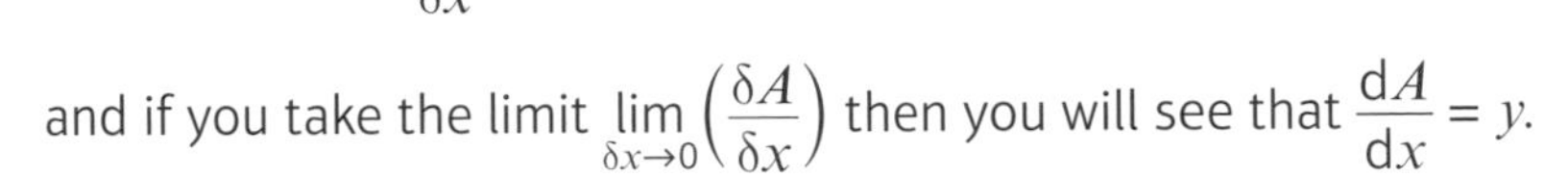

and if you take the limit $\lim\limits_{\delta x \to 0}\left(\dfrac{\delta A}{\delta x}\right)$ then you will see that $\dfrac{\mathrm{d}A}{\mathrm{d}x} = y$.

Now if you know that $\dfrac{\mathrm{d}A}{\mathrm{d}x} = y$, then to find A you have to integrate, giving $A = \int y\,\mathrm{d}x$.

- **The area between a positive curve, the x-axis and the lines $x = a$ and $x = b$ is given by**

  Area = $\displaystyle\int_a^b y\,\mathrm{d}x$

  where $y = \mathrm{f}(x)$ is the equation of the curve.

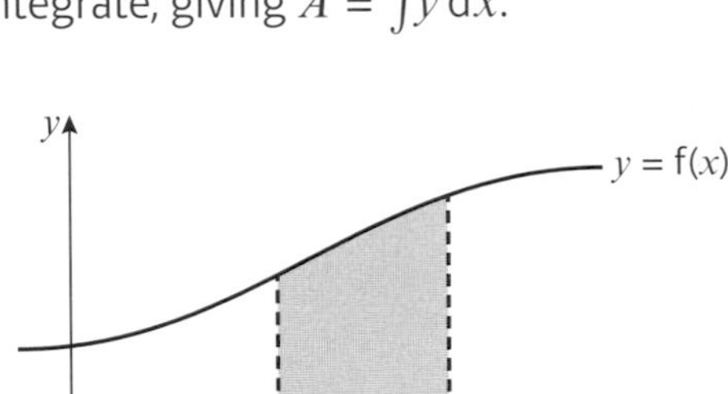

Example 9

Find the area of the finite region between the curve with equation $y = 20 - x - x^2$ and the x-axis.

$y = 20 - x - x^2 = (4 - x)(5 + x)$

Factorise the expression.

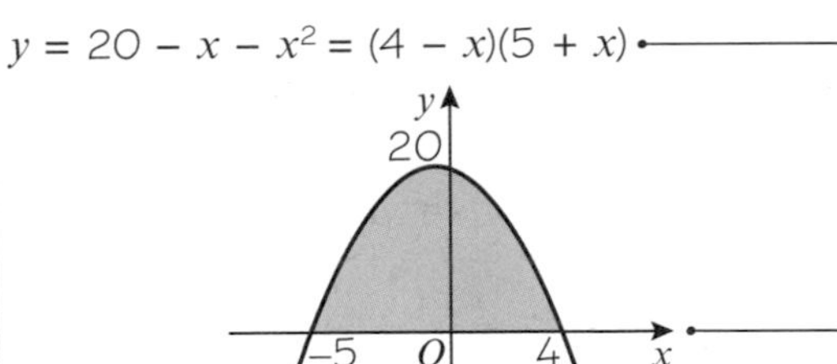

Draw a sketch of the graph. $x = 4$ and $x = -5$ are the points of intersection of the curve and the x-axis.

$$\text{Area} = \int_{-5}^{4}(20 - x - x^2)\mathrm{d}x$$
$$= \left[20x - \frac{x^2}{2} - \frac{x^3}{3}\right]_{-5}^{4}$$
$$= \left(80 - 8 - \frac{64}{3}\right) - \left(-100 - \frac{25}{2} + \frac{125}{3}\right)$$
$$= \frac{243}{2}$$

You don't normally need to give units when you are finding areas on graphs.

Exercise 13E

1 Find the area between the curve with equation $y = f(x)$, the x-axis and the lines $x = a$ and $x = b$ in each of the following cases:

a $f(x) = -3x^2 + 17x - 10$; $a = 1$, $b = 3$

b $f(x) = 2x^3 + 7x^2 - 4x$; $a = -3$, $b = -1$

c $f(x) = -x^4 + 7x^3 - 11x^2 + 5x$; $a = 0$, $b = 4$

d $f(x) = \frac{8}{x^2}$; $a = -4$, $b = -1$

Hint For part **c**, $f(x) = -x(x - 1)^2(x - 5)$

2 The sketch shows part of the curve with equation $y = x(x^2 - 4)$.
Find the area of the shaded region.

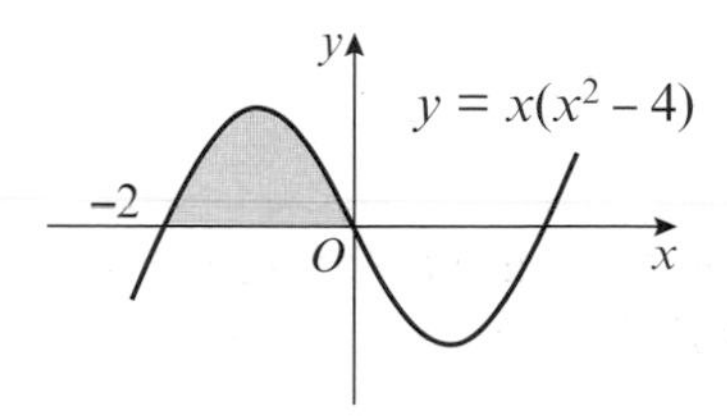

3 The diagram shows a sketch of the curve with equation $y = 3x + \frac{6}{x^2} - 5$, $x > 0$.
The region R is bounded by the curve, the x-axis and the lines $x = 1$ and $x = 3$.
Find the area of R.

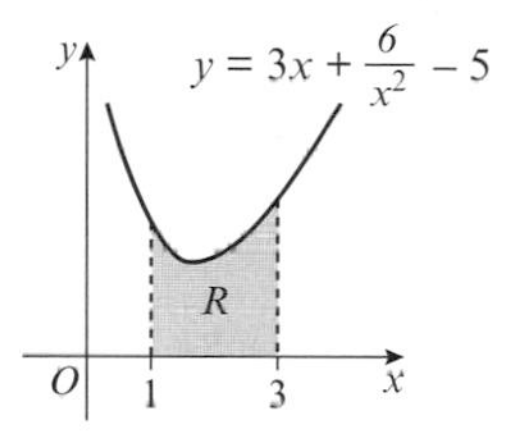

4 Find the area of the finite region between the curve with equation $y = (3 - x)(1 + x)$ and the x-axis.

5 Find the area of the finite region between the curve with equation $y = x(x - 4)^2$ and the x-axis.

(P) 6 Find the area of the finite region between the curve with equation $y = 2x^2 - 3x^3$ and the x-axis.

(P) 7 The shaded area under the graph of the function $f(x) = 3x^2 - 2x + 2$, bounded by the curve, the x-axis and the lines $x = 0$ and $x = k$, is 8. Work out the value of k.

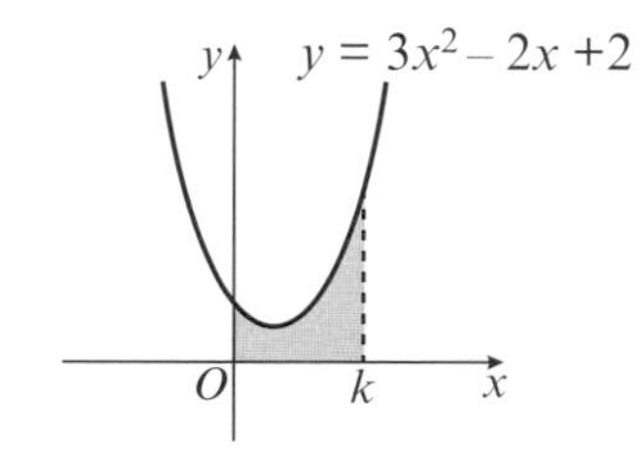

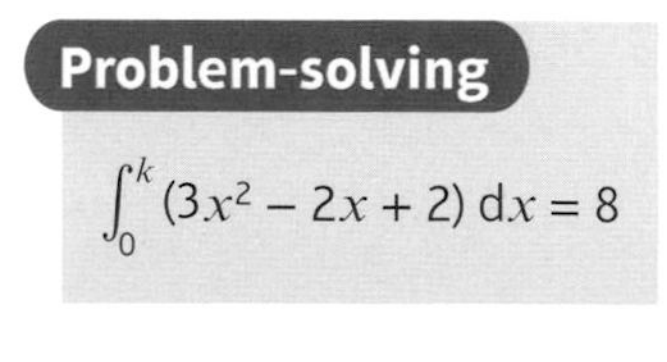

(E) 8 The finite region R is bounded by the x-axis and the curve with equation $y = -x^2 + 2x + 3$, $x \geqslant 0$.
The curve meets the x-axis at points A and B.

a Find the coordinates of point A and point B. **(2 marks)**

b Find the area of the region R. **(4 marks)**

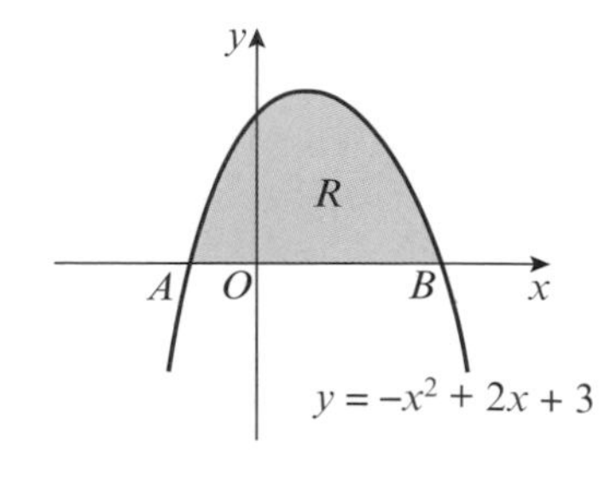

Ⓔ **9** The graph shows part of the curve C with equation $y = x^2(2 - x)$.

The region R, shown shaded, is bounded by C and the x-axis.

Use calculus to find the exact area of R. **(5 marks)**

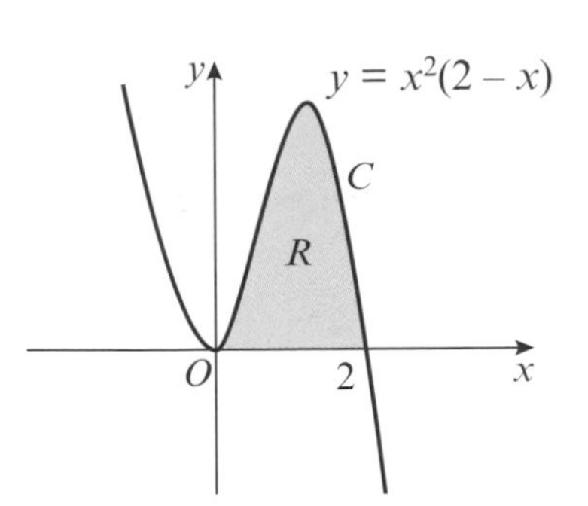

Watch out If a question says "use calculus" then you need to use integration or differentiation, and show clear algebraic working.

13.6 Areas under the x-axis

You need to be careful when you are finding areas below the x-axis.

- **When the area bounded by a curve and the x-axis is below the x-axis, $\int y\,dx$ gives a negative answer.**

Example 10

Find the area of the finite region bounded by the curve $y = x(x - 3)$ and the x-axis.

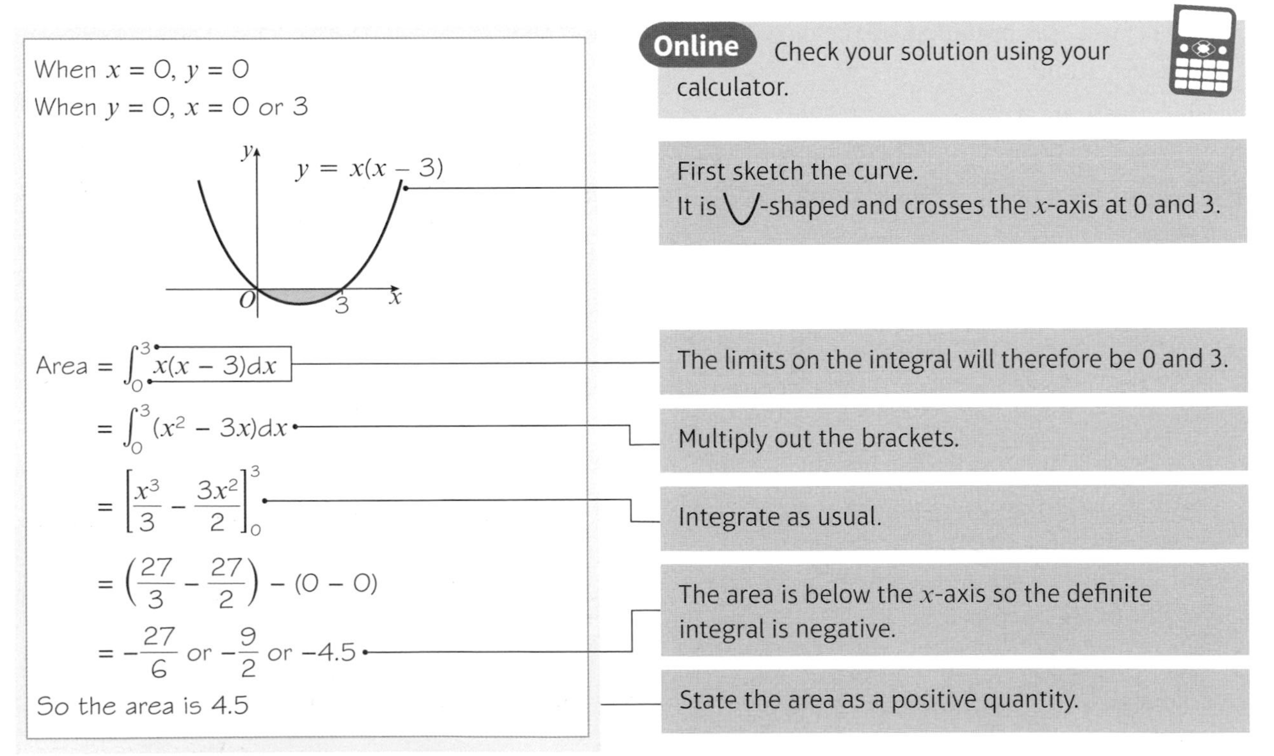

When $x = 0$, $y = 0$

When $y = 0$, $x = 0$ or 3

Area $= \int_0^3 x(x-3)\,dx$

$= \int_0^3 (x^2 - 3x)\,dx$

$= \left[\frac{x^3}{3} - \frac{3x^2}{2}\right]_0^3$

$= \left(\frac{27}{3} - \frac{27}{2}\right) - (0 - 0)$

$= -\frac{27}{6}$ or $-\frac{9}{2}$ or -4.5

So the area is 4.5

Online Check your solution using your calculator.

First sketch the curve. It is ⋃-shaped and crosses the x-axis at 0 and 3.

The limits on the integral will therefore be 0 and 3.

Multiply out the brackets.

Integrate as usual.

The area is below the x-axis so the definite integral is negative.

State the area as a positive quantity.

The following example shows that great care must be taken if you are trying to find an area which straddles the x-axis such as the shaded region.

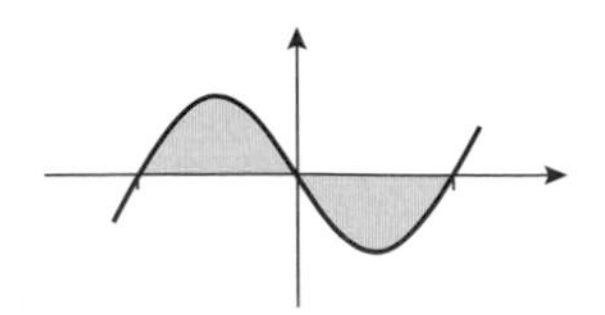

For examples of this type you need to draw a sketch, unless one is given in the question.

Example 11

Sketch the curve with equation $y = x(x - 1)(x + 3)$ and find the area of the finite region bounded by the curve and the x-axis.

When $x = 0$, $y = 0$
When $y = 0$, $x = 0$, 1 or -3

Find out where the curve cuts the axes.

$x \to \infty$, $y \to \infty$
$x \to -\infty$, $y \to -\infty$

Find out what happens to y when x is large and positive or large and negative.

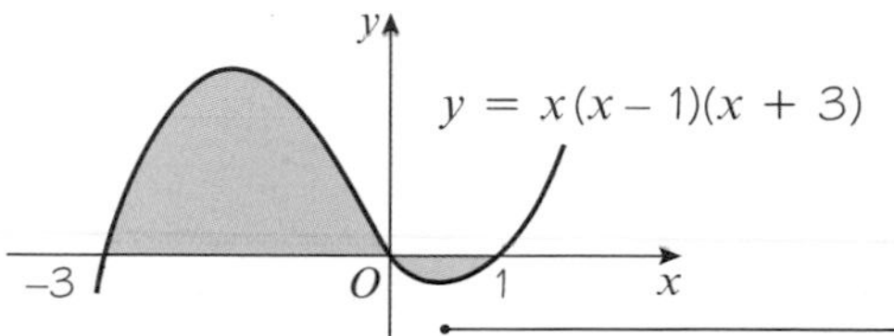

Problem-solving Always draw a sketch, and use the points of intersection with the x-axis as the limits for your integrals.

The area is given by $\int_{-3}^{0} y\,dx - \int_{0}^{1} y\,dx$

Since the area between $x = 0$ and 1 is below the axis the integral between these points will give a negative answer.

Now $\int y\,dx = \int (x^3 + 2x^2 - 3x)dx$

Multiply out the brackets.

$$= \left[\frac{x^4}{4} + \frac{2x^3}{3} - \frac{3x^2}{2}\right]$$

So $\int_{-3}^{0} y\,dx = (0) - \left(\frac{81}{4} - \frac{2}{3} \times 27 - \frac{3}{2} \times 9\right)$

$$= \frac{45}{4}$$

and $\int_{0}^{1} y\,dx = \left(\frac{1}{4} + \frac{2}{3} - \frac{3}{2}\right) - (0)$

$$= -\frac{7}{12}$$

So the area required is $\frac{45}{4} + \frac{7}{12} = \frac{71}{6}$

Watch out If you try to calculate the area as a single definite integral, the positive and negative areas will partly cancel each other out.

Exercise 13F

1 Sketch the following and find the total area of the finite region or regions bounded by the curves and the x-axis:

a $y = x(x + 2)$ b $y = (x + 1)(x - 4)$ c $y = (x + 3)x(x - 3)$

d $y = x^2(x - 2)$ e $y = x(x - 2)(x - 5)$

(E) 2 The graph shows a sketch of part of the curve C with equation $y = x(x + 3)(2 - x)$.

The curve C crosses the x-axis at the origin O and at points A and B.

a Write down the x-coordinates of A and B. **(1 mark)**

The finite region, shown shaded, is bounded by the curve C and the x-axis.

b Use integration to find the total area of the finite shaded region. **(7 marks)**

3 $f(x) = -x^3 + 4x^2 + 11x - 30$

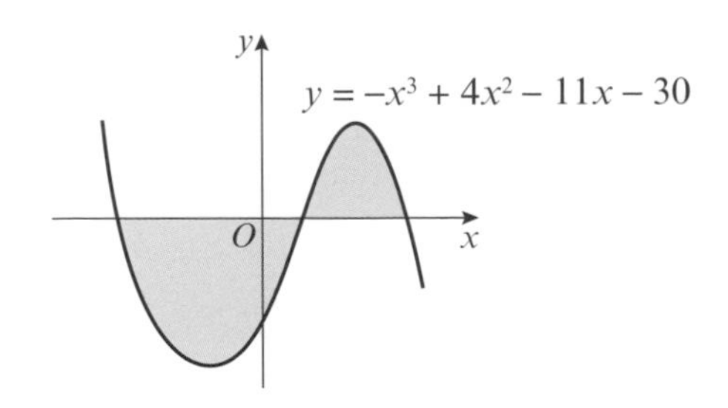

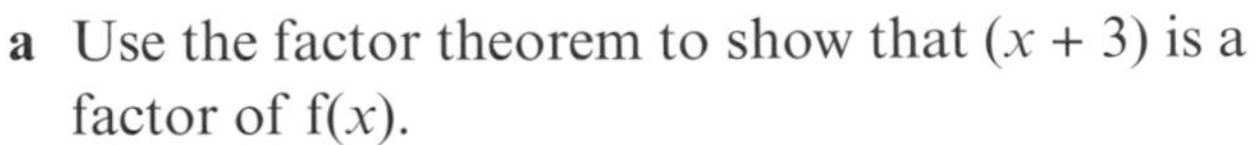
The graph shows a sketch of part of the curve with equation $y = -x^3 + 4x^2 + 11x - 30$.

a Use the factor theorem to show that $(x + 3)$ is a factor of $f(x)$.

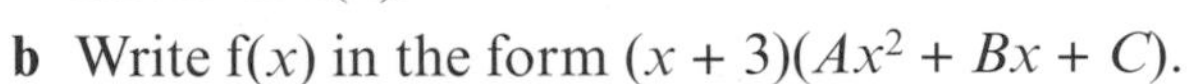
b Write $f(x)$ in the form $(x + 3)(Ax^2 + Bx + C)$.

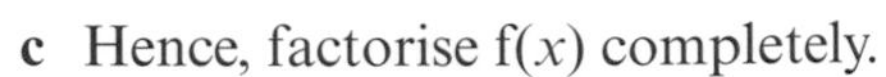
c Hence, factorise $f(x)$ completely.

d Hence, determine the x-coordinates where the curve intersects the x-axis.

e Hence, determine the total shaded area shown on the sketch.

Challenge

1 Given that $f(x) = x(3 - x)$, find the area of the finite region bounded by the x-axis and the curve with equation

a $y = f(x)$ **b** $y = 2f(x)$ **c** $y = af(x)$

d $y = f(x + a)$ **e** $y = f(ax)$.

2 The graph shows a sketch of part of the curve C with equation $y = x(x - 1)(x + 2)$. The curve C crosses the x-axis at the origin O and at point B. The shaded areas above and below the x-axis are equal.

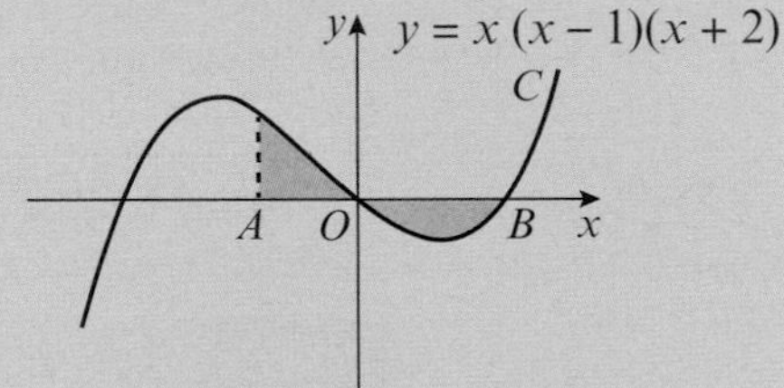

a Show that the x-coordinate of A satisfies the equation

$$(x - 1)^2(3x^2 + 10x + 5) = 0$$

b Hence find the exact coordinates of A, and interpret geometrically the other roots of this equation.

13.7 Areas between curves and lines

- **You can use definite integration together with areas of trapeziums and triangles to find more complicated areas on graphs.**

Example 12

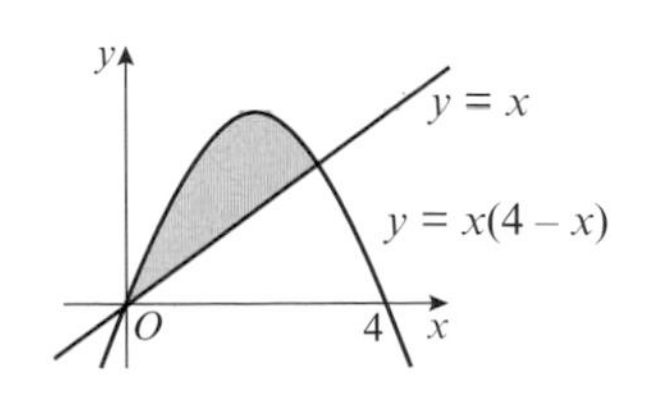

The diagram shows a sketch of part of the curve with equation $y = x(4 - x)$ and the line with equation $y = x$.

Find the area of the region bounded by the curve and the line.

$x(4 - x) = x$
$3x - x^2 = 0$
$x(3 - x) = 0$
$x = 0$ or 3

Area beneath curve $= \int_0^3 (4x - x^2)dx$

$= \left[2x^2 - \frac{x^3}{3}\right]_0^3$

$= 9$

Area beneath triangle $= \frac{1}{2} \times 3 \times 3$

$= \frac{9}{2}$

Shaded area $= 9 - \frac{9}{2} = \frac{9}{2}$

First, find the x-coordinate of the points of intersection of the curve $y = x(4 - x)$ and the line $y = x$.

Shaded area = area beneath curve – area beneath triangle

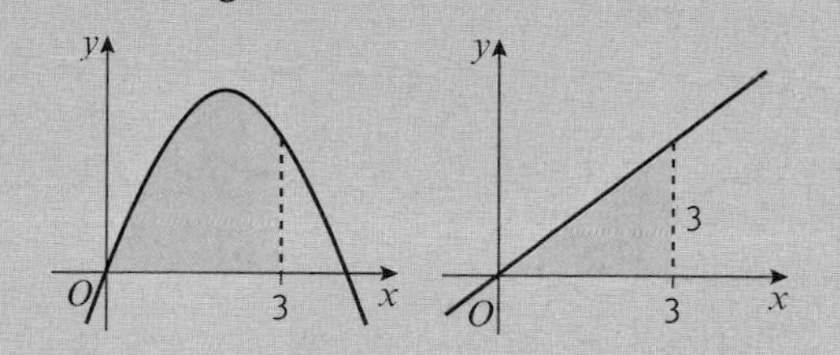

$\left[2x^2 - \frac{x^3}{3}\right]_0^3 = \left(18 - \frac{27}{3}\right) - (0 - 0) = 18 - 9$

Example 13

The diagram shows a sketch of the curve with equation $y = x(x - 3)$ and the line with equation $y = 2x$.

Find the area of the shaded region OAC.

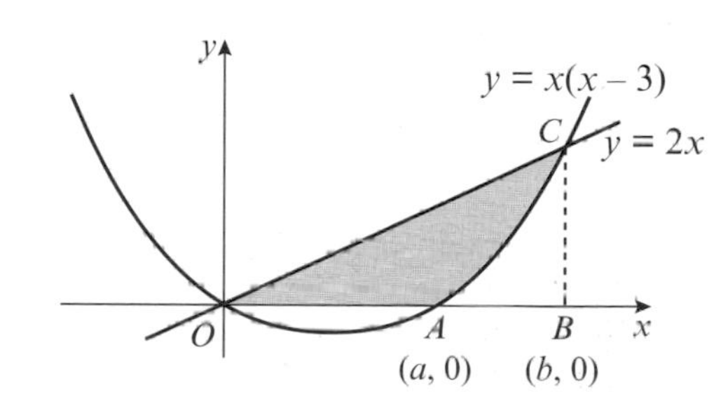

The required area is given by:

Area of triangle $OBC - \int_a^b x(x - 3)dx$

The curve cuts the x-axis at $x = 3$ (and $x = 0$) so $a = 3$.

The curve meets the line $y = 2x$ when $2x = x(x - 3)$.

So $0 = x^2 - 5x$

$0 = x(x - 5)$

$x = 0$ or 5, so $b = 5$

The point C is (5, 10).

Area of triangle $OBC = \frac{1}{2} \times 5 \times 10 = 25$.

Area between curve, x-axis and the line $x = 5$ is

$\int_3^5 x(x - 3)dx = \int_3^5 (x^2 - 3x)dx$

$= \left[\frac{x^3}{3} - \frac{3x^2}{2}\right]_3^5$

Problem-solving

Look for ways of combining triangles, trapeziums and direct integrals to find the missing area.

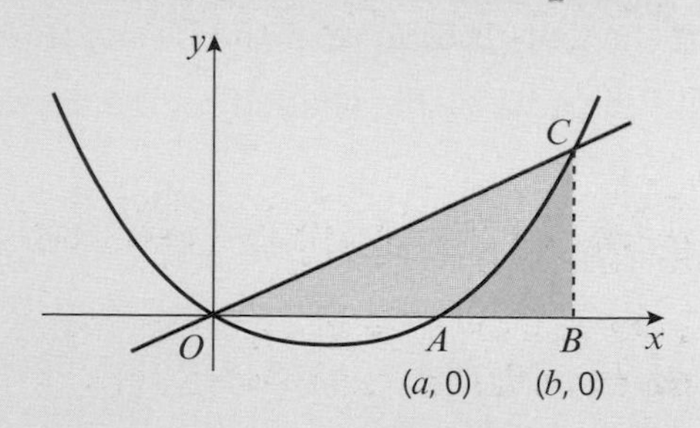

Substituting $x = 5$ into the equation of the line gives $y = 2 \times 5 = 10$.

Work out the definite integral separately. This will help you avoid making errors in your working.

$$= \left(\frac{125}{3} - \frac{75}{2}\right) - \left(\frac{27}{3} - \frac{27}{2}\right)$$

$$= \left(\frac{25}{6}\right) - \left(-\frac{27}{6}\right)$$

$$= \frac{26}{3}$$

Shaded region is therefore $= 25 - \frac{26}{3} = \frac{49}{3}$

Exercise 13G

1 The diagram shows part of the curve with equation $y = x^2 + 2$ and the line with equation $y = 6$. The line cuts the curve at the points A and B.

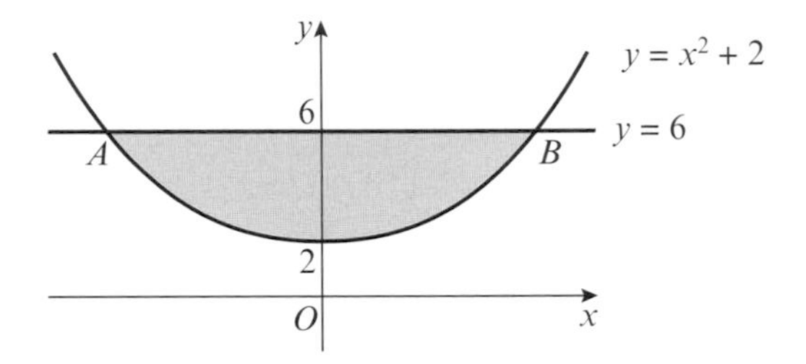

a Find the coordinates of the points A and B.

b Find the area of the finite region bounded by line AB and the curve.

2 The diagram shows the finite region, R, bounded by the curve with equation $y = 4x - x^2$ and the line $y = 3$. The line cuts the curve at the points A and B.

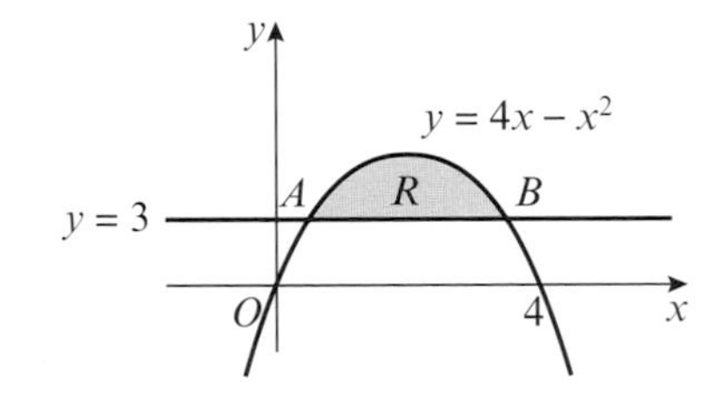

a Find the coordinates of the points A and B.

b Find the area of R.

P 3 The diagram shows a sketch of part of the curve with equation $y = 9 - 3x - 5x^2 - x^3$ and the line with equation $y = 4 - 4x$. The line cuts the curve at the points A $(-1, 8)$ and $B(1, 0)$.

Find the area of the shaded region between AB and the curve.

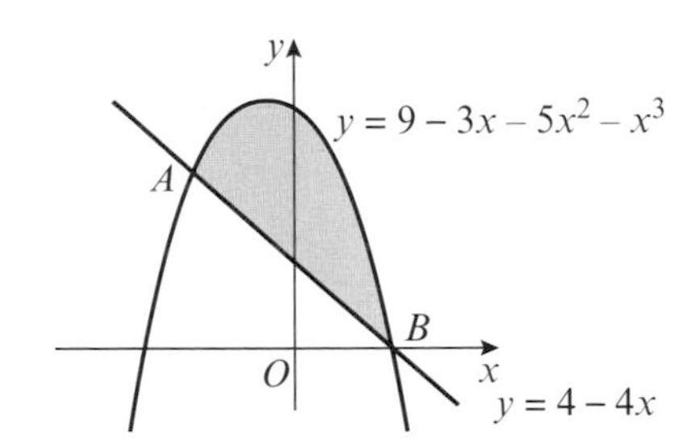

P 4 Find the area of the finite region bounded by the curve with equation $y = (1 - x)(x + 3)$ and the line $y = x + 3$.

5 The diagram shows the finite region, R, bounded by the curve with equation $y = x(4 + x)$, the line with equation $y = 12$ and the y-axis.

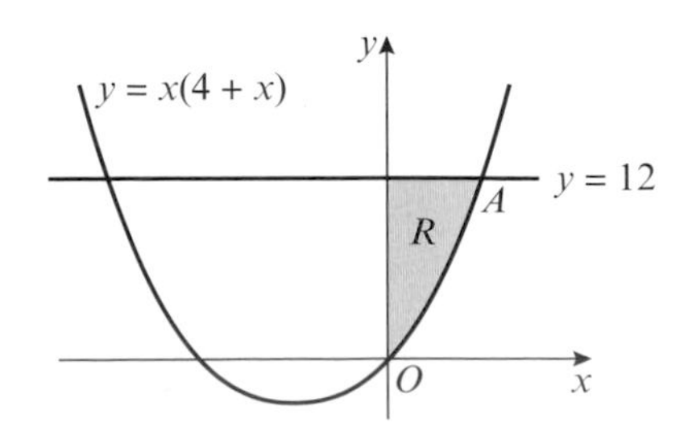

a Find the coordinates of the point A where the line meets the curve.

b Find the area of R.

6 The diagram shows a sketch of part of the curve with equation $y = x^2 + 1$ and the line with equation $y = 7 - x$. The finite region, R_1 is bounded by the line and the curve. The finite region, R_2 is below the curve and the line and is bounded by the positive x- and y-axes as shown in the diagram.

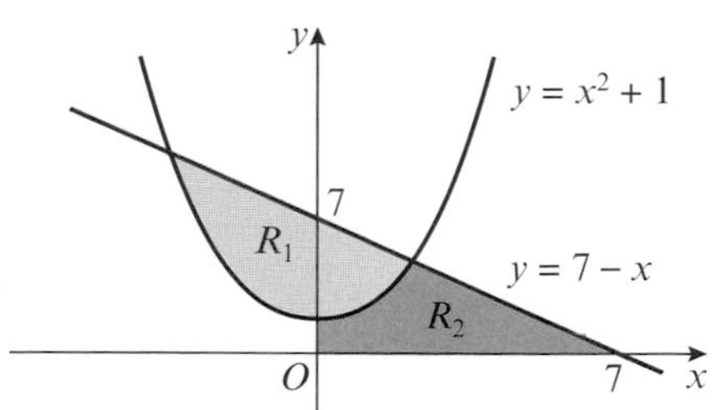

a Find the area of R_1.

b Find the area of R_2.

7 The curve C has equation $y = x^{\frac{2}{3}} - \dfrac{2}{x^{\frac{1}{3}}} + 1$.

a Verify that C crosses the x-axis at the point $(1, 0)$.

b Show that the point $A(8, 4)$ also lies on C.

c The point B is $(4, 0)$. Find the equation of the line through AB.

The finite region R is bounded by C, AB and the positive x-axis.

d Find the area of R.

8 The diagram shows part of a sketch of the curve with equation $y = \dfrac{2}{x^2} + x$. The points A and B have x-coordinates $\frac{1}{2}$ and 2 respectively.

Find the area of the finite region between AB and the curve.

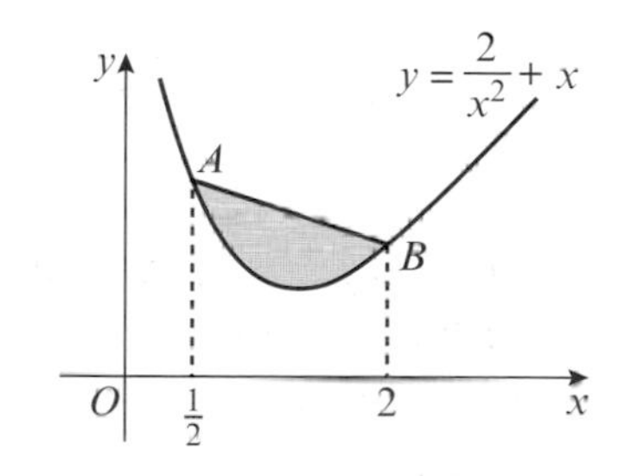

9 The diagram shows part of the curve with equation $y = 3\sqrt{x} - \sqrt{x^3} + 4$ and the line with equation $y = 4 - \frac{1}{2}x$.

a Verify that the line and the curve cross at the point $A(4, 2)$.

b Find the area of the finite region bounded by the curve and the line.

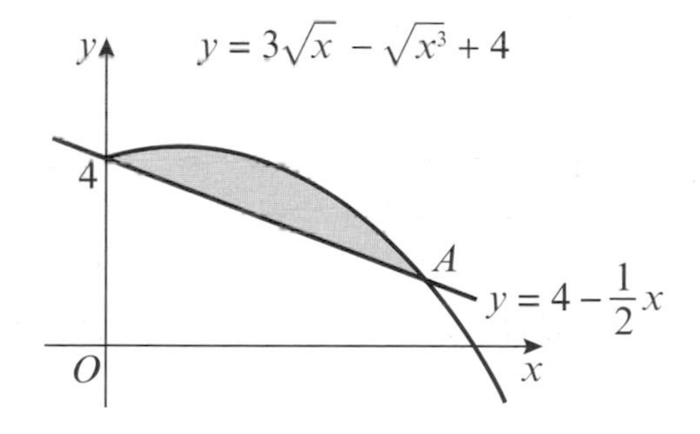

10 The sketch shows part of the curve with equation $y = x^2(x + 4)$. The finite region R_1 is bounded by the curve and the negative x-axis. The finite region R_2 is bounded by the curve, the positive x-axis and AB, where $A(2, 24)$ and $B(b, 0)$.

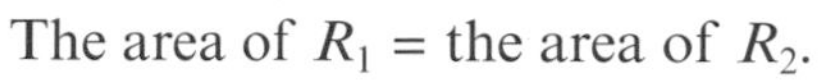
The area of R_1 = the area of R_2.

a Find the area of R_1.

b Find the value of b.

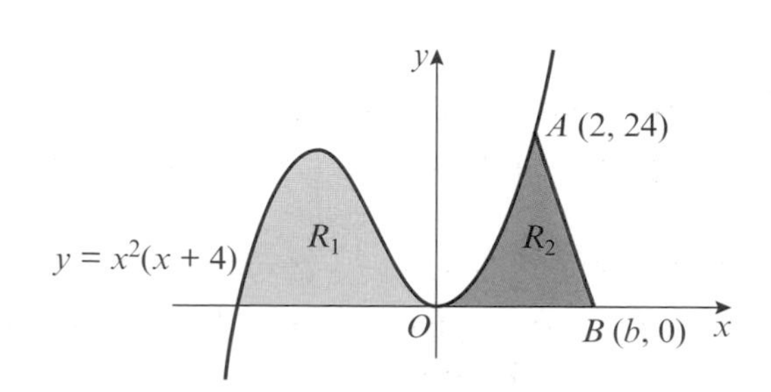

Problem-solving

Split R_2 into two areas by drawing a vertical line at $x = 2$.

11 The line with equation $y = 10 - x$ cuts the curve with equation $y = 2x^2 - 5x + 4$ at the points A and B, as shown.

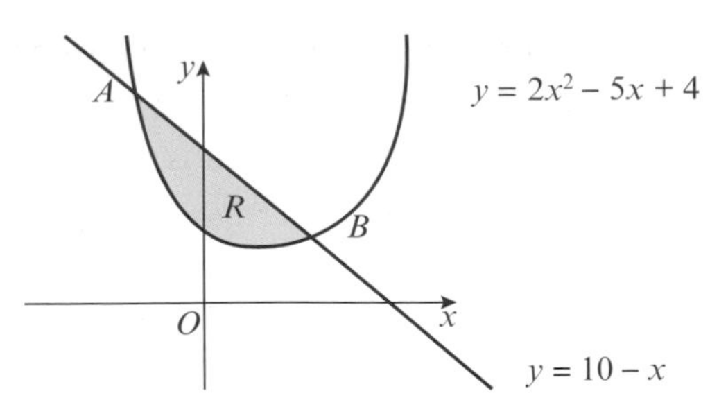

a Find the coordinates of A and the coordinates of B. **(5 marks)**

The shaded region R is bounded by the line and the curve as shown.

b Find the exact area of R. **(6 marks)**

Mixed exercise 13

1 Find:

a $\int (x + 1)(2x - 5)\mathrm{d}x$ **b** $\int (x^{\frac{1}{3}} + x^{-\frac{1}{3}})\mathrm{d}x$

2 The gradient of a curve is given by $f'(x) = x^2 - 3x - \frac{2}{x^2}$. Given that the curve passes through the point (1, 1), find the equation of the curve in the form $y = f(x)$.

3 Find:

a $\int (8x^3 - 6x^2 + 5)\mathrm{d}x$ **b** $\int (5x + 2)x^{\frac{1}{2}}\mathrm{d}x$

(P) **4** Given $y = \frac{(x + 1)(2x - 3)}{\sqrt{x}}$, find $\int y\,\mathrm{d}x$.

(P) **5** Given that $\frac{\mathrm{d}x}{\mathrm{d}t} = (t + 1)^2$ and that $x = 0$ when $t = 2$, find the value of x when $t = 3$.

(E/P) **6** Given that $y^{\frac{1}{2}} = x^{\frac{1}{3}} + 3$:

a show that $y = x^{\frac{2}{3}} + Ax^{\frac{1}{3}} + B$, where A and B are constants to be found. **(2 marks)**

b hence find $\int y\,\mathrm{d}x$. **(3 marks)**

(E/P) **7** Given that $y^{\frac{1}{2}} = 3x^{\frac{1}{4}} - 4x^{-\frac{1}{4}}$ $(x > 0)$:

a find $\frac{\mathrm{d}y}{\mathrm{d}x}$ **(2 marks)**

b find $\int y\,\mathrm{d}x$. **(3 marks)**

(P) **8** $\int \left(\frac{a}{3x^3} - ab\right)\mathrm{d}x = -\frac{2}{3x^2} + 14x + c$

Find the value of a and the value of b.

(P) **9** A rock is dropped off a cliff. The height in metres of the rock above the ground after t seconds is given by the function $f(t)$. Given that $f(0) = 70$ and $f'(t) = -9.8t$, find the height of the rock above the ground after 3 seconds.

(P) 10 A cyclist is travelling along a straight road. The distance in metres of the cyclist from a fixed point after t seconds is modelled by the function $\mathrm{f}(t)$, where $\mathrm{f}'(t) = 5 + 2t$ and $\mathrm{f}(0) = 0$.

a Find an expression for $\mathrm{f}(t)$.

b Calculate the time taken for the cyclist to travel 100 m.

11 The diagram shows the curve with equation $y = 5 + 2x - x^2$ and the line with equation $y = 2$. The curve and the line intersect at the points A and B.

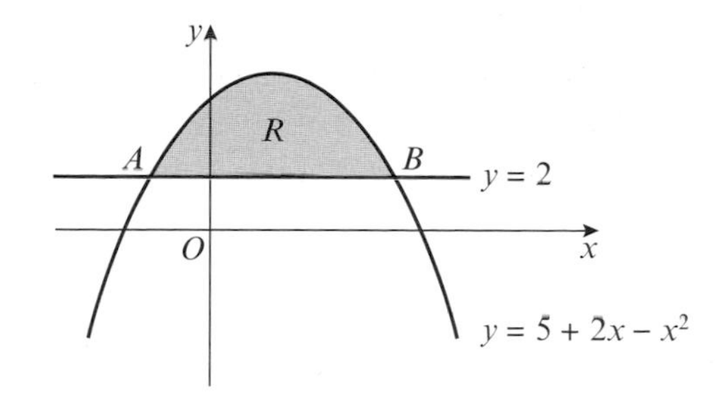

a Find the x-coordinates of A and B.

b The shaded region R is bounded by the curve and the line. Find the area of R.

(E/P) 12 a Find $\int (x^{\frac{1}{2}} - 4)(x^{-\frac{1}{2}} - 1)\,\mathrm{d}x$. **(4 marks)**

b Use your answer to part **a** to evaluate

$$\int_1^4 (x^{\frac{1}{2}} - 4)(x^{-\frac{1}{2}} - 1)\,\mathrm{d}x$$

giving your answer as an exact fraction. **(2 marks)**

(E) 13 The diagram shows part of the curve with equation $y = x^3 - 6x^2 + 9x$. The curve touches the x-axis at A and has a local maximum at B.

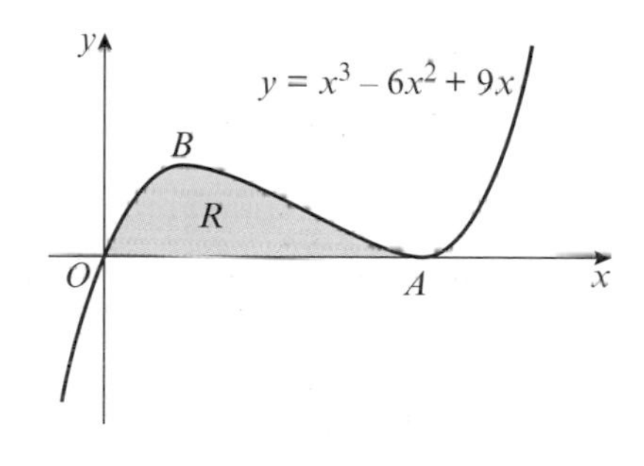

a Show that the equation of the curve may be written as $y = x(x - 3)^2$, and hence write down the coordinates of A. **(2 marks)**

b Find the coordinates of B. **(2 marks)**

c The shaded region R is bounded by the curve and the x-axis. Find the area of R. **(6 marks)**

(E) 14 Consider the function $y = 3x^{\frac{1}{2}} - 4x^{-\frac{1}{2}}$, $x > 0$.

a Find $\frac{\mathrm{d}y}{\mathrm{d}x}$. **(2 marks)**

b Find $\int y\,\mathrm{d}x$. **(3 marks)**

c Hence show that $\int_1^3 y\,\mathrm{d}x = A + B\sqrt{3}$, where A and B are integers to be found. **(2 marks)**

(E/P) 15 The diagram shows a sketch of the curve with equation $y = 12x^{\frac{1}{2}} - x^{\frac{3}{2}}$ for $0 \leqslant x \leqslant 12$.

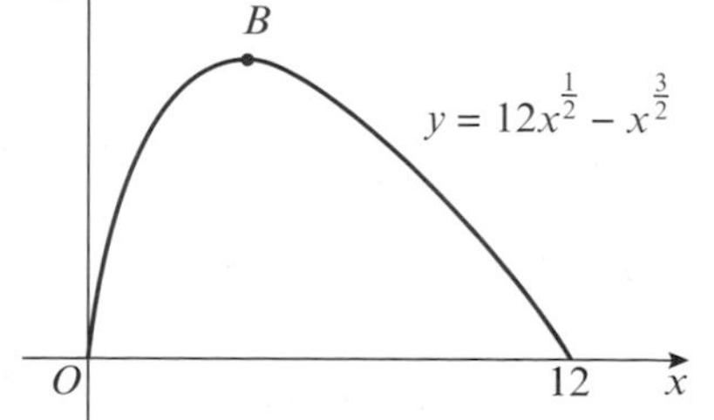

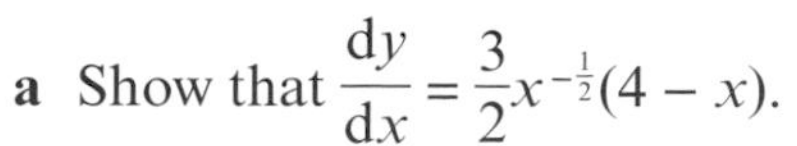

a Show that $\frac{\mathrm{d}y}{\mathrm{d}x} = \frac{3}{2}x^{-\frac{1}{2}}(4 - x)$. **(2 marks)**

b At the point B on the curve the tangent to the curve is parallel to the x-axis. Find the coordinates of the point B. **(2 marks)**

c Find, to 3 significant figures, the area of the finite region bounded by the curve and the x-axis. **(6 marks)**

16 The diagram shows the curve C with equation $y = x(8 - x)$ and the line with equation $y = 12$ which meet at the points L and M.

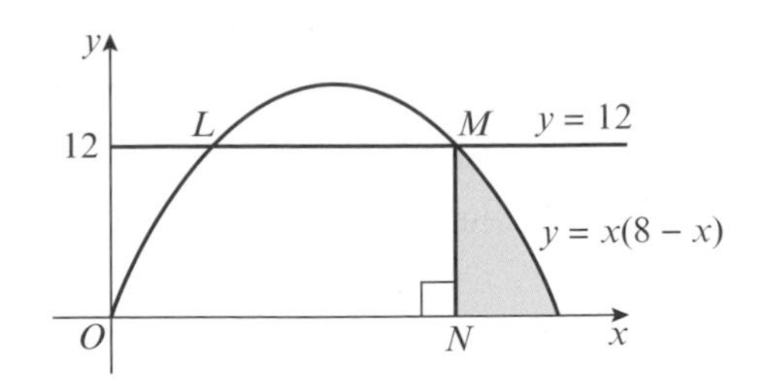

a Determine the coordinates of the point M. **(2 marks)**

b Given that N is the foot of the perpendicular from M on to the x-axis, calculate the area of the shaded region which is bounded by NM, the curve C and the x-axis. **(6 marks)**

17 The diagram shows the line $y = x - 1$ meeting the curve with equation $y = (x - 1)(x - 5)$ at A and C. The curve meets the x-axis at A and B.

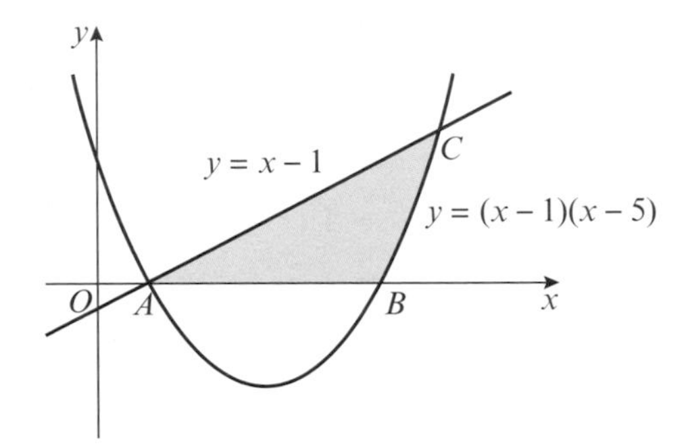

a Write down the coordinates of A and B and find the coordinates of C. **(4 marks)**

b Find the area of the shaded region bounded by the line, the curve and the x-axis. **(6 marks)**

18 The diagram shows part of the curve with equation $y = p + 10x - x^2$, where p is a constant, and part of the line l with equation $y = qx + 25$, where q is a constant. The line l cuts the curve at the points A and B. The x-coordinates of A and B are 4 and 8 respectively. The line through A parallel to the x-axis intersects the curve again at the point C.

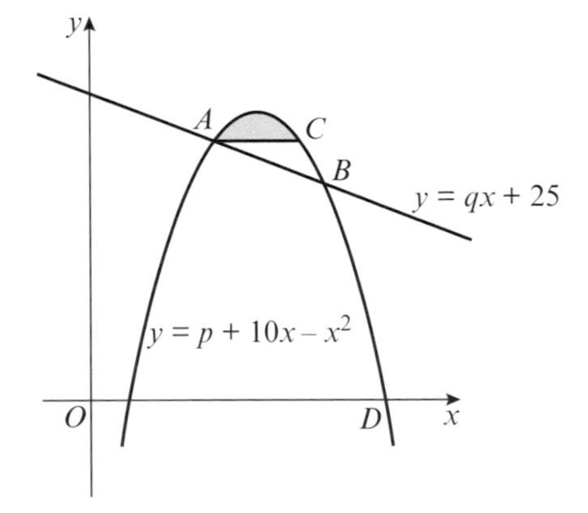

a Show that $p = -7$ and calculate the value of q. **(3 marks)**

b Calculate the coordinates of C. **(2 marks)**

c The shaded region in the diagram is bounded by the curve and the line segment AC. Using integration and showing all your working, calculate the area of the shaded region. **(6 marks)**

(E) **19** Given that $f(x) = \frac{9}{x^2} - 8\sqrt{x} + 4x - 5$, $x > 0$, find $\int f(x)dx$. **(5 marks)**

(E/P) **20** Given that A is constant and $\int_4^9 \left(\frac{3}{\sqrt{x}} - A\right)dx = A^2$ show that there are two possible values for A and find these values. **(5 marks)**

(E/P) **21** $f'(x) = \frac{(2 - x^2)^3}{x^2}$, $x \neq 0$

a Show that $f'(x) = 8x^{-2} - 12 + Ax^2 + Bx^4$, where A and B are constants to be found. **(3 marks)**

b Find $f''(x)$.

Given that the point $(-2, 9)$ lies on the curve with equation $y = f(x)$,

c find $f(x)$. **(5 marks)**

E **22** The finite region S, which is shown shaded, is bounded by the x-axis and the curve with equation $y = 3 - 5x - 2x^2$.

The curve meets the x-axis at points A and B.

a Find the coordinates of point A and point B. **(2 marks)**

b Find the area of the region S. **(4 marks)**

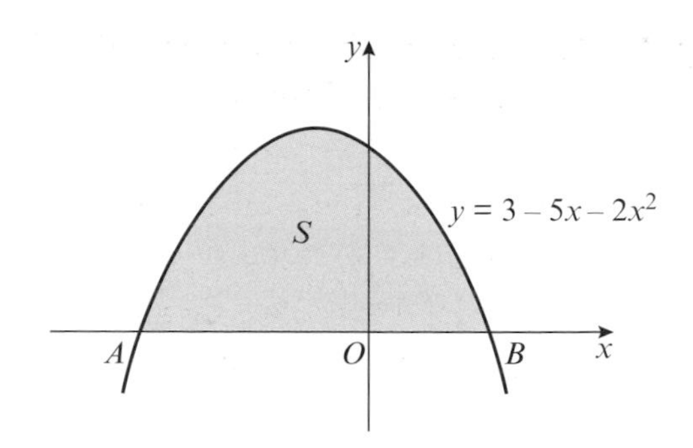

E **23** The graph shows a sketch of part of the curve C with equation $y = (x - 4)(2x + 3)$.

The curve C crosses the x-axis at the points A and B.

a Write down the x-coordinates of A and B. **(1 mark)**

The finite region R, shown shaded, is bounded by C and the x-axis.

b Use integration to find the area of R. **(6 marks)**

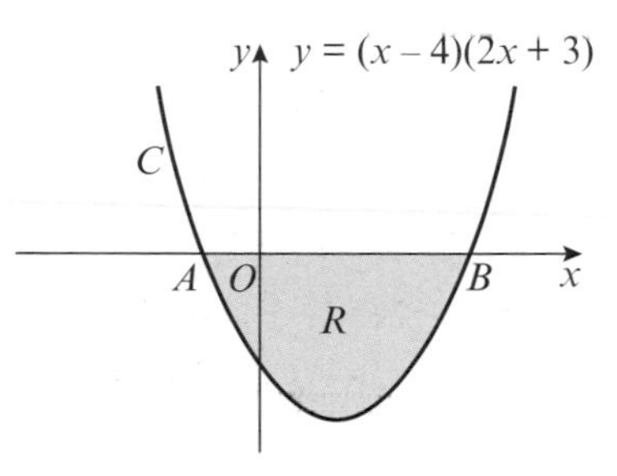

E **24** The graph shows a sketch of part of the curve C with equation $y = x(x - 3)(x + 2)$.

The curve crosses the x-axis at the origin O and the points A and B.

a Write down the x-coordinates of the points A and B. **(1 mark)**

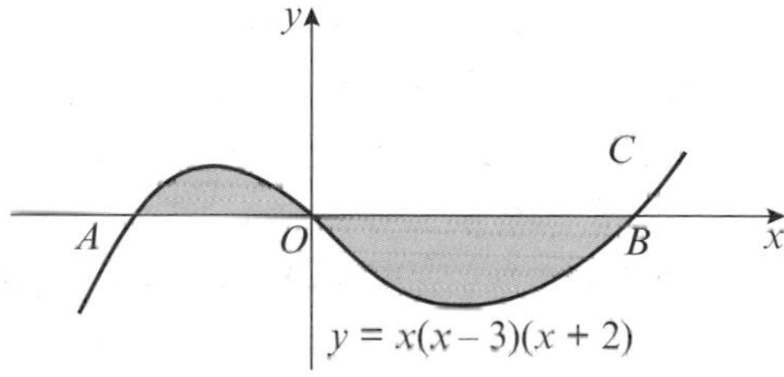

The finite region shown shaded is bounded by the curve C and the x-axis.

b Use integration to find the total area of this region. **(7 marks)**

Challenge

The curve with equation $y = x^2 - 5x + 7$ cuts the curve with equation $y = \frac{1}{2}x^2 - \frac{5}{2}x + 7$. The shaded region R is bounded by the curves as shown.

Find the exact area of R.

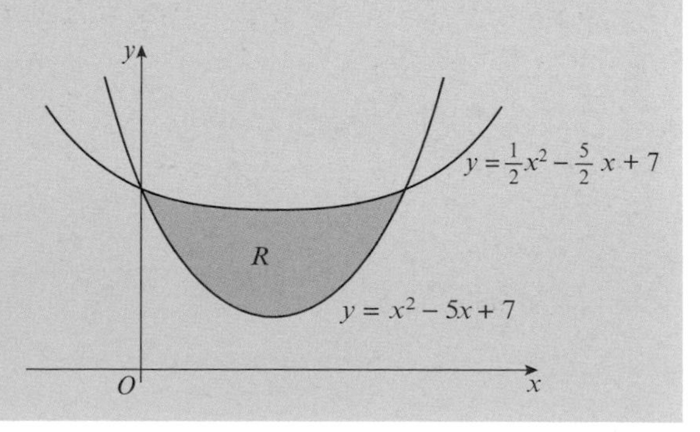

Summary of key points

1 If $\frac{dy}{dx} = x^n$, then $y = \frac{1}{n+1}x^{n+1} + c, n \neq -1$.

Using function notation, if $f'(x) = x^n$, then $f(x) = \frac{1}{n+1}x^{n+1} + c, n \neq -1$.

2 If $\frac{dy}{dx} = kx^n$, then $y = \frac{k}{n+1}x^{n+1} + c, n \neq -1$.

Using function notation, if $f'(x) = kx^n$, then $f(x) = \frac{k}{n+1}x^{n+1} + c, n \neq -1$.

When integrating polynomials, apply the rule of integration separately to each term.

3 $\int f'(x)dx = f(x) + c$

4 $\int(f(x) + g(x))dx = \int f(x)dx + \int g(x)dx$

5 To find the constant of integration, c

- Integrate the function
- Substitute the values (x, y) of a point on the curve, or the value of the function at a given point $f(x) = k$ into the integrated function
- Solve the equation to find c

6 If $f'(x)$ is the derivative of $f(x)$ for all values of x in the interval $[a, b]$, then the definite integral is defined as $\int_a^b f'(x)dx = [f(x)]_a^b = f(b) - f(a)$

7 The area between a positive curve, the x-axis and the lines $x = a$ and $x = b$ is given by

$$\text{Area} = \int_a^b y\,dx$$

where $y = f(x)$ is the equation of the curve.

8 When the area bounded by a curve and the x-axis is below the x-axis, $\int y\,dx$ gives a negative answer.

9 You can use definite integration together with areas of trapeziums and triangles to find more complicated areas on graphs.

Exponentials and logarithms

14

Objectives

After completing this unit you should be able to:

- Sketch graphs of the form $y = a^x$, $y = e^x$, and transformations of these graphs → **pages 312–317**
- Differentiate e^{kx} and understand why this result is important → **pages 314–317**
- Use and interpret models that use exponential functions → **pages 317–319**
- Recognise the relationship between exponents and logarithms → **pages 319–321**
- Recall and apply the laws of logarithms → **pages 321–324**
- Solve equations of the form $a^x = b$ → **pages 324–325**
- Describe and use the natural logarithm function → **pages 326–328**
- Use logarithms to estimate the values of constants in non-linear models → **pages 328–333**

Logarithms are used to report and compare earthquakes. Both the Richter scale and the newer moment magnitude scale use base 10 logarithms to express the size of seismic activity. → **Mixed exercise Q15**

Prior knowledge check

1 Given that $x = 3$ and $y = -1$, evaluate these expressions without a calculator.

a 5^x **b** 3^y **c** 2^{2x-1} **d** 7^{1-y} **e** 11^{x+3y}

← **GCSE Mathematics**

2 Simplify these expressions, writing each answer as a single power.

a $6^8 \div 6^2$ **b** $y^3 \times (y^9)^2$ **c** $\dfrac{2^5 \times 2^9}{2^8}$ **d** $\sqrt{x^8}$

← **Sections 1.1, 1.4**

3 Plot the following data on a scatter graph and draw a line of best fit.

x	1.2	2.1	3.5	4	5.8
y	5.8	7.4	9.4	10.3	12.8

Determine the gradient and intercept of your line of best fit, giving your answers to one decimal place. ← **GCSE Mathematics**

14.1 Exponential functions

Functions of the form $f(x) = a^x$, where a is a constant, are called **exponential functions**. You should become familiar with these functions and the shapes of their graphs.

For an example, look at a table of values of $y = 2^x$.

x	−3	−2	−1	0	1	2	3
y	$\frac{1}{8}$	$\frac{1}{4}$	$\frac{1}{2}$	1	2	4	8

Notation In the expression 2^x, x can be called an **index**, a **power** or an **exponent.**

The value of 2^x tends towards 0 as x decreases, and grows without limit as x increases.

Links Recall that $2^0 = 1$ and that $2^{-3} = \frac{1}{2^3} = \frac{1}{8}$ ← Section 1.4

The graph of $y = 2^x$ is a smooth curve that looks like this:

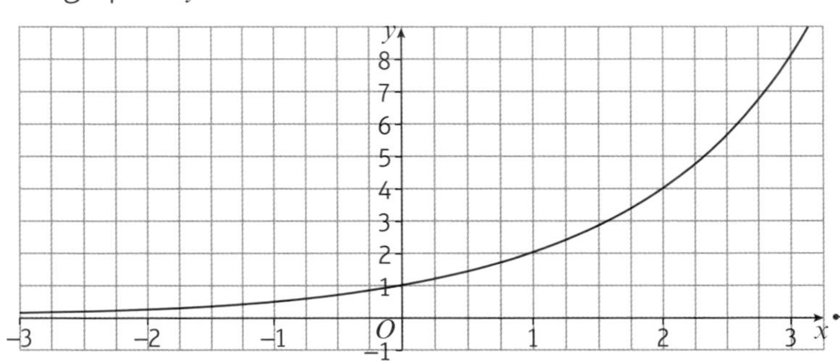

The x-axis is an asymptote to the curve.

Example 1

a On the same axes sketch the graphs of $y = 3^x$, $y = 2^x$ and $y = 1.5^x$.

b On another set of axes sketch the graphs of $y = \left(\frac{1}{2}\right)^x$ and $y = 2^x$.

a For all three graphs, $y = 1$ when $x = 0$.

When $x > 0$, $3^x > 2^x > 1.5^x$.

When $x < 0$, $3^x < 2^x < 1.5^x$.

$a^0 = 1$

Work out the relative positions of the three graphs.

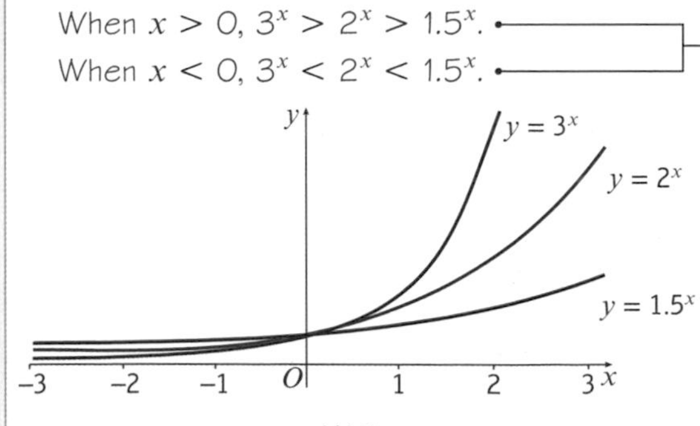

Whenever $a > 1$, $f(x) = a^x$ is an increasing function. In this case, the value of a^x grows without limit as x **increases**, and tends towards 0 as x **decreases**.

b The graph of $y = \left(\frac{1}{2}\right)^x$ is a reflection in the y-axis of the graph of $y = 2^x$.

Since $\frac{1}{2} = 2^{-1}$, $y = \left(\frac{1}{2}\right)^x$ is the same as $y = (2^{-1})^x = 2^{-x}$.

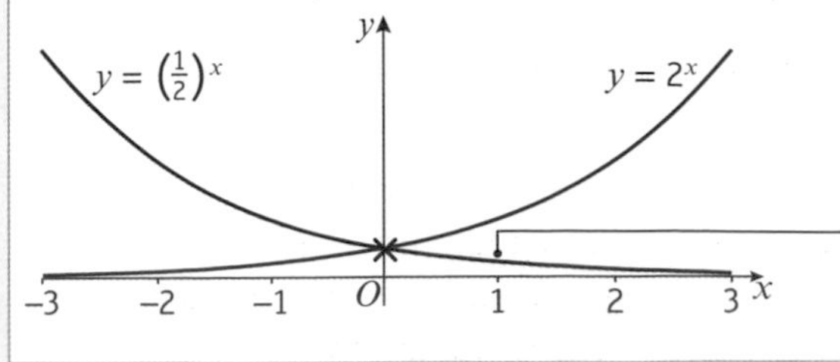

Whenever $0 < a < 1$, $f(x) = a^x$ is a decreasing function. In this case, the value of a^x tends towards 0 as x **increases**, and grows without limit as x **decreases**.

Example 2

Sketch the graph of $y = \left(\frac{1}{2}\right)^{x-3}$. Give the coordinates of the point where the graph crosses the y-axis.

If $f(x) = \left(\frac{1}{2}\right)^x$ then $y = f(x - 3)$.
The graph is a translation of the graph $y = \left(\frac{1}{2}\right)^x$ by the vector $\begin{pmatrix}3\\0\end{pmatrix}$.
The graph crosses the y-axis when $x = 0$.
$y = \left(\frac{1}{2}\right)^{0-3}$
$y = 8$
The graph crosses the y-axis at (0, 8).

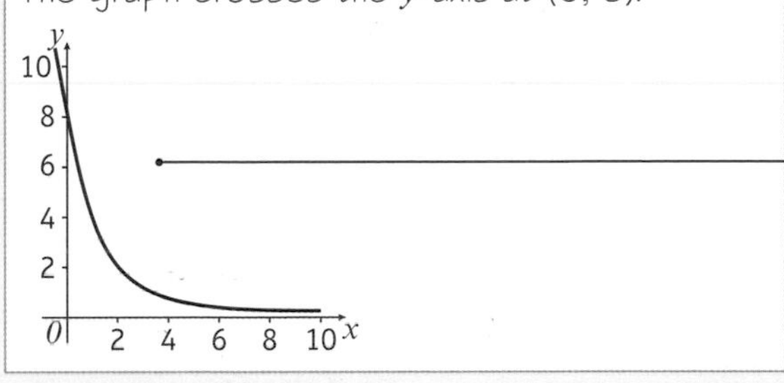

You can also consider this graph as a stretch of the graph $y = \left(\frac{1}{2}\right)^x$

$$y = \left(\frac{1}{2}\right)^{x-3} = \left(\frac{1}{2}\right)^x \times \left(\frac{1}{2}\right)^{-3} = \left(\frac{1}{2}\right)^x \times 8 = 8\left(\frac{1}{2}\right)^x = 8f(x)$$

So the graph of $y = \left(\frac{1}{2}\right)^{x-3}$ is a vertical stretch of the graph of $y = \left(\frac{1}{2}\right)^x$ with scale factor 8.

Problem-solving

If you have to sketch the graph of an unfamiliar function, try writing it as a transformation of a familiar function. ← Section 4.5

Exercise 14A

1 a Draw an accurate graph of $y = (1.7)^x$, for $-4 \leqslant x \leqslant 4$.
b Use your graph to solve the equation $(1.7)^x = 4$.

2 a Draw an accurate graph of $y = (0.6)^x$, for $-4 \leqslant x \leqslant 4$.
b Use your graph to solve the equation $(0.6)^x = 2$.

3 Sketch the graph of $y = 1^x$.

(P) **4** For each of these statements, decide whether it is true or false, justifying your answer or offering a counter-example.
a The graph of $y = a^x$ passes through (0, 1) for all positive real numbers a.
b The function $f(x) = a^x$ is always an increasing function for $a > 0$.
c The graph of $y = a^x$, where a is a positive real number, never crosses the x-axis.

5 The function f(x) is defined as $f(x) = 3^x$, $x \in \mathbb{R}$. On the same axes, sketch the graphs of:
a $y = f(x)$ **b** $y = 2f(x)$ **c** $y = f(x) - 4$ **d** $y = f\left(\frac{1}{2}x\right)$
Write down the coordinates of the point where each graph crosses the y-axis, and give the equations of any asymptotes.

(P) **6** The graph of $y = ka^x$ passes through the points (1, 6) and (4, 48). Find the values of the constants k and a.

Problem-solving

Substitute the coordinates into $y = ka^x$ to create two simultaneous equations. Use division to eliminate one of the two unknowns.

(P) 7 The graph of $y = pq^x$ passes through the points (−3, 150) and (2, 0.048).

a By drawing a sketch or otherwise, explain why $0 < q < 1$.

b Find the values of the constants p and q.

Challenge

Sketch the graph of $y = 2^{x-2} + 5$. Give the coordinates of the point where the graph crosses the y-axis.

14.2 $y = e^x$

Exponential functions of the form $f(x) = a^x$ have a special mathematical property. The graphs of their gradient functions are a similar shape to the graphs of the functions themselves.

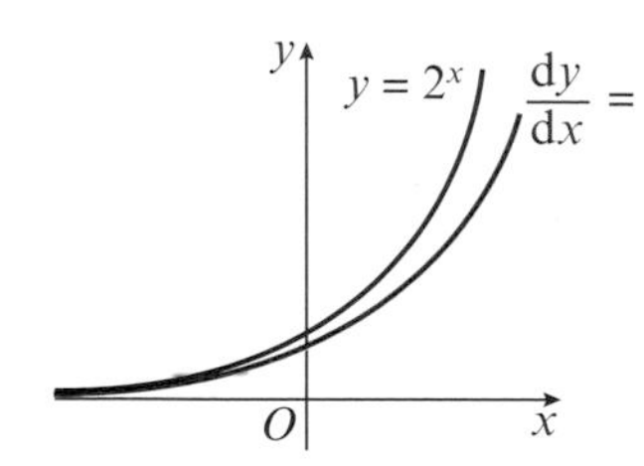

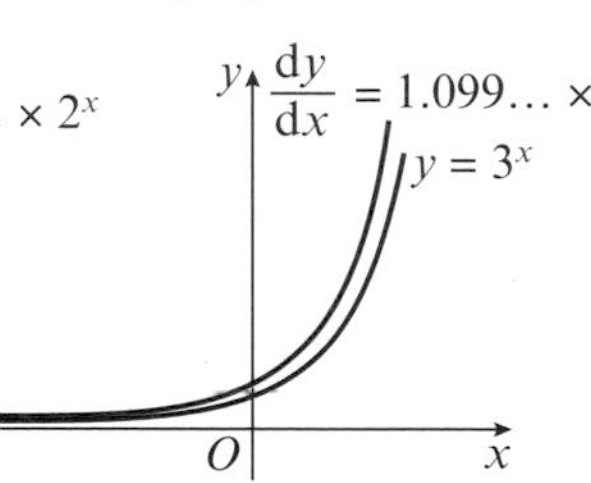

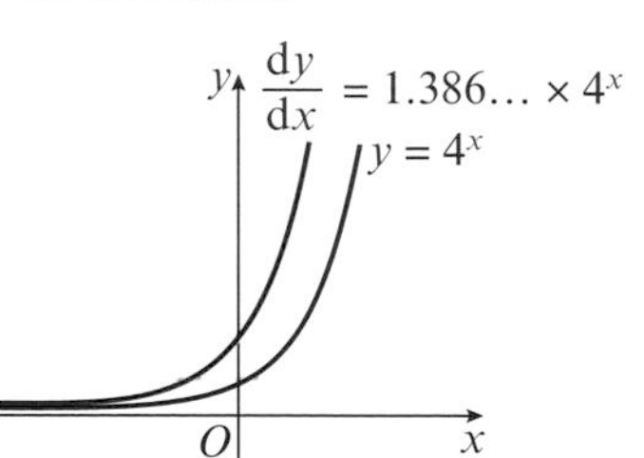

In each case $f'(x) = kf(x)$, where k is a constant. As the value of a increases, so does the value of k.

Something unique happens between $a = 2$ and $a = 3$. There is going to be a value of a where the gradient function is exactly the same as the original function. This occurs when a is approximately equal to 2.71828. The exact value is represented by the letter e. Like π, e is both an important mathematical constant and an irrational number.

Function	Gradient function
$f(x) = 1^x$	$f'(x) = 0 \times 1^x$
$f(x) = 2^x$	$f'(x) = 0.693\ldots \times 2^x$
$f(x) = 3^x$	$f'(x) = 1.099\ldots \times 3^x$
$f(x) = 4^x$	$f'(x) = 1.386\ldots \times 4^x$

Online Explore the relationship between exponential functions and their derivatives using technology.

- **For all real values of x:**
  - **If $f(x) = e^x$ then $f'(x) = e^x$**
  - **If $y = e^x$ then $\frac{dy}{dx} = e^x$**

A similar result holds for functions such as e^{5x}, e^{-x} and $e^{\frac{1}{2}x}$.

- **For all real values of x and for any constant k:**
  - **If $f(x) = e^{kx}$ then $f'(x) = ke^{kx}$**
  - **If $y = e^{kx}$ then $\frac{dy}{dx} = ke^{kx}$**

Example 3

Differentiate with respect to x.

a e^{4x} **b** $e^{-\frac{1}{2}x}$ **c** $3e^{2x}$

a $y = e^{4x}$

$\frac{dy}{dx} = 4e^{4x}$

Use the rule for differentiating e^{kx} with $k = 4$.

b $y = e^{-\frac{1}{2}x}$

$\frac{dy}{dx} = -\frac{1}{2}e^{-\frac{1}{2}x}$

c $y = 3e^{2x}$

$\frac{dy}{dx} = 2 \times 3e^{2x} = 6e^{2x}$

To differentiate ae^{kx} multiply the whole function by k. The derivative is kae^{kx}.

Example 4

Sketch the graphs of the following equations. Give the coordinates of any points where the graphs cross the axes, and state the equations of any asymptotes.

a $y = e^{2x}$ **b** $y = 10e^{-x}$ **c** $y = 3 + 4e^{\frac{1}{2}x}$

a $y = e^{2x}$

When $x = 0$, $y = e^{2\times 0} = 1$ so the graph crosses the y-axis at (0, 1).

The x-axis ($y = 0$) is an asymptote.

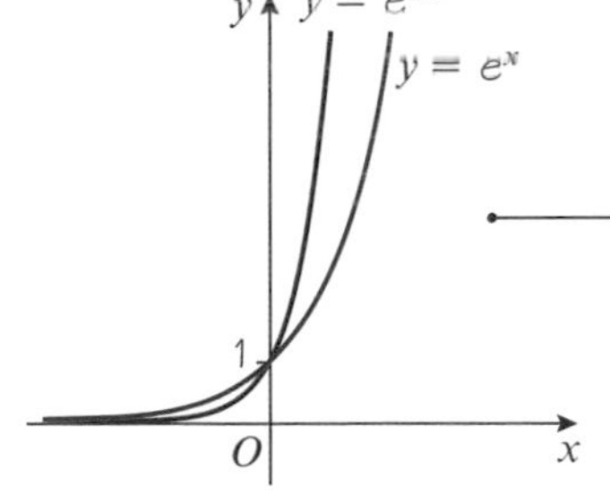

The graph of $y = e^x$ has been shown in purple on this sketch.

This is a stretch of the graph of $y = e^x$, parallel to the x-axis and with scale factor $\frac{1}{2}$

← Section 4.6

b $y = 10e^{-x}$

When $x = 0$, $y = 10e^{-0}$. So the graph crosses the y-axis at (0, 10).

The x-axis ($y = 0$) is an asymptote.

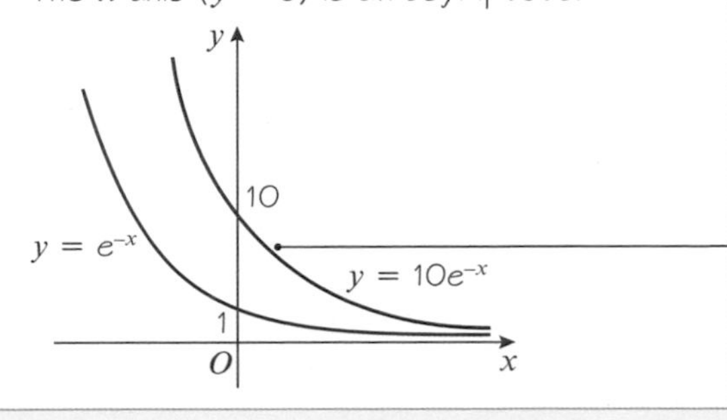

Negative powers of e^x, such as e^{-x} or e^{-4x}, give rise to decreasing functions.

The graph of $y = e^x$ has been reflected in the y-axis and stretched parallel to the y-axis with scale factor 10.

c $y = 3 + 4e^{\frac{1}{2}x}$

When $x = 0$, $y = 3 + 4e^{\frac{1}{2}\times 0} = 7$ so the graph crosses the y-axis at $(0, 7)$.

The line $y = 3$ is an asymptote.

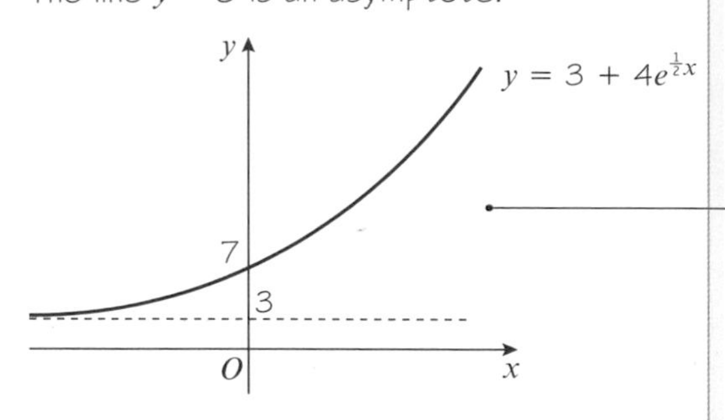

The graph of $y = e^{\frac{1}{2}x}$ has been stretched parallel to the y-axis with scale factor 4 and then translated by $\begin{pmatrix} 0 \\ 3 \end{pmatrix}$.

Problem-solving

If you have to sketch a transformed graph with an asymptote, it is often easier to sketch the asymptote first.

Online Use technology to draw transformations of $y = e^x$.

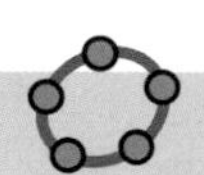

Exercise 14B

1 Use a calculator to find the value of e^x to 5 decimal places when

a $x = 1$ **b** $x = 4$ **c** $x = -10$ **d** $x = 0.2$

2 a Draw an accurate graph of $y = e^x$ for $-4 \leqslant x \leqslant 4$.

b By drawing appropriate tangent lines, estimate the gradient at $x = 1$ and $x = 3$.

c Compare your answers to the actual values of e and e^3.

3 Sketch the graphs of:

a $y = e^{x+1}$ **b** $y = 4e^{-2x}$ **c** $y = 2e^x - 3$

d $y = 4 - e^x$ **e** $y = 6 + 10e^{\frac{1}{2}x}$ **f** $y = 100e^{-x} + 10$

4 Each of the sketch graphs below is of the form $y = Ae^{bx} + C$, where A, b and C are constants. Find the values of A and C for each graph, and state whether b is positive or negative.

a

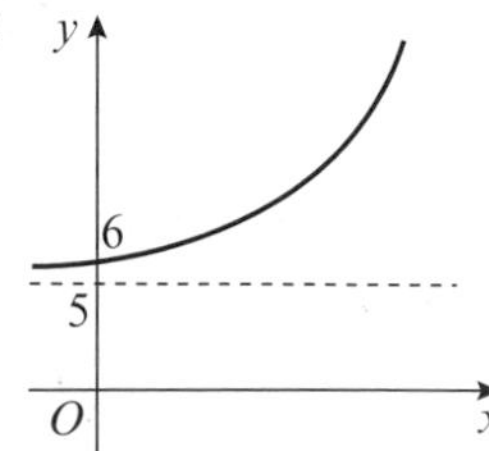

b

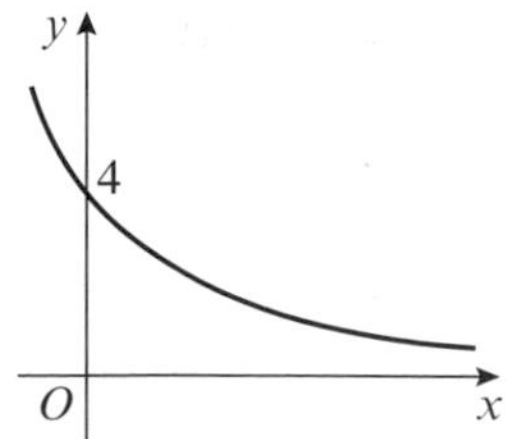

c

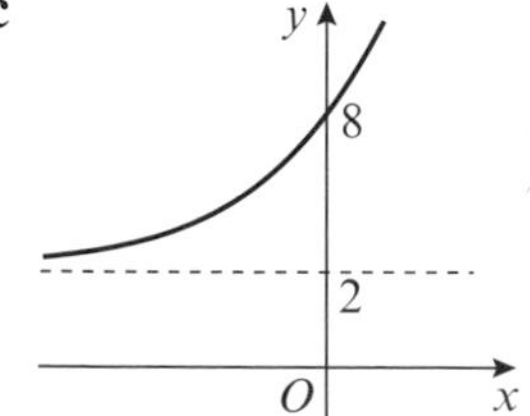

Hint You do not have enough information to work out the value of b, so simply state whether it is positive or negative.

5 Rearrange $f(x) = e^{3x+2}$ into the form $f(x) = Ae^{bx}$, where A and b are constants whose values are to be found. Hence, or otherwise, sketch the graph of $y = f(x)$.

Hint $e^{m+n} = e^m \times e^n$

6 Differentiate the following with respect to x.

a e^{6x} **b** $e^{-\frac{1}{3}x}$ **c** $7e^{2x}$

d $5e^{0.4x}$ **e** $e^{3x} + 2e^x$ **f** $e^x(e^x + 1)$

Hint For part **f**, start by expanding the bracket.

7 Find the gradient of the curve with equation $y = e^{3x}$ at the point where
a $x = 2$ **b** $x = 0$ **c** $x = -0.5$

(P) **8** The function f is defined as $f(x) = e^{0.2x}$, $x \in \mathbb{R}$. Show that the tangent to the curve at the point (5, e) goes through the origin.

14.3 Exponential modelling

You can use e^x to model situations such as population growth, where the rate of **increase** is proportional to the size of the population at any given moment. Similarly, e^{-x} can be used to model situations such as radioactive decay, where the rate of **decrease** is proportional to the number of atoms remaining.

Example 5

The density of a pesticide in a given section of field, P mg/m², can be modelled by the equation

$$P = 160e^{-0.006t}$$

where t is the time in days since the pesticide was first applied.

a Use this model to estimate the density of pesticide after 15 days.

b Interpret the meaning of the value 160 in this model.

c Show that $\frac{dP}{dt} = kP$, where k is a constant, and state the value of k.

d Interpret the significance of the sign of your answer to part **c**.

e Sketch the graph of P against t.

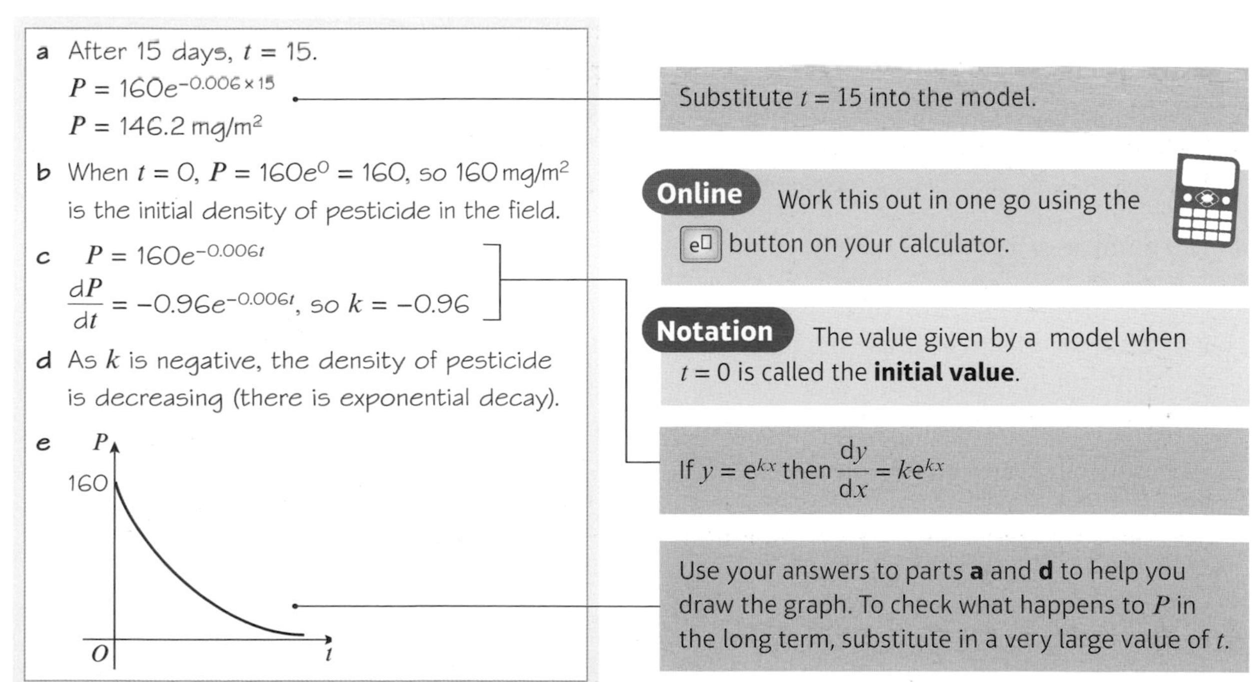

Exercise 14C

1 The value of a car is modelled by the formula

$$V = 20\,000e^{-\frac{t}{12}}$$

where V is the value in £s and t is its age in years from new.

a State its value when new.

b Find its value (to the nearest £) after 4 years.

c Sketch the graph of V against t.

(P) 2 The population of a country is modelled using the formula

$$P = 20 + 10e^{\frac{t}{50}}$$

where P is the population in thousands and t is the time in years after the year 2000.

a State the population in the year 2000.

b Use the model to predict the population in the year 2030.

c Sketch the graph of P against t for the years 2000 to 2100.

d Do you think that it would be valid to use this model to predict the population in the year 2500? Explain your answer.

(P) 3 The number of people infected with a disease is modelled by the formula

$$N = 300 - 100e^{-0.5t}$$

where N is the number of people infected with the disease and t is the time in years after detection.

a How many people were first diagnosed with the disease?

b What is the long term prediction of how this disease will spread?

c Sketch the graph of N against t for $t > 0$.

(P) 4 The number of rabbits, R, in a population after m months is modelled by the formula

$$R = 12e^{0.2m}$$

a Use this model to estimate the number of rabbits after

i 1 month **ii** 1 year

b Interpret the meaning of the constant 12 in this model.

c Show that after 6 months, the rabbit population is increasing by almost 8 rabbits per month.

d Suggest one reason why this model will stop giving valid results for large enough values of t.

Problem-solving

Your answer to part **b** must refer to the context of the model.

E/P **5** On Earth, the atmospheric pressure, p, in bars can be modelled approximately by the formula $p = e^{-0.13h}$ where h is the height above sea level in kilometres.

a Use this model to estimate the pressure at the top of Mount Rainier, which has an altitude of 4.394 km. **(1 mark)**

b Demonstrate that $\frac{dp}{dh} = kp$ where k is a constant to be found. **(2 marks)**

c Interpret the significance of the sign of k in part **b**. **(1 mark)**

d This model predicts that the atmospheric pressure will change by s% for every kilometre gained in height. Calculate the value of s. **(3 marks)**

6 Nigel has bought a tractor for £20 000. He wants to model the depreciation of the value of his tractor, £T, in t years. His friend suggests two models:

Model 1: $T = 20\,000e^{-0.24t}$

Model 2: $T = 19\,000e^{-0.255t} + 1000$

a Use both models to predict the value of the tractor after one year. Compare your results. **(2 marks)**

b Use both models to predict the value of the tractor after ten years. Compare your results. **(2 marks)**

c Sketch a graph of T against t for both models. **(2 marks)**

d Interpret the meaning of the 1000 in model 2, and suggest why this might make model 2 more realistic. **(1 mark)**

14.4 Logarithms

The inverses of exponential functions are called **logarithms**. A relationship which is expressed using an exponent can also be written in terms of logarithms.

- **$\log_a n = x$ is equivalent to $a^x = n$ $(a \neq 1)$**

Notation a is called the base of the logarithm.

Example 6

Write each statement as a logarithm.

a $3^2 = 9$ **b** $2^7 = 128$ **c** $64^{\frac{1}{2}} = 8$

a $3^2 = 9$, so $\log_3 9 = 2$

b $2^7 = 128$, so $\log_2 128 = 7$

c $64^{\frac{1}{2}} = 8$, so $\log_{64} 8 = \frac{1}{2}$

In words, you would say 'the logarithm of 9 to the base 3 is 2'.

Logarithms can take fractional or negative values.

Example 7

Rewrite each statement using a power.

a $\log_3 81 = 4$ **b** $\log_2\left(\frac{1}{8}\right) = -3$

a $\log_3 81 = 4$, so $3^4 = 81$

b $\log_2\left(\frac{1}{8}\right) = -3$, so $2^{-3} = \frac{1}{8}$

Example 8

Without using a calculator, find the value of:

a $\log_3 81$ **b** $\log_4 0.25$ **c** $\log_{0.5} 4$ **d** $\log_a (a^5)$

a $\log_3 81 = 4$ — Because $3^4 = 81$.

b $\log_4 0.25 = -1$ — Because $4^{-1} = \frac{1}{4} = 0.25$.

c $\log_{0.5} 4 = -2$ — Because $0.5^{-2} = \left(\frac{1}{2}\right)^{-2} = 2^2 = 4$.

d $\log_a (a^5) = 5$ — Because $a^5 = a^5$.

You can use your calculator to find logarithms of any base. Some calculators have a specific [log□□] key for this function. Most calculators also have separate buttons for logarithms to the base 10 (usually written as [log] and logarithms to the base e (usually written as [ln]).

Notation Logarithms to the base e are typically called **natural logarithms**. This is why the calculator key is labelled [ln].

Online Use the logarithm buttons on your calculator.

Example 9

Use your calculator to find the following logarithms to 3 decimal places.

a $\log_3 40$ **b** $\log_e 8$ **c** $\log_{10} 75$

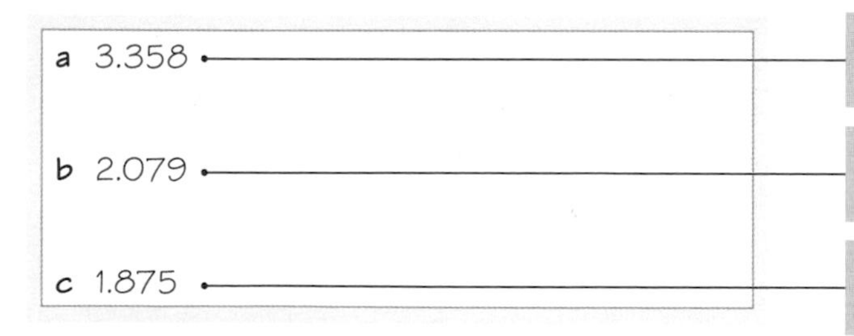

a 3.358 — For part **a** use [log□□].

b 2.079 — For part **b** you can use either [ln] or [log□□].

c 1.875 — For part **c** you can use either [log] or [log□□].

Exercise 14D

1 Rewrite using a logarithm.

a $4^4 = 256$ **b** $3^{-2} = \frac{1}{9}$ **c** $10^6 = 1\,000\,000$

d $11^1 = 11$ **e** $(0.2)^3 = 0.008$

2 Rewrite using a power.

a $\log_2 16 = 4$ **b** $\log_5 25 = 2$ **c** $\log_9 3 = \frac{1}{2}$

d $\log_5 0.2 = -1$ **e** $\log_{10} 100\,000 = 5$

3 Without using a calculator, find the value of

a $\log_2 8$ **b** $\log_5 25$ **c** $\log_{10} 10\,000\,000$ **d** $\log_{12} 12$

e $\log_3 729$ **f** $\log_{10} \sqrt{10}$ **g** $\log_4 (0.25)$ **h** $\log_{0.25} 16$

i $\log_a (a^{10})$ **j** $\log_{\frac{2}{3}} \left(\frac{9}{4}\right)$

4 Without using a calculator, find the value of x for which

a $\log_5 x = 4$ **b** $\log_x 81 = 2$ **c** $\log_7 x = 1$

d $\log_2 (x - 1) = 3$ **e** $\log_3 (4x + 1) = 4$ **f** $\log_x (2x) = 2$

5 Use your calculator to evaluate these logarithms to three decimal places.

a $\log_9 230$ **b** $\log_5 33$ **c** $\log_{10} 1020$ **d** $\log_e 3$

(P) **6 a** Without using a calculator, justify why the value of $\log_2 50$ must be between 5 and 6.

b Use a calculator to find the exact value of $\log_2 50$ to 4 significant figures.

Hint Use corresponding statements involving powers of 2.

7 a Find the values of:

i $\log_2 2$ **ii** $\log_3 3$ **iii** $\log_{17} 17$

b Explain why $\log_a a$ has the same value for all positive values of a ($a \neq 1$).

8 a Find the values of:

i $\log_2 1$ **ii** $\log_3 1$ **iii** $\log_{17} 1$

b Explain why $\log_a 1$ has the same value for all positive values of a ($a \neq 1$).

14.5 Laws of logarithms

Expressions involving more than one logarithm can often be rearranged or simplified. For instance:

$\log_a x = m$ and $\log_a y = n$ — Take two logarithms with the same base

$x = a^m$ and $y = a^n$ — Rewrite these expressions using powers

$xy = a^m \times a^n = a^{m+n}$ — Multiply these powers

$\log_a xy = m + n = \log_a x + \log_a y$ — Rewrite your result using logarithms

This result is one of the **laws of logarithms**.

You can use similar methods to prove two further laws.

- **The laws of logarithms:**
  - $\log_a x + \log_a y = \log_a xy$ **(the multiplication law)**
  - $\log_a x - \log_a y = \log_a \left(\frac{x}{y}\right)$ **(the division law)**
  - $\log_a (x^k) = k \log_a x$ **(the power law)**

Watch out You need to learn these three laws of logarithms, and the special cases below.

- **You should also learn to recognise the following special cases:**
  - $\log_a \left(\frac{1}{x}\right) = \log_a (x^{-1}) = -\log_a x$ **(the power law when $k = -1$)**
  - $\log_a a = 1$ **($a > 0$, $a \neq 1$)**
  - $\log_a 1 = 0$ **($a > 0$, $a \neq 1$)**

Example 10

Write as a single logarithm.

a $\log_3 6 + \log_3 7$ **b** $\log_2 15 - \log_2 3$ **c** $2\log_5 3 + 3\log_5 2$ **d** $\log_{10} 3 - 4\log_{10}\left(\frac{1}{2}\right)$

a $\log_3 (6 \times 7)$
$= \log_3 42$

Use the multiplication law.

b $\log_2 (15 \div 3)$
$= \log_2 5$

Use the division law.

c $2\log_5 3 = \log_5 (3^2) = \log_5 9$
$3\log_5 2 = \log_5 (2^3) = \log_5 8$
$\log_5 9 + \log_5 8 = \log_5 72$

First apply the power law to both parts of the expression.
Then use the multiplication law.

d $4\log_{10}\left(\frac{1}{2}\right) = \log_{10}\left(\frac{1}{2}\right)^4 = \log_{10}\left(\frac{1}{16}\right)$

$\log_{10} 3 - \log_{10}\left(\frac{1}{16}\right) = \log_{10}\left(3 \div \frac{1}{16}\right)$
$= \log_{10} 48$

Use the power law first.
Then use the division law.

Example 11

Write in terms of $\log_a x$, $\log_a y$ and $\log_a z$.

a $\log_a (x^2yz^3)$ **b** $\log_a\left(\frac{x}{y^3}\right)$ **c** $\log_a\left(\frac{x\sqrt{y}}{z}\right)$ **d** $\log_a\left(\frac{x}{a^4}\right)$

a $\log_a (x^2yz^3)$
$= \log_a (x^2) + \log_a y + \log_a (z^3)$
$= 2\log_a x + \log_a y + 3\log_a z$

b $\log_a\left(\frac{x}{y^3}\right)$
$= \log_a x - \log_a (y^3)$
$= \log_a x - 3\log_a y$

c $\log_a\left(\frac{x\sqrt{y}}{z}\right)$
$= \log_a (x\sqrt{y}) - \log_a z$
$= \log_a x + \log_a \sqrt{y} - \log_a z$
$= \log_a x + \frac{1}{2}\log_a y - \log_a z$

Use the power law ($\sqrt{y} = y^{\frac{1}{2}}$).

d $\log_a\left(\frac{x}{a^4}\right)$
$= \log_a x - \log_a (a^4)$
$= \log_a x - 4\log_a a$
$= \log_a x - 4$

$\log_a a = 1$.

Example 12

Solve the equation $\log_{10} 4 + 2\log_{10} x = 2$.

$\log_{10} 4 + 2\log_{10} x = 2$

$\log_{10} 4 + \log_{10} x^2 = 2$ — Use the power law.

$\log_{10} 4x^2 = 2$ — Use the multiplication law.

$4x^2 = 10^2$ — Rewrite the logarithm using powers.

$4x^2 = 100$

$x^2 = 25$

$x = 5$

Watch out $\log_{10} x$ is only defined for **positive** values of x, so $x = -5$ cannot be a solution of the equation.

Example 13

Solve the equation $\log_3 (x + 11) - \log_3 (x - 5) = 2$

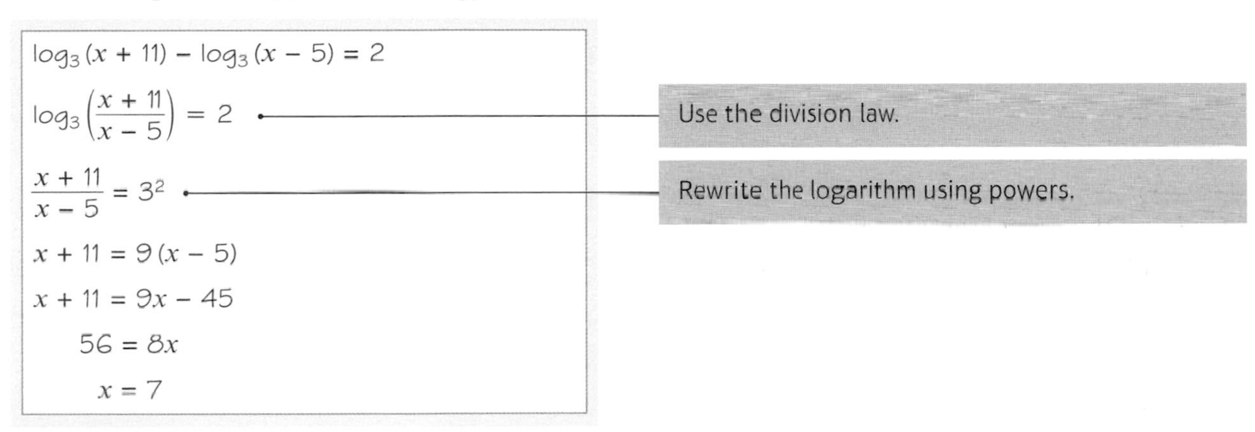

Exercise 14E

1 Write as a single logarithm.

a $\log_2 7 + \log_2 3$ **b** $\log_2 36 - \log_2 4$ **c** $3\log_5 2 + \log_5 10$

d $2\log_6 8 - 4\log_6 3$ **e** $\log_{10} 5 + \log_{10} 6 - \log_{10}\left(\frac{1}{4}\right)$

2 Write as a single logarithm, then simplify your answer.

a $\log_2 40 - \log_2 5$ **b** $\log_6 4 + \log_6 9$ **c** $2\log_{12} 3 + 4\log_{12} 2$

d $\log_8 25 + \log_8 10 - 3\log_8 5$ **e** $2\log_{10} 2 - (\log_{10} 5 + \log_{10} 8)$

3 Write in terms of $\log_a x$, $\log_a y$ and $\log_a z$.

a $\log_a (x^3y^4z)$ **b** $\log_a \left(\frac{x^5}{y^2}\right)$ **c** $\log_a (a^2x^2)$

d $\log_a \left(\frac{x}{z\sqrt{y}}\right)$ **e** $\log_a \sqrt{ax}$

4 Solve the following equations:

a $\log_2 3 + \log_2 x = 2$ **b** $\log_6 12 - \log_6 x = 3$

c $2\log_5 x = 1 + \log_5 6$ **d** $2\log_9 (x + 1) = 2\log_9 (2x - 3) + 1$

Hint Move the logarithms onto the same side if necessary and use the division law.

(P) **5 a** Given that $\log_3 (x + 1) = 1 + 2\log_3 (x - 1)$, show that $3x^2 - 7x + 2 = 0$. **(5 marks)**

b Hence, or otherwise, solve $\log_3 (x + 1) = 1 + 2\log_3 (x - 1)$. **(2 marks)**

(P) **6** Given that a and b are positive constants, and that $a > b$, solve the simultaneous equations

$a + b = 13$

$\log_6 a + \log_6 b = 2$

Problem-solving

Pay careful attention to the conditions on a and b given in the question.

Challenge

By writing $\log_a x = m$ and $\log_a y = n$, prove that $\log_a x - \log_a y = \log_a \left(\frac{x}{y}\right)$.

14.6 Solving equations using logarithms

You can use logarithms and your calculator to solve equations of the form $a^x = b$.

Example 14

Solve the following equations, giving your answers to 3 decimal places.

a $3^x = 20$ **b** $5^{4x-1} = 61$

a $3^x = 20$,

so $x = \log_3 20 = 2.727$

Use the $\log_\square \square$ button on your calculator.

b $5^{4x-1} = 61$, so $4x - 1 = \log_5 61$

$4x = \log_5 61 + 1$

$x = \frac{\log_5 61 + 1}{4}$

$= 0.889$

You can evaluate the final answer in one step on your calculator.

Example 15

Solve the equation $5^{2x} - 12(5^x) + 20 = 0$, giving your answer to 3 significant figures.

$5^{2x} - 12(5^x) + 20$ is a quadratic function of 5^x

An alternative method is to rewrite the equation using the substitution $y = 5^x$: $y^2 - 12y + 20 = 0$.

$(5^x - 10)(5^x - 2) = 0$

$5^x = 10$ or $5^x = 2$

$5^x = 10 \Rightarrow x = \log_5 10 \Rightarrow x = 1.43$

$5^x = 2 \Rightarrow x = \log_5 2 \Rightarrow x = 0.431$

Watch out Solving the quadratic equation gives you two possible values for 5^x. Make sure you calculate both corresponding values of x for your final answer.

You can solve more complicated equations by 'taking logs' of both sides.

- **Whenever $f(x) = g(x)$, $\log_a f(x) = \log_a g(x)$**

Example 16

Find the solution to the equation $3^x = 2^{x+1}$, giving your answer to four decimal places.

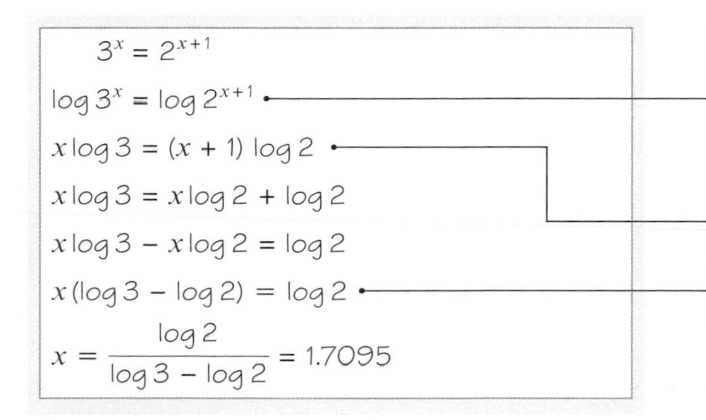

$3^x = 2^{x+1}$

$\log 3^x = \log 2^{x+1}$

$x \log 3 = (x + 1) \log 2$

$x \log 3 = x \log 2 + \log 2$

$x \log 3 - x \log 2 = \log 2$

$x(\log 3 - \log 2) = \log 2$

$x = \dfrac{\log 2}{\log 3 - \log 2} = 1.7095$

This step is called 'taking logs of both sides'. The logs on both sides must be to the **same base**. Here 'log' is used to represent $\log_{10}$.

Use the power law.

Move all the terms in x to one side then factorise.

Exercise 14F

1 Solve, giving your answers to 3 significant figures.

a $2^x = 75$ **b** $3^x = 10$ **c** $5^x = 2$ **d** $4^{2x} = 100$

e $9^{x+5} = 50$ **f** $7^{2x-1} = 23$ **g** $11^{3x-2} = 65$ **h** $2^{3-2x} = 88$

2 Solve, giving your answers to 3 significant figures.

a $2^{2x} - 6(2^x) + 5 = 0$ **b** $3^{2x} - 15(3^x) + 44 = 0$

c $5^{2x} - 6(5^x) - 7 = 0$ **d** $3^{2x} + 3^{x+1} - 10 = 0$

e $7^{2x} + 12 = 7^{x+1}$ **f** $2^{2x} + 3(2^x) - 4 = 0$

g $3^{2x+1} - 26(3^x) - 9 = 0$ **h** $4(3^{2x+1}) + 17(3^x) - 7 = 0$

Hint $3^{x+1} = 3^x \times 3^1 = 3(3^x)$

Problem solving Consider these equations as functions of functions. Part **a** is equivalent to $u^2 - 6u + 5 = 0$, with $u = 2^x$.

(E) **3** Solve the following equations, giving your answers to 3 significant figures where appropriate.

a $3^{x+1} = 2000$ **(2 marks)**

b $\log_5 (x - 3) = -1$ **(2 marks)**

(E/P) **4** **a** Sketch the graph of $y = 4^x$, stating the coordinates of any points where the graph crosses the axes. **(2 marks)**

b Solve the equation $4^{2x} - 10(4^x) + 16 = 0$. **(4 marks)**

Hint Attempt this question without a calculator.

5 Solve the following equations, giving your answers to four decimal places.

a $5^x = 2^{x+1}$ **b** $3^{x+5} = 6^x$ **c** $7^{x+1} = 3^{x+2}$

Hint Take logs of both sides.

14.7 Working with natural logarithms

■ **The graph of $y = \ln x$ is a reflection of the graph $y = e^x$ in the line $y = x$.**

The graph of $y = \ln x$ passes through (1, 0) and does not cross the y-axis.

The y-axis is an asymptote of the graph $y = \ln x$. This means that $\ln x$ is only defined for positive values of x.

As x increases, $\ln x$ grows without limit, but relatively slowly.

You can also use the fact that logarithms are the inverses of exponential functions to solve equations involving powers and logarithms.

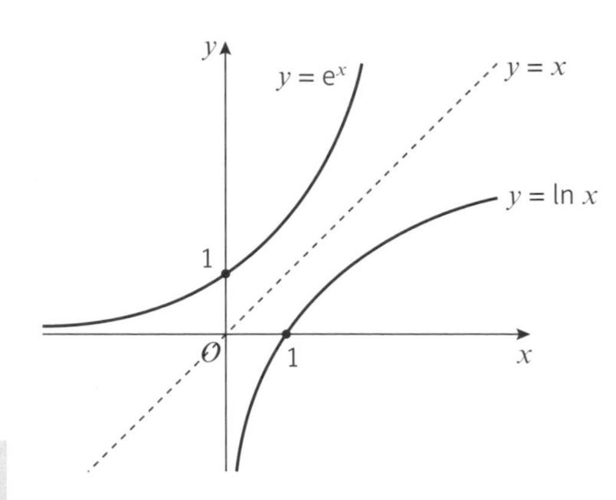

■ $\mathbf{e^{\ln x} = \ln(e^x) = x}$

Notation $\ln x = \log_e x$

Example 17

Solve these equations, giving your answers in exact form.

a $e^x = 5$ **b** $\ln x = 3$

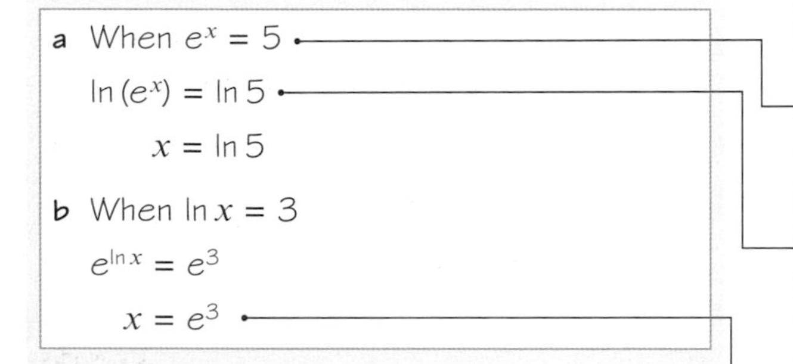
a When $e^x = 5$
$\ln(e^x) = \ln 5$
$x = \ln 5$

b When $\ln x = 3$
$e^{\ln x} = e^3$
$x = e^3$

The inverse operation of raising e to the power x is taking natural logarithms (logarithms to the base e) and vice versa.

You can write the natural logarithm on both sides. $\ln(e^x) = x$

Leave your answer as a logarithm or a power of e so that it is exact.

Example 18

Solve these equations, giving your answers in exact form.

a $e^{2x+3} = 7$ **b** $2\ln x + 1 = 5$ **c** $e^{2x} + 5e^x = 14$

a $e^{2x+3} = 7$
$2x + 3 = \ln 7$
$2x = \ln 7 - 3$
$x = \frac{1}{2}\ln 7 - \frac{3}{2}$

b $2\ln x + 1 = 5$
$2\ln x = 4$
$\ln x = 2$
$x = e^2$

Take natural logarithms of both sides and use the fact that the inverse of e^x is $\ln x$.

Rearrange to make $\ln x$ the subject.

The inverse of $\ln x$ is e^x.

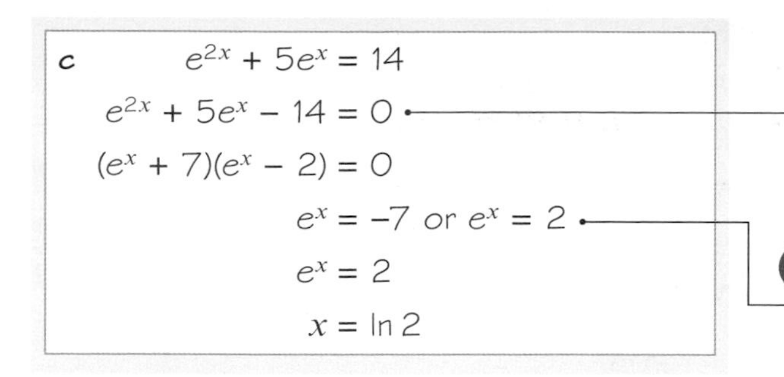

$e^{2x} = (e^x)^2$, so this is a quadratic function of e^x. Start by setting the equation equal to 0 and factorise. You could also use the substitution $u = e^x$ and write the equation as $u^2 + 5u - 14 = 0$.

Watch out e^x is always positive, so you can't have $e^x = -7$. You need to discard this solution.

Exercise 14G

1 Solve these equations, giving your answers in exact form.

a $e^x = 6$ **b** $e^{2x} = 11$ **c** $e^{-x+3} = 20$

d $3e^{4x} = 1$ **e** $e^{2x+6} = 3$ **f** $e^{5-x} = 19$

2 Solve these equations, giving your answers in exact form.

a $\ln x = 2$ **b** $\ln(4x) = 1$ **c** $\ln(2x + 3) = 4$

d $2\ln(6x - 2) = 5$ **e** $\ln(18 - x) = \frac{1}{2}$ **f** $\ln(x^2 - 7x + 11) = 0$

3 Solve these equations, giving your answers in exact form.

a $e^{2x} - 8e^x + 12 = 0$ **b** $e^{4x} - 3e^{2x} = -2$

c $(\ln x)^2 + 2\ln x - 15 = 0$ **d** $e^x - 5 + 4e^{-x} = 0$

e $3e^{2x} + 5 = 16e^x$ **f** $(\ln x)^2 = 4(\ln x + 3)$

Hint All of the equations in question 3 are quadratic equations in a function of x.

Hint First in part **d** multiply each term by e^x.

E/P **4** Find the exact solutions to the equation $e^x + 12e^{-x} = 7$. **(4 marks)**

5 Solve these equations, giving your answers in exact form.

a $\ln(8x - 3) = 2$ **b** $e^{5(x-8)} = 3$ **c** $e^{10x} - 8e^{5x} + 7 = 0$

d $(\ln x - 1)^2 = 4$

E/P **6** Solve $3^x e^{4x-1} = 5$, giving your answer in the form $\frac{a + \ln b}{c + \ln d}$ **(5 marks)**

Hint Take natural logarithms of both sides and then apply the laws of logarithms.

P **7** Officials are testing athletes for doping at a sporting event. They model the concentration of a particular drug in an athlete's bloodstream using the equation $D = 6e^{\frac{-t}{10}}$ where D is the concentration of the drug in mg/l and t is the time in hours since the athlete took the drug.

a Interpret the meaning of the constant 6 in this model.

b Find the concentration of the drug in the bloodstream after 2 hours.

c It is impossible to detect this drug in the bloodstream if the concentration is lower than 3 mg/l. Show that this happens after $t = -10\ln\left(\frac{1}{2}\right)$ and convert this result into hours and minutes.

E/P **8** The graph of $y = 3 + \ln(4 - x)$ is shown to the right.

a State the exact coordinates of point A. **(1 mark)**

b Calculate the exact coordinates of point B. **(3 marks)**

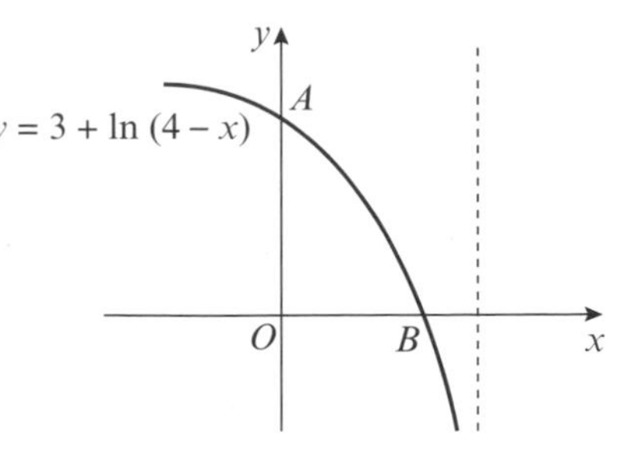

Challenge

The graph of the function $g(x) = Ae^{Bx} + C$ passes through (0, 5) and (6, 10). Given that the line $y = 2$ is an asymptote to the graph, show that $B = \frac{1}{6}\ln\left(\frac{8}{3}\right)$.

14.8 Logarithms and non-linear data

Logarithms can also be used to manage and explore non-linear trends in data.

Case 1: $y = ax^n$

Start with a non-linear relationship ———— $y = ax^n$
Take logs of both sides ($\log = \log_{10}$) ———— $\log y = \log ax^n$
Use the multiplication law ———— $\log y = \log a + \log x^n$
Use the power law ———— $\log y = \log a + n \log x$
Compare this equation to the common form of a straight line, $Y = MX + C$.

$\log y$ variable	=	n constant (gradient)	$\log x$ variable	+	$\log a$ constant (intercept)
Y variable	=	M constant (gradient)	X variable	+	C constant (intercept)

- **If $y = ax^n$ then the graph of $\log y$ against $\log x$ will be a straight line with gradient n and vertical intercept $\log a$.**

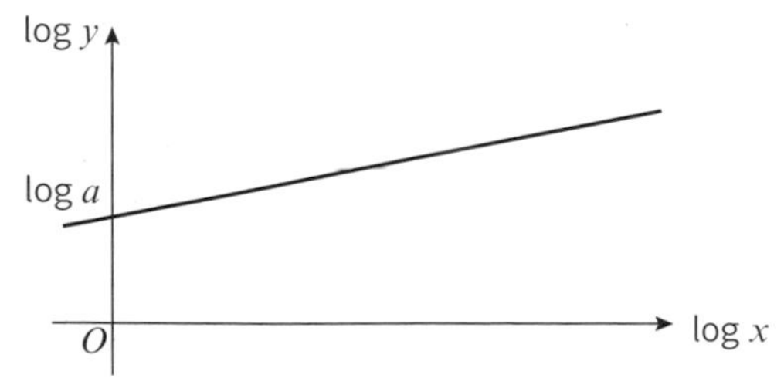

Example 19

The table below gives the rank (by size) and population of the UK's largest cities and districts (London is ranked number 1 but has been excluded as an outlier).

City	Birmingham	Leeds	Glasgow	Sheffield	Bradford
Rank, R	2	3	4	5	6
Population, P (2 s.f.)	1 000 000	730 000	620 000	530 000	480 000

The relationship between the rank and population can be modelled by the formula

$P = aR^n$ where a and n are constants.

a Draw a table giving values of $\log R$ and $\log P$ to 2 decimal places.

b Plot a graph of $\log R$ against $\log P$ using the values from your table and draw a line of best fit.

c Use your graph to estimate the values of a and n to two significant figures.

a

$\log R$	0.30	0.48	0.60	0.70	0.78
$\log P$	6	5.86	5.79	5.72	5.68

b

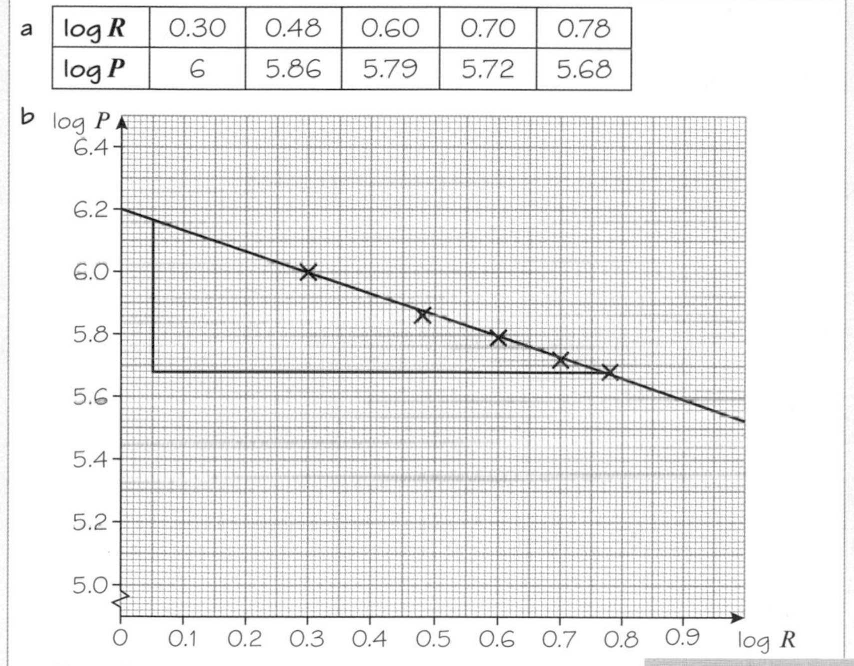

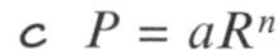

c $P = aR^n$

$\log P = \log a(R^n)$

$\log P = \log a + \log(R^n)$

$\log P = \log a + n \log R$

so the gradient is n and the intercept is $\log a$

Reading the gradient from the graph,

$$n = \frac{5.68 - 6.16}{0.77 - 0.05} = \frac{-0.48}{0.72} = -0.67$$

Reading the intercept from the graph,

$\log a = 6.2$

$a = 10^{6.2} = 1\,600\,000$ (2 s.f.).

Start with the formula given in the question. Take logs of both sides and use the laws of logarithms to rearrange it into a linear relationship between $\log P$ and $\log R$.

The gradient of the line of best fit will give you your value for n.

The vertical intercept will give you the value of $\log a$. You need to raise 10 to this power to find the value of a.

Case 2: $y = ab^x$

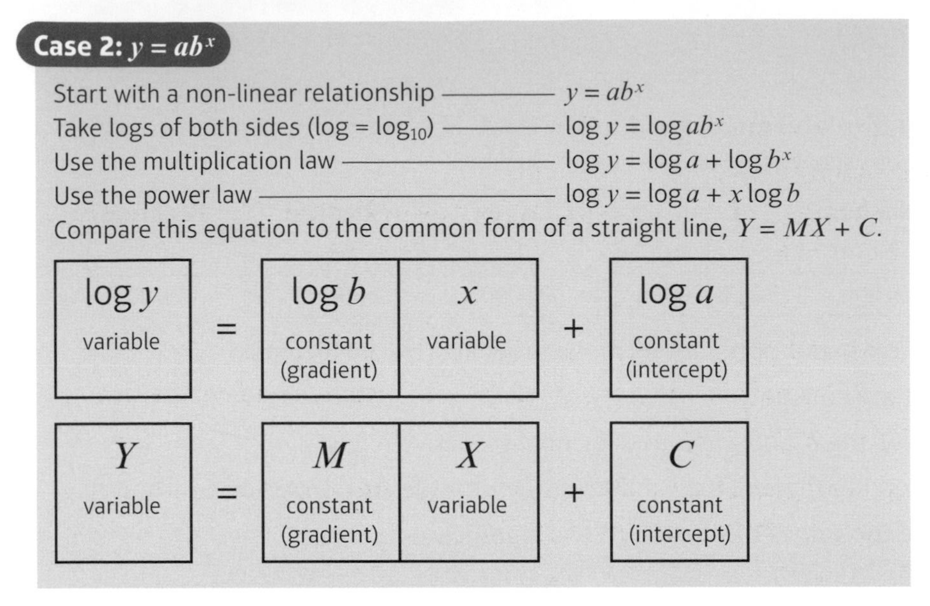

- **If $y = ab^x$ then the graph of $\log y$ against x will be a straight line with gradient $\log b$ and vertical intercept $\log a$.**

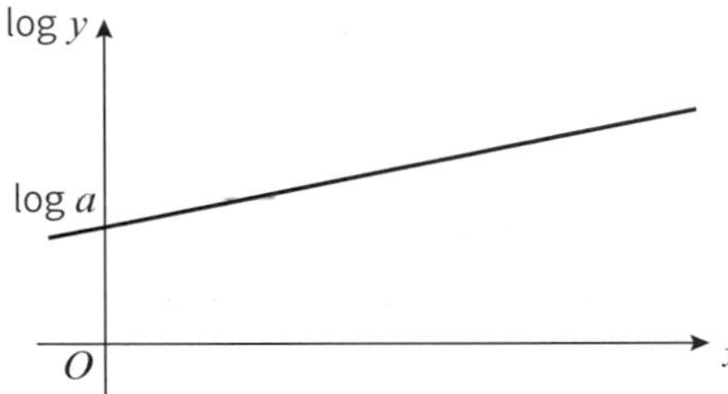

Watch out For $y = ab^x$ you need to plot $\log y$ against x to obtain a linear graph. If you plot $\log y$ against $\log x$ you will **not** get a linear relationship.

Example 20

The graph represents the growth of a population of bacteria, P, over t hours. The graph has a gradient of 0.6 and meets the vertical axis at $(0, 2)$ as shown.

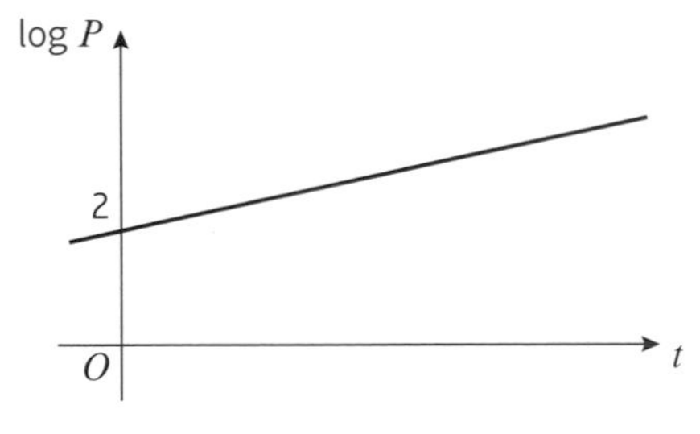

A scientist suggests that this growth can be modelled by the equation $P = ab^t$, where a and b are constants to be found.

a Write down an equation for the line.

b Using your answer to part **a** or otherwise, find the values of a and b, giving them to 3 significant figures where necessary.

c Interpret the meaning of the constant a in this model.

a $\log P = 0.6t + 2$ — $\log P = (\text{gradient}) \times t + (y\text{-intercept})$

b $P = 10^{0.6t + 2}$ — Rewrite the logarithm as a power. An alternative method would be to start with $P = ab^t$ and take logs of both sides, as in Example 19.

$P = 10^{0.6t} \times 10^2$ — Rearrange the equation into the form ab^t. You can use $x^{mn} = (x^m)^n$ to write $10^{0.6t}$ in the form b^t.

$P = 10^2 \times (10^{0.6})^t$

$P = 100 \times 3.98^t$

$a = 100$, $b = 3.98$ (3 s.f.)

c The value of a gives the initial size of the bacteria population.

Exercise 14H

1 Two variables, S and x satisfy the formula $S = 4 \times 7^x$.

a Show that $\log S = \log 4 + x \log 7$.

b The straight line graph of $\log S$ against x is plotted. Write down the gradient and the value of the intercept on the vertical axis.

2 Two variables A and x satisfy the formula $A = 6x^4$.

a Show that $\log A = \log 6 + 4 \log x$.

b The straight line graph of $\log A$ against $\log x$ is plotted. Write down the gradient and the value of the intercept on the vertical axis.

3 The data below follows a trend of the form $y = ax^n$, where a and n are constants.

x	3	5	8	10	15
y	16.3	33.3	64.3	87.9	155.1

a Copy and complete the table of values of $\log x$ and $\log y$, giving your answers to 2 decimal places.

$\log x$	0.48	0.70	0.90	1	1.18
$\log y$	1.21				2.19

b Plot a graph of $\log y$ against $\log x$ and draw in a line of best fit.

c Use your graph to estimate the values of a and n to one decimal place.

4 The data below follows a trend of the form $y = ab^x$, where a and b are constants.

x	2	3	5	6.5	9
y	124.8	424.4	4097.0	30 763.6	655 743.5

a Copy and complete the table of values of x and $\log y$, giving your answers to 2 decimal places.

x	2	3	5	6.5	9
$\log y$	2.10				

b Plot a graph of $\log y$ against x and draw in a line of best fit.

c Use your graph to estimate the values of a and b to one decimal place.

(E) **5** Kleiber's law is an empirical law in biology which connects the mass of an animal, m, to its resting metabolic rate, R. The law follows the form $R = am^b$, where a and b are constants.

The table below contains data on five animals.

Animal	Mouse	Guinea pig	Rabbit	Goat	Cow
Mass, m (kg)	0.030	0.408	4.19	34.6	650
Metabolic rate R (kcal per day)	4.2	32.3	195	760	7637

a Copy and complete this table giving values of $\log R$ and $\log m$ to 2 decimal places. **(1 mark)**

$\log m$	−1.52				
$\log R$	0.62	1.51	2.29	2.88	3.88

b Plot a graph of $\log R$ against $\log m$ using the values from your table and draw in a line of best fit. **(2 marks)**

c Use your graph to estimate the values of a and b to two significant figures. **(4 marks)**

d Using your values of a and b, estimate the resting metabolic rate of a human male with a mass of 80 kg. **(1 mark)**

6 Zipf's law is an empirical law which relates how frequently a word is used, f, to its ranking in a list of the most common words of a language, R. The law follows the form $f = AR^b$, where A and b are constants to be found.

The table below contains data on four words.

Word	'the'	'it'	'well'	'detail'
Rank, R	1	10	100	1000
Frequency per 100 000 words, f	4897	861	92	9

a Copy and complete this table giving values of $\log f$ to 2 decimal places.

$\log R$	0	1	2	3
$\log f$	3.69			

b Plot a graph of $\log f$ against $\log R$ using the values from your table and draw in a line of best fit.

c Use your graph to estimate the value of A to two significant figures and the value of b to one significant figure.

d The word 'when' is the 57th most commonly used word in the English language. A trilogy of novels contains 455 125 words. Use your values of A and b to estimate the number of times the word 'when' appears in the trilogy.

(P) **7** The table below shows the population of Mozambique between 1960 and 2010.

Year	1960	1970	1980	1990	2000	2010
Population, P (millions)	7.6	9.5	12.1	13.6	18.3	23.4

This data can be modelled using an exponential function of the form $P = ab^t$, where t is the time in years since 1960 and a and b are constants.

a Copy and complete the table below.

Time in years since 1960, t	0	10	20	30	40	50
$\log P$	0.88					

b Show that $P = ab^t$ can be rearranged into the form $\log P = \log a + t \log b$.

c Plot a graph of $\log P$ against t using the values from your table and draw in a line of best fit.

d Use your graph to estimate the values of a and b.

e Explain why an exponential model is often appropriate for modelling population growth.

Hint For part **e**, think about the relationship between P and $\frac{dP}{dt}$.

E/P **8** A scientist is modelling the number of people, N, who have fallen sick with a virus after t days.

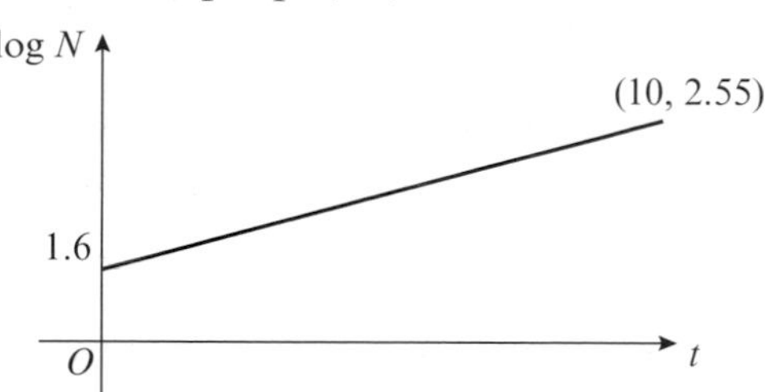

From looking at this graph, the scientist suggests that the number of sick people can be modelled by the equation $N = ab^t$, where a and b are constants to be found.

The graph passes through the points (0, 1.6) and (10, 2.55).

a Write down the equation of the line. **(2 marks)**

b Using your answer to part **a** or otherwise, find the values of a and b, giving them to 2 significant figures. **(4 marks)**

c Interpret the meaning of the constant a in this model. **(1 mark)**

d Use your model to predict the number of sick people to the nearest 100 after 30 days. Give one reason why this might be an overestimate. **(2 marks)**

P **9** A student is investigating a family of similar shapes. She measures the width, w, and the area, A, of each shape. She suspects there is a formula of the form $A = pw^q$, so she plots the logarithms of her results.

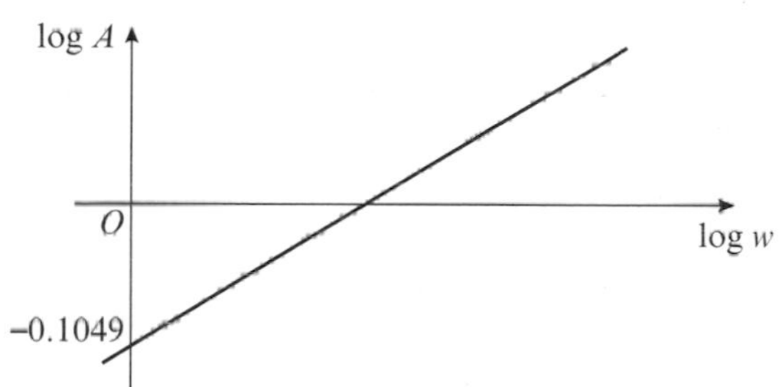

The graph has a gradient of 2 and passes through −0.1049 on the vertical axis.

a Write down an equation for the line.

b Starting with your answer to part **a**, or otherwise, find the exact value of q and the value of p to 4 decimal places.

c Suggest the name of the family of shapes that the student is investigating, and justify your answer.

Hint Multiply p by 4 and think about another name for 'half the width'.

Challenge

Find a formula to describe the relationship between the data in this table.

x	1	2	3	4
y	5.22	4.698	4.2282	3.805 38

Hint Sketch the graphs of $\log y$ against $\log x$ and $\log y$ against x. This will help you determine whether the relationship is of the form $y = ax^n$ or $y = ab^x$.

Mixed exercise 14

1 Sketch each of the following graphs, labelling all intersections and asymptotes.

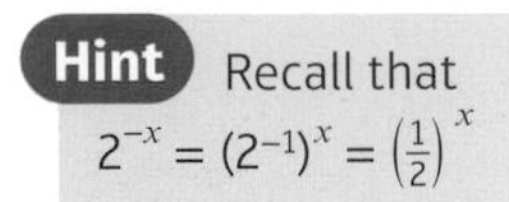

a $y = 2^{-x}$ b $y = 5e^x - 1$ c $y = \ln x$

(P) 2 a Express $\log_a (p^2 q)$ in terms of $\log_a p$ and $\log_a q$.

b Given that $\log_a (pq) = 5$ and $\log_a (p^2 q) = 9$, find the values of $\log_a p$ and $\log_a q$.

(P) 3 Given that $p = \log_q 16$, express in terms of p,

a $\log_q 2$

b $\log_q (8q)$

4 Solve these equations, giving your answers to 3 significant figures.

a $4^x = 23$ b $7^{2x+1} = 1000$ c $10^x = 6^{x+2}$

(E/P) 5 a Using the substitution $u = 2^x$, show that the equation $4^x - 2^{x+1} - 15 = 0$ can be written in the form $u^2 - 2u - 15 = 0$. **(2 marks)**

b Hence solve the equation $4^x - 2^{x+1} - 15 = 0$, giving your answer to 2 decimal places. **(3 marks)**

(E) 6 Solve the equation $\log_2 (x + 10) - \log_2 (x - 5) = 4$. **(4 marks)**

7 Differentiate each of the following expressions with respect to x.

a e^{-x} b e^{11x} c $6e^{5x}$

8 Solve the following equations, giving exact solutions.

a $\ln (2x - 5) = 8$ b $e^{4x} = 5$ c $24 - e^{-2x} = 10$

d $\ln x + \ln (x - 3) = 0$ e $e^x + e^{-x} = 2$ f $\ln 2 + \ln x = 4$

(P) 9 The price of a computer system can be modelled by the formula

$$P = 100 + 850e^{-\frac{t}{2}}$$

where P is the price of the system in £s and t is the age of the computer in years after being purchased.

a Calculate the new price of the system.

b Calculate its price after 3 years, giving your answer to the nearest £.

c When will it be worth less than £200?

d Find its price as $t \to \infty$.

e Sketch the graph showing P against t.

f Comment on the appropriateness of this model.

(P) **10** The points P and Q lie on the curve with equation $y = e^{\frac{1}{2}x}$.
The x-coordinates of P and Q are $\ln 4$ and $\ln 16$ respectively.

a Find an equation for the line PQ.

b Show that this line passes through the origin O.

c Calculate the length, to 3 significant figures, of the line segment PQ.

(E/P) **11** The temperature, T°C, of a cup of tea is given by $T = 55e^{-\frac{t}{8}} + 20 \quad t \geqslant 0$
where t is the time in minutes since measurements began.

a Briefly explain why $t \geqslant 0$. **(1 mark)**

b State the starting temperature of the cup of tea. **(1 mark)**

c Find the time at which the temperature of the tea is 50 °C, giving your answer to the nearest minute. **(3 marks)**

d By sketching a graph or otherwise, explain why the temperature of the tea will never fall below 20 °C. **(2 marks)**

(E) **12** The table below gives the surface area, S, and the volume, V of five different spheres, rounded to 1 decimal place.

S	18.1	50.3	113.1	221.7	314.2
V	7.2	33.5	113.1	310.3	523.6

Given that $S = aV^b$, where a and b are constants,

a show that $\log S = \log a + b \log V$. **(2 marks)**

b copy and complete the table of values of $\log S$ and $\log V$, giving your answers to 2 decimal places. **(1 mark)**

log S					
log V	0.86				

c plot a graph of $\log V$ against $\log S$ and draw in a line of best fit. **(2 marks)**

d use your graph to confirm that $b = 1.5$ and estimate the value of a to one significant figure. **(4 marks)**

(E/P) **13** The radioactive decay of a substance is modelled by the formula $R = 140e^{kt} \quad t \geqslant 0$
where R is a measure of radioactivity (in counts per minute) at time t days, and k is a constant.

a Explain briefly why k must be negative. **(1 mark)**

b Sketch the graph of R against t. **(2 marks)**

After 30 days the radiation is measured at 70 counts per minute.

c Show that $k = c \ln 2$, stating the value of the constant c. **(3 marks)**

(P) **14** The total number of views (in millions) V of a viral video in x days is modelled by

$$V = e^{0.4x} - 1$$

a Find the total number of views after 5 days, giving your answer to 2 significant figures.

b Find $\frac{dV}{dx}$.

c Find the rate of increase of the number of views after 100 days, stating the units of your answer.

d Use your answer to part **c** to comment on the validity of the model after 100 days.

(P) **15** The moment magnitude scale is used by seismologists to express the sizes of earthquakes. The scale is calculated using the formula

$$M = \frac{2}{3}\log_{10}(S) - 10.7$$

where S is the seismic moment in dyne cm.

a Find the magnitude of an earthquake with a seismic moment of 2.24×10^{22} dyne cm.

b Find the seismic moment of an earthquake with
i magnitude 6 **ii** magnitude 7

c Using your answers to part **b** or otherwise, show that an earthquake of magnitude 7 is approximately 32 times as powerful as an earthquake of magnitude 6.

(E/P) **16** A student is asked to solve the equation

$$\log_2 x - \frac{1}{2}\log_2(x + 1) = 1$$

The student's attempt is shown

$$\log_2 x - \log_2\sqrt{x+1} = 1$$
$$x - \sqrt{x+1} = 2^1$$
$$x - 2 = \sqrt{x+1}$$
$$(x-2)^2 = x + 1$$
$$x^2 - 5x + 3 = 0$$
$$x = \frac{5+\sqrt{13}}{2} \quad x = \frac{5-\sqrt{13}}{2}$$

a Identify the error made by the student. **(1 mark)**

b Solve the equation correctly. **(3 marks)**

Challenge

a Given that $y = 9^x$, show that $\log_3 y = 2x$.

b Hence deduce that $\log_3 y = \log_9 y^2$.

c Use your answer to part **b** to solve the equation $\log_3(2 - 3x) = \log_9(6x^2 - 19x + 2)$

Summary of key points

1 For all real values of x:

- If $f(x) = e^x$ then $f'(x) = e^x$
- If $y = e^x$ then $\frac{dy}{dx} = e^x$

2 For all real values of x and for any constant k:

- If $f(x) = e^{kx}$ then $f'(x) = ke^{kx}$
- If $y = e^{kx}$ then $\frac{dy}{dx} = ke^{kx}$

3 $\log_a n = x$ is equivalent to $a^x = n$ $(a \neq 1)$

4 The laws of logarithms:

- $\log_a x + \log_a y = \log_a xy$ (the multiplication law)
- $\log_a x - \log_a y = \log_a \left(\frac{x}{y}\right)$ (the division law)
- $\log_a (x^k) = k \log_a x$ (the power law)

5 You should also learn to recognise the following special cases:

- $\log_a \left(\frac{1}{x}\right) = \log_a (x^{-1}) = -\log_a x$ (the power law when $k = -1$)
- $\log_a a = 1$ $(a > 0, a \neq 1)$
- $\log_a 1 = 0$ $(a > 0, a \neq 1)$

6 Whenever $f(x) = g(x)$, $\log_a f(x) = \log_a g(x)$

7 The graph of $y = \ln x$ is a reflection of the graph $y = e^x$ in the line $y = x$.

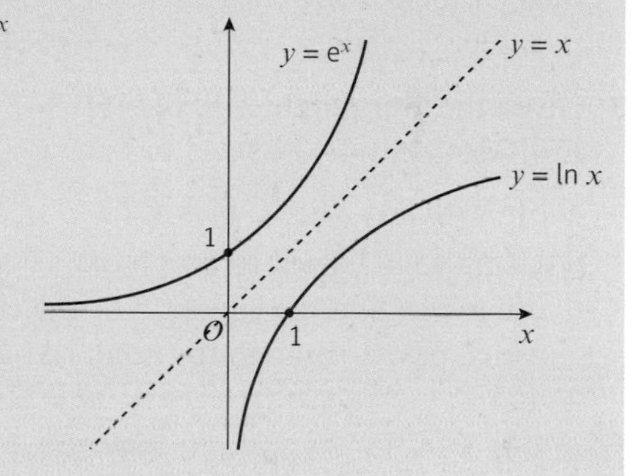

8 $e^{\ln x} = \ln (e^x) = x$

9 If $y = ax^n$ then the graph of $\log y$ against $\log x$ will be a straight line with gradient n and vertical intercept $\log a$.

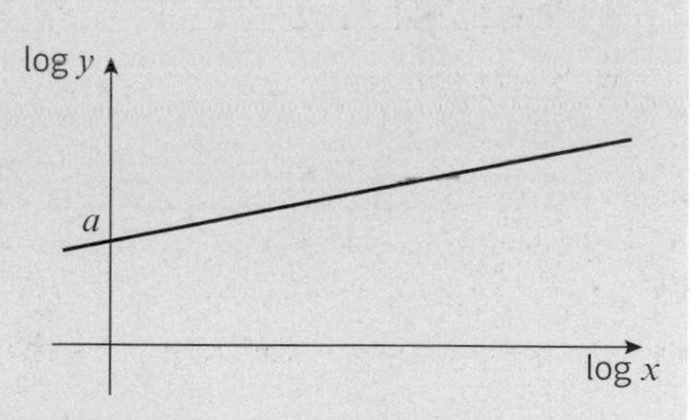

10 If $y = ab^x$ then the graph of $\log y$ against x will be a straight line with gradient $\log b$ and vertical intercept $\log a$.

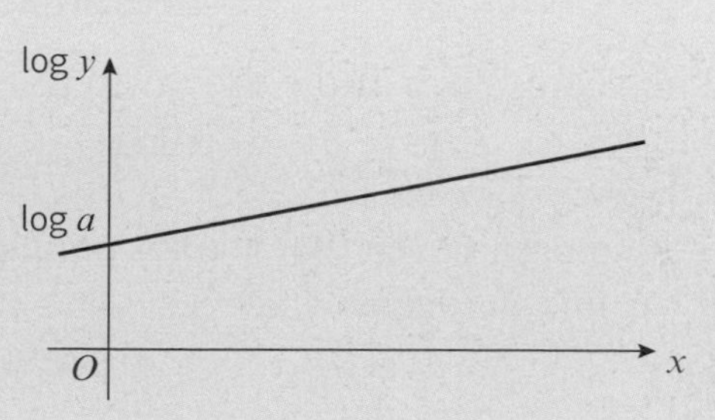

3 Review exercise

(E) **1** The vector $9\mathbf{i} + q\mathbf{j}$ is parallel to the vector $2\mathbf{i} - \mathbf{j}$. Find the value of the constant q. **(2)**

← Section 11.2

(E/P) **2** Given that $|5\mathbf{i} - k\mathbf{j}| = |2k\mathbf{i} + 2\mathbf{j}|$, find the exact value of the positive constant k. **(4)**

← Section 11.3

(E/P) **3** Given the four points $X(9, 6)$, $Y(13, -2)$, $Z(0, -15)$, and $C(1, -3)$,

a Show that $|\overrightarrow{CX}| = |\overrightarrow{CY}| = |\overrightarrow{CZ}|$. **(3)**

b Using your answer to part **a** or otherwise, find the equation of the circle which passes through the points X, Y and Z. **(3)**

← Sections 6.2, 11.4

(E/P) **4** In the triangle ABC, $\overrightarrow{AB} = 9\mathbf{i} + 2\mathbf{j}$ and $\overrightarrow{AC} = 7\mathbf{i} - 6\mathbf{j}$.

a Find $\overrightarrow{BC}$. **(2)**

b Prove that the triangle ABC is isosceles. **(3)**

c Show that $\cos \angle ABC = \frac{1}{\sqrt{5}}$ **(4)**

← Sections 9.1, 11.5

(E/P) **5** The vectors **a**, **b** and **c** are given as $\mathbf{a} = \begin{pmatrix} 8 \\ 23 \end{pmatrix}$, $\mathbf{b} = \begin{pmatrix} -15 \\ x \end{pmatrix}$ and $\mathbf{c} = \begin{pmatrix} -13 \\ 2 \end{pmatrix}$, where x is an integer. Given that $\mathbf{a} + \mathbf{b}$ is parallel to $\mathbf{b} - \mathbf{c}$, find the value of x. **(4)**

← Section 11.2

(E) **6** Two forces, $\mathbf{F}_1$ and $\mathbf{F}_2$, act on a particle.

$\mathbf{F}_1 = 2\mathbf{i} - 5\mathbf{j}$ newtons

$\mathbf{F}_2 = \mathbf{i} + \mathbf{j}$ newtons

The resultant force **R** acting on the particle is given by $\mathbf{R} = \mathbf{F}_1 + \mathbf{F}_2$.

a Calculate the magnitude of **R** in newtons. **(3)**

A third force, $\mathbf{F}_3$ begins to act on the particle, where $\mathbf{F}_3 = k\mathbf{j}$ newtons and k is a positive constant. The new resultant force is given by $\mathbf{R}_{\text{new}} = \mathbf{F}_1 + \mathbf{F}_2 + \mathbf{F}_3$.

b Given that the angle between the line of action of $\mathbf{R}_{\text{new}}$ and the vector **i** is 45 degrees, find the value of k. **(3)**

← Section 11.6

(E/P) **7** A helicopter takes off from its starting position O and travels 100 km on a bearing of 060°. It then travels 30 km due east before landing at point A. Given that the position vector of A relative to O is $(m\mathbf{i} + n\mathbf{j})$ km, find the exact values of m and n. **(4)**

← Sections 10.2, 11.6

(E/P) **8** At the very end of a race, Boat A has a position vector of $(-65\mathbf{i} + 180\mathbf{j})$ m and Boat B has a position vector of $(100\mathbf{i} + 120\mathbf{j})$ m. The finish line has a position vector of $10\mathbf{i}$ km.

a Show that Boat B is closer to the finish line than Boat A. **(2)**

Boat A is travelling at a constant velocity of $(2.5\mathbf{i} - 6\mathbf{j})$ m/s and Boat B is travelling at a constant velocity of $(-3\mathbf{i} - 4\mathbf{j})$ m/s.

b Calculate the speed of each boat. Hence, or otherwise, determine the result of the race. **(4)**

← Section 11.6

(E/P) **9** Prove, from first principles, that the derivative of $5x^2$ is $10x$. **(4)**

← Section 12.2

E **10** Given that $y = 4x^3 - 1 + 2x^{\frac{1}{2}}$, $x > 0$,
find $\frac{dy}{dx}$. **(2)**

← Section 12.5

E/P **11** The curve C has equation
$y = 4x + 3x^{\frac{3}{2}} - 2x^2$, $x > 0$.

a Find an expression for $\frac{dy}{dx}$ **(2)**

b Show that the point $P(4, 8)$ lies on C. **(1)**

c Show that an equation of the normal to C at point P is $3y = x + 20$. **(2)**

The normal to C at P cuts the x-axis at point Q.

d Find the length PQ, giving your answer in simplified surd form. **(2)**

← Section 12.6

E/P **12** The curve C has equation
$y = 4x^2 + \frac{5 - x}{x}$, $x \neq 0$. The point P on C has x-coordinate 1.

a Show that the value of $\frac{dy}{dx}$ at P is 3. **(3)**

b Find an equation of the tangent to C at P. **(3)**

This tangent meets the x-axis at the point $(k, 0)$.

c Find the value of k. **(1)**

← Section 12.6

E/P **13** $f(x) = \frac{(2x + 1)(x + 4)}{\sqrt{x}}$, $x > 0$.

a Show that $f(x)$ can be written in the form $Px^{\frac{3}{2}} + Qx^{\frac{1}{2}} + Rx^{-\frac{1}{2}}$, stating the values of the constants P, Q and R. **(2)**

b Find $f'(x)$. **(3)**

c A curve has equation $y = f(x)$. Show that the tangent to the curve at the point where $x = 1$ is parallel to the line with equation $2y = 11x + 3$. **(3)**

← Section 12.6

E/P **14** Prove that the function $f(x) = x^3 - 12x^2 + 48x$ is increasing for all $x \in \mathbb{R}$. **(3)**

← Section 12.7

E/P **15** The diagram shows part of the curve with equation $y = x + \frac{2}{x} - 3$. The curve crosses the x-axis at A and B and the point C is the minimum point of the curve.

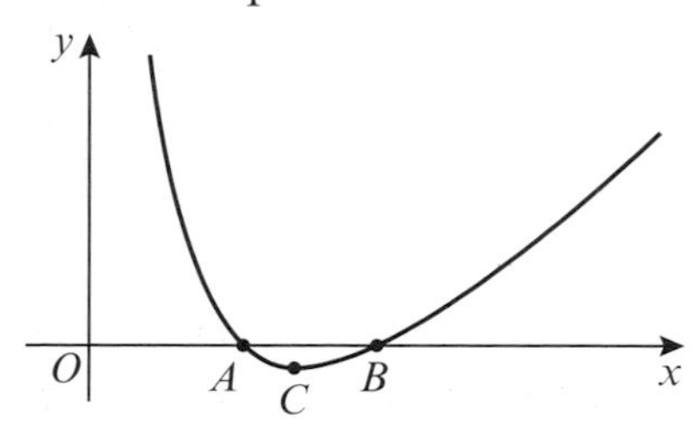

a Find the coordinates of A and B. **(2)**

b Find the exact coordinates of C, giving your answers in surd form. **(4)**

← Section 12.9

E/P **16** A company makes solid cylinders of variable radius r cm and constant volume 128π cm³.

a Show that the surface area of the cylinder is given by $S = \frac{256\pi}{r} + 2\pi r^2$. **(2)**

b Find the minium value for the surface area of the cylinder. **(4)**

← Section 12.11

E **17** Given that $y = 3x^2 + 4\sqrt{x}$, $x > 0$, find

a $\frac{dy}{dx}$ **(2)**

b $\frac{d^2y}{dx^2}$ **(2)**

c $\int y\,dx$ **(3)**

← Sections 12.8, 13.2

E **18** The curve C with equation $y = f(x)$ passes through the point $(5, 65)$.
Given that $f'(x) = 6x^2 - 10x - 12$,

a use integration to find $f(x)$ **(3)**

b hence show that $f(x) = x(2x + 3)(x - 4)$ **(2)**

c sketch C, showing the coordinates of the points where C crosses the x-axis. **(3)**

← Sections 4.1, 13.3

19 Use calculus to evaluate $\int_1^8 (x^{\frac{1}{3}} - x^{-\frac{1}{3}})\,dx$.

← Section 13.4

E/P **20** Given that $\int_0^6 (x^2 - kx)\,dx = 0$, find the value of the constant k. **(3)**

← Section 13.4

E/P **21** The diagram shows a section of the curve with equation $y = -x^4 + 3x^2 + 4$. The curve intersects the x-axis at points A and B. The finite region R, which is shown shaded, is bounded by the curve and the x-axis.

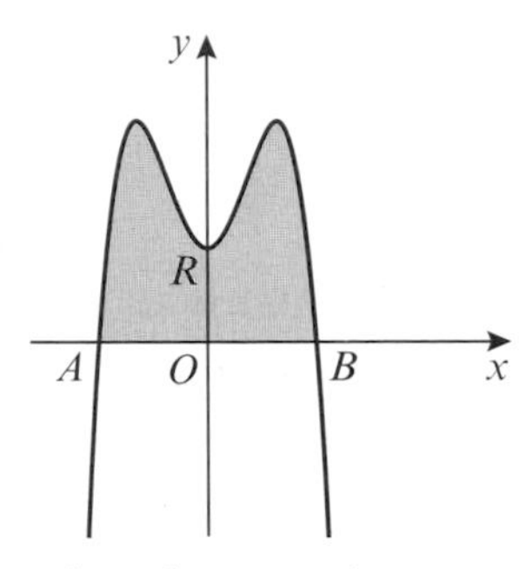

a Show that the equation $-x^4 + 3x^2 + 4 = 0$ only has two solutions, and hence or otherwise find the coordinates of A and B. **(3)**

b Find the area of the region R. **(4)**

← Sections 4.2, 13.5

E **22** The diagram shows the shaded region T which is bounded by the curve $y = (x - 1)(x - 4)$ and the x-axis. Find the area of the shaded region T. **(4)**

← Section 13.6

E/P **23** The diagram shows the curve with equation $y = 5 - x^2$ and the line with equation $y = 3 - x$. The curve and the line intersect at the points P and Q.

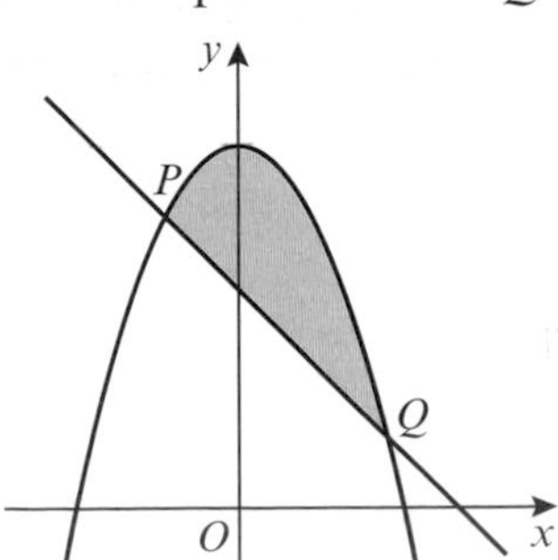

a Find the coordinates of P and Q. **(3)**

b Find the area of the finite region between PQ and the curve. **(6)**

← Section 13.7

E **24** The graph of the function $f(x) = 3e^{-x} - 1$, $x \in \mathbb{R}$, has an asymptote $y = k$, and crosses the x and y axes at A and B respectively, as shown in the diagram.

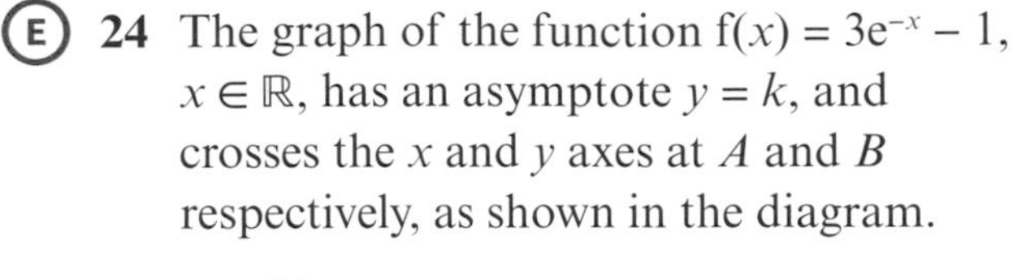

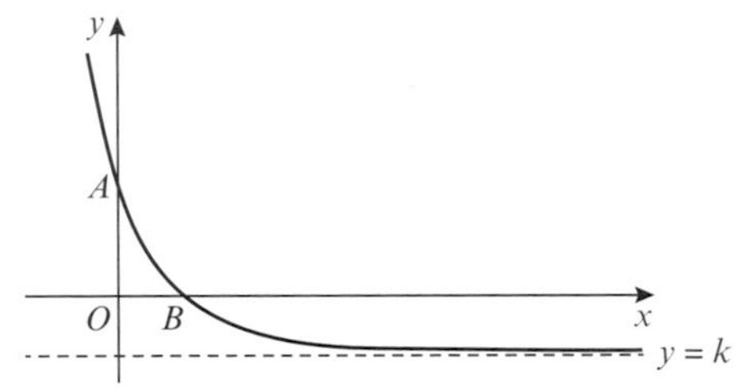

a Write down the value of k and the y-coordinate of A. **(2)**

b Find the exact value of the x-coordinate of B, giving your answer as simply as possible. **(2)**

← Sections 14.2, 14.7

E/P **25** A heated metal ball S is dropped into a liquid. As S cools, its temperature, T°C, t minutes after it enters the liquid, is given by

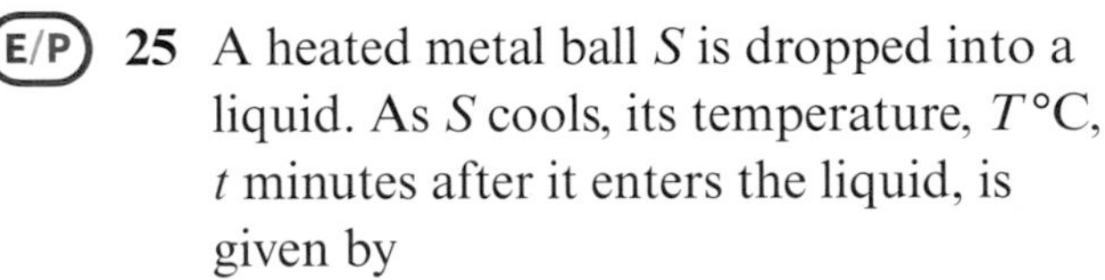

$T = 400e^{-0.05t} + 25, \quad t \geqslant 0.$

a Find the temperature of S as it enters the liquid. **(1)**

b Find how long S is in the liquid before its temperature drops to 300 °C. Give your answer to 3 significant figures. **(3)**

c Find the rate, $\frac{dT}{dt}$, in °C per minute to 3 significant figures, at which the temperature of S is decreasing at the instant $t = 50$. **(3)**

d With reference to the equation given above, explain why the temperature of S can never drop to 20 °C. **(2)**

← Sections 14.3, 14.7

E **26** **a** Find, to 3 significant figures, the value of x for which $5^x = 0.75$. **(2)**

b Solve the equation $2\log_5 x - \log_5 3x = 1$ **(3)**

← Sections 14.5, 14.6

E **27** **a** Solve $3^{2x-1} = 10$, giving your answer to 3 significant figures. **(3)**

b Solve $\log_2 x + \log_2(9 - 2x) = 2$ **(3)**

← Sections 14.5, 14.6

E/P **28** **a** Express $\log_p 12 - (\frac{1}{2}\log_p 9 + \frac{1}{3}\log_p 8)$ as a single logarithm to base p. **(3)**

b Find the value of x in $\log_4 x = -1.5$. **(2)**

← Sections 14.4, 14.5

E/P **29** Find the exact solutions to the equations

a $\ln x + \ln 3 = \ln 6$ **(2)**

b $e^x + 3e^{-x} = 4$ **(4)**

← Section 14.7

30 The table below shows the population of Angola between 1970 and 2010.

Year	Population, P (millions)
1970	5.93
1980	7.64
1990	10.33
2000	13.92
2010	19.55

This data can be modelled using an exponential function of the form $P = ab^t$, where t is the time in years since 1970 and a and b are constants.

a Copy and complete the table below, giving your answers to 2 decimal places. **(1)**

Time in years since 1970, t	$\log P$
0	0.77
10	
20	
30	
40	

b Plot a graph of $\log P$ against t using the values from your table and draw in a line of best fit. **(2)**

c By rearranging $P = ab^t$, explain how the graph you have just drawn supports the assumed model. **(3)**

d Use your graph to estimate the values of a and b to two significant figures. **(4)**

← Section 14.8

Challenge

1 The position vector of a moving object is given by $(\cos\theta)\mathbf{i} + (\sin\theta)\mathbf{j}$, where $0 \leqslant \theta \leqslant 90°$.

a Find the value of θ when the object has a bearing of 090° from the origin.

b Calculate the magnitude of the position vector. ← Sections 10.2, 10.3, 11.3, 11.4

2 The graph of the cubic function $y = f(x)$ has turning points at $(-3, 76)$ and $(2, -49)$.

a Show that $f'(x) = k(x^2 + x - 6)$, where k is a constant.

b Express $f(x)$ in the form $ax^3 + bx^2 + cx + d$, where a, b, c and d are real constants to be found. ← Sections 12.9, 13.3

3 Given that $\int_0^9 f(x)\,dx = 24.2$, state the value of $\int_0^9 (f(x) + 3)\,dx$. ← Sections 4.5, 13.5

4 The functions f and g are defined as $f(x) = x^3 - kx + 1$, where k is a constant, and $g(x) = e^{2x}$, $x \in \mathbb{R}$. The graphs of $y = f(x)$ and $y = g(x)$ intersect at the point P, where $x = 0$.

a Confirm that $f(0) = g(0)$ and hence state the coordinates of P.

b Given that the tangents to the graphs at P are perpendicular, find the value of k.

← Sections 5.3, 14.3

Exam-style practice

Mathematics
AS Level
Paper 1: Pure Mathematics

Time: 2 hours
You must have: Mathematical Formulae and Statistical Tables, Calculator

---

1 **a** Given that $4 = 64^n$, find the value of n. **(1)**

b Write $\sqrt{50}$ in the form $k\sqrt{2}$ where k is an integer to be determined. **(1)**

2 Find the equation of the line parallel to $2x - 3y + 4 = 0$ that passes through the point (5, 6). Give your answer in the form $y = ax + b$ where a and b are rational numbers. **(3)**

3 A student is asked to evaluate the integral $\int_1^2 \left(x^4 - \frac{3}{\sqrt{x}} + 2\right) \mathrm{d}x$

The student's working is shown below

$$\int_1^2 \left(x^4 - \frac{3}{\sqrt{x}} + 2\right)\mathrm{d}x = \int_1^2 (x^4 - 3x^{\frac{1}{2}} + 2)\mathrm{d}x$$
$$= \left[\frac{x^5}{5} - 2x^{\frac{3}{2}} + 2x\right]_1^2$$
$$= \left(\frac{1}{5} - 2 + 2\right) - \left(\frac{32}{5} - 2\sqrt{8} + 4\right)$$
$$= -4.54 \text{ (3 s.f.)}$$

a Identify two errors made by the student. **(2)**

b Evaluate the definite integral, giving your answer correct to 3 significant figures. **(2)**

4 Find all the solutions in the interval $0 \leqslant x \leqslant 180°$ of

$$2\sin^2(2x) - \cos(2x) - 1 = 0$$

giving each solution in degrees. **(7)**

5 A rectangular box has sides measuring x cm, $x + 3$ cm and $2x$ cm.

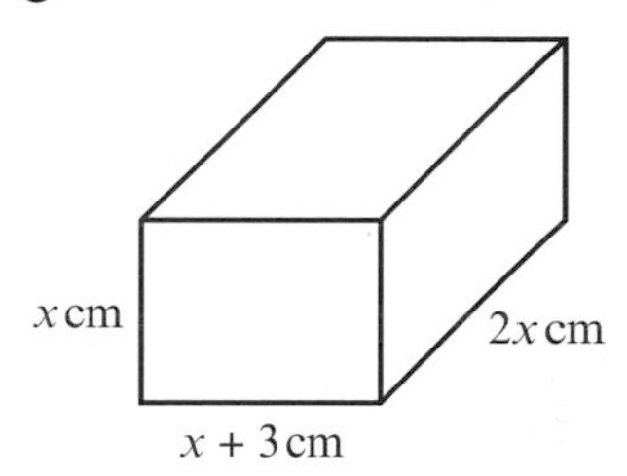

Figure 1

a Write down an expression for the volume of the box. **(1)**

Given that the volume of the box is 980 cm^3,

b Show that $x^3 + 3x^2 - 490 = 0$. **(2)**

c Show that $x = 7$ is a solution to this equation. **(1)**

d Prove that the equation has no other real solutions. **(4)**

6 $f(x) = x^3 - 5x^2 - 2 + \frac{1}{x^2}$

The point P with x-coordinate -1 lies on the curve $y = f(x)$. Find the equation of the normal to the curve at P, giving your answer in the form $ax + by + c = 0$ where a, b and c are positive integers. **(7)**

7 The population, P, of a colony of endangered Caledonian owlet-nightjars can be modelled by the equation $P = ab^t$ where a and b are constants and t is the time, in months, since the population was first recorded.

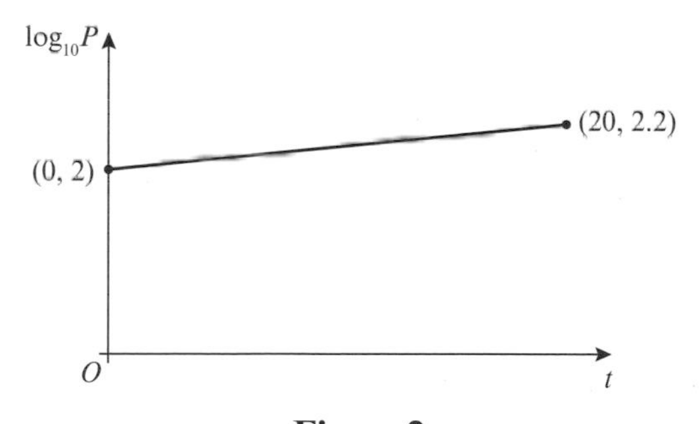

Figure 2

The line l shown in figure 2 shows the relationship between t and $\log_{10}P$ for the population over a period of 20 years.

a Write down an equation of line l. **(3)**

b Work out the value of a and interpret this value in the context of the model. **(3)**

c Work out the value of b, giving your answer correct to 3 decimal places. **(2)**

d Find the population predicted by the model when $t = 30$. **(1)**

8 Prove that $1 + \cos^4 x - \sin^4 x \equiv 2\cos^2 x$. **(4)**

9 Relative to a fixed origin, point A has position vector $6\mathbf{i} - 3\mathbf{j}$ and point B has position vector $4\mathbf{i} + 2\mathbf{j}$.

Find the magnitude of the vector $\overrightarrow{AB}$ and the angle it makes with the unit vector $\mathbf{i}$. **(5)**

10 A triangular lawn ABC is shown in figure 3:

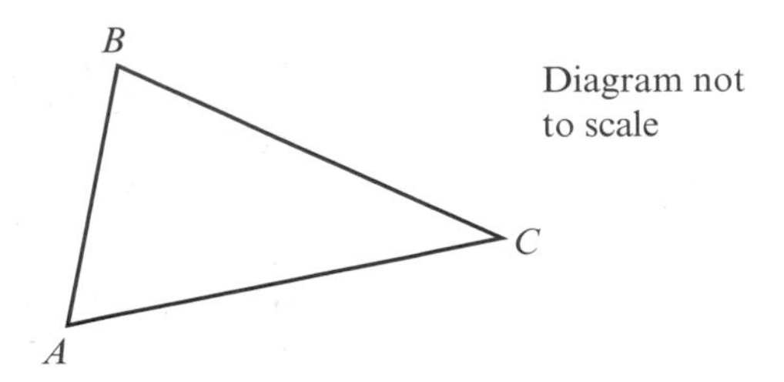

Figure 3

Given that $AB = 7.5\,\text{m}$, $BC = 10.6\,\text{m}$ and $AC = 12.7\,\text{m}$,

a Find angle BAC. **(3)**

Grass seed costs £1.25 per square metre.

b Find the cost of seeding the whole lawn. **(5)**

11 $\text{g}(x) = (x - 2)^2(x + 1)(x - 7)$

a Sketch the curve $y = \text{g}(x)$, showing the coordinates of any points where the curve meets or cuts the coordinate axes. **(4)**

b Write down the roots of the equation $\text{g}(x + 3) = 0$. **(1)**

12 Given that $9^{2x} = 27^{x^2-5}$, find the possible values of x. **(6)**

13 $\text{f}(x) = (1 - 3x)^5$

a Expand $\text{f}(x)$, in ascending powers of x, up to the term in x^2. Give each term in its simplest form. **(3)**

b Hence find an approximate value for 0.97^5. **(2)**

c State, with a reason, whether your approximation is greater or smaller than the true value. **(2)**

14 $\text{f}'(x) = \dfrac{\sqrt{x} - x^2 - 1}{x^2}, x > 0$

a Show that $\text{f}(x)$ can be written as $\text{f}(x) = -\dfrac{x^2 + 2\sqrt{x} - 1}{x} + c$ where c is a constant. **(5)**

Given that $\text{f}(x)$ passes through the point $(3, -1)$,

b find the value of c. Give your answer in the form $p + q\sqrt{r}$ where p, q and r are rational numbers to be found. **(4)**

15 A circle, C, has equation $x^2 + y^2 - 4x + 6y = 12$

a Show that the point $A(5, 1)$ lies on C and find the centre and radius of the circle. **(5)**

b Find the equation of the tangent to C at point A. Give your answer in the form $y = ax + b$ where a and b are rational numbers. **(4)**

c The curve $y = x^2 - 2$ intersects this tangent at points P and Q. Given that O is the origin, find, as a fraction in simplest form, the exact area of the triangle POQ. **(7)**

Answers

CHAPTER 1

Prior knowledge check

1 a $2m^2n + 3mn^2$ b $6x^2 - 12x - 10$
2 a 2^8 b 2^4 c 2^6
3 a $3x + 12$ b $10 - 15x$ c $12x - 30y$
4 a 8 b $2x$ c xy
5 a $2x$ b $10x$ c $\frac{5x}{3}$

Exercise 1A

1 a x^7 b $6x^5$ c k d $2p^2$
e x f y^{10} g $5x^2$ h p^2
i $2a^3$ j $2p$ k $6a^9$ l $3a^2b^3$
m $27x^8$ n $24x^{11}$ o $63a^{12}$ p $32y^6$
q $4a^6$ r $6a^{12}$
2 a $9x - 18$ b $x^2 + 9x$
c $-12y + 9y^2$ d $xy + 5x$
e $-3x^2 - 5x$ f $-20x^2 - 5x$
g $4x^2 + 5x$ h $-15y + 6y^3$
i $-10x^2 + 8x$ j $3x^3 - 5x^2$
k $4x - 1$ l $2x - 4$
m $9d^2 - 2c$ n $13 - r^2$
o $3x^3 - 2x^2 + 5x$ p $14y^2 - 35y^3 + 21y^4$
q $-10y^2 + 14y^3 - 6y^4$ r $4x + 10$
s $11x - 6$ t $7x^2 - 3x + 7$
u $-2x^2 + 26x$ v $-9x^3 + 23x^2$
3 a $3x^3 + 5x^5$ b $3x^4 - x^6$ c $\frac{x^3}{2} - x$
d $4x^2 + \frac{5}{2}$ e $\frac{7x^6}{5} + x$ f $3x^4 - \frac{5x^2}{3}$

Exercise 1B

1 a $x^2 + 11x + 28$
b $x^2 - x - 6$
c $x^2 - 4x + 4$
d $2x^2 + 3x - 2xy - 3y$
e $4x^2 + 11xy - 3y^2$
f $6x^2 - 10xy - 4y^2$
g $2x^2 - 11x + 12$
h $9x^2 + 12xy + 4y^2$
i $4x^2 + 6x + 16xy + 24y$
j $2x^2 + 3xy + 5x + 15y - 25$
k $3x^2 - 4xy - 8x + 4y + 5$
l $2x^2 + 5x - 7xy - 4y^2 - 20y$
m $x^2 + 2x + 2xy + 6y - 3$
n $2x^2 + 15x + 2xy + 12y + 18$
o $13y - 4x + 12 - 4y^2 + xy$
p $12xy - 4y^2 + 3y + 15x + 10$
q $5xy - 20y - 2x^2 + 11x - 12$
r $22y - 4y^2 - 5x + xy - 10$
2 a $5x^2 - 15x - 20$
b $14x^2 + 7x - 70$
c $3x^2 - 18x + 27$
d $x^3 - xy^2$
e $6x^3 + 8x^2 + 3x^2y + 4xy$
f $x^2y - 4xy - 5y$
g $12x^2y + 6xy - 8xy^2 - 4y^2$
h $19xy - 35y - 2x^2y$
i $10x^3 - 4x^2 + 5x^2y - 2xy$
j $x^3 + 3x^2y - 2x^2 + 6xy - 8x$
k $2x^2y + 9xy + xy^2 + 5y^2 - 5y$
l $6x^2y + 4xy^2 + 2y^2 - 3xy - 3y$
m $2x^3 + 2x^2y - 7x^2 + 3xy - 15x$
n $24x^3 - 6x^2y - 26x^2 + 2xy + 6x$
o $6x^3 + 15x^2 - 3x^2y - 18xy^2 - 30xy$
p $x^3 + 6x^2 + 11x + 6$
q $x^3 + x^2 - 14x - 24$
r $x^3 - 3x^2 - 13x + 15$
s $x^3 - 12x^2 + 47x - 60$
t $2x^3 - x^2 - 5x - 2$
u $6x^3 + 19x^2 + 11x - 6$
v $18x^3 - 15x^2 - 4x + 4$
w $x^3 - xy^2 - x^2 + y^2$
x $8x^3 - 36x^2y + 54xy^2 - 27y^3$
3 $2x^2 - xy + 29x - 7y + 24$
4 $4x^3 + 12x^2 + 5x - 6\ \text{cm}^3$
5 $a = 12, b = 32, c = 3, d = -5$

Challenge
$x^4 + 4x^3y + 6x^2y^2 + 4xy^3 + y^4$

Exercise 1C

1 a $4(x + 2)$ b $6(x - 4)$
c $5(4x + 3)$ d $2(x^2 + 2)$
e $4(x^2 + 5)$ f $6x(x - 3)$
g $x(x - 7)$ h $2x(x + 2)$
i $x(3x - 1)$ j $2x(3x - 1)$
k $5y(2y - 1)$ l $7x(5x - 4)$
m $x(x + 2)$ n $y(3y + 2)$
o $4x(x + 3)$ p $5y(y - 4)$
q $3xy(3y + 4x)$ r $2ab(3 - b)$
s $5x(x - 5y)$ t $4xy(3x + 2y)$
u $5y(3 - 4z^2)$ v $6(2x^2 - 5)$
w $xy(y - x)$ x $4y(3y - x)$
2 a $x(x + 4)$ b $2x(x + 3)$
c $(x + 8)(x + 3)$ d $(x + 6)(x + 2)$
e $(x + 8)(x - 5)$ f $(x - 6)(x - 2)$
g $(x + 2)(x + 3)$ h $(x - 6)(x + 4)$
i $(x - 5)(x + 2)$ j $(x + 5)(x - 4)$
k $(2x + 1)(x + 2)$ l $(3x - 2)(x + 4)$
m $(5x - 1)(x - 3)$ n $2(3x + 2)(x - 2)$
o $(2x - 3)(x + 5)$ p $2(x^2 + 3)(x^2 + 4)$
q $(x + 2)(x - 2)$ r $(x + 7)(x - 7)$
s $(2x + 5)(2x - 5)$ t $(3x + 5y)(3x - 5y)$
u $4(3x + 1)(3x - 1)$ v $2(x + 5)(x - 5)$
w $2(3x - 2)(x - 1)$ x $3(5x - 1)(x + 3)$
3 a $x(x^2 + 2)$ b $x(x^2 - x + 1)$
c $x(x^2 - 5)$ d $x(x + 3)(x - 3)$
e $x(x - 4)(x + 3)$ f $x(x + 5)(x + 6)$
g $x(x - 1)(x - 6)$ h $x(x + 8)(x - 8)$
i $x(2x + 1)(x - 3)$ j $x(2x + 3)(x + 5)$
k $x(x + 2)(x - 2)$ l $3x(x + 4)(x + 5)$
4 $(x^2 + y^2)(x + y)(x - y)$
5 $x(3x + 5)(2x - 1)$

Challenge
$(x - 1)(x + 1)(2x + 3)(2x - 3)$

Exercise 1D

1 a x^5 b x^{-2} c x^4 d x^3
e x^5 f $12x^0 = 12$ g $3x^{\frac{1}{2}}$ h $5x$
i $6x^{-1}$ j $x^{\frac{5}{6}}$ k $x^{\frac{17}{6}}$ l $x^{\frac{1}{6}}$

2 a 5 b 729 c 3 d $\frac{1}{16}$
e $\frac{1}{3}$ f $\frac{-1}{125}$ g 1 h 216
i $\frac{125}{64}$ j $\frac{9}{4}$ k $\frac{5}{6}$ l $\frac{64}{49}$

3 a $8x^5$ b $\frac{5}{x^2} - \frac{2}{x^3}$ c $5x^4$
d $\frac{1}{x^2} + 4$ e $\frac{2}{x^3} + \frac{1}{x^2}$ f $\frac{8}{27}x^6$
g $\frac{3}{x} - 5x^2$ h $\frac{1}{3x^2} + \frac{1}{5x}$

4 a 3 b $\frac{16}{\sqrt[3]{x}}$

5 a $\frac{x}{2}$ b $\frac{32}{x^6}$

Exercise 1E

1 a $2\sqrt{7}$ b $6\sqrt{2}$ c $5\sqrt{2}$ d $4\sqrt{2}$
e $3\sqrt{10}$ f $\sqrt{3}$ g $\sqrt{3}$ h $6\sqrt{5}$
i $7\sqrt{2}$ j $12\sqrt{7}$ k $-3\sqrt{7}$ l $9\sqrt{5}$
m $23\sqrt{5}$ n 2 o $19\sqrt{3}$

2 a $2\sqrt{3} + 3$ b $3\sqrt{5} - \sqrt{15}$
c $4\sqrt{2} - \sqrt{10}$ d $6 + 2\sqrt{5} - 3\sqrt{2} - \sqrt{10}$
e $6 - 2\sqrt{7} - 3\sqrt{3} + \sqrt{21}$ f $13 + 6\sqrt{5}$
g $8 - 6\sqrt{3}$ h $5 - 2\sqrt{3}$
i $3 + 5\sqrt{11}$

3 $3\sqrt{3}$

Exercise 1F

1 a $\frac{\sqrt{5}}{5}$ b $\frac{\sqrt{11}}{11}$ c $\frac{\sqrt{2}}{2}$
d $\frac{\sqrt{5}}{5}$ e $\frac{1}{2}$ f $\frac{1}{4}$
g $\frac{\sqrt{13}}{13}$ h $\frac{1}{3}$

2 a $\frac{1 - \sqrt{3}}{-2}$ b $\sqrt{5} - 2$ c $\frac{3 + \sqrt{7}}{2}$
d $3 + \sqrt{5}$ e $\frac{\sqrt{5} + \sqrt{3}}{2}$ f $\frac{(3 - \sqrt{2})(4 + \sqrt{5})}{11}$
g $5(\sqrt{5} - 2)$ h $5(4 + \sqrt{14})$ i $\frac{11(3 - \sqrt{11})}{-2}$
j $\frac{5 - \sqrt{21}}{-2}$ k $\frac{14 - \sqrt{187}}{3}$ l $\frac{35 + \sqrt{1189}}{6}$
m -1

3 a $\frac{11 + 6\sqrt{2}}{49}$ b $9 - 4\sqrt{5}$ c $\frac{44 + 24\sqrt{2}}{49}$
d $\frac{81 - 30\sqrt{2}}{529}$ e $\frac{13 + 2\sqrt{2}}{161}$ f $\frac{7 - 3\sqrt{3}}{11}$

4 $-\frac{7}{4} + \frac{\sqrt{5}}{4}$

Mixed exercise 1

1 a y^8 b $6x^7$ c $32x$ d $12b^9$

2 a $x^2 - 2x - 15$ b $6x^2 - 19x - 7$
c $6x^2 - 2xy + 19x - 5y + 10$

3 a $x^3 + 3x^2 - 4x$ b $x^3 + 6x^2 - 13x - 42$
c $6x^3 - 5x^2 - 17x + 6$

4 a $15y + 12$ b $15x^2 - 25x^3 + 10x^4$
c $16x^2 + 13x$ d $9x^3 - 3x^2 + 4x$

5 a $x(3x + 4)$ b $2y(2y + 5)$
c $x(x + y + y^2)$ d $2xy(4y + 5x)$

6 a $(x + 1)(x + 2)$ b $3x(x + 2)$
c $(x - 7)(x + 5)$ d $(2x - 3)(x + 1)$
e $(5x + 2)(x - 3)$ f $(1 - x)(6 + x)$

7 a $2x(x^2 + 3)$ b $x(x + 6)(x - 6)$
c $x(2x - 3)(x + 5)$

8 a $3x^6$ b 2 c $6x^2$ d $\frac{1}{2}x^{-\frac{1}{3}}$

9 a $\frac{4}{9}$ b $\frac{3375}{4913}$

10 a $\frac{\sqrt{7}}{7}$ b $4\sqrt{5}$

11 a $21\,877$
b $(5x + 6)(7x - 8)$
When $x = 25$, $5x + 6 = 131$ and $7x - 8 = 167$; both 131 and 167 are prime numbers.

12 a $3\sqrt{2} + \sqrt{10}$ b $10 + 2\sqrt{3} - 5\sqrt{5} - \sqrt{15}$
c $24 - 6\sqrt{7} - 4\sqrt{2} + \sqrt{14}$

13 a $\frac{\sqrt{3}}{3}$ b $\sqrt{2} + 1$ c $-3\sqrt{3} - 6$
d $\frac{30 - \sqrt{851}}{-7}$ e $7 - 4\sqrt{3}$ f $\frac{23 + 8\sqrt{7}}{81}$

14 a $b = -4$ and $c = -5$ b $(x + 3)(x - 5)(x + 1)$

15 a $\frac{1}{4}x$ b $256x^{-3}$

16 $\frac{5}{\sqrt{75} - \sqrt{50}} = \frac{1}{\sqrt{3} - \sqrt{2}} = \sqrt{3} + \sqrt{2}$

17 $-36 + 10\sqrt{11}$

18 $x(1 + 8x)(1 - 8x)$

19 $y = 6x + 3$

20 $4\sqrt{3}$

21 $3 - \sqrt{3}$ cm

22 $\frac{4 - 4x^{\frac{1}{2}} + x^1}{x^{\frac{1}{2}}} = 4x^{-\frac{1}{2}} - 4 + x^{\frac{1}{2}}$

23 $\frac{11}{2}$

24 $4x^{\frac{5}{2}} + x^2$, $a = \frac{5}{2}$ $b = 2$

Challenge

a $a - b$

b $\frac{(\sqrt{1} - \sqrt{2}) + (\sqrt{2} - \sqrt{3}) + \ldots + (\sqrt{24} - \sqrt{25})}{-1} = \sqrt{25} - \sqrt{1} = 4$

CHAPTER 2

Prior knowledge check

1 a $x = -5$ b $x = 3$
c $x = 5$ or $x = -5$ d 16 or 0

2 a $(x + 3)(x + 5)$ b $(x + 5)(x - 2)$
c $(3x + 1)(x - 5)$ d $(x - 20)(x + 20)$

3 a
y
O
2
x
−6

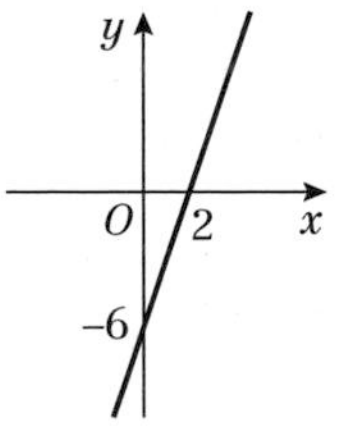

b
y 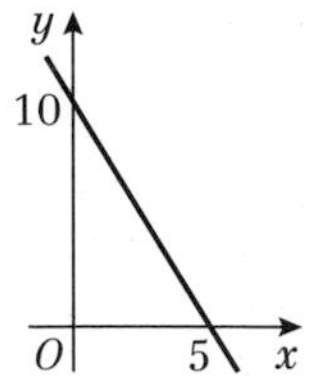
10
O
5
x

3 c

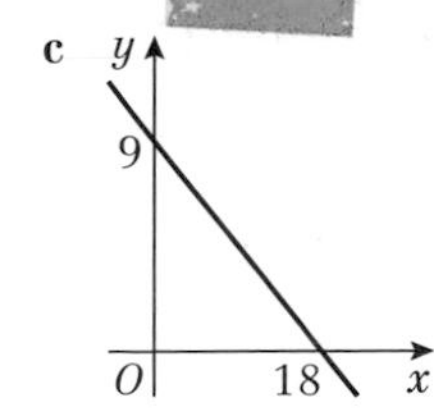

d

4 a $x < 3$ **b** $x \geqslant 9$ **c** $x \leqslant 2.5$ **d** $x > -7$

Exercise 2A

1 a $x = -1$ or $x = -2$ **b** $x = -1$ or $x = -4$
c $x = -5$ or $x = -2$ **d** $x = 3$ or $x = -2$
e $x = 3$ or $x = 5$ **f** $x = 4$ or $x = 5$
g $x = 6$ or $x = -1$ **h** $x = 6$ or $x = -2$

2 a $x = 0$ or $x = 4$ **b** $x = 0$ or $x = 25$
c $x = 0$ or $x = 2$ **d** $x = 0$ or $x = 6$
e $x = -\frac{1}{2}$ or $x = -3$ **f** $x = -\frac{1}{3}$ or $x = \frac{3}{2}$
g $x = -\frac{2}{3}$ or $x = \frac{3}{2}$ **h** $x = \frac{3}{2}$ or $x = \frac{5}{2}$

3 a $x = \frac{1}{3}$ or $x = -2$ **b** $x = 3$ or $x = 0$
c $x = 13$ or $x = 1$ **d** $x = 2$ or $x = -2$
e $x = \pm\sqrt{\frac{5}{3}}$ **f** $x = 3 \pm \sqrt{13}$
g $x = \frac{1 \pm \sqrt{11}}{3}$ **h** $x = 1$ or $x = -\frac{7}{6}$
i $x = -\frac{1}{2}$ or $x = \frac{7}{3}$ **j** $x = 0$ or $x = -\frac{11}{6}$

4 $x = 4$

5 $x = -1$ or $x = -\frac{2}{25}$

Exercise 2B

1 a $x = \frac{1}{2}(-3 \pm \sqrt{5})$ **b** $x = \frac{1}{2}(3 \pm \sqrt{17})$
c $x = -3 \pm \sqrt{3}$ **d** $x = \frac{1}{2}(5 \pm \sqrt{33})$
e $x = \frac{1}{3}(-5 \pm \sqrt{31})$ **f** $x = \frac{1}{2}(1 \pm \sqrt{2})$
g $x = 2$ or $x = -\frac{1}{4}$ **h** $x = \frac{1}{11}(-1 \pm \sqrt{78})$

2 a $x = -0.586$ or $x = -3.41$ **b** $x = 7.87$ or $x = 0.127$
c $x = 0.765$ or $x = -11.8$ **d** $x = 8.91$ or $x = -1.91$
e $x = 0.105$ or $x = -1.90$ **f** $x = 3.84$ or $x = -2.34$
g $x = 4.77$ or $x = 0.558$ **h** $x = 4.89$ or $x = -1.23$

3 a $x = -6$ or $x = -2$ **b** $x = 1.09$ or $x = -10.1$
c $x = 9.11$ or $x = -0.110$ **d** $x = -\frac{1}{2}$ or $x = -2$
e $x = 1$ or $x = -9$ **f** $x = 1$
g $x = 4.68$ or $x = -1.18$ **h** $x = 3$ or $x = 5$

4 Area $= \frac{1}{2}(2x)(x + (x + 10)) = 50\,\text{m}^2$
So $x^2 + 5x - 25 = 0$
Using the quadratic formula:
$x = \frac{1}{2}(-5 \pm 5\sqrt{5})$
Height $= 2x = 5(\sqrt{5} - 1)$ m

Challenge

$x = 13$

Exercise 2C

1 a $(x + 2)^2 - 4$ **b** $(x - 3)^2 - 9$
c $(x - 8)^2 - 64$ **d** $(x + \frac{1}{2})^2 - \frac{1}{4}$
e $(x - 7)^2 - 49$

2 a $2(x + 4)^2 - 32$ **b** $3(x - 4)^2 - 48$
c $5(x + 2)^2 - 20$ **d** $2(x - \frac{5}{4})^2 - \frac{25}{8}$
e $-2(x - 2)^2 + 8$

3 a $2(x + 2)^2 - 7$ **b** $5(x - \frac{3}{2})^2 - \frac{33}{4}$
c $3(x + \frac{1}{3})^2 - \frac{4}{3}$ **d** $-4(x + 2)^2 + 26$
e $-8(x - \frac{1}{8})^2 + \frac{81}{8}$

4 $a = \frac{3}{2}$, $b = \frac{15}{4}$

5 $A = 6$, $B = 0.04$, $C = -10$

Exercise 2D

1 a $x = -3 \pm 2\sqrt{2}$ **b** $x = -6 \pm \sqrt{33}$
c $x = -2 \pm \sqrt{6}$ **d** $x = 5 \pm \sqrt{30}$

2 a $x = \frac{1}{2}(-3 \pm \sqrt{15})$ **b** $x = \frac{1}{5}(-4 \pm \sqrt{26})$
c $x = \frac{1}{8}(1 \pm \sqrt{129})$ **d** $x = \frac{1}{2}(-3 \pm \sqrt{39})$

3 a $p = -7$, $q = -48$
b $(x - 7)^2 = 48$
$x = 7 \pm \sqrt{48} = 7 \pm 4\sqrt{3}$
$r = 7$, $s = 4$

4 $x^2 + 2bx + c = (x + b)^2 - b^2 + c$
$(x + b)^2 = b^2 - c$
$x = -b \pm \sqrt{b^2 - c}$

Challenge

a
$ax^2 + 2bx + c = 0$
$x^2 + \frac{2b}{a}x + \frac{c}{a} = 0$
$\left(x + \frac{b}{a}\right)^2 - \frac{b^2}{a^2} + \frac{c}{a} = 0$
$\left(x + \frac{b}{a}\right)^2 = \frac{b^2 - ac}{a^2}$
$x = -\frac{b}{a} \pm \sqrt{\frac{b^2 - ac}{a^2}}$

b
$ax^2 + bx + c = 0$
$x^2 + \frac{b}{a}x + \frac{c}{a} = 0$
$\left(x + \frac{b}{2a}\right)^2 - \frac{b^2}{4a^2} + \frac{c}{a} = 0$
$\left(x + \frac{b}{2a}\right)^2 = \frac{b^2 - 4ac}{4a^2}$
$x = \frac{-b \pm \sqrt{b^2 - 4ac}}{2a}$

Exercise 2E

1 a 8 **b** 7 **c** 3 **d** 10.5 **e** 0
f 0 **g** 25 **h** 2 **i** 7

2 $a = 4$ or $a = -2$

3 a $\frac{2}{3}$ **b** 2 and -9 **c** -10 and 4
d 12 and -12 **e** 0, -5 and -7 **f** 0, 3 and -8

4 $x = 3$ and $x = 2$

5 $x = 0$, 2.5 and 6

6 a $(x - 1)^2 + 1$
$p = -1$, $q = 1$
b Squared terms are always $\geqslant 0$, so the minimum value is $0 + 1 = 1$

7 a -2 and -1 **b** 2, -2, $2\sqrt{2}$ and $-2\sqrt{2}$
c -1 and $\frac{1}{3}$ **d** $\frac{1}{2}$ and 1
e 4 and 25 **f** 8 and -27

8 a $(3^x - 27)(3^x - 1)$ **b** 0 and 3

Exercise 2F

1 a

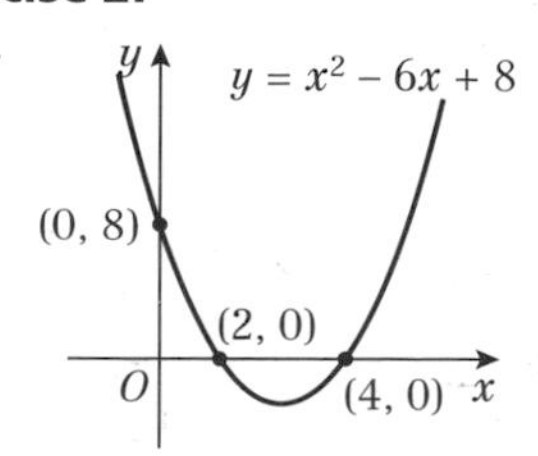

Turning point: $(3, -1)$
Line of symmetry: $x = 3$

b

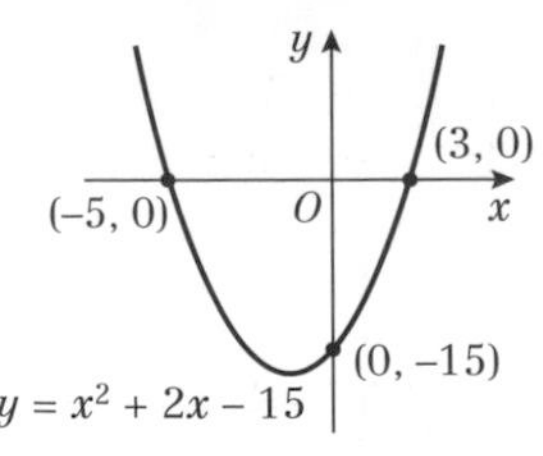

Turning point: $(-1, -16)$
Line of symmetry: $x = -1$

c

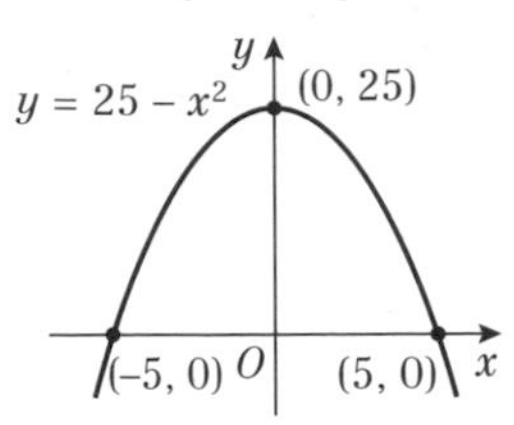

Turning point: $(0, 25)$
Line of symmetry: $x = 0$

d

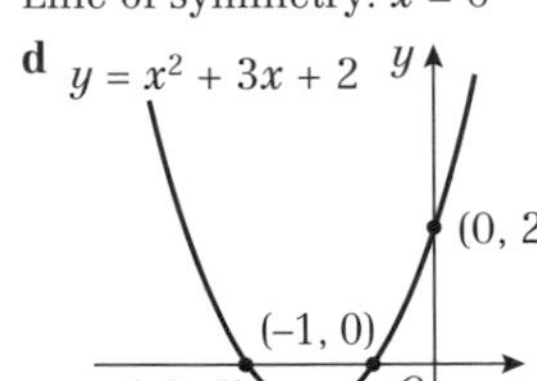

Turning point: $\left(-\frac{3}{2}, -\frac{1}{4}\right)$

Line of symmetry: $x = -\frac{3}{2}$

e

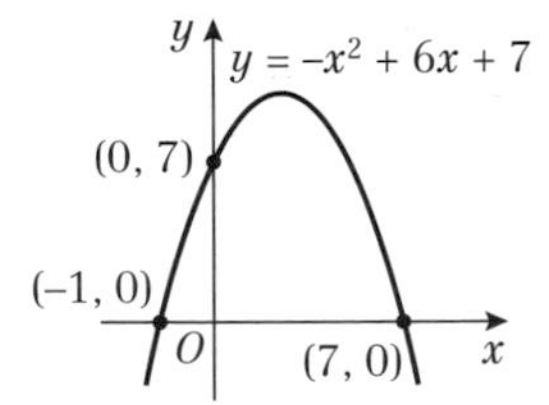

Turning point: $(3, 16)$
Line of symmetry: $x = 3$

f

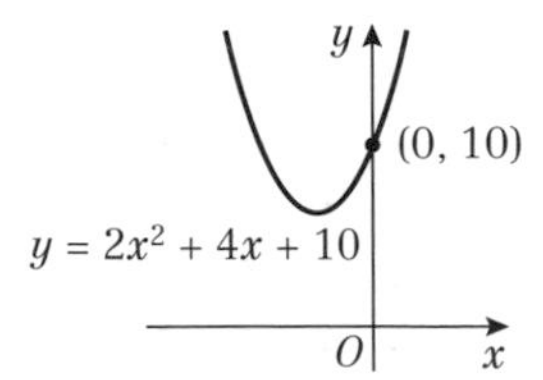

Turning point: $(-1, 8)$
Line of symmetry: $x = -1$

g

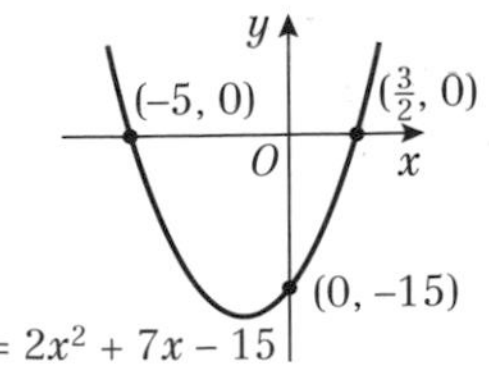

Turning point: $\left(-\frac{7}{4}, -\frac{169}{8}\right)$

Line of symmetry: $x = -\frac{7}{4}$

h

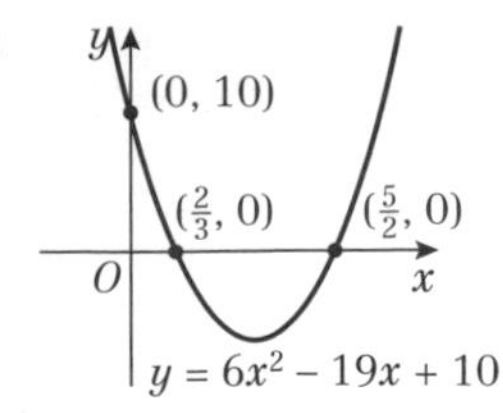

Turning point: $\left(\frac{19}{12}, -\frac{121}{24}\right)$

Line of symmetry: $x = \frac{19}{12}$

i

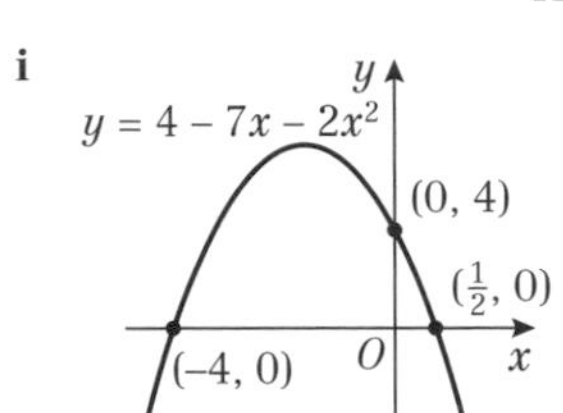

Turning point: $\left(-\frac{7}{4}, \frac{81}{8}\right)$

Line of symmetry: $x = -\frac{7}{4}$

j

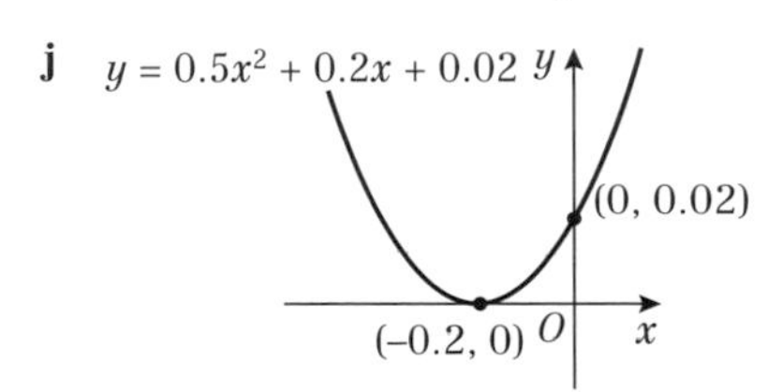

Turning point: $(-0.2, 0)$
Line of symmetry: $x = -0.2$

2 **a** $a = 1, b = -8, c = 15$
b $a = -1, b = 3, c = 10$
c $a = 2, b = 0, c = -18$
d $a = \frac{1}{4}, b = -\frac{3}{4}, c = -1$

3 $a = 3, b = -30, c = 72$

Exercise 2G

1 **a** **i** 52 **ii** −23 **iii** 37
iv 0 **v** −44
b **i** $h(x)$ **ii** $f(x)$ **iii** $k(x)$
iv $j(x)$ **v** $g(x)$

2 $k < 9$

3 $t = \frac{9}{8}$

4 $s = 4$

5 $k > \frac{4}{3}$

6 **a** $p = 6$ **b** $x = -9$

7 **a** $k^2 + 16$
b k^2 is always positive so $k^2 + 16 > 0$

Challenge

a Need $b^2 > 4ac$. If $a, c > 0$ or $a, c < 0$, choose b such that $b > \sqrt{4ac}$. If $a > 0$ and $c < 0$ (or vice versa), then $4ac < 0$, so $4ac < b^2$ for all b.

b Not if one of a or c are negative as this would require b to be the square root of a negative number. Possible if both negative or both positive.

Exercise 2H

1 a The height of the bridge above water level
  b $x = 1103$ and $x = -1103$
  c 2206 m
2 a 21.8 mph and 75.7 mph
  b $A = 39.77, B = 0.01, C = 48.75$
  c 48.75 mph
  d −11 mpg; a negative answer is impossible so this model is not valid for very high speeds
3 a 6 tonnes
  b 39.6 kilograms per hectare.
4 a $M = 40\,000$
  b $r = 400\,000 - 1000(p - 20)^2$
  $A = 400\,000, B = 1000, C = 20$
  c £20

Challenge

a $a = 0.01, b = 0.3, c = -4$
b 36.2 mph

Mixed exercise 2

1 a $y = -1$ or -2 b $x = \frac{2}{3}$ or -5
  c $x = -\frac{1}{5}$ or 3 d $x = \frac{5 \pm \sqrt{7}}{2}$
2 a

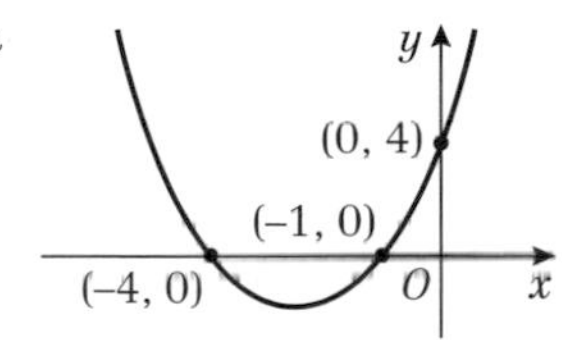

  b

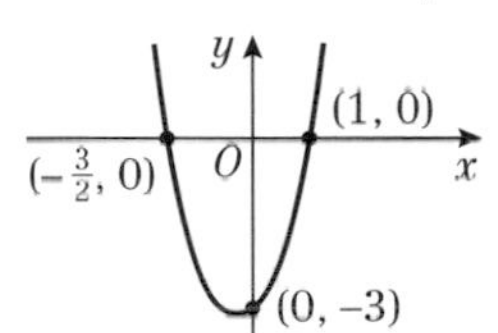

  c

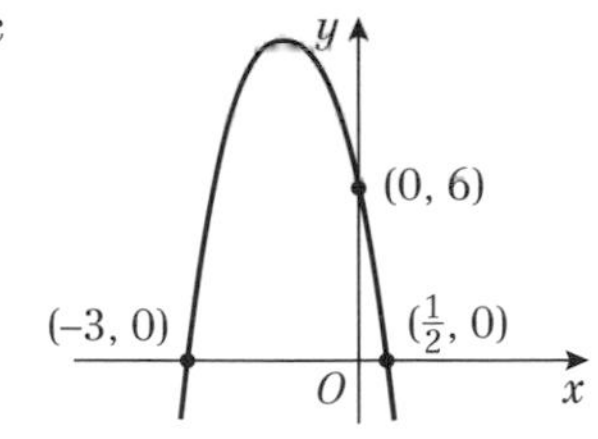

  d

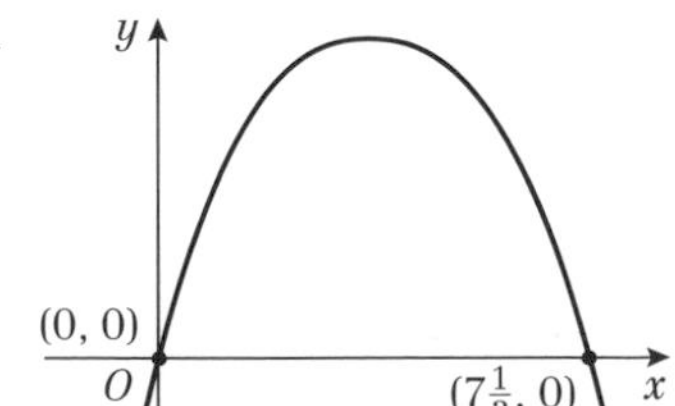

3 a $k = 1$ b $x = 3$ and $x = -2$
4 a $k = 0.0902$ or $k = -11.1$
  b $t = 2.28$ or $t = 0.219$
  c $x = -2.30$ or $x = 1.30$
  d $x = 0.839$ or $x = -0.239$
5 a $(x + 6)^2 - 45$; $p = 1, q = 6, r = -45$
  b $5(x - 4)^2 - 67$; $p = 5, q = -4, r = -67$
  c $-2(x - 2)^2 + 8$; $p = -2, q = -2, r = 8$
  d $2\left(x - \frac{1}{2}\right)^2 - \frac{3}{2}$; $p = 2, q = -\frac{1}{2}, r = -\frac{3}{2}$
6 $k = \frac{1}{5}$
7 a $p = 3, q = 2, r = -7$ b $-2 \pm \sqrt{\frac{7}{3}}$
8 a $f(x) = (2^x - 16)(2^x - 4)$ b 4 and 2
9 $1 \pm \sqrt{13}$
10 $x = -5$ or $x = 4$
11 a 10 m b 1.28 s
  c $h(t) = 10.625 - 10(t - 0.25)^2$
  $A = 10.625, B = 10, C = 0.25$
  d 10.625 m at 0.25 s
12 a $16k^2 + 4$
  b $k^2 \geqslant 0$ for all k, so $16k^2 + 4 > 0$
  c When $k = 0$, $f(x) = 2x + 1$; this is a linear function with only one root
13 1, −1, 2 and −2
14 a $H = 10$
  b $r = 1322.5 - 10(p - 11.5)^2$
  $A = 1322.5, B = 10, C = 11.5$
  c Old revenue is 80 × £15 = £1200; new revenue is £1322.50; difference is £122.50. The best selling price of a cushion is £11.50.

Challenge

a $\frac{b + c}{b} = \frac{b}{c}$
$b^2 - cb - c^2 = 0$
Using quadratic formula: $b = \frac{c + \sqrt{5c^2}}{2}$
So $b : c$ is $\frac{c + \sqrt{5c^2}}{2} : c$
Dividing by c: $\frac{1 + \sqrt{5}}{2} : 1$

b Let $x = \sqrt{1 + \sqrt{1 + \sqrt{1 + \sqrt{1 + \ldots}}}}$
So $x = \sqrt{1 + x} \Rightarrow x^2 - x - 1 = 0$
Using quadratic formula: $x = \frac{1 + \sqrt{5}}{2}$

CHAPTER 3

Prior knowledge check

1 a $A \cap B = \{1, 2, 4\}$ b $(A \cup B)' = \{7, 9, 11, 13\}$
2 a $5\sqrt{3}$ b $\sqrt{5} + 2\sqrt{2}$
3 a graph **ii** b graph **iii** c graph **i**

Exercise 3A

1 a $x = 4, y = 2$ b $x = 1, y = 3$
  c $x = 2, y = -2$ d $x = 4\frac{1}{2}, y = -3$
  e $x = -\frac{2}{3}, y = 2$ f $x = 3, y = 3$
2 a $x = 5, y = 2$ b $x = 5\frac{1}{2}, y = -6$
  c $x = 1, y = -4$ d $x = 1\frac{3}{4}, y = \frac{1}{4}$
3 a $x = -1, y = 1$ b $x = 4, y = -4$
  c $x = 0.5, y = -2.5$
4 a $3x + ky = 8$ (1); $x - 2ky = 5$ (2)
  (1) × 2: $6x + 2ky = 16$ (3)
  (2) + (3) $7x = 21$ so $x = 3$
  b −2
5 $p = 3, q = 1$

Exercise 3B

1 a $x = 5, y = 6$ or $x = 6, y = 5$
  b $x = 0, y = 1$ or $x = \frac{4}{5}, y = -\frac{3}{5}$
  c $x = -1, y = -3$ or $x = 1, y = 3$

d $a = 1, b = 5$ or $a = 3, b = -1$
e $u = 1\frac{1}{2}, v = 4$ or $u = 2, v = 3$
f $x = -1\frac{1}{2}, y = 5\frac{3}{4}$ or $x = 3, y = -1$

2 a $x = 3, y = \frac{1}{2}$ or $x = 6\frac{1}{3}, y = -2\frac{5}{6}$
b $x = 4\frac{1}{2}, y = 4\frac{1}{2}$ or $x = 6, y = 3$
c $x = -19, y = -15$ or $x = 6, y = 5$

3 a $x = 3 + \sqrt{13}, y = -3 + \sqrt{13}$ or $x = 3 - \sqrt{13}$, $y = -3 - \sqrt{13}$
b $x = 2 - 3\sqrt{5}, y = 3 + 2\sqrt{5}$ or $x = 2 + 3\sqrt{5}, y = 3 - 2\sqrt{5}$

4 $x = -5, y = 8$ or $x = 2, y = 1$

5 a $3x^2 + x(2 - 4x) + 11 = 0$
$3x^2 + 2x - 4x^2 + 11 = 0$
$x^2 - 2x - 11 = 0$
b $x = 1 + 2\sqrt{3}, y = -2 - 8\sqrt{3}$
$x = 1 - 2\sqrt{3}, y = -2 + 8\sqrt{3}$

6 a $k = 3, p = -2$
b $x = -6, y = -23$

Challenge

$y = x + k$
$x^2 + (x + k)^2 = 4$
$x^2 + x^2 + 2kx + k^2 - 4 = 0$
$2x^2 + 2kx + k^2 - 4 = 0$ for one solution $b^2 - 4ac = 0$
$4k^2 - 4 \times 2(k^2 - 4) = 0$
$4k^2 - 8k^2 + 32 = 0$ $4k^2 = 32$ $k^2 = 8$ $k = \pm 2\sqrt{2}$

Exercise 3C

1 a i

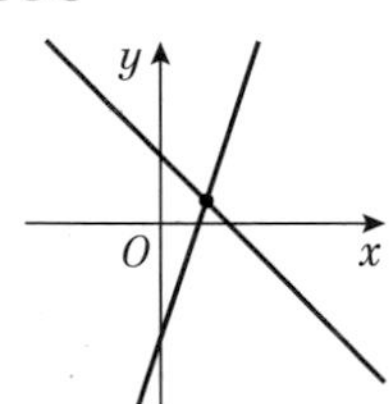

ii (2, 1)

b i

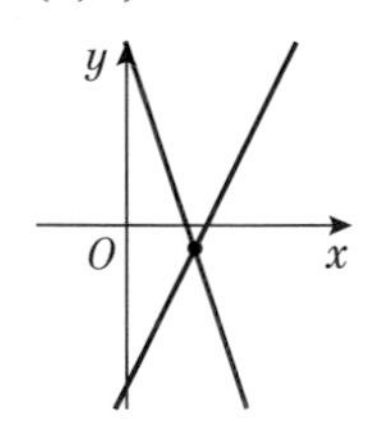

ii (3, −1)

c i

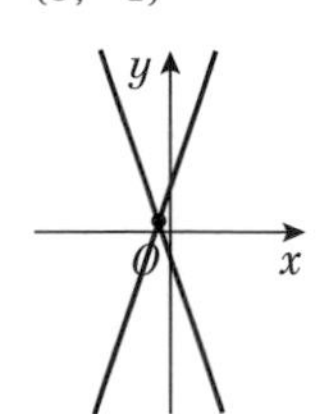

ii (−0.5, 0.5)

2 a

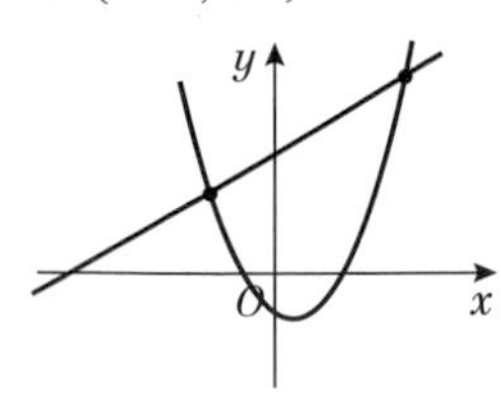

b (3.5, 9) and (−1.5, 4)

3 a

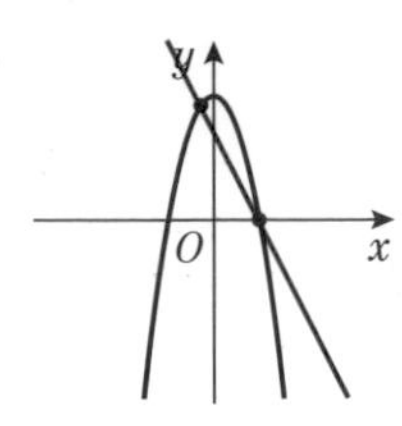

b (−1, 8) and (3, 0)

4 a

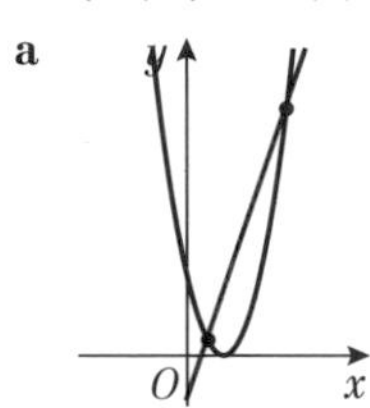

b (6, 16) and (1, 1)

5 (−11, −15) and (3, −1)

6 $(-1\frac{1}{6}, -4\frac{1}{2})$ and (2, 5)

7 a 2 points b 1 point c 0 points

8 a $y = 2x - 1$
$x^2 + 4k(2x - 1) + 5k = 0$
$x^2 + 8kx - 4k + 5k = 0$ $x^2 + 8kx + k = 0$
b $k = \frac{1}{16}$ c $x = -\frac{1}{4}, y = -\frac{3}{2}$

9 If swimmer reaches the bottom of the pool
$0.5x^2 - 3x = 0.3x - 6$
$0.5x^2 - 3.3x + 6 = 0$
$b^2 - 4ac = (-3.3)^2 - 4 \times 0.5 \times 6 = -1.11$
negative so no points of intersection and diver does not reach the bottom of the pool

Exercise 3D

1 a $x < 4$ b $x \geqslant 7$ c $x > 2\frac{1}{2}$ d $x \leqslant -3$
e $x < 11$ f $x < 2\frac{3}{5}$ g $x > -12$ h $x < 1$
i $x \leqslant 8$ j $x > 1\frac{1}{7}$

2 a $x \geqslant 3$ b $x < 1$ c $x \leqslant -3\frac{1}{4}$ d $x < 18$
e $x > 3$ f $x \geqslant 4\frac{2}{5}$ g $x < 4$ h $x > -7$
i $x \leqslant -\frac{1}{2}$ j $x \geqslant \frac{3}{4}$ k $x \geqslant -\frac{10}{3}$ l $x \geqslant \frac{9}{11}$

3 a $\{x: x > 2\frac{1}{2}\}$ b $\{x: 2 < x < 4\}$
c $\{x: 2\frac{1}{2} < x < 3\}$ d No values
e $x = 4$ f $\{x: x < 1.2\} \cup \{x: x > 2.2\}$
g $\left\{x: x \leqslant -\frac{2}{3}\right\} \cup \left\{x: x \geqslant \frac{3}{2}\right\}$

Challenge

$p = -1, q = 4, r = 6$

Exercise 3E

1 a $3 < x < 8$ b $-4 < x < 3$
c $x < -2, x > 5$ d $x \leqslant -4, x \geqslant -3$
e $-\frac{1}{2} < x < 7$ f $x < -2, x > 2\frac{1}{2}$
g $\frac{1}{2} \leqslant x \leqslant 1\frac{1}{2}$ h $x < \frac{1}{3}, x > 2$
i $-3 < x < 3$ j $x < -2\frac{1}{2}, x > \frac{2}{3}$
k $x < 0, x > 5$ l $-1\frac{1}{2} \leqslant x \leqslant 0$

2 a $-5 < x < 2$ b $x < -1, x > 1$
c $\frac{1}{2} < x < 1$ d $-3 < x < \frac{1}{4}$

3 a $\{x: 2 < x < 4\}$ b $\{x: x > 3\}$
c $\{x: -\frac{1}{4} < x < 0\}$ d No values

e $\{x: -5 < x < -3\} \cup \{x: x > 4\}$

f $\{x: -1 < x < 1\} \cup \{x: 2 < x < 3\}$

4 **a** $x < 0$ or $x > 2$ **b** $x < 0$ or $x > 0.8$

c $x < -1$ or $x > 0$ **d** $x < 0$ or $x > 0.5$

e $x < -\frac{1}{5}$ or $x > \frac{1}{5}$ **f** $x \leqslant -\frac{2}{3}$ or $x \geqslant 3$

5 **a** $-2 < k < 6$ **b** $p \leqslant -8$ or $p \geqslant 0$

6 $\{x: x < -2\} \cup \{x: x > 7\}$

7 **a** $\{x: x < \frac{2}{3}\}$ **b** $\{x: -\frac{1}{2} < x < 3\}$

c $\{x: -\frac{1}{2} < x < \frac{2}{3}\}$

8 $x < 3$ or $x > 5.5$

9 No real roots $b^2 - 4ac < 0$ $(-2k)^2 - 4 \times k \times 3 < 0$

$4k^2 - 12k = 0$ when $k = 0$ and $k = 3$

solution $0 \leqslant k < 3$

note when $k = 0$ equation gives $3 = 0$

Exercise 3F

1 **a** $P(3.2, -1.8)$ **b** $x < 3.2$

2 **a** **i**

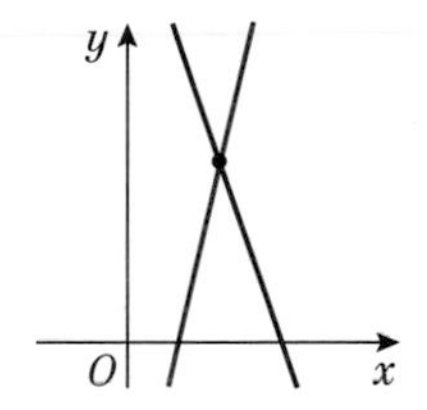

ii (4, 5) **iii** $x \leqslant 4$

b **i**

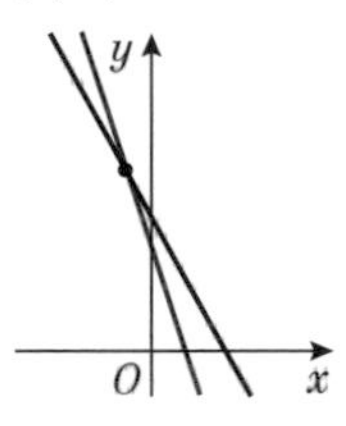

ii (−3, 23) **iii** $x \geqslant -3$

c **i**

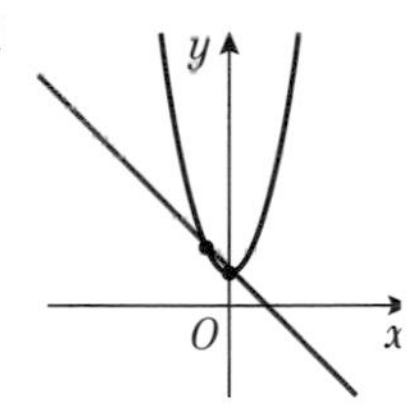

ii (−2, 9), (0, 5) **iii** $-2 \leqslant x \leqslant 0$

d **i**

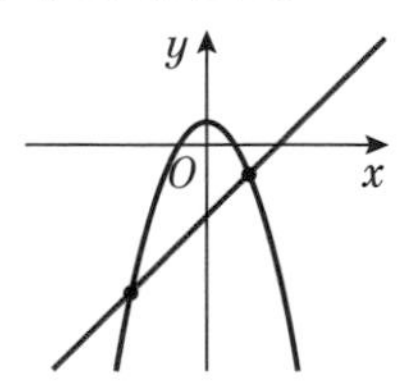

ii (−5, −22), (3, −6) **iii** $x \leqslant -5$ or $x \geqslant 3$

e **i**

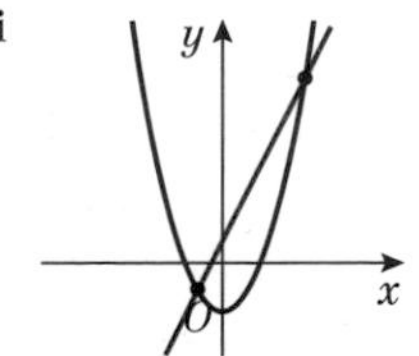

ii (−2, −1), (9, 76) **iii** $-2 \leqslant x \leqslant 9$

f **i**

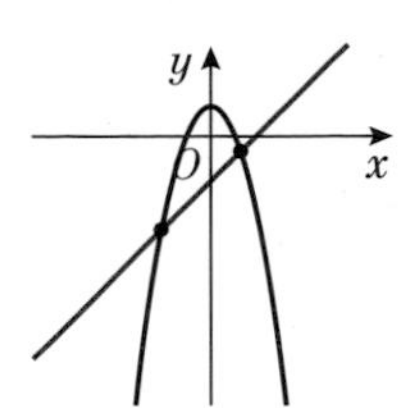

ii (−5, −18), (3, −2) **iii** $x \leqslant -5$ or $x \geqslant 3$

3 **a** $-1 < x < 2$ **b** $0.5 < x < 3$

c $x < 0.5$ or $x > 3$ **d** $x < 0$ or $x > 2$

e $1 < x < 3$ **f** $x < -1$ or $x > -0.75$

Challenge

a (−1.5, −3.75), (6, 0)

b $\{x: -1.5 < x < 6\}$

Exercise 3G

1

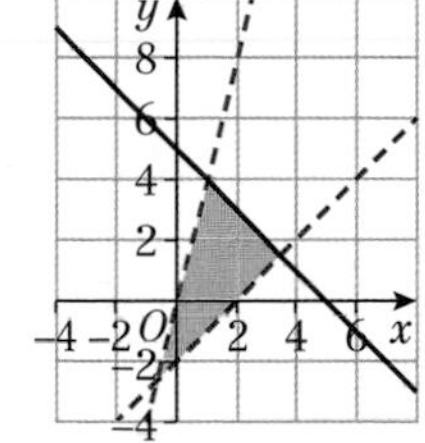

2

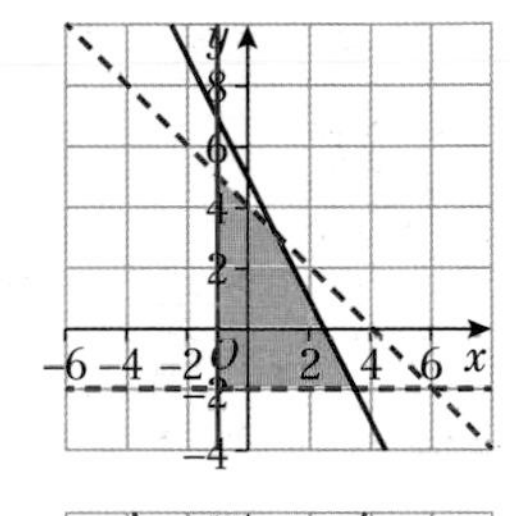

3

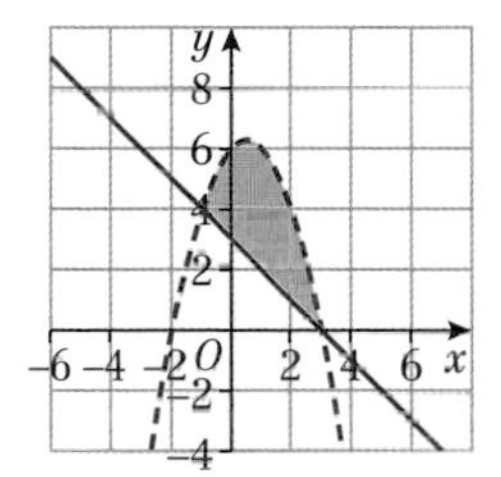

4

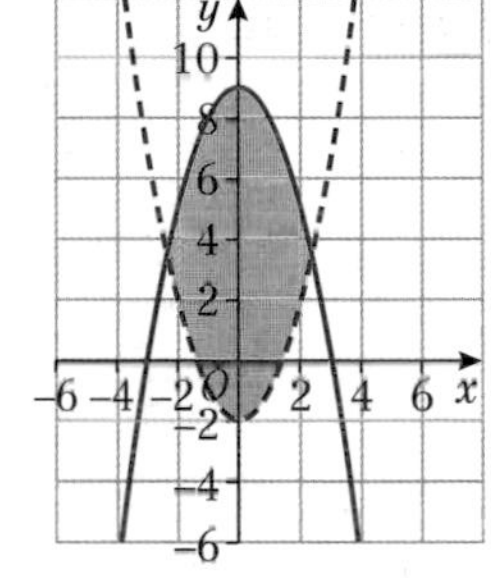

5

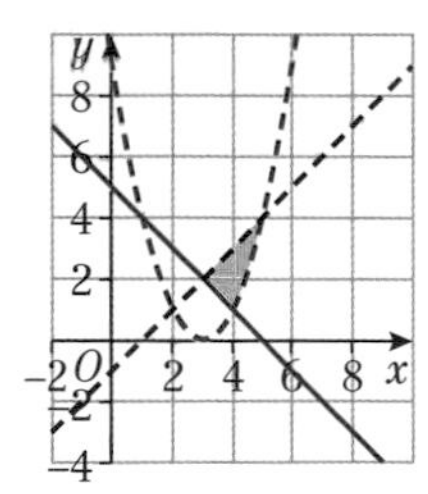

6 **a** (1, 6), (3, 4), (1, 2)

b $x \geqslant 1, y \leqslant 7 - x, y \geqslant x + 1$

7 $y < 2 - 5x - x^2, 2x + y \geqslant 0, x + y \leqslant 4$

8 **a**

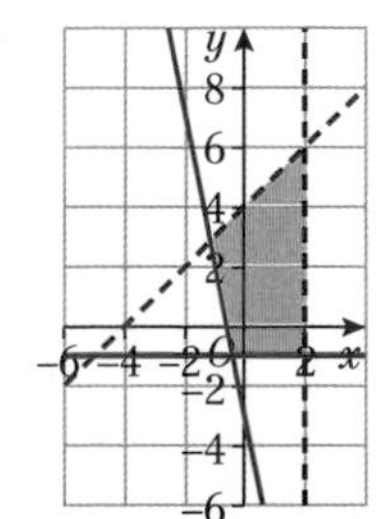

b $(-\frac{7}{6}, \frac{17}{6})$, (2, 6), (2, −1), (−0.4, −1)

c (−0.4, −1) **d** $\frac{941}{60}$

Mixed exercise 3

1 **a** $4kx - 2y = 8$
$4kx + 3y = -2$
$-5y = 10$
$y = -2$

b $x = \frac{1}{k}$

2 $x = -4, y = 3\frac{1}{2}$

3 **a** Substitute $x = 1 + 2y$ into $3xy - y^2 = 8$

b $(3, 1)$ and $\left(-\frac{11}{5}, -\frac{8}{5}\right)$

4 **a** Substitute $y = 2 - x$ into $x^2 + xy - y^2 = 0$

b $x = 3 \pm \sqrt{6}, y = -1 \pm \sqrt{6}$

5 **a** $3^x = (3^2)^{y-1} = 3^{2y-2} \Rightarrow x = 2y - 2$

b $x = 4, y = 3$ and $x = -2\frac{2}{3}, y = -\frac{1}{3}$

6 $x = -1\frac{1}{2}, y = 2\frac{1}{4}$ and $x = 4, y = -\frac{1}{2}$

7 **a** $k = -2$ **b** $(-1, 2)$

8 Yes, the ball will hit the ceiling

9 **a** $\{x: x > 10\frac{1}{2}\}$ **b** $\{x: x < -2\} \cup \{x: x > 7\}$

10 $3 < x < 4$

11 **a** $x = -5, x = 4$ **b** $\{x: x < -5\} \cup \{x: x > 4\}$

12 **a** $x < 2\frac{1}{2}$ **b** $\frac{1}{2} < x < 5$

c $0 < x < 4$ **d** $\frac{1}{2} < x < 2\frac{1}{2}$

13 $1 \leqslant x \leqslant 8$

14 $k \leqslant 3\frac{1}{5}$

15 $b^2 < 4ac$ so $16k^2 < -40k$
$8k(2k + 5) < 0$ so $-\frac{5}{2} < k < 0$

16 **a**

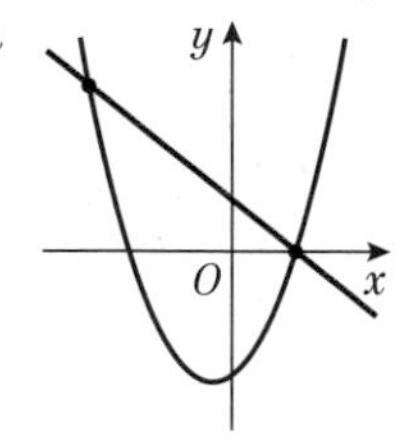

b $(-7, 20), (3, 0)$ **c** $x < -7, x > 3$

17 $\frac{1}{4}(-1 - \sqrt{185}) < x < \frac{1}{4}(-1 + \sqrt{185})$

18

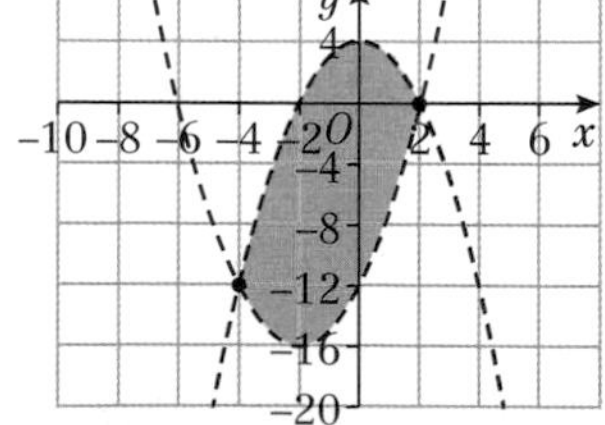

19 **a**

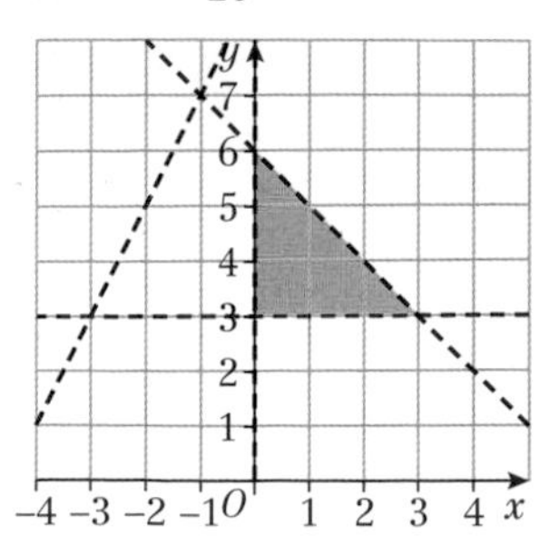

b $\frac{9}{2}$

Challenge

1 $0 < x < 1.6$ **2** $-2 < k < 7$

CHAPTER 4

Prior knowledge check

1 **a** $(x + 5)(x + 1)$ **b** $(x - 3)(x - 1)$

2 **a**

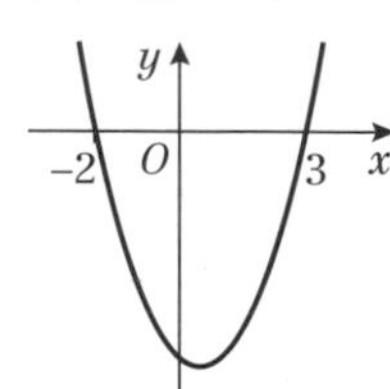

b

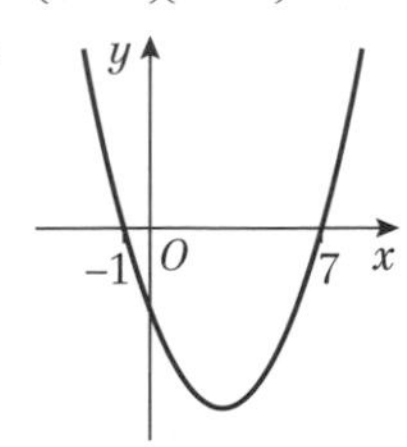

3 **a**

x	-2	-1.5	-1	-0.5	0
y	-12	-6.875	-4	-2.625	-2

x	0.5	1	1.5	2
y	-1.375	0	2.875	8

b

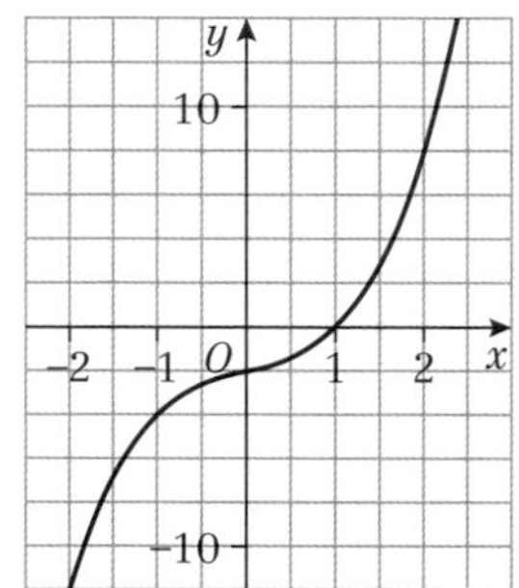

4 **a** $x = 2, y = 4$ **b** $x = 1, y = 1$

Exercise 4A

1 **a**

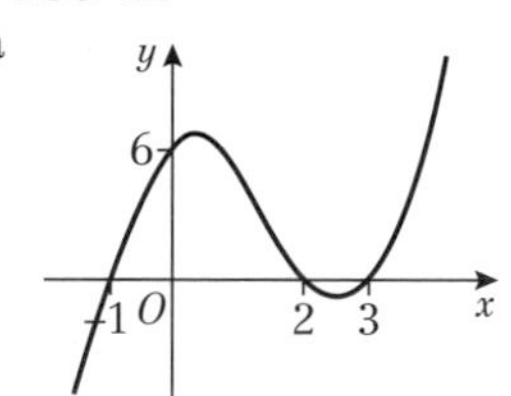

b

c

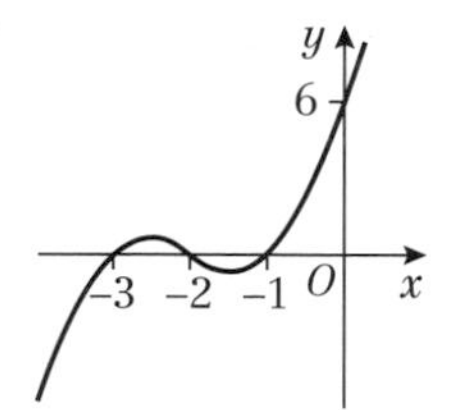

d

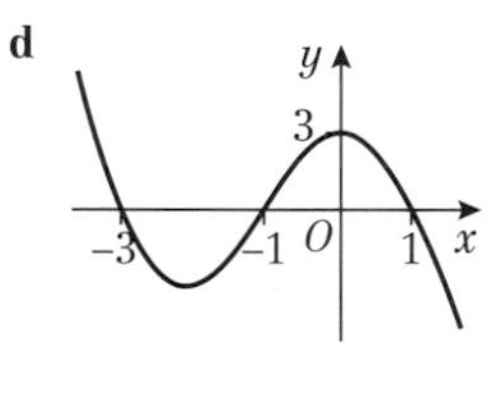

e

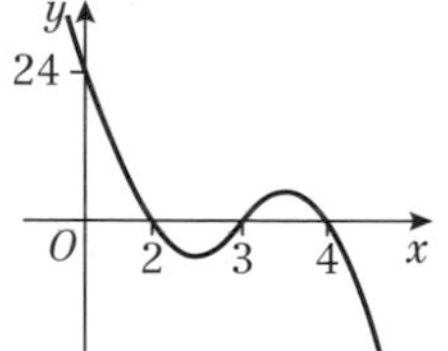

f

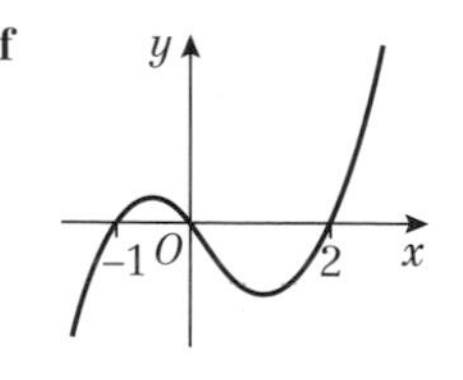

g

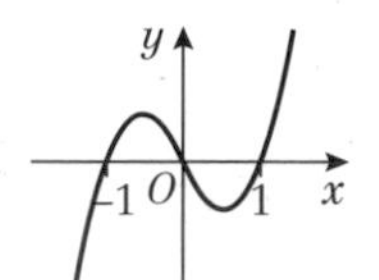

h

i

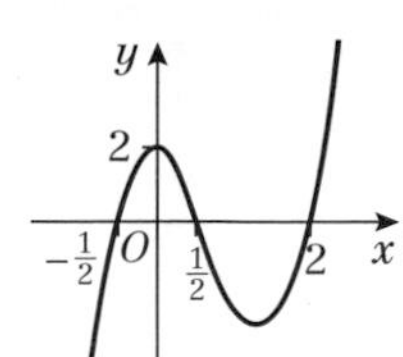

j

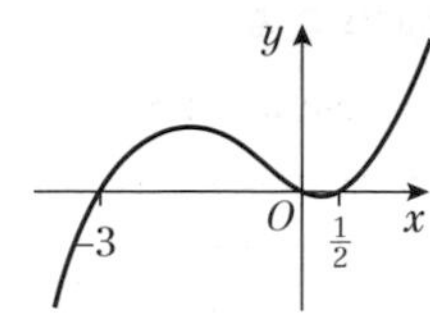

2 a

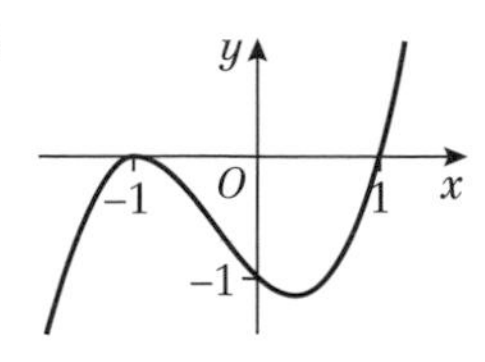

b

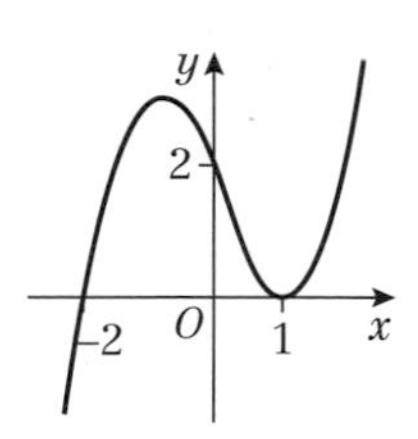

c

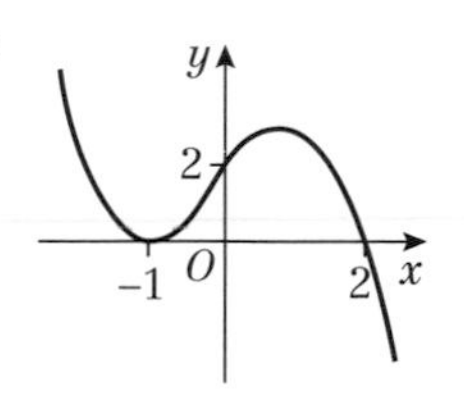

d

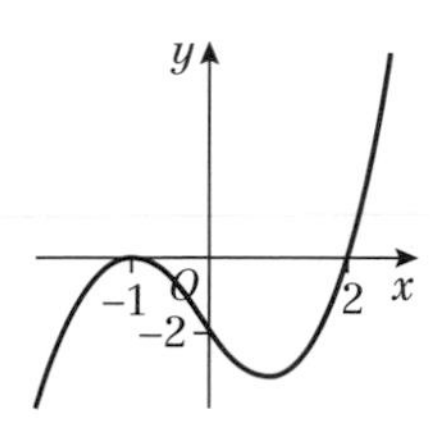

e

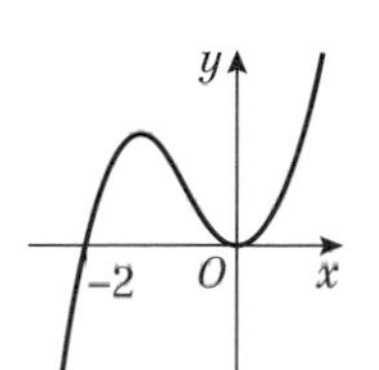

f

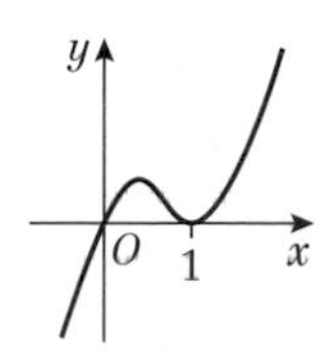

g

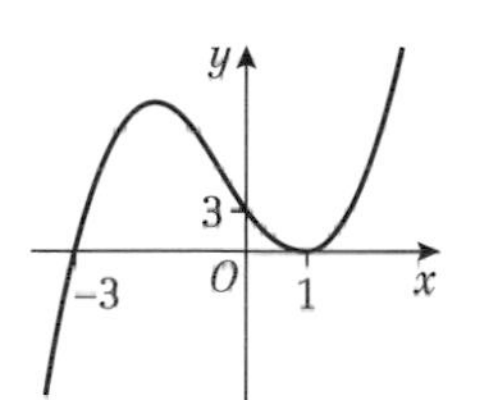

h

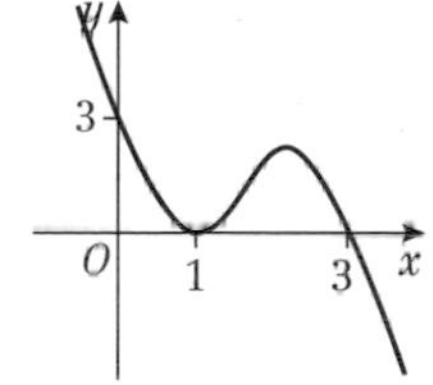

i

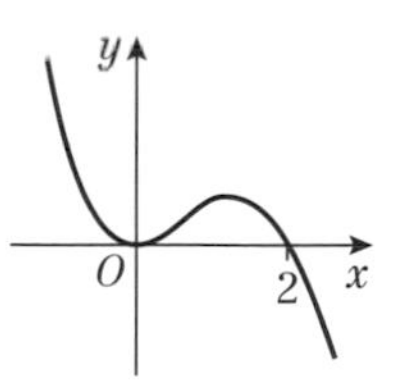

j

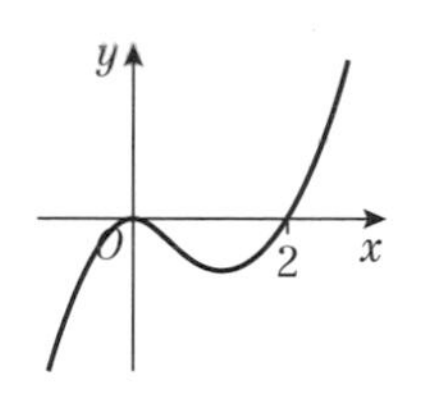

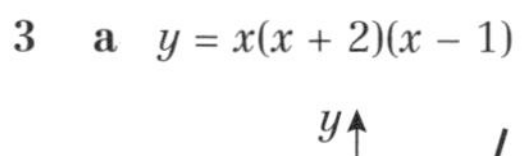

3 a $y = x(x + 2)(x - 1)$

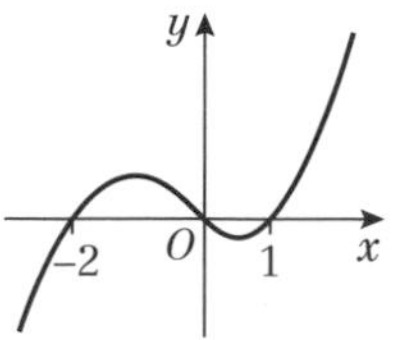

b $y = x(x + 4)(x + 1)$

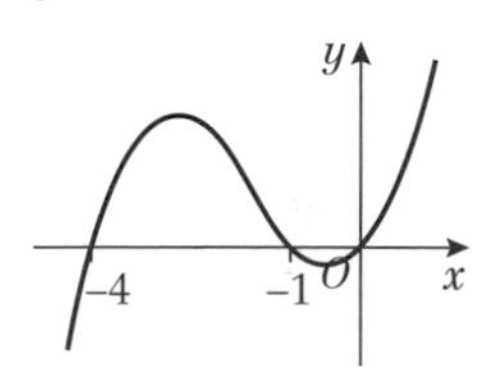

c $y = x(x + 1)^2$

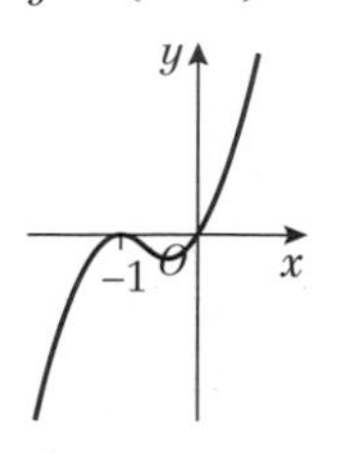

d $y = x(x + 1)(3 - x)$

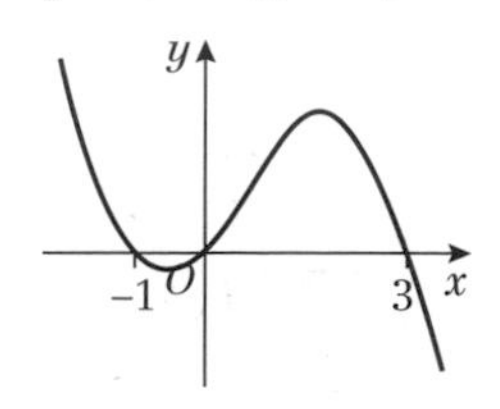

e $y = x^2(x - 1)$

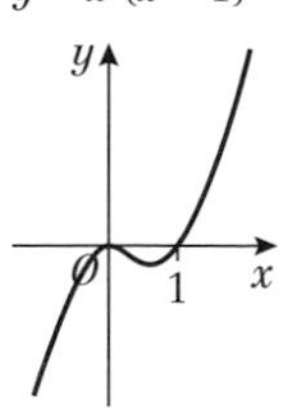

f $y = x(1 - x)(1 + x)$

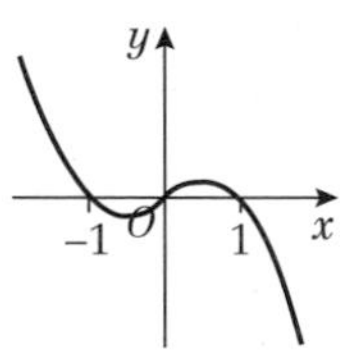

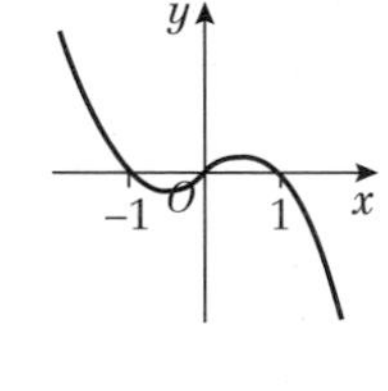

g $y = 3x(2x - 1)(2x + 1)$

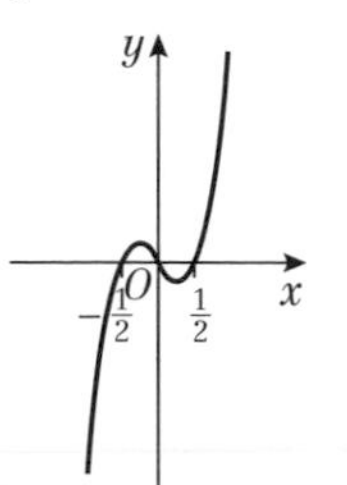

h $y = x(x + 1)(x - 2)$

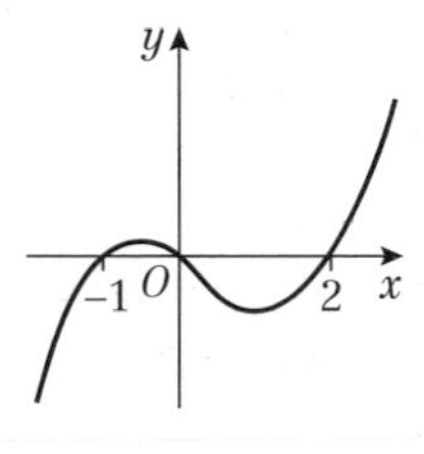

i $y = x(x - 3)(x + 3)$

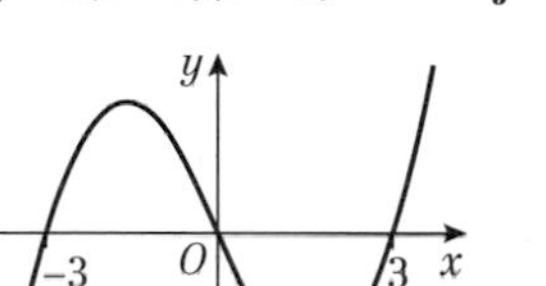

j $y = x^2(x - 9)$

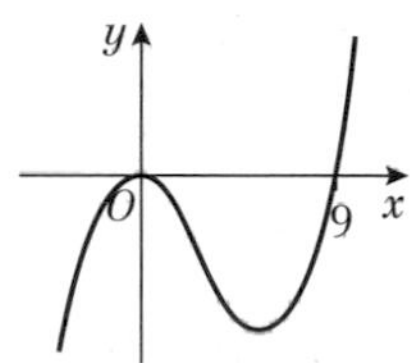

4 a

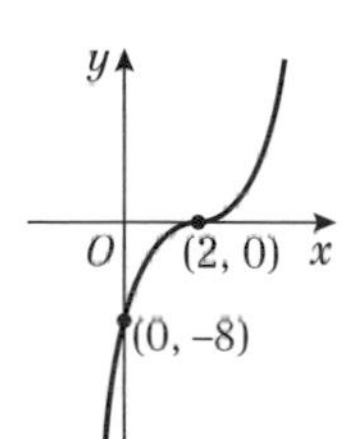

b

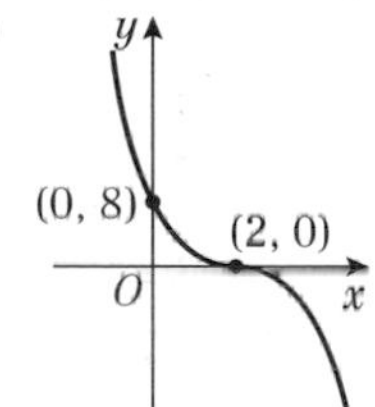

c

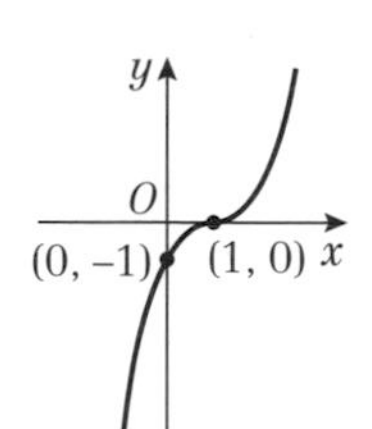

d

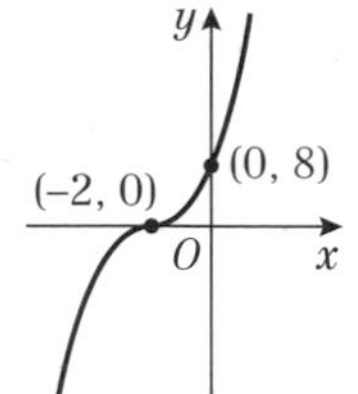

e

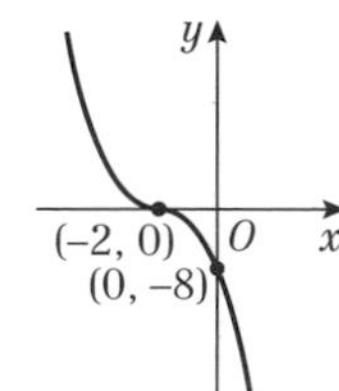

f

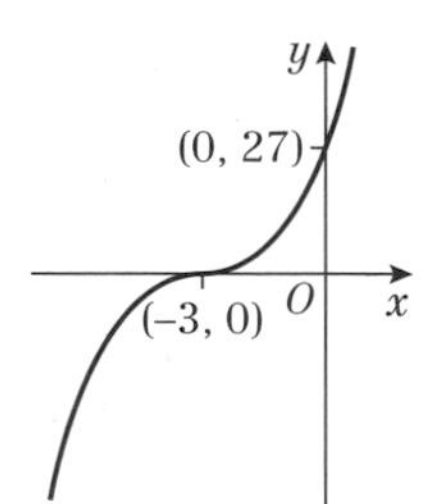

g

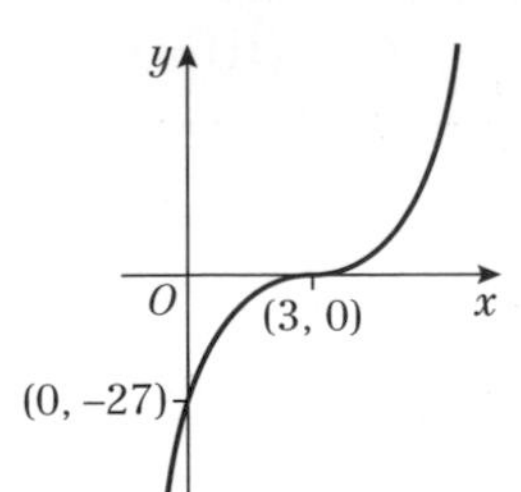

h

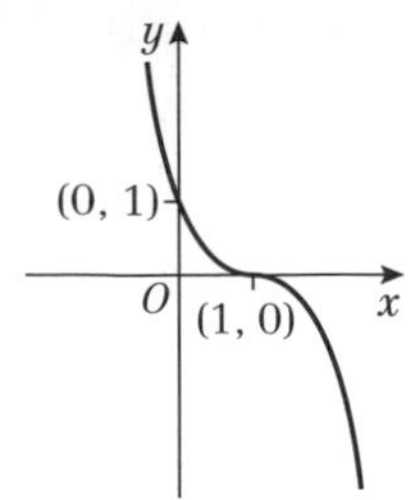

i

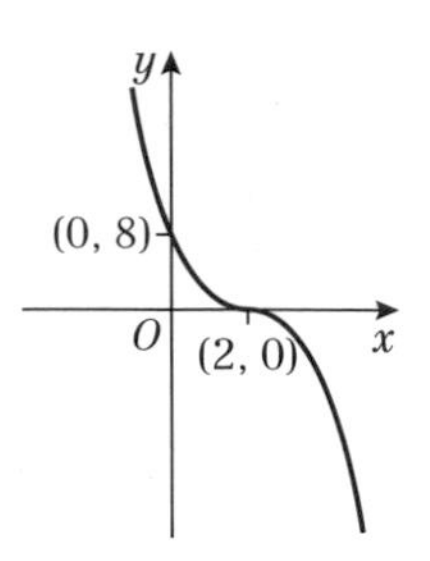

j

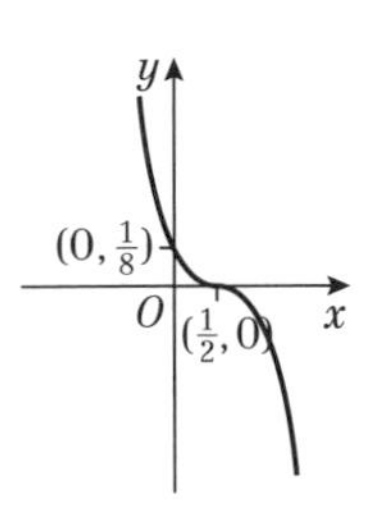

5 **a** $b = 4, c = 1, d = -6$ **b** $(0, -6)$

6 $a = \frac{1}{3}, b = -\frac{4}{3}, c = \frac{1}{3}, d = 2$

7 **a** $x(x^2 - 12x + 32)$ **b** $x(x - 8)(x - 4)$

c

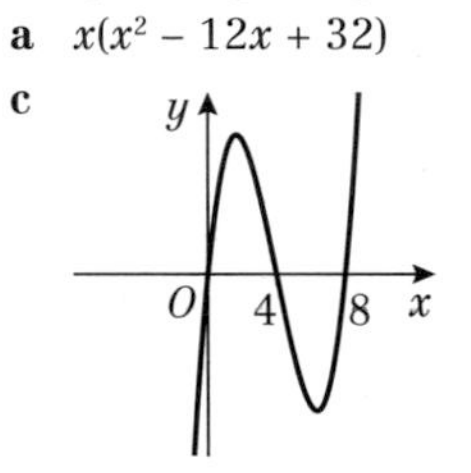

Exercise 4B

1 **a**

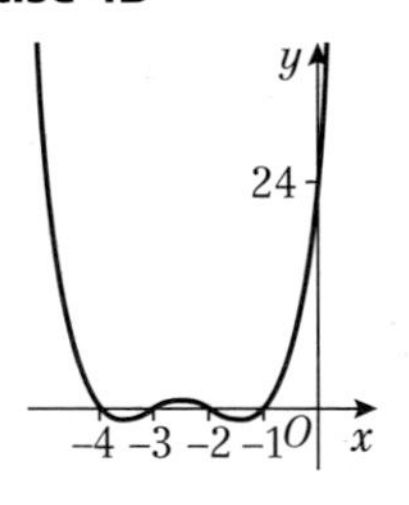

b

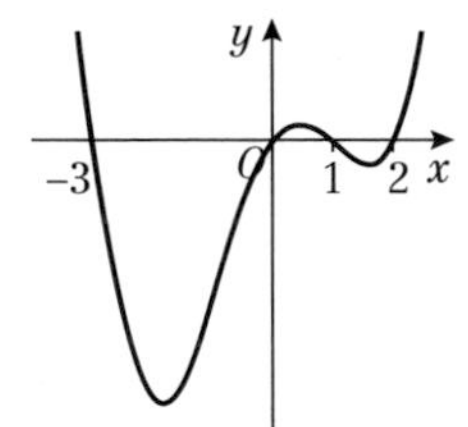

c

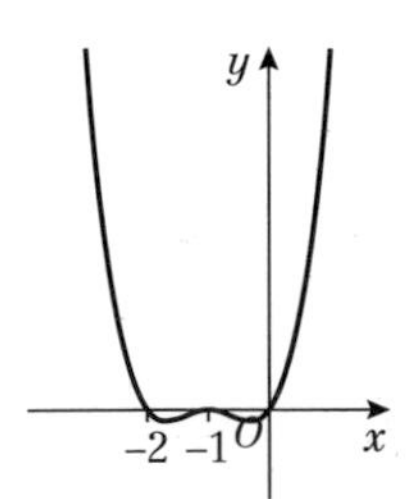

d

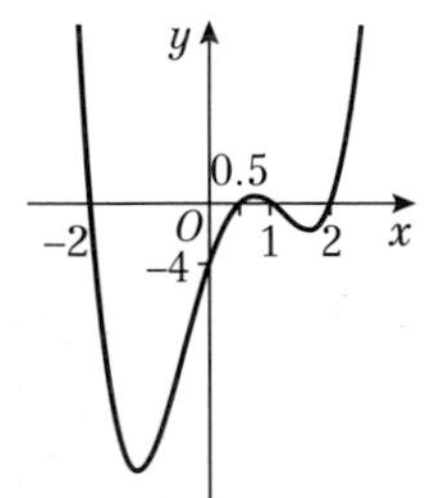

e

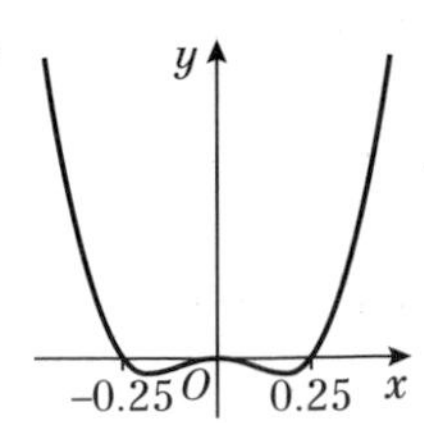

f

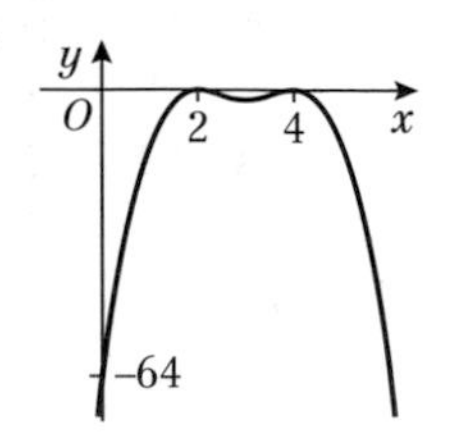

g

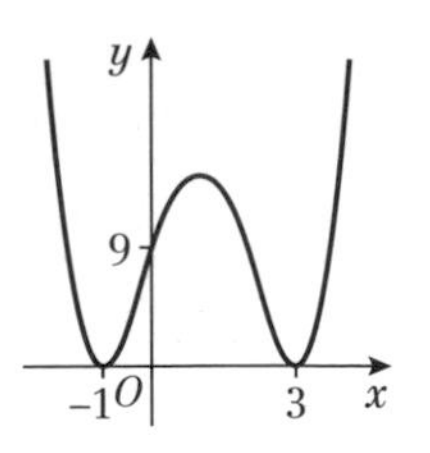

h

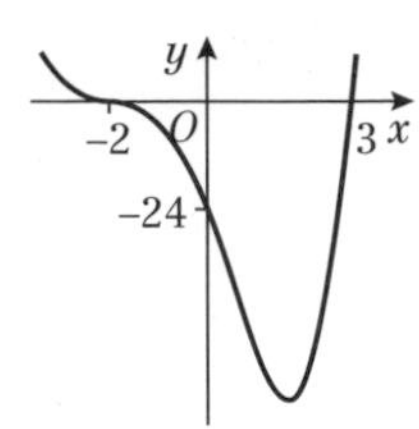

i

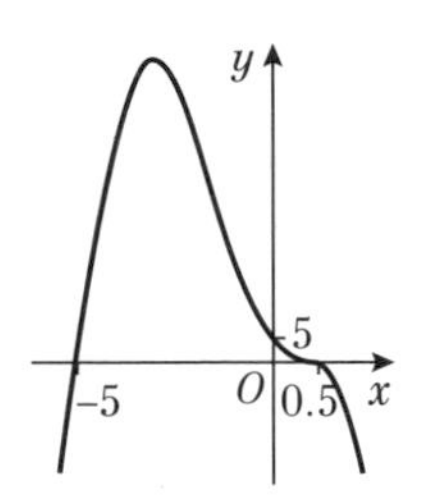

j

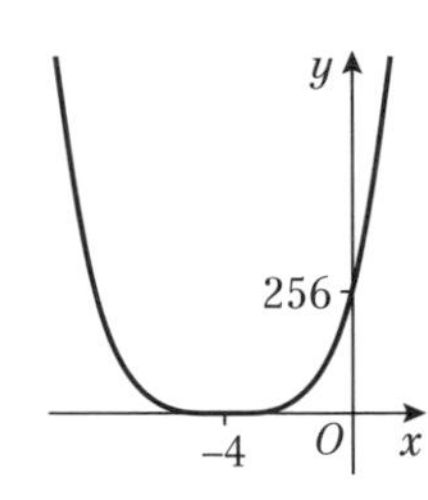

2 **a**

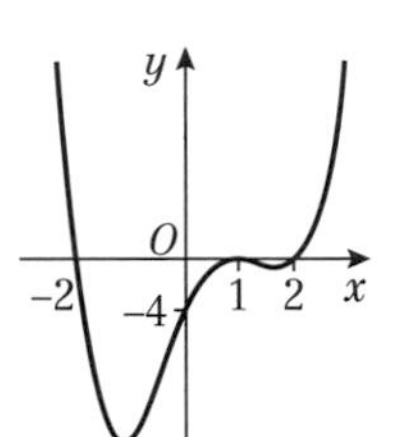

b

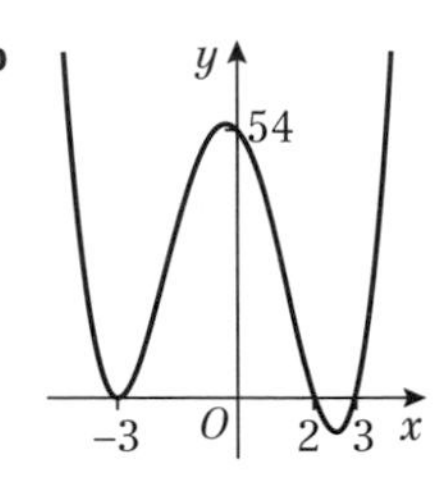

c

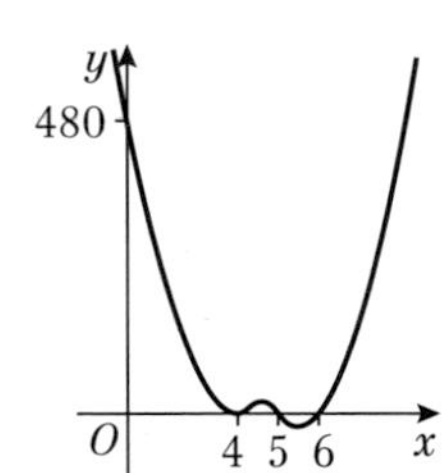

d

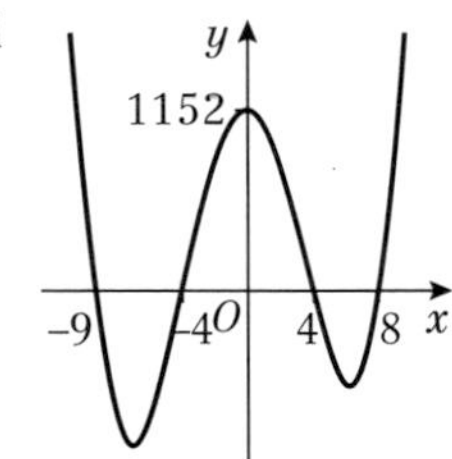

3 **a** $(0, 12)$

b $b = -2, c = -7, d = 8, e = 12$

4

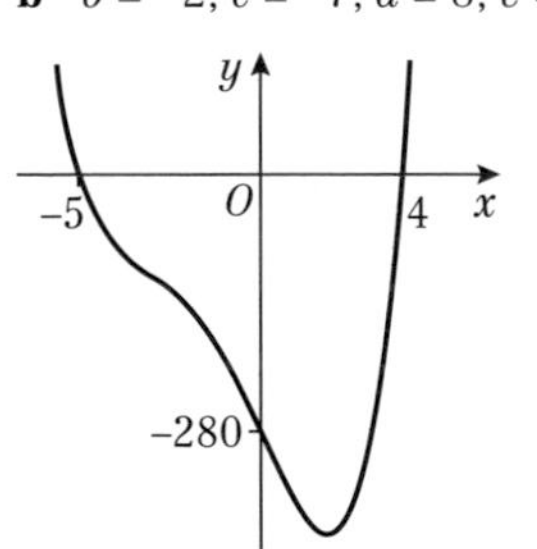

Challenge

$a = \frac{1}{3}, b = -\frac{4}{3}, c = -\frac{2}{3}, d = 4, e = 3$

Exercise 4C

1 a

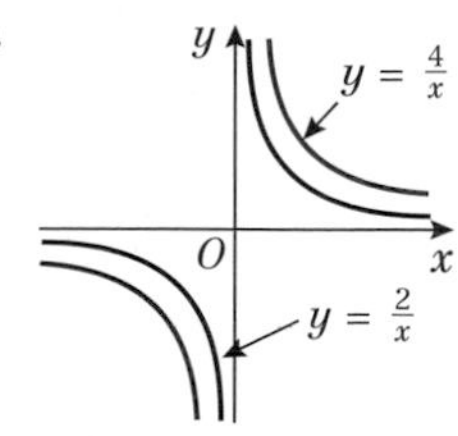

b

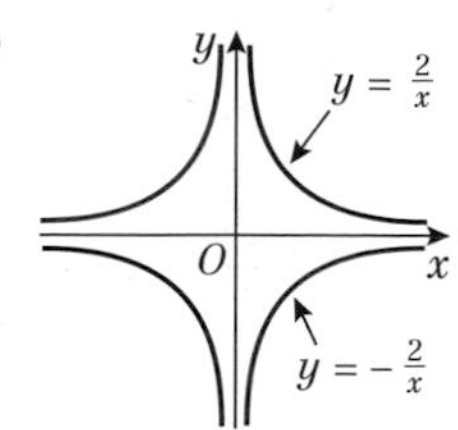

c

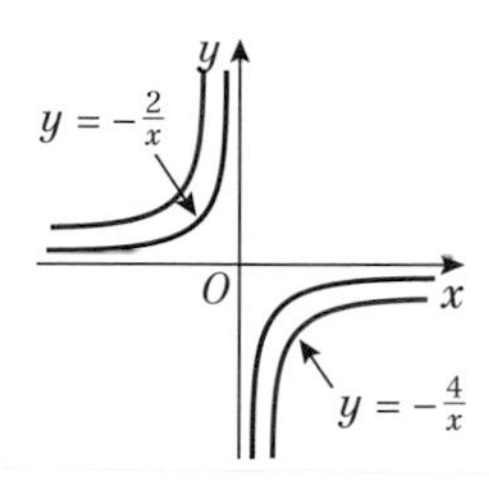

d

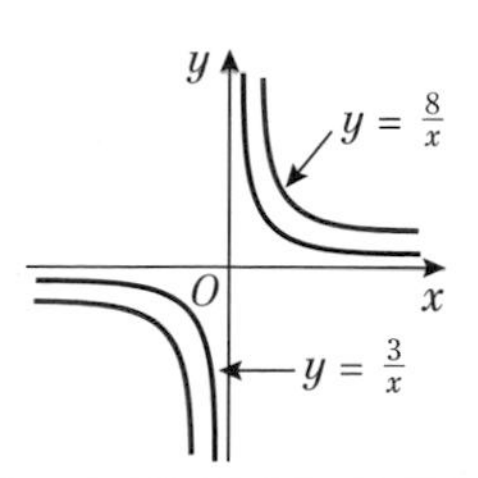

e

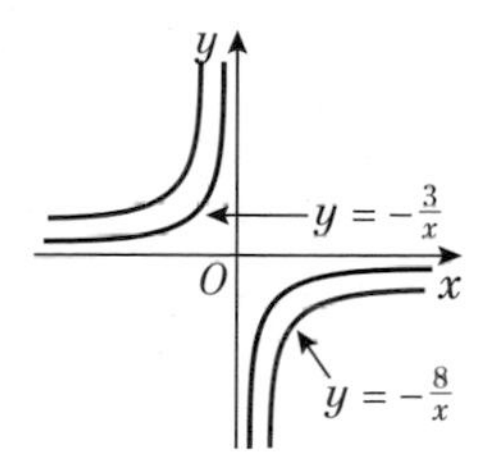

2 a

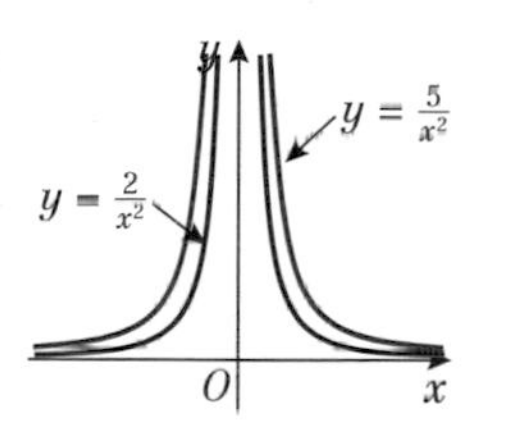

b

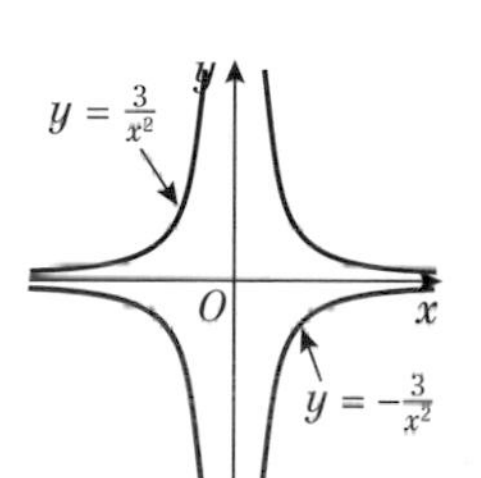

c

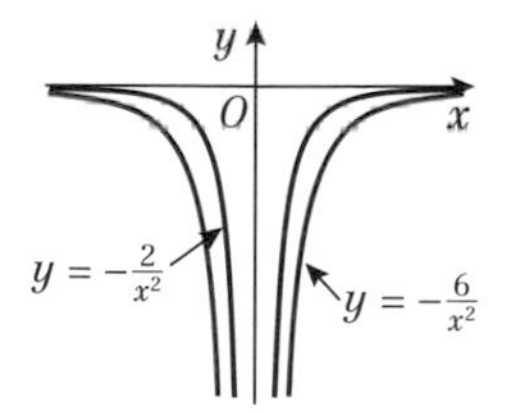

Exercise 4D

1 a i

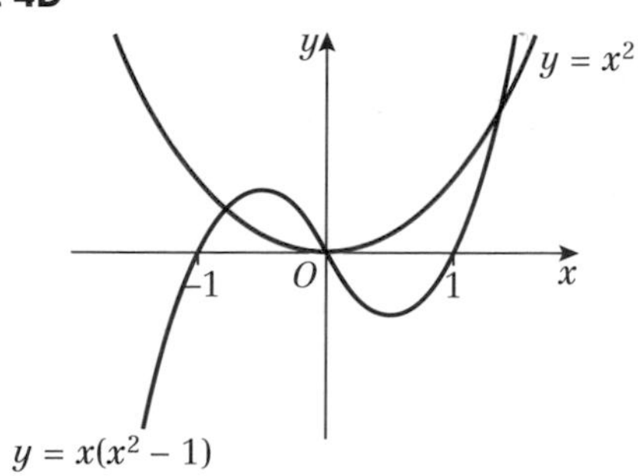

ii 3 **iii** $x^2 = x(x^2 - 1)$

b i

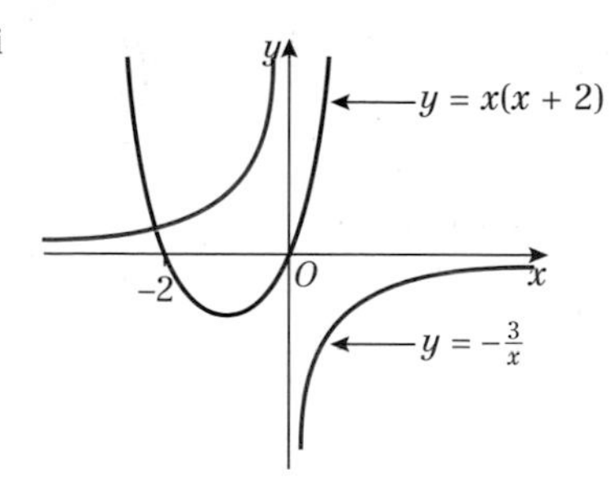

ii 1 **iii** $x(x + 2) = -\frac{3}{x}$

c i

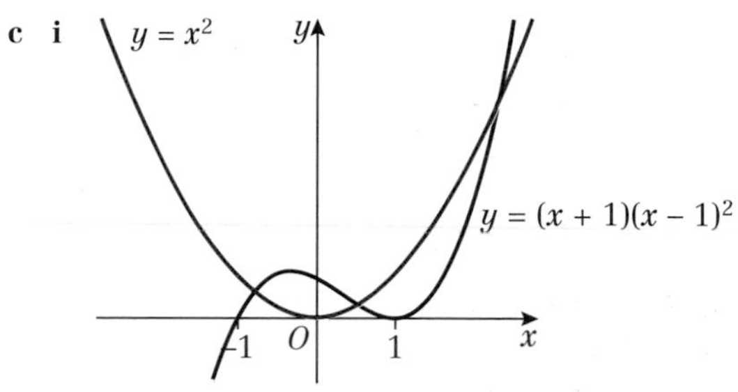

ii 3 **iii** $x^2 = (x + 1)(x - 1)^2$

d i

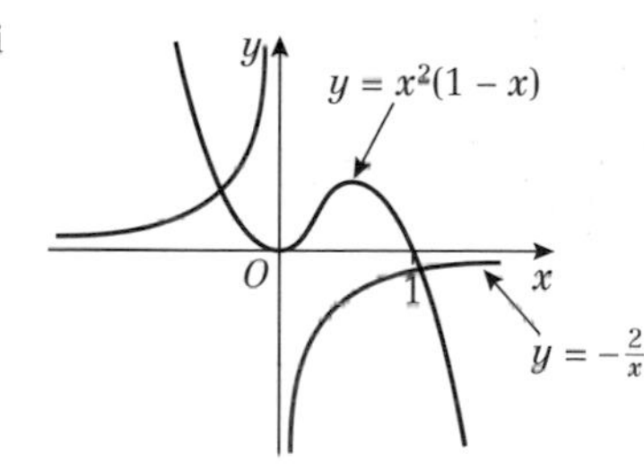

ii 2 **iii** $x^2(1 - x) = -\frac{2}{x}$

e i

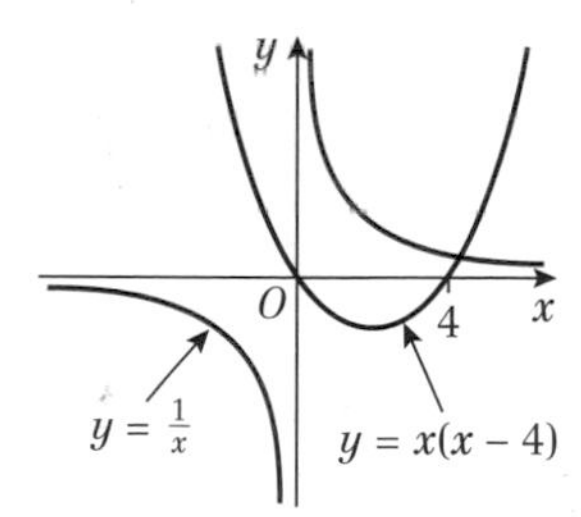

ii 1 **iii** $x(x - 4) = \frac{1}{x}$

f i

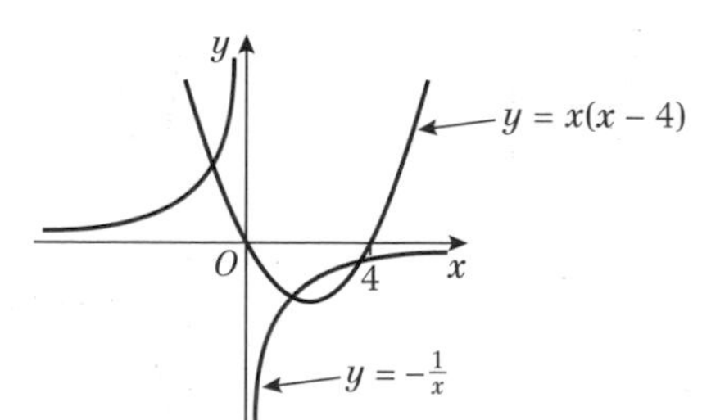

ii 3 **iii** $x(x - 4) = -\frac{1}{x}$

g **i**

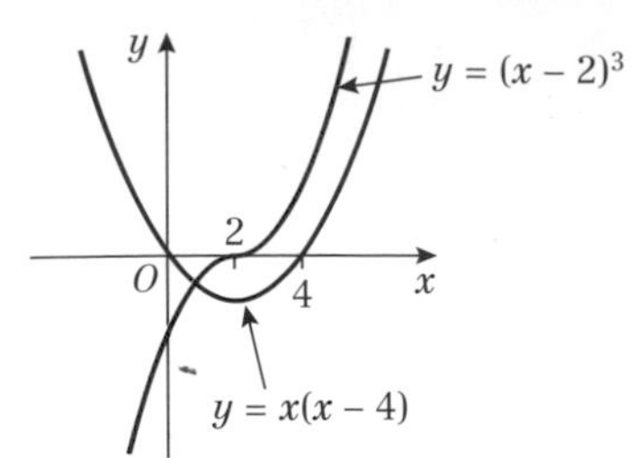

ii 1 **iii** $x(x-4) = (x-2)^3$

h **i**

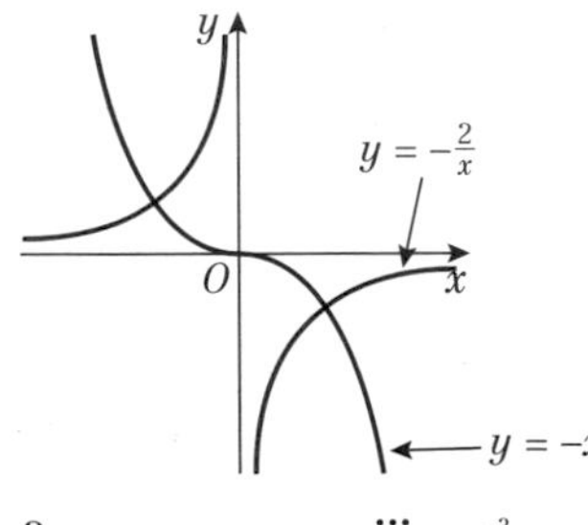

ii 2 **iii** $-x^3 = -\frac{2}{x}$

i **i**

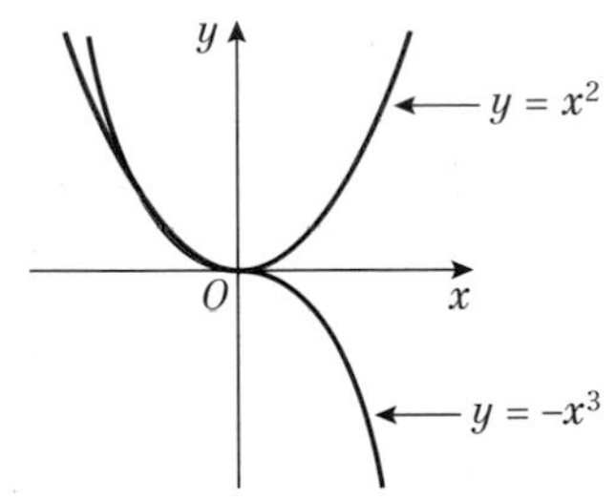

ii 2 **iii** $-x^3 = x^2$

j **i**

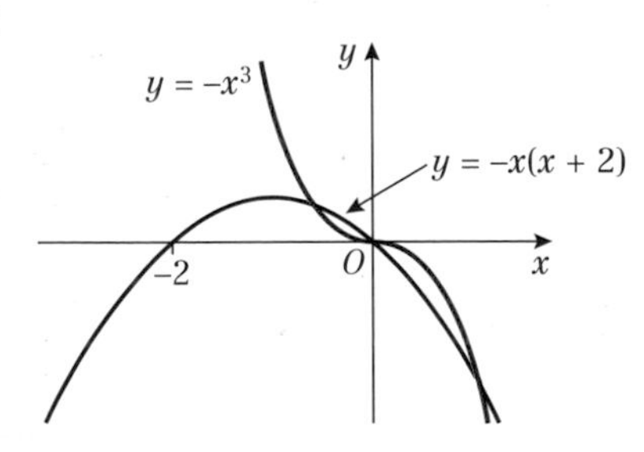

ii 3 **iii** $-x^3 = -x(x+2)$

k **i**

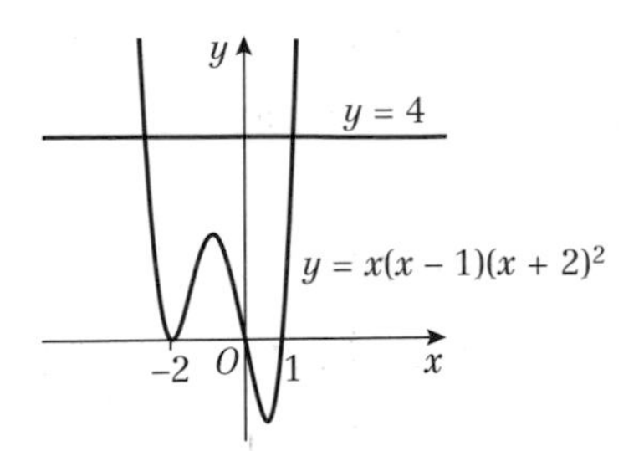

ii 2 **iii** $x(x-1)(x+2)^2 = 4$

l **i**

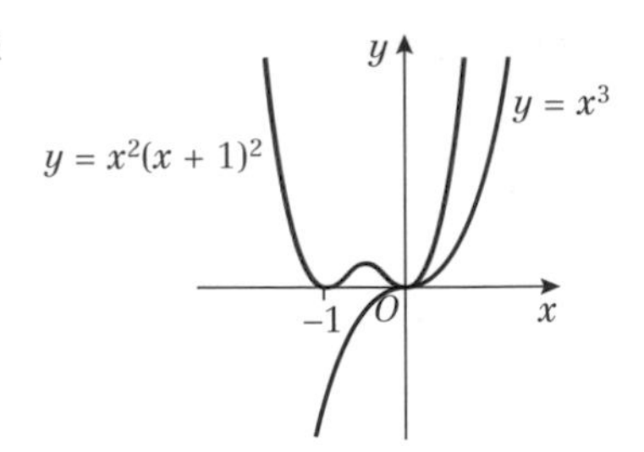

ii 1 **iii** $x^3 = x^2(x+1)^2$

2 **a**

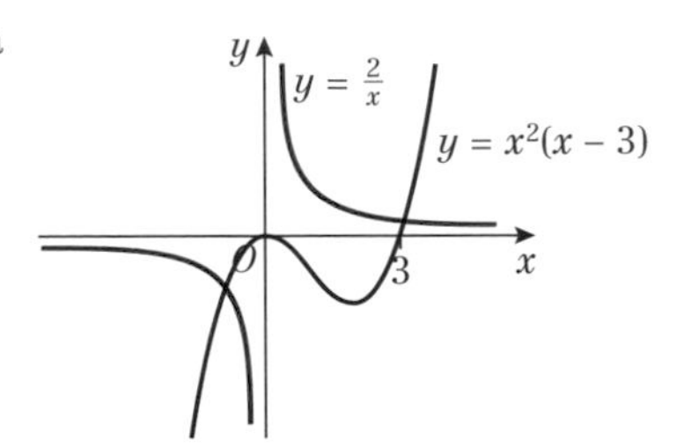

b Only 2 intersections

3 **a**

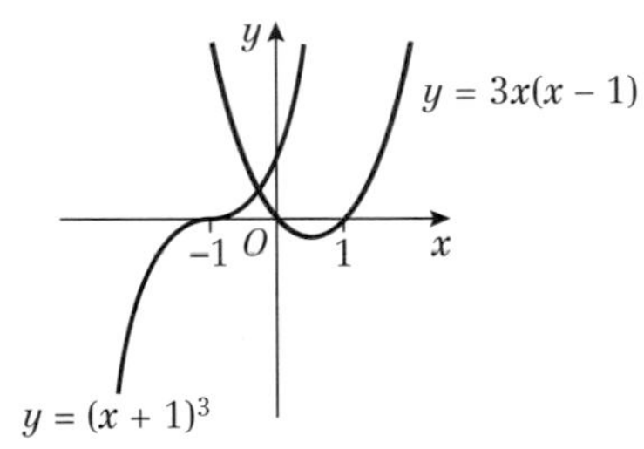

b Only 1 intersection

4 **a**

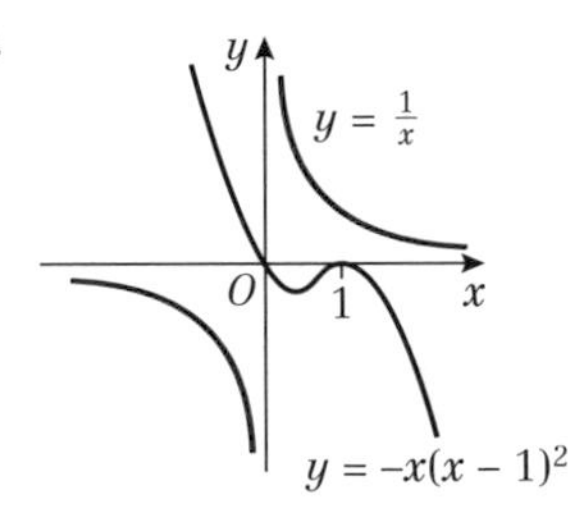

b Graphs do not intersect

5 **a**

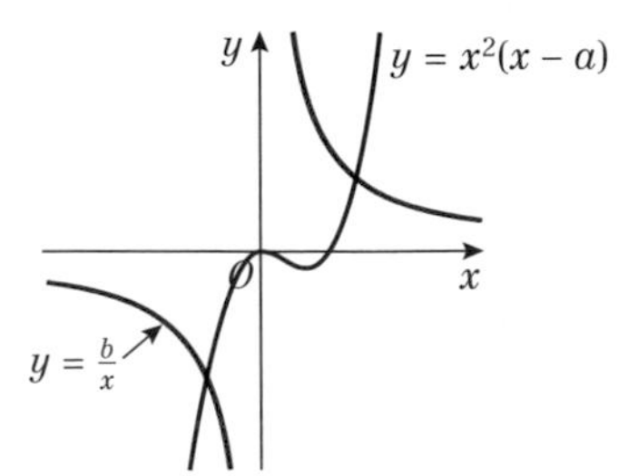

b 2; the graphs cross in two places so there are two solutions.

6 **a**

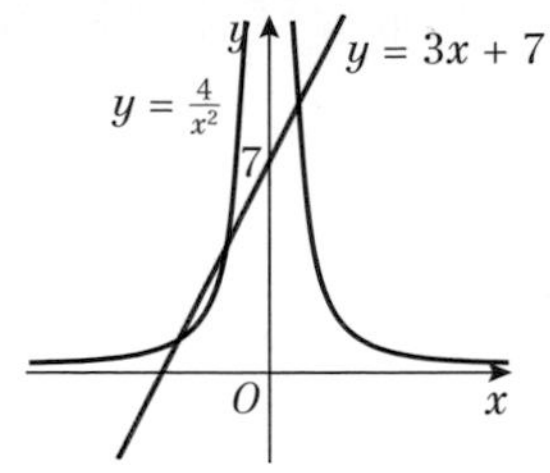

b 3

c Expand brackets and rearrange.

d $(-2, 1)$, $(-1, 4)$, $(\frac{2}{3}, 9)$

7 **a**

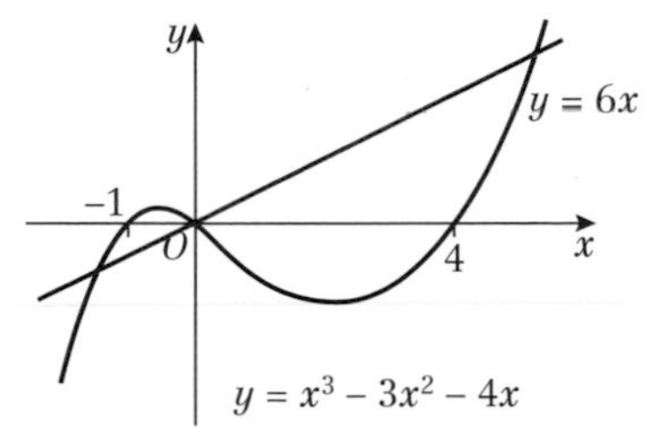

b $(0, 0)$; $(-2, -12)$; $(5, 30)$

8 **a**

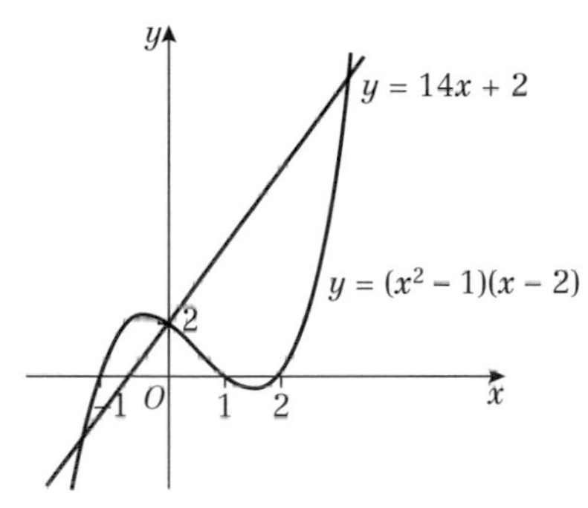

b $(0, 2)$; $(-3, -40)$; $(5, 72)$

9 **a**

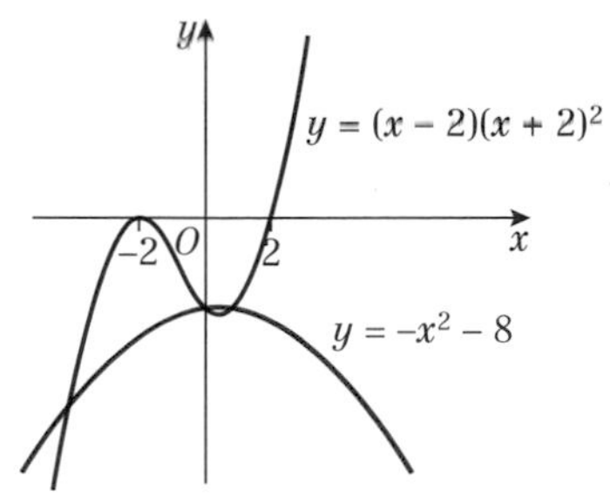

b $(0, -8)$; $(1, -9)$; $(-4, -24)$

10 **a**

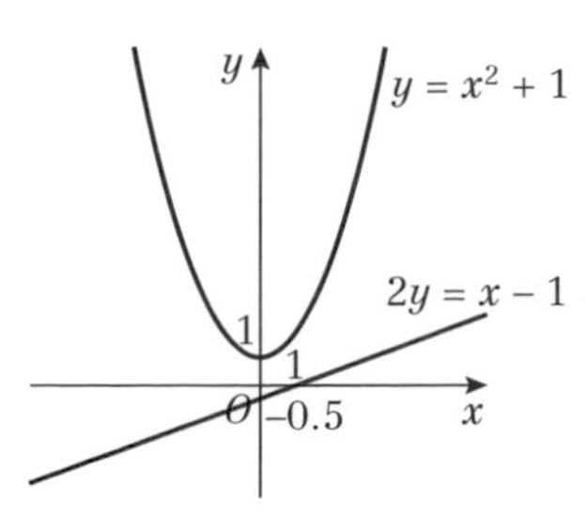

b Graphs do not intersect.

c $a < -\frac{7}{16}$

11 **a**

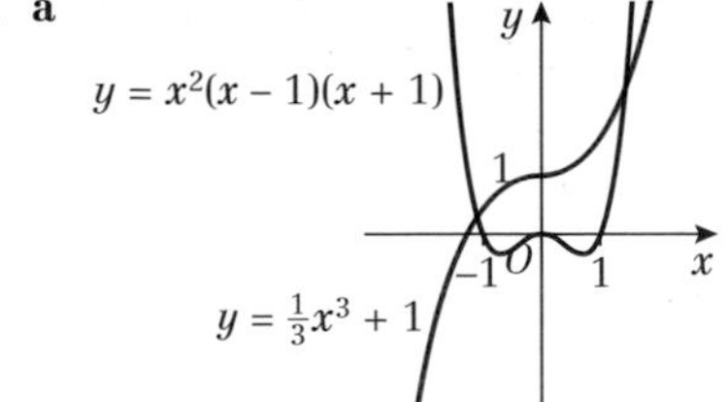

b 2

Exercise 4E

1 **a** **i** **ii**

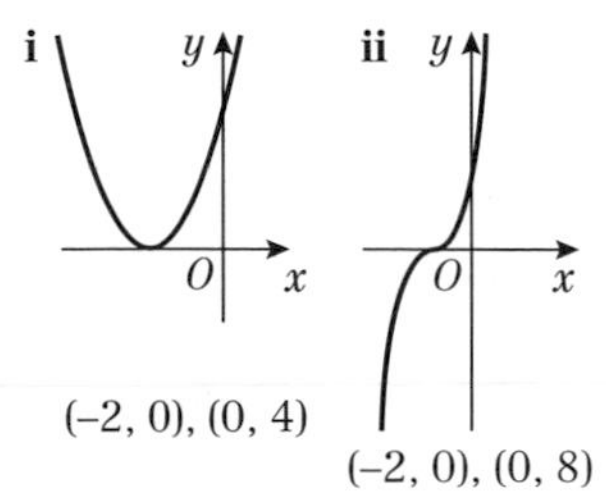

$(-2, 0)$, $(0, 4)$ $\quad$ $(-2, 0)$, $(0, 8)$

iii

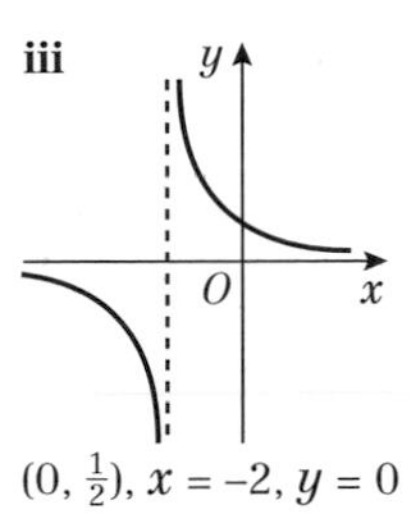

$(0, \frac{1}{2})$, $x = -2$, $y = 0$

b **i** **ii**

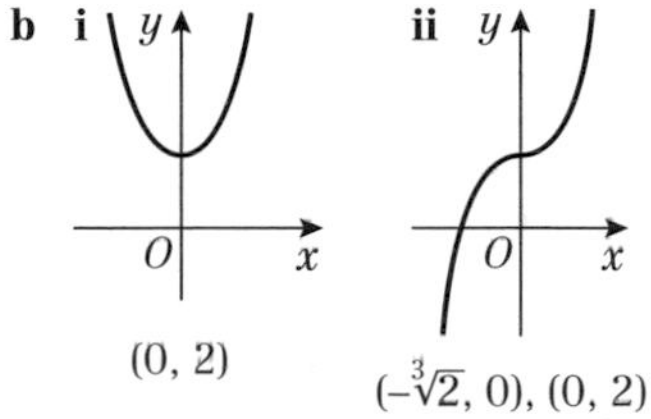

$(0, 2)$ $\quad$ $(-\sqrt[3]{2}, 0)$, $(0, 2)$

iii

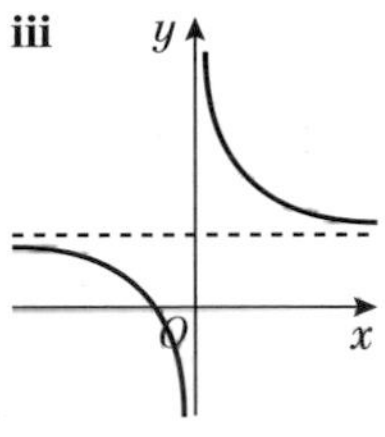

$(-\frac{1}{2}, 0)$, $y = 2$, $x = 0$

c **i**

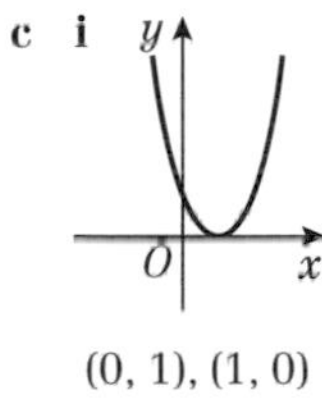

$(0, 1)$, $(1, 0)$

ii

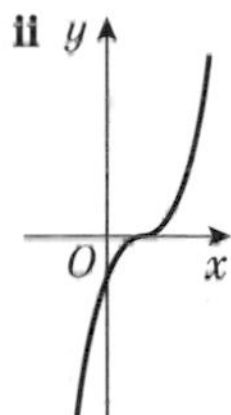

$(0, -1)$, $(1, 0)$

iii

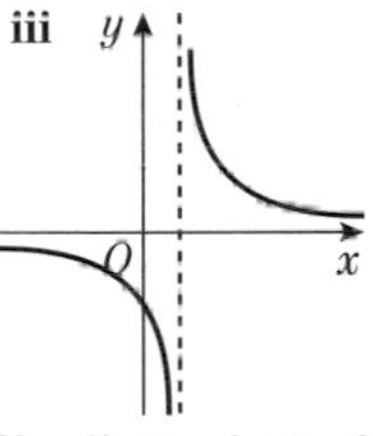

$(0, -1)$, $x = 1$, $y = 0$

d **i**

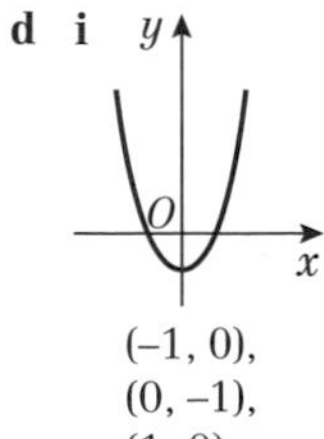

$(-1, 0)$, $(0, -1)$, $(1, 0)$

ii

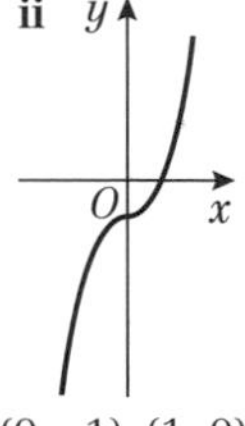

$(0, -1)$, $(1, 0)$

iii

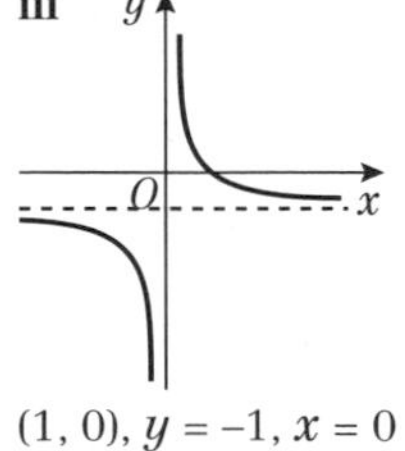

$(1, 0)$, $y = -1$, $x = 0$

e **i**

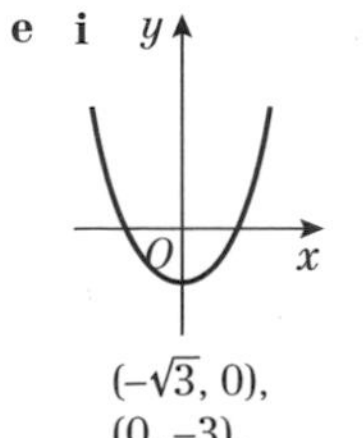

$(-\sqrt{3}, 0)$, $(0, -3)$, $(\sqrt{3}, 0)$

ii

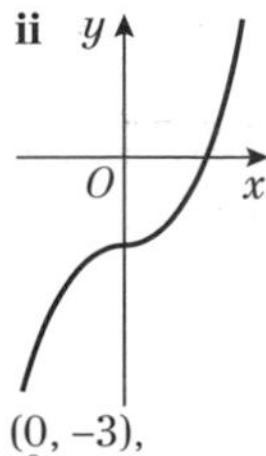

$(0, -3)$, $(\sqrt[3]{3}, 0)$

iii

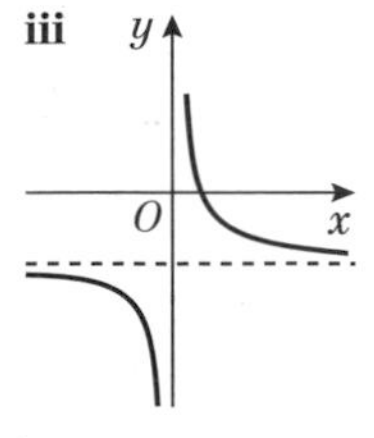

$(\frac{1}{3}, 0)$, $y = -3$, $x = 0$

f **i** $(0, 9), (3, 0)$

ii $(0, -27), (3, 0)$

iii $(0, -\frac{1}{3}), x = 3, y = 0$

2 **a**

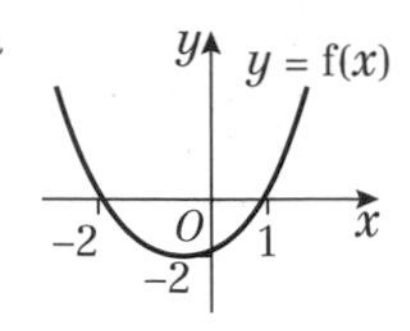

b **i**

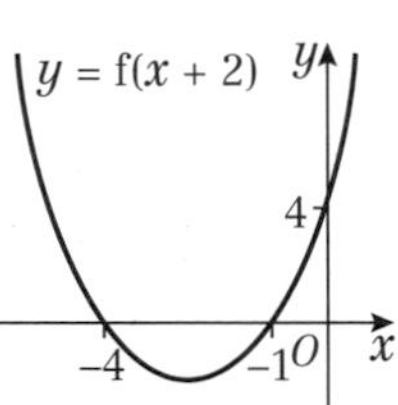

ii

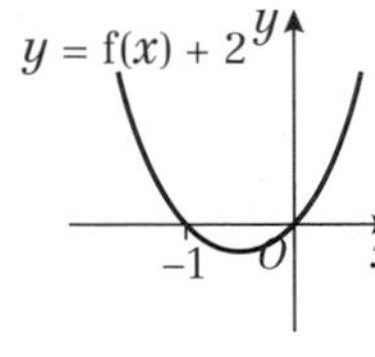

c $f(x + 2) = (x + 1)(x + 4)$; $(0, 4)$
$f(x) + 2 = (x - 1)(x + 2) + 2$; $(0, 0)$

3 **a**

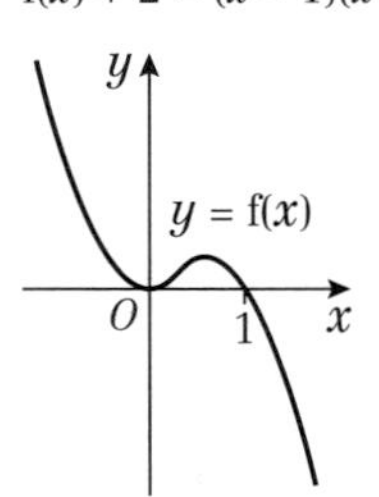

b

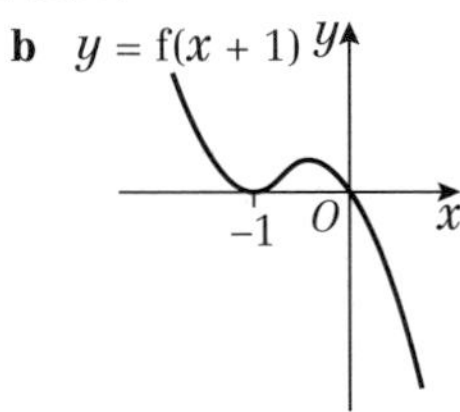

c $f(x + 1) = -x(x + 1)^2$; $(0, 0)$

4 **a**

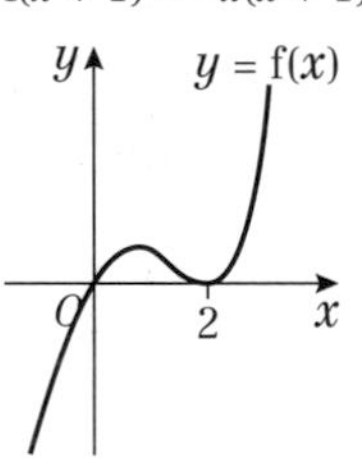

b

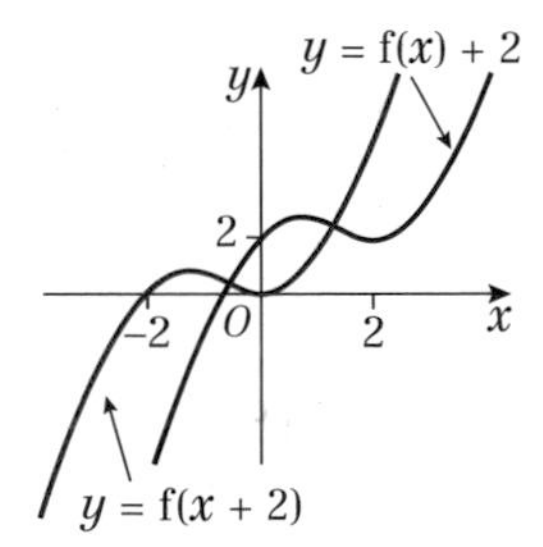

c $f(x + 2) = (x + 2)x^2$; $(0, 0)$; $(-2, 0)$

5 **a**

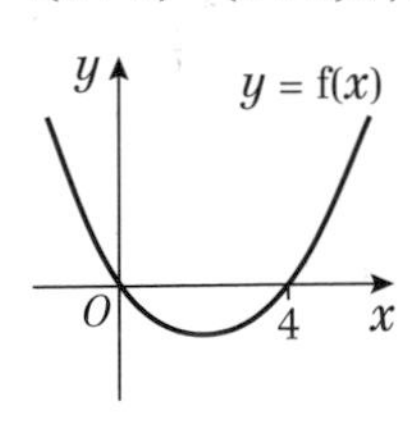

b

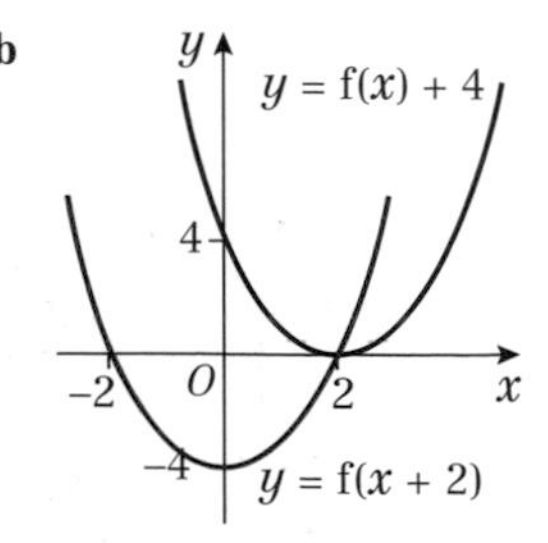

c $f(x + 2) = (x + 2)(x - 2)$; $(2, 0)$; $(-2, 0)$; $(0, -4)$
$f(x) + 4 = (x - 2)^2$; $(2, 0)$; $(0, 4)$

6 **a**

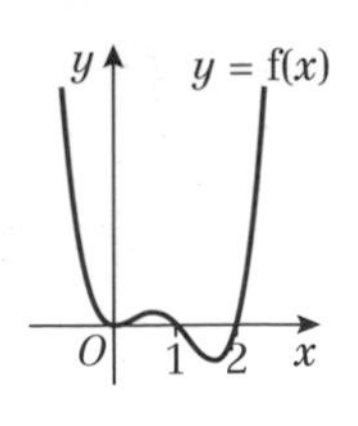

b

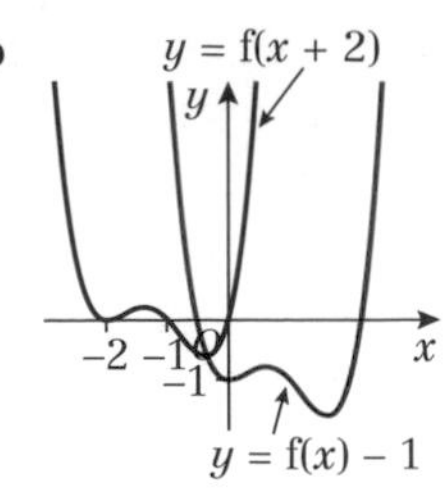

7 **a** $(6, -1)$ **b** $(4, 2)$

8 $y = \frac{1}{x - 4}$

9

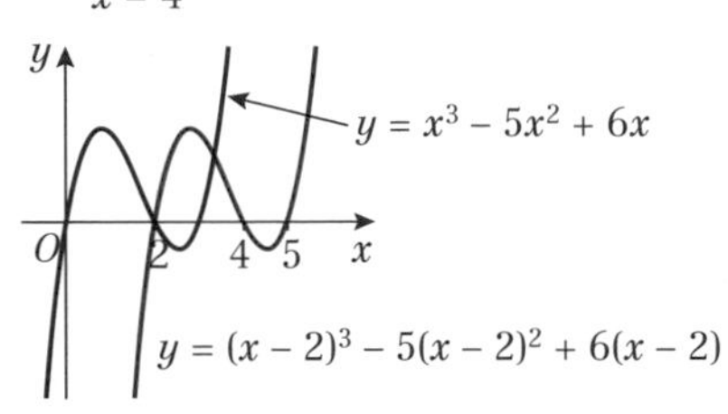

10

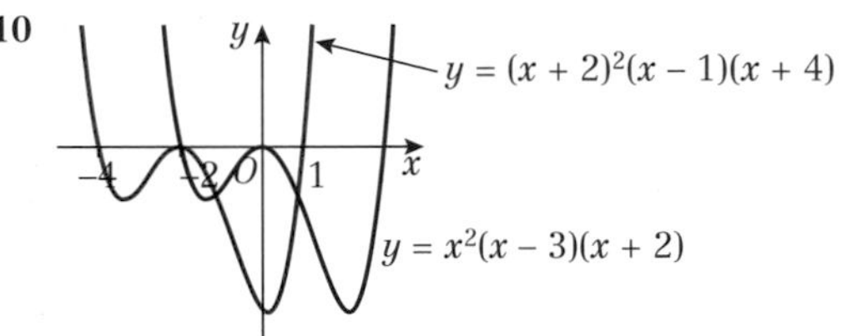

11 **a**

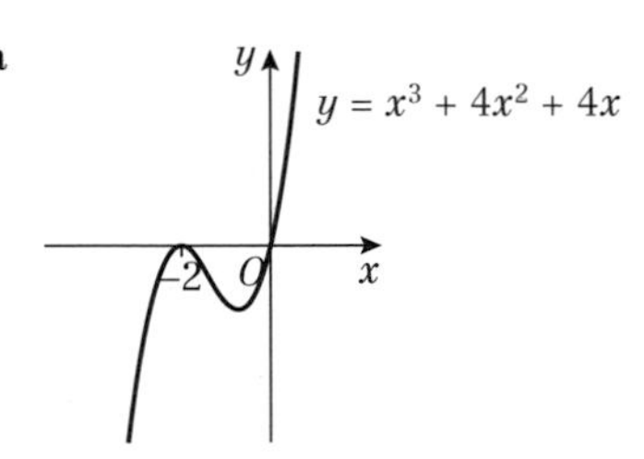

b -1 or 1

12 **a**

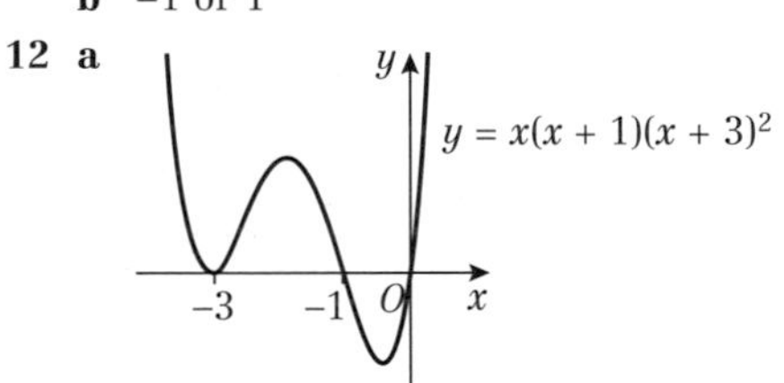

b -2, -3 or -5

Challenge

1 $(3, 2)$

2 **a** $(-7, -12)$ **b** $f(x - 2) + 1$

Exercise 4F

1 **a** **i** y f(2x) f(x) O x

ii y f(2x) f(x) O x

iii

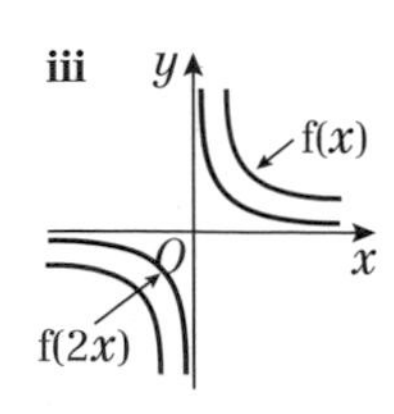

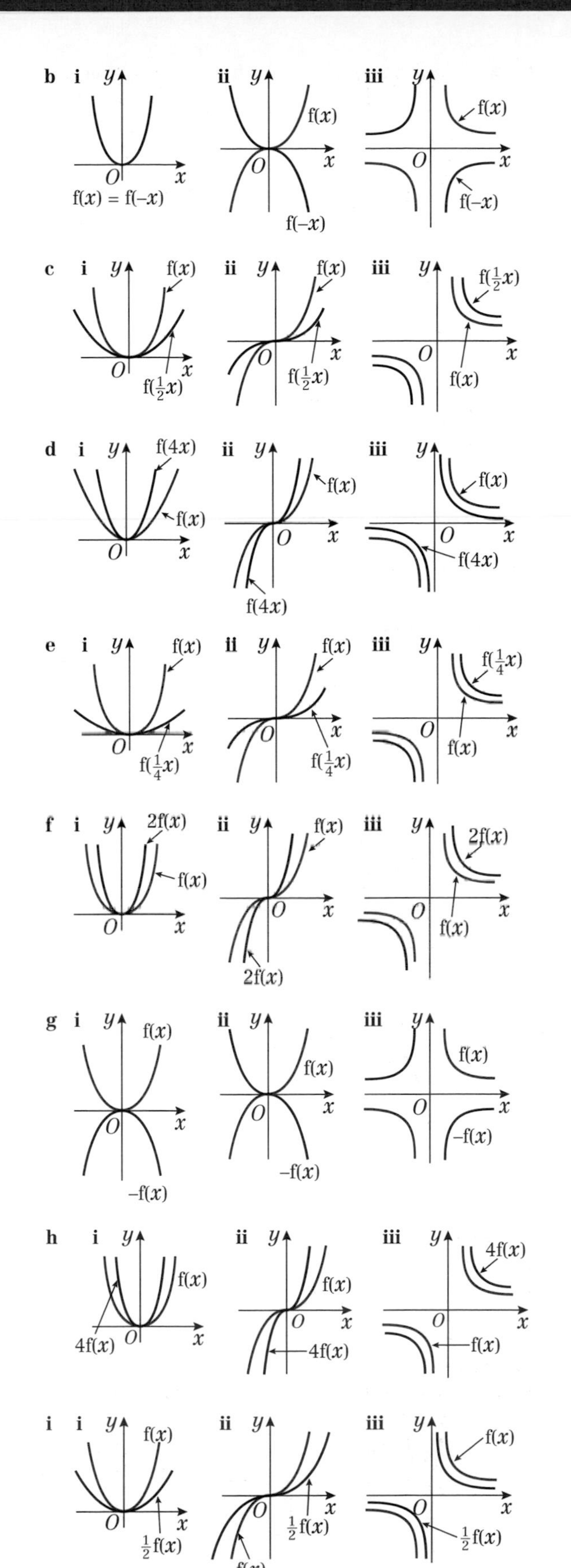
b i ii iii
f(x) = f(−x)
f(x)
f(−x)
c i ii iii
f(x)
f($\frac{1}{2}$x)
d i ii iii
f(4x)
f(x)
e i ii iii
f(x)
f($\frac{1}{4}$x)
f i ii iii
2f(x)
f(x)
g i ii iii
f(x)
−f(x)
h i ii iii
4f(x)
f(x)
i i ii iii
f(x)
$\frac{1}{2}$f(x)

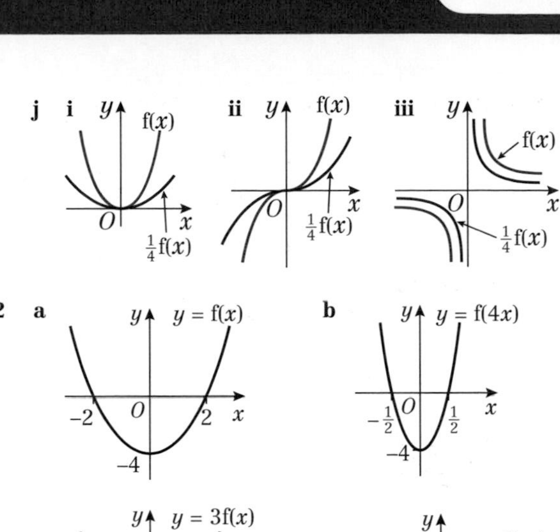
j i ii iii
f(x)
$\frac{1}{4}$f(x)
2 a
y = f(x)
−2
2
−4
b
y = f(4x)
$-\frac{1}{2}$
$\frac{1}{2}$
−4

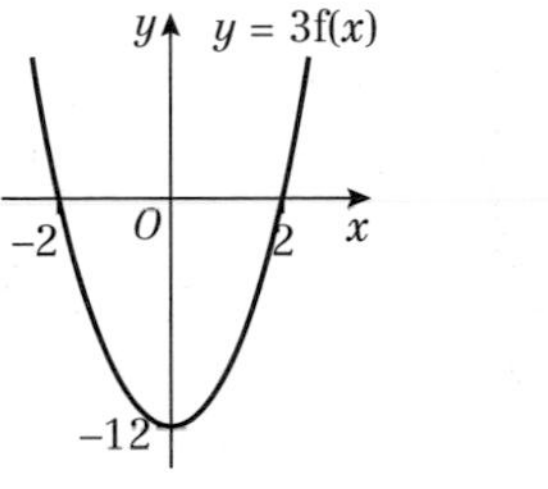
y = 3f(x)
−2
2
−12

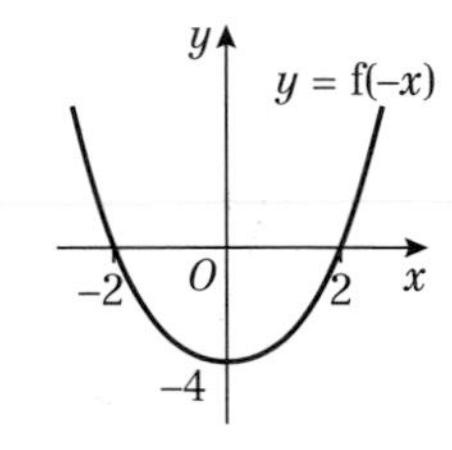
y = f(−x)
−2
2
−4

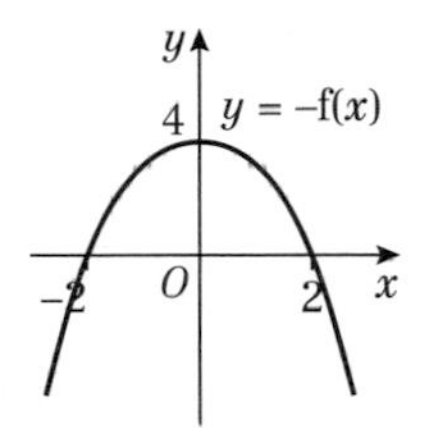
y = −f(x)
4
−2
2

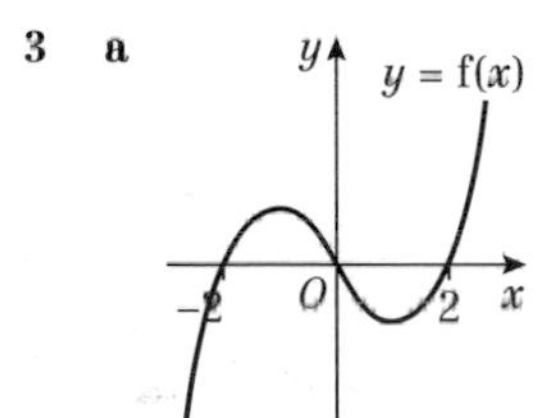
3 a
y = f(x)
−2
2

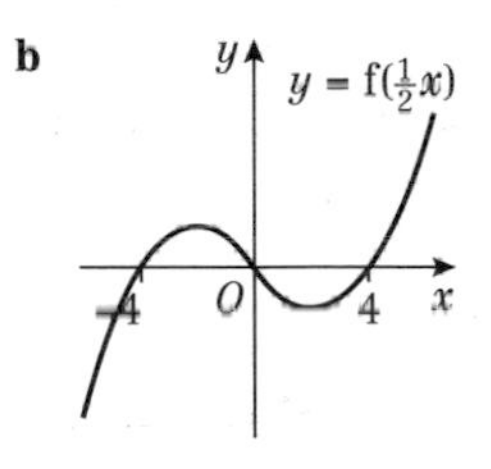
b
y = f($\frac{1}{2}$x)
−4
4

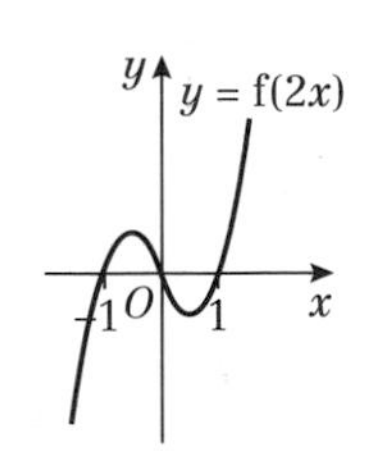
y = f(2x)
−1
1

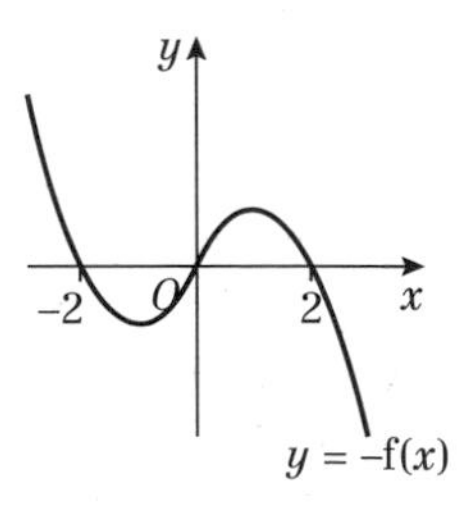
−2
2
y = −f(x)

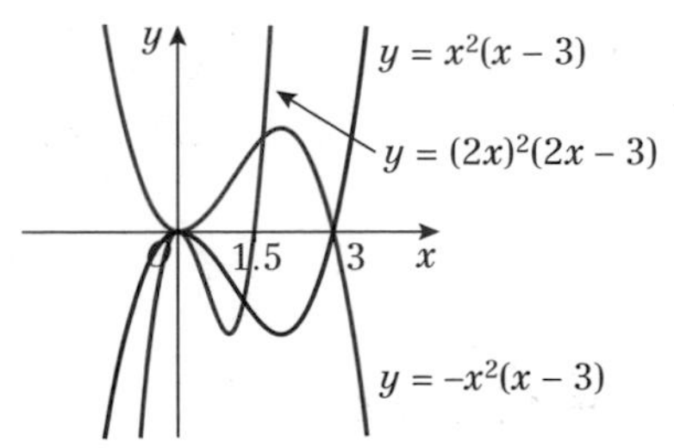
4
y = x²(x − 3)
y = (2x)²(2x − 3)
1.5
3
y = −x²(x − 3)

5

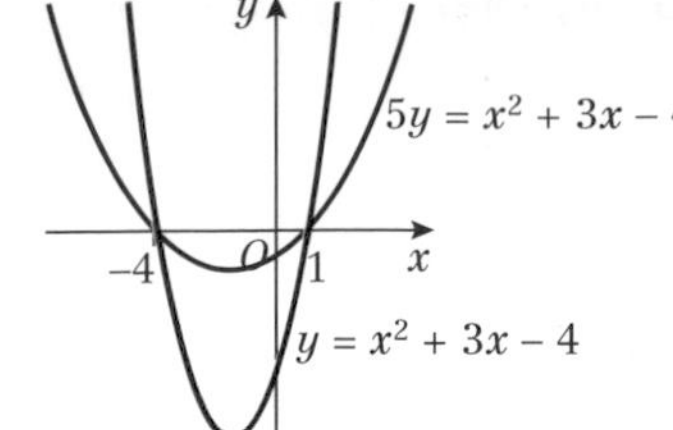

6

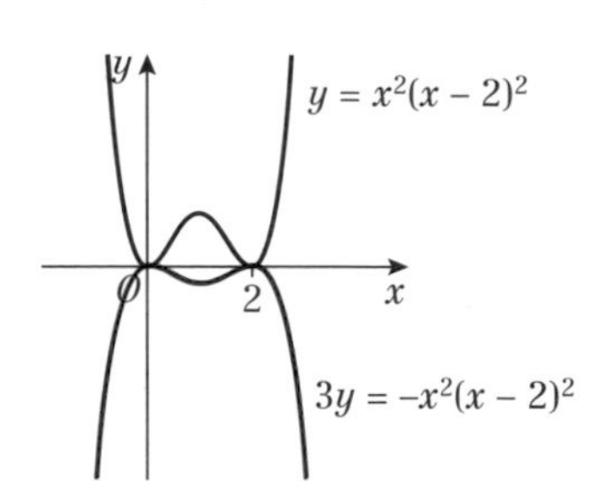

7 **a** $(1, -3)$ **b** $(2, -12)$

8 $(-4, 8)$

9 **a**

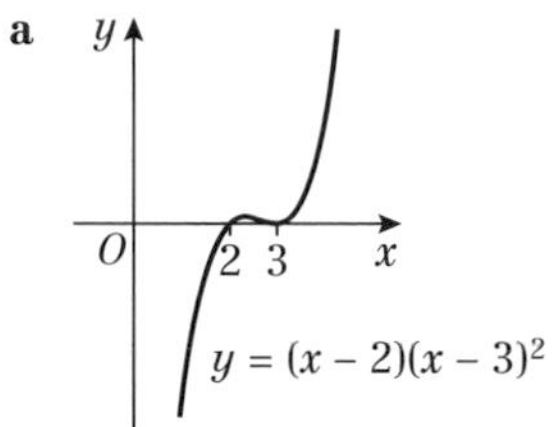

b 2 and 3

Challenge

1 $(2, -2)$

2 $\frac{1}{4}\mathrm{f}(\frac{1}{2}x)$

Exercise 4G

1 **a**

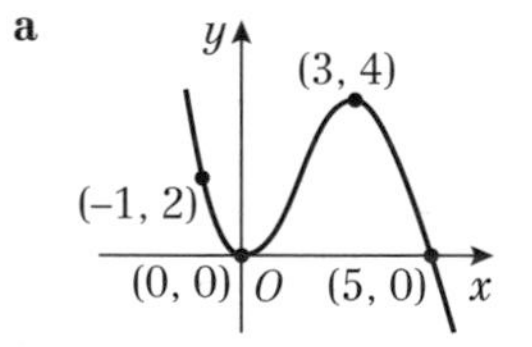

b

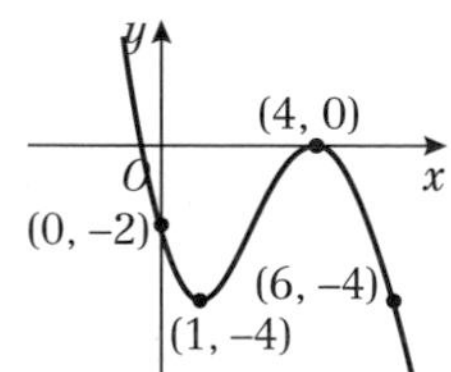

c

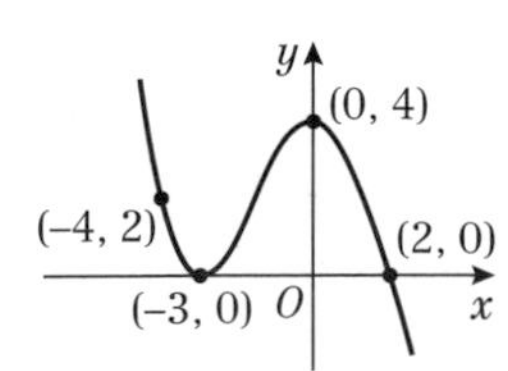

d

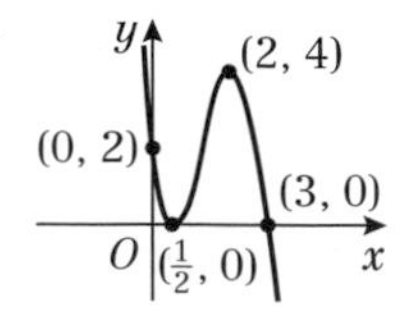

e

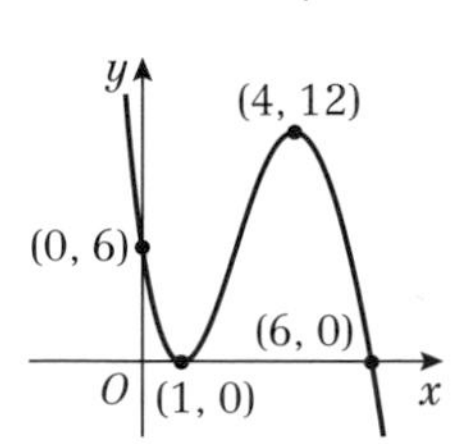

f

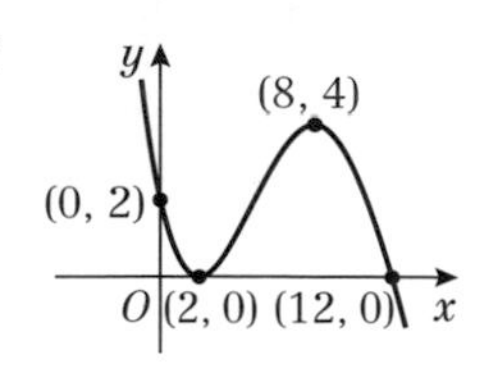

g

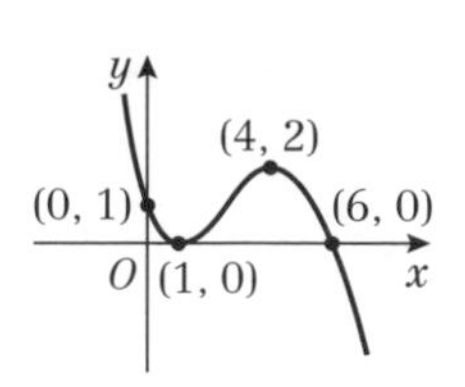

h

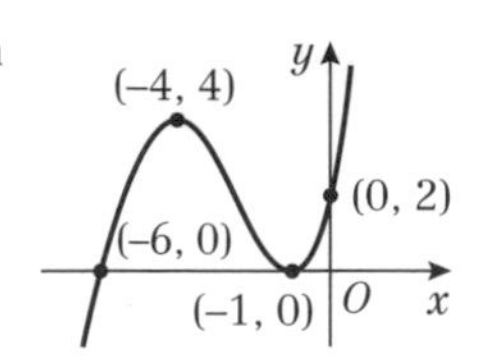

2 **a** $y = 4$, $x = 1$, $(0, 2)$

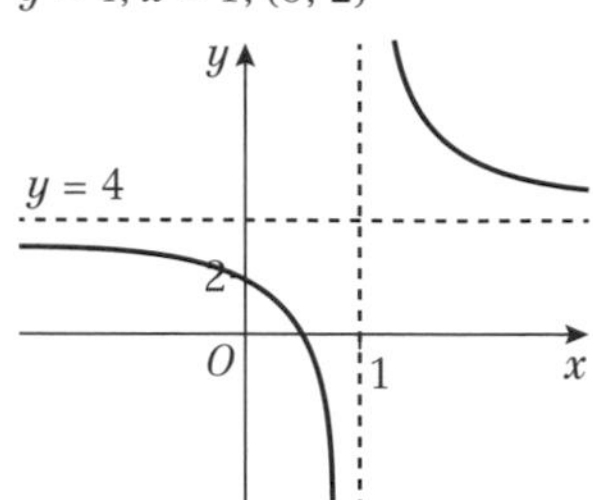

b $y = 2$, $x = 0$, $(-1, 0)$

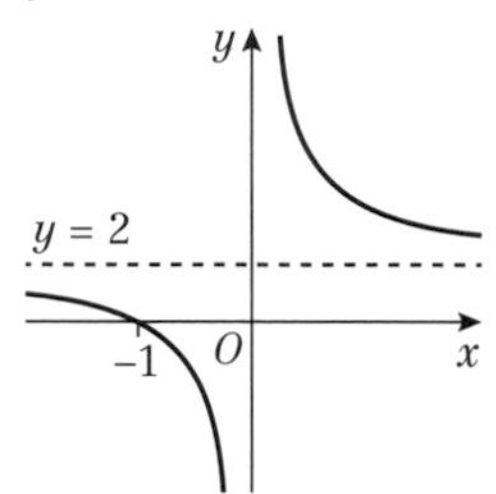

c $y = 4$, $x = 1$, $(0, 0)$

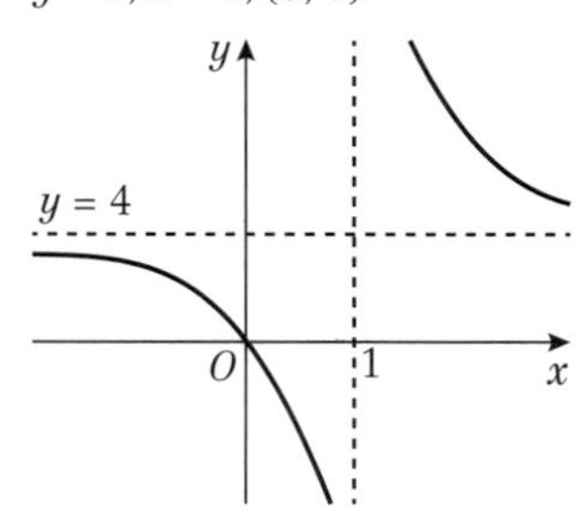

d $y = 0$, $x = 1$, $(0, -2)$

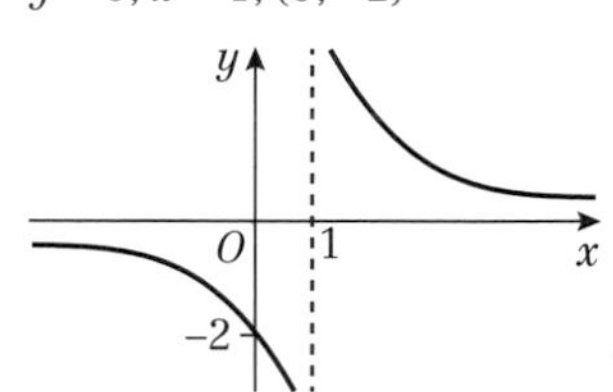

e $y = 2$, $x = \frac{1}{2}$, $(0, 0)$

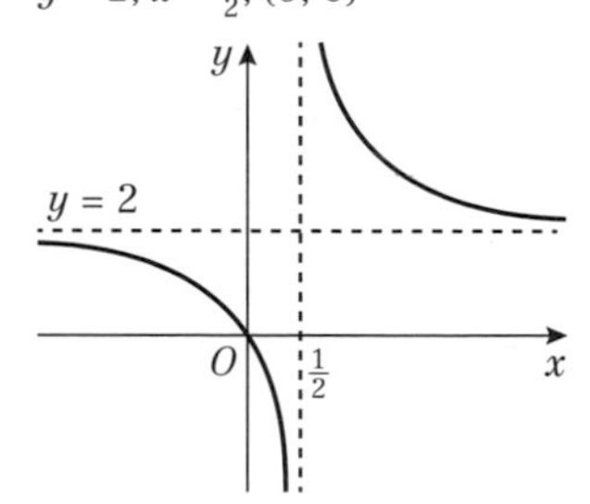

f $y = 2$, $x = 2$, $(0, 0)$

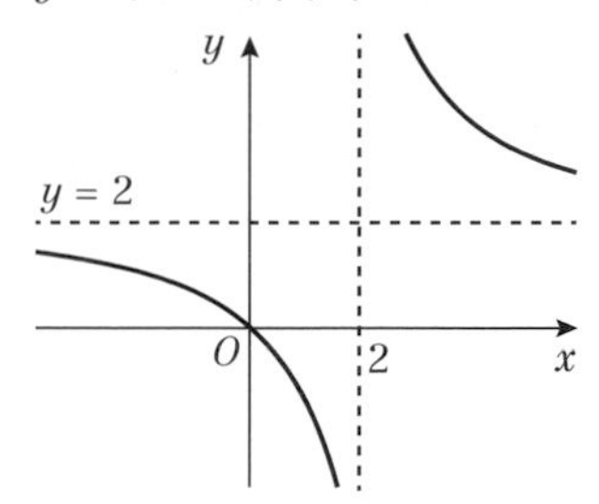

g $y = 1$, $x = 1$, $(0, 0)$

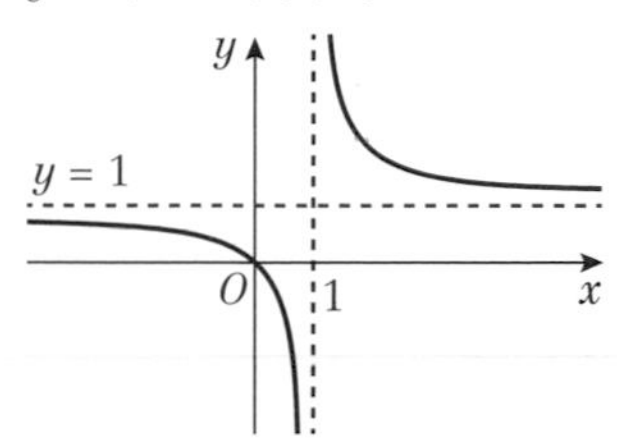

h $y = -2$, $x = 1$, $(0, 0)$

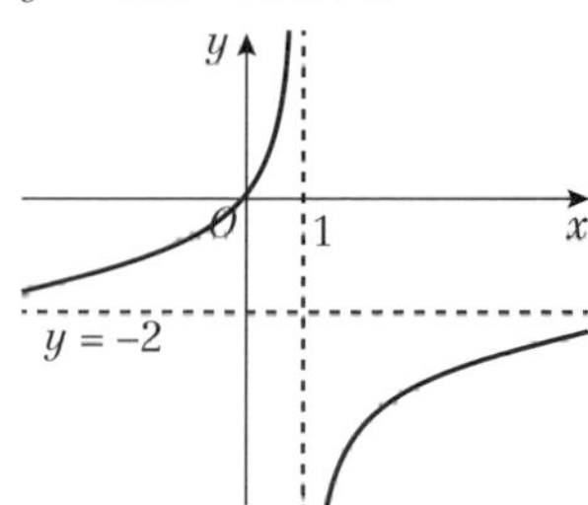

3 a $A(-2, -6)$, $B(0, 0)$, $C(2, -3)$, $D(6, 0)$

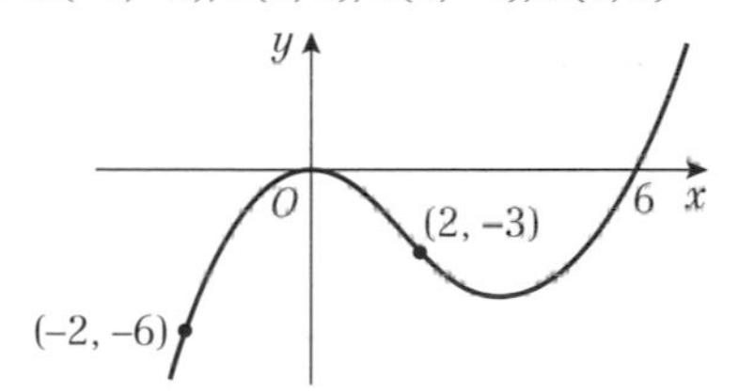

b $A(-4, 0)$, $B(-2, 6)$, $C(0, 3)$, $D(4, 6)$

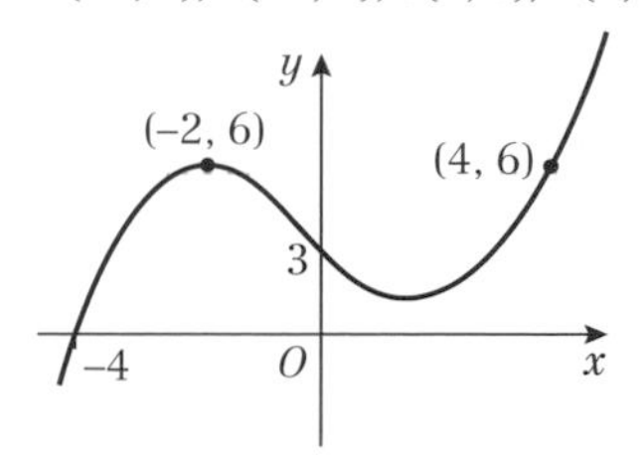

c $A(-2, -6)$, $B(-1, 0)$, $C(0, -3)$, $D(2, 0)$

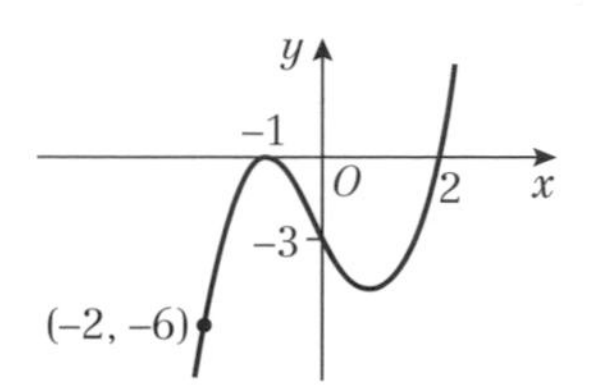

d $A(-8, -6)$, $B(-6, 0)$, $C(-4, -3)$, $D(0, 0)$

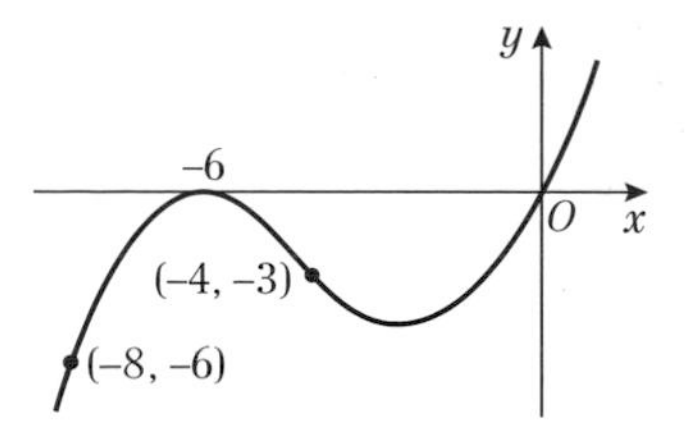

e $A(-4, -3)$, $B(-2, 3)$, $C(0, 0)$, $D(4, 3)$

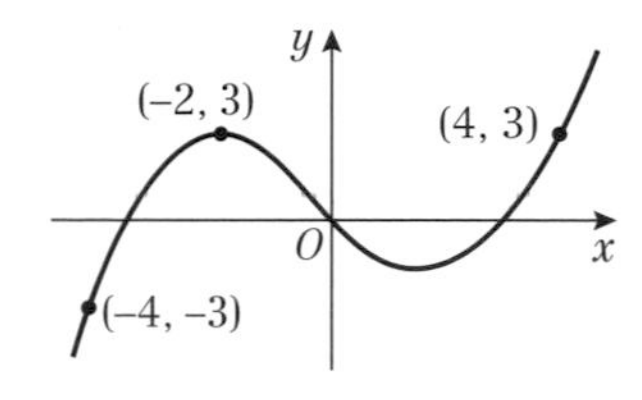

f $A(-4, -18)$, $B(-2, 0)$, $C(0, -9)$, $D(4, 0)$

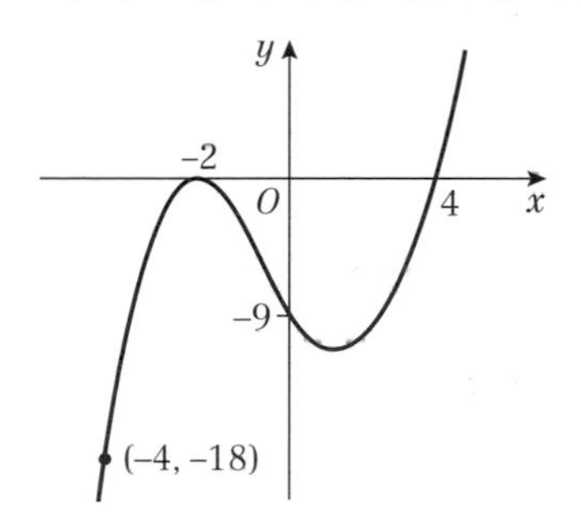

g $A(-4, -2)$, $B(-2, 0)$, $C(0, -1)$, $D(4, 0)$

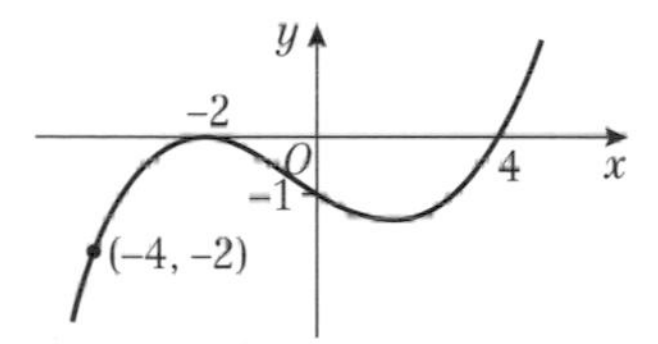

h $A(-16, -6)$, $B(-8, 0)$, $C(0, -3)$, $D(16, 0)$

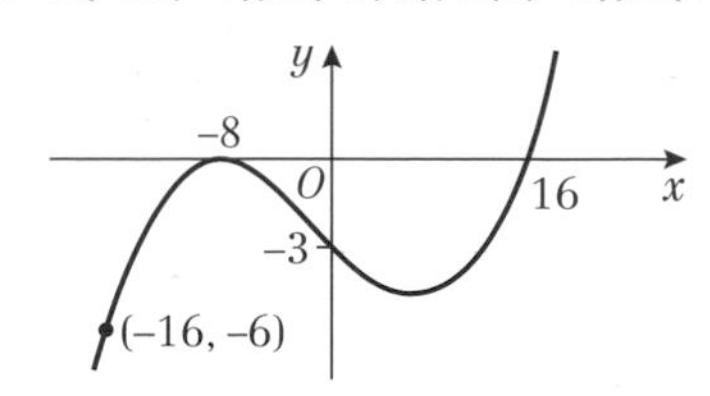

i $A(-4, 6)$, $B(-2, 0)$, $C(0, 3)$, $D(4, 0)$

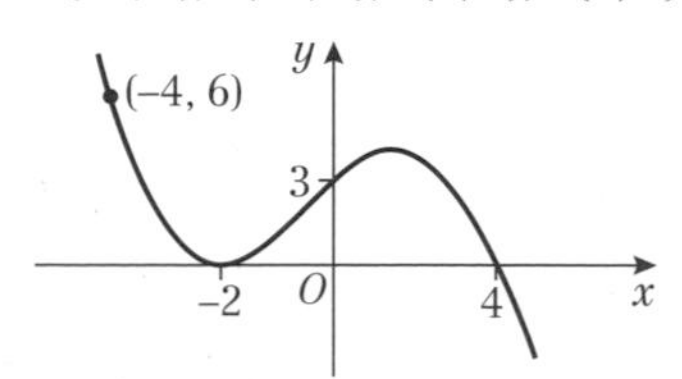

j $A(4, -6)$, $B(2, 0)$, $C(0, -3)$, $D(-4, 0)$

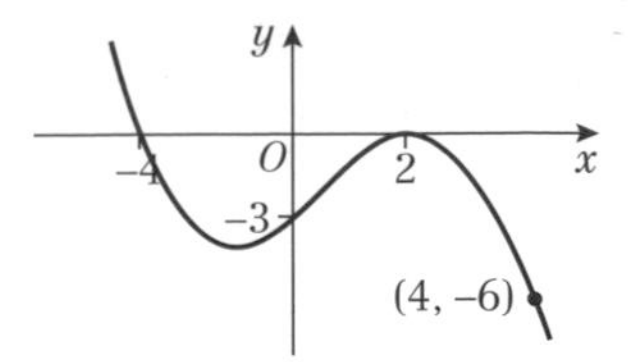

4 a i $x = -2$, $y = 0$, $(0, 2)$

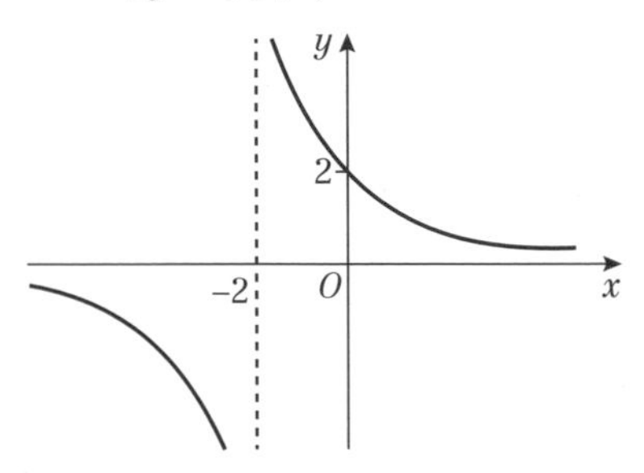

ii $x = -1$, $y = 0$, $(0, 1)$

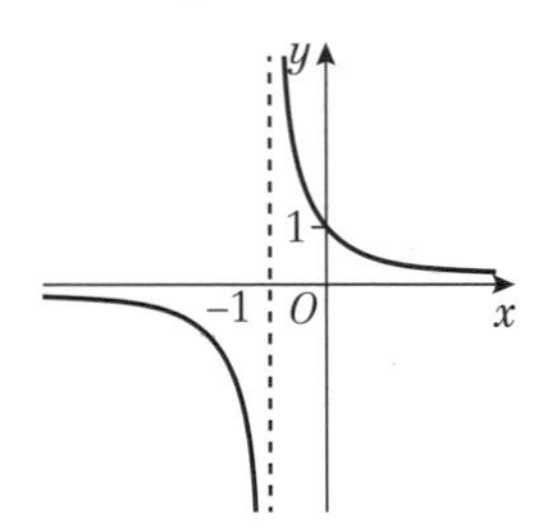

iii $x = 0$, $y = 0$

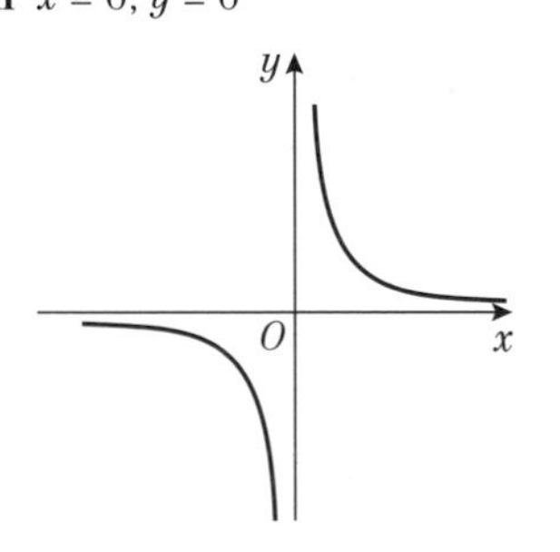

iv $x = -2$, $y = -1$, $(0, 0)$

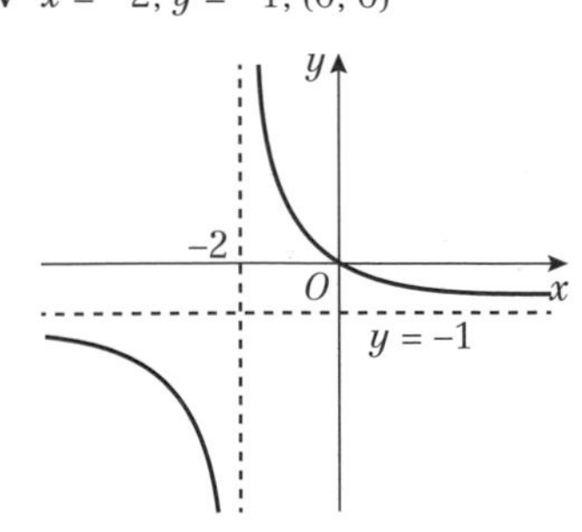

v $x = 2$, $y = 0$, $(0, 1)$

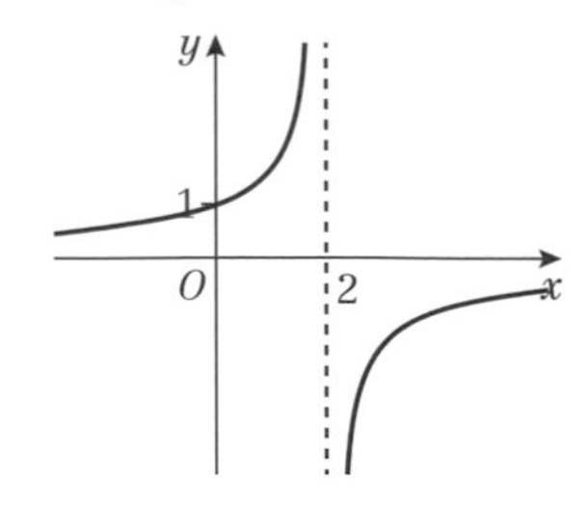

vi $x = -2$, $y = 0$, $(0, -1)$

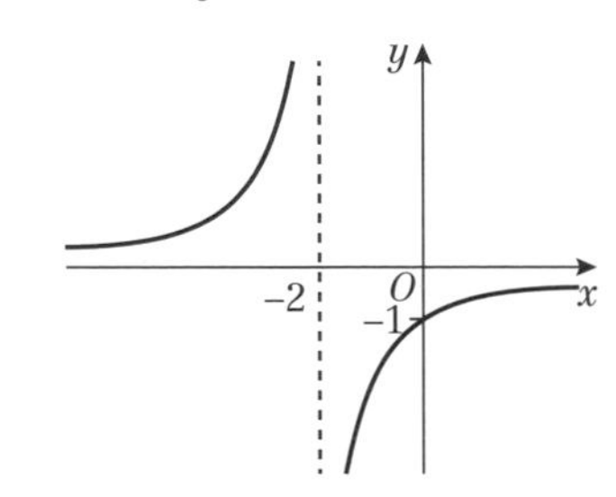

b $f(x) = \dfrac{2}{x + 2}$

5 a $\frac{1}{2}$

b i $(6, 1)$ **ii** $(2, 3)$ **iii** $(2, -3.5)$

6 a $A(-1, -2)$ $B(0, 0)$ $C(1, 0)$ $D(2, -2)$

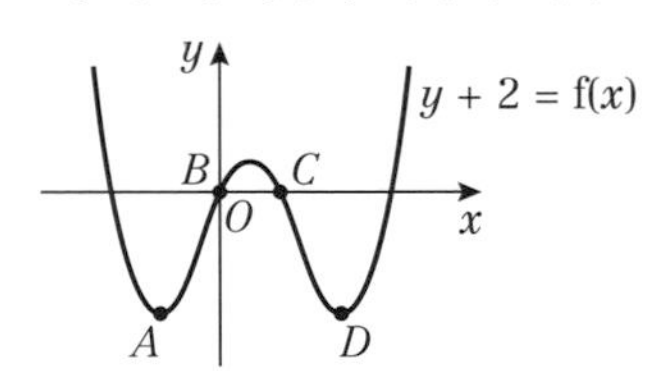

b $A(-1, 0)$ $B(0, 4)$ $C(1, 4)$ $D(2, 0)$

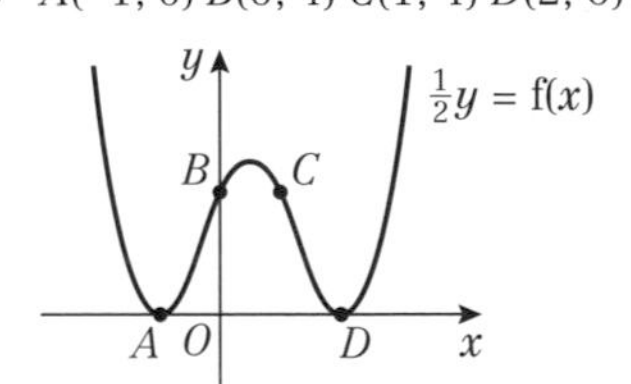

c $A(-1, 3)$ $B(0, 5)$ $C(1, 5)$ $D(2, 3)$

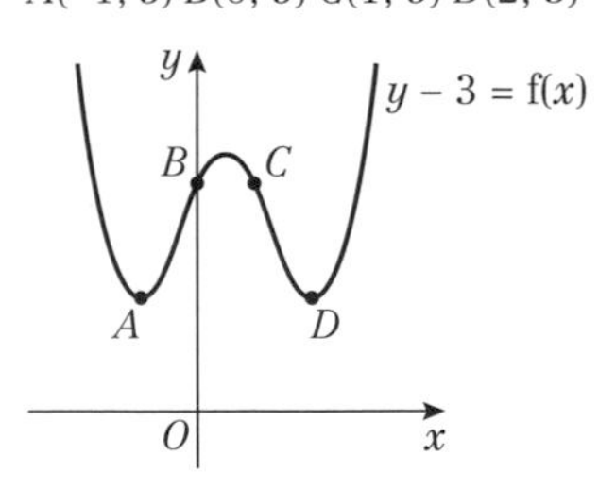

d $A(-1, 0)$ $B(0, \frac{2}{3})$ $C(1, \frac{2}{3})$ $D(2, 0)$

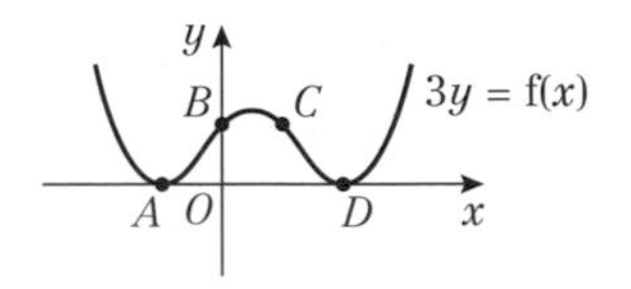

e $A(-1, 0.5)$ $B(0, 1.5)$ $C(1, 1.5)$ $D(2, 0.5)$

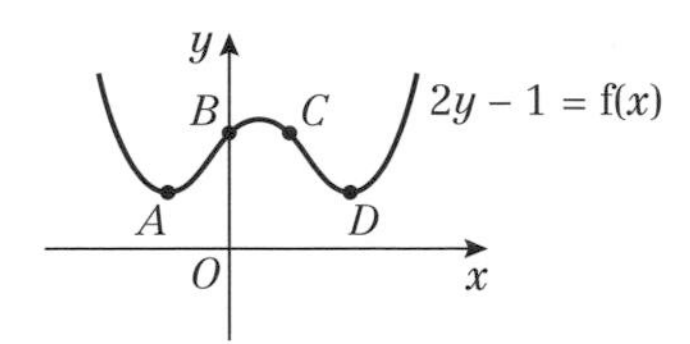

Mixed exercise 4

1 a

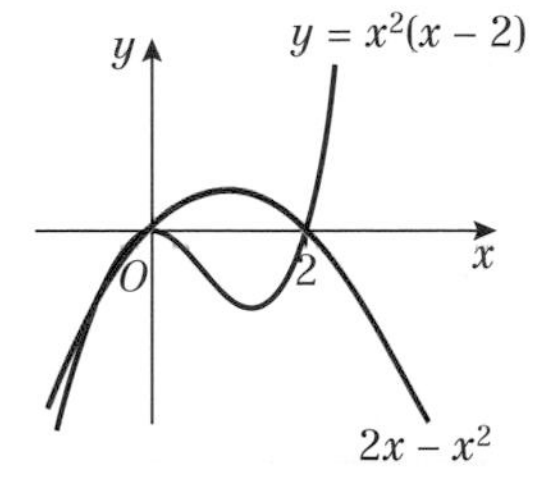

b $x = 0, -1, 2$; points $(0, 0)$, $(2, 0)$, $(-1, -3)$

2 a

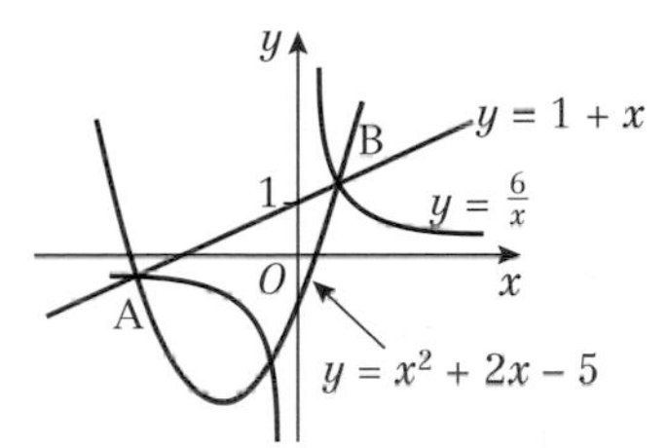

b $A(-3, -2)$, $B(2, 3)$

c $y = x^2 + 2x - 5$

3 a

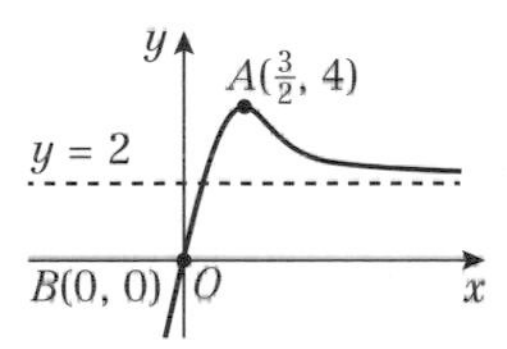

b

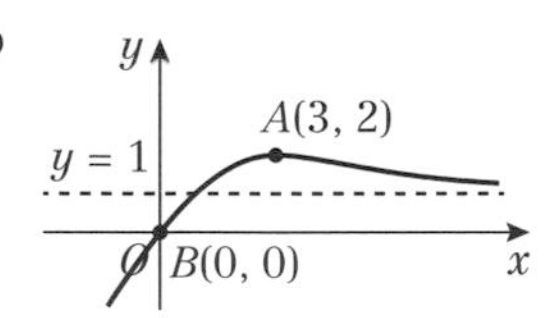

c

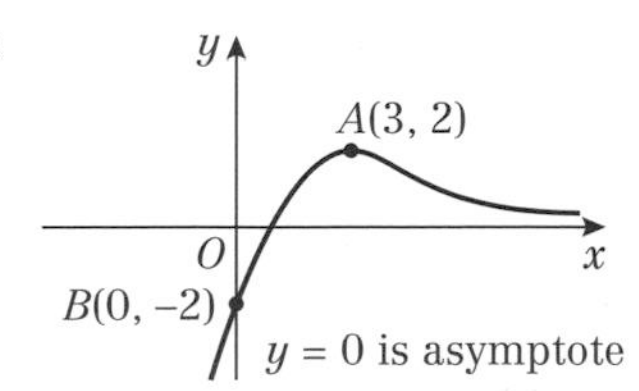

d

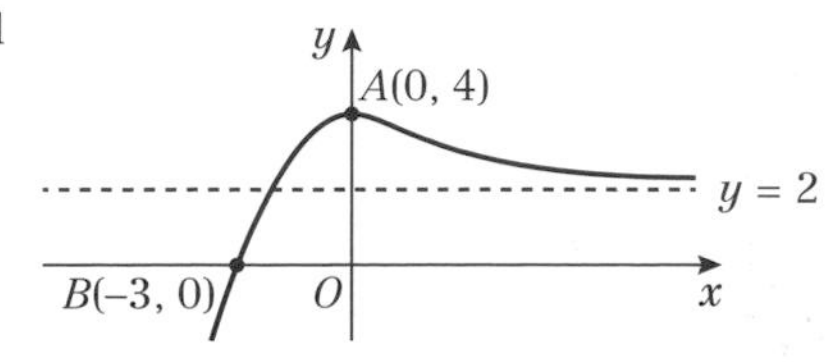

e

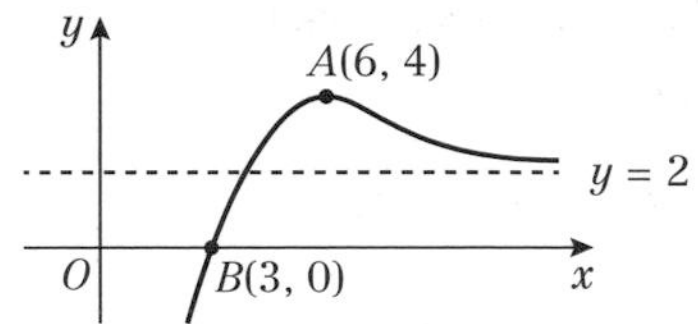

f

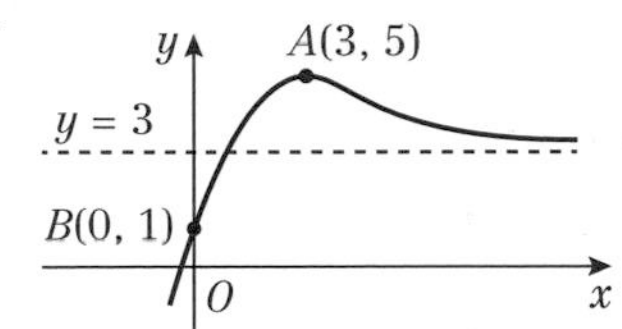

4 a $x = -1$ at A, $x = 3$ at B

5 a, b

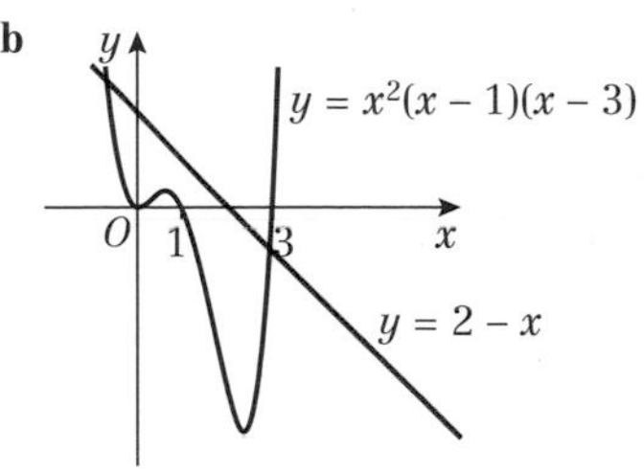

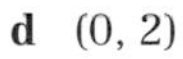

c 2 **d** $(0, 2)$

6 a

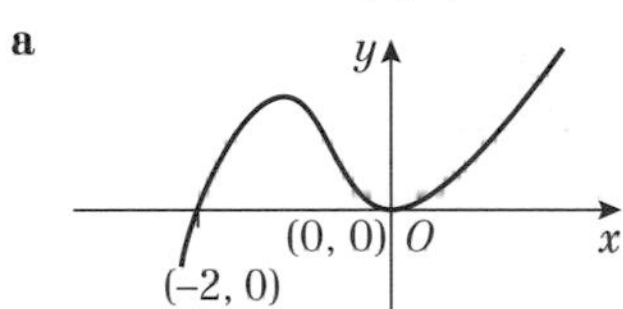

b

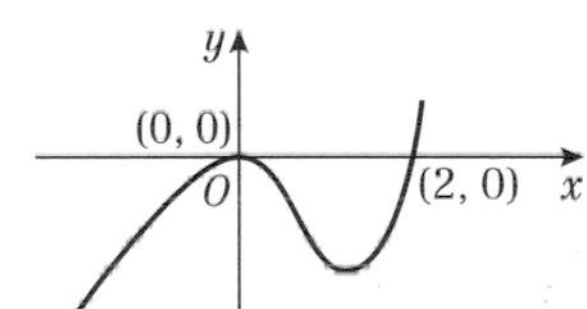

7 a $y = x^2 - 4x + 3$

b i

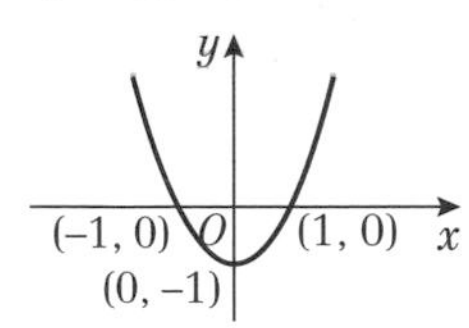

ii

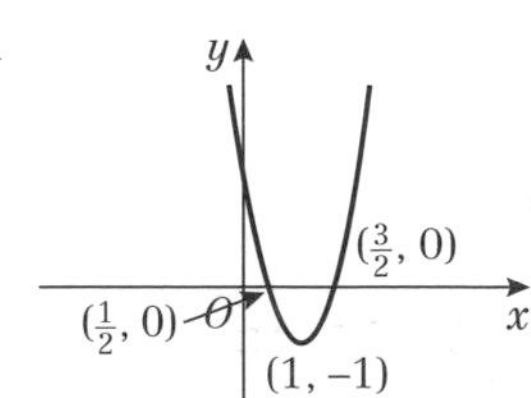

8 a $(0, 2)$ **b** -2 **c** $-1, 1, 2$

9 a i $(\frac{4}{3}, 3)$ **ii** $(4, 6)$ **iii** $(9, 3)$

iv $(4, -3)$ **v** $(4, -\frac{1}{2})$

b $f(2x)$, $f(x + 2)$

c i $f(x - 4) + 3$ **ii** $2f(\frac{1}{2}x)$

10 a

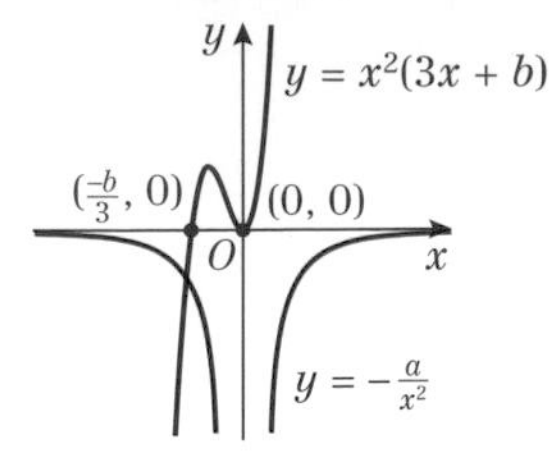

b 1; only one intersection of the two curves

11 a $x(x - 3)^2$

b

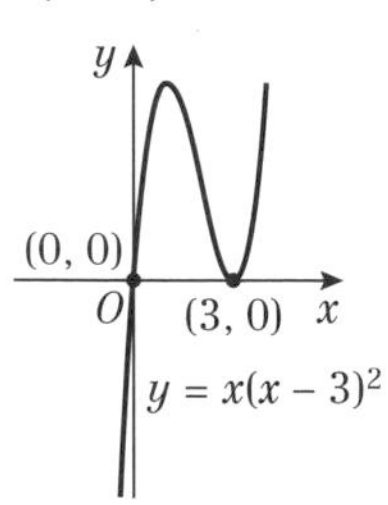

c −4 and −7

12 a

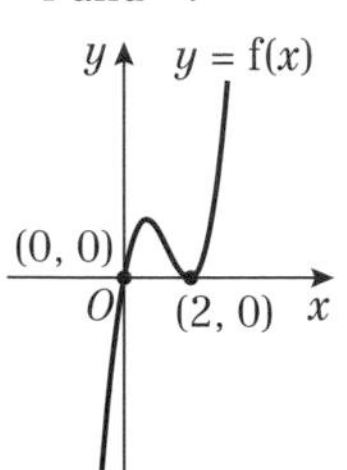

b

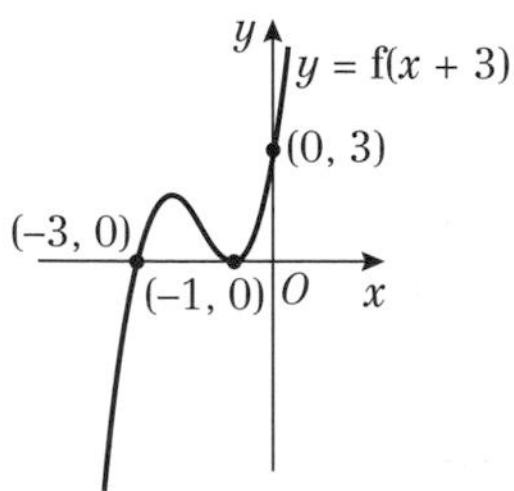

13 a Asymptotes at $x = 0$ and $y = -2$

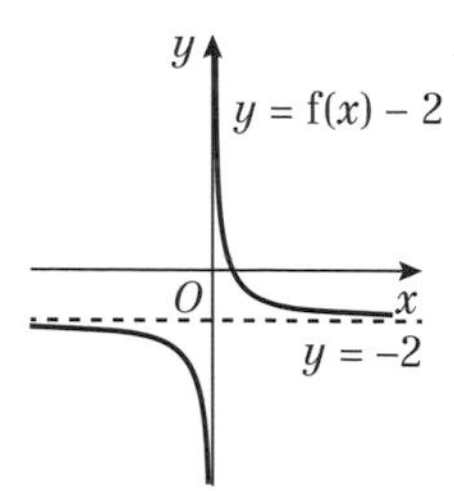

b $(\frac{1}{2}, 0)$

c

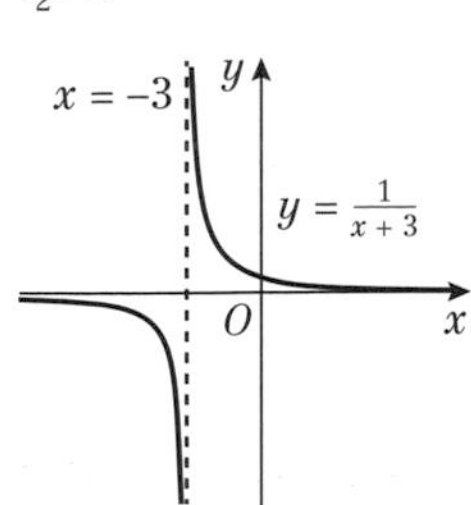

d Asymptotes at $y = 0$ and $x = -3$; intersection at $(0, \frac{1}{3})$

Challenge

$(6 - c, -4 - d)$

Review exercise 1

1 a 2 **b** $\frac{1}{4}$

2 a 625 **b** $\frac{4}{3}x^{\frac{2}{3}}$

3 a $4\sqrt{5}$ **b** $21 - 8\sqrt{5}$

4 a 13 **b** $8 - 2\sqrt{3}$

5 a $1 + 2\sqrt{k}$ **b** $1 + 6\sqrt{k}$

6 a $25x^{-4}$ **b** x^2

7 $8 + 8\sqrt{2}$

8 $1 - 2\sqrt{2}$

9 a $(x - 8)(x - 2)$ **b** $y = 1, y = \frac{1}{3}$

10 a $a = -4, b = -45$ **b** $x = 4 \pm 3\sqrt{5}$

11 4.19 (3 s.f.)

12 a The height of the athlete's shoulder is 1.7 m

b 2.16 s (3 s.f.)

c $6.7 - 5(t - 1)^2$

d 6.7 m after 1 second

13 a $(x - 3)^2 + 9$

b P is (0, 18), Q is (3, 9)

c $x = 3 + 4\sqrt{2}$

14 a $k = 2$

b

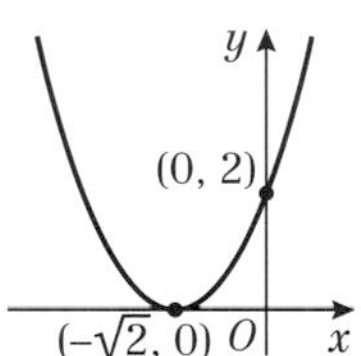

15 a $x^3(x^3 - 8)(x^3 + 1)$ **b** −1, 0, 2

16 a $a = 5, b = 11$

b $(x + 5)^2 = -11$, so no real roots

c $k = 25$

d

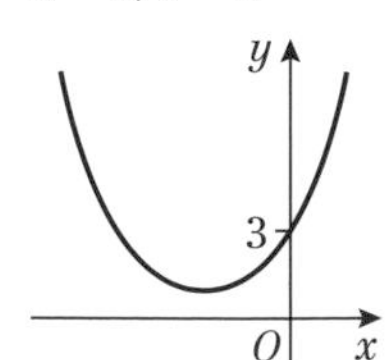

17 a $a = 1, b = 2$

b

c discriminant = −8, so no real roots

d $-2\sqrt{3} < k < 2\sqrt{3}$

18 a Substitute $y = x - 4$ into $2x^2 - xy = 8$ and rearrange.

b $x = -2 \pm 2\sqrt{3}, y = -6 \pm 2\sqrt{3}$

19 a $x > \frac{1}{4}$

b $x < \frac{1}{2}$ or $x > 3$

c $\frac{1}{4} < x < \frac{1}{2}$ or $x > 3$

20 $-2(x + 1) = x^2 - 5x + 2$

$x^2 - 3x + 4 = 0$

The discriminant of this is $-7 < 0$, so no real solutions.

Online Full worked solutions are available in SolutionBank.

21 a $x = \frac{7}{2}, y = -2; x = -3, y = 11$

b $x < -3$ or $x > 3\frac{1}{2}$

22 a Different real roots, discriminant > 0 so $k^2 - 4k - 12 > 0$

b $k < -2$ or $k > 6$

23 $x < -5$ or $x > -2$

24 a, b

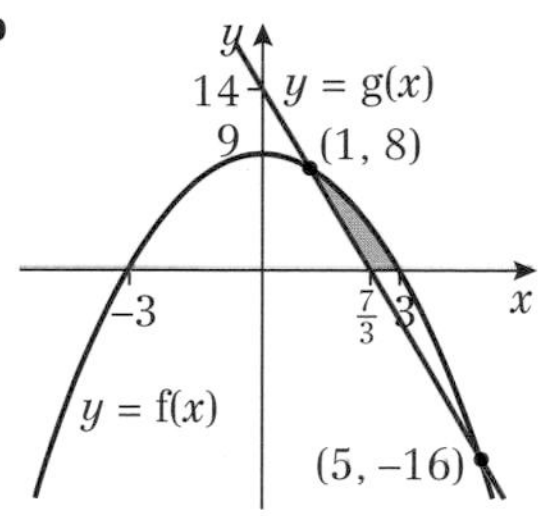

25 a $x(x-2)(x+2)$

b

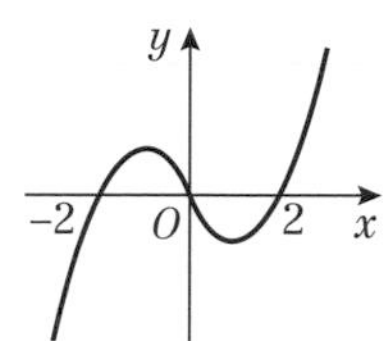

c

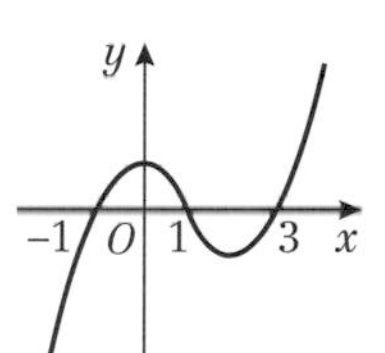

26 a

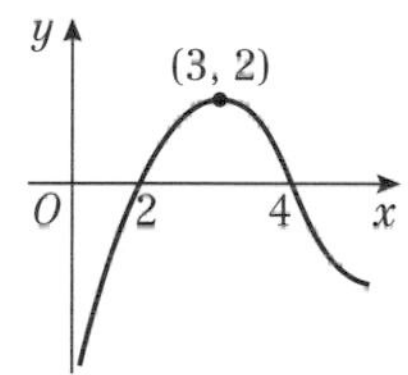

(2, 0) (4, 0) and (3, 2)

b

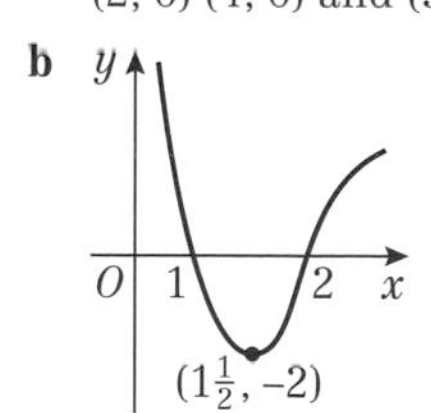

(1, 0) (2, 0) and $(1\frac{1}{2}, -2)$

27 a

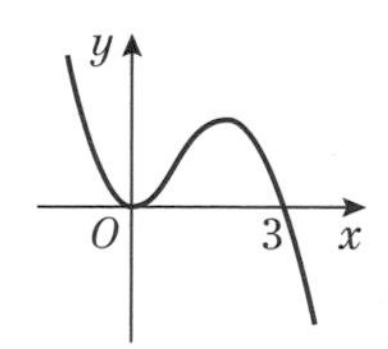

(0, 0) and (3, 0)

b

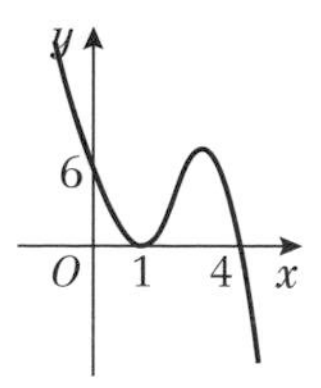

(1, 0) (4, 0) and (0, 6)

c

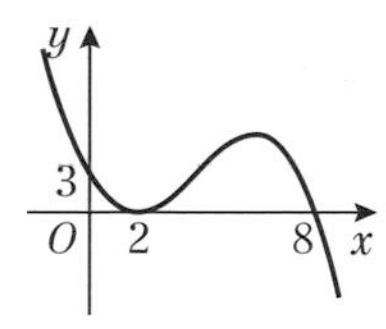

(2, 0) (8, 0) and (0, 3)

28 a

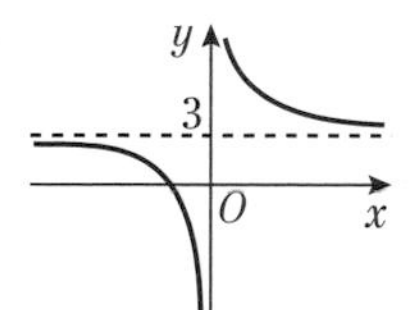

Asymptotes: $y = 3$ and $x = 0$

b $(-\frac{1}{3}, 0)$

29 a 0.438, 1, 4, 4.56

b y; y = t(x); 0.438; 8; O; 1; 4; 4.56; x

30 a (6, 8) **b** (9, −8) **c** (6, −4)

31 a

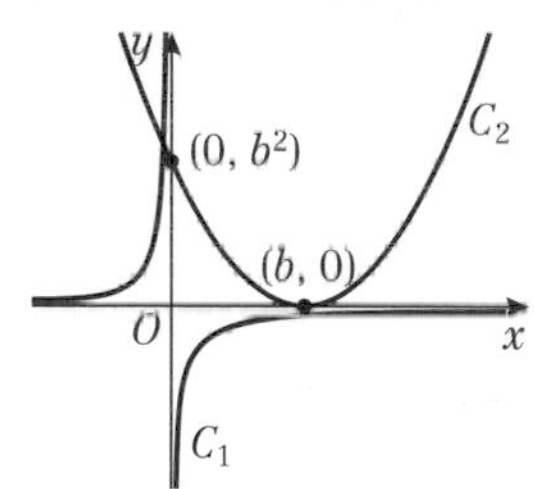

b 1

32 a

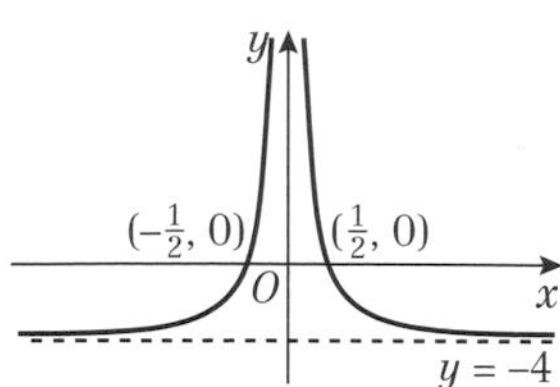

b $-\frac{1}{2}, \frac{1}{2}$

Challenge

1 a $x = 1, x = 9$ **b** $x = 0, x = 2$

2 $\sqrt{2}$ cm, $3\sqrt{2}$ cm

3 $3x^3 + x^2 - x = 2x(x-1)(x+1)$

$3x^3 + x^2 - x = 2x^3 - 2x$

$x^3 + x^2 + x = 0$

$x(x^2 + x + 1) = 0$

The discriminant of the bracket is $-3 < 0$ so this contributes no real solutions.

The only solution is when $x = 0$ at (0, 0).

4 −3, 3

CHAPTER 5

Prior knowledge check

1 **a** $(-2, -1)$ **b** $\left(\frac{9}{19}, \frac{26}{19}\right)$ **c** $(7, 3)$

2 **a** $4\sqrt{5}$ **b** $10\sqrt{2}$ **c** $5\sqrt{5}$

3 **a** $y = 5 - 2x$ **b** $y = \frac{2}{5}x - \frac{9}{5}$ **c** $y = \frac{3}{7}x + \frac{12}{7}$

Exercise 5A

1 **a** $\frac{1}{2}$ **b** $\frac{1}{6}$ **c** $-\frac{3}{5}$ **d** 2
e -1 **f** $\frac{1}{2}$ **g** $\frac{1}{2}$ **h** 8
i $\frac{2}{3}$ **j** -4 **k** $-\frac{1}{3}$ **l** $-\frac{1}{2}$
m 1 **n** $\frac{q^2 - p^2}{q - p} = q + p$

2 7

3 12

4 $4\frac{1}{3}$

5 $2\frac{1}{4}$

6 $\frac{1}{4}$

7 26

8 -5

9 Gradient of AB = gradient of BC = 0.5; point B is common

10 Gradient of AB = gradient of BC = −0.5; point B is common

Exercise 5B

1 **a** -2 **b** -1 **c** 3 **d** $\frac{1}{3}$
e $-\frac{2}{3}$ **f** $\frac{5}{4}$ **g** $\frac{1}{2}$ **h** 2
i $\frac{1}{2}$ **j** $\frac{1}{2}$ **k** -2 **l** $-\frac{3}{2}$

2 **a** 4 **b** -5 **c** $-\frac{2}{3}$ **d** 0
e $\frac{7}{5}$ **f** 2 **g** 2 **h** -2
i 9 **j** -3 **k** $\frac{3}{2}$ **l** $-\frac{1}{2}$

3 **a** $4x - y + 3 = 0$ **b** $3x - y - 2 = 0$
c $6x + y - 7 = 0$ **d** $4x - 5y - 30 = 0$
e $5x - 3y + 6 = 0$ **f** $7x - 3y = 0$
g $14x - 7y - 4 = 0$ **h** $27x + 9y - 2 = 0$
i $18x + 3y + 2 = 0$ **j** $2x + 6y - 3 = 0$
k $4x - 6y + 5 = 0$ **l** $6x - 10y + 5 = 0$

4 $(3, 0)$

5 $(0, 0)$

6 $(0, 5)$, $(-4, 0)$

7 **a** $\frac{1}{3}$ **b** $x - 3y + 15 = 0$

8 **a** $-\frac{2}{5}$ **b** $2x + 5y - 10 = 0$

9 $ax + by + c = 0$
$by = -ax - c$
$y = \left(-\frac{a}{b}\right)x - \left(\frac{c}{b}\right)$

10 $a = 6$, $c = 10$

11 $P(3,0)$

12 **a** -16 **b** -27

Challenge

Gradient $= -\frac{a}{b}$; y-intercept $= a$. So $y = -\frac{a}{b}x + a$

Rearrange to give $ax + by - ab = 0$

Exercise 5C

1 **a** $y = 2x + 1$ **b** $y = 3x + 7$ **c** $y = -x - 3$
d $y = -4x - 11$ **e** $y = \frac{1}{2}x + 12$ **f** $y = -\frac{2}{3}x - 5$
g $y = 2x$ **h** $y = -\frac{1}{2}x + 2b$

2 **a** $y = 4x - 4$ **b** $y = x + 2$ **c** $y = 2x + 4$
d $y = 4x - 23$ **e** $y = x - 4$ **f** $y = \frac{1}{2}x + 1$
g $y = -4x - 9$ **h** $y = -8x - 33$ **i** $y = \frac{6}{5}x$
j $y = \frac{2}{7}x + \frac{5}{14}$

3 $5x + y - 37 = 0$

4 $y = x + 2$, $y = -\frac{1}{6}x - \frac{1}{3}$, $y = -6x + 23$

5 $a = 3$, $c = -27$

6 $a = -4$, $b = 8$

Challenge

a $m = \frac{(y_2 - y_1)}{(x_2 - x_1)}$

b $y - y_1 = \frac{(y_2 - y_1)}{(x^2 - y_1)}(x - x_1)$

$\frac{(y - y_1)}{(y_2 - y_1)} = \frac{(x - x_1)}{(x_2 - x_1)}$

c $y = \frac{3}{7}x + \frac{52}{7}$

Exercise 5D

1 $y = 3x - 6$

2 $y = 2x + 8$

3 $2x - 3y + 24 = 0$

4 $-\frac{1}{5}$

5 $(-3, 0)$

6 $(0, 1)$

7 $(0, 3\frac{1}{2})$

8 $y = \frac{2}{5}x + 3$

9 $2x + 3y - 12 = 0$

10 $\frac{8}{5}$

11 $y = \frac{4}{3}x - 4$

12 $6x + 15y - 10 = 0$

13 $y = -\frac{4}{5}x + 4$

14 $x - y + 5 = 0$

15 $y = -\frac{3}{8}x + \frac{1}{2}$

16 $y = 4x + 13$

Exercise 5E

1 **a** Parallel **b** Not parallel **c** Not parallel

2 r: $y = \frac{4}{5}x + 3.2$, s: $y = \frac{4}{5}x - 7$
Gradients equal therefore lines are parallel.

3 Gradient of $AB = \frac{3}{5}$, gradient of $BC = -\frac{7}{2}$, gradient of $CD = \frac{3}{5}$, gradient of $AD = \frac{10}{3}$. The quadrilateral has a pair of parallel sides, so it is a trapezium.

4 $y = 5x + 3$

5 $2x + 5y + 20 = 0$

6 $y = -\frac{1}{2}x + 7$

7 $y = \frac{2}{3}x$

8 $4x - y + 15 = 0$

Exercise 5F

1 **a** Perpendicular **b** Parallel
c Neither **d** Perpendicular
e Perpendicular **f** Parallel
g Parallel **h** Perpendicular
i Perpendicular **j** Parallel
k Neither **l** Perpendicular

2 $y = -\frac{1}{6}x + 1$

3 $y = \frac{8}{3}x - 8$

4 $y = -\frac{1}{3}x$

5 $y = -\frac{1}{3}x + \frac{13}{3}$

6 $y = -\frac{3}{2}x + \frac{17}{2}$

7 $3x + 2y - 5 = 0$

8 $7x - 4y + 2 = 0$

9 l has gradient $-\frac{1}{3}$ and n has gradient 3. Gradients are negative reciprocals, therefore lines perpendicular.

10 AB: $y = -\frac{1}{2}x + 4\frac{1}{2}$, CD: $y = -\frac{1}{2}x - \frac{1}{2}$, AD: $y = 2x + 7$, BC: $y = 2x - 13$. Two pairs of parallel sides and lines with gradients 2 and $-\frac{1}{2}$ are perpendicular, so $ABCD$ is a rectangle.

11 **a** $A(\frac{7}{5}, 0)$ **b** $55x - 25y - 77 = 0$

12 $-\frac{9}{4}$

Exercise 5G

1 **a** 10 **b** 13 **c** 5 **d** $\sqrt{5}$
e $\sqrt{106}$ **f** $\sqrt{113}$

2 Distance between A and $B = \sqrt{50}$ and distance between B and $C = \sqrt{50}$ so the lines are congruent.

3 Distance between P and $Q = \sqrt{74}$ and distance between Q and $R = \sqrt{73}$ so the lines are not congruent.

4 $x = -8$ or $x = 6$

5 $y = -2$ or $y = 16$

6 **a** Both lines have gradient 2.
b $y = -\frac{1}{2}x + \frac{23}{2}$ or $x + 2y - 23 = 0$
c $\left(\frac{29}{5}, \frac{43}{5}\right)$
d $\frac{7\sqrt{5}}{5}$

7 $P\left(-\frac{3}{5}, \frac{29}{5}\right)$ or $P(3, -5)$

8 **a** $AB = \sqrt{178}$, $BC = 3$ and $AC = \sqrt{205}$. All sides are different lengths, therefore the triangle is a scalene triangle.
b $\frac{39}{2}$ or 19.5

9 **a** $A(2, 11)$
b $B\left(\frac{41}{4}, 0\right)$
c $\frac{451}{8}$

10 **a** $\left(\frac{5}{2}, 0\right)$ **b** $(-5, 0)$
c $(-10, -10)$ **d** $\frac{75}{2}$

11 **a** $y = \frac{1}{2}x - \frac{9}{2}$ **b** $y = -2x + 8$
c $T(0, 8)$ **d** $RS = 2\sqrt{5}$ and $TR = 5\sqrt{5}$
e 25

12 **a** $x + 4y - 52 = 0$ **b** $A(0,13)$
c $B(4, 12)$ **d** 26

Exercise 5H

1 **a** **i** $k = 50$ **ii** $d = 50t$
b **i** $k = 0.3$ or £0.30 **ii** $C = 0.3t$
c **i** $k = \frac{3}{5}$ **ii** $p = \frac{3}{5}t$

2 **a** not linear

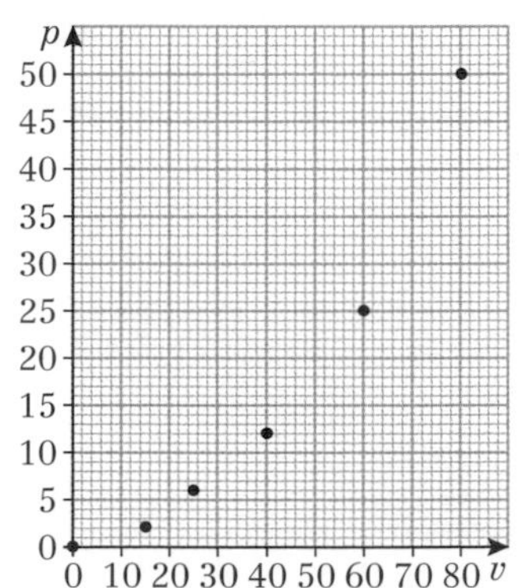

b linear

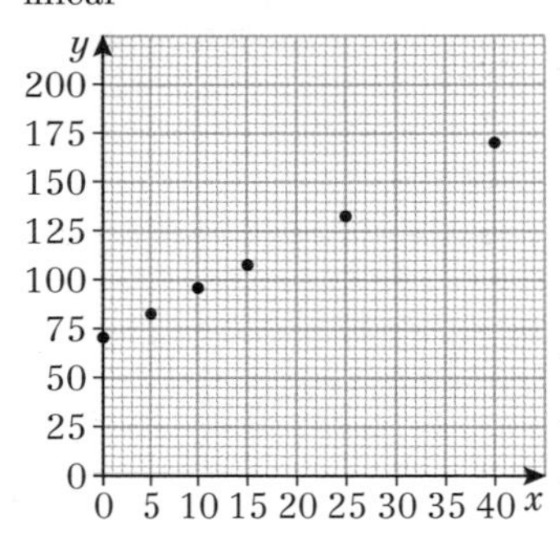

c not linear

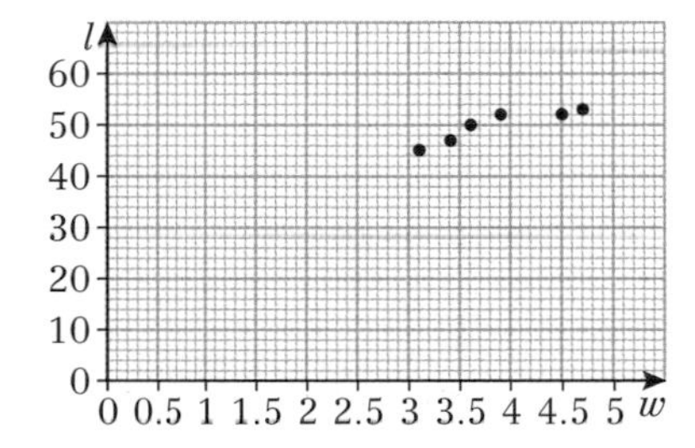

3 **a**

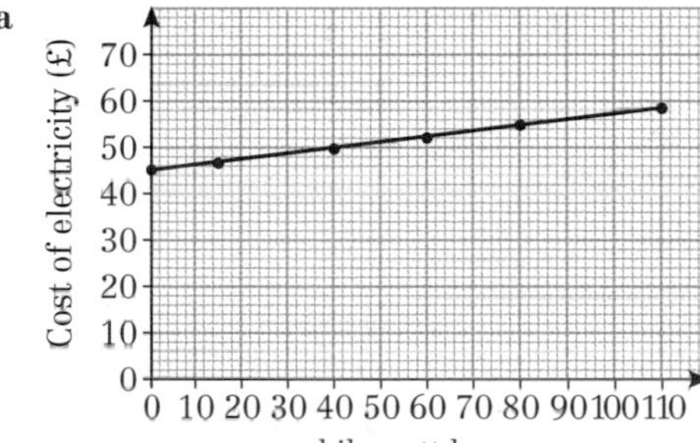

b The data forms a straight line, so a linear model is appropriate.
c $E = 0.12h + 45$
d a = £0.12 = cost of 1 kilowatt hour of electricity, b = £45 = fixed electricity costs (per month or per quarter)
e £52.80

4 **a**

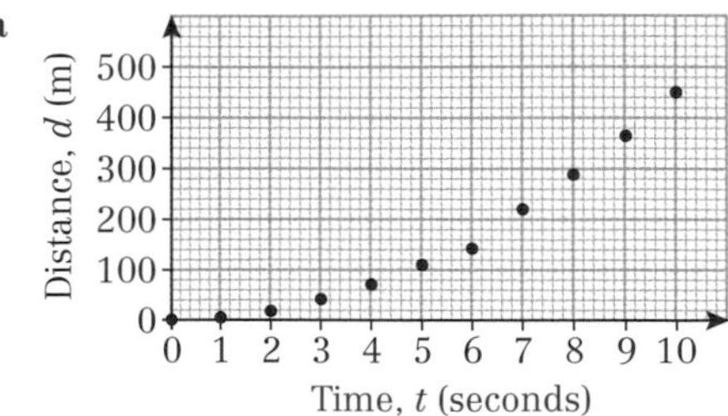

b The data does not follow a straight line. There is a definite curve to the points on the graph.

5 **a** $C = 350d + 5000$
b $a = 350$ = daily fee charged by the website designer. $b = 5000$ = initial cost charged by the website designer.
c 24 days

6 a $F = 1.8C + 32$ or $F = \frac{9}{5}C + 32$
b $a = 1.8$ = increase in Fahrenheit temperature when the Celsius temperature increases by 1°C.
$b = 32$ temperature in Fahrenheit when temperature in Celsius is 0°.
c 38.5°C
d −40°C
7 a $n = 750t + 17\,500$
b The increase in the number of homes receiving the internet will be the same each year.
8 a All the points lie close to the straight line shown.
b $h = 4f + 69$
c 175 cm
9 a

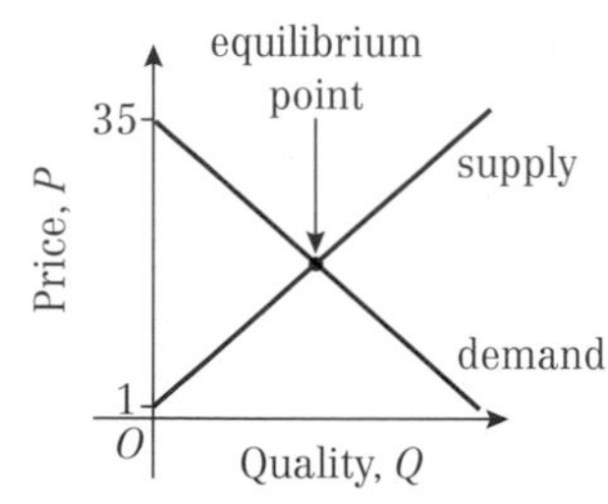

b $Q = 24, P = 17$

Mixed exercise 5

1 a $y = -\frac{5}{12}x + \frac{11}{6}$ b −22
2 a $\frac{2k-2}{8-k} = \frac{1}{3}$ therefore $7k = 14$, $k = 2$
b $y = \frac{1}{3}x + \frac{1}{3}$
3 a $L_1 = y = \frac{1}{7}x + \frac{12}{7}$, $L_2 = y = -x + 12$
b (9, 3)
4 a $y = \frac{3}{2}x - \frac{3}{2}$ b (3, 3)
5 $11x - 10y + 19 = 0$
6 a $y = -\frac{1}{2}x + 3$ b $y = \frac{1}{4}x + \frac{9}{4}$
7 Gradient $= \frac{3 + 4\sqrt{3} - 3\sqrt{3}}{2 + \sqrt{3} - 1} = \frac{3 + \sqrt{3}}{1 + \sqrt{3}} = \sqrt{3}$
$y = \sqrt{3}x + c$ and $A(1, 3\sqrt{3})$, so $c = 2\sqrt{3}$
Equation of line is $y = \sqrt{3}x + 2\sqrt{3}$
When $y = 0$, $x = -2$, so the line meets the x-axis at (−2, 0)
8 a $y = -3x + 14$ b (0, 14)
9 a $y = -\frac{1}{2}x + 4$ b Students own work.
c (1, 1). Note: equation of line n: $y = -\frac{1}{2}x + \frac{3}{2}$
10 20
11 a $2x + y = 20$ b $y = \frac{1}{3}x + \frac{4}{3}$
12 a $\frac{1}{2}$ b 6 c $2x + y - 16 = 0$
d 10
13 a $7x + 5y - 18 = 0$ b $\frac{162}{35}$
14 a

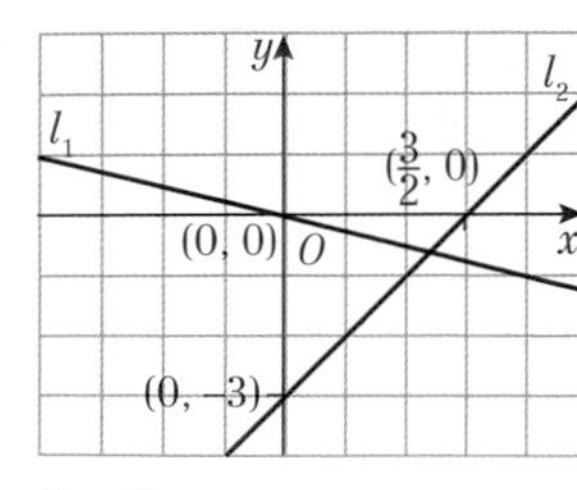

b $\left(\frac{4}{3}, -\frac{1}{3}\right)$ c $12x - 3y - 17 = 0$
15 a $x + 2y - 16 = 0$
b $y = -\frac{2}{3}x$
c $C(-48, 32)$
d Slope of OA is $\frac{3}{2}$. Slope of OC is $-\frac{2}{3}$. Lines are perpendicular.
e $OA = 2\sqrt{13}$ and $OC = 16\sqrt{13}$
f Area = 208
16 a $d = \sqrt{50a^2} = 5a\sqrt{2}$ b $5\sqrt{2}$
c $15\sqrt{2}$ d $25\sqrt{2}$
17 a $d = \sqrt{10x^2 - 28x + 26}$
b $B\left(-\frac{6}{5}, -\frac{18}{5}\right)$ and $C(4, 12)$
c $y = -\frac{1}{3}x + \frac{14}{3}$
d $\left(\frac{7}{5}, \frac{21}{5}\right)$
e 20.8
18 a gradient = 10.5
b $C = 10.5P - 10755$
c When the oil production increases by 1 million tonnes, the carbon dioxide emissions increase by 10.5 million tonnes.
d The model is not valid for small values of P, as it is not possible to have a negative amount of carbon dioxide emissions. It is always dangerous to extrapolate beyond the range on the model in this way.

Challenge
1 130
2 $\left(\frac{78}{19}, \frac{140}{19}\right)$
3 $\left(a, \frac{a(c-a)}{b}\right)$

CHAPTER 6

Prior knowledge check

1 a $(x + 5)^2 + 3$ b $(x - 3)^2 - 8$
c $(x - 6)^2 - 36$ d $(x + \frac{7}{2})^2 - \frac{49}{4}$
2 a $y = \frac{9}{4}x - 6$ b $y = -\frac{1}{2}x - \frac{3}{2}$
c $y = \frac{4}{3}x + \frac{10}{3}$
3 a $b^2 - 4ac = -7$ No real solutions
b $b^2 - 4ac = 89$ Two real solutions
c $b^2 - 4ac = 0$ One real solution
4 $y = -\frac{5}{6}x - \frac{3}{2}$

Exercise 6A

1 a (5, 5) b (6, 4) c (−1, 4) d (0, 0)
e (2, 1) f $(-8, \frac{3}{2})$ g $(4a, 0)$ h $\left(-\frac{u}{2}, -v\right)$
i $(2a, a - b)$ j $(3\sqrt{2}, 4)$ k $(2\sqrt{2}, \sqrt{2} + 3\sqrt{3})$
2 $a = 10, b = 1$
3 $(\frac{3}{2}, 7)$
4 $\left(\frac{3a}{5}, \frac{b}{4}\right)$
5 a $(\frac{3}{2}, 3)$ or (1.5, 3) b $y = 2x$, $3 = 2 \times 1.5$
6 a $\left(\frac{1}{8}, \frac{5}{3}\right)$ b $\frac{2}{3}$
7 Centre is $\left(3, -\frac{7}{2}\right)$. $3 - 2\left(-\frac{7}{2}\right) - 10 = 0$
8 (10, 5)
9 $(-7a, 17a)$

10 $p = 8, q = 7$
11 $a = -2, b = 4$

Challenge

a $p = 9, q = -1$
b $y = -x + 13$
c AC: $y = -x + 8$. Lines have the same slope, so they are parallel.

Exercise 6B

1 a $y = 2x + 3$ b $y = -\frac{1}{3}x + \frac{47}{3}$ c $y = \frac{5}{2}x - 25$
d $y = 3$ e $y = -\frac{3}{4}x + \frac{37}{8}$ f $x = 9$
2 $y = -x + 7$
3 $2x - y - 8 = 0$
4 a $y = -\frac{5}{3}x - \frac{13}{3}$ b $y = 3x - 8$ c $\left(\frac{11}{14}, -\frac{79}{14}\right)$
5 $q = -\frac{5}{4}, b = -\frac{189}{8}$

Challenge

a PR: $y = -\frac{5}{2}x + \frac{9}{4}$
PQ: $y = -\frac{1}{4}x + \frac{33}{8}$
RQ: $y = 2x + 6$
b $\left(-\frac{5}{6}, \frac{13}{3}\right)$

Exercise 6C

1 a $(x - 3)^2 + (y - 2)^2 = 16$
b $(x + 4)^2 + (y - 5)^2 = 36$
c $(x - 5)^2 + (y + 6)^2 = 12$
d $(x - 2a)^2 + (y - 7a)^2 = 25a^2$
e $(x + 2\sqrt{2})^2 + (y + 3\sqrt{2})^2 = 1$
2 a $(-5, 4)$, 9 b $(7, 1)$, 4
c $(-4, 0)$, 5 d $(-4a, -a)$, $12a$
e $(3\sqrt{5}, -\sqrt{5})$, $3\sqrt{3}$
3 a $(4 - 2)^2 + (8 - 5)^2 = 4 + 9 = 13$
b $(0 + 7)^2 + (-2 - 2)^2 = 49 + 16 = 65$
c $7^2 + (-24)^2 = 49 + 576 = 625 = 25^2$
d $(6a - 2a)^2 + (-3a + 5a)^2 = 16a^2 + 4a^2 = 20a^2$
e $(\sqrt{5} - 3\sqrt{5})^2 + (-\sqrt{5} - \sqrt{5})^2 = (-2\sqrt{5})^2 + (-2\sqrt{5})^2$
$= 20 + 20 = 40 = (2\sqrt{10})^2$
4 $(x - 8)^2 + (y - 1)^2 = 25$
5 $(x - \frac{3}{2})^2 + (y - 4)^2 = \frac{65}{4}$
6 $\sqrt{5}$
7 a $r = 2$
b Distance $PQ = PR = RQ = 2\sqrt{3}$, three equal length sides therefore triangle is equilateral.
8 a $(x - 2)^2 + y^2 = 15$
b Centre (2, 0) and radius $= \sqrt{15}$
9 a $(x - 5)^2 + (y + 2)^2 = 49$
b Centre (5, −2) and radius = 7
10 a Centre (1, −4), radius 5
b Centre (−6, 2), radius 7
c Centre (11, 3), radius $3\sqrt{10}$
d 10 Centre (−2.5, 1.5), radius $\frac{5\sqrt{2}}{2}$
e Centre (2 ,−2), radius $\sqrt{6.5}$
11 a Centre (−6, −1)
b $k > -37$
12 $Q(-13, 28)$
13 $k = -2$ and $k = 8$

Challenge

1 $k = 3$, $(x - 3)^2 + (y - 2)^2 = 50$
$k = 5$, $(x - 5)^2 + (y - 2)^2 = 50$
2 $(x + f)^2 - f^2 + (y + g)^2 - g^2 + c = 0$
So $(x + f)^2 + (y + g)^2 = f^2 + g^2 - c$
Circle with centre $(-f, -g)$ and radius $\sqrt{f^2 + g^2 - c}$.

Exercise 6D

1 (7, 0), (−5, 0)
2 (0, 2), (0, −8)
3 (6, 10), (−2, 2)
4 (4, −9), (−7, 2)
5 $2x^2 - 24x + 79 = 0$ has no real solutions, therefore lines do not intersect circle.
6 a $b^2 - 4ac = 64 - 4 \times 1 \times 16 = 0$. So there is only one point of intersection.
b (4, 7)
7 a (0, −2), (4, 6) b midpoint of AB is (2, 2)
8 a 13 b $p = 1$ or 5
9 a $A(5, 0)$ and $B(-3, -8)$ (or vice-versa)
b $y = -x - 3$
c $(4, -7)$ is a solution to $y = -x - 3$.
d 20
10 a Substitute $y = kx$ to give
$(k^2 + 1)x^2 - (12k + 10)x + 57 = 0$
$b^2 - 4ac > 0$, $-84k^2 + 240k - 128 > 0$,
$21k^2 - 60k + 32 < 0$
b $0.71 < k < 2.15$
Exact answer is $\frac{10}{7} - \frac{2\sqrt{57}}{21} < k < \frac{10}{7} + \frac{2\sqrt{57}}{21}$
11 $k < \frac{8}{17}$
12 $k = -20 \pm 2\sqrt{105}$

Exercise 6E

1 a $3\sqrt{10}$
b Gradient of radius = 3, gradient of line $= -\frac{1}{3}$, gradients are negative reciprocals and therefore perpendicular.
2 a $(x - 4)^2 + (y - 6)^2 = 73$ b $3x + 8y + 13 = 0$
3 a $y = -2x - 1$
b Centre of circle (1, −3) satisfies $y = -2x - 1$.
4 a $y = \frac{1}{2}x - 3$
b Centre of circle (2, −2) satisfies $y = \frac{1}{2}x - 3$
5 a (−7, −6) satisfies $x^2 + 18x + y^2 - 2y + 29 = 0$
b $y = \frac{2}{7}x - 4$ c $R(0, -4)$ d $\frac{53}{2}$
6 a (0, −17), (17, 0)
b 144.5
7 $y = 2x + 27$ and $y = 2x - 13$
8 a $p = 4, p = -6$
b (3, 4) and (3, −6)
9 a $(x - 11)^2 + (y + 5)^2 = 100$
b $y = \frac{3}{4}x - \frac{3}{4}$
c $A(8 - 4\sqrt{3}, -1 - 3\sqrt{3})$ and $B(8 + 4\sqrt{3}, -1 + 3\sqrt{3})$
d $10\sqrt{3}$
10 a $y = 4x - 22$
b $a = 5$
c $(x - 5)^2 + (y + 2)^2 = 34$
d $A(5 + \sqrt{2}, -2 + 4\sqrt{2})$ and $B(5 - \sqrt{2}, -2 - 4\sqrt{2})$

11 a $P(-2, 5)$ and $Q(4, 7)$
b $y = 2x + 9$ and $y = -\frac{1}{2}x + 9$
c $y = -3x + 9$
d $(0, 9)$

Challenge

1 $y = \frac{1}{2}x - 2$
2 a $\angle CPR = \angle CQR = 90°$ (Angle between tangent and radius)
$CP = CQ = \sqrt{10}$ (Radii of circle)
$CR = \sqrt{(6-2)^2 + (-1-1)^2} = \sqrt{20}$
So using Pythagoras' Theorem,
$PR = QR = \sqrt{20 - 10} = \sqrt{10}$
4 equal sides and two opposite right–angles, so $CPRQ$ is a square
b $y = \frac{1}{3}x - 3$ and $y = -3x + 17$

Exercise 6F

1 a $WV^2 = WU^2 + UV^2$
b $(2, 3)$
c $(x-2)^2 + (y-3)^2 = 41$
2 a $AC^2 = AB^2 + BC^2$
b $(x-5)^2 + (y-2)^2 = 25$
c 15
3 a i $y = \frac{3}{2}x + \frac{21}{2}$ **ii** $y = -\frac{2}{3}x + 4$
b $(-3, 6)$
c $(x+3)^2 + (y-6)^2 = 169$
4 a i $y = \frac{1}{3}x + \frac{10}{3}$ **ii** $x = -1$
b $(x+1)^2 + (y-3)^2 = 125$
5 $(x-3)^2 + (y+4)^2 = 50$
6 a $AB^2 + BC^2 = AC^2$
$AB^2 = 400, BC^2 = 100, AC^2 = 500$
b $(x+2)^2 + (y-5)^2 = 125$
c $D(8, 0)$ satisfies the equation of the circle.
7 a $AB = BC = CD = DA = \sqrt{50}$
b 50
c $(3,6)$
8 a $DE^2 = b^2 + 6b + 13$
$EF^2 = b^2 + 10b + 169$
$DF^2 = 200$
So $b^2 + 6b + 13 + b^2 + 10b + 169 = 200$
$(b+9)(b-1) = 0$; as $b > 0$, $b = 1$
b $(x+5)^2 + (y+4)^2 = 50$
9 a Centre $(-1, 12)$ and radius $= 13$
b Use distance formula to find $AB = 26$. This is twice radius, so AB is the diameter. Other methods possible.
c $C(-6, 0)$

Mixed exercise 6

1 a $C(3, 6)$
b $r = 10$
c $(x-3)^2 + (y-6)^2 = 100$
d P satisfies the equation of the circle.
2 $(0-5)^2 + (0+2)^2 = 5^2 + 2^2 = 29 < 30$ therefore point is inside the circle
3 a Centre $(0, -4)$ and radius $= 3$
b $(0, -1)$ and $(0, -7)$
c Students' own work. Equation $x^2 = -7$ has no real solutions.
4 a $P(8, 8)$, $(8+1)^2 + (8-3)^2 = 9^2 + 5^2 = 81 + 25 = 106$
b $\sqrt{106}$
5 a All points satisfy $x^2 + y^2 = 1$, therefore all lie on circle.
b $AB = BC = CA$
6 a $k = 1, k = -\frac{2}{5}$
b $(x-1)^2 + (y-3)^2 = 13$
7 Substitute $y = 3x - 9$ into the equation
$x^2 + px + y^2 + 4y = 20$
$x^2 + px + (3x-9)^2 + 4(3x-9) = 20$
$10x^2 + (p-42)x + 25 = 0$
Using the discriminant: $(p-42)^2 - 1000 < 0$
$42 - 10\sqrt{10} < p < 42 + 10\sqrt{10}$
8 $(x-2)^2 + (y+4)^2 = 20$
9 a $2\sqrt{29}$ **b** 12
10 $(-1, 0), (11, 0)$
11 The values of m and n are $7 - \sqrt{105}$ and $7 + \sqrt{105}$.
12 a $a = 6$ and $b = 8$ **b** $y = -\frac{4}{3}x + 8$ **c** 24
13 a $p = 0, q = 24$ **b** $(0, 49), (0, -1)$
14 $x + y + 10 = 0$
15 60
16 l_1: $y = -4x + 12$ and l_2: $y = -\frac{8}{19}x + 12$
17 a $y = \frac{1}{3}x + \frac{8}{3}$
b $(x+2)^2 + (y-2)^2 = 50$
c 20
18 a $P(-3, 1)$ and $Q(9, -7)$
b $y = \frac{3}{2}x + \frac{11}{2}$ and $y = \frac{3}{2}x - \frac{41}{2}$
19 a $y = -4x + 6$ and $y = \frac{1}{4}x + 6$
b $P(-4, 5)$ and $Q(1, 2)$
c 17
20 a $P(5, 16)$ and $Q(13, 8)$
b l_2: $y = \frac{1}{7}x + \frac{107}{7}$ and l_3: $y = 7x - 83$
c l_4: $y = x + 3$
d All 3 equations have solution $x = \frac{43}{3}, y = \frac{52}{3}$ so $R(\frac{43}{3}, \frac{52}{3})$
e $\frac{200}{3}$
21 a $(4,0), (0,12)$
b $(2,6)$
c $(x-2)^2 + (y-6)^2 = 40$
22 a $q = 4$
b $(x + \frac{5}{2})^2 + (y-2)^2 = \frac{65}{4}$
23 a $RS^2 + ST^2 = RT^2$
b $(x-2)^2 + (y+2)^2 = 61$
24 $(x-1)^2 + (y-3)^2 = 34$
25 a i $y = -4x - 4$ **ii** $x = -2$
b $(x+2)^2 + (y-4)^2 = 34$

Challenge

a $x + y - 14 = 0$
b $P(7, 7)$ and $Q(9, 5)$
c 10

CHAPTER 7

Prior knowledge check

1 a $15x^7$ **b** $\frac{x}{3y}$
2 a $(x-6)(x+4)$ **b** $(3x-5)(x-4)$
3 a 8567 **b** 1652
4 a $y = 1 - 3x$ **b** $y = \frac{1}{2}x - 7$
5 a $(x-1)^2 - 21$ **b** $2(x+1)^2 + 13$

Exercise 7A

1 a $4x^3 + 5x - 7$ **b** $2x^4 + 9x^2 + x$
c $-x^3 + 4x + \frac{6}{x}$ **d** $7x^4 - x^2 - \frac{4}{x}$

e $4x^3 - 2x^2 + 3$ f $3x - 4x^2 - 1$

g $\frac{7x^2}{5} - \frac{x^3}{5} - \frac{2}{5x}$ h $2x - 3x^3 + 1$

i $\frac{x^7}{2} - \frac{9x^3}{2} + 2x^2 - \frac{3}{x}$ j $3x^8 + 2x^5 - \frac{4x^3}{3} + \frac{2}{3x}$

2 a $x + 3$ b $x + 4$ c $x + 3$
d $x + 7$ e $x + 5$ f $x + 4$
g $\frac{x - 4}{x - 3}$ h $\frac{x + 2}{x + 4}$ i $\frac{x + 4}{x - 6}$
j $\frac{2x + 3}{x - 5}$ k $\frac{2x - 3}{x + 1}$ l $\frac{x - 2}{x + 2}$
m $\frac{2x + 1}{x - 2}$ n $\frac{x + 4}{3x + 1}$ o $\frac{2x + 1}{2x - 3}$

3 $a = 1, b = 4, c = -2$

Exercise 7B

1 a $(x + 1)(x^2 + 5x + 3)$ b $(x + 4)(x^2 + 6x + 1)$
c $(x + 2)(x^2 - 3x + 7)$ d $(x - 3)(x^2 + 4x + 5)$
e $(x - 5)(x^2 - 3x - 2)$ f $(x - 7)(x^2 + 2x + 8)$

2 a $(x + 4)(6x^2 + 3x + 2)$ b $(x + 2)(4x^2 + x - 5)$
c $(x + 3)(2x^2 - 2x - 3)$ d $(x - 6)(2x^2 - 3x - 4)$
e $(x + 6)(-5x^2 + 3x + 5)$ f $(x - 2)(-4x^2 + x - 1)$

3 a $x^3 + 3x^2 - 4x + 1$ b $4x^3 + 2x^2 - 3x - 5$
c $-3x^3 + 3x^2 - 4x - 7$ d $-5x^4 + 2x^3 + 4x^2 - 3x + 7$

4 a $x^3 + 2x^2 - 5x + 4$ b $x^3 - x^2 + 3x - 1$
c $2x^3 + 5x + 2$ d $3x^4 + 2x^3 - 5x^2 + 3x + 6$
e $2x^4 - 2x^3 + 3x^2 + 4x - 7$ f $4x^4 - 3x^3 - 2x^2 + 6x - 5$
g $5x^3 + 12x^2 - 6x - 2$ h $3x^4 + 5x^3 + 6$

5 a $x^2 - 2x + 5$ b $2x^2 - 6x + 1$
c $-3x^2 - 12x + 2$

6 a $x^2 + 4x + 12$ b $2x^2 - x + 5$
c $-3x^2 + 5x + 10$

7 Divide $x^3 + 2x^2 - 5x - 10$ by $(x + 2)$ to give $(x^2 - 5)$. So $x^3 + 2x^2 - 5x - 10 = (x + 2)(x^2 - 5)$.

8 a -8 b -7 c -12

9 $f(1) = 3 - 2 + 4 = 5$

10 $f(-1) = 3 + 8 + 10 + 3 - 25 = -1$

11 $(x + 4)(5x^2 - 20x + 7)$

12 $3x^2 + 6x + 4$

13 $x^2 + x + 1$

14 $x^3 - 2x^2 + 4x - 8$

15 14

16 a -200 b $(x + 2)(x - 7)(3x + 1)$

17 a i 30 ii 0 b $x = -3, x = -4, x = 1$

18 a $a = 1, b = 2, c = -3$
b $f(x) = (2x - 1)(x + 3)(x - 1)$
c $x = 0.5, x = -3, x = 1$

19 a $a = 3, b = 2, c = 1$
b Quadratic has no real solutions so only $\frac{1}{4}$ is a solution

Exercise 7C

1 a $f(1) = 0$ b $f(-3) = 0$ c $f(4) = 0$

2 $(x - 1)(x + 3)(x + 4)$

3 $(x + 1)(x + 7)(x - 5)$

4 $(x - 5)(x - 4)(x + 2)$

5 $(x - 2)(2x - 1)(x + 4)$

6 a $(x + 1)(x - 5)(x - 6)$ b $(x - 2)(x + 1)(x + 2)$
c $(x - 5)(x + 3)(x - 2)$

7 a i $(x - 1)(x + 3)(2x + 1)$ ii

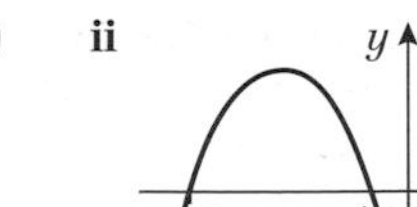

b i $(x - 3)(x - 5)(2x - 1)$ ii

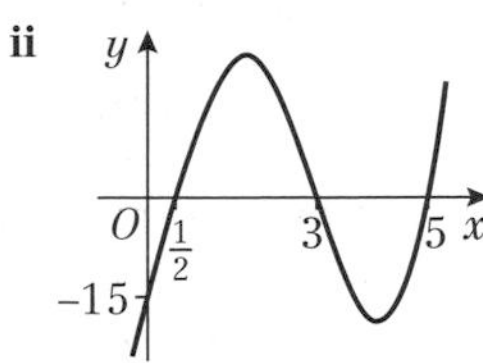

c i $(x + 1)(x + 2)(3x - 1)$ ii

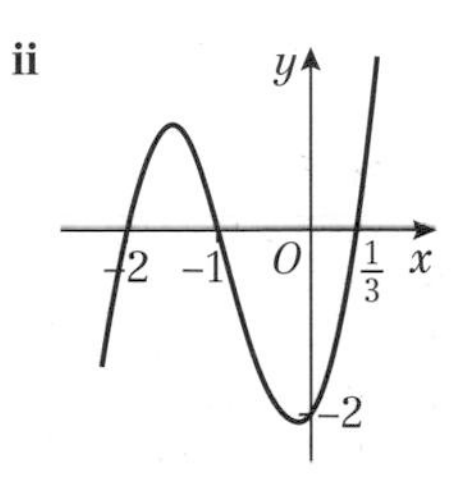

d i $(x + 2)(2x - 1)(3x + 1)$ ii

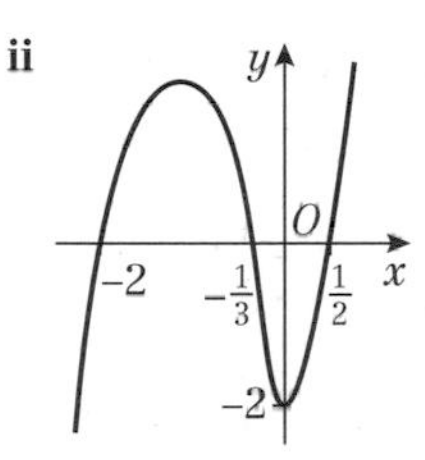

e i $(x - 2)(2x - 5)(2x + 3)$ ii

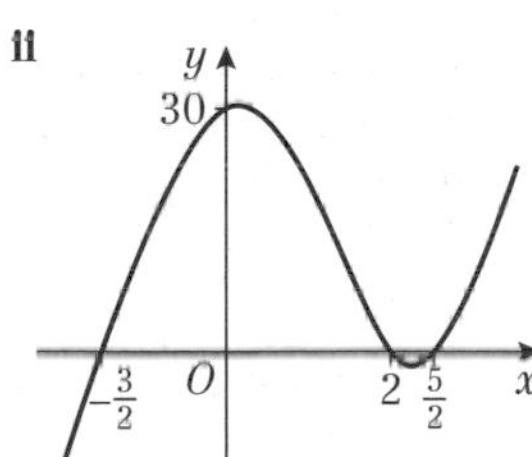

8 2

9 -16

10 $p = 3, q = 7$

11 $c = 2, d = 3$

12 $g = 3, h = -7$

13 a $f(4) = 0$
b $f(x) = (x - 4)(3x^2 + 6)$
For $3x^2 + 6 = 0$, $b^2 - 4ac = -72$ so there are no real roots. Therefore, 4 is the only real root of $f(x) = 0$.

14 a $f(-2) = 0$ b $(x + 2)(2x + 1)(2x - 3)$
c $x = -2, x = -\frac{1}{2}$ and $x = 1\frac{1}{2}$

15 a $f(2) = 0$ b $x = 0, x = 2, x = -\frac{1}{3}$ and $x = \frac{1}{3}$

Challenge

a $f(1) = 2 - 5 - 42 - 9 + 54 = 0$
$f(-3) = 162 + 135 - 378 + 27 + 54 = 0$

b $2x^4 - 5x^3 - 42x^2 - 9x + 54$
$= (x - 1)(x + 3)(x - 6)(2x + 3)$
$x = 1$ $x = -3, x = 6, x = -1.5$

Exercise 7D

1 $n^2 - n = n(n-1)$
If n is even, $n - 1$ is odd and even × odd = even
If n is odd, $n - 1$ is even and odd × even = even

2 $\frac{x}{(1+\sqrt{2})} \times \frac{(1-\sqrt{2})}{(1-\sqrt{2})} = \frac{x(1-\sqrt{2})}{(1-2)} = \frac{x - x\sqrt{2}}{-1} = x\sqrt{2} - x$

3 $(x + \sqrt{y})(x - \sqrt{y}) = x^2 - x\sqrt{y} + x\sqrt{y} - y = x^2 - y$

4 $(2x-1)(x+6)(x-5) = (2x-1)(x^2+x-30)$
$= 2x^3 + x^2 - 61x + 30$

5 LHS $= x^2 + bx$, using completing the square,
$\left(x + \frac{b}{2}\right)^2 - \left(\frac{b}{2}\right)^2$

6 $x^2 + 2bx + c = 0$, using completing the square
$(x+b)^2 + c - b^2 = 0$
$(x+b)^2 = b^2 - c$
$x + b = \pm\sqrt{b^2 - c}$
$x = -b \pm \sqrt{b^2 - c}$

7 $\left(x - \frac{2}{x}\right)^3 = \left(x - \frac{2}{x}\right)\left(x^2 - 4 + \frac{4}{x^2}\right) = x^3 - 6x + \frac{12}{x} - \frac{8}{x^3}$

8 $\left(x^3 - \frac{1}{x}\right)(x^{\frac{3}{2}} + x^{\frac{-5}{2}}) = x^{\frac{9}{2}} + x^{\frac{1}{2}} - x^{\frac{1}{2}} - x^{-\frac{7}{2}} = x^{\frac{9}{2}} - x^{-\frac{7}{2}}$
$= x^{\frac{1}{2}}\left(x^4 - \frac{1}{x^4}\right)$

9 $3n^2 - 4n + 10 = 3\left[n^2 - \frac{4}{3}n + \frac{10}{3}\right] = 3\left[\left(n - \frac{2}{3}\right)^2 + \frac{10}{3} - \frac{4}{9}\right]$
$= 3\left(n - \frac{2}{3}\right)^2 + \frac{26}{3}$
The minimum value is $\frac{26}{3}$ so $3n^2 - 4n + 10$ is always positive.

10 $-n^2 - 2n - 3 = -[n^2 + 2n + 3] = -[(n+1)^2 + 3 - 1]$
$= -(n+1)^2 - 2$
The maximum value is -2 so $-n^2 - 2n - 3$ is always negative.

11 $x^2 + 8x + 20 = (x+4)^2 + 4$
The minimum value is 4 so $x^2 + 8x + 20$ is always greater than or equal to 4.

12 $kx^2 + 5kx + 3 = 0$, $b^2 - 4ac < 0$, $25k^2 - 12k < 0$, $k(25k - 12) < 0$, $0 < k < \frac{12}{25}$.
When $k = 0$ there are no real roots, so $0 \leqslant k < \frac{12}{25}$

13 $px^2 - 5x - 6 = 0$, $b^2 - 4ac > 0$, $25 + 24p > 0$, $p > -\frac{25}{24}$

14 Gradient $AB = -\frac{1}{2}$, gradient $BC = 2$,
Gradient AB × gradient $BC = -\frac{1}{2} \times 2 = -1$,
so AB and BC are perpendicular.

15 Gradient $AB = 3$, gradient $BC = \frac{1}{4}$, gradient $CD = 3$, gradient $AD = \frac{1}{4}$
Gradient AB = gradient CD so AB and CD are parallel.
Gradient BC = gradient AD so BC and AD are parallel.

16 Gradient $AB = \frac{1}{3}$, gradient $BC = 3$, gradient $CD = \frac{1}{3}$, gradient $AD = 3$
Gradient AB = gradient CD so AB and CD are parallel.
Gradient BC = gradient AD so BC and AD are parallel.
Length $AB = \sqrt{10}$, $BC = \sqrt{10}$, $CD = \sqrt{10}$ and $AD = \sqrt{10}$, so all four sides are equal.

17 Gradient $AB = -3$, gradient $BC = \frac{1}{3}$,
Gradient AB × gradient $BC = -3 \times \frac{1}{3} = -1$, so AB and BC are perpendicular.
Length $AB = \sqrt{40}$, $BC = \sqrt{40}$, $AB = BC$

18 $(x-1)^2 + y^2 = k$, $y = ax$, $(x-1)^2 + a^2x^2 = k$,
$x^2(1 + a^2) - 2x + 1 - k = 0$
$b^2 - 4ac > 0$, $k > \frac{a^2}{1 + a^2}$.

19 $x = 2$. There is only one solution so the line $4y - 3x + 26 = 0$ only touches the circle in one place so is the tangent to the circle.

20 Area of square $= (a+b)^2 = a^2 + 2ab + b^2$
Shaded area $= 4\left(\frac{1}{2}ab\right)$
Area of smaller square: $a^2 + 2ab + b^2 - 2ab$
$= a^2 + b^2 = c^2$

Challenge

1 The equation of the circle is $(x-3)^2 + (y-5)^2 = 25$ and all four points satisfy this equation.

2 $2k + 1 = 1 \times (2k+1) = ((k+1) - k)((k+1) + k) = (k+1)^2 - k^2$

Exercise 7E

1 3, 4, 5, 6, 7 and 8 are not divisible by 10

2 3, 5, 7, 11, 13, 17, 19, 23 are prime numbers. 9, 15, 21, 25 are the product of two prime numbers.

3 $1^2 + 2^2 = 5$, $2^2 + 3^2 =$ odd, $3^2 + 4^2 =$ odd, $4^2 + 5^2 =$ odd, $5^2 + 6^2 =$ odd, $6^2 + 7^2 =$ odd, $7^2 + 8^2 = 113$

4 $(3n)^3 = 27n^3 = 9n(3n^2)$ which is a multiple of 9
$(3n+1)^3 = 27n^3 + 27n^2 + 9n + 1 = 9n(3n^2 + 3n + 1) + 1$
which is one more than a multiple of 9
$(3n+2)^3 = 27n^3 + 54n^2 + 36n + 8 = 9n(3n^2 + 6n + 4) + 8$
which is one less than a multiple of 9

5 **a** For example, when $n = 2$, $2^4 - 2 = 14$, 14 is not divisible by 4.
b Any square number
c For example, when $n = \frac{1}{2}$
d For example, when $n = 1$

6 **a** Assuming that x and y are positive
b e.g. $x = 0$, $y = 0$

7 $(x+5)^2 \geqslant 0$ for all real values of x, and
$(x+5)^2 + 2x + 11 = (x+6)^2$, so $(x+6)^2 \geqslant 2x + 11$

8 If $a^2 + 1 \geqslant 2a$ (a is positive, so multiplying both sides by a does not reverse the inequality), then $a^2 - 2a + 1 \geqslant 0$, and $(a-1)^2 \geqslant 0$, which we know is true.

9 **a** $(p+q)^2 = p^2 + 2pq + q^2 = (p-q)^2 + 4pq$
$(p-q)^2 \geqslant 0$ since it is a square, so $(p+q)^2 \geqslant 4pq$
$p > 0, q > 0 \Rightarrow p + q > 0 \Rightarrow p + q \geqslant \sqrt{4pq}$
b e.g. $p = q = -1$: $p + q = -2$, $\sqrt{4pq} = 2$

10 **a** Starts by assuming the inequality is true: i.e. negative ⩾ positive
b e.g. $x = y = -1$: $x + y = -2$, $\sqrt{x^2 + y^2} = \sqrt{2}$
c $(x+y)^2 = x^2 + 2xy + y^2 > x^2 + y^2$ since $x > 0$, $y > 0 \Rightarrow 2xy > 0$
As $x + y > 0$, can take square roots: $x + y \geqslant \sqrt{x^2 + y^2}$

Mixed exercise 7

1 **a** $x^3 - 7$ **b** $\frac{x+4}{x-1}$ **c** $\frac{2x-1}{2x+1}$

2 $3x^2 + 5$

3 $2x^2 - 2x + 5$

4 **a** When $x = 3$, $2x^3 - 2x^2 - 17x + 15 = 0$
b $A = 2$, $B = 4$, $C = -5$

5 **a** When $x = 2$, $x^3 + 4x^2 - 3x - 18 = 0$
b $p = 1$, $q = 3$

6 $(x-2)(x+4)(2x-1)$

7 7

8 **a** $p = 1$, $q = -15$ **b** $(x+3)(2x-5)$

9 **a** $r = 3$, $s = 0$ **b** $x(x+1)(x+3)$

10 **a** $(x-1)(x+5)(2x+1)$ **b** $-5, -\frac{1}{2}, 1$

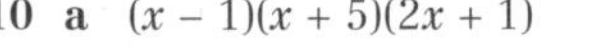

11 a When $x = 2$, $x^3 + x^2 - 5x - 2 = 0$

b $2, -\frac{3}{2} \pm \frac{\sqrt{5}}{2}$

12 $\frac{1}{2}, 3$

13 a When $x = -4$, $f(x) = 0$

b $x = -4$, $x = 1$ and $x = 5$

14 a $f(\frac{2}{3}) = 0$, therefore $(3x - 2)$ is a factor of $f(x)$
$a = 2$, $b = 7$ and $c = 3$

b $(3x - 2)(2x + 1)(x + 3)$

c $x = \frac{2}{3}, -\frac{1}{2}, -3$

15 $\frac{x - y}{(\sqrt{x} - \sqrt{y})} \times \frac{(\sqrt{x} + \sqrt{y})}{(\sqrt{x} + \sqrt{y})} = \frac{(x - y)(\sqrt{x} + \sqrt{y})}{x - y} = \sqrt{x} + \sqrt{y}$

16 $n^2 - 8n + 20 = (n - 4)^2 + 4$, 4 is the minimum value so $n^2 - 8n + 20$ is always positive

17 Gradient $AB = \frac{1}{2}$, gradient $BC = -2$, gradient $CD = \frac{1}{2}$, gradient $AD = -2$
AB and BC, BC and CD, CD and AD and AB and AD are all perpendicular
Length $AB = \sqrt{5}$, $BC = \sqrt{5}$, $CD = \sqrt{5}$ and $AD = \sqrt{5}$, all four sides are equal

18 1 + 3 = even, 3 + 5 = even, 5 + 7 = even, 7 + 9 = even

19 For example when $n = 6$

20 $\left(x - \frac{1}{x}\right)(x^{\frac{4}{3}} + x^{\frac{-2}{3}}) = x^{\frac{7}{3}} + x^{\frac{1}{3}} - x^{\frac{1}{3}} - x^{-\frac{5}{3}} = x^{\frac{1}{3}}\left(x^2 - \frac{1}{x^2}\right)$

21 RHS $= (x + 4)(x - 5)(2x + 3) = (x + 4)(2x^2 - 7x - 15)$
$= 2x^3 + x^2 - 43x - 60 =$ LHS

22 $x^2 - kx + k = 0$, $b^2 - 4ac = 0$, $k^2 - 4k = 0$, $k(k - 4) = 0$, $k = 4$.

23 The distance between opposite edges
$= 2\sqrt{\sqrt{3}^2 - \left(\frac{\sqrt{3}}{2}\right)^2} = 2\sqrt{\frac{9}{4}} = 3$ which is rational.

24 a $(2n + 2)^2 - (2n)^2 = 8n + 4 = 4(2n + 1)$ is always divisible by 4.

b Yes, $(2n + 1)^2 - (2n - 1)^2 = 8n$ which is always divisible by 4.

25 a The assumption is that x is positive

b $x = 0$

Challenge

1 a Perimeter of inside square $= 4\left(\sqrt{\left(\frac{1}{2}\right)^2 + \left(\frac{1}{2}\right)^2}\right) = \frac{4}{\sqrt{2}} = 2\sqrt{2}$

Perimeter of outside square = 4
Circumference of circle $= \pi \times 1^2 = \pi$
therefore $2\sqrt{2} < \pi < 4$.

b Perimeter of inside hexagon = 3

Perimeter of outside hexagon $= 6 \times \frac{\sqrt{3}}{3} = 2\sqrt{3}$,
therefore $3 < \pi < 2\sqrt{3}$

2 $ax^3 + bx^2 + cx + d \div (x - p) = ax^2 + (b + ap)x + (c + bp + ap^2)$ with remainder $d + cp + bp^2 + ap^3$
$f(p) = ap^3 + bp^2 + cp + d = 0$, which matches the remainder, so $(x - p)$ is a factor of $f(x)$.

CHAPTER 8

Prior knowledge check

1 a $4x^2 - 12xy + 9y^2$ **b** $x^3 - 3x^2y + 3xy^2 - y^3$
c $8 + 12x + 6x^2 + x^3$

2 a $-8x^3$ **b** $\frac{1}{81x^4}$ **c** $\frac{4}{25}x^2$ **d** $\frac{27}{x^3}$

3 a $5\sqrt{x}$ **b** $\frac{1}{16\sqrt[3]{x^2}}$ **c** $\frac{10}{3\sqrt{x}}$ **d** $\frac{16\sqrt[3]{x^4}}{81}$

Exercise 8A

1 a 4th row **b** 16th row
c $(n + 1)$th row **d** $(n + 5)$th row

2 a $x^4 + 4x^3y + 6x^2y^2 + 4xy^3 + y^4$
b $p^5 + 5p^4q + 10p^3q^2 + 10p^2q^3 + 5pq^4 + q^5$
c $a^3 - 3a^2b + 3ab^2 - b^3$
d $x^3 + 12x^2 + 48x + 64$
e $16x^4 - 96x^3 + 216x^2 - 216x + 81$
f $a^5 + 10a^4 + 40a^3 + 80a^2 + 80a + 32$
g $81x^4 - 432x^3 + 864x^2 - 768x + 256$
h $16x^4 - 96x^3y + 216x^2y^2 - 216xy^3 + 81y^4$

3 a 16 **b** −10 **c** 8 **d** 1280
e 160 **f** −2 **g** 40 **h** −96

4 $1 + 9x + 30x^2 + 44x^3 + 24x^4$

5 $8 + 12y + 6y^2 + y^3$, $8 + 12x - 6x^2 - 11x^3 + 3x^4 + 3x^5 - x^6$

6 ±3

7 $\frac{5}{2}, -1$

8 $12p$

9 $500 + 25X + \frac{X^2}{2}$

Challenge

$\frac{3}{4}$

Exercise 8B

1 a 24 **b** 362 880 **c** 720 **d** 210

2 a 6 **b** 15 **c** 20 **d** 5
e 45 **f** 126

3 a 5005 **b** 120 **c** 184 756 **d** 1140
e 2002 **f** 8568

4 $a = {}^4C_1$, $b = {}^5C_2$, $c = {}^6C_2$, $d = {}^6C_3$

5 330

6 a 120, 210 **b** 960

7 a 286, 715 **b** 57 915

8 0.1762 to 4 decimal places. Whilst it seems a low probability, there is more chance of the coin landing on 10 heads than any other number of heads.

9 a ${}^nC_1 = \frac{n!}{1!(n-1)!}$

$= \frac{1 \times 2 \times \ldots \times (n-2) \times (n-1) \times n}{1 \times 1 \times 2 \times \ldots \times (n-3) \times (n-2) \times (n-1)} = n$

b ${}^nC_2 = \frac{n!}{2!(n-2)!}$

$= \frac{1 \times 2 \times \ldots \times (n-2) \times (n-1) \times n}{1 \times 2 \times 1 \times 2 \times \ldots \times (n-3) \times (n-2)} = \frac{n(n-1)}{2}$

10 $a = 37$

11 $p = 17$

Challenge

a ${}^{10}C_3 = \frac{10!}{3!7!} = 120$ and ${}^{10}C_7 = \frac{10!}{7!3!} = 120$

b ${}^{14}C_5 = \frac{14!}{5!9!} = 2002$ and ${}^{14}C_9 = \frac{14!}{9!5!} = 2002$

c The two answers for part **a** are the same and the two answers for part **b** are the same.

d ${}^nC_r = \frac{n!}{r!(n-r)!}$ and ${}^nC_{n-r} = \frac{n!}{(n-r)!r!}$, therefore ${}^nC_r = {}^nC_{n-r}$

Exercise 8C

1 a $1 + 4x + 6x^2 + 4x^3 + x^4$
b $81 + 108x + 54x^2 + 12x^3 + x^4$
c $256 - 256x + 96x^2 - 16x^3 + x^4$

d $x^6 + 12x^5 + 60x^4 + 160x^3 + 240x^2 + 192x + 64$
e $1 + 8x + 24x^2 + 32x^3 + 16x^4$
f $1 - 2x + \frac{3}{2}x^2 - \frac{1}{2}x^3 + \frac{1}{16}x^4$

2 **a** $1 + 10x + 45x^2 + 120x^3$
b $1 - 10x + 40x^2 - 80x^3$
c $1 + 18x + 135x^2 + 540x^3$
d $256 - 1024x + 1792x^2 - 1792x^3$
e $1024 - 2560x + 2880x^2 - 1920x^3$
f $2187 - 5103x + 5103x^2 - 2835x^3$

3 **a** $64x^6 + 192x^5y + 240x^4y^2 + 160x^3y^3$
b $32x^5 + 240x^4y + 720x^3y^2 + 1080x^2y^3$
c $p^8 - 8p^7q + 28p^6q^2 - 56p^5q^3$
d $729x^6 - 1458x^5y + 1215x^4y^2 - 540x^3y^3$
e $x^8 + 16x^7y + 112x^6y^2 + 448x^5y^3$
f $512x^9 - 6912x^8y + 41\,472x^7y^2 - 145\,152x^6y^3$

4 **a** $1 + 8x + 28x^2 + 56x^3$
b $1 - 12x + 60x^2 - 160x^3$
c $1 + 5x + \frac{45}{4}x^2 + 15x^3$
d $1 - 15x + 90x^2 - 270x^3$
e $128 + 448x + 672x^2 + 560x^3$
f $27 - 54x + + 36x^2 - 8x^3$
g $64 - 576x + 2160x^2 - 4320x^3$
h $256 + 256x + 96x^2 + 16x^3$
i $128 + 2240x + 16\,800x^2 + 70\,000x^3$

5 $64 - 192x + 240x^2$

6 $243 - 810x + 1080x^2$

7 $x^5 + 5x^3 + 10x + \frac{10}{x} + \frac{5}{x^3} + \frac{1}{x^5}$

Challenge

a $(a + b)^4 = a^4 + 4a^3b + 6a^2b^2 + 4ab^3 + b^4$
$(a - b)^4 = a^4 - 4a^3b + 6a^2b^2 - 4ab^3 + b^4$
$(a + b)^4 - (a - b)^4 = 8a^3b + 8ab^3 = 8ab(a^2 + b^2)$

b $82\,896 = 2^4 \times 3 \times 11 \times 157$

Exercise 8D

1 **a** 90 **b** 80 **c** −20
d 1080 **e** 120 **f** −4320
g 1140 **h** −241 920 **i** −2.5
j 354.375 **k** −224 **l** 3.90625

2 $a = \pm\frac{1}{2}$

3 $b = -2$

4 $1, \frac{5 \pm \sqrt{105}}{8}$

5 **a** $p = 5$ **b** −10 **c** −80

6 **a** $5^{30} + 5^{29} \times 30px + 5^{28} \times 435p^2x^2$
b $p = 10$

7 **a** $1 + 10qx + 45q^2x^2 + 120q^3x^3$
b $q = \pm 3$

8 **a** $1 + 11px + 55p^2x^2$
b $p = 7, q = 2695$

9 **a** $1 + 15px + 105p^2x^2$
b $p = -\frac{5}{7}, q = 10\frac{5}{7}$

10 $\frac{q}{p} = 2.1$

Challenge

a 314 928 **b** 43 750

Exercise 8E

1 **a** $1 - 0.6x + 0.15x^2 - 0.02x^3$
b 0.941 48

2 **a** $1024 + 1024x + 460.8x^2 + 122.88x^3$
b 1666.56

3 $(1 - 3x)^5 = 1^5 + \binom{5}{1}1^4(-3x)^1 + \binom{5}{2}1^3(-3x)^2 = 1 - 15x + 90x^2$
$(2 + x)(1 - 3x)^5 = (2 + x)(1 - 15x + 90x^2)$
$= 2 - 30x + 180x^2 + x - 15x^2 + 90x^3 \approx 2 - 29x + 165x^2$

4 $a = 162, b = 135, c = 0$

5 **a** $1 + 16x + 112x^2 + 448x^3$
b $x = 0.01, 1.02^8 \approx 1.171\,648$

6 **a** $1 - 150x + 10875x^2 - 507500x^3$
b 0.860 368
c 0.860 384, 0.0019%

7 **a** $59\,049 - 39\,366x + 11\,809.8x^2$
b Substitute $x = 0.1$ into the expansion.

8 **a** $1 - 15x + 90x^2 - 270x^3$
b $(1 + x)(1 - 3x)^5 \approx (1 + x)(1 - 15x) \approx 1 - 14x$

9 **a** So that higher powers of p can be ignored as they tend to 0
b $1 - 200p + 19\,900p^2$
c $p = 0.000417$ (3 s.f.)

Mixed exercise 8

1 **a** 455, 1365 **b** 3640

2 $a = 28$

3 **a** 0.0148 **b** 0.000 000 000 034 9 **c** 0.166

4 **a** $p = 16$ **b** 270 **c** −1890

5 $A = 8192, B = -53\,248, C = 159\,744$

6 **a** $1 - 20x + 180x^2 - 960x^3$
b 0.817 04, $x = 0.01$

7 **a** $1024 - 15\,360x + 103\,680x^2 - 414\,720x^3$
b 880.35

8 **a** $81 + 216x + 216x^2 + 96x^3 + 16x^4$
b $81 - 216x + 216x^2 - 96x^3 + 16x^4$
c 1154

9 **a** $n = 8$ **b** $\frac{35}{8}$

10 **a** $81 + 1080x + 5400x^2 + 12\,000x^3 + 10\,000x^4$
b 1 012 054 108 081, $x = 100$

11 **a** $1 + 24x + 264x^2 + 1760x^3$ **b** 1.268 16, $x = 0.01$
c 1.268 241 795 **d** 0.006 45% (3 sf)

12 $x^5 - 5x^3 + 10x - \frac{10}{x} + \frac{5}{x^3} - \frac{1}{x^5}$

13 **a** $\binom{n}{2}(2k)^{n-2} = \binom{n}{3}(2k)^{n-3}$

$$\frac{n!(2k)^{n-2}}{2!(n-2)!} = \frac{n!(2k)^{n-3}}{3!(n-3)!}$$

$$\frac{2k}{n-2} = \frac{1}{3}$$

So $n = 6k + 2$

b $\frac{4096}{729} + \frac{2048}{81}x + \frac{1280}{27}x^2 + \frac{1280}{27}x^3$

14 **a** $64 + 192x + 240x^2 + 160x^3 + 60x^4 + 12x^5 + x^6$
b $k = 1560$

15 **a** $k = 1.25$ **b** 3500

16 **a** $A = 64, B = 160, C = 20$ **b** $x = \pm\sqrt{\frac{3}{2}}$

17 **a** $p = 1.5$ **b** 50.625

18 672

19 **a** $128 + 448px + 672p^2x^2$
b $p = 5, q = 16\,800$

20 **a** $1 - 12px + 66p^2x^2$
b $p = -1\frac{1}{11}, q = 13\frac{1}{11}$

21 a $128 + 224x + 168x^2$
 b Substitute $x = 0.1$ into the expansion.

22 $k = \frac{1}{2}$

Challenge

1 $540 - 405p = 0$, $p = \frac{4}{3}$

2 -4704

CHAPTER 9

Prior knowledge check

1 a 3.10 cm b 9.05 cm

2 a 25.8° b 77.2°

3 a graph of $x^2 + 3x$ b graph of $(x + 2)^2 + 3(x + 2)$
 c graph of $x^2 + 3x - 3$ b graph of $(0.5x)^2 + 3(0.5x)$

Exercise 9A

1 a 3.19 cm b 1.73 cm ($\sqrt{3}$ cm) c 9.85 cm
 d 4.31 cm e 6.84 cm f 9.80 cm

2 a 108(.2)° b 90° c 60°
 d 52.6° e 137° f 72.2°

3 192 km

4 11.2 km

5 128.5° or 031.5° (Angle BAC = 48.5°)

6 302 yards (301.5...)

7 Using the cosine rule $\frac{5^2 + 4^2 - 6^2}{2 \times 5 \times 4} = \frac{1}{8}$

8 Using the cosine rule $\frac{2^2 + 3^2 - 4^2}{2 \times 2 \times 3} = -\frac{1}{4}$

9 ACB = 22.3°

10 ABC = 108(.4)°

11 104(.48)°

12 x = 4.4 cm

13 x = 42 cm

14 a $y^2 = (5 - x)^2 + (4 + x)^2 - 2(5 - x)(4 + x)\cos 120°$
 $= 25 - 10x + x^2 + 16 + 8x + x^2 - 2(20 + x - x^2)\left(-\frac{1}{2}\right)$
 $= x^2 - x + 61$
 b Minimum $AC^2 = 60.75$; it occurs for $x = \frac{1}{2}$

15 a $\cos\angle ABC = \frac{x^2 + 5^2 - (10 - x)^2}{2x \times 5}$
 $= \frac{20x - 75}{10x} = \frac{4x - 15}{2x}$
 b 3.5

16 65.3°

17 a 28.7 km b 056.6°

Exercise 9B

1 a 15.2 cm b 9.57 cm c 8.97 cm d 4.61 cm

2 a $x = 84°$, $y = 6.32$
 b $x = 13.5$, $y = 16.6$
 c $x = 85°$, $y = 13.9$
 d $x = 80°$, $y = 6.22$ (isosceles triangle)
 e $x = 6.27$, $y = 7.16$
 f $x = 4.49$, $y = 7.49$ (right-angled)

3 a 36.4° b 35.8° c 40.5° d 130°

4 a 48.1° b 45.6° c 14.8° d 48.7°
 e 86.5° f 77.4°

5 a 1.41 cm ($\sqrt{2}$ cm) b 1.93 cm

6 QPR = 50.6°, PQR = 54.4°

7 a $x = 43.2°$, $y = 5.02$ cm b $x = 101°$, $y = 15.0$ cm
 c $x = 6.58$ cm, $y = 32.1°$ d $x = 54.6°$, $y = 10.3$ cm
 e $x = 21.8°$, $y = 3.01$ f $x = 45.9°$, $y = 3.87°$

8 a 6.52 km b 3.80 km

9 a 7.31 cm b 1.97 cm

10 a 66.3° b 148 m

11 Using the sine rule, $x = \frac{4\sqrt{2}}{2 + \sqrt{2}}$; rationalising
 $x = \frac{4\sqrt{2}(2 - \sqrt{2})}{2} = 4\sqrt{2} - 4 = 4(\sqrt{2} - 1)$.

12 a 36.5 m
 b That the angles have been measured from ground level

Exercise 9C

1 a 70.5°, 109° (109.5°)
 b

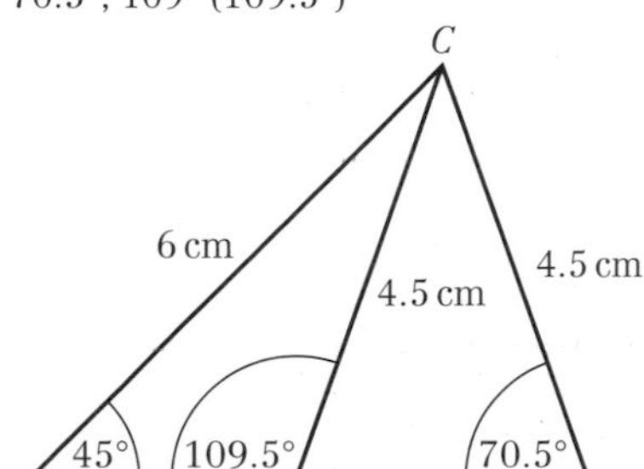

2 a $x = 74.6°$, $y = 65.4°$
 $x = 105°$, $y = 34.6°$
 b $x = 59.8°$, $y = 48.4$ cm
 $x = 120°$, $y = 27.3$cm
 c $x = 56.8°$, $y = 4.37$ cm
 $x = 23.2°$, $y = 2.06$ cm

3 a 5 cm (ACB = 90°) b 24.6°
 c 45.6°, 134(.4)°

4 2.97 cm

5 In one triangle ABC = 101° (100.9°); in the other BAC = 131° (130.9°)

6 a 62.0° b The swing is symmetrical

Exercise 9D

1 a 23.7 cm² b 4.31 cm² c 20.2 cm²

2 a x = 41.8° or 138(.2)°
 b x = 26.7° or 153(.3)°
 c x = 60° or 120°

3 275(.3) m (third side = 135.3 m)

4 3.58

5 a Area $= \frac{1}{2}(x + 2)(5 - x)\sin 30°$
 $= \frac{1}{2}(10 + 3x - x^2) \times \frac{1}{2}$
 $= \frac{1}{4}(10 + 3x - x^2)$
 b Maximum $A = 3\frac{1}{16}$, when $x = 1\frac{1}{2}$

6 a $\frac{1}{2}x(5 + x)\sin 150° = \frac{15}{4}$
 $\frac{1}{2}(5x + x^2) \times \frac{1}{2} = \frac{15}{4}$
 $5x + x^2 = 15$
 $x^2 + 5x - 15 = 0$
 b 2.11

Exercise 9E

1 a $x = 37.7°$, $y = 86.3°$, $z = 6.86$
 b $x = 48°$, $y = 19.5$, $z = 14.6$
 c $x = 30°$, $y = 11.5$, $z = 11.5$
 d $x = 21.0°$, $y = 29.0°$, $z = 8.09$
 e $x = 93.8°$, $y = 56.3°$, $z = 29.9°$
 f $x = 97.2°$, $y = 41.4°$, $z = 41.4°$

g $x = 45.3°, y = 94.7°, z = 14.7$
or $x = 135°, y = 5.27°, z = 1.36$
h $x = 7.07, y = 73.7°, z = 61.3°$
or $x = 7.07, y = 106°, z = 28.7°$
i $x = 49.8°, y = 9.39, z = 37.0°$

2 a $ACB = 32.4°, ABC = 108°, AC = 15.1$ cm
Area = 41.3 cm^2
b $BAC = 41.5°, ABC = 28.5°, AB = 9.65$ cm
Area = 15.7 cm^2

3 a 8 km b 060°

4 107 km

5 12 km

6 a 5.44 b 7.95 c 36.8°

7 a $AB + BC > AC \Rightarrow x + 6 > 7 \Rightarrow x > 1$;
$AC + AB > BC \Rightarrow 11 > x + 2 \Rightarrow x < 9$
b i $x = 6.08$ from $x^2 = 37$
Area = 14.0 cm^2
ii $x = 7.23$ from $x^2 - 4(\sqrt{2} - 1)x - (29 + 8\sqrt{2}) = 0$
Area = 13.1 cm^2

8 a $x = 4$ b 4.68 cm^2

9 $AC = 1.93$ cm

10 a $AC^2 = (2 - x)^2 + (x + 1)^2 - 2(2 - x)(x + 1)\cos 120°$
$= (4 - 4x + x^2) + (x^2 + 2x + 1) - 2(-x^2 + x + 2)\left(-\frac{1}{2}\right)$
$= x^2 - x + 7$
b $\frac{1}{2}$

11 $4\sqrt{10}$

12 $AC = 1\frac{2}{3}$ cm and $BC = 6\frac{1}{3}$ cm
Area = 5.05 cm^2

13 a 61.3° b 78.9 cm^2

14 a $DAB = 136.3°, BCD = 50.1°$
b 13.1 m^2
c 5.15 m

15 34.2 cm^2

Exercise 9F

1

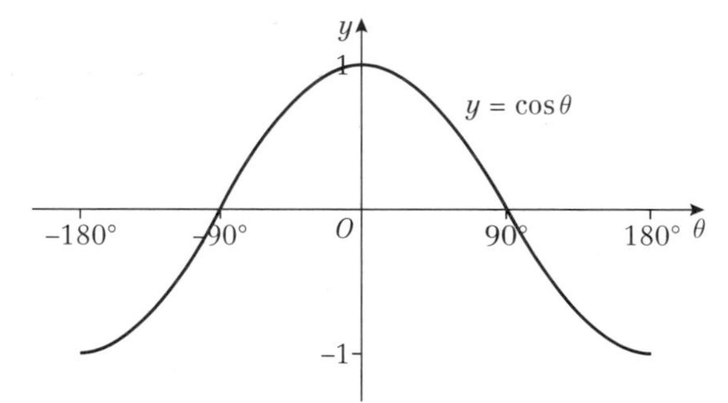

2

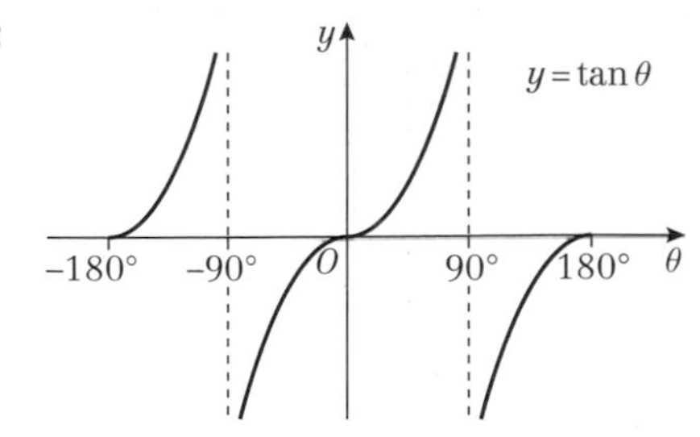

3

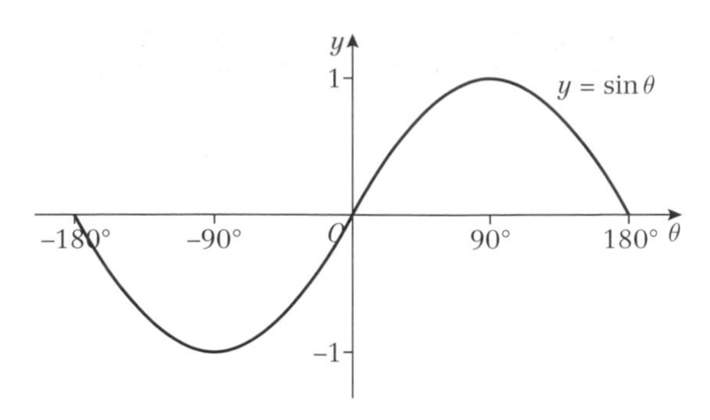

4 a −30°
b i −120° ii −60°, 120°
c i 135° ii −45°, −135°

Exercise 9G

1 a i 1, $x = 0°$ ii −1, $x = 180°$
b i 4, $x = 90°$ ii −4, $x = 270°$
c i 1, $x = 0°$ ii −1, $x = 180°$
d i 4, $x = 90°$ ii 2, $x = 270°$
e i 1, $x = 270°$ ii −1, $x = 90°$
f i 1, $x = 30°$ ii −1, $x = 90°$

2

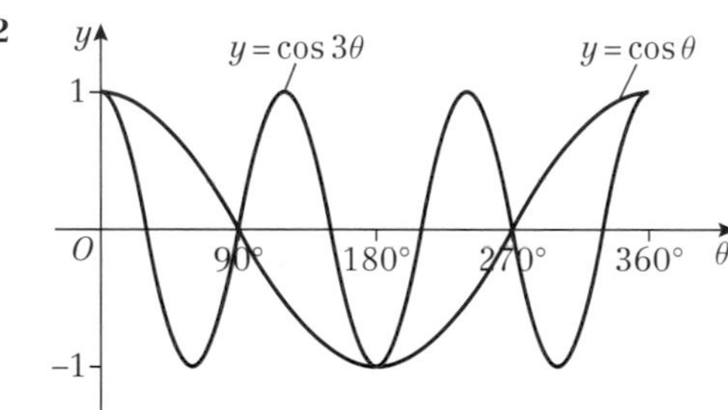

3 a The graph of $y = -\cos\theta$ is the graph of $y = \cos\theta$ reflected in the θ-axis

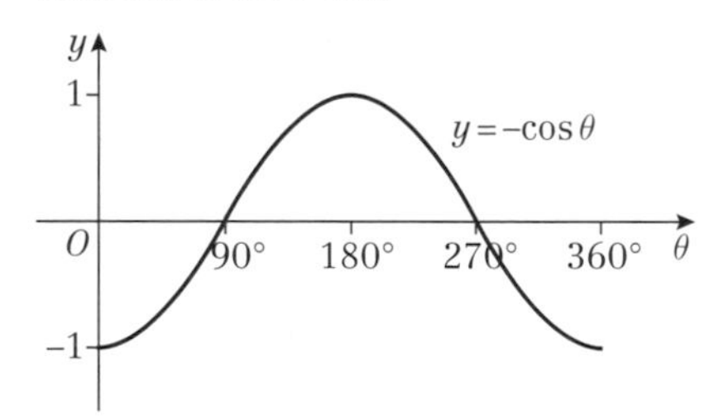

Meets θ-axis at (90°, 0), (270°, 0)
Meets y-axis at (0°, −1)
Maximum at (180°, 1)
Minimum at (0°, −1) and (360°, −1)

b The graph of $y = \frac{1}{3}\sin\theta$ is the graph of $y = \sin\theta$ stretched by a scale factor $\frac{1}{3}$ in the y direction.

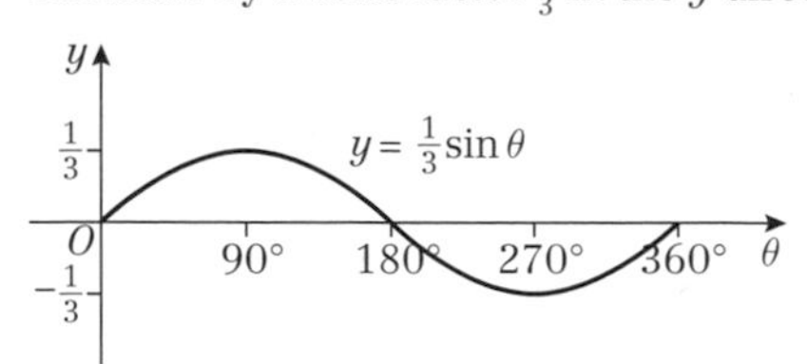

Meets θ-axis at (0°, 0), (180°, 0), (360°, 0)
Meets y-axis at (0°, 0)
Maximum at $(90°, \frac{1}{3})$
Minimum at $(270°, -\frac{1}{3})$

c The graph of $y = \sin\frac{1}{3}\theta$ is the graph of $y = \sin\theta$ stretched by a scale factor 3 in the θ direction.

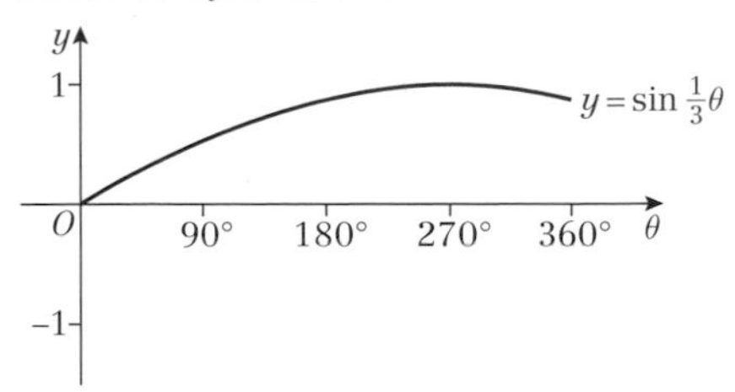

Only meets axis at origin
Maximum at (270°, 1)

d The graph of $y = \tan(\theta - 45°)$ is the graph of $\tan\theta$ translated by 45° in the positive θ direction.

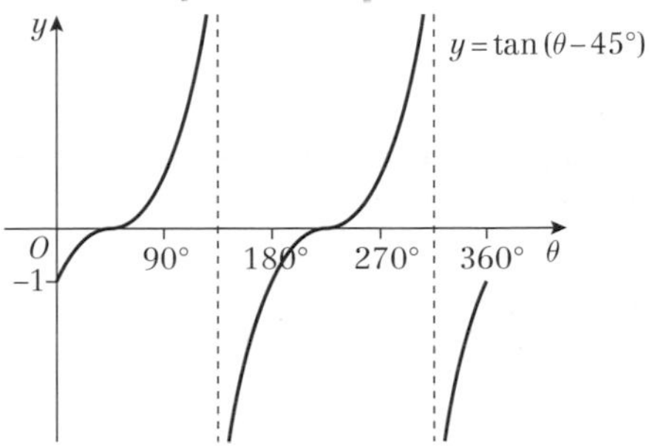

Meets θ-axis at (45°, 0), (225°, 0)
Meets y-axis at (0°, −1)
(Asymptotes at $\theta = 135°$ and $\theta = 315°$)

4 a This is the graph of $y = \sin\theta$ stretched by scale factor −2 in the y-direction (i.e. reflected in the θ-axis and scaled by 2 in the y-direction).

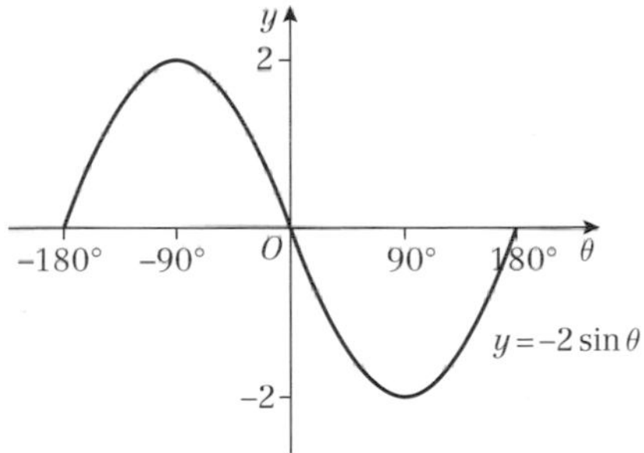

Meets θ-axis at (−180°, 0), (0, 0), (180, 0)
Maximum at (−90°, 2)
Minimum at (90°, −2).

b This is the graph of $y = \tan\theta$ translated by 180° in the negative θ direction.

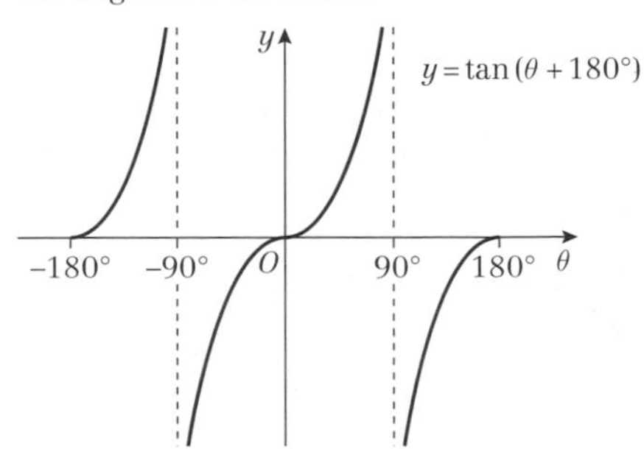

As $\tan\theta$ has a period of 180°
$\tan(\theta + 180°) = \tan\theta$
Meets θ-axis at (−180°, 0), (0, 0), (180°, 0)
Meets y-axis at (0, 0)

c This is the graph of $y = \cos\theta$ stretched by scale factor $\frac{1}{4}$ horizontally.

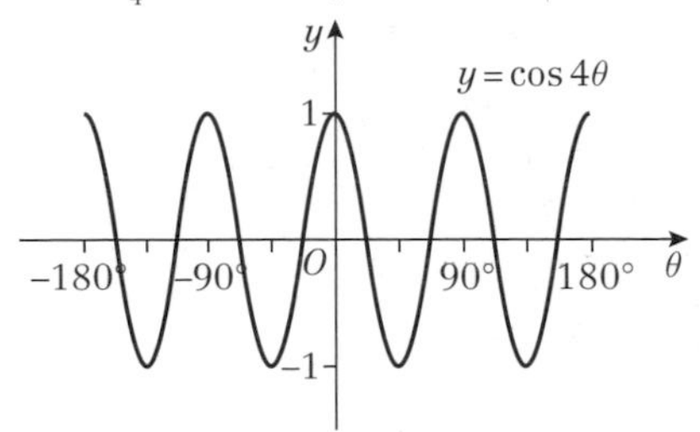

Meets θ-axis at $(-157\frac{1}{2}°, 0)$, $(-112\frac{1}{2}°, 0)$, $(-67\frac{1}{2}°, 0)$, $(-22\frac{1}{2}°, 0)$, $(22\frac{1}{2}°, 0)$, $(67\frac{1}{2}°, 0)$, $(112\frac{1}{2}°, 0)$, $(157\frac{1}{2}°, 0)$
Meets y-axis at (0, 1)
Maxima at (−180°, 1), (−90°, 1), (0, 1), (90°, 1), (180°, 1)
Minima at (−135°, −1), (−45°, −1), (45°, −1), (135°, −1)

d This is the graph of $y = \sin\theta$ reflected in the y-axis. (This is the same as $y = -\sin\theta$.)

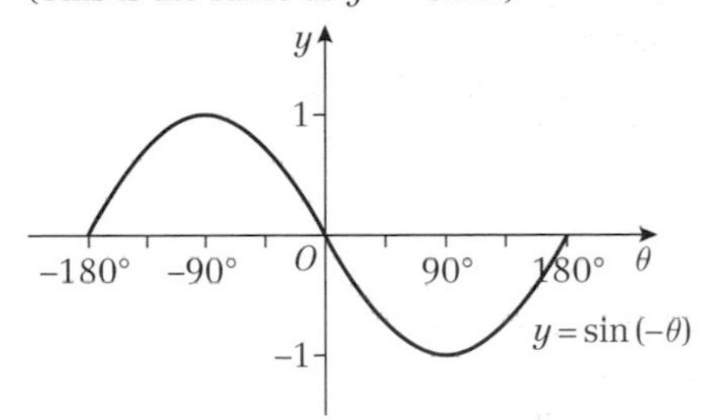

Meets θ-axis at (−180°, 0), (0°, 0), (180°, 0)
Maximum at (−90°, 1)
Minimum at (90°, −1)

5 a Period = 720°

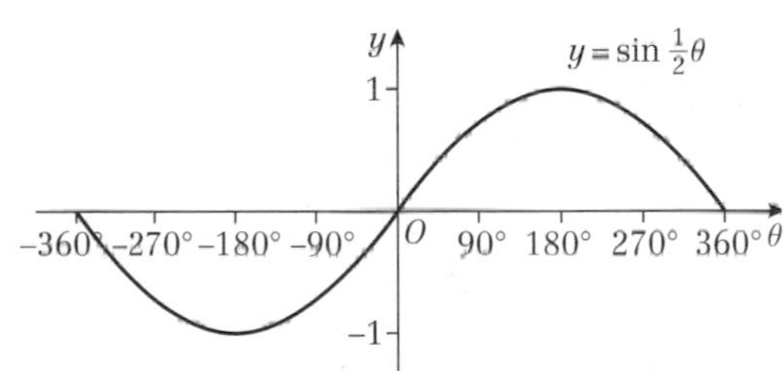

b Period = 360°

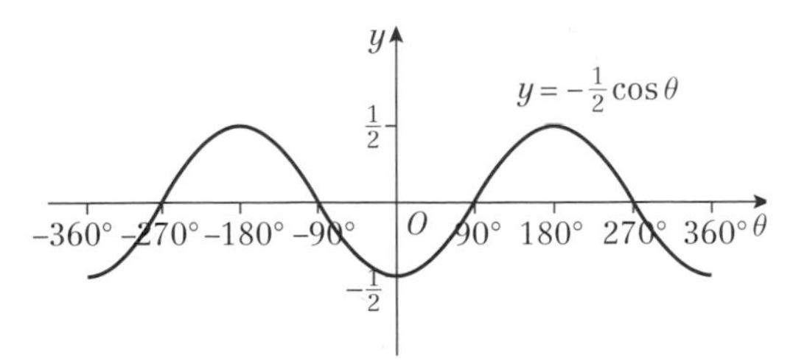

c Period = 180°

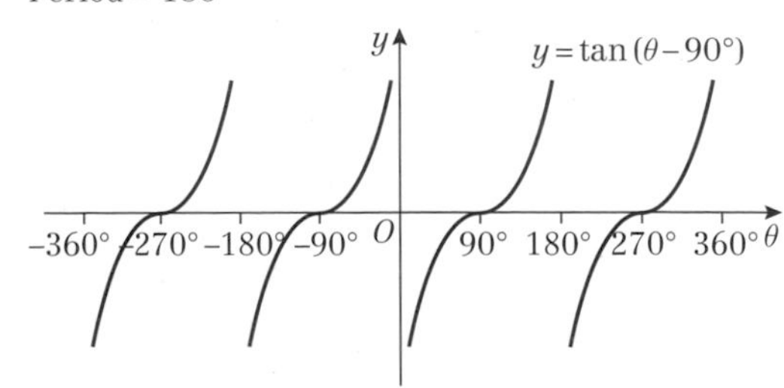

d Period = 90°

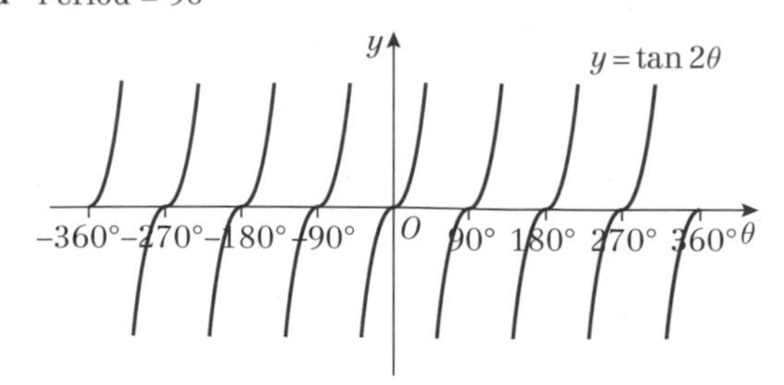

6 a i $y = \cos(-\theta)$ is a reflection of $y = \cos\theta$ in the y-axis, which is the same curve, so $\cos\theta = \cos(-\theta)$.

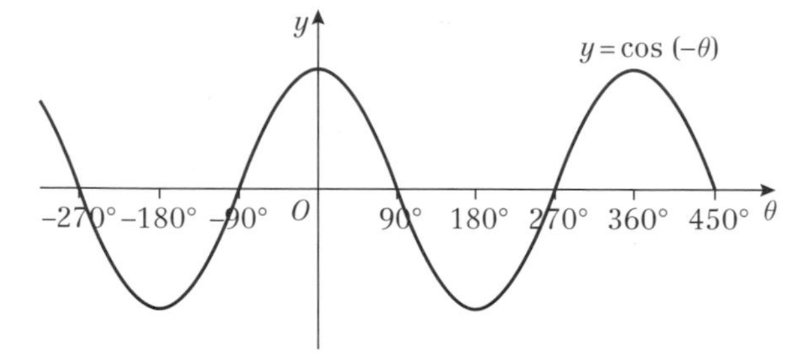

ii $y = \sin(-\theta)$ is a reflection of $y = \sin\theta$ in the y-axis.

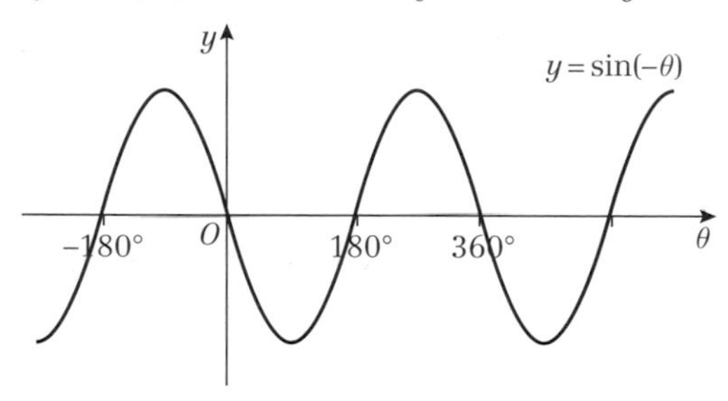

$y = -\sin(-\theta)$ is a reflection of $y = \sin(-\theta)$ in the θ-axis, which is the graph of $y = \sin\theta$, so $-\sin(-\theta) = \sin\theta$.

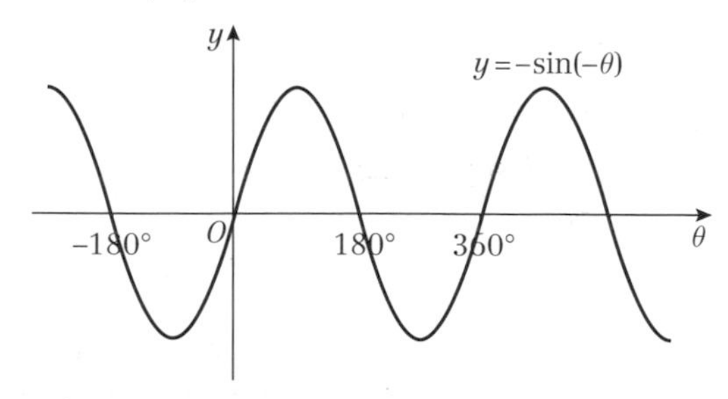

iii $y = \sin(\theta - 90°)$ is the graph of $y = \sin\theta$ translated by 90° to the right, which is the graph of $y = -\cos\theta$, so $\sin(\theta - 90°) = -\cos\theta$.

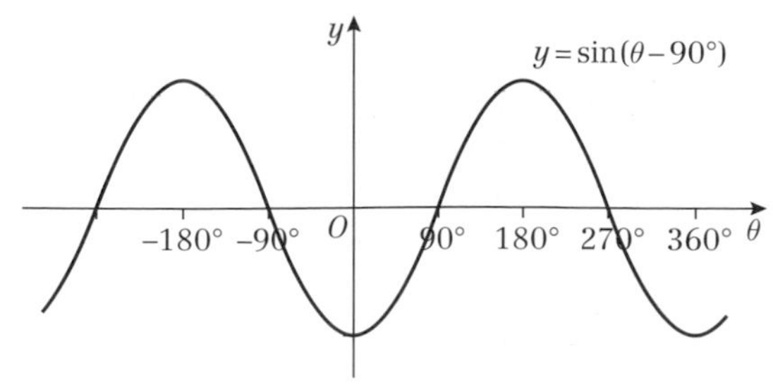

b $\sin(90° - \theta)$
$= -\sin(-(90° - \theta)) = -\sin(\theta - 90°)$
using (a) (ii)
$= -(-\cos\theta)$ using (a) (iii)
$= \cos\theta$

c Using (a)(i) $\cos(90° - \theta) = \cos(-(90° - \theta))$
$= \cos(\theta - 90°)$, but $\cos(\theta - 90°) = \sin\theta$,
so $\cos(90° - \theta) = \sin\theta$

7 a (–300°, 0), (–120°, 0), (60°, 0), (240°, 0)

b $\left(0°, \frac{\sqrt{3}}{2}\right)$

8 a $k = 60°$

b Yes – the graph of $y = \sin\theta$ repeats every 360°, so e.g. $k = 420°$.

9 a

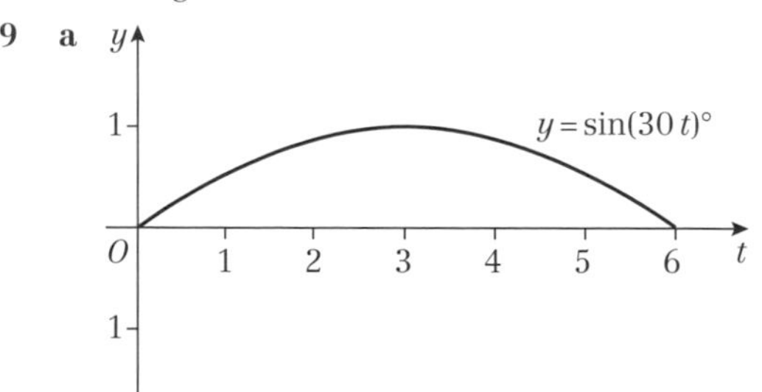

b Between 1 pm and 5 pm

Mixed exercise 9

1 a 155° **b** 13.7 cm

2 a $x = 49.5°$, area = 1.37 cm^2

b $x = 55.2°$, area = 10.6 cm^2

c $x = 117°$, area = 6.66 cm^2

3 6.50 cm^2

4 a 50.9 cm^2 **b** 12.0 cm^2

5 a 5 **b** $\frac{25\sqrt{3}}{2}$ cm^2

6 area $= \frac{1}{2}ab\sin C$

$1 = \frac{1}{2} \times 2\sqrt{2}\sin C$

$\frac{1}{\sqrt{2}} = \sin C \Rightarrow C = 45°$

Use the cosine rule to find the other side:

$x^2 = 2^2 + (\sqrt{2})^2 - 2 \times 2\sqrt{2}\cos C \Rightarrow x = \sqrt{2}$ cm

So the triangle is isosceles, with two 45° angles, thus is also right-angled.

7 a $AC = \sqrt{5}$, $AB = \sqrt{18}$, $BC = \sqrt{5}$

$$\cos\angle ACB = \frac{AC^2 + BC^2 - AB^2}{2 \times AC \times BC}$$

$$= \frac{5 + 5 - 18}{2 \times \sqrt{5} \times \sqrt{5}}$$

$$= -\frac{8}{10} = -\frac{4}{5}$$

b $1\frac{1}{2}$ cm^2

8 a 4 **b** $\frac{15\sqrt{3}}{4}$ (6.50) cm^2

9 a 1.50 km **b** 241° **c** 0.789 km^2

10 359 m^2

11 35.2 m

12 a A stretch of scale factor 2 in the x direction.

b A translation of +3 in the y direction.

c A reflection in the x-axis.
d A translation of -20 in the x direction.

13 a

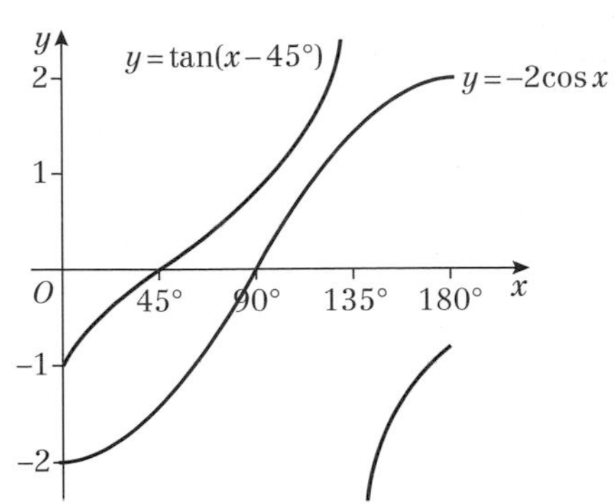

b There are no solutions.

14 a 300° **b** (30°, 1) **c** 60° **d** $\frac{\sqrt{3}}{2}$

15 a $p = 5$

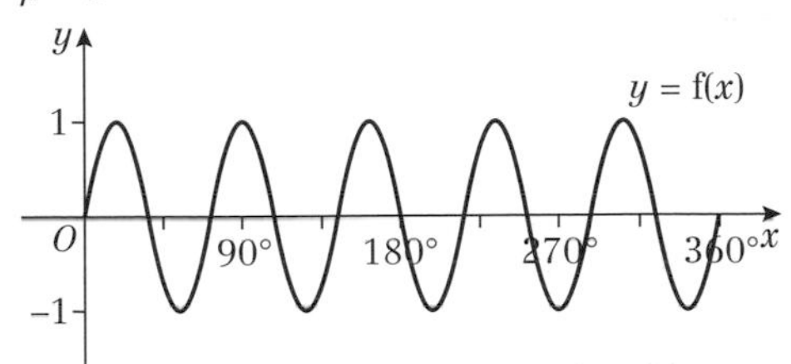

b 72°

16 a The four shaded regions are congruent.

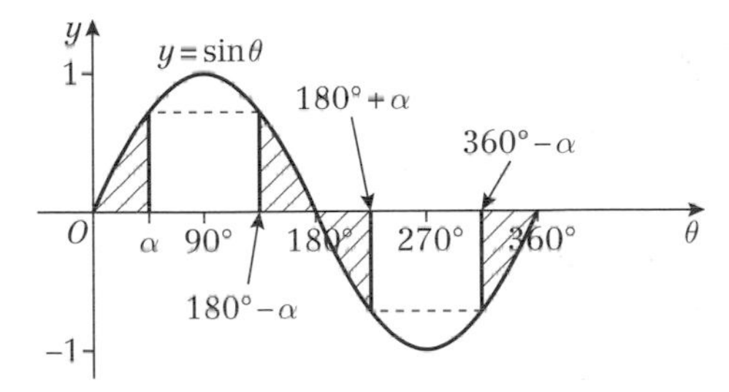

b $\sin\alpha$ and $\sin(180° - \alpha)$ have the same y value, (call it k)
so $\sin\alpha = \sin(180° - \alpha)$
$\sin(180° + \alpha)$ and $\sin(360° - \alpha)$ have the same y value, (which will be $-k$)
so $\sin\alpha = \sin(180° - \alpha)$
$= -\sin(180° + \alpha) = -\sin(360° - \alpha)$

17 a

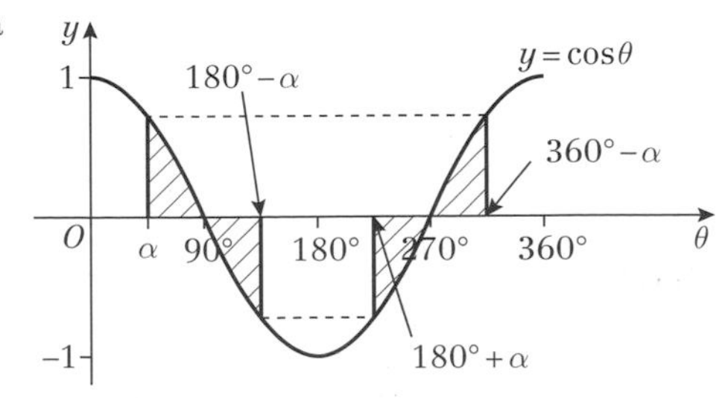

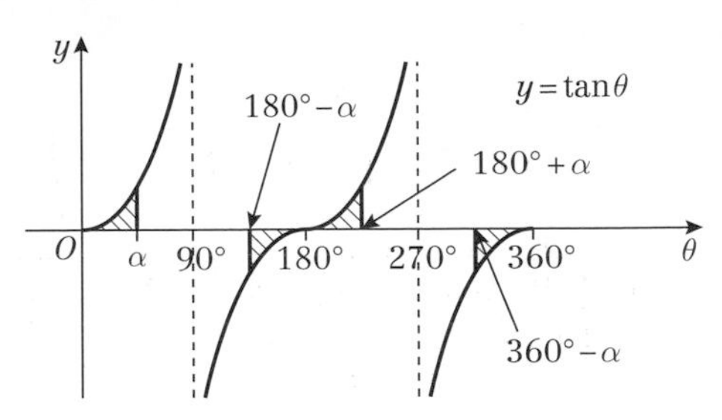

b i From the graph of $y = \cos\theta$, which shows four congruent shaded regions, if the y value at α is k, then y at $180° - \alpha$ is $-k$, y at $180° + \alpha = -k$ and y at $360° - \alpha = +k$
so $\cos\alpha = -\cos(180° - \alpha)$
$= -\cos(180° + \alpha) = \cos(360° - \alpha)$

ii From the graph of $y = \tan\theta$, if the y value at α is k, then at $180° - \alpha$ it is $-k$, at $180° + \alpha$ it is $+k$ and at $360° - \alpha$ it is $-k$,
so $\tan\alpha = -\tan(180° - \alpha)$
$= +\tan(180° + \alpha) = -\tan(360° - \alpha)$

18 a

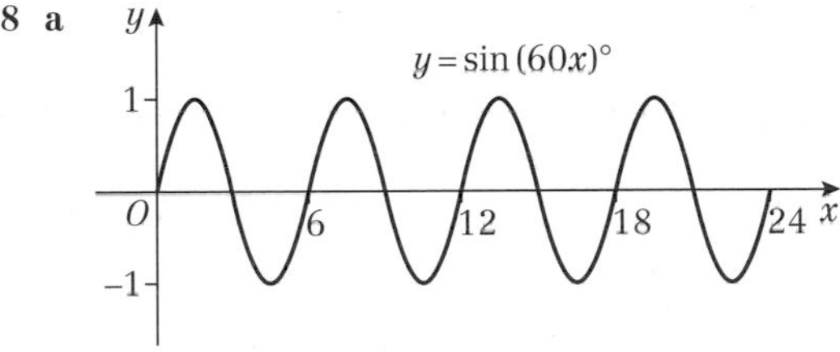

b 4
c The dunes may not all be the same height.

Challenge
Using the sine rule:

$$\sin(180° - \angle ADB - \angle AEB) = \frac{5\left(\frac{1}{\sqrt{5}}\right)}{\sqrt{10}} = \frac{1}{\sqrt{2}}$$

$180° - \angle ADB - \angle AEB = 135°$ (obtuse)
so $\angle ADB + \angle AEB = 45° = \angle ACB$

CHAPTER 10

Prior knowledge check

1 a

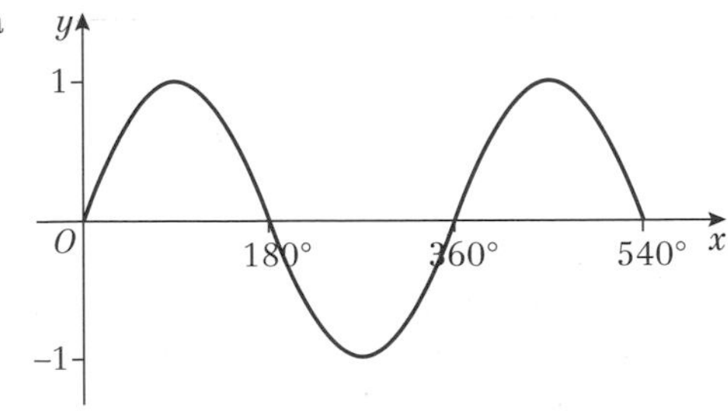

b 4
c 143.1°, 396.9°, 503.1°

2 a 57.7° **b** 73.0°

3 a $x = 11$ **b** $x = \frac{9}{4}$ **c** $x = -44.4°$

4 a $x = 1$ or $x = 3$
b $x = 1$ or $x = -9$
c $x = \frac{3 \pm \sqrt{65}}{4}$

Exercise 10A

1 a
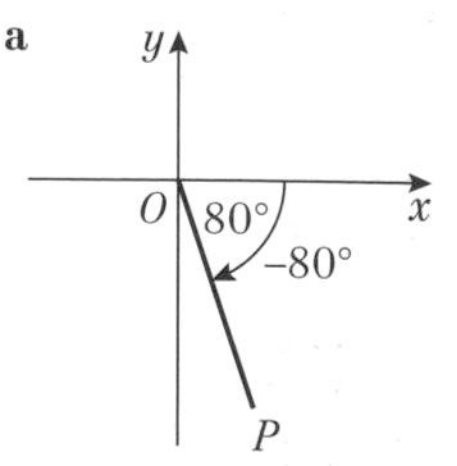

b
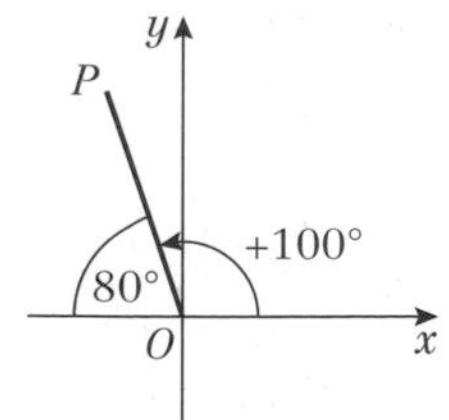

c
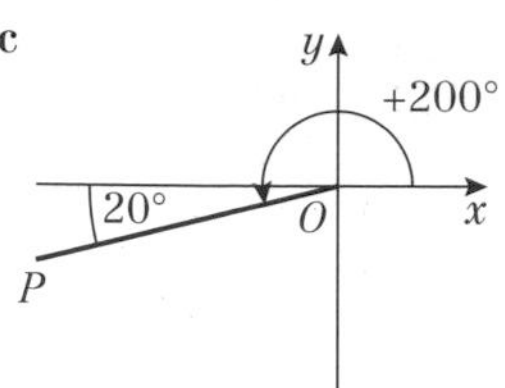

d
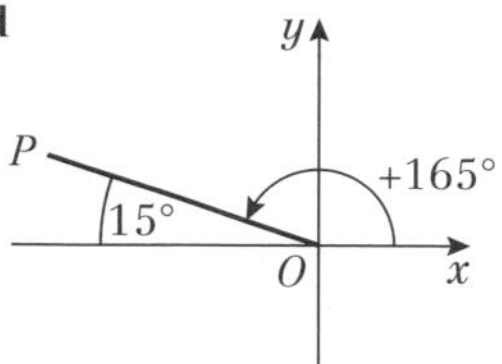

e
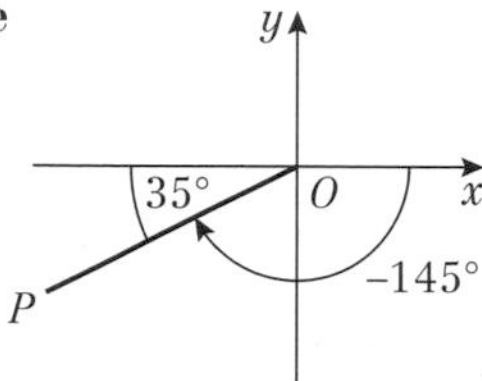

f
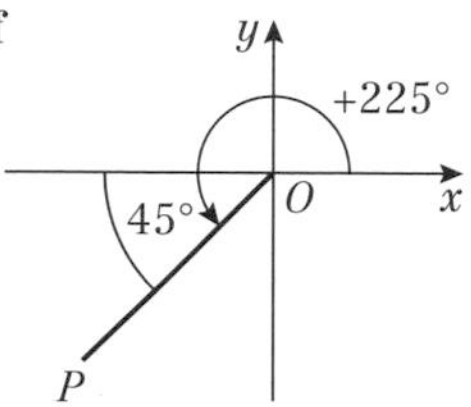

g
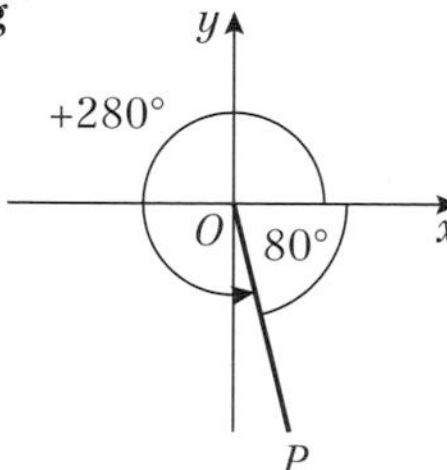

h
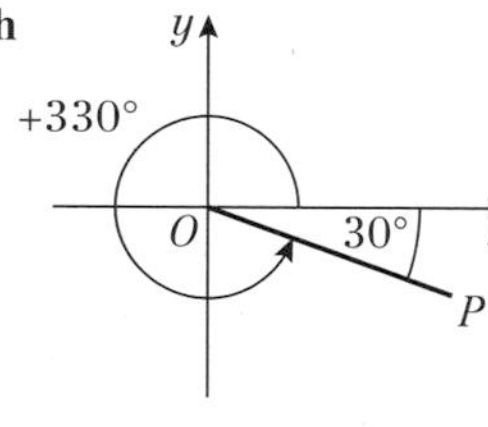

i
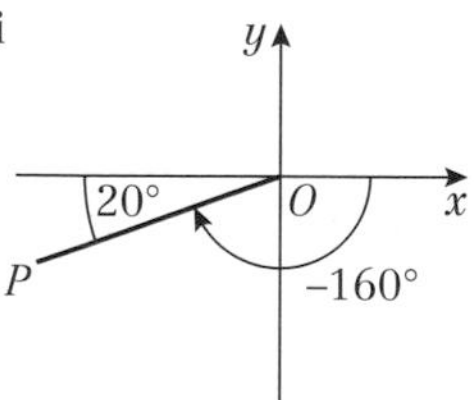

j
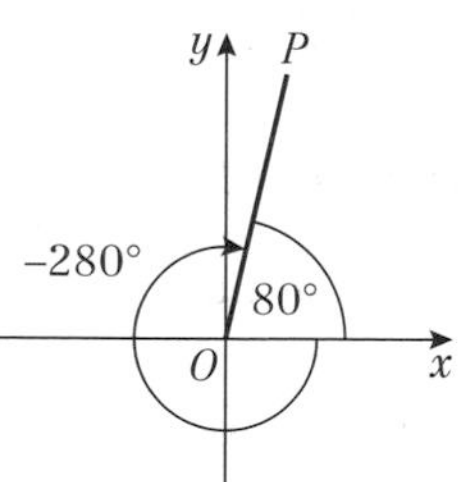

2 **a** First **b** Second **c** Second
d Third **e** Third

3 **a** –1 **b** 1 **c** 0 **d** –1 **e** –1
f 0 **g** 0 **h** 0 **i** 0 **j** 0

4 **a** $-\sin 60°$ **b** $-\sin 80°$ **c** $\sin 20°$
d $-\sin 60°$ **e** $\sin 80$ **f** $-\cos 70°$
g $-\cos 80°$ **h** $\cos 50°$ **i** $-\cos 20°$
j $-\cos 5°$ **k** $-\tan 80°$ **l** $-\tan 35°$
m $-\tan 30°$ **n** $\tan 5°$ **o** $\tan 60°$

5 **a** $-\sin\theta$ **b** $-\sin\theta$ **c** $-\sin\theta$
d $\sin\theta$ **e** $-\sin\theta$ **f** $\sin\theta$
g $-\sin\theta$ **h** $-\sin\theta$ **i** $\sin\theta$

6 **a** $-\cos\theta$ **b** $-\cos\theta$ **c** $\cos\theta$
d $-\cos\theta$ **e** $\cos\theta$ **f** $-\cos\theta$
g $-\tan\theta$ **h** $-\tan\theta$ **i** $\tan\theta$
j $\tan\theta$ **k** $-\tan\theta$ **l** $\tan\theta$

Challenge

a
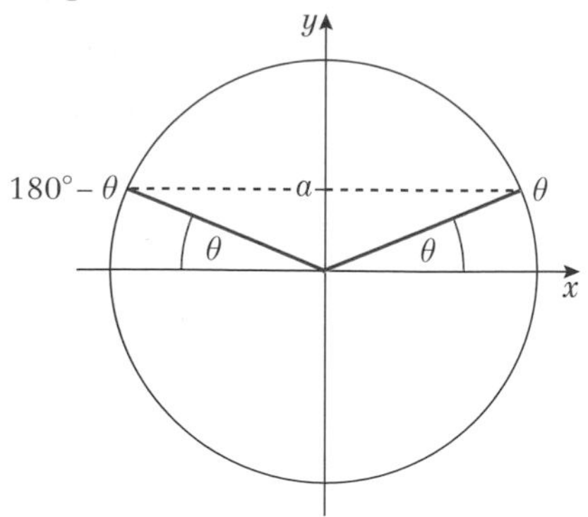

$\sin\theta = \sin(180° - \theta) = a$

b
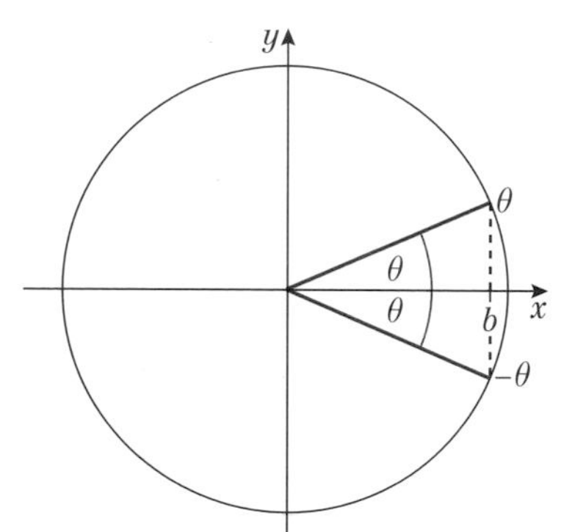

$\cos\theta = \cos(-\theta) = b$

c
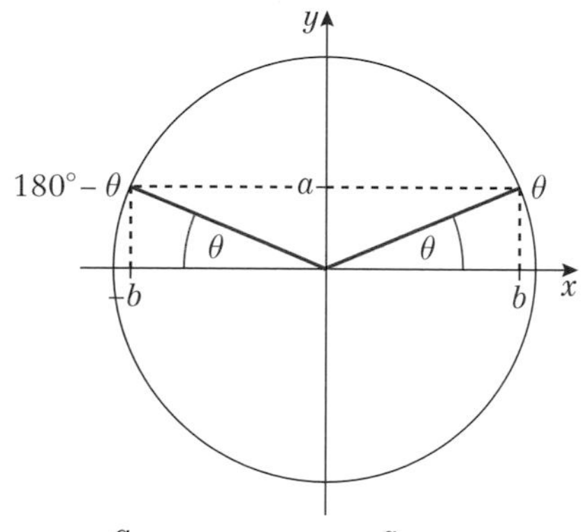

$\tan\theta = \frac{a}{b}$; $\tan(180° - \theta) = \frac{a}{-b} = -\tan\theta$

Exercise 10B

1 **a** $\frac{\sqrt{2}}{2}$ **b** $-\frac{\sqrt{3}}{2}$ **c** $-\frac{1}{2}$ **d** $\frac{\sqrt{3}}{2}$
e $\frac{\sqrt{3}}{2}$ **f** $-\frac{1}{2}$ **g** $\frac{1}{2}$ **h** $-\frac{\sqrt{2}}{2}$
i $-\frac{\sqrt{3}}{2}$ **j** $-\frac{\sqrt{2}}{2}$ **k** –1 **l** –1
m $\frac{\sqrt{3}}{3}$ **n** $-\sqrt{3}$ **o** $\sqrt{3}$

Challenge

a **i** $\sqrt{3}$ **ii** 2 **iii** $\sqrt{2+\sqrt{3}}$ **iv** $\sqrt{2+\sqrt{3}} - \sqrt{2}$
b 15°
c **i** $\frac{\sqrt{2+\sqrt{3}} - \sqrt{2}}{2}$ **ii** $\frac{\sqrt{2+\sqrt{3}}}{2}$

Exercise 10C

1 a $\sin^2 \frac{\theta}{2}$ b 5 c $-\cos^2 A$
d $\cos\theta$ e $\tan x$ f $\tan 3A$
g 4 h $\sin^2\theta$ i 1

2 $1\frac{1}{2}$

3 $3\tan y$

4 a $1 - \sin^2\theta$ b $\frac{\sin^2\theta}{1-\sin^2\theta}$ c $\sin\theta$
d $\frac{1-\sin^2\theta}{\sin\theta}$ e $1 - 2\sin^2\theta$

5 (One outline example of a proof is given)

a $\text{LHS} = \sin^2\theta + \cos^2\theta + 2\sin\theta\cos\theta$
$= 1 + 2\sin\theta\cos\theta$
$= \text{RHS}$

b $\text{LHS} = \frac{1-\cos^2\theta}{\cos\theta} = \frac{\sin^2\theta}{\cos\theta} = \sin\theta \times \frac{\sin\theta}{\cos\theta}$
$= \sin\theta\tan\theta = \text{RHS}$

c $\text{LHS} = \frac{\sin x}{\cos x} + \frac{\cos x}{\sin x} = \frac{\sin^2 x + \cos^2 x}{\sin x\cos x}$
$= \frac{1}{\sin x\cos x} = \text{RHS}$

d $\text{LHS} = \cos^2 A - (1 - \cos^2 A) = 2\cos^2 A - 1$
$= 2(1 - \sin^2 A) - 1 = 1 - 2\sin^2 A = \text{RHS}$

e $\text{LHS} = (4\sin^2\theta - 4\sin\theta\cos\theta + \cos^2\theta)$
$+ (\sin^2\theta + 4\sin\theta\cos\theta + \cos^2\theta)$
$= 5(\sin^2\theta + \cos^2\theta) = 5 = \text{RHS}$

f $\text{LHS} = 2 - (\sin^2\theta - 2\sin\theta\cos\theta + \cos^2\theta)$
$= 2(\sin^2\theta + \cos^2\theta) - (\sin^2\theta - 2\sin\theta\cos\theta + \cos^2\theta)$
$= \sin^2\theta + 2\sin\theta\cos\theta + \cos^2\theta$
$= (\sin\theta + \cos\theta)^2 = \text{RHS}$

g $\text{LHS} = \sin^2 x(1 - \sin^2 y) - (1 - \sin^2 x)\sin^2 y$
$= \sin^2 x - \sin^2 y = \text{RHS}$

6 a $\sin\theta = \frac{5}{13}$, $\cos\theta = \frac{12}{13}$
b $\sin\theta = \frac{4}{5}$, $\tan\theta = -\frac{4}{3}$
c $\cos\theta = \frac{24}{25}$, $\tan\theta = -\frac{7}{24}$

7 a $-\frac{\sqrt{5}}{3}$ b $-\frac{2\sqrt{5}}{5}$

8 a $-\frac{\sqrt{3}}{2}$ b $\frac{1}{2}$

9 a $-\frac{\sqrt{7}}{4}$ b $-\frac{\sqrt{7}}{3}$

10 a $x^2 + y^2 = 1$
b $4x^2 + y^2 = 4$ $\left(\text{or } x^2 + \frac{y^2}{4} = 1\right)$
c $x^2 + y = 1$
d $x^2 = y^2(1 - x^2)$ $\left(\text{or } x^2 + \frac{x^2}{y^2} = 1\right)$
e $x^2 + y^2 = 2$ $\left(\text{or } \frac{(x+y)^2}{4} + \frac{(x-y)^2}{4} = 1\right)$

11 a Using cosine rule: $\cos B = \frac{8^2 + 12^2 - 10^2}{2 \times 8 \times 12} = \frac{9}{16}$
b $\frac{\sqrt{175}}{16}$

12 a Using sine rule: $\sin Q = \frac{\sin 30}{6} \times 8 = \frac{2}{3}$
b $-\frac{\sqrt{5}}{3}$

Exercise 10D

1 a $-63.4°$ b $116.6°, 296.6°$

2 a $66.4°$ b $66.4°, 113.6°, 246.4°, 293.6°$

3 a $270°$ b $60°, 240°$
c $60°, 300°$ d $15°, 165°$
e $140°, 220°$ f $135°, 315°$
g $90°, 270°$ h $230°, 310°$

4 a $45.6°, 134.4°$ b $135°, 225°$
c $132°, 228°$ d $229°, 311°$
e $8.13°, 188°$ f $61.9°, 242°$
g $105°, 285°$ h $41.8°, 318°$

5 a $30°, 210°$ b $135°, 315°$
c $53.1°, 233°$ d $56.3°, 236°$
e $54.7°, 235°$ f $148°, 328°$

6 a $-120°, -60°, 240°, 300°$ b $-171°, -8.63°$
c $-144°, 144°$ d $-327°, -32.9°$
e $150°, 330°, 510°, 690°$ f $251°, 431°$

7 a $\tan x$ should be $\frac{2}{3}$
b Squaring both sides creates extra solutions
c $-146.3°, 33.7°$

8 a

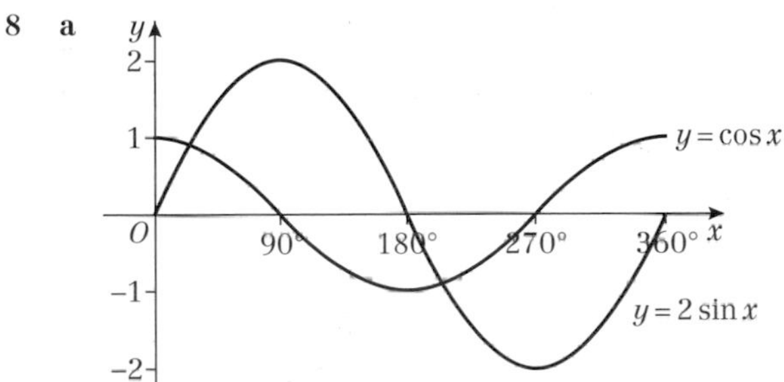

b 2 c $26.6°, 206.6°$

9 $71.6°, 108.4°, 251.6°, 288.4°$

10 a $4\sin^2 x - 3(1 - \sin^2 x) = 2$.
Rearrange to get $7\sin^2 x = 5$
b $57.7°, 122.3°, 237.7°, 302.3°$

11 a $2\sin^2 x + 5(1 - \sin^2 x) = 1$.
Rearrange to get $3\sin^2 x = 4$
b $\sin x > 1$

Exercise 10E

1 a $0°, 45°, 90°, 135°, 180°, 225°, 270°, 315°, 360°$
b $60°, 180°, 300°$
c $22\frac{1}{2}°, 112\frac{1}{2}°, 202\frac{1}{2}°, 292\frac{1}{2}°$
d $30°, 150°, 210°, 330°$
e $300°$
f $225°, 315°$

2 a $90°, 270°$ b $50°, 170°$ c $165°, 345°$
d $250°, 310°$ e $16.9°, 123°$

3 a $11.2°, 71.2°, 131.2°$ b $6.3°, 186.3°, 366.3°$
c $37.0°, 127.0°$ d $-150°, 30°$

4 a $10°, 130°$ b $71.6°, 108.4°$

5 a

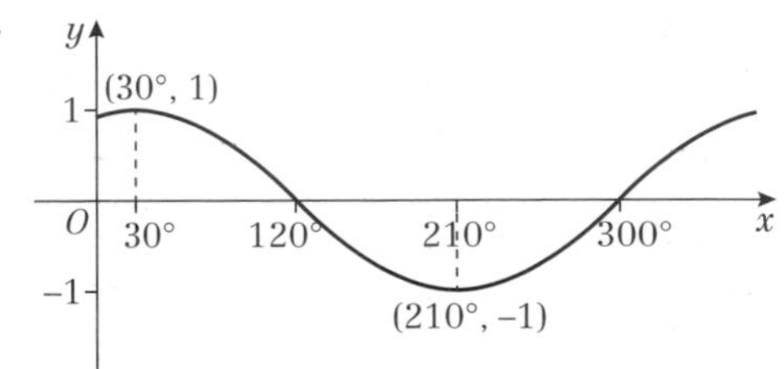

b $\left(0°, \frac{\sqrt{3}}{2}\right)$, (120°, 0), (300°, 0)

c 86.6°, 333.4°

6 a 0.75

b 18.4°, 108.4°, 198.4°, 288.4°

7 a 2.5

b No: increasing k will bring another 'branch' of the tan graph into place.

Challenge

25°, 65°, 145°

Exercise 10F

1 a 60°, 120°, 240°, 300°

b 45°, 135°, 225°, 315°

c 0°, 180°, 199°, 341°, 360°

d 77.0°, 113°, 257°, 293°

e 60°, 300°

f 204°, 336°

g 30°, 60°, 120°, 150°, 210°, 240°, 300°, 330°

2 a ±45°, ±135° b −180°, −117°, 0°, 63.4°, 180°

c ±114° d 0°, ±75.5°, ±180°

3 a 72°, 144° b 0°, 60°

c No solutions in range

4 a ±41.8°, ±138° b 38.2°, 142°

5 60°, 75.5°, 284.5°, 300°

6 48.2°, 131.8°, 228.2°, 311.8°

7 $2\cos^2 x + \cos x - 6 = (2\cos x - 3)(\cos x + 2)$

There are no solutions to $\cos x = -2$ or to $\cos x = \frac{3}{2}$

8 a $1 - \sin^2 x = 2 - \sin x$

Rearrange to get $\sin^2 x - \sin x + 1 = 0$

b The equation has no real roots as $b^2 - 4ac < 0$

9 a $p = 1$, $q = 5$

b 72.8°, 129.0°, 252.8°, 309.0°, 432.8°, 489.0°

Challenge

1 −180°, −60°, 60°, 180°

2 0°, 90°, 180°, 270°, 360°

Mixed exercise 10

1 a −cos 57° b −sin 48° c +tan 10°

2 a 0 b $-\frac{\sqrt{2}}{2}$ c −1

d $\sqrt{3}$ e −1

3 Using $\sin^2 A = 1 - \cos^2 A$, $\sin^2 A = 1 - \left(-\sqrt{\frac{7}{11}}\right)^2 = \frac{4}{11}$.

Since angle A is obtuse, it is in the second quadrant and sin is positive, so $\sin A = \frac{2}{\sqrt{11}}$.

Then $\tan A = \frac{\sin A}{\cos A} = \frac{2}{\sqrt{11}} \times \left(-\sqrt{\frac{11}{7}}\right) = -\frac{2}{\sqrt{7}} = -\frac{2}{7}\sqrt{7}$.

4 a $-\frac{\sqrt{21}}{5}$ b $-\frac{2}{5}$

5 a $\cos^2\theta - \sin^2\theta$ b $\sin^4 3\theta$ c 1

6 a 1 b $\tan y = \frac{4 + \tan x}{2\tan x - 3}$

7 a LHS $= (1 + 2\sin\theta + \sin^2\theta) + \cos^2\theta$

$= 1 + 2\sin\theta + 1$

$= 2 + 2\sin\theta$

$= 2(1 + \sin\theta) =$ RHS

b LHS $= \cos^4\theta + \sin^2\theta$

$= (1 - \sin^2\theta)^2 + \sin^2\theta$

$= 1 - 2\sin^2\theta + \sin^4\theta + \sin^2\theta$

$= (1 - \sin^2\theta) + \sin^4\theta$

$= \cos^2\theta + \sin^4\theta =$ RHS

8 a No solutions: $-1 \leqslant \sin\theta \leqslant 1$

b 2 solutions: $\tan\theta = -1$ has two solutions in the interval.

c No solutions: $2\sin\theta + 3\cos\theta > -5$ so $2\sin\theta + 3\cos\theta + 6$ can never be equal to 0.

d No solutions: $\tan^2\theta = -1$ has no real solutions.

9 a $(4x - y)(y + 1)$ b 14.0°, 180°, 194°

10 a $3\cos 3\theta$ b 16.1, 104, 136, 224, 256, 344

11 a $2\sin 2\theta = \cos 2\theta \Rightarrow \frac{2\sin 2\theta}{\cos 2\theta} = 1$

$\Rightarrow 2\tan 2\theta = 1 \Rightarrow \tan 2\theta = 0.5$

b 13.3°, 103.3°, 193.3°, 283.3°

12 a 225°, 345°

b 22.2°, 67.8°, 202.2°, 247.8°

13 30°, 150°, 210°

14 0°, 131.8°, 228.2°, 360°

15 a Found additional solutions after dividing by three rather than before. Not applied the full interval for solutions.

b −350°, −310°, −230°, −190°, −110°, −70°, 10°, 50°, 130°, 170°, 250°, 290°

16 a

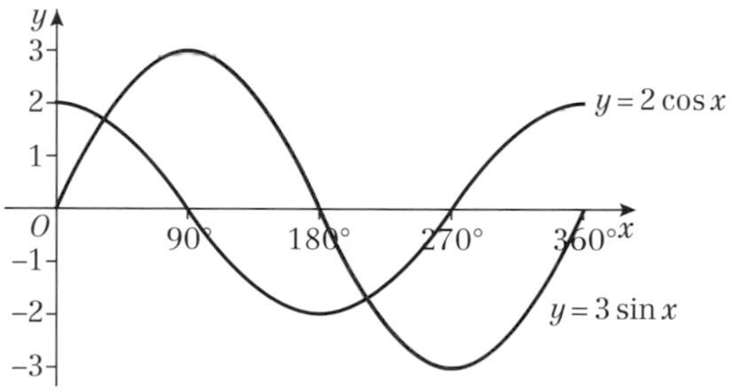

b 2 c 33.7°, 213.7°

17 a $\frac{9}{11}$ b $\frac{\sqrt{40}}{11}$

18 a Using sine rule: $\sin Q = \sin 45° \times \frac{6}{5} = \frac{\sqrt{2}}{2} \times \frac{6}{5} = \frac{3\sqrt{2}}{5}$

b $-\frac{\sqrt{7}}{5}$

19 a $3\sin^2 x - (1 - \sin^2 x) = 2$.

Rearrange to give $4\sin^2 x = 3$.

b −120°, −60°, 60°, 120°

20 −318.2°, −221.8°, 41.8°, 138.2°

Challenge

45°, 54.7°, 125.3°, 135°, 225°, 234.7°, 305.3°, 315°

Review exercise 2

1 $x + 3y - 22 = 0$

2 $x - 3y - 21 = 0$

3 4, −2.5

4 a 0.45

b $l = 0.45h$

c The model may not be valid for young people who are still growing.

5 a $y = -\frac{1}{3}x + 4$ b C is (3, 3) c 15

6 $3\sqrt{5}$
7 $(-6, 0)$
8 $(x + 3)^2 + (y - 8)^2 = 10$
9 a $(x - 3)^2 + (y + 1)^2 = 20$ ($a = 3$, $b = -1$, $r = \sqrt{20}$)
b Centre $(3, -1)$, radius $\sqrt{20}$
10 a $(3, 5)$ and $(4, 2)$
b $\sqrt{10}$
11 $0 < r < \sqrt{\frac{2}{5}}$
12 a $(x - 1)^2 + (y - 5)^2 = 58$
b $7y - 3x + 26 = 0$
13 a $AB = \sqrt{32}$; $BC = \sqrt{8}$; $AC = \sqrt{40}$; $AC^2 = AB^2 + BC^2$
b AC is a diameter of the circle.
c $(x - 5)^2 + (y - 2)^2 = 10$
14 $a = 3, b = -2, c = -8$
15 a $2(\frac{1}{2})^3 - 7(\frac{1}{2})^2 - 17(\frac{1}{2}) + 10 = 0$
b $(2x - 1)(x - 5)(x + 2)$
c

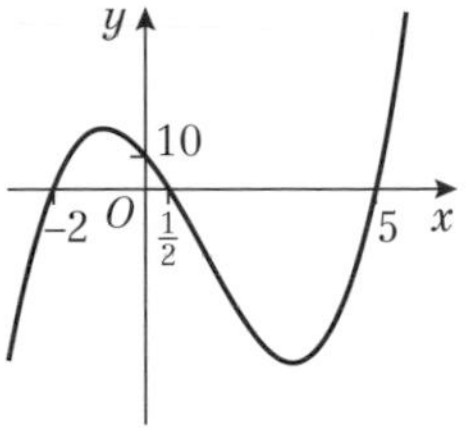

16 a 24
b $(x - 3)(3x - 2)(x + 4)$
17 a $g(3) = 3^3 - 13(3) + 12 = 0$
b $(x - 3)(x + 4)(x - 1)$
18 a $a = 0, b = 0$
b $a > 0, b > 0$
19 a $5^2 = 24 + 1$; $7^2 = 2(24) + 1$; $11^2 = 5(24) + 1$; $13^2 = 7(24) + 1$; $17^2 = 12(24) + 1$; $19^2 = 15(24) + 1$
b $3(24) + 1 = 73$ which is not a square of a prime number
20 a $(x - 5)^2 + (y - 4)^2 = 3^2$
b $\sqrt{41}$
c Sum of radii $= 3 + 3 < \sqrt{41}$ so circles do not touch
21 a $1 - 20x + 180x^2 - 960x^3 + \dots$
b 0.817
22 $a = 2, b = 19, c = 70$
23 4
24 $\sqrt{10}$ cm
25 a $\cos 60° = \frac{1}{2} = (5^2 + (2x - 3)^2 - (x + 1)^2) \div 2(5)(2x - 3)$
$5(2x - 3) = (25 + 4x^2 - 12x + 9 - x^2 - 2x - 1)$
$0 = 3x^2 - 24x + 48$
$x^2 - 8x + 16 = 0$
b 4
c 10.8 cm²
26 a 11.93 km
b 100.9°
27 a $AB = BC = 10$ cm, $AC = 6\sqrt{10}$ cm
b 143.1°
28 19.4 cm²
29 a $(x - 5)^2 + (y - 2)^2 = 25$
b 6
c $XY = \sqrt{90}$; $YZ = \sqrt{20}$; $XZ = \sqrt{98}$
$\cos XYZ = (20 + 90 - 98) \div (2 \times \sqrt{20} \times \sqrt{90})$
$\cos XYZ = 12 \div 60\sqrt{2} = \sqrt{2} \div 10$
30 a

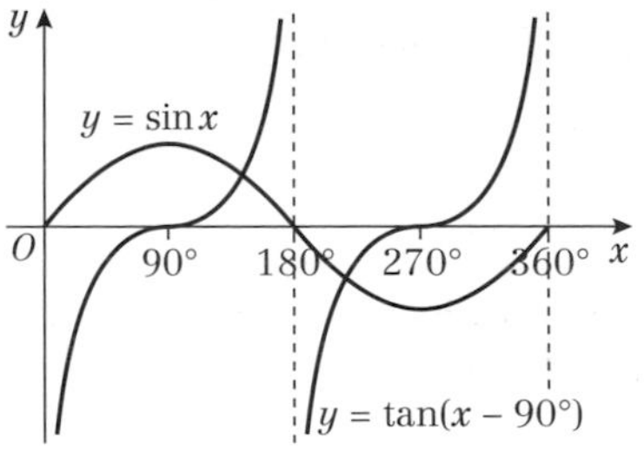

b 2
31 a $(-225, 0)$, $(-45, 0)$, $(135, 0)$ and $(315, 0)$
b $\left(0, \frac{\sqrt{2}}{2}\right)$
32 Area of triangle $= \frac{1}{2} \times s \times s \times \sin 60° = \frac{\sqrt{3}}{4}s^2$
Area of square $= s^2$
Total surface area $= 4 \times \left(\frac{\sqrt{3}}{4}s^2\right) + s^2 = (\sqrt{3} + 1)s^2$ cm²
33 a 1
b 45°, 225°
34 30°, 150°, 210°, 330°
35 90°, 150°
36 a $2(1 - \sin^2 x) = 4 - 5\sin x$
$2 - 2\sin^2 x = 4 - 5\sin x$
$2\sin^2 x - 5\sin x + 2 = 0$
b $x = 30°, 150°$
37 72.3°, 147.5°, 252.3°, 327.5°
38 0°, 78.5°, 281.5°, 360°
39 $\cos^2 x(\tan^2 x + 1)$
$= \cos^2 x\left(\frac{\sin^2 x}{\cos^2 x} + 1\right)$
$= \sin^2 x + \cos^2 x = 1$

Challenge

1 a 160
b $\left(-\frac{28}{3}, 0\right)$
2 The equation of the second circle is $(x + 4)^2 + (y - 5)^2 = 10^2$ so it has the same centre but a larger radius
3 $$\binom{n}{k} + \binom{n}{k+1} = \frac{n!}{k!(n-k)!} + \frac{n!}{(k+1)!(n-k-1)!}$$
$$= \frac{n!(k+1)}{(k+1)!(n-k)!} + \frac{n!(n-k)}{(k+1)!(n-k)!}$$
$$= \frac{n!((k+1) + (n-k))}{(k+1)!(n-k)!}$$
$$= \frac{n!(n+1)}{(k+1)!(n-k)!}$$
$$= \frac{(n+1)!}{(k+1)!(n-k)!}$$
$$= \binom{n+1}{k+1}$$
4 0°, 30°, 150°, 180°, 270°, 360°

CHAPTER 11

Prior knowledge check

1 a $\binom{4}{2}$ b $\binom{5}{-2}$ c $\binom{-1}{-3}$
2 a $\frac{7}{9}$ b $\frac{2}{9}$ c $\frac{7}{2}$
3 a 123.2° b 13.6 c 5.3 d 21.4°

Exercise 11A

1 **a**

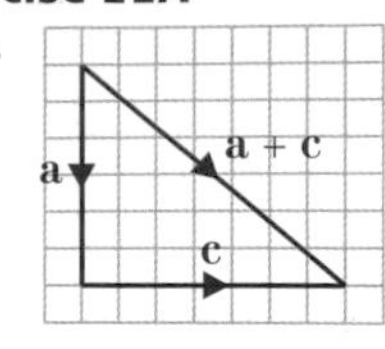

b

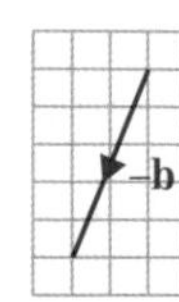

c

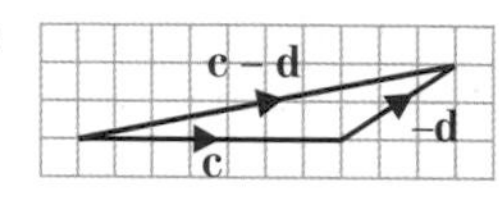

d

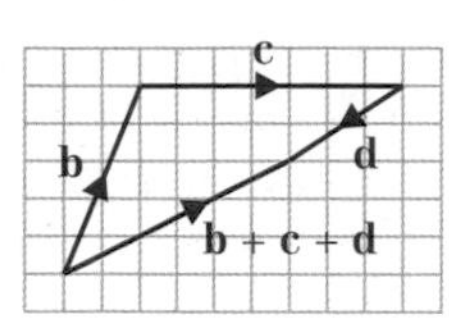

e

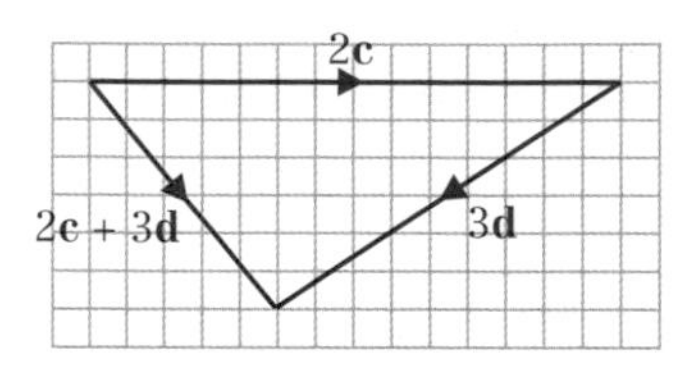

f

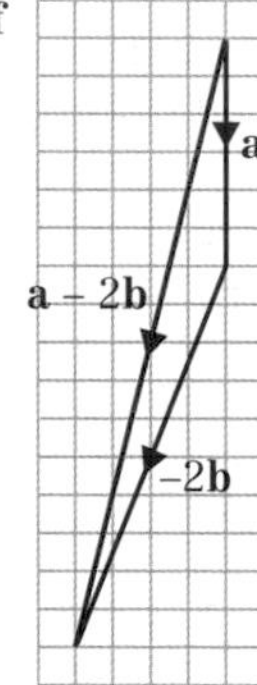

g

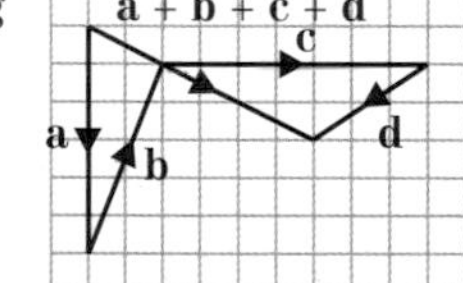

2 **a** 2**b** **b** **d** **c** **b**
d 2**b** **e** **d** + **b** **f** **d** + **b**
g –2**d** **h** –**b** **i** 2**d** + **b**
j –**b** + 2**d** **k** –**b** + **d** **l** –**d** – **b**

3 **a** 2**m** **b** 2**p** **c** **m**
d **m** **e** **p** + **m** **f** **p** + **m**
g **p** + 2**m** **h** **p** – **m** **i** –**m** –**p**
j –2**m** + **p** **k** –2**p** + **m** **l** –**m** – 2**p**

4 **a** **d** – **a** **b** **a** + **b** + **c**
c **a** + **b** – **d** **d** **a** + **b** + **c** – **d**

5 **a** 2**a** + 2**b** **b** **a** + **b** **c** **b** – **a**

6 **a** **b** **b** **b** – 3**a** **c** **a** – **b**
d 2**a** – **b**

7 **a** $\overrightarrow{OB} = \mathbf{a} + \mathbf{b}$ **b** $\overrightarrow{OP} = \frac{5}{8}(\mathbf{a} + \mathbf{b})$ **c** $\overrightarrow{AP} = \frac{5}{8}\mathbf{b} - \frac{3}{8}\mathbf{a}$

8 **a** Yes ($\lambda = 2$) **b** Yes ($\lambda = 4$) **c** No
d Yes ($\lambda = -1$) **e** Yes ($\lambda = -3$) **f** No

9 **a** **i** **b** – **a** **ii** $\frac{1}{2}\mathbf{a}$ **iii** $\frac{1}{2}\mathbf{b}$ **iv** $\frac{1}{2}\mathbf{b} - \frac{1}{2}\mathbf{a}$
b $\overrightarrow{BC} = \mathbf{b} - \mathbf{a}$, $\overrightarrow{PQ} = \frac{1}{2}(\mathbf{b} - \mathbf{a})$ so PQ is parallel to BC.

10 **a** **i** 2**b** **ii** **a** – **b**
b $\overrightarrow{AB} = 2\mathbf{b}$, $\overrightarrow{OC} = 3\mathbf{b}$ so AB is parallel to OC.

11 1.2

Exercise 11B

1 $\mathbf{v}_1$: $8\mathbf{i}$, $\begin{pmatrix}8\\0\end{pmatrix}$ $\mathbf{v}_2$: $9\mathbf{i} + 3\mathbf{j}$, $\begin{pmatrix}9\\3\end{pmatrix}$ $\mathbf{v}_3$: $-4\mathbf{i} + 2\mathbf{j}$, $\begin{pmatrix}-4\\2\end{pmatrix}$
$\mathbf{v}_4$: $3\mathbf{i} + 5\mathbf{j}$, $\begin{pmatrix}3\\5\end{pmatrix}$ $\mathbf{v}_5$: $-3\mathbf{i} - 2\mathbf{j}$, $\begin{pmatrix}-3\\-2\end{pmatrix}$ $\mathbf{v}_6$: $-5\mathbf{j}$, $\begin{pmatrix}0\\-5\end{pmatrix}$

2 **a** $8\mathbf{i} + 12\mathbf{j}$ **b** $\mathbf{i} + 1.5\mathbf{j}$ **c** $-4\mathbf{i} + \mathbf{j}$
d $10\mathbf{i} + \mathbf{j}$ **e** $-2\mathbf{i} + 11\mathbf{j}$ **f** $-2\mathbf{i} - 10\mathbf{j}$
g $14\mathbf{i} - 7\mathbf{j}$ **h** $-8\mathbf{i} + 9\mathbf{j}$

3 **a** $\begin{pmatrix}45\\35\end{pmatrix}$ **b** $\begin{pmatrix}4\\0.5\end{pmatrix}$ **c** $\begin{pmatrix}12\\3\end{pmatrix}$
d $\begin{pmatrix}-1\\16\end{pmatrix}$ **e** $\begin{pmatrix}-21\\-29\end{pmatrix}$ **f** $\begin{pmatrix}10\\2\end{pmatrix}$

4 **a** $\lambda = 5$ **b** $\mu = -\frac{3}{2}$

5 **a** $\lambda = \frac{1}{3}$ **b** $\mu = -1$
c $s = -1$ **d** $t = -\frac{1}{17}$

6 **i** – **j**

7 **a** $\overrightarrow{AC} = 5\mathbf{i} - 4\mathbf{j} = \begin{pmatrix}5\\-4\end{pmatrix}$ **b** $\overrightarrow{AP} = 3\mathbf{i} - \frac{12}{5}\mathbf{j} = \begin{pmatrix}3\\-\frac{12}{5}\end{pmatrix}$
c $\overrightarrow{OP} = 5\mathbf{i} + \frac{8}{5}\mathbf{j} = \begin{pmatrix}5\\\frac{8}{5}\end{pmatrix}$

8 $j = 4$, $k = 11$

9 $p = 3$, $q = 2$

10 **a** $p = 5$ **b** $8\mathbf{i} - 12\mathbf{j}$

Exercise 11C

1 **a** 5 **b** 10 **c** 13
d 4.47 (3 s.f.) **e** 5.83 (3 s.f.) **f** 8.06 (3 s.f.)
g 5.83 (3 s.f.) **h** 4.12 (3 s.f.)

2 **a** $\sqrt{26}$ **b** $5\sqrt{2}$ **c** $\sqrt{101}$

3 **a** $\frac{1}{5}\begin{pmatrix}4\\3\end{pmatrix}$ **b** $\frac{1}{13}\begin{pmatrix}5\\-12\end{pmatrix}$
c $\frac{1}{25}\begin{pmatrix}-7\\24\end{pmatrix}$ **d** $\frac{1}{\sqrt{10}}\begin{pmatrix}1\\-3\end{pmatrix}$

4 **a** 53.1° above **b** 53.1° below
c 67.4° above **d** 63.4° above

5 **a** 149° to the right **b** 29.7° to the right
c 31.0° to the left **d** 104° to the left

6 **a** $\frac{15\sqrt{2}}{2}\mathbf{i} + \frac{15\sqrt{2}}{2}\mathbf{j}$, $\begin{pmatrix}\frac{15\sqrt{2}}{2}\\\frac{15\sqrt{2}}{2}\end{pmatrix}$ **b** $7.52\mathbf{i} + 2.74\mathbf{j}$, $\begin{pmatrix}7.52\\2.74\end{pmatrix}$
c $18.1\mathbf{i} - 8.45\mathbf{j}$, $\begin{pmatrix}18.1\\-8.45\end{pmatrix}$ **d** $\frac{5\sqrt{3}}{2}\mathbf{i} - 2.5\mathbf{j}$, $\begin{pmatrix}\frac{5\sqrt{3}}{2}\\-2.5\end{pmatrix}$

7 **a** $|3\mathbf{i} + 4\mathbf{j}| = 5$, 53.1° above

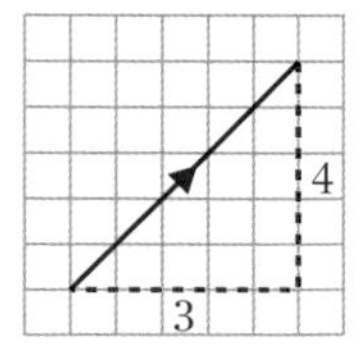

b $|2\mathbf{i} - \mathbf{j}| = \sqrt{5}$, 26.6° below

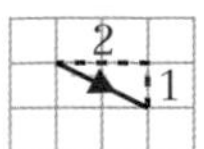

c $|-5\mathbf{i} + 2\mathbf{j}| = \sqrt{29}$, 158.2° above

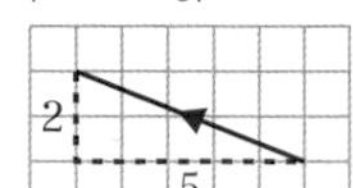

8 $k = \pm 6$

9 $p = \pm 8$, $q = 6$

Online Full worked solutions are available in SolutionBank.

10 a $36.9°$ b $33.7°$ c $70.6°$
11 a $67.2°$ b 19.0

Challenge
Possible solution:

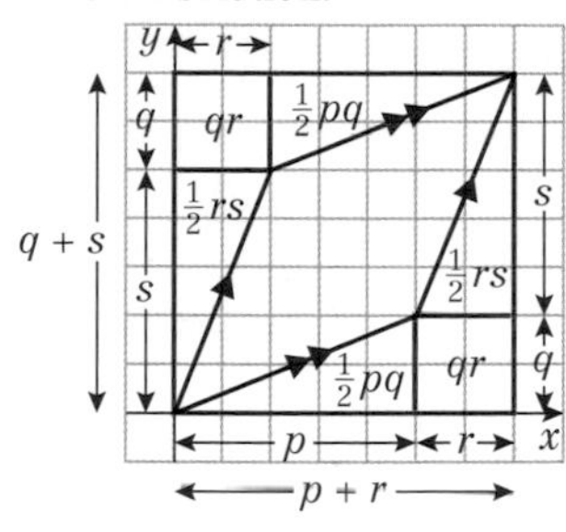

Area of parallelogram = area of large rectangle − 2(area of small rectangle) − 2 (area triangle 1) − 2(area triangle 2)

Area of parallelogram $= (p + r)(q + s) - 2qr - 2(\frac{1}{2}pq) - 2(\frac{1}{2}rs) = ps - qr$

Exercise 11D

1 a i $\overrightarrow{OA} = 3\mathbf{i} - \mathbf{j}$, $\overrightarrow{OB} = 4\mathbf{i} + 5\mathbf{j}$, $\overrightarrow{OC} = -2\mathbf{i} + 6\mathbf{j}$
ii $\mathbf{i} + 6\mathbf{j}$ iii $-5\mathbf{i} + 7\mathbf{j}$
b i $\sqrt{40} = 2\sqrt{10}$ ii $\sqrt{37}$ iii $\sqrt{74}$

2 a $-\mathbf{i} + 5\mathbf{j}$ or $\begin{pmatrix}-1\\5\end{pmatrix}$
b i 5 ii $\sqrt{13}$ iii $\sqrt{26}$

3 a $-\mathbf{i} - 9\mathbf{j}$ or $\begin{pmatrix}-1\\-9\end{pmatrix}$
b i $\sqrt{82}$ ii 5 iii $\sqrt{61}$

4 a $-2\mathbf{a} + 2\mathbf{b}$ b $-3\mathbf{a} + 2\mathbf{b}$ c $-2\mathbf{a} + \mathbf{b}$

5 $\begin{pmatrix}7\\9\end{pmatrix}$ or $\begin{pmatrix}9\\3\end{pmatrix}$

6 a $2\mathbf{i} + 8\mathbf{j}$ b $2\sqrt{17}$

7 $\frac{3\sqrt{5}}{5}$

Challenge
$\overrightarrow{OB} = 2\mathbf{i} + 3\mathbf{j}$ or $\overrightarrow{OB} = \frac{46}{13}\mathbf{i} + \frac{9}{13}\mathbf{j}$

Exercise 11E

1 $\overrightarrow{XY} = \mathbf{b} - \mathbf{a}$ and $\overrightarrow{YZ} = \mathbf{c} - \mathbf{b}$, so $\mathbf{b} - \mathbf{a} = \mathbf{c} - \mathbf{b}$.
Hence $\mathbf{a} + \mathbf{c} = 2\mathbf{b}$.

2 a i $2\mathbf{r}$ ii $\mathbf{r}$
b Sides of triangle OAB are twice the length of sides of triangle PAQ and angle A is common to both SAS.

3 a $\frac{2}{3}\mathbf{a} + \frac{1}{3}\mathbf{b}$
b $\overrightarrow{AN} = \frac{1}{3}(\mathbf{b} - \mathbf{a})$, $\overrightarrow{AB} = \mathbf{b} - \mathbf{a}$, $\overrightarrow{NB} = \frac{2}{3}(\mathbf{b} - \mathbf{a})$
so $AN : NB = 1 : 2$.

4 a $\frac{3}{5}\mathbf{a} + \frac{2}{5}\mathbf{c}$
b $\overrightarrow{AP} = -\mathbf{a} + \frac{3}{5}\mathbf{a} + \frac{2}{5}\mathbf{c} = \frac{2}{5}(\mathbf{c} - \mathbf{a})$,
$\overrightarrow{PC} = \mathbf{c} - (\frac{3}{5}\mathbf{a} + \frac{2}{5}\mathbf{c}) = \frac{3}{5}(\mathbf{c} - \mathbf{a})$ so $AP : PC = 2 : 3$

5 a $\sqrt{26}$ b $2\sqrt{2}$ c $3\sqrt{2}$
d $\angle BAC = 56°$, $\angle ABC = 34°$, $\angle ACB = 90°$

6 a $\overrightarrow{OR} = \mathbf{a} + \frac{1}{3}(\mathbf{b} - \mathbf{a}) = \frac{2}{3}\mathbf{a} + \frac{1}{3}\mathbf{b}$,
$\overrightarrow{OS} = 3\overrightarrow{OR} = 3(\frac{2}{3}\mathbf{a} + \frac{1}{3}\mathbf{b}) = 2\mathbf{a} + \mathbf{b}$
b $\overrightarrow{TP} = \overrightarrow{TO} + \overrightarrow{OP} = \mathbf{a} + \mathbf{b}$, $\overrightarrow{PS} = \overrightarrow{PO} + \overrightarrow{OS} = -\mathbf{a} + 2\mathbf{a} + \mathbf{b} = \mathbf{a} + \mathbf{b}$
$\overrightarrow{TP}$ is parallel (and equal) to $\overrightarrow{PS}$ and they have a point, P, in common so T, P and S lie on a straight line.

Challenge:
a $\overrightarrow{PR} = \mathbf{b} - \mathbf{a}$, $\overrightarrow{PX} = j(\mathbf{b} - \mathbf{a}) = -j\mathbf{a} + j\mathbf{b}$
b $\overrightarrow{ON} = \mathbf{a} + \frac{1}{2}\mathbf{b}$, $\overrightarrow{PX} = -\mathbf{a} + k(\mathbf{a} + \frac{1}{2}\mathbf{b}) = (k - 1)\mathbf{a} + \frac{1}{2}k\mathbf{b}$
c Coefficients of $\mathbf{a}$ and $\mathbf{b}$ must be the same in both expressions for $\overrightarrow{PX}$
Coefficients of $\mathbf{a}$: $k - 1 = -j$; Coefficients of $\mathbf{b}$: $j = \frac{1}{2}k$
d Solving simultaneously gives $j = \frac{1}{3}$ and $k = \frac{2}{3}$
e $\overrightarrow{PX} = \frac{1}{3}\overrightarrow{PR}$.
By symmetry, $\overrightarrow{PX} = \overrightarrow{YR} = \overrightarrow{XY}$, so ON and OM divide PR into 3 equal parts.

Exercise 11F

1 a $5\,\text{m s}^{-1}$ b $25\,\text{km h}^{-1}$
c $5.39\,\text{m s}^{-1}$ d $8.06\,\text{cm s}^{-1}$
2 a 50 km b 51.0 m
c 4.74 km d 967 cm
3 a $5\,\text{m s}^{-1}$, 75 m b $5.39\,\text{m s}^{-1}$, 16.2 m
c $5.39\,\text{km h}^{-1}$, 16.2 km d $13\,\text{km h}^{-1}$, 6.5 km
4 $(2.8\mathbf{i} - 1.6\mathbf{j})\,\text{m s}^{-2}$
5 a $54.5°$ b $0.3\sqrt{74} = 2.58\,\text{N}$
6 a $26.6°$ below $\mathbf{i}$
b $\mathbf{R} = (3 + p)\mathbf{i} + (q - 4)\mathbf{j}$, $3 + p = 2\lambda$ and $q - 4 = -\lambda \Rightarrow \lambda = 4 - q$
$3 + p = 2(4 - q) \Rightarrow 3 + p = 8 - 2q$ so $p + 2q = 5$
c $|\mathbf{R}| = 2\sqrt{5}$ newtons
7 a $10\mathbf{i} - 100\mathbf{j}$ metres b $109.4°$
c $1700\,\text{m}^2$
8 a $\sqrt{41} = 64.0\,\text{km}$ b $321.3°$
c $\overrightarrow{AB} = 4\mathbf{i} - 5\mathbf{j}$, $\mathbf{v} = 2(4\mathbf{i} - 5\mathbf{j})$ so the boat is travelling directly towards the buoy.
d $2\sqrt{41} = 12.8\,\text{km h}^{-1}$ e 30 minutes

Mixed exercise 11

1 a $2\sqrt{10}$ newtons b $18°$ to the left
2 a $108°$ b $9.49\,\text{km h}^{-1}$
3 a $9.85\,\text{m s}^{-1}$ b 59.1 m
c The model ignores friction and air resistance.
The model will become less accurate as t increases.
4 a $\mathbf{b} - \frac{3}{5}\mathbf{a}$ b $\mathbf{b} - 4\mathbf{a}$ c $\frac{8}{5}\mathbf{a} - \mathbf{b}$ d $3\mathbf{a} - \mathbf{b}$
5 1.25
6 a $\begin{pmatrix}12\\-1\end{pmatrix}$ b $\begin{pmatrix}-18\\5\end{pmatrix}$ c $\begin{pmatrix}49\\13\end{pmatrix}$
7 a $3\mathbf{i} - 2\mathbf{j}$ b $32.5°$ c 10.5
8 a $p = -1.5$ b $\mathbf{i} - 1.5\mathbf{j}$
9 a i $\frac{1}{17}(8\mathbf{i} + 15\mathbf{j})$ ii $61.9°$ above
b i $\frac{1}{25}(24\mathbf{i} - 7\mathbf{j})$ ii $16.3°$ below
c i $\frac{1}{41}(-9\mathbf{i} + 40\mathbf{j})$ ii $102.7°$ above
d i $\frac{1}{\sqrt{13}}(3\mathbf{i} - 2\mathbf{j})$ ii $33.7°$ below
10 $p = 8.6$, $q = 12.3$
11 ± 6
12 a $\frac{3}{5}\mathbf{a} + \frac{2}{5}\mathbf{b}$ b $\frac{2}{5}\mathbf{b}$
c $\overrightarrow{AB} = \mathbf{b} - \mathbf{a}$, $\overrightarrow{AN} = \frac{2}{5}(\mathbf{b} - \mathbf{a})$ so $AN : NB = 2 : 3$

13 **a** 18.4° below

b $\mathbf{R} = (4 + p)\mathbf{i} + (-5 + q)\mathbf{j}$, $4 + p = 3\lambda$ and $-5 + q = -\lambda$
$4 + p = 3(q - 5)$ so $p + 3q = 11$

c $2\sqrt{10} = 6.32$ newtons

14 $\dfrac{\sqrt{193}}{2} = 6.95\,\text{m s}^{-1}$

Challenge

$\overrightarrow{OB} = \frac{3}{2}\mathbf{i} + \frac{5}{2}\mathbf{j}$ or $\frac{99}{34}\mathbf{i} + \frac{5}{34}\mathbf{j}$

CHAPTER 12

Prior knowledge check

1 **a** 5 **b** $-\frac{2}{3}$ **c** $\frac{1}{3}$

2 **a** x^{10} **b** $x^{\frac{2}{3}}$ **c** x^{-1} **d** $x^{\frac{3}{4}}$

3 **a** $y = \frac{1}{2}x - 2$ **b** $y = -\frac{1}{2}x + 8\frac{1}{2}$ **c** $y = -\frac{1}{4}x + 7\frac{1}{2}$

4 $y = -\frac{1}{2}x$

Exercise 12A

1 **a**

x-coordinate	−1	0	1	2	3
Estimate for gradient of curve	−4	−2	0	2	4

b Gradient = $2p - 2$ **c** 1

2 **a** $\sqrt{1 - 0.6^2} = \sqrt{0.64} = 0.8$

b Gradient = −0.75

c **i** −1.21 (3 s.f.) **ii** −1 **iii** −0.859 (3 s.f.)

d As other point moves closer to A, gradient tends to −0.75.

3 **a** **i** 7 **ii** 6.5 **iii** 6.1 **iv** 6.01 **v** $h + 6$

b Gradient of tangent = 6

4 **a** **i** 9 **ii** 8.5 **iii** 8.1 **iv** 8.01 **v** $8 + h$

b Gradient of tangent = 8

Exercise 12B

1 **a** $$f'(2) = \lim_{h\to0}\frac{f(2 + h) - f(2)}{h} = \lim_{h\to0}\frac{(2 + h)^2 - 2^2}{h} = \lim_{h\to0}\frac{4h + h^2}{h} = \lim_{h\to0}(4 + h) = 4$$

b $$f'(-3) = \lim_{h\to0}\frac{f(-3 + h) - f(-3)}{h} = \lim_{h\to0}\frac{(-3 + h)^2 - 3^2}{h} = \lim_{h\to0}\frac{-6h + h^2}{h} = \lim_{h\to0}(-6 + h) = -6$$

c $$f'(0) = \lim_{h\to0}\frac{f(0 + h) - f(0)}{h} = \lim_{h\to0}\frac{h^2 - 0^2}{h} = \lim_{h\to0}h = 0$$

d $$f'(50) = \lim_{h\to0}\frac{f(50 + h) - f(50)}{h} = \lim_{h\to0}\frac{(50 + h)^2 - 50^2}{h} = \lim_{h\to0}\frac{100h + h^2}{h} = \lim_{h\to0}(100 + h) = 100$$

2 **a** $$f'(x) = \lim_{h\to0}\frac{f(x + h) - f(x)}{h} = \lim_{h\to0}\frac{(x + h)^2 - x^2}{h} = \lim_{h\to0}\frac{2xh + h^2}{h} = \lim_{h\to0}(2x + h)$$

b As $h \to 0$, $f'(x) = \lim_{h\to0}(2x + h) = 2x$

3 **a** $$g = \lim_{h\to0}\frac{(-2 + h)^3 - (-2)^3}{h} = \lim_{h\to0}\frac{-8 + 3(-2)^2h + 3(-2)h^2 + h^3 + 8}{h} = \lim_{h\to0}\frac{12h - 6h^2 + h^3}{h} = \lim_{h\to0}(12 - 6h + h^2)$$

b $g = 12$

4 **a** $$\text{Gradient of } AB = \frac{(-1 + h)^3 - 5(-1 + h) - 4}{(-1 + h) - (-1)} = \frac{-1 + 3h - 3h^2 + h^3 + 5 - 5h - 4}{h} = \frac{h^3 - 3h^2 - 2h}{h} = h^2 - 3h - 2$$

b gradient = −2

5 $$\frac{dy}{dx} = \lim_{h\to0}\frac{6(x + h) - 6x}{h} = \lim_{h\to0}\frac{6h}{h} = 6$$

6 $$\frac{dy}{dx} = \lim_{h\to0}\frac{4(x + h)^2 - 4x^2}{h} = \lim_{h\to0}\frac{8xh + 4h^2}{h} = \lim_{h\to0}(8x + 4h) = 8x$$

7 $$\frac{dy}{dx} = \lim_{h\to0}\frac{a(x + h)^2 - ax^2}{h} = \lim_{h\to0}\frac{(a - a)x^2 + 2axh + ah^2}{h} = \lim_{h\to0}\frac{2axh + ah^2}{h} = \lim_{h\to0}(2ax + ah) = 2ax$$

Challenge

a $$f'(x) = \lim_{h\to0}\frac{\frac{1}{x + h} - \frac{1}{x}}{h} = \lim_{h\to0}\frac{x - (x + h)}{xh(x + h)} = \lim_{h\to0}\frac{-1}{x(x + h)} = \lim_{h\to0}\frac{-1}{x^2 + xh}$$

b $$f'(x) = \lim_{h\to0}\frac{-1}{x(x + h)} = \frac{-1}{x^2 + xh} = \frac{-1}{x^2 + 0} = -\frac{1}{x^2}$$

Exercise 12C

1 **a** $7x^6$ **b** $8x^7$ **c** $4x^3$ **d** $\frac{1}{3}x^{-\frac{2}{3}}$
e $\frac{1}{4}x^{-\frac{3}{4}}$ **f** $\frac{1}{3}x^{-\frac{2}{3}}$ **g** $-3x^{-4}$ **h** $-4x^{-5}$
i $-2x^{-3}$ **j** $-5x^{-6}$ **k** $-\frac{1}{2}x^{-\frac{3}{2}}$ **l** $-\frac{1}{3}x^{-\frac{4}{3}}$
m $9x^8$ **n** $5x^4$ **o** $3x^2$ **p** $-2x^{-3}$
q 1 **r** $3x^2$

2 **a** $6x$ **b** $54x^8$ **c** $2x^3$ **d** $5x^{-\frac{3}{4}}$
e $\frac{15}{2}x^{\frac{1}{4}}$ **f** $-10x^{-2}$ **g** $6x^2$ **h** $-\dfrac{1}{2x^5}$
i $x^{-\frac{3}{2}}$ **j** $\frac{15}{2}\sqrt{x}$

3 **a** $\frac{3}{4}$ **b** $\frac{1}{2}$ **c** 3 **d** 2

4 $\dfrac{dy}{dx} = \dfrac{3}{2}\sqrt{\dfrac{x}{2}}$

Exercise 12D

1 **a** $4x - 6$ **b** $x + 12$ **c** $8x$ **d** $16x + 7$
e $4 - 10x$

2 **a** 12 **b** 6 **c** 7 **d** $2\frac{1}{2}$
e −2 **f** 4

3 4, 0

4 (−1, −8)

5 1, −1

6 6, −4

7 a, b

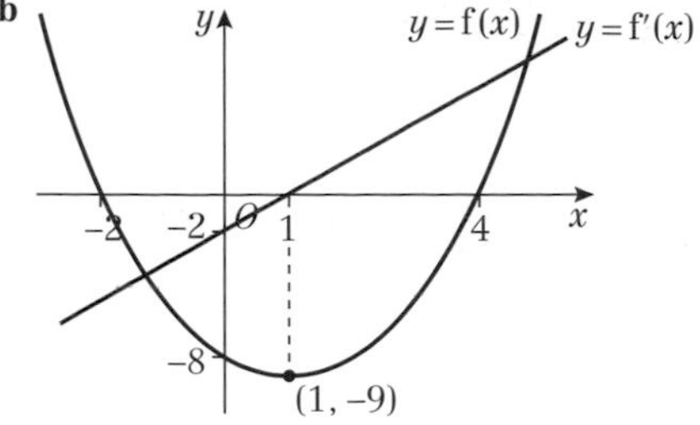

c At the turning point, the gradient of $y = f(x)$ is zero, i.e. $f'(x) = 0$.

Exercise 12E

1 **a** $4x^3 - x^{-2}$ **b** $10x^4 - 6x^{-3}$ **c** $9x^{\frac{1}{2}} - x^{-\frac{3}{2}}$

2 **a** 0 **b** $11\frac{1}{2}$

3 **a** $(2\frac{1}{2}, -6\frac{1}{4})$ **b** $(4, -4)$ and $(2, 0)$
c $(16, -31)$ **d** $(\frac{1}{2}, 4)$ $(-\frac{1}{2}, -4)$

4 **a** $x^{-\frac{1}{2}}$ **b** $-6x^{-3}$ **c** $-x^{-4}$
d $\frac{4}{3}x^3 - 2x^2$ **e** $\frac{1}{2}x^{-\frac{1}{2}} - 6x^{-4}$ **f** $\frac{1}{3}x^{-\frac{2}{3}} - \frac{1}{2}x^{-2}$
g $-3x^{-2}$ **h** $3 + 6x^{-2}$ **i** $5x^{\frac{3}{2}} + \frac{3}{2}x^{-\frac{1}{2}}$
j $3x^2 - 2x + 2$ **k** $12x^3 + 18x^2$ **l** $24x - 8 + 2x^{-2}$

5 **a** 1 **b** $\frac{2}{9}$ **c** -4 **d** 4

6 $-\frac{3}{4}\sqrt{2}$

7 **a** $512 - 2304x + 4608x^2$
b $f'(x) \approx \frac{d}{dx}(512 - 2304x + 4608x^2)$
$= -2304 + 2 \times 4608x$
$= 9216x - 2304$

Exercise 12F

1 **a** $y + 3x - 6 = 0$ **b** $4y - 3x - 4 = 0$
c $3y - 2x - 18 = 0$ **d** $y = x$
e $y = 12x + 14$ **f** $y = 16x - 22$

2 **a** $7y + x - 48 = 0$ **b** $17y + 2x - 212 = 0$

3 $(1\frac{2}{9}, 1\frac{8}{9})$

4 $y = -x$, $4y + x - 9 = 0$; $(-3, 3)$

5 $y = -8x + 10$, $8y - x - 145 = 0$

6 $\left(-\frac{3}{4}, \frac{9}{8}\right)$

Challenge
L has equation $y = 12x - 8$.

Exercise 12G

1 **a** $x \geqslant -\frac{4}{3}$ **b** $x \leqslant \frac{2}{3}$ **c** $x \leqslant -2$
d $x \leqslant 2, x \geqslant 3$ **e** $x \in \mathbb{R}$ **f** $x \in \mathbb{R}$
g $x \geqslant 0$ **h** $x \geqslant 6$

2 **a** $x \leqslant 4.5$ **b** $x \geqslant 2.5$
c $x \geqslant -1$ **d** $-1 \leqslant x \leqslant 2$
e $-3 \leqslant x \leqslant 3$ **f** $-5 \leqslant x < 0, 0 < x \leqslant 5$
g $0 < x \leqslant 9$ **h** $-2 \leqslant x \leqslant 0$

3 $f'(x) = -6x^2 - 3$
$x^2 \geqslant 0$ for all $x \in \mathbb{R}$, so $-6x^2 - 3 \leqslant 0$ for all $x \in \mathbb{R}$.
$\therefore f(x)$ is decreasing for all $x \in \mathbb{R}$.

4 **a** Any $p \geqslant 2$
b No. Can be any $p \geqslant 2$.

Exercise 12H

1 **a** $24x + 3$, 24
b $15 - 3x^{-2}$, $6x^{-3}$
c $\frac{9}{2}x^{-\frac{1}{2}} + 6x^{-3}$, $-\frac{9}{4}x^{-\frac{3}{2}} - 18x^{-4}$
d $30x + 2$, 30
e $-3x^{-2} - 16x^{-3}$, $6x^{-3} + 48x^{-4}$

2 Acceleration $= \frac{3}{4}t^{-\frac{1}{2}} + \frac{3}{2}t^{-\frac{5}{2}}$

3 $\frac{3}{2}$

4 $-\frac{1}{2}$

Exercise 12I

1 **a** -28 **b** -17 **c** $-\frac{1}{5}$

2 **a** 10 **b** 4 **c** 12.25

3 **a** $\left(-\frac{3}{4}, -\frac{9}{4}\right)$ minimum
b $\left(\frac{1}{2}, 9\frac{1}{4}\right)$ maximum
c $\left(-\frac{1}{3}, 1\frac{5}{27}\right)$ maximum, $(1, 0)$ minimum
d $(3, -18)$ minimum, $\left(-\frac{1}{3}, \frac{14}{27}\right)$ maximum
e $(1, 2)$ minimum, $(-1, -2)$ maximum
f $(3, 27)$ minimum
g $\left(\frac{9}{4}, -\frac{9}{4}\right)$ minimum
h $(2, -4\sqrt{2})$ minimum
i $(\sqrt{6}, -36)$ minimum, $(-\sqrt{6}, -36)$ minimum, $(0, 0)$ maximum

4 **a**

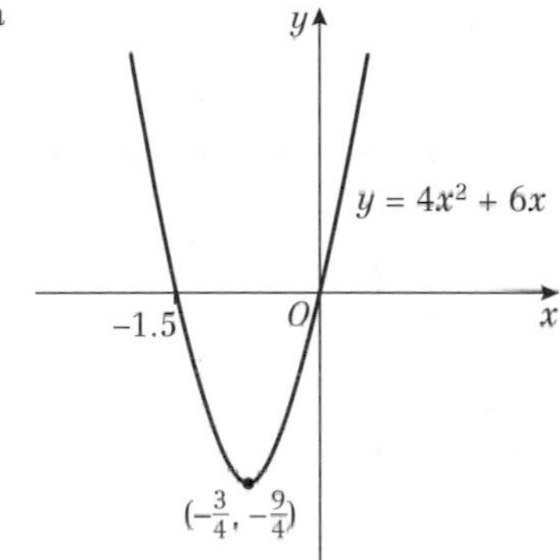

b

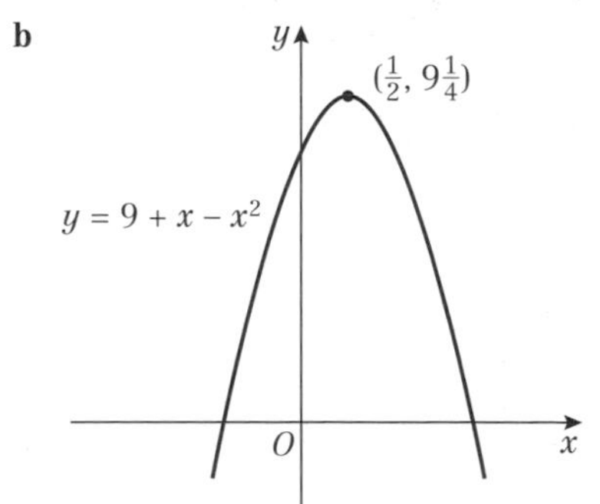

c

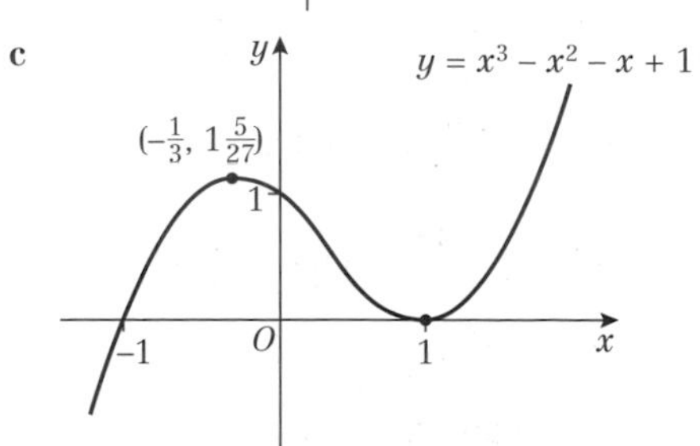

d

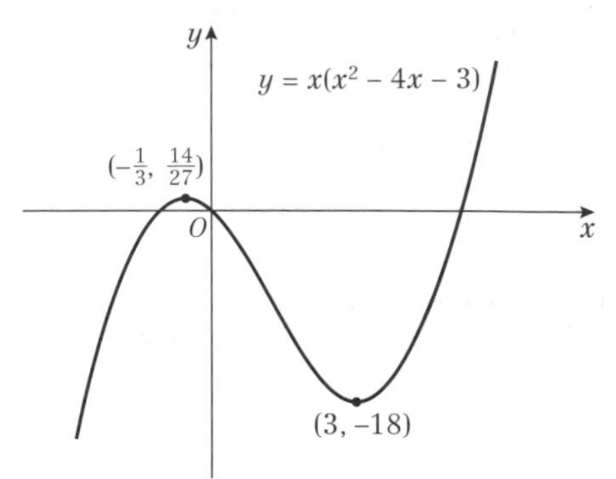

5 (1, 1) inflection (gradient is positive either side of point)

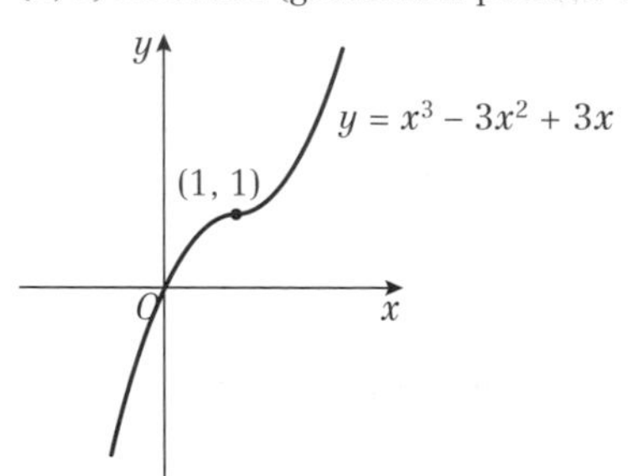

6 Maximum value is 27; $f(x) \leqslant 27$

7 **a** (1, −3): minimum, (−3, −35): minimum, $\left(-\frac{1}{4}, \frac{357}{256}\right)$: maximum

b

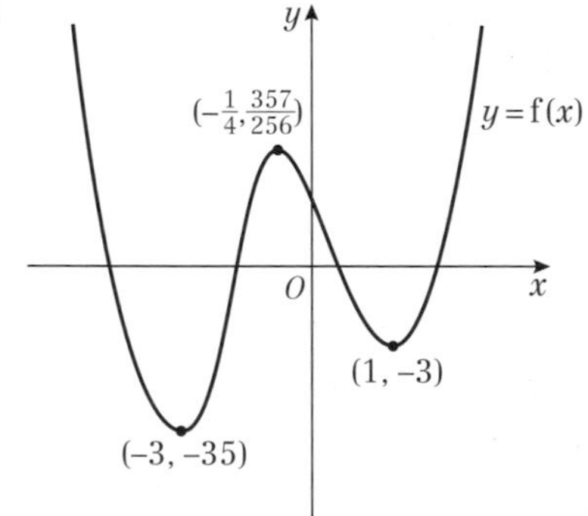

Exercise 12J

1 **a**

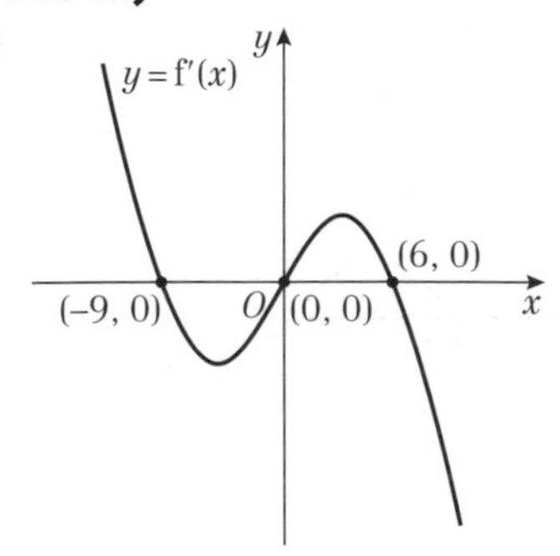

b

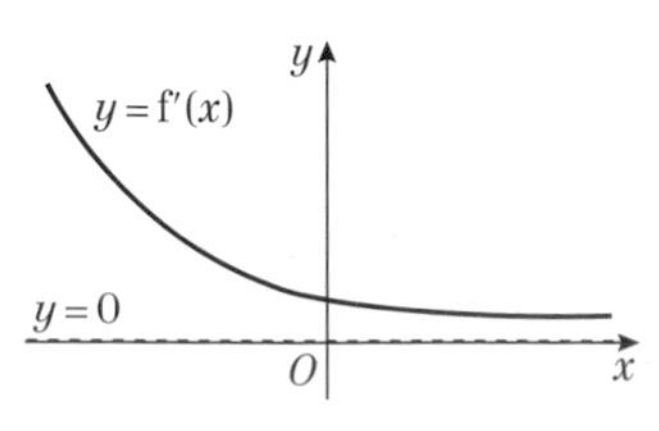

c

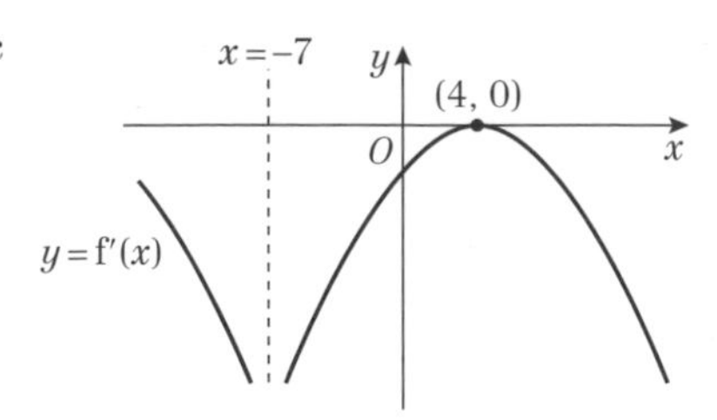

d

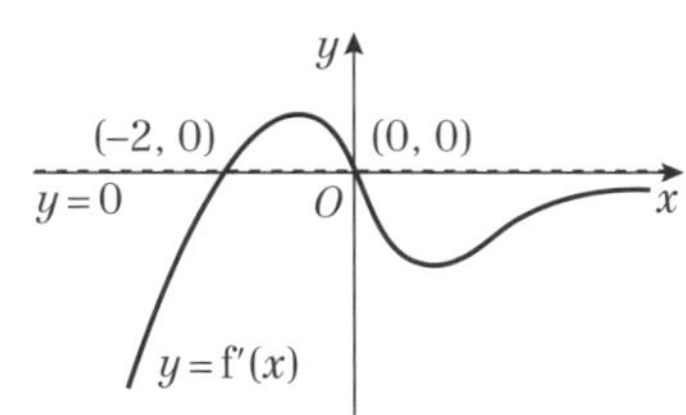

e

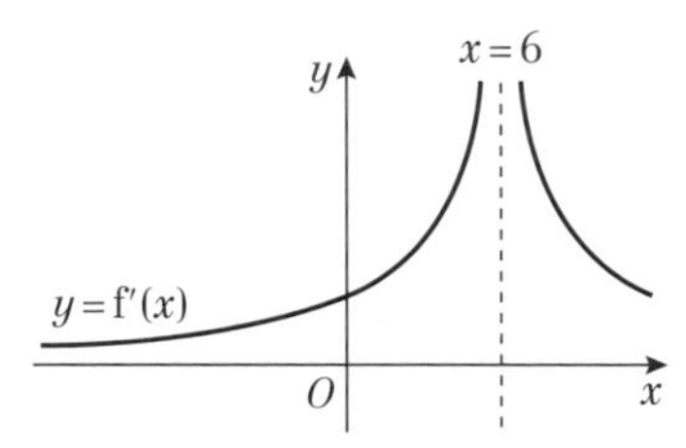

f

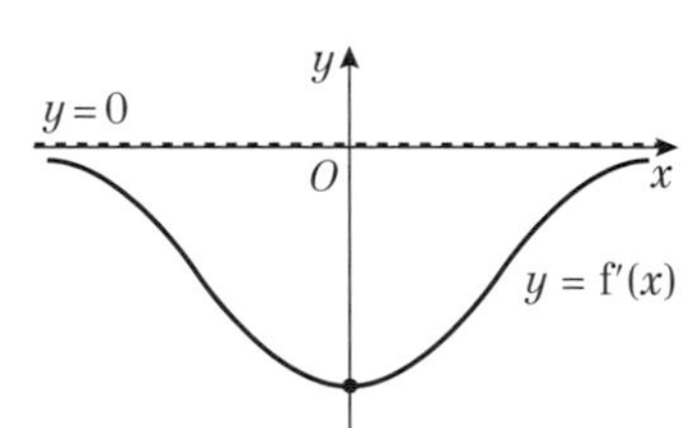

2 **a** $f(x) = x^3 - 7x^2 + 8x + 16$
$f'(x) = 3x^2 - 14x + 8 = (3x - 2)(x - 4)$

b

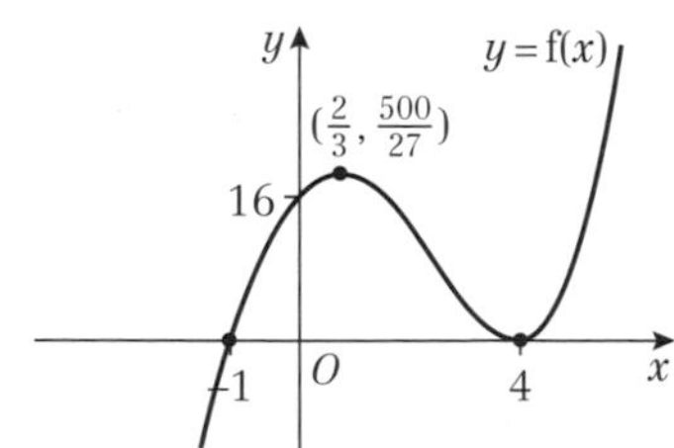

c (4, 0), $\left(\frac{2}{3}, 0\right)$ and (0, 8)

d

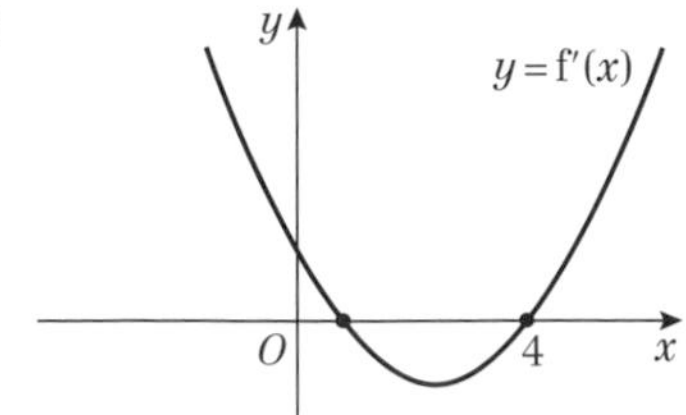

Exercise 12K

1 $2t - 3$ 2 2π 3 $-\frac{4}{3}$

4 $48\pi\,\text{cm}^3$ per cm 5 $18\,\text{m}\,\text{s}^{-1}$

6 a Let x = width of garden.
$x + 2y = 80 \Rightarrow x = 80 - 2y$
Area $A = xy = y(80 - 2y)$

b $20\,\text{m} \times 40\,\text{m}$, $800\,\text{m}^2$

7 a $2\pi r^2 + 2\pi rh = 600\pi \Rightarrow h = \dfrac{300 - r^2}{r}$
$V = \pi r^2 h = \pi r(300 - r^2) = 300\pi r - \pi r^3$

b $2000\pi\,\text{cm}^3$

8 a Let θ = angle of sector.
$\pi r^2 \times \dfrac{\theta}{360} = 100 \Rightarrow \theta = \dfrac{36\,000}{\pi r^2}$
$P = 2r + 2\pi r \times \dfrac{\theta}{360} = 2r + \dfrac{200\pi r}{\pi r^2}$
$= 2r + \dfrac{200}{r}$
$\theta < 2\pi \Rightarrow$ Area $< \pi r^2$, so $\pi r^2 > 100$
$\therefore r > \sqrt{\dfrac{100}{\pi}}$

b 40 cm

9 a Let h = height of rectangle.
$P = \pi r + 2r + 2h = 40 \Rightarrow 2h = 40 - 2r - \pi r$
$A = \frac{\pi}{2}r^2 + 2rh = \frac{\pi}{2}r^2 + r(40 - 2r - \pi r)$
$= 40r - 2r^2 - \frac{\pi}{2}r^2$

b $\dfrac{800}{4 + \pi}\,\text{cm}^2$

10 a $18x + 14y = 1512 \Rightarrow y = \dfrac{1512 - 18x}{14}$
$A = 12xy = 12x\left(\dfrac{1512 - 18x}{14}\right)$
$= 1296x - \dfrac{108x^2}{7}$

b $27\,216\,\text{mm}^2$

Mixed exercise 12

1 $f'(x) = \lim\limits_{h \to 0} \dfrac{10(x + h)^2 - 10x^2}{h} = \lim\limits_{h \to 0} \dfrac{20xh + 10h^2}{h}$
$= \lim\limits_{h \to 0} (20x + 10h) = 20x$

2 a y-coordinate of $B = (\delta x)^3 + 3(\delta x)^2 + 6\delta x + 4$
Gradient $= \dfrac{((\delta x)^3 + 3(\delta x)^2 + 6\delta x + 4) - 4}{(1 + \delta x) - 1}$
$= \dfrac{(\delta x)^3 + 3(\delta x)^2 + 6\delta x}{(\delta x)} = (\delta x)^2 + 3\delta x + 6$

b 6

3 $4, 11\frac{3}{4}, 17\frac{25}{27}$

4 $2, 2\frac{2}{3}$

5 $(2, -13)$ and $(-2, 15)$

6 a $1 - \dfrac{9}{x^2}$ b $x = \pm 3$

7 $\frac{3}{2}x^{-\frac{1}{2}} + 2x^{-\frac{3}{2}}$

8 a $\dfrac{dy}{dx} = 6x^{-\frac{1}{2}} - \frac{3}{2}x^{\frac{1}{2}} = \frac{3}{2}x^{-\frac{1}{2}}(4 - x)$ b $(4, 16)$

9 a $x + x^{\frac{3}{2}} - x^{-\frac{1}{2}} - 1$ b $1 + \frac{3}{2}x^{\frac{1}{2}} + \frac{1}{2}x^{-\frac{3}{2}}$ c $4\frac{1}{16}$

10 $6x^2 + \frac{1}{2}x^{-\frac{1}{2}} - 2x^{-2}$

11 $a = 1, b = -4, c = 5$

12 a $3x^2 - 10x + 5$

b i $\frac{1}{3}$ ii $y = 2x - 7$ iii $\frac{7}{2}\sqrt{5}$

13 $y = 9x - 4$ and $9y + x = 128$

14 a $\left(\frac{4}{5}, -\frac{2}{5}\right)$ b $\frac{1}{5}$

15 P is $(0, -1)$, $\dfrac{dy}{dx} = 3x^2 - 4x - 4$
Gradient at $P = -4$, so L is $y = -4x - 1$.
$-4x - 1 = x^3 - 2x^2 - 4x - 1 \Rightarrow x^2(x - 2) = 0$
$x = 2 \Rightarrow y = -9$, so Q is $(2, -9)$
Distance $PQ = \sqrt{(2 - 0)^2 + (-9 - (-1))^2} = \sqrt{68} = 2\sqrt{17}$

16 a $x = 4, y = 20$

b $\dfrac{d^2y}{dx^2} = \dfrac{3}{4}x^{-\frac{1}{2}} + 96x^{-3}$
At $x = 4$, $\dfrac{d^2y}{dx^2} = \dfrac{15}{8} > 0$
$(4, 20)$ is a local minimum.

17 $(1, -11)$ and $\left(\frac{7}{3}, -\frac{329}{27}\right)$

18 a $7\frac{31}{32}$

b $f'(x) = \left(x - \dfrac{1}{x}\right)^2 \geqslant 0$ for all values of x

19 $(1, 4)$

20 a $(1, 33)$ maximum, $(2, 28)$ and $(-1, 1)$ minimum

b
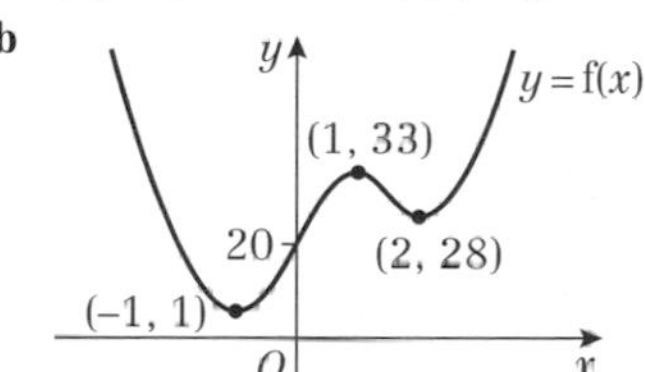

21 a $\dfrac{250}{x^2} - 2x$ b $(5, 125)$

22 a $P\left(x, 5 - \frac{1}{2}x^2\right)$
$OP^2 = (x - 0)^2 + \left(5 - \frac{1}{2}x^2 - 0\right)^2$
$= \frac{1}{4}x^4 - 4x^2 + 25$

b $x = \pm 2\sqrt{2}$ or $x = 0$

c When $x = \pm 2\sqrt{2}$, $f''(x)$ so minimum
When $x = \pm 2\sqrt{2}$, $y = 9$ so $OP = 3$

23 a $3 + 5(3) + 3^2 - 3^3 = 0$ therefore C on curve

b A is $(-1, 0)$; B is $\left(\frac{5}{3}, 9\frac{13}{27}\right)$

24
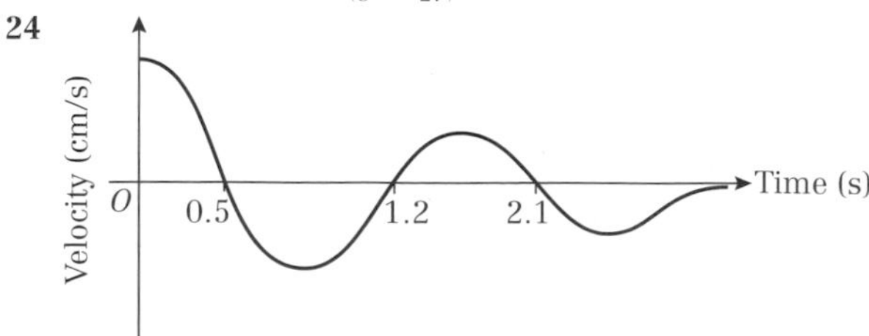

25 $\dfrac{10}{3}, \dfrac{2300\pi}{27}$

26 $\dfrac{dA}{dx} = 4\pi x - \dfrac{2000}{x^2}$
$\dfrac{dA}{dx} = 0$: $4\pi x = \dfrac{2000}{x^2} \rightarrow x^3 = \dfrac{2000}{4\pi} = \dfrac{500}{\pi}$

27 a $y = 1 - \dfrac{x}{2} - \dfrac{\pi x}{4}$

b $R = xy + \frac{\pi}{2}\left(\frac{x}{2}\right)^2$

$= x\left(1 - \frac{x}{2} - \frac{\pi x}{4}\right) + \frac{\pi x^2}{8}$

$= x - \frac{x^2}{2} - \frac{\pi x^2}{4} + \frac{\pi x^2}{8}$

$= \frac{x}{8}(8 - 4x - \pi x)$

c $\frac{2}{4 + \pi}$ m² (0.280 m²)

28 a $SA = \pi x^2 + 2\pi x + \pi x^2 + 2\pi xh = 80\pi$

$h = \frac{40 - x - x^2}{x}$

$V = \pi x^2 h = \pi x^2\left(\frac{40 - x - x^2}{x}\right)$

$= \pi(40x - x^2 - x^3)$

b $\frac{10}{3}$ **c** $\frac{d^2V}{dx^2} < 0 \therefore$ maximum

d $\frac{2300\pi}{27}$ **e** $22\frac{2}{9}\%$

29 a Length of short sides $= \frac{x}{\sqrt{2}}$

Area $= \frac{1}{2} \times$ base $\times$ height

$= \frac{1}{2}\left(\frac{x^2}{2}\right) = \frac{1}{4}x^2$ m²

b Let l be length of EF.

$\frac{1}{4}x^2 l = 4000 \Rightarrow l = \frac{16\,000}{x^2}$

$S = 2\left(\frac{1}{4}x^2\right) + \frac{2xl}{\sqrt{2}}$

$= \frac{1}{2}x^2 + \frac{32\,000x}{\sqrt{2}x^2} = \frac{x^2}{2} + \frac{16\,000\sqrt{2}}{x}$

c $x = 20\sqrt{2}$, $S = 1200$ m² **d** $\frac{d^2S}{dx^2} > 0$

Challenge

a $x^7 + 7x^6h + 21x^5h^2 + 35x^4h^3$

b $\frac{d}{dx}(x^7) = \lim_{h\to0}\frac{(x + h)^7 - x^7}{h} = \lim_{h\to0}\frac{7x^6h + 21x^5h^2 + 35x^4h^3}{h}$

$= \lim_{h\to0}(7x^6 + 21x^5h + 35x^4h^2) = 7x^6$

CHAPTER 13

Prior knowledge check

1 a $x^{\frac{5}{2}}$ **b** $2x^{\frac{3}{2}}$ **c** $x^{\frac{5}{2}} - \sqrt{x}$ **d** $x^{-\frac{3}{2}} + 4x$

2 a $6x^2 + 3$ **b** $x - 1$ **c** $3x^2 + 2x$ **d** $-\frac{1}{x^2} - 3x^2$

3 a

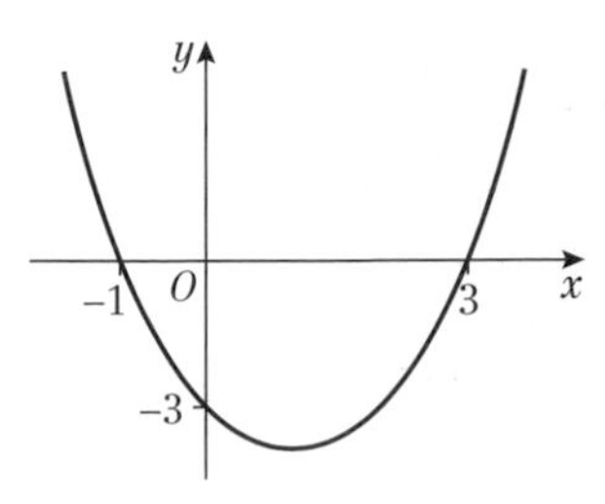

b

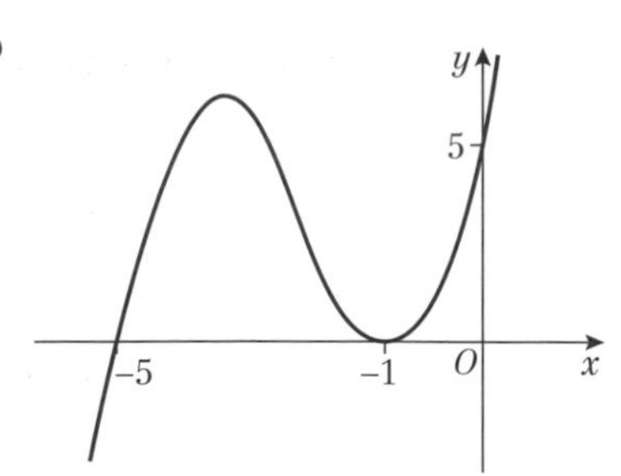

Exercise 13A

1 a $y = \frac{1}{6}x^6 + c$ **b** $y = 2x^5 + c$
c $y = x^{-1} + c$ **d** $y = 2x^{-2} + c$
e $y = \frac{3}{5}x^{\frac{5}{3}} + c$ **f** $y = \frac{8}{3}x^{\frac{3}{2}} + c$
g $y = -\frac{2}{7}x^7 + c$ **h** $y = 2x^{\frac{1}{2}} + c$
i $y = -10x^{-\frac{1}{2}} + c$ **j** $y = \frac{9}{2}x^{\frac{4}{3}} + c$
k $y = 3x^{12} + c$ **l** $y = 2x^{-7} + c$
m $y = -9x^{\frac{1}{3}} + c$ **n** $y = -5x + c$
o $y = 3x^2 + c$ **p** $y = \frac{10}{3}x^{0.6} + c$

2 a $y = \frac{1}{4}x^4 - 3x^{\frac{1}{2}} + 6x^{-1} + c$ **b** $y = x^4 + 3x^{\frac{1}{3}} + x^{-1} + c$
c $y = 4x + 4x^{-3} + 4x^{\frac{1}{2}} + c$ **d** $y = 3x^{\frac{5}{3}} - 2x^5 - \frac{1}{2}x^{-2} + c$
e $y = 4x^{-\frac{1}{3}} - 3x + 4x^2 + c$ **f** $y = x^5 + 2x^{-\frac{1}{2}} + 3x^{-4} + c$

3 a $f(x) = 6x^2 - 3x^{-\frac{1}{2}} + 5x + c$ **b** $f(x) = x^6 - x^{-6} + x^{-\frac{1}{6}} + c$
c $f(x) = x^{\frac{1}{2}} + x^{-\frac{1}{2}} + c$ **d** $f(x) = 2x^5 - 4x^{-2} + c$
e $f(x) = 3x^{\frac{2}{3}} - 6x^{-\frac{2}{3}} + c$
f $f(x) = 3x^3 - 2x^{-2} + \frac{1}{2}x^{\frac{1}{2}} + c$

4 $y = \frac{4x^3}{3} + 6x^2 + 9x + c$

5 $f(x) = -3x^{-1} + 4x^{\frac{3}{2}} + \frac{x^2}{2} - 4x + c$

Challenge

$y = -\frac{12}{7x^{\frac{7}{2}}} - \frac{4}{5x^{\frac{5}{2}}} + \frac{3}{2x^2} + \frac{1}{x} + c$

Exercise 13B

1 a $\frac{x^4}{4} + c$ **b** $\frac{x^8}{8} + c$
c $-x^{-3} + c$ **d** $\frac{5x^3}{3} + c$

2 a $\frac{1}{5}x^5 + \frac{1}{2}x^4 + c$ **b** $\frac{x^4}{2} - \frac{x^3}{3} + \frac{5x^2}{2} + c$
c $2x^{\frac{5}{2}} - x^3 + c$

3 a $-4x^{-1} + 6x^{\frac{1}{2}} + c$ **b** $-6x^{-1} - \frac{2}{3}x^{\frac{3}{2}} + c$
c $-4x^{-\frac{1}{2}} + \frac{x^3}{3} - 2x^{\frac{1}{2}} + c$

4 a $x^4 + x^{-3} + rx + c$ **b** $\frac{1}{2}x^2 + 2x^{\frac{1}{2}} - 2x^{-\frac{1}{2}} + c$
c $\frac{px^5}{5} + 2qx - 3x^{-1} + c$

5 a $t^3 + t^{-1} + c$ **b** $\frac{2}{3}t^3 + 6t^{-\frac{1}{2}} + t + c$
c $\frac{p}{4}t^4 + q^2t + pr^3t + c$

6 a $x^2 - \frac{3}{x} + c$ **b** $\frac{4}{3}x^3 + 6x^2 + 9x + c$
c $\frac{4}{5}x^{\frac{5}{2}} + 2x^{\frac{3}{2}} + c$

Online Full worked solutions are available in SolutionBank.

7 a $\frac{1}{3}x^3 + 2x - \frac{1}{x} + c$ b $\frac{1}{2}x^2 + \frac{8}{3}x^{\frac{3}{2}} + 4x + c$
c $2x^{\frac{1}{2}} + \frac{4}{3}x^{\frac{3}{2}} + c$

8 a $\frac{3}{5}x^{\frac{5}{3}} - \frac{2}{x^2} + c$ b $-\frac{1}{x^2} - \frac{1}{x} + 3x + c$
c $\frac{1}{4}x^4 - \frac{1}{3}x^3 + \frac{3}{2}x^2 - 3x + c$ d $\frac{8}{5}x^{\frac{5}{2}} + \frac{8}{3}x^{\frac{3}{2}} + 2x^{\frac{1}{2}} + c$
e $3x + 2x^{\frac{1}{2}} + 2x^3 + c$ f $\frac{2}{5}x^{\frac{5}{2}} + 3x^2 + 6x^{\frac{3}{2}} + c$

9 a $-\frac{A}{x} - 3x + c$ b $\frac{2}{3}\sqrt{P}\,x^{\frac{3}{2}} - \frac{1}{x^2} + c$
c $-\frac{p}{x} + \frac{2qx^{\frac{3}{2}}}{3} + rx + c$

10 $-\frac{6}{x} + \frac{8x^{\frac{3}{2}}}{3} - \frac{3x^2}{2} + 2x + c$

11 $2x^4 + 3x^2 - 6x^{\frac{1}{2}} + c$

12 a $(2 + 5\sqrt{x})^2 = 4 + 10\sqrt{x} + 10\sqrt{x} + 25x = 4 + 20\sqrt{x} + 25x$
b $4x + \frac{40x^{\frac{3}{2}}}{3} + \frac{25x^2}{2} + c$

13 $\frac{x^6}{2} - 8x^{\frac{1}{2}} + c$

14 $p = -4$, $q = -2.5$

15 a $1024 - 5120x + 11\,520x^2$
b $1024x - 2560x^2 + 3840x^3 + c$

Exercise 13C

1 a $y = x^3 + x^2 - 2$ b $y = x^4 - \frac{1}{x^2} + 3x + 1$
c $y = \frac{2}{3}x^{\frac{3}{2}} + \frac{1}{12}x^3 + \frac{1}{3}$ d $y = 6\sqrt{x} - \frac{1}{2}x^2 - 4$
e $y = \frac{1}{3}x^3 + 2x^2 + 4x + \frac{2}{3}$ f $y = \frac{2}{5}x^{\frac{5}{2}} + 6x^{\frac{1}{2}} + 1$

2 $f(x) = \frac{1}{2}x^4 + \frac{1}{x} + \frac{1}{2}$

3 $y = 1 - \frac{2}{\sqrt{x}} - \frac{3}{x}$

4 $f(x) = 3x^3 + 2x^2 - 3x - 2$

5 $y = 6x^{\frac{1}{2}} - \frac{4x^{\frac{5}{2}}}{5} + \frac{118}{5}$

6 a $p = \frac{1}{2}$, $q = 1$ b $y = 4x^{\frac{3}{2}} + \frac{5x^2}{2} - \frac{421}{2}$

7 a $f(t) = 10t - \frac{5t^2}{2}$ b $7\frac{1}{2}$

8 a $f(t) = -4.9t^2 + 35$ b 23.975 m
c 35 m d 2.67 seconds
e e.g. the ground is flat

Challenge

1 $f_2(x) = \frac{x^3}{3}$; $f_4(x) = \frac{x^4}{12}$ b $\frac{x^{n+1}}{3 \times 4 \times 5 \times \ldots \times (n+1)}$

2 $f_2(x) = x + 1$; $f_3(x) = \frac{1}{2}x^2 + x + 1$; $f_4(x) = \frac{1}{6}x^3 + \frac{1}{2}x^2 + x + 1$

Exercise 13D

1 a $152\frac{1}{4}$ b $48\frac{2}{5}$ c $5\frac{1}{3}$ d 2

2 a $5\frac{1}{4}$ b 10 c $11\frac{5}{6}$ d $60\frac{1}{2}$

3 a $16\frac{2}{3}$ b $46\frac{1}{2}$ c $\frac{11}{14}$ d $2\frac{1}{2}$

4 $A = -7$ or 4

5 28

6 $-8 + 8\sqrt{3}$

7 $k = \frac{25}{4}$

8 450 m

Challenge
$k = 2$

Exercise 13E

1 a 22 b $36\frac{2}{3}$ c $48\frac{8}{15}$ d 6

2 4 3 6 4 $10\frac{2}{3}$

5 $21\frac{1}{3}$ 6 $\frac{4}{81}$ 7 $k = 2$

8 a $(-1, 0)$ and $(3, 0)$ b $10\frac{2}{3}$

9 $1\frac{1}{3}$

Exercise 13F

1 a $1\frac{1}{3}$ b  $20\frac{5}{6}$

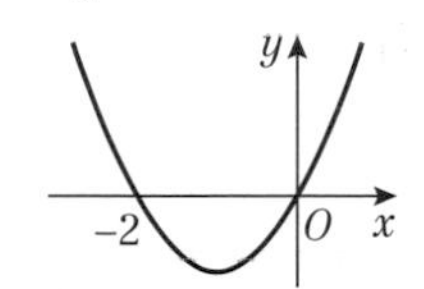

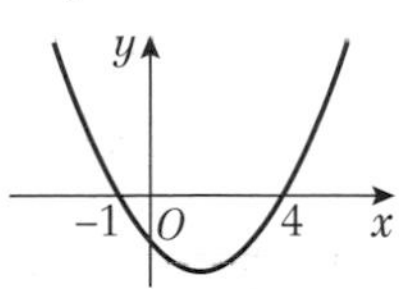

c $40\frac{1}{2}$ d $1\frac{1}{3}$

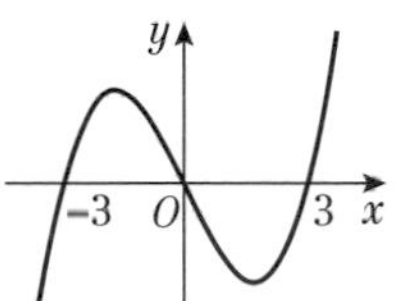

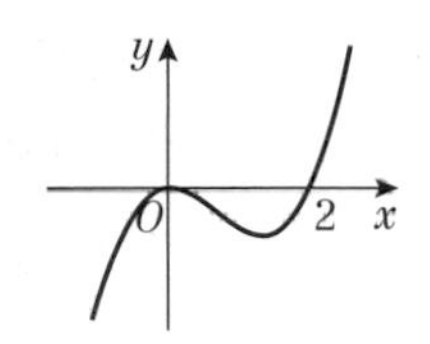

e $21\frac{1}{12}$

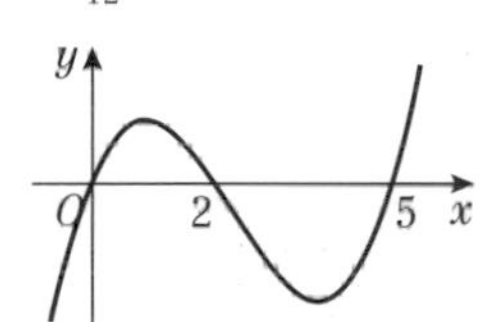

2 a $(-3, 0)$ and $(2, 0)$ b $21\frac{1}{12}$

3 a $f(-3) = 0$
b $f(x) = (x + 3)(-x^2 + 7x - 10)$
c $f(x) = (x + 3)(x - 5)(2 - x)$
d $(-3, 0)$, $(2, 0)$ and $(5, 0)$
e $143\frac{5}{6}$

Challenge

1 a  $4\frac{1}{2}$ b 9 c 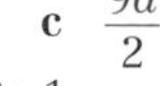 $\frac{9a}{2}$ d $4\frac{1}{2}$ e $\frac{9}{2a}$

2 a B has x-coordinate 1
$\int_0^1 (x^3 + x^2 - 2x)\,dx = \left[\frac{1}{4}x^4 + \frac{1}{3}x^3 - x^2\right]_0^1$
$= \frac{1}{4} + \frac{1}{3} - 1 = -\frac{5}{12}$
So area under x-axis is $\frac{5}{12}$
Area above x-axis is
$\left(\frac{1}{4}0^4 + \frac{1}{3}0^3 - 0^2\right) - \left(\frac{1}{4}x^4 + \frac{1}{3}x^3 - x^2\right) = \frac{5}{12}$
So the x-coordinate of a satisfies
$3x^4 + 4x^3 - 12x^2 + 5 = 0$
Then use the factor theorem twice to get
$(x - 1)^2(3x^2 + 10x + 5) = 0$

b A has coordinates $\left(\frac{-5+\sqrt{10}}{3}, \frac{-80+37\sqrt{10}}{27}\right)$

The roots at 1 correspond to point B.

The root $\frac{-5-\sqrt{10}}{3}$ gives a point on the curve to the left of -2 below the x-axis, so cannot be A.

Exercise 13G

1 a $A(-2, 6)$, $B(2, 6)$ **b** $10\frac{2}{3}$
2 a $A(1, 3)$, $B(3, 3)$ **b** $1\frac{1}{3}$
3 $6\frac{2}{3}$
4 4.5
5 a (2, 12) **b** $13\frac{1}{3}$
6 a $20\frac{5}{6}$ **b** $17\frac{1}{6}$
7 a, b Substitute into equation for y
c $y = x - 4$ **d** $8\frac{3}{5}$
8 $3\frac{3}{8}$
9 a Substitute $x = 4$ into both equations
b 7.2
10 a $21\frac{1}{3}$ **b** $2\frac{5}{9}$
11 a $(-1, 11)$ and $(3, 7)$ **b** $21\frac{1}{3}$

Mixed exercise 13

1 a $\frac{2}{3}x^3 - \frac{3}{2}x^2 - 5x + c$ **b** $\frac{3}{4}x^{\frac{4}{3}} + \frac{3}{2}x^{\frac{2}{3}} + c$
2 $\frac{1}{3}x^3 - \frac{3}{2}x^2 + \frac{2}{x} + \frac{1}{6}$
3 a $2x^4 - 2x^3 + 5x + c$ **b** $2x^{\frac{5}{2}} + \frac{4}{3}x^{\frac{3}{2}} + c$
4 $\frac{4}{5}x^{\frac{5}{2}} - \frac{2}{3}x^{\frac{3}{2}} - 6x^{\frac{1}{2}} + c$
5 $x = \frac{1}{3}t^3 + t^2 + t - 8\frac{2}{3}$; $x = 12\frac{1}{3}$
6 a $A = 6, B = 9$ **b** $\frac{3}{5}x^{\frac{5}{3}} + \frac{9}{2}x^{\frac{4}{3}} + 9x + c$
7 a $\frac{9}{2}x^{-\frac{1}{2}} - 8x^{-\frac{3}{2}}$ **b** $6x^{\frac{3}{2}} + 32x^{\frac{1}{2}} - 24x + c$
8 $a = 4, b = -3.5$
9 25.9 m
10 a $f(t) = 5t + t^2$ **b** 7.8 seconds
11 a $-1,3$ **b** $10\frac{2}{3}$
12 a $-\frac{2x^{\frac{3}{2}}}{3} + 5x - 8\sqrt{x} + c$ **b** $\frac{7}{3}$
13 a (3, 0) **b** (1, 4) **c** $6\frac{3}{4}$
14 a $\frac{3}{2}x^{-\frac{1}{2}} + 2x^{-\frac{3}{2}}$ **b** $2x^{\frac{3}{2}} - 8x^{\frac{1}{2}} + c$ **c** $A = 6, B = -2$
15 a $\frac{dy}{dx} = 6x^{-\frac{1}{2}} - \frac{3}{2}x^{\frac{1}{2}} = \frac{3}{2}x^{-\frac{1}{2}}(4 - x)$
b (4,16) **c** 133 (3 sf)
16 a (6,12) **b** $13\frac{1}{3}$
17 a $A(1, 0)$, $B(5, 0)$, $C(6, 5)$ **b** $10\frac{1}{6}$
18 a $q = -2$ **b** $C(6,17)$ **c** $1\frac{1}{3}$
19 $-\frac{9}{x} - \frac{16x^{\frac{3}{2}}}{3} + 2x^2 - 5x + c$
20 $A = -6$ or 1
21 a $f'(x) = \frac{(2 - x^2)(4 - 4x^2 + x^4)}{x^2} = 8x^{-2} - 12 + 6x^2 - x^4$
b $f''(x) = -16x^{-3} + 12x - 4x^3$
c $f(x) = -\frac{8}{x} - 12x + 2x^3 - \frac{x^5}{5} - \frac{47}{5}$
22 a $(-3, 0)$ and $(\frac{1}{2}, 0)$ **b** $14\frac{7}{24}$
23 a $(-\frac{3}{2}, 0)$ and $(4, 0)$ **b** $55\frac{11}{24}$
24 a -2 and 3 **b** $21\frac{1}{12}$

Challenge

$10\frac{5}{12}$

CHAPTER 14

Prior knowledge check

1 a 125 **b** $\frac{1}{3}$ **c** 32 **d** 49 **e** 1
2 a 6^6 **b** y^{21} **c** 2^6 **d** x^4
3 gradient 1.5, intercept 4.1

Exercise 14A

1 a

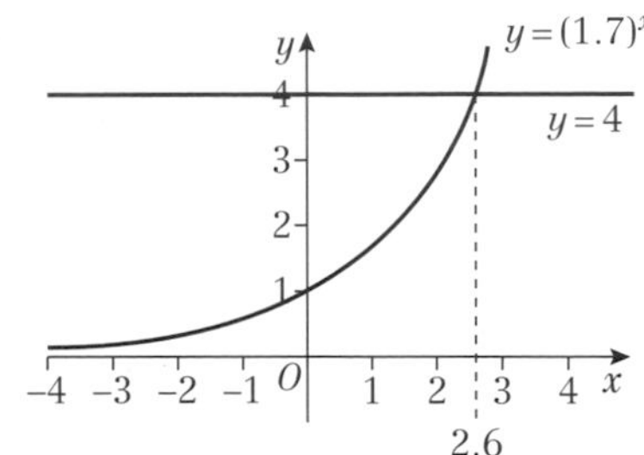

b $x \approx 2.6$

2 a

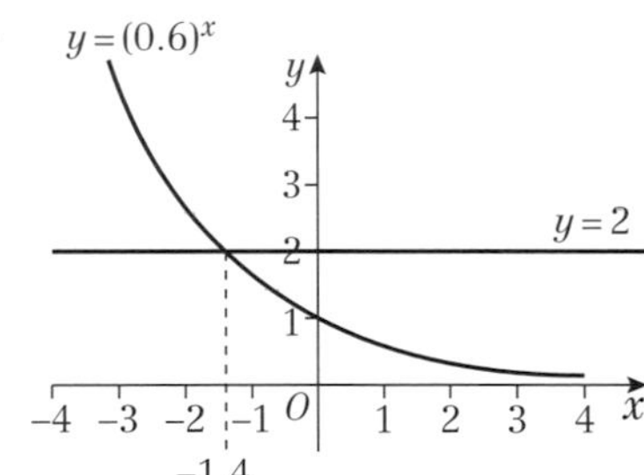

b $x \approx -1.4$

3

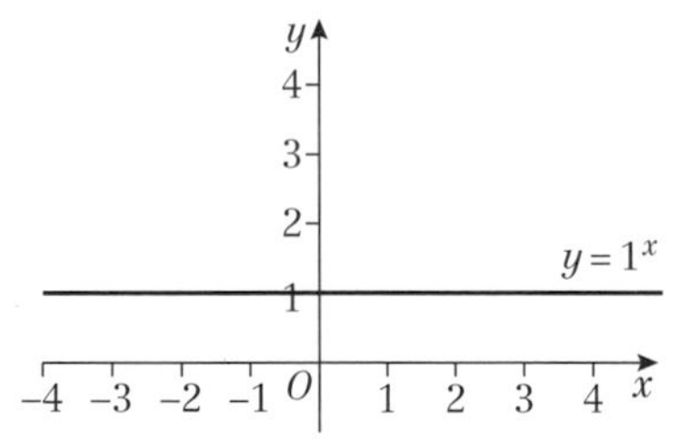

4 a True, because $a^0 = 1$ whenever a is positive
b False, for example when $a = \frac{1}{2}$
c True, because when a is positive, $a^x > 0$ for all values of x

5

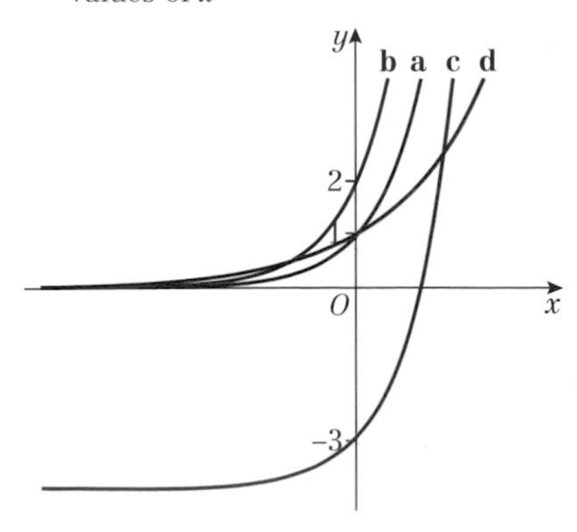

6 $k = 3$, $a = 2$
7 a As x increases, y decreases
b $p = 1.2$, $q = 0.2$

Challenge

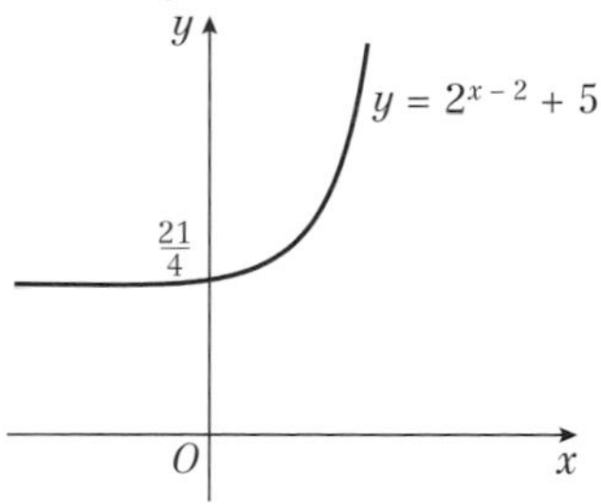

Exercise 14B

1 a 2.718 28 b 54.598 15 c 0.000 04 d 1.221 40
2 a

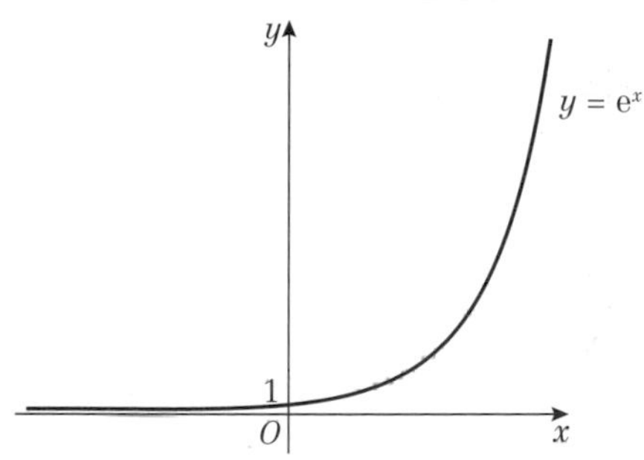

b Student's own answers
c e = 2.71828…
e^3 = 20.08553…

3 a

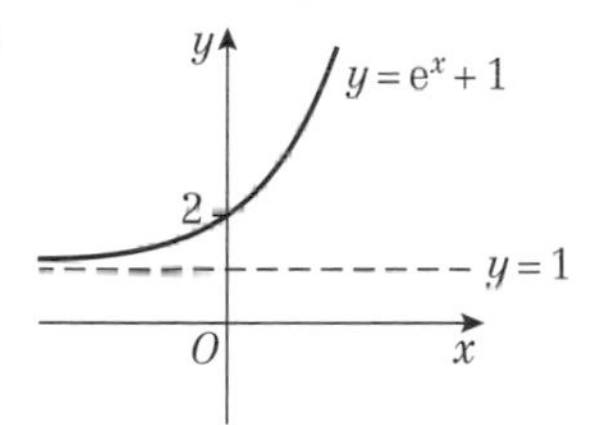

b

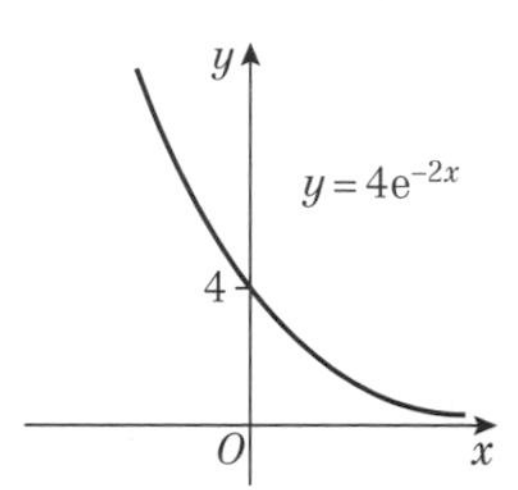

c

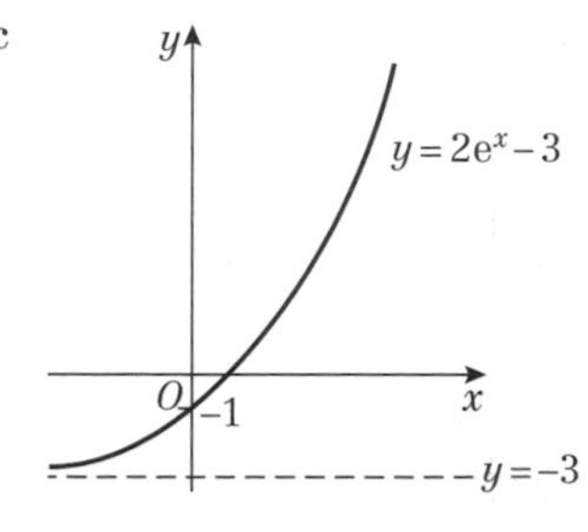

d

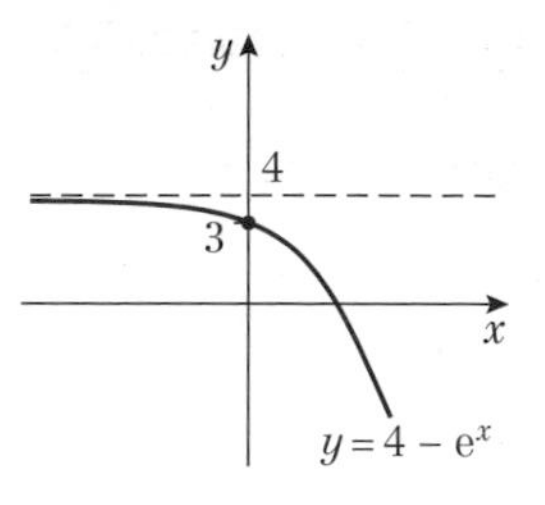

e

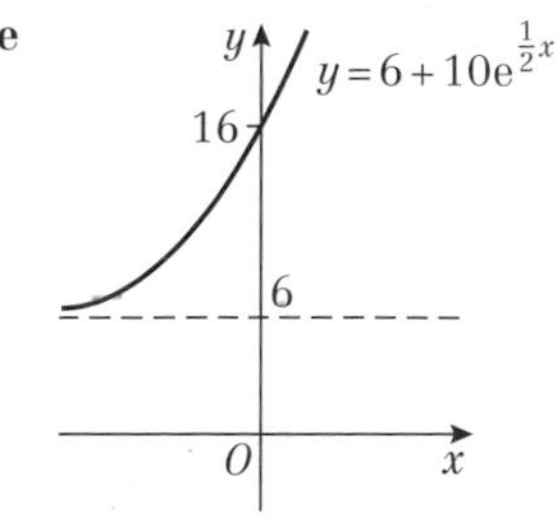

f

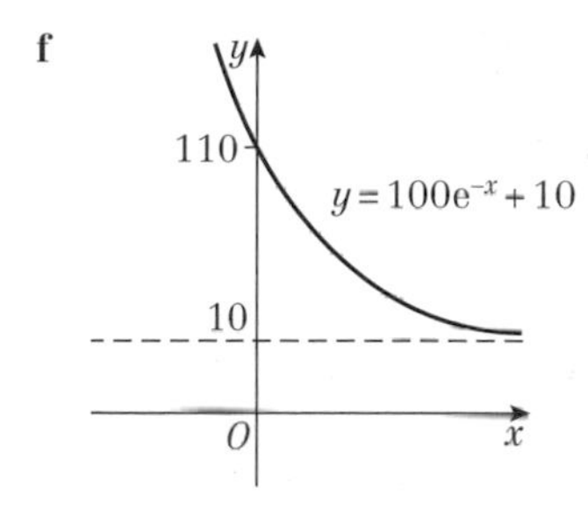

4 a $A = 1$, $C = 5$, b is positive
b $A = 4$, $C = 0$, b is negative
c $A = 6$, $C = 2$, b is positive
5 $A = e^2$, $b = 3$

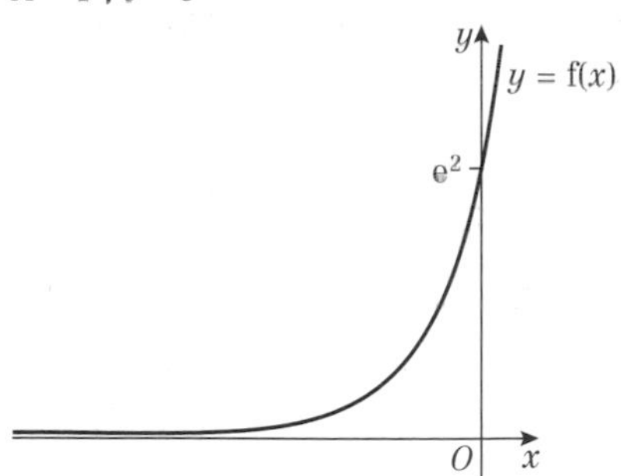

6 a $6e^{6x}$ b $-\frac{1}{3}e^{-\frac{1}{3}x}$ c $14e^{2x}$
d $2e^{0.4x}$ e $3e^{3x} + 2e^x$ f $2e^{2x} + e^x$
7 a $3e^6$ b 3 c $3e^{-1.5}$
8 $f'(x) = 0.2e^{0.2x}$
The gradient of the tangent when $x = 5$ is
$f'(5) = 0.2e^1 = 0.2e$.
The equation of the tangent is therefore $y = (0.2e)x + c$.
At (5, e), $e = 0.2e \times 5 + c$, so $c = 0$ and when $x = 0$, $y = 0$.

Exercise 14C

1 a £20 000 b £14 331
c

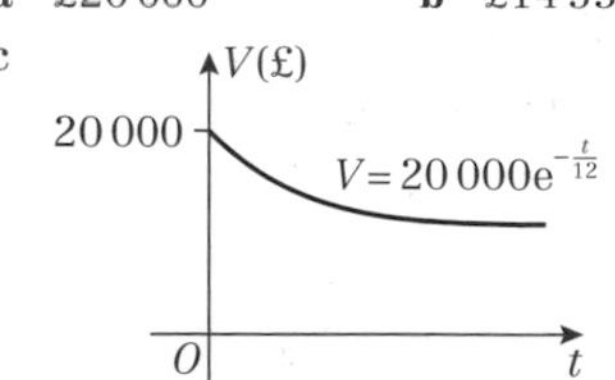

2 a 30 000 b 38 221

c

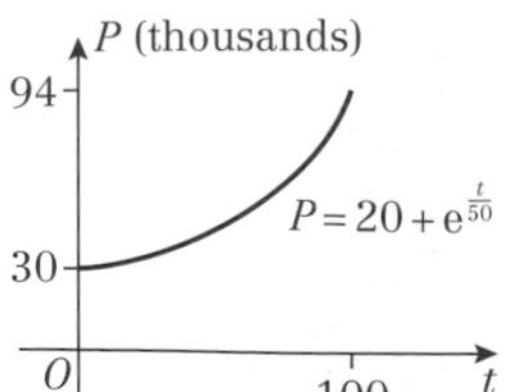

d Model predicts population of the country to be over 200 million, this is highly unlikely and by 2500 new factors are likely to affect population growth. Model not valid for predictions that far into the future.

3 a 200

b Disease will infect up to 300 people.

c

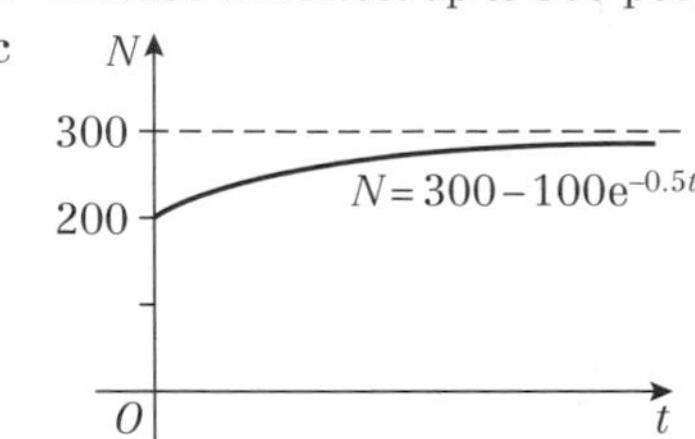

4 a i 15 rabbits ii 132 rabbits

b The initial number of rabbits

c $\frac{dR}{dm} = 2.4\,e^{0.2m}$

When $m = 6$, $\frac{dR}{dm} = 7.97 \approx 8$

d The rabbits may begin to run out of food or space

5 a 0.565 bars

b $\frac{dp}{dh} = -0.13e^{-0.13h} = -0.13p$, $k = -0.13$

c The atmospheric pressure decreases exponentially as the altitude increases

d 12%

6 a Model 1: £15 733
Model 2: £15 723 Similar results

b Model 1: £1814
Model 2: £2484 Model 2 predicts a larger value

c

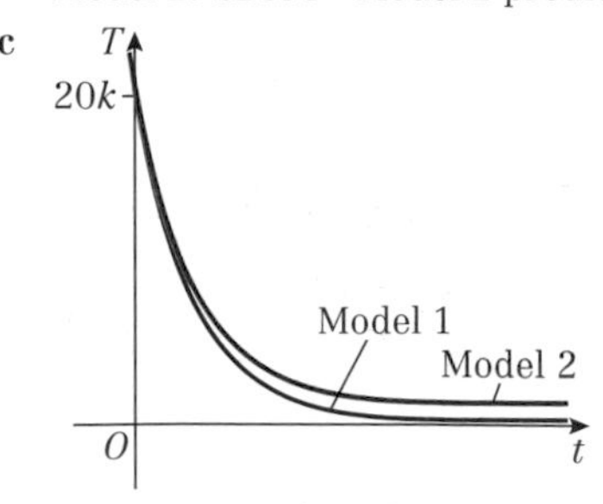

d In Model 2 the tractor will always be worth at least £1000. This could be the value of the tractor as scrap metal.

Exercise 14D

1 a $\log_4 256 = 4$ b $\log_3 \frac{1}{9} = -2$
c $\log_{10} 1\,000\,000 = 6$ d $\log_{11} 11 = 1$
e $\log_{0.2} 0.008 = 3$

2 a $2^4 = 16$ b $5^2 = 25$
c $9^{\frac{1}{2}} = 3$ d $5^{-1} = 0.2$
e $10^5 = 100\,000$

3 a 3 b 2 c 7 d 1
e 6 f $\frac{1}{2}$ g −1 h −2
i 10 j −2

4 a 625 b 9 c 7 d 9
e 20 f 2

5 a 2.475 b 2.173 c 3.009 d 1.099

6 a $5 = \log_2 32 < \log_2 50 < \log_2 64 = 6$
b 5.644

7 a i 1 ii 1 iii 1 b $a^1 \equiv a$

8 a i 0 ii 0 iii 0 b $a^0 \equiv 1$

Exercise 14E

1 a $\log_2 21$ b $\log_2 9$ c $\log_5 80$
d $\log_6 \left(\frac{64}{81}\right)$ e $\log_{10} 120$

2 a $\log_2 8 = 3$ b $\log_6 36 = 2$ c $\log_{12} 144 = 2$
d $\log_8 2 = \frac{1}{3}$ e $\log_{10} \frac{1}{10} = -1$

3 a $3\log_a x + 4\log_a y + \log_a z$
b $5\log_a x - 2\log_a y$
c $2 + 2\log_a x$
d $\log_a x - \frac{1}{2}\log_a y - \log_a z$
e $\frac{1}{2} + \frac{1}{2}\log_a x$

4 a $\frac{4}{3}$ b $\frac{1}{18}$ c $\sqrt{30}$ d 2

5 a $\log_3 (x + 1) - 2\log_3 (x - 1) = 1$

$\log_3 \left(\frac{x + 1}{(x - 1)^2}\right) = 1$

$\frac{x + 1}{(x - 1)^2} = 3$

$x + 1 = 3(x-1)^2$

$x + 1 = 3(x^2 - 2x + 1)$

$3x^2 - 7x + 2 = 0$

b $x = 2$

6 $a = 9$, $b = 4$

Challenge

$\log_a x = m$ and $\log_a y = n$

$x = a^m$ and $y = a^n$

$x \div y = a^m \div a^n = a^{m-n}$

$\log_a \left(\frac{x}{y}\right) = m - n = \log_a x - \log_a y$

Exercise 14F

1 a 6.23 b 2.10 c 0.431
d 1.66 e −3.22 f 1.31
g 1.25 h −1.73

2 a 0, 2.32 b 1.26, 2.18 c 1.21
d 0.631 e 0.565, 0.712 f 0
g 2 h −1

3 a 5.92 b 3.2

4 a (0, 1)

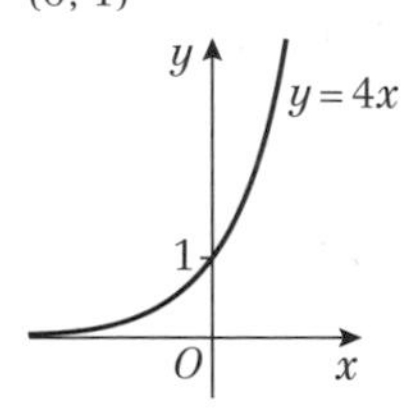

b $\frac{1}{2}, \frac{3}{2}$

5 a 0.7565 b 7.9248 c 0.2966

Exercise 14G

1 a $\ln 6$ b $\frac{1}{2}\ln 11$ c $3 - \ln 20$
d $\frac{1}{4}\ln\left(\frac{1}{3}\right)$ e $\frac{1}{2}\ln 3 - 3$ f $5 - \ln 19$

2 a e^2 b $\frac{e}{4}$ c $\frac{1}{2}e^4 - \frac{3}{2}$
d $\frac{1}{6}(e^{\frac{5}{2}} + 2)$ e $18 - e^{\frac{1}{2}}$ f 2, 5

3 a $\ln 2, \ln 6$ b $\frac{1}{2}\ln 2, 0$ c e^3, e^{-5}
d $\ln 4, 0$ e $\ln 5, \ln\left(\frac{1}{3}\right)$ f e^6, e^{-2}

4 $\ln 3, 2\ln 2$

5 a $\frac{1}{8}(e^2 + 3)$ b $\frac{1}{5}(\ln 3 + 40)$ c $\frac{1}{5}\ln 7, 0$
d e^3, e^{-1}

6 $\dfrac{1 + \ln 5}{4 + \ln 3}$

7 a The initial concentration of the drug in mg/l
b 4.91 mg/l
c $3 = 6e^{-\frac{t}{10}}$
$\frac{1}{2} = e^{-\frac{t}{10}}$
$\ln\left(\frac{1}{2}\right) = -\dfrac{t}{10}$
$t = -10\ln\left(\frac{1}{2}\right) = 6.931... = 6$ hours 56 minutes

8 a $(0, 3 + \ln 4)$ b $(4 - e^{-3})$

Challenge
As $y = 2$ is an asymptote, $C = 2$.
Substituting (0, 5) gives $5 = Ae^0 + 2$, so A is 3.
Substituting (6, 10) gives $10 = 3e^{6B} + 2$.
Rearranging this gives $B = \frac{1}{6}\ln\left(\frac{8}{3}\right)$.

Exercise 14H

1 a $\log S = \log(4 \times 7^x)$
$\log S = \log 4 + \log 7^x$
$\log S = \log 4 + x\log 7$
b gradient $\log 7$, intercept $\log 4$

2 a $\log A = \log(6x^4)$
$\log A = \log 6 + \log x^4$
$\log A = \log 6 + 4\log x$
b gradient 4, intercept $\log 6$

3 a Missing values 1.52, 1.81, 1.94
b

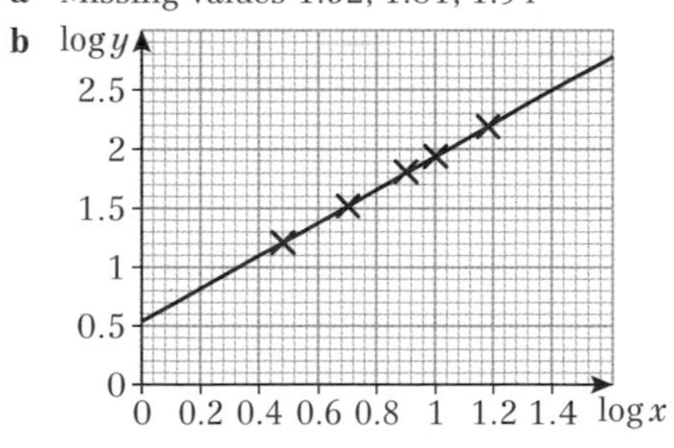

c Approximately $a = 3.5$, $n = 1.4$

4 a Missing values 2.63, 3.61, 4.49, 5.82
b

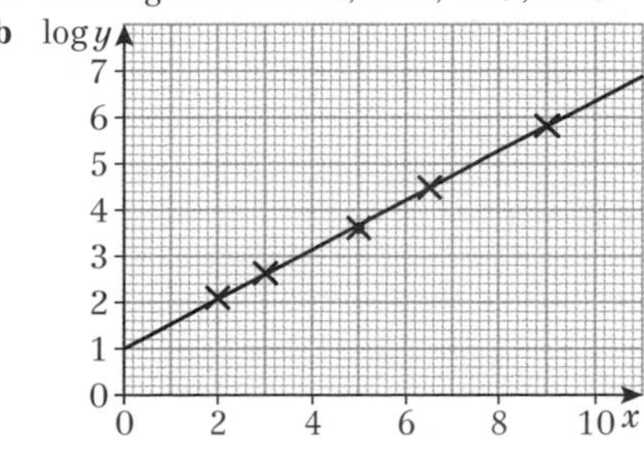

c Approximately $b = 3.4$, $a = 10$

5 a Missing values −0.39, 0.62, 1.54, 2.81
b

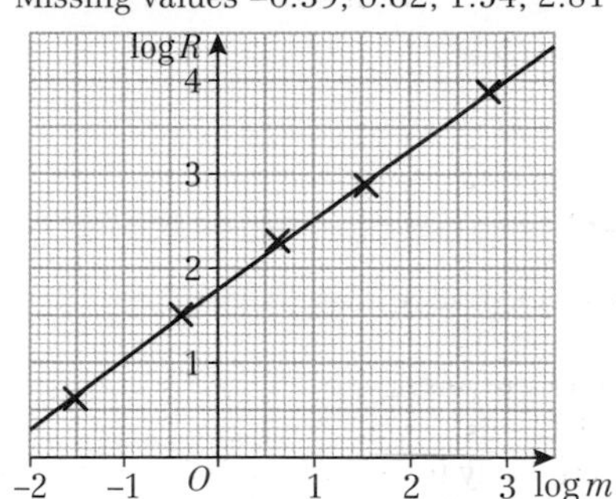

c Approximately $a = 60$, $b = 0.75$
d Approximately 1,600 kcal per day (2 s.f.)

6 a Missing values 2.94, 1.96, 0.95
b

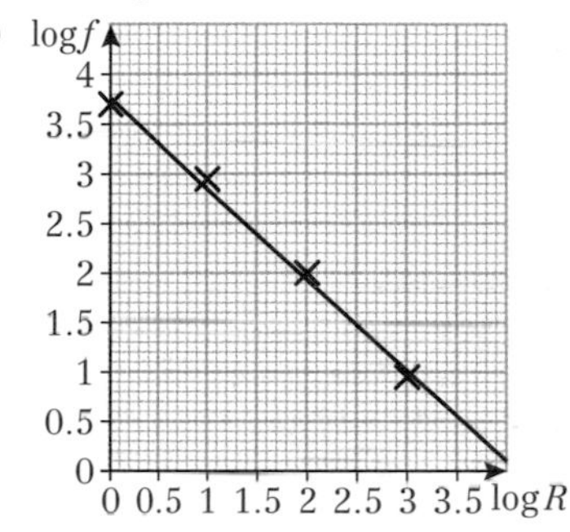

c Approximately $A = 5800$, $b = -0.9$
d Approximately 690 times

7 a Missing values 0.98, 1.08, 1.13, 1.26, 1.37
b $P = ab^t$
$\log P = \log(ab^t)$
$\log P = \log a + \log b^t$
$\log P = \log a + t\log b$
c

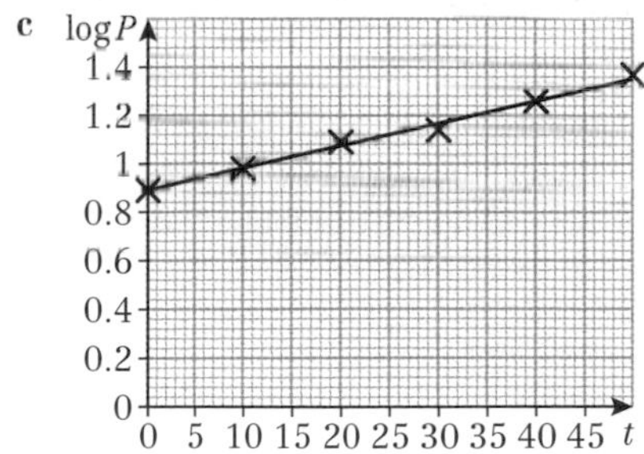

d Approximately $a = 7.6$, $b = 1.0$
e The rate of growth is often proportional to the size of the population

8 a $\log N = 0.095t + 1.6$
b $a = 40$, $b = 1.2$
c The initial number of sick people
d 9500 people. After 30 days people may start to recover, or the disease may stop spreading as quickly.

9 a $\log A = 2\log w - 0.1049$
b $q = 2$, $p = 0.7854$
c Circles: p is approximately one quarter π, and the width is twice the radius, so $A = \frac{\pi}{4}w^2 = \frac{\pi}{4}(2r)^2 = \pi r^2$.

Challenge
$y = 5.8 \times 0.9^x$

Mixed exercise 14

1 a

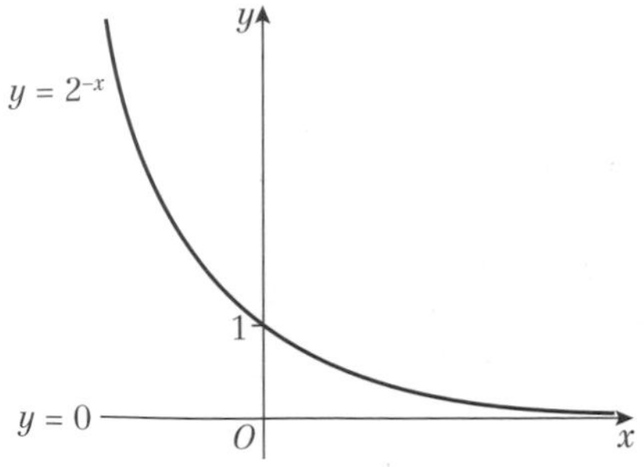

b

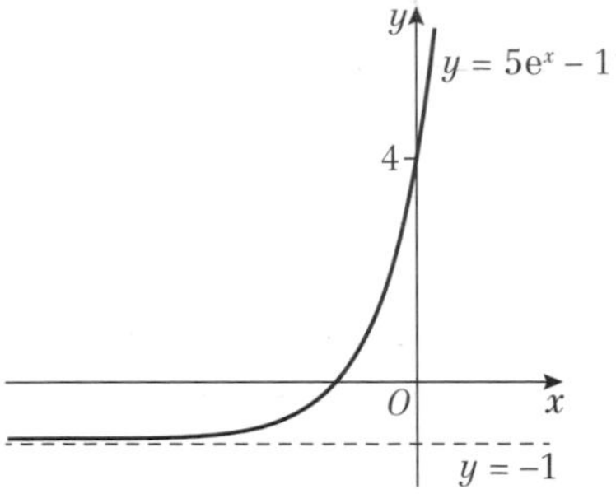

c

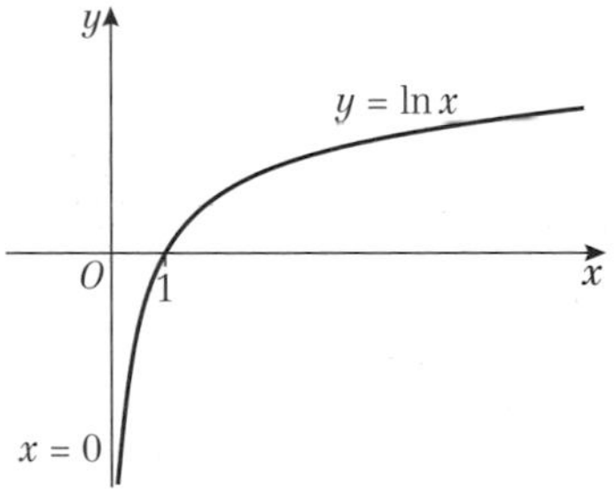

2 a $2\log_a p + \log_a q$ **b** $\log_a p = 4$, $\log_a q = 1$

3 a $\frac{1}{4}p$ **b** $\frac{3}{4}p + 1$

4 a 2.26 **b** 1.27 **c** 7.02

5 a $4^x - 2^{x+1} - 15 = 0$
$2^{2x} - 2 \times 2^x - 15 = 0$
$(2^x)^2 - 2 \times 2^x - 15 = 0$
$u^2 - 2u - 15 = 0$

b 2.32

6 $x = 6$

7 a $-e^{-x}$ **b** $11e^{11x}$ **c** $30e^{5x}$

8 a $\dfrac{e^8 + 5}{2}$ **b** $\dfrac{\ln 5}{4}$ **c** $-\frac{1}{2}\ln 14$ **d** $\dfrac{3 + \sqrt{13}}{2}$

e 0 **f** $\dfrac{e^4}{2}$

9 a £950 **b** £290 **c** 4.28 years **d** £100

e

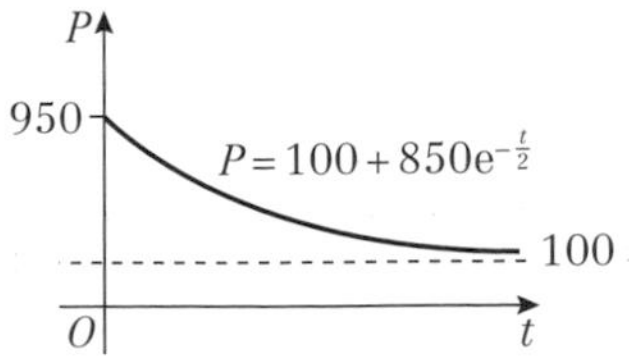

f A good model. The computer will always be worth something

10 a $y = \left(\dfrac{2}{\ln 4}\right)x$

b (0, 0) satisfies the equation of the line.

c 2.43

11 a We cannot go backwards in time

b 75°C

c 5 minutes

d The exponential term will always be positive, so the overall temperature will be greater than 20°C.

12 a $S = aV^b$
$\log S = \log(aV^b)$
$\log S = \log a + \log(V^b)$
$\log S = \log a + b\log V$

b

$\log S$	1.26	1.70	2.05	2.35	2.50
$\log V$	0.86	1.53	2.05	2.49	2.72

c

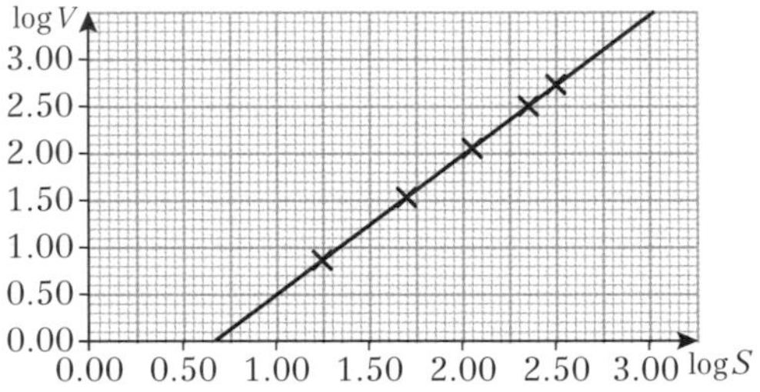

d The gradient is approximately 1.5; $a \approx 0.09$

13 a The model concerns decay, not growth

b

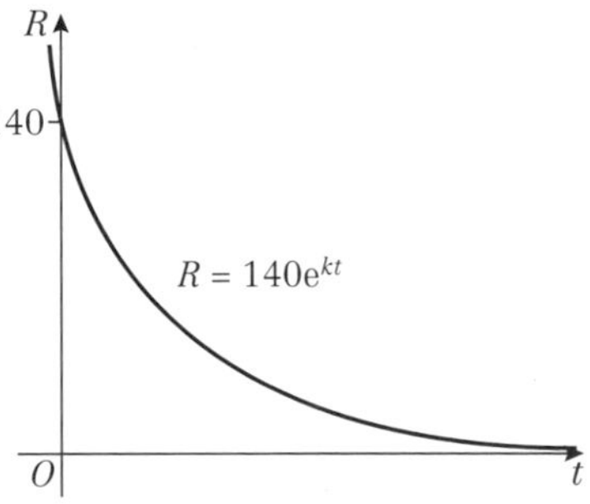

c $70 = 140e^{30k}$
$\frac{1}{2} = e^{30k}$
$\ln\left(\frac{1}{2}\right) = 30k$
$k = \frac{1}{30}\ln\left(\frac{1}{2}\right)$
$k = -\frac{1}{30}\ln(2)$, so $c = -\frac{1}{30}$

14 a 6.4 million views

b $\dfrac{dV}{dx} = 0.4e^{0.4x}$

c 9.42×10^{16} new views per day

d This is too big, so the model is not valid after 100 days

15 a 4.2

b i 1.12×10^{25} dyne cm

ii 3.55×10^{26} dyne cm

c divide **b ii** by **b i**

16 a They exponentiated the two terms on RHS separately rather than combining them first.

b $x = 2 + 2\sqrt{2}$

Challenge

a $y = 9^x = 3^{2x}$, $\log_3(y) = 2x$
b $y^2 = (9^x)^2 = 9^{2x}$, $\log_9(y^2) = 2x$
c $x = -\frac{1}{3}$ or $x = -2$

Review exercise 3

1 -4.5
2 $\sqrt{7}$
3 **a** All equal to $\sqrt{145}$
b $(x-1)^2 + (y+3)^2 = 145$
4 **a** $-2\mathbf{i} - 8\mathbf{j}$
b $|\overrightarrow{AB}| = |\overrightarrow{AC}| = \sqrt{85}$
c $|\overrightarrow{BC}| = \sqrt{68}$

$\cos \angle ABC = \dfrac{85 + 68 - 85}{2 \times \sqrt{85} \times \sqrt{68}} = \dfrac{1}{\sqrt{5}}$

5 12
6 **a** 5N **b** 7
7 $m = 50\sqrt{3} + 30$, $n = 50$
8 **a** $\sqrt{(-75)^2 + 180^2} = 195 > 150 = \sqrt{90^2 + 120^2}$
b Boat A: 6.5 m/s; Boat B: 5 m/s; Both boats arrive at the same time – it is a tie.
9 $\lim_{h \to 0} \dfrac{5(x+h)^2 - 5x^2}{h}$

$= \lim_{h \to 0} \dfrac{5x^2 + 10xh + 5h^2 - 5x^2}{h}$

$= \lim_{h \to 0} \dfrac{10xh + 5h^2}{h}$

$= \lim_{h \to 0} 10x + 5h$

$= 10x$

10 $\dfrac{dy}{dx} = 12x^2 + x^{-\frac{1}{2}}$
11 **a** $\dfrac{dy}{dx} = 4 + \frac{9}{2}x^{\frac{1}{2}} - 4x$
b Substitute $x = 4$ into equation for C
c Gradient of tangent $= -3$ so gradient of normal $= \frac{1}{3}$
Substitute (4, 8) into $y = \frac{1}{3}x + c$
Rearrange $y = \frac{1}{3}x + \frac{20}{3}$
d $PQ = 8\sqrt{10}$
12 **a** $\dfrac{dy}{dx} = 8x - 5x^{-2}$, at P this is 3
b $y = 3x + 5$
c $k = -\frac{5}{3}$
13 **a** $P = 2$, $Q = 9$, $R = 4$
b $3x^{\frac{1}{2}} + \frac{9}{2}x^{-\frac{1}{2}} - 2x^{-\frac{3}{2}}$
c When $x = 1$, $f'(x) = 5\frac{1}{2}$, gradient of $2y = 11x + 3$ is $5\frac{1}{2}$, so it is parallel with tangent
14 $f'(x) = 3x^2 - 24x + 48 = 3(x-4)^2 > 0$
15 **a** $A(1,0)$ and $B(2,0)$
b $(\sqrt{2}, 2\sqrt{2} - 3)$
16 **a** $V = \pi r^2 h = 128\pi$, so $h = \dfrac{128}{r^2}$

$S = 2\pi rh + 2\pi r^2 = \dfrac{256\pi}{r} + 2\pi r^2$

b 96π cm^2
17 **a** $\dfrac{dy}{dx} = 6x + 2x^{-\frac{1}{2}}$
b $\dfrac{d^2y}{dx^2} = 6 - x^{-\frac{3}{2}}$
c $x^3 + \frac{8}{3}x^{\frac{3}{2}} + c$
18 **a** $2x^3 - 5x^2 - 12x$
b $2x^3 - 5x^2 - 12x$
$= x(2x^2 - 5x - 12)$
$= x(2x + 3)(x - 4)$
c

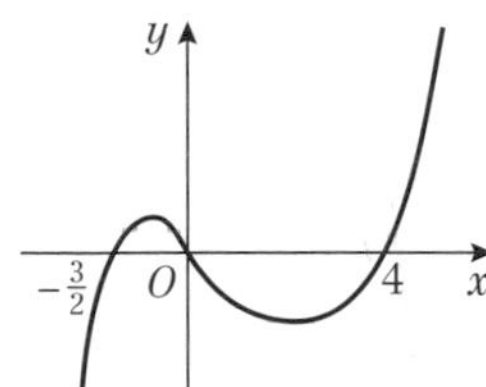

19 $6\frac{3}{4}$
20 4
21 **a** $-x^4 + 3x^2 + 4 = (-x^2 + 4)(x^2 + 1)$; $x^2 + 1 = 0$ has no real solutions; so solutions are $A(-2, 0)$, $B(2, 0)$
b 19.2
22 $4\frac{1}{2}$
23 **a** $P(-1, 4)$, $Q(2, 1)$
b 4.5
24 **a** $k = -1$, $A(0, 2)$
b $\ln 3$
25 **a** 425 °C **b** 7.49 minutes **c** 1.64 °C/minute
d The temperature can never go below 25 °C, so cannot reach 20 °C.
26 **a** -0.179 **b** $x = 15$
27 **a** $x = 1.55$ **b** $x = 4$ or $x = \frac{1}{2}$
28 **a** $\log_p 2$ **b** 0.125
29 **a** $x = 2$ **b** $x = \ln 3$ or $x = \ln 1 = 0$
30 **a** Missing values 0.88, 1.01, 1.14 and 1.29
b

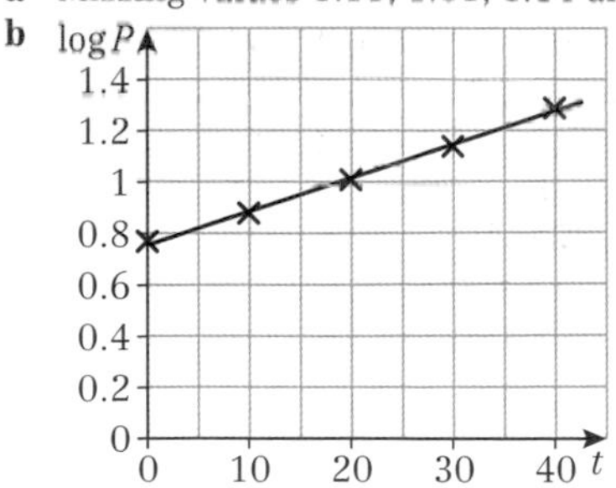

c $P = ab^t$
$\log P = \log(ab^t) = \log a + t \log b$
This is a linear relationship. The gradient is $\log b$ and the intercept is $\log a$.
d $a = 5.9$, $b = 1.0$

Challenge

1 **a** 0
b 1
2 **a** $f'(-3) = f'(2) = 0$, so $f'(x) = k(x + 3)(x - 2)$ $= k(x^2 + x - 6)$; there are no other factors as $f(x)$ is cubic
b $2x^3 + 3x^2 - 36x - 5$
3 51.2
4 **a** $f(0) = 0^3 - k(0) + 1 = 1$; $g(0) = e^{2(0)} = 1$; $P(0, 1)$
b $\frac{1}{2}$

Exam-style practice

1 a $\frac{1}{3}$ b $5\sqrt{2}$

2 $y = \frac{2}{3}x + \frac{8}{3}$

3 a error 1: $= -\dfrac{3}{\sqrt{x}} = -3x^{-\frac{1}{2}}$, not $-3x^{\frac{1}{2}}$

error 2: $\left[\dfrac{x^5}{5} - 2x^{\frac{3}{2}} + 2x\right]_1^2 = \left(\dfrac{32}{5} - 2\sqrt{8} + 4\right) - \left(\dfrac{1}{5} - 2 + 2\right)$

not $\left(\dfrac{1}{5} - 2 + 2\right) - \left(\dfrac{32}{5} - 2\sqrt{8} + 4\right)$

b 5.71 (3 s.f.)

4 $x = 30°, 90°, 150°$

5 a $2x^2(x + 3)$

b $2x^2(x + 3) = 980 \Rightarrow 2x^3 + 6x^2 - 980 = 0$
$\Rightarrow x^3 + 3x^2 - 490 = 0$

c For $f(x) = x^3 + 3x^2 - 490 = 0$, $f(7) = 0$, so $x - 7$ is a factor of $f(x)$ and $x = 7$ is a solution.

d Equation becomes $(x - 7)(x^2 + 10x + 70) = 0$
Quadratic has discriminant $10^2 - 4 \times 1 \times 70 = -180$
So the quadratic has no real roots, and the equation has no more real solutions.

6 $x + 15y + 106 = 0$

7 a $\log_{10} P = 0.01t + 2$

b 100, initial population

c 1.023

d Accept answers from 195 to 200

8 $1 + \cos^4 x - \sin^4 x \equiv 1 + (\cos^2 x + \sin^2 x)(\cos^2 - \sin^2 x)$
$\equiv (1 - \sin^2 x) + \cos^2 x \equiv 2\cos^2 x$

9 Magnitude $= \sqrt{29}$, angle $= 112°$ (3 s.f.)

10 a 56.5° (3 s.f.) b £49.63

11 a

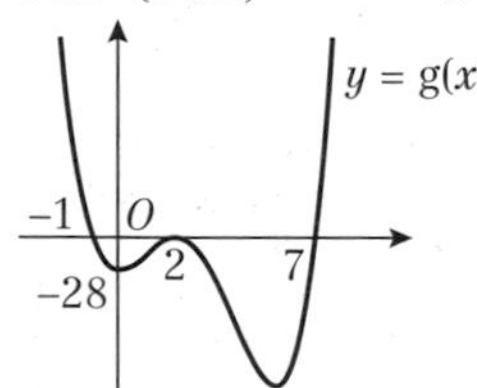

b $-4, -1, 4$

12 $x = 3$ or $-\frac{5}{3}$

13 a $1 - 15x + 90x^2$

b 0.859

c Greater: The next term will be subtracting from this, and future positive terms will be smaller.

14 a $f(x) = \int (x^{-\frac{3}{2}} - 1 - x^{-2})\,dx = -2x^{-\frac{1}{2}} - x + x^{-1} + c$

$= -\dfrac{x^2 + 2\sqrt{x} - 1}{x} + c$

b $c = \frac{5}{3} + \frac{2}{3}\sqrt{3}$

15 a $5^2 + 1^2 - 4 \times 5 + 6 \times 1 = 12$, so $(5, 1)$ lies on C.
Centre $= (2, -3)$, radius $= 5$

b $y = -\frac{3}{4}x + \frac{19}{4}$

c $12\frac{15}{32}$

Index